# Computer Algebra Handbook

Springer
*Berlin*
*Heidelberg*
*New York*
*Hong Kong*
*London*
*Milan*
*Paris*
*Tokyo*

Johannes Grabmeier
Erich Kaltofen
Volker Weispfenning (Editors)

# Computer Algebra Handbook

Foundations · Applications · Systems

 Springer

*Editors:*
Johannes Grabmeier
University of Applied Sciences
Edlmairstrasse 6 + 8
94469 Deggendorf, Germany
e-mail: Johannes.grabmeier@fh-deggendorf.de

Erich Kaltofen
North Carolina State University
Department of Mathematics
Raleigh, NC 27695-8205, USA
e-mail: kaltofen@math.ncsu.edu

Volker Weispfenning
University of Passau
Department of Mathematics
Innstrasse 33
94032 Passau, Germany
e-mail: weispfen@fmi.uni-passau.de

Library of Congress Cataloging-in-Publication Data applied for

A catalog record for this book is available from the Library of Congress.

Bibliographic information published by Die Deutsche Bibliothek
Die Deutsche Bibliothek lists this publication in the Deutsche Nationalbibliografie;
detailed bibliographic data is available in the Internet at http://dnb.ddb.de

---

Mathematics Subject Classification (2000): Explicit machine computation and programs (XX-04) in the fields 11-XX – 20-XX, 33-XX – 40-XX, 70-XX – 86-XX, 97-XX

---

ISBN 3-540-65466-6 Springer-Verlag Berlin Heidelberg New York

This work is subject to copyright. All rights are reserved, whether the whole or part of the material is concerned, specifically the rights of translation, reprinting, reuse of illustrations, recitation, broadcasting, reproduction on microfilm or in any other way, and storage in data banks. Duplication of this publication or parts thereof is permitted only under the provisions of the German Copyright Law of September 9, 1965, in its current version, and permission for use must always be obtained from Springer-Verlag. Violations are liable for prosecution under the German Copyright Law.

*Please note:* The software is protected by copyright. The publisher and the authors accept no legal responsibility for any damage caused by improper use of the instructions and programs contained in this book and the CD-ROM. Although the software has tested with extreme care, errors in the software cannot excluded. Decompiling, disassembling, reverse engineering or in any way changing the program is expressly forbidden. For more details concerning the conditions of use and warranty we refer to *License Agreement* on the CD-ROM (license.txt).

Springer-Verlag Berlin Heidelberg New York
a member of BertelsmannSpringer Science+Business Media GmbH
http://www.springer.de
© Springer-Verlag Berlin Heidelberg 2003
Printed in Germany

The use of general descriptive names, registered names, trademarks etc. in this publication does not imply, even in the absence of a specific statement, that such names are exempt from the relevant protective laws and regulations and therefore free for general use.

Cover design: *Erich Kirchner, Heidelberg*
Printed on acid-free paper    SPIN 10521668    46/3142ck-5 4 3 2 1 0

# Foreword

*Two ideas lie gleaming on the jeweler's velvet. The first is the calculus, the second, the algorithm. The calculus and the rich body of mathematical analysis to which it gave rise made modern science possible; but it has been the algorithm that has made possible the modern world.*

—David Berlinski, *The Advent of the Algorithm*

First there was the concept of integers, then there were symbols for integers: |, ||, |||, ||||, ||||| (what might be called a sticks and stones representation); I, II, III, IV, V (Roman numerals); 1, 2, 3, 4, 5 (Arabic numerals), etc. Then there were other concepts with symbols for them and algorithms (sometimes) for manipulating the new symbols. Then came collections of mathematical knowledge (tables of mathematical computations, theorems of general results). Soon after algorithms came devices that provided assistance for carrying out computations. Then mathematical knowledge was organized and structured into several related concepts (and symbols): logic, algebra, analysis, topology, algebraic geometry, number theory, combinatorics, etc. This organization and abstraction lead to new algorithms and new fields like universal algebra. But always our symbol systems reflected and influenced our thinking, our concepts, and our algorithms.

In the latter half of the twentieth century a powerful new computational device, the electronic computer, was stirred into the mix. These devices stimulated work on new algorithms, new ways of representing algorithms (more symbol systems) as programs written in various languages devised for use with the new computing devices, even validating (at least for the present) the formal notation of algorithm provided earlier by the logicians (Church's thesis and related concepts). New ways of representing mathematical knowledge in electronic form followed: data bases, electronic documents with new ways of representing mathematical concepts/symbols (TeX, for example) and more recently MathML that promises even more powerful and general ways of electronically storing, manipulating, and displaying mathematical knowledge. There are many other advances along these lines that could be mentioned. But the over arching question is: Where will all this lead?

Will we one day have a giant interacting web of mathematical machines capable of executing all known mathematical algorithms, of finding, retrieving, and organizing mathematical knowledge, of using heuristic procedures to answer mathematical questions when purely algorithmic methods do not exist or are unknown, of verifying mathematical proofs, of generating new conjectures and in some cases proving or finding counterexamples of conjectures (both those generated by machines and those generated by humans)?

In the last century giant strides have been made toward a grand vision of computational mathematics, in the broadest sense of this phrase. An important aspect of this progress has been the research on the application of computers to (mostly) exact mathematical computation—a field that we now call *computer algebra*—a field that brings together Berlinski's role of the calculus and the algorithm in a far-reaching manner. Important new data structures and algorithms, incredibly fast computational processors, large memories, new software systems, and applications to a myriad of problems in a multitude of disciplines have been the building blocks used by the computer algebraists in this fascinating, productive, and ever-widening journey.

This volume, this *Computer Algebra Handbook*, is an impressive snapshot taken with a wide-angle lens of the state of computer algebra research and applications in the last decade of the twentieth century. As with all snapshots of dynamic objects it can reveal a moment in time that reflects only a part of the past and give some indication of the future. But this volume, with the wide-angle lens of two hundred authors, gives a picture of the field that should be valuable to all current researchers and to generations of future researchers. So, gentle reader, I recommend this volume and all its concepts, symbols, and algorithms to you.

Bob Caviness<br>
Newark, Delaware<br>
(October 2001)

# Editorial Remarks

The German special interest group for computer algebra (in German: *Fachgruppe für Computeralgebra*) in 1993 published a report "Computeralgebra in Deutschland Bestandsaufnahme, Möglichkeiten, Perspektiven," (in English "Computer Algebra in Germany Status, Possibilities, Perspectives"). The book was written in the German language and contained contributions mostly from German persons working in computer algebra. In 1995 Grabmeier and Weispfenning, who had edited the German report, approached Kaltofen with the idea to produce a translated and expanded version of their book in the English language. The "Computer Algebra Handbook" is the product of that effort. The outline of the original Report was retained, but the Handbook became about twice as long. The pre-dominantly German authorship of the contributions in the Handbook remains as a legacy from the German Report.

The Handbook describes computer algebra in three main chapters: Chapter 2 surveys the methodology of our discipline, Chapter 3 applications and Chapter 4 software. Chapter 1 attempts a definition of what computer algebra is, and Chapter 5 rounds out the Handbook with lists of activities and publications. The Handbook has over two hundred contributing authors, whose efforts and patience we thankfully acknowledge. Consequently, the individual sections are quite diverse in style and focus. One goal of our editorial work was to preserve this diversity. A reader can obtain a homogeneous presentation of computer algebra from any of the excellent books on the subjects cited in Chapter 5 that have been written by small author teams.

The Handbook is to serve as a reference to computer algebra. The bibliography is accordingly large, with over 2,100 entries. However, it is simply impossible to account for all important work in our field of considerable breadth, quantity, and active research. Therefore, the omission of a reference to one's work or software should in no way be taken as a value judgement. We intend to maintain a current bibliography on the web and invite everyone to send us reference to their omitted or future work.

The Handbook contains descriptions of software that may no longer be actively used. We believe it an important archival task to maintain a record of such systems. In the fast-moving world of computer software, significant ideas may simply fall by the wayside as program source code is removed from the storage media. In the attached compact disk we attempt to stem the tide by providing data files that may contain such information. This is just a beginning, and we believe that publishers of journals and proceedings in the discipline ought to follow our example. Like in physics, the experiments with our software must be repeatable.

As the great Donald Knuth has told us, the three tasks: (i) publishing a paper, (ii) publishing a book, (iii) writing a large computer program, are of increasing level of difficulty. Perhaps before and after (iii) one can place the tasks of writing a book and a program with 200 authors. Grants from the National Science Foundation (USA) have supported Kaltofen in part during his work on this book, which is gratefully acknowledged. We thank our colleagues Friedrich W. Hehl,

Cologne for editing section 3.1 Applications in Physics, Wolfram Koepf, Kassel for editing section 3.6 Computer Algebra in Education and Werner M. Seiler, Mannheim for editing section 2.11 Symbolic Methods for Differential Equations. Our work was assisted by several individuals who we would like to name. Markus Hitz, now at North Georgia College and State University, translated Chapter 2 of the German Report into English, layed out the production directory structure that we followed, and performed several other tasks. Ilias Kotsireas, now at the University of Western Ontario, helped us in processing the email of the final updates. And last but not least, our sincere thanks go to Martin Peters, our editor at Springer Verlag, and his staff members Ruth Allewelt and Claudia Kehl, who gave their enthusiastic support and showed great patience with our decidedly slow progress.

Now, we wish the reader enjoyment with the contents of our book.

Johannes Grabmeier<br>
Erich Kaltofen<br>
Volker Weispfenning<br>
(October 2001)

# Table of Contents

| | |
|---|---|
| Foreword | v |
| Editorial Remarks | vii |
| Table of Contents | ix |
| List of Contributing Authors | xv |
| **1 Development, Characterization, Prospects** | 1 |
|   1.1 Historical Remarks | 1 |
|   1.2 General Characterization | 1 |
|   1.3 Impact on Education | 2 |
|   1.4 Impact on Research | 4 |
|   1.5 Computer Algebra – Today and Tomorrow | 6 |
|     1.5.1 Today | 6 |
|     1.5.2 Outlook | 7 |
| **2 Topics of Computer Algebra** | 11 |
|   2.1 Exact Arithmetic | 11 |
|     2.1.1 Long Integer Arithmetic | 11 |
|     2.1.2 Arithmetic with Polynomials, Rational Functions and Power Series | 13 |
|     2.1.3 Euclid's Algorithm and Continued Fractions | 16 |
|     2.1.4 Modular Arithmetic and the Chinese Remainder Theorem | 17 |
|     2.1.5 Computations with Algebraic Numbers | 18 |
|     2.1.6 Real Algebraic Numbers | 19 |
|     2.1.7 $p$-adic Numbers and Approximations | 20 |
|     2.1.8 Finite Fields | 21 |
|   2.2 Algorithms for Polynomials and Power Series | 23 |
|     2.2.1 The Division Algorithm | 23 |
|     2.2.2 Factorization of Polynomials | 24 |
|     2.2.3 Absolute Factorization of Polynomials | 26 |
|     2.2.4 Polynomial Decomposition | 26 |
|     2.2.5 Gröbner Bases | 28 |
|     2.2.6 Standard Bases | 32 |
|     2.2.7 Characteristic Sets | 32 |
|     2.2.8 Algorithmic Invariant Theory | 33 |
|   2.3 Linear Algebra | 36 |
|     2.3.1 Linear Systems | 36 |
|     2.3.2 Algorithms for Matrix Canonical Forms | 38 |
|   2.4 Constructive Methods of Number Theory | 41 |
|     2.4.1 Primality Tests | 41 |
|     2.4.2 Integer Factorization | 44 |
|     2.4.3 Algebraic Number Fields and Algebraic Function Fields | 45 |
|     2.4.4 Galois Groups | 47 |
|     2.4.5 Rational Points on Elliptic Curves | 48 |
|     2.4.6 Geometry of Numbers | 50 |
|   2.5 Algorithms of Commutative Algebra and Algebraic Geometry | 51 |
|     2.5.1 Algorithms for Polynomial Ideals and Their Varieties | 51 |

|  |  | 2.5.2 Singularities of Varieties | 54 |
|---|---|---|---|
|  |  | 2.5.3 Real Algebraic Geometry | 55 |
|  | 2.6 | Algorithmic Aspects of the Theory of Algebras | 57 |
|  |  | 2.6.1 Structure Constants | 58 |
|  |  | 2.6.2 Generators and Relations, Swapping and G-Algebras | 58 |
|  |  | 2.6.3 Monad Algebras, Path Algebras and Generalizations | 59 |
|  |  | 2.6.4 Finite-Dimensional Lie Algebras | 60 |
|  |  | 2.6.5 Non-commutative Gröbner Bases | 60 |
|  |  | 2.6.6 Structural Issues and Classification | 63 |
|  |  | 2.6.7 Identities | 63 |
|  |  | 2.6.8 Computational Aspects in the Representation Theory of Quivers and Path Algebras | 64 |
|  | 2.7 | Computational Group Theory | 65 |
|  |  | 2.7.1 A Crash Course in Group Theory | 66 |
|  |  | 2.7.2 Describing Groups | 67 |
|  |  | 2.7.3 A Brief History | 69 |
|  |  | 2.7.4 Permutation Groups | 71 |
|  |  | 2.7.5 Matrix Groups | 74 |
|  |  | 2.7.6 Black Box Groups | 75 |
|  |  | 2.7.7 Abelian Groups | 76 |
|  |  | 2.7.8 Polycyclic Groups | 76 |
|  |  | 2.7.9 Finitely Presented Groups | 78 |
|  |  | 2.7.10 Group-Theoretic Software | 83 |
|  |  | 2.7.11 Another Perspective | 83 |
|  | 2.8 | Algorithms of Representation Theory | 84 |
|  |  | 2.8.1 Ordinary Representation Theory | 84 |
|  |  | 2.8.2 Modular Representation Theory | 85 |
|  |  | 2.8.3 Generic Character Tables | 87 |
|  |  | 2.8.4 Summary of Systems | 88 |
|  | 2.9 | Algebraic Methods for Constructing Discrete Structures | 89 |
|  | 2.10 | Summation and Integration | 91 |
|  |  | 2.10.1 Definite Summation and Hypergeometric Identities | 91 |
|  |  | 2.10.2 Symbolic Integration | 94 |
|  | 2.11 | Symbolic Methods for Differential Equations | 96 |
|  |  | 2.11.1 Introduction | 96 |
|  |  | 2.11.2 Differential Galois Theory | 97 |
|  |  | 2.11.3 Lie Symmetries | 98 |
|  |  | 2.11.4 Painlevé Theory | 99 |
|  |  | 2.11.5 Completion | 102 |
|  |  | 2.11.6 Differential Ideal Theory | 104 |
|  |  | 2.11.7 Dynamical Systems | 105 |
|  |  | 2.11.8 Numerical Analysis | 108 |
|  | 2.12 | Symbolic/Numeric Methods | 109 |
|  |  | 2.12.1 Computer Analysis | 109 |
|  |  | 2.12.2 Algorithms for Computing Validated Results | 110 |

|       | 2.12.3 Hybrid Methods ............................................. 112 |
| ----- | --------------------------------------------------------------- |
| 2.13  | Algebraic Complexity Theory .................................. 125 |
| 2.14  | Coding Theory and Cryptography ............................. 128 |
|       | 2.14.1 Coding Theory ............................................. 128 |
|       | 2.14.2 Quantum Coding Theory ................................ 130 |
|       | 2.14.3 Cryptography .............................................. 131 |
| 2.15  | Algorithmic Methods in Universal Algebra and Logic ........... 132 |
|       | 2.15.1 Term Rewriting Systems ................................. 132 |
|       | 2.15.2 Decision Procedures and Quantifier Elimination Methods for Algebraic Theories ............................... 137 |
| 2.16  | Knowledge Representation and Abstract Data Types ............ 140 |
|       | 2.16.1 Mathematical Knowledge Representation and Expert Systems ................................................ 140 |
|       | 2.16.2 Abstract Data Types ..................................... 142 |
| 2.17  | On the Design of Computer Algebra Systems ................... 143 |
|       | 2.17.1 Memory Management .................................... 143 |
|       | 2.17.2 Program Verification and Abstract Data Types .......... 144 |
|       | 2.17.3 The Concept of Types ................................... 144 |
|       | 2.17.4 Genericity ................................................ 145 |
|       | 2.17.5 Modularization ........................................... 145 |
|       | 2.17.6 Parallel Implementation ................................. 145 |
|       | 2.17.7 Continuing Development of Computer Algebra Systems ... 146 |
| 2.18  | Parallel Computer Algebra Systems ............................. 146 |
|       | 2.18.1 Parallel Architectures and Operating Systems Supports ... 146 |
|       | 2.18.2 Parallel Execution: Mapping and Scheduling ............. 147 |
|       | 2.18.3 Parallelism Expression and Languages ................... 149 |
| 2.19  | Interfaces and Standardization .................................. 150 |
|       | 2.19.1 Interfaces to Word Processors ........................... 150 |
|       | 2.19.2 Graphics .................................................. 150 |
|       | 2.19.3 Interfaces to Numerical Software ........................ 150 |
|       | 2.19.4 User Interfaces ........................................... 152 |
|       | 2.19.5 General Problem-Solving Environments ................. 152 |
|       | 2.19.6 Standardisation .......................................... 153 |
|       | 2.19.7 MathML ................................................. 154 |
| 2.20  | Hardware Implementation of Computer Algebra Algorithms ..... 161 |

3 Applications of Computer Algebra ................................. 163
   3.1 Physics ............................................................. 163
      3.1.1 Elementary Particle Physics ............................. 164
      3.1.2 Gravity .................................................... 172
      3.1.3 'Central Configurations' in the Newtonian N-Body Problem of Celestial Mechanics ........................ 176
      3.1.4 CA-Systems for Differential Geometry and Applications ... 180
      3.1.5 Differential Equations in Physics ....................... 187
   3.2 Mathematics ...................................................... 195
      3.2.1 Computer Algebra in Group Theory .................... 196

- 3.2.2 The Tangent Cone Algorithm and Applications in the Theory of Singularities .............................. 197
- 3.2.3 Automatic Theorem Proving in Geometry .............. 201
- 3.2.4 Homological Algebra ..................................... 207
- 3.2.5 Study of Differential Structures on Quantum Groups ...... 212
- 3.2.6 Orthogonal Polynomials and Computer Algebra .......... 214
- 3.2.7 Computer Algebra in Symmetric Bifurcation Theory ...... 215
- 3.2.8 Symbolic-Numeric Treatment of Equivariant Systems of Equations ......................................... 216
- 3.3 Computer Science ............................................... 217
  - 3.3.1 Computer Algebra in Computer Science ................. 217
  - 3.3.2 Decomposable Structures, Generating Functions and Average-Case of Algorithms ........................... 219
  - 3.3.3 Telecommunication Management Networks .............. 221
- 3.4 Engineering ..................................................... 221
  - 3.4.1 Computer Algebra, a Modern Research Tool for Engineering ............................................ 221
  - 3.4.2 Critical Load Computations for Jet Engines ............. 226
  - 3.4.3 Audio Signal Processing ............................... 227
  - 3.4.4 Robotics ............................................. 229
  - 3.4.5 Computer Aided Design and Modelling .................. 234
- 3.5 Chemistry ...................................................... 242
  - 3.5.1 Computer Algebra in Chemistry and Crystallography ..... 242
  - 3.5.2 Chemical Reaction Systems ............................ 243
- 3.6 Computer Algebra in Education ................................. 244
  - 3.6.1 New Hand-Held Computer Symbolic Algebra Tools in Mathematics Education ............................. 245
  - 3.6.2 The Dutch Perspective ................................ 247
  - 3.6.3 Computer Algebra in Teaching and Learning Mathematics: Experiences at the University of Plymouth, England .............................................. 250
  - 3.6.4 The Educational Use of Computer Algebra Systems at the University of Illinois ............................ 253
  - 3.6.5 Mathematics Education from a MATHEMATICA Perspective ............................................. 254
  - 3.6.6 Visualization: Courseware for Mathematics Education .... 256
- 4 Computer Algebra Systems ........................................ 261
  - 4.1 General Purpose Systems ..................................... 261
    - 4.1.1 AXIOM ............................................. 261
    - 4.1.2 Aldor .............................................. 265
    - 4.1.3 DERIVE and the TI-92 .............................. 271
    - 4.1.4 Macsyma .......................................... 283
    - 4.1.5 MAGMA ........................................... 295
    - 4.1.6 Maple ............................................. 308
    - 4.1.7 *Mathematica* ..................................... 314

|        | 4.1.8  | *MuPAD* ............................................. 321 |
|--------|--------|------------------------------------------------------|
|        | 4.1.9  | REDUCE ............................................. 333 |
| 4.2    | Special Purpose Systems ................................. 345 | |
|        | 4.2.1  | Algebraic Combinatorics Environment (ACE) ............ 345 |
|        | 4.2.2  | Building Nonassociative Algebras With Albert .......... 346 |
|        | 4.2.3  | ALGEB ............................................... 348 |
|        | 4.2.4  | AMORE .............................................. 348 |
|        | 4.2.5  | BERGMAN ............................................ 349 |
|        | 4.2.6  | CANNES / PARCAN .................................... 351 |
|        | 4.2.7  | CARAT ............................................... 354 |
|        | 4.2.8  | CASA ................................................ 356 |
|        | 4.2.9  | CHEVIE .............................................. 359 |
|        | 4.2.10 | C-Meataxe ............................................ 363 |
|        | 4.2.11 | CoCoA ............................................... 364 |
|        | 4.2.12 | CREP ................................................ 368 |
|        | 4.2.13 | The Desir Project and Its Continuation ................. 370 |
|        | 4.2.14 | DISCRETA: A Tool for Constructing $t$-Designs ......... 372 |
|        | 4.2.15 | FELIX ............................................... 375 |
|        | 4.2.16 | *Fermat* .............................................. 380 |
|        | 4.2.17 | FOXBOX and Other Blackbox Systems .................. 383 |
|        | 4.2.18 | GAP ................................................. 385 |
|        | 4.2.19 | GiNaC ............................................... 391 |
|        | 4.2.20 | Kan/sm1 ............................................. 392 |
|        | 4.2.21 | KANT V4 ............................................ 396 |
|        | 4.2.22 | LiDIA ............................................... 403 |
|        | 4.2.23 | Lie .................................................. 408 |
|        | 4.2.24 | LIE .................................................. 411 |
|        | 4.2.25 | A Brief Introduction to Macaulay 2 ..................... 411 |
|        | 4.2.26 | MAS ................................................. 421 |
|        | 4.2.27 | MASYCA ............................................. 428 |
|        | 4.2.28 | MOC ................................................. 429 |
|        | 4.2.29 | NTL: A Library for Doing Number Theory .............. 430 |
|        | 4.2.30 | PARI ................................................ 431 |
|        | 4.2.31 | PARSAC ............................................. 434 |
|        | 4.2.32 | QUOTPIC ............................................. 436 |
|        | 4.2.33 | ReDuX ............................................... 437 |
|        | 4.2.34 | REPTILES  A Program for Interactively Generating Periodic Tilings ...................................... 438 |
|        | 4.2.35 | SAC-1, Aldes/SAC-2, Saclib ........................... 439 |
|        | 4.2.36 | SciNapse: Software that Writes PDE Software .......... 440 |
|        | 4.2.37 | SENAC ............................................... 441 |
|        | 4.2.38 | SIMATH - Algorithms in Number Theory ............... 442 |
|        | 4.2.39 | SINGULAR – A Computer Algebra System for Polynomial Computations ............................... 445 |
|        | 4.2.40 | SymbMath ........................................... 451 |

                4.2.41  SYMMETRICA .......................................... 452
                4.2.42  Theorema: Computation and Deduction in Natural Style .. 453
                4.2.43  THEORIST—a User Interface for Symbolic Algebra ........ 454
        4.3  Packages ..................................................... 459
                4.3.1  ANU Polycyclic Quotient Programs ..................... 459
                4.3.2  AREP ................................................. 461
                4.3.3  CALI ................................................. 463
                4.3.4  CLN .................................................. 464
                4.3.5  CRACK, LIEPDE, APPLYSYM and CONLAW ............ 465
                4.3.6  DIMSYM ............................................... 468
                4.3.7  EinS .................................................. 469
                4.3.8  *FeynArts* and *FormCalc* ................................. 469
                4.3.9  FeynCalc – Tools and Tables for Elementary Particle
                         Physics .............................................. 471
                4.3.10 GRAPE ................................................ 473
                4.3.11 Recognising Matrix Groups over Finite Fields ........... 474
                4.3.12 MOLGEN .............................................. 476
                4.3.13 ORME ................................................. 477
                4.3.14 Ratappr .............................................. 477
                4.3.15 TTC: Tools of Tensor Calculus ......................... 480
5  Meetings and Publications ........................................ 485
        5.1  Conferences and Proceedings .................................. 485
        5.2  Books on Computer Algebra .................................. 490
Cited References ..................................................... 493
Subject Index ........................................................ 623
Index for Authors' Contributions ..................................... 635

# List of Contributing Authors

See the author index in the back of this book for the pages where the contributions appear.

**Akers, Robert L.** USA akers@scicomp.com
**Apel, Joachim** Leipzig D apel@mathematik.uni-leipzig.de
**Backelin, Jörgen** Stockholm S joeb@matematik.su.se
**Balfagón, Alberto** Llull ES abalf@fletxa.iqs.url.es
**Bauer, Christian** Mainz D Christian.Bauer@Uni-Mainz.DE
**Baumann, Gerd** Ulm D Gerd.Baumann@physik.uni-ulm.de
**Becker, Eberhard** Dortmund D becker@math.uni-dortmund.de
**Behnke, Kurt** Düsseldorf D kurt.behnke@edd.ericsson.se, eddkube@edd.ericsson.se
**Belabas, Karim** Orsay F Karim.Belabas@math.u-psud.fr
**Berry, John** Plymouth GB JBerry@plymouth.ac.uk
**Beth, Thomas** Karlsruhe D EISS_Office@ira.uka.de
**Betten, Anton** Bayreuth D Anton.Betten@uni-bayreuth.de
**Bonadio, Allen** San Francisco USA
**Boston, Nigel** Urbana-Champaign USA boston@math.uiuc.edu
**Breuer, Thomas** Aachen D thomas.breuer@math.rwth-aachen.de
**Bronstein, Manuel** Sophia Antipolis F Manuel.Bronstein@sophia.inria.fr
**Buchberger, Bruno** Linz A Bruno.Buchberger@risc.uni-linz.ac.at
**Buchmann, Johannes** Darmstadt D buchmann@cdc.informatik.th-darmstadt.de
**Bündgen, Reinhard** Böblingen D BUENDGEN@de.ibm.com
**Bürgisser, Peter** Paderborn D pbuerg@math.uni-paderborn.de
**Calmet, Jaques** Karlsruhe D calmet@ira.uka.de
**Cannon, John** Sydney Au john@maths.usyd.edu.au
**Capani, Antonio** Genova I cap@ideal.dima.unige.it
**Clausen, Michael** Bonn D clausen@leon.cs.uni-bonn.de
**Cohen, Arjeh M.** Eindhoven NL amc@win.tue.nl
**Cojocaru, Svetlana** Chisinau MD sveta@math.md
**Conrad, Marc** Saarbrücken D marc@math.uni-sb.de
**Corless, Robert M.** London, Ontario CA Rob.Corless@uwo.ca
**Degen, Wendelin** Stuttgart D degen@mathematik.uni-stuttgart.de
**Delgado Friedrichs, Olaf** Bielefeld D delgado@mathematik.uni-bielefeld.de
**Dentzer, Ralf** Heidelberg D ralf.dentzer@sap-ag.de

**Dewar, Michael** Oxford GB miked@nag.co.uk
**Di Crescenzo, Claire** Grenoble F Claire.Dicrescenzo@imag.fr
**Dolzmann, Andreas** Passau D dolzmann@uni-passau.de
**Dräxler, Peter** Bielefeld D draexler@mathematik.uni-bielefeld.de
**Dress, Andreas W.M.** Bielefeld D dress@mathematik.uni-bielefeld.de
**Drijvers, Paul** Utrecht NL pauld@fi.ruu.nl
**Eck, Hagen** Walldorf D Hagen.Eck@sap-ag.de
**Egner, Sebastian** Eindhoven NL egner@natlab.research.philips.com
**Fachgruppe Computeralgebra, (Steering Committee)** D
**Fleischer, Jochem** Bielefeld D FLEISCHER@physik.uni-bielefeld.de
**Ford, David** Montreal CA ford@cs.concordia.ca
**Fortenbacher, Albrecht** Berlin D a.fortenbacher@fhtw-berlin.de
**Fowler, David** Lincoln USA dfowler@unlinfo.unl.edu
**Frink, Alexander** Mainz D Alexander.Frink@Uni-Mainz.DE
**Gatermann, Karin** Berlin D gatermann@math.fu-berlin.de
**Gathen, Joachim von zur** Paderborn D gathen@uni-paderborn.de
**Gautier, Thierry** F Thierry.Gautier@inrialpes.fr
**Geck, Meinolf** Paris F geck@desargues.univ-lyon1.fr
**Geiselmann, Willi** Karlsruhe D geiselma@ira.uka.de
**Gerdt, Vladimir P.** Dubna R gerdt@jinr.ru
**Gerhard, Jürgen** Paderborn D jngerhar@mupad.de
**Giesbrecht, Mark** London, Ontario CA mwg@csd.uwo.ca
**Giusti, Marc** Palaiseau F Marc.Giusti@ariana.gage.polytechnique.fr
**Gloor, Oliver** Bern CH og@amrhein.ch
**Gonzalez-Vega, Laureano** Cantabria ES gvega@matesco.unican.es
**Gräbe, Hans-Gert** Leipzig D graebe@informatik.uni-leipzig.de
**Grabmeier, Johannes** Deggendorf D johannes.grabmeier@fh-deggendorf.de
**Graham, Ted** Plymouth GB egraham@plymouth.ac.uk
**Grassl, Markus** Karlsruhe D grassl@ira.uka.de
**Grayson, Daniel R.** Urbana-Champaign USA dan@math.uiuc.edu
**Greuel, Gert-Martin** Kaiserslautern D greuel@mathematik.uni-kl.de
**Gutierrez, Jaime** Cantabria ES gutierrez@ccucvx.unican.es
**Hahn, Thomas** Karlsruhe D hahn@particle.physik.uni-karlsruhe.de;
    hahn@feynarts.de
**Haible, Bruno** Gentilly F haible@ilog.fr
**Hantzschmann, Karl** Rostock D kh@informatik.uni-rostock.de
**Head, Alan** Melbourne AUS head@artemis.mst.csiro.au

**Heckenberger, Istvan** Leipzig D heckenbe@mathematik.uni-leipzig.de
**Hehl, Friedrich W.** Cologne D hehl@thp.uni-koeln.de
**Heinicke, Christian** Cologne D chh@thp.uni-koeln.de
**Hemmecke, Ralf** Linz A hemmecke@risc.uni-linz.ac.at
**Hereman, Willy** Boulder USA whereman@mines.edu
**Hillgarter, Erik** Linz A Erik.Hillgarter@risc.uni-linz.ac.at
**Hiss, Gerhard** Aachen D Gerhard.Hiss@Math.RWTH-Aachen.DE
**Homann, Karsten** München D karsten.homann@pn.siemens.de
**Hong, Hoon** Raleigh USA hong@math.ncsu.edu
**Huang, Weiguang** Sydney AUS w.huang@unsw.edu.au
**Hulpke, Alexander** Columbus USA ahulpke@math.ohio-state.edu
**Huson, Daniel H.** Rockville USA
**Jacobs, David P.** Clemson USA dpj@CLEMSON.EDU
**Jaén, Xavier** Catalan ES Xavier.Jaen@upc.es
**Jebelean, Tudor** Linz A Tudor.Jebelean@risc.uni-linz.ac.at
**Jerie, Michael** AUS m.jerie@latrobe.edu.au
**Jung, Francoise** Grenoble F Francoise.Jung@imag.fr
**Kaltofen, Erich** Raleigh USA kaltofen@eos.ncsu.edu
**Kant, Elaine** USA kant@scicomp.com
**Kemper, Gregor** Heidelberg D Gregor.Kemper@IWR.Uni-Heidelberg.De
**Kerber, Adalbert** Bayreuth D kerber@uni-bayreuth.de
**Klappenecker, Andreas** College Station USA andreask@math.tamu.edu
**Klaus, Uwe** Leipzig D uklaus@uklaus.deuklaus@hgb-leipzig.de
**Klioner, Sergei** St. Petersburg RU klioner@Rcs1.urz.tu-dresden.de
**Klüners, Jürgen** Heidelberg D juergen@maths.usyd.edu.au
**Koepf, Wolfram** Kassel D koepf@mathematik.uni-kassel.de
**Kohnert, Axel** Bayreuth D kohnert@uni-bayreuth.de
**Kotsireas, Ilias** Waterloo CA Ilias.Kotsireas@orcca.on.ca
**Kovaćs, Peter** Berlin D kovacs@cs.tu-berlin.de
**Kozen, Dexter** Cornell USA kozen@cs.cornell.edu
**Kreckel, Richard** Mainz D Richard.Kreckel@Uni-Mainz.DE
**Kredel, Heinz** Mannheim D kredel@rz.uni-mannheim.de
**Küblbeck, Sepp** Backnang D
**Küchlin, Wolfgang** Tübingen D kuechlin@informatik.uni-tuebingen.de
**Kurth, Frank** Bonn D kurth@cs.uni-bonn.de
**Kutzler, Bernd** Linz AT kutzler@eunet.at
**Lambe, Larry A.** Rutgers, Bangor USA llambe@caip.rutgers.edu

**Landau, Susan** Amherst USA susan.landau@east.sun.com
**Laßner, Wolfgang** Senftenberg D lassner@informatik.fh-lausitz.de
**Laue, Reinhard** Bayreuth D laue@btm2xg.mat.uni-bayreuth.de
**Laushnyk, Oksana** Lviv UKR lviv@utel.com.ua
**Lazic, Dejan** Karlsruhe D lazic@ira.uka.de
**Leedham-Green, Charles R.** London GB C.R.Leedham-Green@qmw.ac.uk
**Leeuwen, Marc A. A. van** Poirtiers F maavl@mathlabo.univ-poitiers.fr
**Lescanne, Pierre** Vandœuvre-les-Nancy F lescanne@loria.fr
**Lewis, Robert H.** New York USA rlewis@murray.fordham.edu
**Loos, Rüdiger** Tübingen D loos@informatik.uni-tuebingen.de
**Lübeck, Frank** Aachen F Frank.Luebeck@Math.RWTH-Aachen.DE
**Lux, Klaux** Tucson USA klux@math.arizona.edu
**Macsyma Inc., submitted by R. H. Berman** Arlington USA
 info@macsyma.com
**Madlener, Klaus E.** Kaiserslautern D madlener@informatik.uni-kl.de
**Matzat, B. Heinrich** Heidelberg D matzat@iwr.uni-heidelberg.de
**Melenk, Herbert** Berlin D melenk@zib.de
**Mertig, Rolf** Amsterdam NL info@mertig.com
**Müller-Quade, Jörn** Tokyio JP muellerq@hideki.iis.u-tokyo.ac.jp
**Müller, Volker** Yogyakarta ID vmueller@ukdw.ac.id
**Newman, M.F.** AUS newman@wintermute.anu.edu.au
**Nickel, Werner** Darmstadt D nickel@mathematik.tu-darmstadt.de
**Niemeyer, Alice C.** Western Australia AUS alice@maths.uwa.edu.au
**Niesi, Gianfranco** Genova I niesi@dima.unige.it
**Niklasch, Gerhard** München D gerhard.niklasch@okay.net
**Nörenberg, Rainer** Essen D noerenbg@exp-math.uni-essen.de
**Nückel, Armin** Darmstadt D nueckel@aol.com
**O'Brien, Eamonn. A.** Auckland AUS obrien@wintermute.anu.edu.au
**Obukhov, Yuri N.** Moscow RU general@elnet.msk.ru yo@thp.uni-koeln.de
**Ollivier, Francois** Palaiseau F francois.ollivier@gage.polytechnique.fr
**Opgenorth, Jürgen** Aachen D juergen@momo.math.rwth-aachen.de
**Pardo, Luis Miguel** Palaiseau, Santander ES pardo@zmat.usc.es
**Paule, Peter** Linz A peter.paule@risc.uni-linz.ac.at
**Pesch, Michael** Passau D pesch@unicorn.fmi.uni-passau.de
**Pfahler, Thomas** Darmstadt D pfahler@cdc.Informatik.TU-Darmstadt.DE
**Pfister, Gerhard** Kaiserslautern D pfister@mathematik.uni-kl.de
**Pflügel, Eckhard** GB epfluegel@mpc-data.co.uk

**Plesken, Wilhelm** Aachen D plesken@momo.math.rwth-aachen.de
**Pohst, Michael E.** Berlin D pohst@math.tu-berlin.de
**Popov, Bogdan** Lviv UKR bogdan@popov.lviv.ua
**Praeger, Cheryl E.** Western Australia GB praeger@maths.uwa.edu.au
**Prince, Geoff** AUS g.prince@latrobe.edu.au
**Püschel, Markus** Pittsburgh USA pueschel@earthlink.net
**Punjani, Minaz** London GB M.Punjani@uk.ac.ulcc
**Rabuka, Scott** Waterloo CA srabuka@maplesoft.com
**Randall, Curt** USA randall@scicomp.com
**Rees, Sarah** Newcastle GB Sarah.Rees@newcastle.ac.uk
**Reinert, Birgit** Kaiserslautern D reinert@informatik.uni-kl.de
**Robbiano, Lorenzo** Genua I robbiano@dima.unige.it
**Roch, Jean-Louis** F JeanLouis.Roch@imag.fr
**Roesner, Karl G.** Darmstadt D karo@tollmien.mechanik.tu-darmstadt.de
**Roggenbach, Markus** Bremen D roba@informatik.uni-bremen.de
**Roy, Marie-Francoise** Rennes F Marie-Francoise.Coste-Roy@univ-rennes1.fr
**Rump, Siegfried M.** Hamburg D rump@tu-harburg.de
**Saunders, B. David** Delaware USA saunders@mail.eecis.udel.edu
**Schäfer-Lorinser, Frank** Darmstadt D lorinser@tzd.telekom.de
**Schmitt, Susanne** Saarbrücken D susanne@math.uni-sb.de
**Schönemann, Hans** Kaiserslautern D hannes@mathematik.uni-kl.de
**Schreiner, Wolfgang** Linz A Wolfgang.Schreiner@risc.uni-linz.ac.at
**Schrüfer, Eberhard** St. Augustin D eberhard.schruefer@gmd.de
**Schüler, Axel** Leipzig D schueler@mathematik.uni-leipzig.de
**Schulz, Tilman** Aachen D tilman@momo.math.rwth-aachen.de
**Schupp, Sibylle** Troy USA schupp@cs.rpi.edu
**Schwarz, Fritz** St. Augustin D Fritz.Schwarz@gmd.de
**Seiler, Werner** Mannheim D seiler@euler.math.uni-mannheim.de
**Sharp, Jenny** Plymouth GB jsharp@plymouth.ac.uk
**Shokrollahi, M. Amin** Murray Hill USA amin@research.bell-labs.com
**Sims, Charles C.** Rutgers USA sims@math.rutgers.edu
**Soicher, Leonard H.** London GB L.H.Soicher@qmw.ac.uk
**Sorgatz, Andreas** Paderborn D Andreas.Sorgatz@sciface.com
**Steinberg, Stan** Albuquerque USA stanly@math.unm.edu,
    steinberg@scicomp.com
**Steinhauser, Matthias** Hamburg D msteinh@mail.desy.de
**Stillman, Michael E.** Cornell USA mike@math.cornell.edu

**Storjohann, Arne** Zürich CH storjoha@inf.ethz.ch
**Strehl, Volker** Erlangen D strehl@informatik.uni-erlangen.de
**Sturm, Thomas** Passau D sturm@uni-passau.de
**Takayama, Nobuki** Kobe JP taka@math.s.kobe-u.ac.jp
**Tertychniy, Sergey I.** Moscow RU bpt97@mail.ru
**Thomas, Wolfgang** Kiel D thomas@informatik.rwth-aachen.de
**Townend, Stewart** Plymouth GB stownend@plymouth.ac.uk
**Ufnarovski, Victor** Lund S aet018a@tninet.se
**Ulmer, Felix** Rennes F ulmer@univ-rennes1.fr
**Unger, William** Sydney Au billu@maths.usyd.edu.au
**Veigneau, Sebastien** Noisy-le-Grand F ACE@univ-mlv.fr
**Vermaseren, Jos** Amsterdam NL t68@nikhef.nl
**Villard, Gilles** Lyon F Gilles.Villard@ens-lyon.fr
**Waits, Bert K.** Ohio USA waitsb@math.ohio-state.edu
**Wassermann, Alfred** Bayreuth D Alfred.Wassermann@uni-bayreuth.de
**Watkins, Anthony** Plymouth GB ajwatkins@plymouth.ac.uk
**Watt, Stephen M.** London, Ontrario CA watt@csd.uwo.ca
**Weiglein, Georg** Genf CH Georg.Weiglein@cern.ch
**Weispfenning, Volker** Passau D weispfen@uni-passau.de
**Wildanger, Klaus** Berlin D
**Windsteiger, Wolfgang** Linz A Wolfgang.Windsteiger@risc.uni-linz.ac.at
**Winkler, Franz** Linz A Franz.Winkler@risc.uni-linz.ac.at
**Wolf, Thomas W.** London GB T.Wolf@qmw.ac.uk
**Young, Robert L.** USA ryoung@scicomp.com
**Zayer, Jörg** Darmstadt D joerg.zayer@cosmosDirekt.dej.zayer@ids-scheer.de
**Zimmer, Horst G.** Saarbrücken D zimmer@arabella.math.uni-sb.de

# 1 Computer Algebra – Historical Development, Characterization, and Prospects

## 1.1 Historical Remarks

Historically, the terms *algebra* and *algorithm* originate from the same source: the book *Kitab al muhtasar fi hisab al-gabr w'al-muqabalah* by the Persian scientist Abu Ja'far Mohammed ibn Musa *al-Khorezmi* (cf. [Zemanek 1981] or [Ifrah 1991]) – compiled at the academy of science in Bagdad in the ninth century – uses *al-gabr* and *muqabalah* to describe symbolic transformations and term reductions respectively, which are performed to solve algebraic equations. *Algorithmic* manipulation of symbolic algebraic expressions remained to be the major task of algebra about until the end of the nineteenth century. At the beginning of the twentieth century, this method was amended and shadowed by developments in *abstract algebra*. There, the main interest is focussed on formal investigations of algebraic structures derived from axioms. This structural algebraic method has taken over considerable parts of mathematics nowadays.

During the past decades, algorithmic aspects of algebra became prevailing again. In particular, the accelerated development of computers and digital data processing made it feasible to automate manipulation of formulas and symbolic computations. This renewed interest into methods of algebra and discrete mathematics generated novel applications within mathematics itself as well as in natural sciences and in engineering, where before numerical methods almost exclusively had been used.

Within this setting, *computer algebra* evolved as a discipline, linking algorithmic and abstract algebra in its most general sense to methods of computer science, and providing a new methodical tool in the border area between applied mathematics and computer science. Computer algebra has its theoretical roots in algorithmic oriented mathematics of the nineteenth and the beginning twentieth century, as well as in algorithmic methods of logic, which were developed during the first half of the twentieth century. The final initiative was sparked by the need of physicists and mathematicians for extensive symbolic computations which could not be conducted by hand anymore.

Beginning in the forties, initially small improvised systems for the solution of particular tasks were implemented. Further insight into the various interrelations and dependencies among individual methods lead to the development of systems on a larger scale with a wider range of applications. As a consequence, utilization of computer algebra was, and still is on the rise.

Steering committee of the German special interest group (Fachgruppe) of computer algebra

## 1.2 General Characterization

What exactly does computer algebra mean? In the following, we try to characterize the notion of computer algebra in form of theses:

Computer Algebra is a subject of science devoted to methods for solving mathematically formulated problems by symbolic algorithms, and to implementation of these algorithms in software and hardware. It is based on the exact finite representation of finite or infinite mathematical objects and structures, and allows for symbolic and abstract manipulation by a computer. Structural mathematical knowledge is used during the design as well as for verification and complexity analysis of the respective algorithms. Therefore computer algebra can be effectively employed for answering questions from various areas of computer science and mathematics, as well as natural sciences and engineering, provided they can be expressed in a mathematical model.

Algorithms of computer algebra typically surpass basic number arithmetic. They extend to computations involving specifically represented algebraic objects like indeterminates, elementary functions or permutations, as well as logical variables being able to serve as parameter. Only that way computation in many algebraic structures becomes feasible, e.g., in groups, number fields, Lie algebras, or rings of differential operators. Symbolic parameters allow the uniform treatment of whole classes of problems in a generic way, which can reduce the cost of the solution process. In addition, instabilities caused by inaccurate input data – which are a well known problem in classic numerical computing – can be detected and contained.

Elaborate methods of computer science and profound results of abstract mathematics find their way into the design, verification, and complexity analysis of computer algebra algorithms. Therefore, computer algebra continues traditional algorithmic computing, and incorporates elements of more recent structural mathematics at the same time, transforming them by means of computer science into a powerful tool for modern research and technology.

Although computer algebra prevailed many areas of mathematics and computer science, it is by no means a substitute for creativity and mathematical knowledge. Just as little should computer algebra systems considered to be universal mathematical problem solvers. However, it can assist in tapping the vast mathematical resources, and is therefore an essential tool of experimental mathematics as well as an integral part of scientific computing. In its approach, computer algebra differs from numerical computing, though it can improve accuracy and efficiency of numerical methods. Methods and results of computer algebra left their mark on a number of computer algebra systems, see 4; however, they are not limited to the design and generation of such systems. However, not every software development on a symbolic base in general, such as text processing, compiler design, and computer linguistics, can be added to the list of computer algebra applications.

## 1.3 Impact on Education

Just the mere existence of computer algebra systems as a computational tool has a profound impact on teaching. Among other things, standard problems

## 1.3.0 Impact on Education

in high school mathematics can be solved by the push of a button nowadays. More and more students take advantage of specially priced computer algebra packages, or have access to some computer algebra system on a high school or campus network. The pace of change is accelerated by the fact that in the past few years hand-held computer algebra tools were developed by Casio and Texas Instruments (TI-89, TI-92, CASIO CFX-9970G and CASIO Algebra FX2.0) which are increasingly used in high school education. A few of these worldwide activities are described in Section 3.6.

Now, there is the opportunity to rid the curriculum of technical ballast for theoretically lesser inclined users of mathematics (engineers, computer scientists, economists, medics, biologists, etc.). This enables to teach more application oriented material, which is usually more extensive but also more interesting, and there is more time to explore more general and conceptual aspects of mathematics. Both can considerably improve comprehension of mathematical methods, their potential and their limitations, by non-mathematicians. On the other hand, there is the danger that computation by hand and related skills are not acquired to the same extent as it used to be before the introduction of the computer to the classroom. This could go so far that future generations of mathematicians and users of mathematics could become incapable of solving even simple problems without the help of a computer. For example, in the era of numerical calculators, many users of mathematics might not know how to compute decimal approximations of square roots, although they can easily find these numbers using a calculator. That particular tendency should be counteracted whenever possible.

A typical computer algebra curriculum is usually divided into two parts. Courses on the first level impart the necessary skills for proper utilization of computer algebra systems. For general purpose systems (described in Section 4.1), there exists a number of textbooks (cf. [Jenks and Sutor 1992] for AXIOM, [Char et al. 1991a] for MAPLE, [Wolfram 1996] for MATHEMATICA, etc.), which can be consulted for preparation of introductory courses. Courses are offered on a regular basis to interested parties of the scientific community as well as of the industrial sector, partly by universities, partly by user groups, and other organizations. Furthermore, in the meantime many textbooks exist for mathematics courses, which teach basic mathematical knowledge and the use of computer algebra systems at the same time. For special purpose systems, generally no easily accessible introductory texts—other than technical manuals—are available. However for some of those systems (e.g., CAYLEY, GAP, KANT V4, SIMATH, Singular), sometimes the developers themselves offer workshops or weekend courses to get to know the system and for practice. Workshops that deal with the use of some of the general purpose systems or hand-held devices are also scheduled on a regular basis.

The second level addresses anyone who is required to have an in-depth knowledge of the structure of computer algebra systems, especially of their underlying algorithms, in order to be able to use them in critical applications. For this area one has to refer to special literature which is still insufficiently covered by textbooks. Another source of information are special lectures and supporting computer algebra laboratories where algorithms are mathematically derived and analyzed. Many universities already offer such courses on a regular basis, however they should eventually become part of the standard curriculum. Exemplary topics could include "Fast Arithmetic" [Knuth 1998; Geddes et al. 1992], "Irreducibility and Factorization" [Koblitz 1987; Cohen 1996b], "Gröbner Bases with Applications" [Geddes et al. 1992; Cox et al. 1992; Becker et al. 1993], "Algorithms of Algebraic Number Theory" [Pohst and Zassenhaus 1989; Cohen 1996b], "Algorithmic Methods in Group Theory" [Johnson 1990; Butler 1991; Sims 1994], "Symbolic Summation" [Graham et al. 1994; Petkovšek et al. 1996; Koepf 1998a], "Symbolic Treatment of Differential Equations" [Fakler 1997a], and "Symbolic Integration" [Geddes et al. 1992; Bronstein 1997]. A selection of other textbooks, like [Davenport et al. 1989] and [Mignotte 1992], is compiled in Section 5.2.

Wolfram Koepf (Kassel)

## 1.4 Impact on Research

Computer algebra systems have significantly influenced scientific research in many fields, among them mathematics, computer science, physics, chemistry, engineering. They have supplemented the well-established numerical packages by new tools aimed at problems that require exact answers, or a closed-form analysis of the dependence of a problem class on certain parameters. They offer a convenient environment for high-level programming of specialized algorithms involving complicated data-types. Researchers can profit from a wide variety of specialized computer algebra systems and special purpose add-on packages to popular general-purpose systems. Thus the use of computer algebra systems has become a standard tool for experiments in advanced research that help to find, support, or refute conjectures. They have become absolutely indispensible in the design and update of tables of mathematical, physical or other scientific objects.

The role of computer algebra systems in the sciences and in engineering can be quite diverse:

First, it can support straight-forward activities, like carrying out a manipulation of mathematical expressions on one of our systems, for testing conjectures, deriving solution properties, or simply for recording the steps in a scratchpad-like fashion on a computer possibly avoiding hand-calculation mistakes. We anticipate that the trend will spread where one publishes, along with one's written paper, the worksheets or notebooks which store the computations peformed on the computer algebra systems.

Second, the application can be intricate, like designing and building a computer algebra system for a specialized task in one's research area. The SCHOONSCHIP system (see section 3.1.1.1) for performing high energy physics calculation is a successful example. The Nobel Prize in physics was awarded in 1999 to its author, and as the citation of the Nobel Foundation reads [www.nobel.se/

announcement-99/physics99.html] "At the end of the 1960s ... [Martinus J. G.] Veltman had developed the Schoonschip computer program which, using symbols, performed algebraic simplifications of the complicated expressions that all quantum field theories result in when quantitative calculations are performed. ... With the help of Veltman's computer program [Gerardus] 't Hooft's partial results were now verified and together they worked out a calculation method in detail."

Third, it can be "in-between," taking a scientific computational task at hand and studying how current algorithms and their implementations can be customized and improved to solve the problem. A scenario for such activity is the application of the Gröbner basis algorithms to solve the algebraic constraints on a wavelet-based filter for wireless communication; or a Smith normal form computation on a large sparse integer matrix in order to investigate the homology groups of large simplical complexes in topology. The mode of research is best conducted by the scientists or engineers teaming up with experts in the symbolic computation discipline, and the resulting software then can become available to the larger community. Those teams may embrace both the symbolic and the numeric methodology, as a satisfactory solution of the problems at hand may be obtainable only by combination (hybrid) of both (see section 2.12.3).

In the following we review some specific major contributions of computer algebra to research in mathematics, computer science, physics and engineering. For applications in other fields of research compare the section 3.5.

**Mathematics.** Computer algebra supported experiments have been used to support or refute well-known conjectures and to gain additional insight in these problems, such as the conjecture of Birch and Swinnerton-Dyer, or the Zassenhaus conjecture on group rings. They have been applied in treating a large number of residual special cases in proofs of general theorems (four-colour problem, diophantine equations). In a similar spirit they are used to do large systematic calculations and classifications, such as the computation of generic character tables.

They have contributed significantly to establish and update large mathematical databases such as integral tables, tables of special functions, and the atlas of finite groups.

Finally they have contributed to a new upswing in algorithmic methods not only as tools, but as new objects worthy of mathematical study. This concerns both the design, verification, and complexity analysis of computer algebra algorithms, as well as non-algorithmic structural mathematics. In fact an algorithmic approach to a classical problem may lead to a significant refinement of classical structure theory irrespective of algorithmic considerations. Typical examples are the theory of Gröbner bases, standard bases and involutive bases in commutative and non-commutative algebras and in differential algebra (see the sections 2.5, 2.11).

**Computer Science.** The goals of computer algebra constituted a major challenge for the development and refinement of a wide range of computer science tools and methods, among them list processing, abstract data types, parametric polymorphism, constraint logic programming, parallel computing at various levels of granularity, dynamical memory management, user interfaces and interfaces between systems, mathematical knowledge representation and management (see the sections 2.18, 2.16, 2.17, 2.19, 3.3). Thus the inherent needs of computer algebra provide an ideal test ground for the practical relevance, power and efficiency of these methods. Despite the significant progress made in the past, most of these problems are still far from a satisfactory solution. This concerns in particular user interfaces and interfaces between systems, mathematical knowledge representation and management. These problems are closely tied to the problem of reusability of efficient special purpose software in a larger environment.

**Physics.** Physics is one of the oldest application field of computer algebra; in fact many of the early computer algebra systems were designed or strongly influences by physicists in order to suit their research needs, in particular the need to handle highly complicated mathematical relations with parameters in an exact way. Among these early systems are FORMAC, SCHOONSCHIP, CAMAL, MACSYMA, REDUCE, of which the latter two evolved to a general purpose systems. In addition many current special purpose systems and packages are designed by physicists (see the sections 4.2 and 4.3).

Nowadays computer algebra is used in almost all areas of physics as a standard tool. The section 3.1 describes some typical applications of computer algebra methods and software in various research areas of physics.

**Engineering.** While the use of computer algebra in engineering is not as universally established as in physics, it is meanwhile recognized as a valuable tool supplementing and extending numerical software. A prominent example is the inverse kinematic problem and the path planning problem in robotics (section 3.4.4). These are typical instances of
parametric problems, where the symbolic-algebraic approach may offer a closed form solution - and thus deeper insights - in comparison to purely numeric solutions of specific problem instances. Similar advantages may occur in Computer Aided Geometric Design, where computer algebra methods greatly increase the quality and flexibility of solutions 3.4.5. Other engineering areas, where computer algebra methods have a major impact include mechanics, flow dynamics, thermodynamics and combustion, as well as audio signal processing (sections 3.4.1, 3.4.2, 3.4.3).

Erich Kaltofen (Raleigh, USA)
Volker Weispfenning (Passau, Germany)

## 1.5 Computer Algebra – Today and Tomorrow

### 1.5.1 Today

Similarly as numerical methods and packages have grown into standard tools in all areas of natural sciences and engineering, as well as in economy, methods and systems of computer algebra have come in widespread use and acceptance. In particular physicists – once the first who required and developed systems for symbolic computations – now use computer algebra systems in a selfevident way as a standard tool. In particular the systems MAPLE and MATHEMATICA have a very large user base and are used in almost every university throughout the world with an increasing tendency of being used also in industry and business companies.

The majority of computer algebra systems contains a comprehensive set of well-established algorithms, and will give access to a wealth of acquired knowledge organized in databases. Improved, or even new, algorithms will be distributed relatively fast through updates of computer algebra systems, making them accessible to a large number of users. Through these algorithms a considerable part of constructive mathematics is available as an intelligent tool together with simple directions for use (*black box*), therefore allowing for the solution of more and more complex problems.

User interfaces have become much more convenient for the user. Now they also allow the non-experts to simply click to develop complicated formulae, to access mathematical knowledge through simple queries, and to solve standard mathematical problems. A recent direction is the integration of user interface and browser technology.

### 1.5.2 Outlook

Prospects for development of particular subjects, systems and applications of computer algebra will be discussed in the next three chapters both from a theoretical and a technical point of view and under the sign of applications. Here, we would like to give an outline of more general anticipated tendencies.

One of the most important impact of computer algebra and its systems is seen in their increased usage in education both in schools and at universities. A lot of effort is taken to develop courses in many different subjects based on computer algebra systems. Hence, the subsection 1.3 is devoted to this highly relevant topic for the future of computer algebra.

The impact of computer algebra to research is discussed in subsection 1.4. For top-quality research in scientific computing, it was always necessary to develop special purpose algorithms for many problems, and to carry out problem dependent implementations due to the limitations of existing systems. This brings up the desire to increase the applicability of computer algebra systems for such tasks, and to make them more attractive for particularly complex applications from other areas. To that end, a series of research activities in computer algebra will have to be promoted, or created in the first place. For a great number of further tasks new or faster algorithms have to be found, more proficient implementations, special purpose systems, or even hardware solutions have to be developed. Whereby, special attention has to be paid to optimal utilization of existing hardware and computing environment. I.e., besides the plain problem of running time and memory space, other questions like being able to split up, to parallelize, and the requirements for distributed memory will become more important.

For the future development of computer algebra systems the following tendencies become manifest: on one hand, we expect the growth of general purpose systems to continue, their efficiency and ease of use to increase at the same time. On the other hand, it can be forseen that despite progress in the hardware sector, these systems will generally not be able to provide the required power for complex computing tasks. For that purpose, special software packages, or specialized computer algebra systems will have to be designed, and used. In order to keep the necessary work at a minimum, it becomes imperative to think about means of standardization and the opening up of computer algebra systems to facilitate the integration of customized program parts, and to allow data transfer between systems. The OpenMath initiative – for a detailed description see section 2.19.6 – is devoted to this idea and attempts to create a standard for communication between computer algebra system. Furthermore, it is desirable to have at least basic algorithms in standardized form, maybe even implemented on the processor level, available as a platform for developers of special systems and software packages. Examples in this direction are packages for long integer arithmetic. At the same time, exchange of program parts originating from systems based on the same platform should be made easier.

Furthermore, computer algebra systems will increasingly have access to numerical packages and integrate them into the system. Conversely, computer algebra functions should be callable from other programs, the same way one can link FORTRAN routines from libraries nowadays. Most of symbolic libraries are still devoted to a specific computer algebra system. However, we see the development of system independent libraries of data, data types and programs as well as stand-alone programs using symbolic code.

It is also obvious that computer algebra will be not only accessible via systems and algorithms. The increasing need of cryptographic techniques will require more and more basic methods of computer algebra to be used directly or indirectly by a large amount of people communicating via the Internet and in particular for electronic commerce. Another example of implicit usage of computer algebra techniques are error correcting codes used, e.g., in all storage devices.

The rapidly exploding Internet success will also influence computer algebra. We expect that computer algebra systems are employing the network computing facilities in an Internet environment and using languages like JAVA. There are already examples of computer algebra servers in the Internet devoted to certain topics of computer algebra, where each user can depose specific problems to be solved there. Computer algebra plugins will allow standard browser to be an interface for solving symbolic problems.

Most of the discussed facts make mathematical knowledge directly or indirectly accessible in an easy and natural way by means of computer algebra to a large number of people. To support this process of enlarging the user base of mathematical knowledge by means of computer algebra, mathematicians have to take over a growing number of more general consulting positions, and get involved in mathematical modeling of complicated problems. There will be special consulting for computer algebra, ranging from advice on selecting a system, from introductory courses to development of application packages and using computer algebra related internet facilities.

Another requirement is the comparison of systems and algorithms under various points of view. The variety of topics and themes of computer algebra will not allow to have standard benchmarks as in numerics and number crunching. Nevertheless, to develop benchmarks for different subthemes, for different systems for the solution of certain problem classes, to compare the effectivity of implementation of algorithms implemented differently on different platforms, test suites and parametrized problem settings, etc. are important tasks to be solved in the future.

Furthermore, the problem of registration and the refereeing of computer algebra systems will become more pressing. Specifically, non-experts should be given the means to check whether wrong data was published due to an internal error of some computer algebra system. To that end, users would have to keep a record of versions of the computer algebra systems they use, and to cite them in publications accordingly. Updates and known errors should be communicated to registered users without delay, and publicly announced in order to allow screening of results obtained by the corresponding version of the system. Officially released versions of computer algebra systems must be kept available for an extended period of time to allow for verification later on.

Continued development and maintenance of systems adhering to the above mentioned standards will take up resources to an extent which will make financing viable only through cost sharing by users, e.g., through service or license fees. Experimental and specialized packages and systems, on the other hand, will be essentially free of charge. However for such systems, one cannot expect to be entitled for further support services.

Johannes Grabmeier and B. Heinrich Matzat (Heidelberg)

# 2 Topics of Computer Algebra

In the following we are going to discuss the main points of emphasis and the central topics of computer algebra. A quick glance at the section headers already provides an impression of the wide spectrum of areas and trends involved in computer algebra. This variety makes it all the more difficult to structure the field in a comprehensive and universally accepted manner. We have chosen to order the topics roughly as follows: the first part (section 2.1 until the beginning of section 2.4) deals with *basic algorithms and methods of computer algebra*, followed by *computer algebra in special areas of mathematics* (sections 2.4 through 2.15). The sections at the end of the chapter – covering *Computer Science aspects of computer algebra* (2.13 to 2.20) – partially overlap with each other. In general the transitions are fluid. Therefore, the sections are intentionally numbered in consecutive order not implying any ranking with regard to importance.

The editors

## 2.1 Exact Arithmetic

### 2.1.1 Long Integer Arithmetic

About two thirds or 480 pages of Knuth' [Knuth 1998] volume on *Seminumerical algorithms* are devoted to Chapter 4, entitled *Arithmetic*. He states *Research on seminumerical algorithms continues at a phenomenal rate*. They are called *seminumerical because they lie on the borderline between numeric and symbolic calculation. Each algorithm not only computes the answer to a numerical problem, it also is intended to blend well with the internal operations of a digital computer.*

This is in particular true for integer arithmetic, one of the fundamental building blocks of computer algebra systems.

One kind of integer arithmetic is based on finite precision arithmetic – which still can lead to several thousand bytes per integer – while most implementations today allow for arbitrary precision, limited only by the size of the machine memory. Even for simple computer algebra algorithms, estimates for the size of resulting integers are difficult to obtain. As a consequence, long integers have to be implemented as dynamic data structures. Representations include polynomials in some fixed radix (cf. 2.1.2, for example the largest unsigned integer + 1 of the machine arithmetic, or sometimes a residue number system where the moduli are bound by the word size (cf. 2.1.4). Dynamic memory management is obviously a crucial part of any computer algebra system, in contrast to systems based on numerical methods.

Many other algebraic data types are based on integers. Implementations of exact rational arithmetic rely on efficient GCD algorithms (cf. 2.1.3) 2.1.3) for integers of arbitrary size. The complexity of the basic rational operations is dominated by the complexity of the integer GCD algorithm. A practically improved GCD algorithm is described by [Weber 1995]. Gaussian integers or Gaussian rationals are another example for applications of integer arithmetic in related algebraic domains. Even GCD algorithms on univariate polynomials over the integers can be mapped to long integer GCD computation.

Real numbers can be represented as intervals given by rational upper and lower bounds. Due to the fact that intervals allow only ring operations, a *binary rational* arithmetic, where all denominators are a power of two, suffices. Closely related to binary rational numbers are floating-point numbers in base two with mantisses and exponents of arbitrary length. For example, this representation was chosen for *bigfloats* in several systems mentioned below. In arithmetic there is no strong division line between computer algebra and numerics; the famous book *Numerical Recipes in C* [Flannery et al. 1992] describes in section 20.6 *Arithmetic at Arbitrary Precision*.

Some computer algebra systems and most stand alone packages for arithmetic provide non-classical algorithms for integer arithmetic; for example the Karatsuba algorithm for multiplication, or the Schönhage-Strassen algorithm and Winograd's algorithm can be found in some implementations. Fast multiplication algorithms lead to fast other arithmetical algorithms in particular division. Exact division was improved by Jebelean [Jebelean 1993], Schönhage and Vetter [Schönhage and Vetter 1994].

In general, exact arithmetic imposes performance penalties compared to inexact methods; therefore analysis of old and new algorithms is an important major research topic in this area. It is still unkown to date how far the complexity of the Schönhage-Strassen algorithm differs from the optimum considering the lack of non-trivial lower bounds for multiplication (see 2.11).

Efficient arithmetic packages are freely available:

- apfloat [Tommila 1998], a high precision package for very long integers,
- BigNum by [Serpette et al. 1990] and others,
- Haible's CLN [Haible 1998] class library for long numbers,
- [Vetter et al. 1994] Schönhage's Fast Algorithms,
- Leyland's freelip [Leyland 1997] extends lip,
- the GNU multiple precision package [Granlund 1996] by T. Granlund,
- Arndt's Hfloat [Arndt 1990],
- Dentzer's libI [Dentzer 1993],
- A. K. Lenstra's lip package [Lenstra 1995],
- Shamus' MIRACL [Shamus Software 1998] (free for educational purposes),
- Baile's MPFUN[Bailey 1990],
- Shoup's NTL [Shoup 1998],
- Cohen's PARI [Batut et al. 2000],
- Wedeniwski's Piologie [Wedeniwski 1996],

to name a few. These packages *are intended to blend well with the internal operations of a digital computer*, for example, the gmp provides for many machines assembler routines for the critical inner loops, as Schoenhage does for his *Fast Algorithms*[Vetter et al. 1994]. In afloat extensive care is spent to pipe and cache based optimizations for processors prevalent in workstations and PCs today. On such machines a branch may be much more expensive than a multiplication or division. This effectively invalidates the traditional MIX scale for concrete algorithm analysis.

Fast arithmetic is, next to much improved algorithms, a necessary prerequisite for many common research projects, for example to factor large integers with the elliptic curve method [Lenstra Jr. 1987] or to compute $\pi$ to millions of digits in less than an hour.

In [Riordan 1994] one can find an (older) list of packages, starting with one of the first ones, Brent's [Brent 1981] Fortran based package. Today, most packages are written in C, newer ones in C++ like [Tommila 1998], [Haible 1998], [Wedeniwski 1996], one to our knowledge in Java [Systemics Ltd. 1996]. Fast techniques in C++ are exploited in Blitz++ [Veldhuizen 1998], not yet applied to multiprecision arithmetic. Another new method allows for the integration of modules from different sources, even different languages. A promising technique, found in the process of C++ standardization, is the traits technique [Myers 1996]. The LiDIA [LiDIA-Group 1997] number theory package, written in C++, can use different arithmetics like [Dentzer 1993], [Leyland 1997], [Granlund 1996], [Lenstra 1995], in its kernel, written in C, by an abstract interface realized by C-macros. Different arithmetical packages are also used in the research paper by Erlingsson, Kaltofen and Musser [Erlingsson et al. 1996] which exploits C++-genericity at compile-, link- and runtime; interesting enough, it contains a *null* arithmetic which does not do any arithmetic but counts the number of arithemtical operations. [Zimmermann 1998] gives a comparison of three different arithmetical packages. He keeps also a record of challenges and achievements on problems where fast multiprecision arithmetic is crucial.

Surprisingly, in the current standard of C, and hence the new standard of C++, the integer remainder (for negative integers) is not standardized. Other gaps are wrapping on integer overflow and notifications. Boute [Boute 1992] discusses problems of *div* and *mod* definitions in programming languages. Recently, the problem was satisfactorily addressed in the standarization of Language Independent Arithmetic [International Standard 1994], [International Standard 1995] and the new C-Standard draft [International Standard 1990] follows these standards and closes the gaps – in one of the possible ways.

Additional reference: [Vuillemin 1990].

Rüdiger Loos (Tübingen)

## 2.1.2 Arithmetic with Polynomials, Rational Functions and Power Series

Traditionally, computing with univariate polynomials over a commutative ring with unity is similar to long integer arithmetic. Here polynomials are represented by their coefficient list whereas integers are given by their binary (bit) expansion. In practical applications, polynomials are often sparsely populated, i.e. many of the coefficients are zero. In this case, the internal representation consists of lists of pairs (index, non-zero coefficient) to encode only non-zero terms. Both approaches leads to algorithms conceptually based on a linear algebra/vector space framework, and addition, subtraction, and multiplication are readily implemented.

Remark however that usual convolution for polynomial multiplication is less efficient than *Fast Fourier Transform* (FFT). Convolution is the bilinear form associated to the syntactical enconding of polynomials as a coefficient list, whereas FFT recalls that a polynomial is a function and, hence, it takes advantage of the semantical features of the polynomials to multiply (FFT evaluates polynomials at sufficiently many points and then interpolates to get the coefficients of the product polynomial). Besides power series expansions, other data structures are required for efficient computation with polynomial functions,

A basic alternative example is that of *straight-line programs* (SLPs). They were introduced by A. M. Ostrowsky in the mid-fifties (cf. [Ostrowski 1954]) to model semi-numerical computations and polynomial evaluation with floating point arithmetics. A straight-line program for polynomial evaluation is a directed acyclic graph whose nodes are labelled gates which perform arithmetic operations involving precomputed results (see section 2.13 below for more detailed discussions). As straight-line programs evaluate polynomials, they can also be used to represent polynomials [von zur Gathen 1985; Kaltofen 1988]. In fact, straight-line programs are closer to the functional nature of polynomials and hence they seem to be closer to the semantical features of concrete polynomials. For instance, the determinant of a generic $n \times n$ matrix is a sparse polynomial with $n!$ non-zero terms, whereas a straight-line program encoding of such determinantal polynomial requires no more than $O(n^4)$ memory units. For a related better result see [Kaltofen 1992a].

Of course, addition, subtraction, and multiplication of polynomials and rational functions given by straight-line program encoding become an easy task. The same holds for FFT or even for probabilistic zero tests versus coefficient comparisons (cf. [Zippel 1979], [Schwartz 1980], [Heintz and Schnorr 1980], [Giusti and Heintz 1993], or [Krick and Pardo 1996]).

Besides basic arithmetics, some higher level operations are of interest: euclidean division, GCD, and factoring (provided that the ground ring allows unique factorization). For more detailed comments about the classical setting see Subsections 2.2.1 and 2.2.2 below. Efficient differentiation of polynomials given by straight-line program encoding follows from the methods described in [Baur and Strassen 1983]. There is, however, a significant open problem in straight-line program manipulation. Up to now no procedure is known to perform Euclidean division of two polynomials given by SLPs and such that its running time is polynomial-time in the size of the program. Note that the degrees can be exponential in the length, and this problem is probably NP-hard (see [Plaisted 1978]). As a matter of fact, resultants are easy to compute when one of the polynomials has some good intrinsic properties (very few complex zeros, for instance). In this case a resultant is not a remainder any more, but a minimal polynomial vanishing on all common zeros. Moreover resultants can then be computed in very efficient running time for straight-line program encoding of the inputs.

## 2.1.2 Arithmetic with Polynomials, Rational Functions and Power Series

An outstanding advance in univariate polynomial computations was the Lenstra-Lenstra-Lovasz (LLL) basis reduction procedure (see [Lenstra et al. 1982], and Subsection 2.2.2 below). This method, coming from the early eighties, explores the connections between approximate computations and exact computations. Namely, from an approximate factor (with respect to some non-archimedean norm) the LLL procedure outputs an exact factor of a given polynomial. This yields polynomial time factoring procedures and reveals a hidden connection between numerical analysis and symbolic computing, which has not been explored in depth (see [Castro et al. 2001] for instance). The LLL algorithm was also used by E. Kaltofen as a soubroutine to design efficient factoring algorithms dealing with straight-line program encoding [Kaltofen 1989, 1992b]. Even more, the LLL basis reduction procedure allows a numerical analysis (also diophantine approximation) approach to factoring. Given a univariate polynomial with rational coefficients in straight-line program encoding the task of computing only irreducible factors with rational coefficients of bounded precision (i.e. where the bit length of the coefficients of the irreducible factor is bounded by some given bound) can be performed in time which depends only on the degree, the size of the data structure and the given precision (cf. [Kannan et al. 1984], [Castro et al. 2001], or [Lenstra 1984]).

The search of lower complexity bounds for polynomial evaluation procedures, one of the main goals of Algebraic Complexity Theory, has not been dropped off (cf. [Baur 1997], [Aldaz et al. 2000], and Subsection 2.13 below). Recent advances have shown, for instance, hidden connections between polynomials that are hard-to-compute and transcendental function theory (cf. [Aldaz et al. 1998a], [Aldaz et al. 1998b], [Aldaz et al. 2001]). Moreover, some new combinatorial techniques to get examples of polynomials that require big straight-line programs have been found and, by the way, new examples of transcendental holomorphic functions are available.

As for multivariate polynomials given in either dense or sparse encoding, two main data structures are currently used: either *recursive presentation* (as univariate polynomial whose coefficients are polynomials in the remaining variables) or in *distributive representation* (as sum of monomials, each consisting of a coefficient and a product of powers of the variables). Recursive algorithms (such as factoring and resultant computations) favor the first representation, whereas the distributive representation is used whenever single terms have to be manipulated (e.g., flexible choice of the term ordering in Gröbner bases computations, cf. Subsections 2.2.5 and 2.2.6 below).

Rational functions are represented using techniques analogous to those used to compute with rational numbers (cf. 2.1.1 and 2.1.3).

Due to their infinitary nature, formal power series , Laurent series and Puiseux series cannot be finitely represented in general. Exceptions are series whose coefficients are given by an explicit formula or by some recursion (as, for instance, algebraic power series given by a Newton-Hensel Lifting procedure and a non-archimedean approximate zero). In that case, series can be stored in a computer algebra system as finite list of (already computed) coefficients, together with a function for computing more coefficients on demand. This so-called *lazy evaluation* technique is, e.g., implemented in AXIOM (*streams*, cf. [Burge and Watt 1987]). An alternative method uses truncation to some fixed (previously chosen) order, and consequently polynomial arithmetic. Inversion, and therefore division, of power series (provided their constant term is invertible), Laurent series, or Puiseux series respectively, can be based on the same data structures, for example by computing the coefficients recursively.

M. Giusti (Palaiseau), L.M. Pardo (Palaiseau, Santander)

### 2.1.3 Euclid's Algorithm and Continued Fractions

Besides basic operations, computation of the greatest common divisor (GCD) is by far the most important task for any integer arithmetic. The *Euclidean Algorithm* computes GCDs by iterated division with remainder. Optionally, one can obtain a representation of the GCD as a linear combination $d = \text{GCD}(a, b) = u \cdot a + v \cdot b$ ("Extended Euclidean Algorithm"), as well as the continued fraction expansion for $a/b$. One can view this as constructing explicitly the free rank one subgroup $d\mathbb{Z}$ of $\mathbb{Z}$ from the pair of generators $a, b \in \mathbb{Z}$ (and thus as the simplest case of a Hermite Normal Form algorithm; cf. [Cohen 1996b, 2.4–2.7]).

*Continued Fraction Expansions* exist for every real number; they generate sequences of rational approximations which are optimal in some sense (cf. [Perron 1954], or any standard text on Number Theory). Applications are the extraction of the numerator and denominator of a rational number from a given decimal approximation, factorization of integers [Morrison and Brillhart 1975], and the computation of units in real quadratic number fields — which in turn is equivalent to the solution of certain diophantine equations.

More generally, continued fraction methods can successfully be applied to classes of diophantine equations whose solutions are related to rational approximations of algebraic irrationalities, or of quotients of the logarithms of two algebraic numbers. The text by [de Weger 1989] contains an introduction to this subject, also covering multidimensional continued fraction algorithms and, as an alternative, lattice basis reduction (using LLL). In dimension 2, both of these reduce to the basic Extended Euclidean Algorithm.

The efficiency of many computer algebra algorithms is greatly influenced by the running time of the Euclidean algorithm. Therefore any acceleration of this algorithm is of particular interest. An idea by Lehmer allows the almost exclusive use of divisions involving only "small" numbers (in the range of the machine word length), which leads to significant improvements in practical computations.

A further speed-up can be achieved by a binary variant of the Euclidean algorithm which uses comparison, addition, subtraction, and shift (equivalent to division or multiplication by a power of 2) only. This method is especially suitable for hardware implementations in extremely long registers (in the range of several thousand bits). In this particular model, multiplication, and computation of the extended GCD all belong to the same complexity class (cf. [Wagstaff and Smith 1987]).

A comprehensive presentation with a complexity analysis of most variants of the Euclidean algorithm can be found in [Knuth 1981b, Ch. 4.5].

Additional reference: [Vuillemin 1990].

Gerhard Niklasch (München) and Ralf Dentzer (Heidelberg)

## 2.1.4 Modular Arithmetic and the Chinese Remainder Theorem

Modular arithmetic (arithmetic of residue classes modulo a natural number $m$) is performed on representatives of the congruence classes. For the complexity analysis, one chooses a special system of representatives like $\{0,\ldots,m-1\}$. Modular multiplication then requires one multiplication and one division with remainder (of an integer by $m$); for odd $m$, the division can be avoided. Residue classes which are relatively prime to $m$ can be inverted in the ring $\mathbb{Z}/m\mathbb{Z}$; the Euclidean algorithm will compute $1 = \text{GCD}(a,m) = u \cdot a + v \cdot m$, where $u$ is the inverse of $a$ modulo $m$.

If $m = m_1 \cdots m_r$ is the product of $r$ pairwise coprime integers $m_1,\ldots,m_r$, then the *Chinese Remainder Theorem* allows one to recover any number in the range $0,\ldots,m-1$ from its residues mod $m_i$: given residue classes modulo $m_i$ with representatives $a_i$, for $i = 1,\ldots,r$, there is a unique residue class $a+m\mathbb{Z}$ modulo $m$ with representative $a$ such that $a + m_i\mathbb{Z} = a_i + m_i\mathbb{Z}$ for each $i = 1,\ldots,r$. (In other words, the set $a + m\mathbb{Z}$ is the intersection of the sets $a_i + m_i\mathbb{Z}$.) Problems having a solution modulo $m$ can now be split into $r$ smaller problems, whose solutions can be combined through an application of the theorem. In particular, the method can be used for long integer arithmetic, by performing computations modulo sufficiently many "small" powers of prime numbers (each in the range of a computer word). The actual implementation of the Chinese Remainder Theorem is based on the Euclidean algorithm.

There are several generalizations and variants of this procedure. From any identity in the polynomial ring $\mathbb{Z}[X_1,\ldots,X_n]$, one can derive an identity over the ring of residue classes $\mathbb{Z}/m\mathbb{Z}$ by reducing each coefficient modulo $m$. In this way, problems over the integers are transformed into problems over a finite ring, in which they can often be solved considerably faster (and with bounded memory requirements). Afterwards, the modular solutions have to be lifted back to $\mathbb{Z}$, or it has to be shown that a solution over $\mathbb{Z}$ does not exist. A classic example where this strategy is useful is the factorization of polynomials in $\mathbb{Z}[X]$ (see section 2.2.2).

Modular arithmetic in a more general sense also includes computations with algebraic numbers (treated as residue classes of $\mathbb{Z}[X]$ modulo an irreducible polynomial, together with an integer denominator if required). It also includes arithmetic in finite fields of prime power order, which can be constructed as residue class rings of $\mathbb{F}_p[X]$ (cf. section 2.1.8). These topics are discussed extensively in [Pohst and Zassenhaus 1989] and in [Lidl and Niederreiter 1986], respectively, as well as in [Mignotte 1992, Ch. 6] and [Cohen 1996b, sections 4.2 and 3.2, 3.4]. Applications are numerous — let us only mention the *Fast Dis-*

*crete Fourier Transform* (FDFT), which in turn is the basis of asymptotically fast methods for the multiplication of large integers.

Gerhard Niklasch (München) and Ralf Dentzer (Heidelberg)

### 2.1.5 Computations with Algebraic Numbers

Fundamental to algebraic numbers is the issue of representation. The natural method of presenting an algebraic number $\alpha$ as a root of an irreducible polynomial $f(x)$ over Q leaves ambiguous exactly which of the conjugates is meant. For many applications, however, we are concerned only with a single root, and this representation suffices. When conjugacy issues do arise — for example if one seeks to extend the field $Q(\alpha) \simeq Q[x]/(f(x))$ by another root of $f(x)$ not already in the field — one can specify the additional root by the irreducible factor of $f(x)$ to which the root belongs. In this case, the new field $Q(\alpha, \beta) \simeq Q[z]/(h(z))$ where $\alpha$ satisfies the irreducible polynomial $f(x)$ over Q, $\beta$ satisfies $g(y)$ over $Q[x]/(f(x))$, and $h(z) = \text{Resultant}_x(f(x), g(y + cx, x))$ for a suitable choice of $c$ [Trager 1976; Landau 1985]; see also 2.4.3. The classic theory of resultants, described in [van der Waerden 1966] [Zippel 1993], captures the arithmetic operations one might seek to perform on polynomials.

Alternate forms of representation include writing the root $\alpha$ as a sum of basis elements $\theta_j$ in $K$, a number field; in this case arithmetic operations are straightforward linear algebra calculations in the $\theta_j$. One may also represent $\alpha$ as a sum of numerical approximations of its conjugates; computations — addition, subtraction, multiplication, and division — are all clear, but one must take care of roundoff errors. Cohen's lovely book has an excellent discussion of these various representations [Cohen 1996b].

Radicals present intriguing problems for symbolic computation. For example, the nested radical $\alpha = \sqrt{5 + 2\sqrt{6}}$ gives rise to the basis $\{1, \alpha, \alpha^2, \alpha^3\}$ for the field $Q(\sqrt{5 + 2\sqrt{6}})$ over Q. This basis has pleasing algebraic properties, but to the human user it is considerably less intuitive than an alternative: $\{1, \sqrt{2}, \sqrt{3}, \sqrt{6}\}$. (That the two describe the same field is clear once one checks that $\sqrt{5 + 2\sqrt{6}} = \sqrt{2} + \sqrt{3}$.) Early work on simplification presented methods to denest radicals of a special form [Zippel 1985; Hopcroft et al. 1985]. There are algorithms for the general case that rely on extensions of Q containing all roots of unity — not a particularly efficient computational approach — while other theorems depend upon adding specific roots of unity to Q [Landau 1992; Horng and Huang 1999]. But this method sometimes introduces imaginary numbers into the simplification of nested real radicals, which is not altogether satisfying. Using approximations, Blömer showed how to avoid this difficulty in the case of simplifying depth 2 nested real radicals [Blömer 1992].

Determining the sign of an expression containing radicals continues to be difficult, which is to say, current methods to determine whether an expression with $k$ input roots is positive or negative take time exponential in $k$. By contrast, there are fast methods to determine if the sum is zero [Blömer 1991, 1998].

Sign-detecting algorithms depend on root-separation bounds, and recent work on such bounds for low-degree polynomials has proved useful in computational geometry, which is typically interested in intersections of straight lines and curves of degree 2. Numerical calculations have been the computation of choice in this area, but these sometimes run into round-off error difficulties. Meanwhile symbolic computation, though infinitely precise, takes much more time, and should only be used when numerical computations are likely to fail. Improved separation bounds show when to trade off between the two approaches in certain computations [Schirra et al. 1997]. The interaction between numeric and symbolic computation looks as if it will bear fruit not only in computational geometry but also in other aspects of symbolic algebra; see 2.12.3.

Susan Landau * (Amherst, USA)

---

* Partially supported by NSF grant CDA-9753055.

### 2.1.6 Real Algebraic Numbers

Real algebraic numbers are real roots of a polynomial with integer coefficients. Real algebraic points are points whose coordinates are real algebraic numbers.

A real algebraic number $x$ can be characterized in two ways:

1. by a polynomial $P$ and an isolating interval $(a, b)$, such that $x$ is the only real root of $P$ in $(a, b)$,
2. by a polynomial $P$ and its Thom encoding, i.e. the signs of the successive derivatives of $P$ at $x$. This information characterizes uniquely the number $x$ [Coste and Roy 1988].

Note that we do not require the polynomial $P$ to be irreducible.

The basic operations required on real algebraic numbers are: to decide whether two numbers are equal, which of two numbers is the larger one, to determine the sign of a polynomial at a real algebraic number.

The two main tools to characterize the real roots of a univariate polynomial are

1. Descartes's rule: Let $P = a_1 X^{d_1} + \ldots + a_k X^{d_k}$ be a univariate polynomial in $\mathbb{R}[X]$ with $d_1 > \ldots > d_k$ and all $a_i \neq 0$. We write $V(P)$ for the number of sign changes in $a_1, \ldots, a_k$ and $\text{pos}(P)$ for the number of positive real roots of $P$ counted with multiplicity. Then
   (a) $\text{pos}(P) \leq V(P)$,
   (b) $V(P) - \text{pos}(P)$ is even.
   Descartes's rule does not in general give complete information on the number of positive real roots of $P$. It gives a complete answer though whenever $V(P)$ is 0 or 1 or when we know in advance that all the roots of the polynomials are real (for example if $P$ is the characteristic polynomial of a symmetric matrix), since, in this special case, $\text{pos}(P) = V(P)$. There is however a method for finding isolating intervals based on Descartes's rule, Uspensky's method in [Uspensky 1948; Collins and Loos 1982a].

2. Sturm-Sylvester theorem (or improvements using subresultants). The signed Euclidean remainder sequence of $P$ and $P'Q$ is defined by

$$S_0(P,Q) = P, S_1(P,Q) = P'Q,$$

$$S_{i+1}(P,P'Q) = -\mathrm{rem}(S_{i-1}(P,P'Q), S_i(P,P'Q))$$

for $0 \leq i \leq k$ where $k$ is the least number such that $S_{k+1} = 0$. Then $S_k = \gcd(P, P'Q)$. The difference between the number of sign variations of the sequence $(S_0(a), \ldots, S_k(a))$ and the number of sign variations of the sequence $(S_0(b), \ldots, S_k(b))$ equals the difference between the number of real roots of $P$ in $(a,b)$ where $Q$ is positive and the the number of real roots of $P$ in $(a,b)$ where $Q$ is negative. Sturm-Sylvester (and variants [Roy 1996; Gonzalez-Vega et al. 1998b]) give isolation techniques based on dichotomy as well as methods for determining the non empty sign conditions realized by a family of polynomials at the zeroes of a polynomial. Applied to the derivatives of $P$, this gives the Thom encodings of the real roots.

Real algebraic numbers play an important role in all the algorithms of real algebraic geometry (see 2.5.3). For example

1. in order to compute the topology of a real algebraic curve [Gonzalez-Vega and Kahoui 1996],
2. in order to decide whether a semi-algebraic set is empty. Though the answer sought for is yes or no, the known algorithms for this problem produce, when the answer is yes, at least a real algebraic point in every connected component of the set [Roy 1996].

In the algorithms of real algebraic geometry, it is convenient, for complexity reasons, to make perturbations of the initial equations, using infinitesimals, in order to ensure some convenient geometric properties, as for example smoothness. It is thus important to deal with "real roots of equations with coefficients in $Z[\varepsilon]$", which are no real numbers but elements of the real closed field of real algebraic Puiseux series. Isolation intervals are not available in this situation, since the field is not archimedean, while methods based on Thom encodings are still valid.

Marie-Françoise Roy (Rennes)

### 2.1.7 $p$-adic Numbers and Approximations

In analogy to the construction of the reals from the rational numbers by "completion" based on the absolute value, there exists a completion $\mathbb{Q}_p$ of the rationals for any prime $p$, defined by the $p$-adic norm $|\ |_p$. This norm is discrete and non-archemedian. It has certain advantages over the absolute value making computations in $\mathbb{Q}_p$ more efficient than computing with reals. For example, a series of $p$-adic numbers is convergent if just the sequence of values converges to zero in the $p$-adic norm, and if there exists an entire rational number in every

$|\ |_p$-neighborhood of $x$, for every $x \in \mathbb{Q}_p$ with $|x|_p \leq 1$. The *p-adic numbers* defined in this manner can be represented by formal power series with finitely many terms of negative order (*Laurent Series*). Their arithmetic is analogous to power series arithmetic ([Knuth 1981b, Ch. 4.7]), with the sole exception that there will be carries which propagate from left to right (as opposed to right to left in integer arithmetic).

In a similar way, like one would represent real numbers by floating point numbers with arbitrary long mantissas, elements of $\mathbb{Q}_p$ can be approximated to arbitrary precision by truncated series $R = \sum_{i=M}^{N} a_i p^i$, where $a_i \in \{0, 1, \ldots, p-1\}$, and $M, N \in \mathbb{Z}$, $N \geq M$. The quality of that approximation is measured in the $p$-adic norm $|\ |_p$. Working with approximations $r$ to $x \in \mathbb{Q}_p$ can be interpreted as constructing numbers $r \in \frac{1}{p^M}\mathbb{Z}$ which are congruent to $x$ modulo $p^N$ for sufficiently large $N$. the most important tool for lifting a congruence modulo $p^N$ to one modulo $p^{N+1}$ is Hensel's lemma, which is the $p$-adic analogue of Newton's method. Compared to Newton's method it has the adavantage that it will always converge as soon as the starting conditions are satisfied. In particular, approximations of roots of rational numbers can be computed that way, provided that those roots belong to the field $\mathbb{Q}_p$.

The book by Koblitz [Koblitz 1984] is a good survey of this special area. The introduction of $p$-adic numbers into computer algebra systems started a few of years ago and is now under thorough investigation.

Additional reference: [Vuillemin 1990].

Michael E. Pohst (Berlin)

### 2.1.8 Finite Fields

In the last few decades, computing with finite fields $\mathbb{F}$ has found extensive application in coding theory and cryptography (see also section 2.14). Arithmetic in $\mathbb{F}$ is particularly simple in the case of residue class fields $\mathbb{F}_p \cong \mathbb{Z}/p\mathbb{Z}$. Arithmetic in those fields is reduced to modular computations (for computing inverses, the extended Euclidian algorithm is required, see 2.2.1 and 2.1.3). The realization of finite fields $\mathbb{F}_q$, whose cardinality is a power of a prime number $q = p^n$, is more difficult. $\mathbb{F}_q$ is generated by a normalized, irreducible polynomial $f \in \mathbb{F}_p$ of degree $n$. Field arithmetic can then be implemented as polynomial arithmetic modulo $f$. The probability to find such a polynomial by random choice is relatively high, namely $1/n$. However, one will need some irreducibility test for verfication, e.g., Berlekamp's algorithm (see 2.2.2). Deterministic methods for finding generating polynomials are, in general, recursive and rather complicated [Pohst and Zassenhaus 1989].

Additionally, one can take advantage of the fact that the multiplicative group of $\mathbb{F}_q$ is cyclic. To that end, one has to first find a generating element $\zeta$ (a so-called *primitive root*) of $\mathbb{F}_q \setminus \{0\}$ as a representative for some residue class in $\mathbb{F}_p[x]/(f)$, a process which can be quite costly. Once $\zeta$ is determined, one is able to precompute an index table consisting of residue class representatives of all

$\alpha \in \mathbb{F}_q$ and their exponent $ind(\alpha)$, defined by $\zeta^{ind(\alpha)} = \alpha$. At least for small $q$, basic arithmetic can be performed very easily and fast that way: multiplication by adding indices, addition by adding the representatives of the residue classes. The storage requirement can be further reduced, by keeping only exponents in memory. Addition can be expressed in terms of the generating element as $\zeta^i + \zeta^j = \zeta^i(1 + \zeta^{j-i})$, which can be accomplished through table lookup of *Zech* or *Jacobi logarithms* (defined by $\zeta^{Z(k)} = 1 + \zeta^k$).

For large $q$, $\mathbb{F}_q$ is considered as vector space over $\mathbb{F}_p$. Arithmetic is reduced to multiplication of basis elements $\{1, \beta, \ldots, \beta^{n-1}\}$, where $\beta$ is the residue class of $x$, which can be implemented through a multiplication table. This table also plays an important role in the *normal basis representation* of finite fields. If $\beta$ has the property that its $p$ powers form a basis $(\beta^{(p^i)}|0 \leq i \leq n-1)$ of $\mathbb{F}_p$ then this basis can be used for the representation of all elements and arithmetic in $\mathbb{F}_q$. One advantage is the fact that the Frobenius automorphism $\varphi : \mathbb{F}_q \to \mathbb{F}_q : \alpha \mapsto \alpha^p$ can be realized as cyclic translation of the coordinates. A lot of research is done on the construction of normal bases which produce multiplication tables with only a few – for *optimal* normal bases, just $2n - 1$ – non-zero entries; see also section 2.20. What is still missing, however, is a fast method for establishing an isomorphism between two finite fields which are given in different representations.

Different constructions of finite fields have different advantages and disadvantages regarding running time and memory requirements. Therefore any choice will greatly depend on the particular application. A thourough description of various methods and implementations can be found in [Grabmeier and Scheerhorn 1992].

For the embedding of different finite fields into larger ones, J.H. Conway proposed to use *norm compatible* irreducible polynomials for the construction of the field. These so-called *Conway polynomials* are, in general, difficult to obtain. However, once they are computed, the actual embedding can easily be accomplished through multiplication by a natural number. For an overview of these ideas, and an analogous theory based on *trace compatible* polynomials, as well as constructions of the algebraic closure of finite fields, we refer to [Scheerhorn 1992, 1993].

Important applications of finite fields are the factorization of integers and polynomials (the algorithm by Berlekamp and Cantor-Zassenhaus); see section 2.2.2. Another profound problem is the computation of discrete logarithms, i.e., to find a natural number $x$ which satisfies the equation $a^x = b$ for given $a, b \in \mathbb{F}_q$, if there exists such a solution at all. An overview of the problem can be found in the book by Odlyzko [Odlyzko 1984]. More recent results with algorithms basing on sieve methods (in principle derivations of the number field sieve) are discussed in a survey article by O. Schirokauer, D. Weber, and Th. Denny [Schirokauer et al. 1996].

Finally, the book [Lidl and Niederreiter 1986] by Lidl and Niederreiter is a standard reference for finite fields. Computational aspects are treated in the book [Shparlinski 1992] by Shparlinski. Additional reference: [Vuillemin 1990].

Michael E. Pohst (Berlin) and Johannes Grabmeier (Heidelberg)

## 2.2 Algorithms for Polynomials and Power Series

### 2.2.1 The Division Algorithm

In the ring $K[X]$ of univariate polynomials over a field $K$, most symbolic algorithmic methods are based on the *division algorithm* for polynomials.

The algorithm produces by successive reduction steps for a given pair $(f, g) \in K[X]^2$ with $\deg(g) > 0$ a uniquely determined pair $(q, r) \in K[X]^2$ such that $f = qg + r$ and ($r = 0$ or $\deg(r) < \deg(g)$). Iterative application of the division algorithm leads to the *Euclidian algorithm* which returns a *greatest common divisor* (GCD) $d \in K[X]$ of $f$ and $g$; the algorithm can be augmented to the *extended Euclidian algorithm* (cf. section 2.1.3) that returns in addition the cofactors of $f$ and $g$ in the representation of $d$ as a linear combination of $f$ and $g$. The Euclidian algorithm is also the basic ingredient for the squarefree factorization of polynomials (see section 2.2.2). Through multiplication of the dividend $f$ by an appropriate power of the highest coefficient $c$ of the divisor $g$, the division algorithm can also be performed in a *fraction-free* form. This so-called *pseudo-division with remainder* is applicable over any integral domain $K$, and also furnishes the GCD of $f$ and $g$ over a field $K$ up to a constant factor, if applied iteratively. For univariate polynomials $f, g$ with indeterminate coefficients, or multivariate polynomials construed as such, it yields, up to a change in sign and a powers of the highest order coefficients, the *resultant* of $f$ and $g$. In the special case where $g = f'$ the resultant becomes the *discriminant* of $f$. See [Loos 1983] and [Knuth 1981b] for an overview of variants and optimized versions of the iterated pseudo-division algorithm, especially the *subresultant algorithm* that replaces iterated division with explicit determinantal expressions. With regard to the complexity, we also refer to section 2.13.

Expressed in the language of the theory of ideals, the Euclidian algorithm computes a generating element – namely the GCD of $f_1, \ldots, f_m$ – for the ideal $I = Id(f_1, \ldots, f_m)$, when applied iteratively to $f_1, \ldots, f_m \in K[X]$. By virtue of this representation of $I$ as *principal ideal*, we can derive algorithms for the basic operations on ideals of $K[X]$ (such as product, intersection, quotient, determination of radicals). Geometrically, one obtains an algorithmic criterium – taking into account Hilbert's Nullstellensatz – for deciding whether there exists a common zero of $f_1, \ldots, f_m$ in the algebraic closure $\overline{K}$ of $K$ at which additional polynomials $g_1, \ldots, g_k \in K[X]$ do not vanish.

In the ring of univariate formal power series over a field the corresponding problems become almost trivial: a power series $f$ divides another power series $g$ if and only if its order $ord(f)$ divides $ord(g)$. Therefore, the GCD of a set of power series is simply an element from this set which has minimal order. In terms of algorithms, power series are infinite objects, allowing computation of the quotient of two power series to arbitrary, yet finite precision. For practical implementations, power series can be realized as *streams*, see section 2.1.2.

For multivariate polynomials the computation of the unique normal form of a polynomial wrt. a Gröbner basis generalizes the division algorithm (cf. section 2.2.5). Similarly for multivariate power series and standard bases (cf. section 2.2.6).

Volker Weispfenning (Passau)

### 2.2.2  Factorization of Polynomials

Since the time of Carl Friedrich Gauß, we know that a polynomial (in several variables) over a field factors (essentially) uniquely into a product of irreducible polynomials. The question how this decomposition can be done efficiently is the starting point for one of the most appealing and most successful areas of computer algebra.

In the following, we briefly discuss the computation of greatest common divisors (gcds), squarefree factorization, and partial fraction decomposition. We continue with an overview on factorization of polynomials in one variable over finite fields, over the rationals, and over algebraic number fields. Finally, we look at polynomials in several variables. An extensive exposition of this topic, including references, can be found in the three survey articles by Kaltofen [Kaltofen 1982, 1990, 1992b], and also in [von zur Gathen and Panario 2001].

The extended Euclidean algorithm for univariate polynomials over a field has many important applications: the Chinese remainder theorem, interpolation, inversion in extension rings, linear recurrence sequences, linear algebra for sparse matrices à la Krylov, and the decoding of BCH-codes (Berlekamp-Massey algorithm). The theory of subresultants provides us with theoretical insights, strategies for avoiding excessively large intermediate expressions, and above all, important modular algorithms for polynomials in one or several variables over $\mathbb{Q}$. The details are explained, e.g., in [von zur Gathen and Gerhard 1999].

The squarefree part $f_1 \cdots f_r$ of a polynomial $f = f_1^{e_1} \cdots f_r^{e_r} \in F[x]$, where $F$ is a field and $f_1, \ldots, f_r$ are irreducible and pairwise non-associated, is $f/\gcd(f, f')$. In fields of characteristic $p > 0$, one additionally has to take some $p$-th roots. By iteration of this procedure, one obtains the *squarefree decomposition* $f = \mathrm{lc}(f) g_1^1 \cdot g_2^2 \cdots g_n^n$, where the $g_i \in F[X]$ are monic, squarefree, and pairwise relatively prime. This squarefree decomposition is used in the integration of rational functions $f/g$ via the partial fraction decomposition; see section 2.10.2. A full-scale partial fraction decomposition, however, requires the complete factorization of the denominator $g$.

The most basic task of factorization is the case of a univariate polynomial of degree $n$ over a finite field $\mathbb{F}_q$ with $q$ elements, where $q$ is the power of a prime number. Berlekamp [Berlekamp 1967, 1970] devised a deterministic algorithm with running time (meaning number of operations in $\mathbb{F}_q$) $O^{\sim}(n^3 + nq)$, where the *soft* $O^{\sim}$ indicates that $\log n$ factors are not accounted for. A milestone on the

### 2.2.2 Factorization of Polynomials

way to fast methods is his probabilistic algorithm with polynomial running time $O\tilde{\ }(n^3+n\log q)$. This was the first major application of this concept in computer algebra, and until today, no deterministic method running in polynomial time is known for this problem. An important idea was introduced by the algorithm of Cantor and Zassenhaus [Cantor and Zassenhaus 1981] requiring $O\tilde{\ }(n^2 \log q)$ operations. The (expected) running time of these algorithms depends on two parameters: the degree $n$ and the length $\log q$ of the representation of a field element. Among the many algorithms known today, there are several which are the fastest in a certain range of inputs: [Huang and Pan 1998] for very small $q$, say $\log q < n^{0.0017}$, [Kaltofen and Shoup 1998] for $\log q$ up to $n^{0.466}$ with a running time of $O(n^{1.806}(\log q)^{0.416})$ (see [von zur Gathen and Gerhard 1999]), then [von zur Gathen and Shoup 1992] for $\log q \leq n^{1.376}$ with a running time of $O\tilde{\ }(n^2+n\log q)$. For larger $q$, Berlekamp's algorithm has not been improved. For large fields $\mathbb{F}_q$ of small characteristic, say $q$ a power of 2, a faster algorithm is in [Kaltofen and Shoup 1997]. Some of these algorithms can be found in textbooks: [Knuth 1997], [Geddes et al. 1992], [von zur Gathen and Gerhard 1999].

A typical computer algebra system like MAPLE factors polynomials of degree $n \approx 200$ over a field with $q \approx 2^{200}$ elements within a day running on an average workstation. The power of recent developments is illustrated by the factorization at degree 1024 over a prime field with about $2^{1024}$ elements in 50 hours ([Kaltofen and Shoup 1998]) and of polynomials over $\mathbb{F}_2$ of degree more than one million ([Bonorden et al. 2001]).

For polynomials over $\mathbb{Q}$ or over algebraic number fields the method of choice is factoring modulo a prime $p$ and *Hensel lifting*, which was introduced by Zassenhaus [Zassenhaus 1969] into computer algebra. Since an irreducible polynomial might become reducible modulo $p$, one has to determine which modular factors together form the image of a true irreducible factor. The method of *factor combination*, based on exhaustive search, has exponential running time for certain inputs. The method of *short vectors in lattices* by Lenstra et al. (see [Lenstra et al. 1982] and 2.4.6) has guaranteed polynomial running time, but is slower on many inputs. Schönhage proposed very fast algorithms for numerically factoring polynomials over $\mathbb{R}$ and $\mathbb{C}$ by computing root approximations [Schönhage 1987].

For multivariate polynomials the choice of data structures is a central issue. For a small number of variables, a variant of *Hensel lifting* can be used, for many variables one has to switch to the *sparse representation* of the polynomials. The most promising concepts are effective versions of Hilbert's irreducibility theorem, and the representation by *arithmetic circuits*. Kaltofen's factoring algorithm [Kaltofen 1989] based on this model is one of the highlights in polynomial factorization. An even more efficient data structure for multivariate polynomials is via a *black box*, , into which one feeds the appropriate number of field elements and obtains the value of the polynomials at those points. Kaltofen and Trager [Kaltofen and Trager 1990] solve the factorization problem in this data structure, also in random polynomial time.

Joachim von zur Gathen (Paderborn)

### 2.2.3 Absolute Factorization of Polynomials

The absolutely irreducible factors of a polynomial are those factors with coefficients in the algebraic closure of the coefficient field that remain irreducible over that algebraic closure. For example, for a univariate polynomial with rational coefficients the absolutely irreducible factors are the linear factors that determine the complex roots. Multivariate polynomials of degree more than one over the rational numbers can remain irreducible over the complex numbers, for example, $x^3 - y^2$.

The problem of computing arbitrary high precision approximations to the complex roots of a rational polynomial is a classical mathematical task. Many very efficient numerical algorithms are known. Algorithms that yield all the roots on all input polynomials are known at least since Routh's stability criterion, and the challenge of investigating the efficiency of "infallible" methods was begun in the early 1980s. We shall only cite the papers [Pan 1997; Kirrinnis 1998; Bini and Fiorentino 2000] and point to recent ISSAC Proceedings (see section 5.1) for current work and additional pointers to the literature.

In the multivariate case, the problem of computing factors over algebraically closed fields constitutes an exponential case of the Berlekamp-Zassenhaus algorithm (see the previous section 2.2.2). The projected univariate polynomial factors completely and for irreducible inputs exponential many combinations need to be checked. Polynomial-time algorithms were discovered in the mid-1980s (see [Kaltofen 1995b] and the cited literature there), which includes the representation of the algebraic coefficients of the factors. A different idea, based on a partial differential equation from [Ruppert 1986, 1999], has yielded an efficient and practical algorithm [Gao 2001] that can also be made into a contender for factoring a multivariate polynomial over the coefficient field itself.

Algorithms for factoring multivariate polynomials approximately over the complex numbers are discussed in section 2.12.3. We conclude by observing that the approach of determining suitable sums of powers of roots, first introduced in [Sasaki et al. 1991] for the multivariate factoring problem over complex numbers and suggested for univariate polynomials over the rationals in [Sasaki and Sasaki 1993], is also an ingredient in the new lattice-based method [van Hoeij 2001] for the hard univariate cases over the rational numbers.

Erich Kaltofen (Raleigh, USA)

### 2.2.4 Polynomial Decomposition

*Polynomial decomposition* is the problem of representing a given polynomial $f(x)$ as a functional composition $g(h(x))$ of polynomials of smaller degree. Polynomial decomposition is useful in simplifying the representation of field extensions of high degree and is provided as a primitive by many major symbolic algebra systems. Decomposition problems for multivariate polynomials and for rational and algebraic functions are also of interest.

The theory of polynomial decomposition was initiated by Ritt in 1922 [Ritt 1922] and further developed in [Fried and MacRae 1969; Dorey and Whaples

### 2.2.4 Polynomial Decomposition

1974]. A survey was presented in [Schinzel 1982]. Decompositions of $f(x) \in F[x]$ are intimately related to the intermediate fields between $F(f)$ and $F(x)$ [Dorey and Whaples 1974; Ritt 1922].

The problem breaks into two cases, the *tame* and the *wild*. The tame case is when the characteristic of $F$ does not divide the degree of $g$, including characteristic 0. In the tame case, the problem is rational—that is, if $f$ decomposes over an extension of $F$, then if decomposes over $F$ [Schinzel 1982]—and maximal decompositions are unique up to insertion of linear decomposition factors and commutativity of powers of $x$ and Chebyshev polynomials [Ritt 1922]. Both facts may fail in the wild case [Dorey and Whaples 1974; Schinzel 1982].

The first analyzed algorithms for decomposition of polynomials were provided by Barton and Zippel [Barton and Zippel 1985] and Alagar and Thanh [Alagar and Thanh 1985], who considered polynomials over fields of characteristic zero. Both solutions involved polynomial factorization and took exponential time. For some time, the problem of finding nontrivial decompositions was considered to be computationally hard; a cryptographic protocol was based on its supposed intractability [Cade 1985].

Kozen and Landau [Kozen and Landau 1989a] discovered the first polynomial-time algorithms in the tame case that do not require factorization. The time bounds were $O(n^3)$, $O(n^2)$ if $F$ supports an FFT. A similar algorithm was discovered independently by Gutierrez et al. [Gutierrez et al. 1989]. Kozen and Landau also gave efficient NC algorithms (parallel polylog time on polynomially many processors) with a time bound of $O(\log^2 n)$. In the wild case, they reduced the problem to factorization and gave an $O(n^{\log n})$ algorithm for the decomposition of irreducible polynomials over general fields admitting a polynomial-time factorization algorithm and an NC algorithm for irreducible polynomials over finite fields. The sequential bounds in the tame case were improved by von zur Gathen [von zur Gathen 1990a] to $O(n \log^2 n \log \log n)$, $O(n \log^2 n)$ if $F$ supports an FFT. He also gave an improved algorithm for the wild case, yielding a polynomial-time reduction to factorization, and observed undecidability over sufficiently general fields [von zur Gathen 1990b].

Several extensions and variations have been investigated. Dickerson [Dickerson 1987] and von zur Gathen [von zur Gathen 1990a] investigated the decomposition of multivariate polynomials. Gutierrez [Gutierrez 1991] gave a polynomial decomposition algorithm over factorial domains. Von zur Gathen and Weiss [von zur Gathen and Weiss 1995] gave an exponential method for finding decompositions of the form $f(x) = g(h(x), k(x))$ for homogeneous $g(y, z)$. Casperson et al. [Casperson et al. 1996] gave an algorithm that, given an irreducible $f(x)$, finds $g(x)$ and $h(x)$ such that $f(x)$ divides $g(h(x))$. The latter two algorithms were generalized and improved by Klüners and Pohst [Klüners and Pohst 1997] and Klüners [Klüners 1999]. Binder [Binder 1995] gave a fast method to compute the Lüroth generator of the union field of two polynomials.

Hommenl and Kovács [Hommenl and Kovács 1992] defined and investigated the *sine-cosine decomposition problem*: determine whether a given bivariate polynomial $f(s, c)$ has a decomposition of the form $f(s, c) = g(h(s, c)) \bmod s^2 + c^2 - 1$.

This problem has applications in robot kinematics. Gutierrez and Recio [Gutierrez and Recio 1998] gave a cubic-time algorithm that does not require factorization. It is based on approximating a root of $f(s,c)$ mod $s^2 + c^2 - 1$.

Zippel [Zippel 1991] presented a polynomial-time algorithm to decompose rational functions over any field admitting efficient polynomial factorization. His approach uncovers a strong relation to Lüroth's theorem. Alonso et al. [Alonso et al. 1995b] gave two exponential-time algorithms for rational function decomposition that compare favorably with Zippel's in practice. They also presented several applications: faithful reparametrization of unfaithfully parameterized curves, computing intermediate fields in a simple purely transcendental extension of $F$, and providing a birationality test for subfields of $F(x)$.

Kozen, Landau, and Zippel [Kozen et al. 1996] addressed the decomposition problem for algebraic functions. They uncovered a connection to univariate resultants over algebraic function fields and showed that the decomposition problem essentially asks whether some power of a given irreducible bivariate polynomial $f(x,z)$ can be expressed as the resultant with respect to $y$ of two bivariate polynomials $g(x,y)$, $h(y,z)$. They determined necessary and sufficient conditions for the existence of a nontrivial decomposition and classified all such decompositions up to isomorphism. They also gave an exponential-time algorithm for finding a nontrivial decomposition if one exists.

Jaime Gutierrez (Cantabria) and Dexter Kozen (Cornell)

### 2.2.5 Gröbner Bases

The ring $R = K[X]$ of univariate polynomials over a field is a Euclidean domain; the *Euclidean algorithm* consisting of an iterated division with remainder computes the *greatest common divisor* $d$ of a finite set $F$ polynomials. The ideal generated by $F$ coincides with the principal ideal generated by $d$, and the *extended Euclidean algorithm* provides a representation of $d$ as linear combination of the polynomials in $F$. As a consequence the *variety* of $F$, i. e. the set of all common zeros of polynomials in $F$ in some extension field of $K$, coincides with the variety of $d$. This reduces the study of univariate polynomial varieties to the study of the zeros of a single polynomial [Weispfenning 1995] (see also section 2.2.1).

By way of contrast rings $R = K[X_1, \ldots, X_n]$ of multivariate polynomials over fields are no longer principal ideal domains for $n \geq 2$. In these rings, the construction of *Gröbner bases* takes over the role of the Euclidian algorithm for univariate polynomials as an algorithmic method for the study ideals in $R$, and varieties polynomial sets in $R$ in extension fields of $K$. The method was established by B. Buchberger 1965 (see [Buchberger 1965, 1970]), and experienced rapid development starting in the seventies. Nowadays, it is one of the most important fundamental algorithms of computer algebra [Buchberger 1985b; Cox et al. 1992; Mishra 1993; Becker et al. 1998; Adams and Loustaunau 1994b; Froeberg 1997; Cox et al. 1998a].

### 2.2.5 Gröbner Bases

Every non–zero polynomial in $R = K[X_1, \ldots, X_n]$ is a finite sum of monomials $at$, where $0 \neq a \in K$ and $t$ is a *term*, i. e. a power–product of the indeterminates $X_i$. A linear order "<" on the multiplicative monoid $T$ of terms is called an (admissible) *term order* if 1 is the smallest term and multiplication of terms is monotone wrt. the order. Term orders can be characterized completely via certain real matrices [Robbiano 1985; Weispfenning 1987; Hong and Weispfenning 1999]. Any term order is a well–order on $T$ and induces a well–quasi–order on $R$. In the theory of standard bases one considers also more general term orders, where 1 is not required to be the smallest term. These orders are in general not well–orders on $T$ (compare section 2.2.6). After choosing a fixed term order < on $T$, one can define the *reduction* of a polynomial $f$ by a set $P$ of other polynomials in complete analogy to the division of univariate polynomials. Iterative application of such reductions will eventually produce a *normal form* of $f$ with respect to $P$. In general normal forms are not unique.

A finite set $G \subseteq R$ is a *Gröbner basis* (of the ideal $I$ generated by $G$) iff the highest term of every polynomial $0 \neq f \in I$ is a multiple of the highest term of a polynomial in $G$. Roughly speaking, this means that the cancellation of high monomials in linear combinations of polynomials from $G$ is insignificant. An equivalent condition requires that normal forms of arbitrary polynomials in $R$ wrt. reduction relative to $G$ are unique. Given two polynomials $f, g \in R$, there is an obvious "minimal" linear combination $h$ of $f$ and $g$ with monomial coefficients, where a cancellation of highest monomials occurs; this polynomial $h$ is called the *S–polynomial* of $f$ and $g$.

In 1965, Buchberger showed how to construct a Gröbner basis $G$ from a given finite ideal basis $F$. His algorithm adds iteratively non–zero normal forms of *S-polynomials* (*critical pairs*) of pairs of polynomials in $F$ to $F$. The termination of the process is guaranteed by *Dickson's lemma* or alternatively by the *Hilbert basis theorem*. Buchberger's algorithm leaves a lot of lee–way concerning the order in which S-polynomials are considered and reduced to normal form. Moreover there are a–priori criteria that recognize certain S-polynomials as superfluous, since their normal form is guaranteed to be zero at the present or a later stage of the algorithm by theoretical reasons.

Numerous schemes have been proposed and tested in order to speed up Gröbner basis computations in general or for special types of input systems (see e. g. [Giovine et al. 1991; Traverso 1996]). In the worst case, the total degree of the polynomials in an ideal basis may grow doubly exponentially during the computation of a Gröbner basis [Winkler 1984; Möller and Mora 1984]; nevertheless Gröbner basis computations can in principle always be performed in exponential work space [Mayr 1989; Kühnle and Mayr 1996]. On the other hand Gröbner basis computations are exponential space hard, since they solve the ideal membership problem for multivariate polynomial ideals [Mayr and Meyer 1982]. A complexity problem arising frequently in practice is the occurence of very large rational coefficients in Gröbner basis computations over the field of rational numbers. Over finite field this the coefficient size is of course bounded from the outset. This fact has stimulated multiple modular representation possibly together with a numeric representation of rational coefficients in Gröbner basis computations [Winkler 1988; Traverso 1988; Pauer 1992; Gräbe 1993]. Another information that may speed up Gröbner basis computations is the knowledge of the Hilbert function of the ideal [Traverso 1996].

*Reduced Gröbner bases* are Gröbner bases, that are interreduced as polynomial sets and consist of monic polynomials. A reduced Gröbner basis $H$ for a polynomial ideal $I$ can be obtained from an arbitrary Gröbner basis $G$ of $I$ with respect to the same term order by iterative interreduction of the polynomials in $G$. In contrast to an arbitrary Gröbner basis $g$ for $I$, a reduced Gröbner basis of $I$ is uniquely determined by $I$ and the term order.

The choice of the term order can be of decisive importance for the speed of a Gröbner basis computation. As a rule of thumb, term orders refining the total degree quasi–order tend to perform better than purely lexicographical term orders. This empirical observation has stimulated the search for efficient algorithms converting a Gröbner basis for a given (e. g. total degree term order) into another (e. g. lexicographical) term order that may be required by certain applications. This conversion can be achieved either by one big step [Faugere et al. 1993] or more efficiently in several smaller steps by "Gröbner walk" algorithms [Collart et al. 1997; Amrhein et al. 1997a; Amrhein and Gloor 1998]. *Gröbner walk algorithms* use the important fact that a given polynomial ideal has only finitely many different reduced Gröbner bases with respect to all term orders. This leads to the concepts of the *Gröbner fan* and of a *Universal Gröbner basis* [Schwartz 1988; Weispfenning 1989a; Schemmel 1987; Mora and Robbiano 1988; Mall 1997] of a polynomial ideal.

Gröbner bases (sometimes in combination with other polynomial algorithms, like factorization) allow for the solution of a host of algorithmic problems related to multivariate polynomials and their zeroes (in algebraically closed fields). We mention a few (compare [Möller and Mora 1986; Becker et al. 1998]): the problem of ideal membership, inclusion among ideals, computation of intersection and quotient of ideals, computation of the radical of an ideal, and decomposition into primary ideals, a generalized form of the Chinese remainder theorem [Becker and Weispfenning 1991] , computation of the dimension and of the Hilbert function of an ideal, and the vector space dimension of the corresponding residue class ring $\bar{R}$, explicit construction of a basis of $\bar{R}$, as well as of its structure constants, computation of a generating system for the module of syzygies of an ideal basis, solution of systems of linear equations over $\bar{R}$, membership in a polynomial subalgebra, arithmetic in finite algebraic field extensions, transformation of systems of algebraic equations into triangular form, allowing an exact, symbolic solution in a suitable extension of the base field (cf. section 2.5). Applications in the realm of logic include word problems, first-order logic and geometric theorem proving [Becker et al. 1998; Kapur and Narendran 1985; Kutzler and Stifter 1986; Kutzler 1988; Wang 1998], compare also section 3.2.3. Gröbner bases for binomial ideals have applications to linear integer programming [Pasqualina and Carlo 1991; Sturmfels 1995; Hosten and Thomas 1998]. A variant of the Buchberger algorithm leads to an analogue of Gröbner bases for subalgebras instead of ideals called *SAGBI bases* [Robbiano and Sweedler 1988; Kapur and Madlener 1989]. In general SAGBI bases for finitely generated polynomial algebras are infinite.

Buchberger's Gröbner basis method for multivariate polynomials over a field is implemented in most general purpose computer algebra systems, as well as in many special purpose systems like CoCoA, FELIX, MACAULAY, and MAS. It has been extended to a number of other rings, and to finitely generated modules over such rings, like rings of multivariate polynomials over principal ideal domains. Large parts of the Gröbner basis theory have also been extended to non-commutative polynomials of various kinds (see section 2.6.5).

For multivariate polynomials with parametric coefficients Gröbner bases are in general not stable under specialization of the parameters in some field. Conditions for stability under specializations have been studied in [Bayer et al. 1991; Gianni 1987; Becker 1994; Kalkbrener 1997]. A more subtle construction yields for every ideal of multivariate parametric polynomials a finite *comprehensive Gröbner basis* that remains stable under arbitrary specialization of the parameters in any field. Comprehensive Gröbner basis have many applications in geometric elimination theory, in particular to quantifier elimination in algebraically closed fields (see section 2.5).

*Involutive bases* provide an alternative approach to the construction of Gröbner bases [Zharkov and Blinkov 1993; Zharkov 1996; Apel 1995, 1998b; Gerdt 1994, 1998b,a,c, 1999; Gerdt et al. 1999]. The method is derived from the algebraic theory of partial differential equations, where it was developed in order to construct integrability conditions (compare sections 2.11.2 and 2.11.5). Instead of adding normal forms of S-polynomials as in the Buchberger algorithm, involutive bases are constructed by adding normal forms of multiples of polynomials by certain variables. The set of variables is split into multiplicative and non-multiplicative variables; this can be done in many different ways, either globally or relative to finite sets of terms, such that certain axioms are satisfied. Each specific splitting is called an *involutive division*. Examples are Janet, Pommaret and Thomas division. Reduction is performed similar as in the Buchberger algorithm except that the cofactors have to be multiplicative; hence it is called multiplicative reduction. A given ideal basis is extended to an involutive basis by adding successively multiplicatively reduced normal forms of non-multiplicative multiples of polynomials in the current basis, and by performing multiplicative interreduction. Termination is not always guaranteed; for example in the case of Pommaret division the ideal has to be zero-dimensional. For other involutive divisions termination can be shown for arbitrary ideals. Involutive bases are always Gröbner bases wrt. the given term order, but in general they are not reduced Gröbner bases. For global involutive divisions minimal involutive bases are uniquely determined by the ideal and the term order [Gerdt 1998c]. The redundancy of involutive bases carries additional information that can be used e. g. for fast computation of the Hilbert function of the ideal [Apel 1998b].

Volker Weispfenning (Passau)

## 2.2.6 Standard Bases

In rings $K[[X_1,\ldots,X_n]]$ of formal power series over a field $K$, the notion of a *standard basis* is an equivalent to the concept of a Gröbner basis in the polynomial ring $K[X_1,\ldots,X_n]$. In this context, the lowest order terms of the power series take over the role of the highest terms of polynomials (with respect to a fixed not well–founded term ordering). In 1964, Hironaka gave a non-constructive proof for the existence of Gröbner bases and standard bases for certain term orderings. As for Gröbner bases standard bases can be characterized by the condition that normal forms of arbitrary power series in $K[[X_1,\ldots,X_n]]$ relative to a standard basis are unique. Given two polynomials Approximate standard bases can be computed in a similar way as Gröbner bases with respect to a total degree term ordering, starting from a given finite ideal basis $F$ up to arbitrary precision (defined in the natural topology on $K[[X_1,\ldots,X_n]]$) [Becker 1990; Ebner-Altunay 1991]. In case of an ideal basis $F$ consisting of polynomials only, an exact polynomial standard basis can be computed by Mora's *tangent cone algorithm* and extensions thereof [Mora 1982; Mora et al. 1992; Gräbe 1994; Greuel and Pfister 1996, 1998]. The computation is performed in the localization of the polynomial ring with respect to the multiplicative set of all polynomials with non–vanishing absolute term. Alternatively the computation of a standard basis can be achieved via homogenization in a polynomial ring [Lazard 1983; Becker et al. 1998; Gräbe 1994]. Standard bases are of great importance for algorithmic classification of singularities in algebraic varieties (see section 2.5.2).

Volker Weispfenning (Passau)

## 2.2.7 Characteristic Sets

Characteristic sets were introduced by Ritt in 1950 as an important tool for structural investigation of algebraic differential equations (cf. sections 2.11.2 and 2.11.5). Starting in 1984, Wu transferred the algorithmic aspects of the method to algebraic equations with the goal to find an efficient algorithm for automatic theorem proving in geometry [Wu 1984a,b, 1986]. Since then, this approach has been the subject of intensive investigations [Wang and Gao 1987; Wang 1995; Chou 1988, 1990; Chou and Yiao-Shan 1990a; Mishra 1993; Kapur and Mundy 1988; Kapur 1998]. Let $F$ a set of polynomials from the ring $R$ of multivariate polynomials over a field. A *characteristic set* of $F$ is a minimal ascending subset of $F$ in the sense of Ritt [Ritt 1950]. Existence of a characteristic set follows from the fact that the respective ordering among ascending chains is well-founded. If $F$ is finite then a characteristic subset can be easily derived algorithmically.

Wu's algorithm consists of a completion procedure for finite characteristic sets and successive splitting of the resulting sets into finitely many parts. During the completion step, characteristic sets are complemented by remainders of polynomials from $F$ under pseudo-division by suitably chosen variables. The splitting step involves testing of all cases of possible vanishing of initials of the polynomials used in pseudo-division.

The result can be interpreted as splitting of the set $V$ of zeroes of $F$ into finitely many, pairwise disjoint boolean combinations of varieties (*constructible sets*), which for many practical purposes is sufficient to reveal all the relevant information on $V$. In general however, the algorithm cannot decide whether a given polynomial $g$ vanishes on $V$ or not. It yields only a semi-decision method. Despite this imperfection, the method is often more efficient for practical investigations in geometrical applications than corresponding Gröbner basis methods [Wu 1984a,b, 1986; Wang and Gao 1987; Wang 1995; Chou 1988, 1990; Chou and Yiao-Shan 1990a,b; Kutzler 1988; Kapur 1986a; Kapur and Mundy 1988; Kapur and Lakshman 1992; Kapur 1998], compare also section 3.2.3. It may, however, fail for geometrical problems that depend on the real numbers as the coordinate field [Pasqualina and Carlo 1995]. The method can be supplemented by successive factorization of polynomials from the characteristic set over the algebraic extension field defined by the previous polynomials of the characteristic set. That way, the amount of information provided by the method increases while its efficiency considerably decreases.

The text book [Cox et al. 1992] and the appendix of [Becker et al. 1998] give an overview of the method, [Chou 1988; Mishra 1993] contain more detailled descriptions.

Volker Weispfenning (Passau)

## 2.2.8 Algorithmic Invariant Theory

Invariant theory is a classical branch of mathematics which reached a bloom at about the turn of the century and was associated to names like Gordan, Young, and Hilbert, to name just a few famous ones. In fact, much of the algebraic work of Hilbert and his colleagues was inspired by invariant theory. After having proved his Finiteness Theorem on the finite generation of invariant rings, Hilbert pronounced the subject dead and sent it to a near dormant state for several decades. However, in the last few years invariant theory has experienced a remarkable renaissance, which it owes (among other factors) to the emergence of new algorithmic methods and the availability of better software and hardware. The new interest in the subject is illustrated by the appearance of three new books [Sturmfels 1993; Benson 1993; Smith 1995], one of which focuses expressly on algorithmic aspects. As further reading on invariant theory I recommend [Kraft et al. 1987; Kraft 1984; Mumford et al. 1994; Popov and Vinberg 1994], and [Springer 1977].

The standard setting of invariant theory is as follows. We consider a group $G \leq \mathrm{GL}_n(K)$ of invertible matrices over a field $K$. Then $G$ acts on the polynomial ring $K[x_1, \ldots, x_n]$ by linear transformations of the variables. The invariant ring $K[x_1, \ldots, x_n]^G$ consists of all polynomials which are fixed by the action of $G$. Equivalently, a polynomial is an invariant if it is constant on each orbit of the action of $G$ on $K^n$. A classical goal of invariant theory is to find a (finite) set of invariants which generate $K[x_1, \ldots, x_n]^G$ as an algebra over $K$. It is clear that the invariant ring contains comprehensive information about the geometry of the $G$-orbits, and that properties of points which are stable under the $G$-action can usually be expressed in terms of invariants.

A typical example is the case that $G$ is the symmetric group $S_n$ acting by permutations of the $x_i$. It is well known that the invariant ring is generated by the elementary symmetric polynomials. Another example is $G = \mathrm{SO}_2(\mathbf{R})$, the special orthogonal group. Here clearly $x_1^2 + x_2^2$ is an invariant, and every invariant can only depend on the distance of a point from the origin, so it is quite plausible (and indeed true) that $K[x_1, x_2]^G$ is generated by $x_1^2 + x_2^2$. We can also take $G$ to be the subgroup of $\mathrm{SO}_2(\mathbf{R})$ of order 4 generated by a rotation about 90 degrees. In that case the invariant is generated by three invariants, among which there is an algebraic relation.

The question whether the invariant ring is finitely generated has become known as Hilbert's 14th problem. It is now known that the general answer is "no", but the theorem of Hilbert and Nagata [Hilbert 1890; Nagata 1965] says that the invariant ring is finitely generated if $G$ is a reductive group. Important examples of reductive groups include all finite groups and the classical groups over $\mathbf{C}$.

Apart from being an interesting subject in its own right, invariant theory has applications in many fields. Examples include:

1. equivariant dynamics [Sattinger 1979; Gatermann 1996b; Worfolk 1994], also see Section 3.2.7,
2. solving systems of algebraic equations with symmetries [Worfolk 1994; Sturmfels 1993, Section 2.6], also see Section 3.2.8
3. theoretical physics and chemistry [Jarić et al. 1984; Ischtwan and Collins 1991],
4. coding theory [Sloane 1977],
5. image processing (no reference available at the moment),
6. combinatorics [Stanley 1979; Dress and Wenzel 1991],
7. incidence geometry [Sturmfels 1993, Section 3.4],
8. cohomology of finite groups [Adem and Milgram 1994].

Moreover, modular invariant theory (the case that the field $K$ has positive characteristic) has been a very active area of research in the last few years and continues to offer many unresolved, interesting questions (see, for example, [Smith 1997]). In most applications, the computation of a generating system of invariant is the starting point, and often the algebraic relations ("syzygies") between the generators are required as well. In the first two points of the above list, the idea is to choose coordinates which are invariants. Then a $G$-orbit of points in the original coordinates shrinks into just one point in the new coordinates, which often renders calculations feasible that would otherwise be out of the reach of computation.

There has been a considerable interest in developing efficient algorithms to compute invariant rings for finite groups $G$ [Sturmfels 1993; Kemper 1996; Kemper and Steel 1999]. Here the algorithms become much more intricate in the modular case, but this is the case that is most interesting for structural investigations and for applications in cohomology theory. For the special case of permutation groups, an interesting algorithm which is independent of the ground field (or ring) $K$ was given in [Göbel 1995]. Quite recently Harm Derksen gave an algorithm which applies to linearly reductive groups [Derksen 1999]. All algorithms use Gröbner basis methods and linear algebra as main ingredients.

I will finish this survey by giving a list of some computer algebra packages that are devoted to invariant theory, ordered chronologically (to my best knowledge).

1. The Invariant Package of MAS [Göbel 1997]. A package for invariants of permutation groups written by Manfred Göbel in MAS (see Section 4.2.26). The core procedure performs the reduction of invariants given in [Göbel 1995]. The Invariant Package is included in the standard distribution of MAS.
2. INVAR [Kemper 1993]. A Maple package for the computation of invariant rings of finite groups and their properties. An older version is part of the Maple share library which is shipped with the standard distribution of Maple. A new version also covers the modular case and can be obtained by anonymous ftp from the site ftp.iwr.uni-heidelberg.de under /pub/kemper/INVAR2/. For more information, please contact Gregor.Kemper@iwr.uni-heidelberg.de.
3. FINVAR [Heydtmann 1997]. A package based on Singular (see Section 4.2.39) written by Agnes E. Heydtmann. The scope of the package is roughly the same as that of INVAR. For more information, please contact decker@math.uni-sb.de.
4. The Magma package for invariants [Kemper and Steel 1999]. Algorithms for invariant theory of finite groups have quite recently become part of the standard distribution of Magma (see Section 4.1.5). They were implemented by Allan Steel with the collaboration of the author. In my opinion, this is the fastest and most sophisticated implementation available at the moment. Extensions of the functionalities are under development.
5. SYMMETRY [Gatermann and Guyard 1996]. A Maple package written by K. Gatermann and F. Guyard for invariants and equivariants of finite and continuous groups. The focus lies on the computation of invariants and equivariants of compact Lie groups. See Section 3.2.7 for further information on the topic.
6. INVAR. A Singular library written by Gerhard Pfister for the computation of invariants of the additive group $G_a$ over $\mathbf{C}$. Different actions of $G_a$ are given by their differential vector field, and also non-degree-preserving actions are dealt with. This package is part of the standard distribution of Singular.
7. Harm Derksen is developing an implementation of his algorithm for invariants of linearly reductive groups [Derksen 1999] in Singular. For more information, please contact hderksen@math.mit.edu.

Gregor Kemper (Heidelberg)

## 2.3 Linear Algebra

### 2.3.1 Linear Systems

Many computer algebra systems include facilities for the solution of linear systems. Also some recent algorithmic developments exist that have not been extensively implemented. The emphasis in computer algebra is primarily on the exact solution of systems over such domains as

1. the integers (Diophantine systems)
2. the field of rational numbers
3. a finite field
4. the rational numbers extended by some algebraic or transcendental elements

We discuss solution methods based in Gaussian elimination, determinant expansions, and the black box methods.

**2.3.1.1 Elimination Methods** Basic Gaussian elimination works over any field, The values computed in elimination methods are minors or quotients of minors of the original matrix. Thus there can be a severe "intermediate expression size swell" when the entry field is infinite and exact arithmetic is used. McClellan [McClellan 1973] addressed this problem with a method which carries out the computation in a number of homomorphic images (entries in an integral subring of the field), and then uses the Chinese remainder algorithm or interpolation.

Another approach is to work in one homomorphic image and then construct successively better approximations by Hensel lifting until the exact result is obtained, [Moenck and Carter 1979; Dixon 1982; Mulders and Storjohann 1999]. One may view Hensel lifting as an exact variant of the numerical Gauss-Seidel method, with the amazing effect that the complexity is cubic in the dimension. Thus, for example, the exact solution of a system of linear equations with rational numbers as entries has asymptotically the same complexity as computing the solution by classical Gaussian elimination and in floating point arithmetic.

In some cases non-modular methods work well. Bareiss [Bareiss 1968] developed an approach based on determinant identities of Sylvester. The divisions are exact and quotients need not be stored, saving substantial storage and simplification time. Improvements have been made for sparse systems to avoid some fill-in and unnecessary steps. See, for example, [Sasaki and Murao 1982], [Lee and Saunders 1995].

**2.3.1.2 Determinant-Based Methods** Consider a a row operation on an integer matrix, $R_i \leftarrow aR_k + bR_i$. Because of the products, if $a$ and $b$ are the same size as the entries, generally the integer sizes will double. If the matrix is very sparse, the number of nonzero entries on the row will also nearly double. This problem is compounded if the entries themselves are sparse objects, as is typical if they are multivariate polynomials. If elimination is used, there is the fill-in and, worse, the growth of expressions can be exponential in the dimensions of the sub-matrices involved.

In such cases, it may be more efficient to exploit minor expansion. This situation was analyzed vis-à-vis determinant computation by Gentleman and Johnson, [Gentleman and Johnson 1974] or [Gentleman and Johnson 1976]. In extreme cases, nothing improves on direct use of the formula $\det(A) = \sum_\sigma (-1)^{\text{sgn}(\sigma)} \prod_i a_{i,\sigma i}$.

**2.3.1.3 Diophantine Solutions** For dense methods see the discussion of Hermite form computation in 2.3.2 and of the Lenstra, Lenstra, Lovász lattice basis reduction in 2.4.6. A recent development is [Mulders and Storjohann 1999]. Additional citations are to be found in the bibliography in [Giesbrecht 1997], where a method suitable for sparse matrices is also given.

**2.3.1.4 Black Box Methods** In numerical linear system solving a distinction is made between iterative methods and direct methods. For solving the linear system $Ax = b$, elimination is direct, exploiting and manipulating the internal structure of the matrix $A$. On the other hand, iterative methods generally treat the matrix as a black box (a term coined in [Kaltofen and Trager 1990]), a representation of the linear transformation on vectors, $u \to Au$. Black box methods are well suited to situations where the cost of the matrix-vector product $Au$ is relatively small. This is the case for sparse matrices and for certain structured matrices such as Toeplitz and Vandermonde matrices. For them the matrix-vector product can be done rapidly using FFT (Fast Fourier Transform) techniques.

In a watershed paper, Wiedemann [Wiedemann 1986] opened up the use of iterative methods in exact system solving by developing and analyzing the Krylov space method, with projection to recursive sequences in the entry domain and use of the Berlekamp/Massey algorithm to compute the minimal polynomial. Since then adaptation of the Lanczos and conjugate gradient methods to exact computation has been done [LaMacchia and Odlyzko 1991; Lambert 1996; Eberly and Kaltofen 1997; Teitelbaum 1998]. Lambert [Lambert 1996] discusses the relations among these approaches and shows their essential equivalence. Kaltofen [Kaltofen 1992a] applies the Wiedemann approach to computing the determinant and characteristic polynomial of a dense matrix without divisions. The algorithm is asymptotically about a $\sqrt{n}$ factor slower than the matrix multiplication algorithm that is employed, but no divisions are used. The [Kaltofen 1992a] result can be modified to obtain an algorithm of low bit complexity for computing integer and polynomial determinants. Kaltofen and Villard [to be published] have employed blocked methods (see below) and obtain a complexity of $O(n^{2.698})$ (or $O(n^{3+1/3})$ when using a cubic matrix multiplication procedure) for both the division free determinant problem of $n \times n$ matrices and the bit complexity of the determinant, for instance when the entries are fixed length integers.

For singular systems a number of issues arise. One may ask what constitutes a solution. To provide a basis for the solution plane is asking too much in the context of a sparse system. The basis may require an amount of space quadratic in the input size. One might as well use elimination. Instead Kaltofen and Saunders [Kaltofen and Saunders 1991] provide a method to rapidly obtain a random sample of the solution manifold. For this algorithm random preconditioners are necessary. The Beneš network preconditioners in [Wiedemann 1986] or the Toeplitz preconditioners in [Kaltofen and Saunders 1991] may be used (see [Chen et al. 2001] for further preconditioners). In the process the rank of the matrix is determined. Another approach to the rank of a sparse matrix is found in [Mulmuley 1987].

Possibly inconsistent systems then present yet another problem. Because of the random preconditioners, when a solution to a singular system is not found, it is desirable to determine whether it is due to inconsistency of the input system or to bad luck in the preconditioners. This may be resolved by computing an inconsistency certificate [Giesbrecht et al. 1998].

Coppersmith [Coppersmith 1994] described a block matrix version of the Wiedemann algorithm for linear system solution over GF(2) in the context of large integer factorization. In a number of papers [Kaltofen 1995a; Villard 1997b,c, 1998] this has been extended and improved. Similarly, block Lanczos methods have been developed [Coppersmith 1993; Montgomery 1995].

**2.3.1.5 Structured Systems** Many of the fast algorithms developed for matrices and linear systems with a special structure, e.g., Toeplitz, Vandermonde, and Hilbert matrices, are applicable in the exact and symbolic setting. We refer our reader to the excellent survey by Olshevsky [Olshevsky 2001] and the forthcoming book by Pan [Pan 2001] on methods for structured systems.

**2.3.1.6 Parallel Algorithms** For dense matrices, processor efficient algorithms are known [Kaltofen and Pan 1991, 1992; Villard 2000b]. For black box matrices, the block methods allow for use of parallel computations. Further development of this in [Kaltofen and Lobo 1999a].

Erich Kaltofen (North Carolina State U.),
B. David Saunders (U. Delaware)

## 2.3.2 Algorithms for Matrix Canonical Forms

Canonical forms of matrices are fundamental tools in various domains of algebra. The fact that they are very difficult to compute by means of fixed-precision reals makes them particularly amenable to computer algebra. For a matrix $A$ over a principal ideal domain R, the *Hermite form* $H = UA$ is triangular and the *Smith form* $S = VAW$ is diagonal. Both forms are obtained by unimodular transformations $U, V$ and $W$ over R. They are canonical representatives of the equivalence classes of matrices under unimodular pre-multiplication (Hermite), and unimodular pre- and post-multiplication (Smith). For a square matrix $A$ over a field K, the *Frobenius form* over K and the *Jordan form* over an algebraic extension of K, are block diagonal matrices similar to $A$. These canonical forms capture and display algebraic and geometric invariants of the matrices under similarity transformations. The classical theory offers constructive definitions of these forms [Gantmacher 1966; Newman 1972]. Considerable recent activity has produced new algorithms for all these forms which are practically fast and rigourously analysed.

**2.3.2.1 The Hermite and the Smith Canonical Forms.** The Hermite form is a generalization of the reduced row echelon form of a matrix over a field to a matrix over a principal ideal domain R. An important application is to solving systems of equations over R, see 2.3.1. The Smith form of an integer matrix is a basic tool of abelian group theory, see 2.7.7.

To control intermediate expression swell, many recent algorithms [Hafner and McCurley 1991; Storjohann 1996, 2000; Storjohann and Labahn 1996] follow the approach of [Domich et al. 1987; Iliopoulos 1989] and work modulo the determinant $d$ of a maximal rank minor of the input matrix. The algorithms over $\mathbb{Z}$ in [Storjohann 1996; Storjohann and Labahn 1996] are based on matrix multiplication. In addition to the form itself, unimodular transforming matrices may be required. Since these are in general non-unique, the goal is to produce ones with good size bounds on the entries. The currently fastest algorithms for recovering such transforms for an input matrix with arbitrary shape and rank profile are described in [Storjohann 2000]. Lattice reduction can be used to produce transforms with possibly much smaller entries, but at an increased cost [Havas et al. 1998]. Heuristic strategies for controlling expression swell have also been developed, see for example [Havas and Majewski 1997].

The Hermite form has been generalized in [Howell 1986] to matrices over a principal ideal ring which may have zero divisors. Algorithms appear in [Buchmann and Neis 1996; Mulders and Storjohann 1998; Storjohann 2000]. See also [Cohen 1996c].

The above algorithms using the modulo $d$ approach are adaptable also to matrices over $\mathsf{K}[x]$ with $\mathsf{K}$ a finite field. Recently, a faster (non-modular) algorithm for Hermite form over $\mathsf{K}[x]$ has been proposed [Mulders and Storjohann 2000]. However, over $\mathbb{Q}[x]$, both the degrees of intermediate polynomials and the sizes of rational number coefficients need to be bounded. Computing the Hermite form over $\mathbb{Q}[x]$ was first shown to be in $\mathcal{P}$ in [Kannan 1985]. The algorithm in [Labhalla et al. 1995] reduces the computation of the Hermite form over $\mathsf{K}[x]$ to the problem of triangularizing a large matrix over $\mathsf{K}$. This leads to good bounds for the degrees of intermediate polynomials and for the sizes of rational number coefficients when $\mathsf{K} = \mathbb{Q}$. A fraction free method for computing the Hermite form over $\mathbb{Q}[x]$ is derived in [Beckermann et al. 1999]. In [Beckermann et al. 1999] and [Mulders and Storjohann 2000] analogous results are obtained for column reduced and Popov forms. The approach of [Labhalla et al. 1995] is adapted in [Villard 1995] to show that computing transforming matrices for the Smith form over $\mathbb{Q}[x]$ is in $\mathcal{P}$. A practical sequential algorithm for computing the Smith form (without transforms) over $\mathbb{Q}[x]$ is given in [Storjohann and Labahn 1997].

For the problem of computing the determinant or Smith form of a dense integer matrix without transforms, randomization can be used to get asymptotically faster algorithms. The algorithm for Smith form over $\mathbb{Z}$ in [Giesbrecht 1995a] has an improved space complexity compared to previous algorithms. A combination [Kaltofen 2000] of [Kaltofen 1992a] and [Giesbrecht 1996] or the recent algorithm in [Eberly et al. 2000] achieve better asymptotic costs.

**2.3.2.2 The Frobenius and the Jordan Canonical Forms.** The Frobenius and Jordan forms effectively display all similarity invariants of a matrix. Once known, it is easy to determine matrix similarity, find the minimal polynomial of a matrix, and to quickly exponentiate and evaluate polynomials at that matrix [Giesbrecht 1995b].

Computing the Frobenius form of an $n \times n$ matrix $A$ over a field K was first accomplished efficiently in [Ozello 1987] with an algorithm requiring $O(n^4)$ operations in K. Essentially, a tower of maximal Krylov subspaces are built which lead to a construction for the transforming matrix. The cost is reduced to $O(n^3)$ field operations in [Storjohann 1998] (see also [Storjohann and Villard 2000]). Using a related approach to [Ozello 1987] and randomization, the cost of computing the Frobenius form is linked to that of matrix multiplication in [Giesbrecht 1995b], where an algorithm requiring $O(\log n)$ multiplications of $n \times n$ matrices is presented (see also [Eberly 2000]). Deterministically, the characteristic polynomial and a (non-canonical) shift-Hessenberg form can be computed in this same time [Keller-Gehrig 1985]. Recently, the Frobenius form problem has been reduced to computing $O(\log \log n)$ shift-Hessenberg forms of $n \times n$ matrices [Storjohann 2000], thus showing that also deterministically the Frobenius form can be computed in essentially the time required for matrix multiplication.

The more refined Jordan form comes at the expense working symbolically in an extension of the ground field K. Even if the eigenvalues are not known, most of the information given by the form – including the dimensions of the blocks – can be computed over an arbitrary field. The cost is alleviated through dynamic evaluation for explicit computations with algebraic numbers [Duval 1994], or through factor refinement [Kaltofen et al. 1990] to find the best possible annihilating polynomials over K for the eigenvalues. Efficient algorithms for the Frobenius and Jordan forms of matrices over the integers, rationals and number fields are proposed and analysed in [Giesbrecht 1994a; Giesbrecht and Storjohann 2000; Ozello 1987].

**2.3.2.3 Sparse Matrices.** The paradigm of black-box matrices as discussed in 2.3.1 may be successfully employed for computing matrix normal forms. The immediate relevance of computing the minimal polynomial of a matrix $A$ by iterative methods [Kaltofen and Saunders 1991; Wiedemann 1986], is that the companion matrix of the minimal polynomial appears as the largest block of the Frobenius form of $A$. Recently, in [Villard 2000a], it has been shown how to compute the remaining blocks in the Frobenius form quickly by means of randomized perturbations of the input matrix.

The black-box paradigm may also be employed to compute the Smith form of sparse integer matrices. In [Giesbrecht 1996, 2001] the Smith form of an $n \times n$ integer matrix $A$ is computed with high probability from the minimal polynomials of $O(\log n)$ non-derogatory matrices obtained from randomized Toeplitz pre-conditionings of $A$. Asymptotically this is the fastest known method to compute the Smith form of a sparse integer matrix.

**2.3.2.4 Fast Parallel Algorithms.** Fast parallel algorithms to compute the canonical forms are known under the arithmetic PRAM model [Karp and Ramachandran 1990] over a field K. By linearization, the problem of computing the Hermite form over K[$x$] reduces to solving a linear system of equations over K and is thus in $\mathcal{NC}$ [Kaltofen et al. 1987; Villard 1996].

Randomization techniques – analogous to those used in sequential algorithms – may then be applied in parallel to avoid the inherently sequential aspects of the elimination processes. This is done in [Kaltofen et al. 1990] for the Smith form and in [Giesbrecht 1995b] for the Frobenius form, placing the problems in $\mathcal{RNC}$ [Kaltofen et al. 1990]. The latter algorithm can be implemented in a processor-efficient manner [Giesbrecht 1995b].

A different approach based on the use of the embedded kernels $\ker(A - xI)^k$ leads to deterministic algorithms. Using algebraic numbers, a symbolic Jordan form is computed from the kernels [Roch and Villard 1996]. By gathering its blocks together appropriately, the Smith and Frobenius forms may be deduced. These problems are shown in [Villard 1997a] to be in $\mathcal{NC}$. The existing theoretical links between the forms have been used to obtain similar algorithms, either with respect to similarity or to unimodular equivalence.

Mark Giesbrecht (London), Arne Storjohann (Zurich)<br>and Gilles Villard (Lyon)

## 2.4 Constructive Methods of Number Theory

### 2.4.1 Primality Tests

An important problem of algorithmic number theory is to decide whether a natural number $n$ is prime or not. A common procedure is to apply a probabilistic *primality test* first. It either certifies that $n$ is composite, or indicates that $n$ is prime with high probability. In the latter case one uses a deterministic primality test to show that $n$ actually is a prime number.

A first method uses *Fermat's little theorem*: Let $n$ be prime. Then for all $a \in \mathbb{Z}$ with $\gcd(a, n) = 1$, the congruence $a^{n-1} \equiv 1 \bmod n$ holds.

The theorem leads to the following, so-called *Fermat test*: We choose a random integer $0 < a < n$, and compute the GCD of this number and $n$. If the result is greater than 1, then we found a divisor of $n$, therefore $n$ is composite. Otherwise, we compute $a^{n-1} \bmod n$, which can be done efficiently by *binary exponentiation*. If the result is different from 1 then $n$ has to be composite. However, the method does not yield a divisor of $n$. On the other hand, when the result is 1, we have to repeat this step by selecting another integer $a$. Provided that $n$ is composite, the procedure "almost always" terminates after a few iterations. Unfortunately, this is not true for any composite number. For the so-called *Carmichael numbers* - which are composite - the congruence $a^{n-1} \equiv 1 \bmod n$ is true for every $0 < a < n$ with $\gcd(a, n) = 1$. The smallest Carmichael number is $561 = 3 \cdot 11 \cdot 17$.

There is a whole class of similar methods which allow to decide whether a given number $n \neq 2$ is composite or "probably prime". We assume in these methods that $a$ is chosen randomly in $\{2, \ldots, n-1\}$ and is coprime with $n$. A first improvement over the Fermat's method is the test by *Solovay-Strassen*: We test whether $a^{(n-1)/2} \equiv \left(\frac{a}{n}\right)$ modulo $n$, where $\left(\frac{a}{n}\right)$ denotes the Jacobi-Kronecker symbol which is quite easy to compute (see, e.g. [Cohen 1996b; Bach and Shallit 1996]). In fact, if this congruence does not hold, then we have proven that $n$ is composite. Moreover, at most $n/2$ values of $a$ will fulfil this equation for composite $n$, in other words the probability that we can prove the compositeness of $n$ is greater than $\frac{1}{2}$. If $n$ passes this test for, say 20 different values of $a$, then we know that $n$ is probably prime with a probability greater than $1 - (\frac{1}{2})^{20}$.

The *Miller-Rabin* primality test improves the Fermat test by checking for nontrivial square roots of 1, modulo $n$, when computing the modular exponentiation, since the existence of a number $x \neq \pm 1$ such that $x^2 = 1 \bmod n$ proves the compositeness of $n$. A single Miller-Rabin test fails with probability less than $\frac{1}{4}$. Hence fewer trials have to be made to ensure a given probability, and since this test does not require the computation of a Jacobi-Kronecker symbol, this test completely supersedes the Solovay-Strassen test and is part of almost any number theory software package (such as LiDIA 4.2.22). As a remark we note that neither of the latter two tests admits the possibility of analogues to Carmichael numbers.

The solution of the second problem, namely to prove that a number which is "presumably prime" is in fact prime, turns out to be far more difficult, for all practical purposes. There are two applicable methods, one is based on Jacobi sums (see [Cohen and Lenstra 1984], [Mihalescu 1998]), another one uses elliptic curves (see [Morain 1998]). The latter method will now be described in more detail:

A theoretical result concerning the solution of our problem is the following theorem by *Pocklington*: Let $s > \sqrt{n}$ divide $n - 1$. If for all divisors $q$ of $s$, there exists a number $a$ with $a^{n-1} \equiv 1 \bmod n$ and $\gcd(a^{(n-1)/q}, n) = 1$, then $n$ is prime.

In practical applications however, we encounter the problem of having to find a "large" divisor of $n - 1$, which often is a rather difficult task. Like in factorization, we can obliterate this difficulty by replacing the prime residue class group $(\mathbb{Z}/n\mathbb{Z})^*$ by the point set of an elliptic curve.

An elliptic curve $E$ over a field $K$ can be defined by an equation of the following (affine) type:

$$E: \quad y^2 + a_1 xy + a_3 = x^3 + a_2 x^2 + a_4 x + a_6,$$

where the parameters $a_1, a_2, a_3, a_4, a_6$ lie in $K$. The word "elliptic" stems from the theory of elliptic integrals, which for instance arise in the context of computing the length of an arc of an ellipse. The non-empty set of (projective) points which are on the curve, together with the unique point $\mathcal{O}$ at infinity, is called the point set of $E$ and is denoted $E(K)$. With simple formulas for arithmetic operations, $E(K)$ forms a finite abelian (additive) group; the zero element is the point $\mathcal{O}$. A geometric interpretation of the addition law can be given as follows: If $P$ and $Q$ are points in $E(K)$, then the line from $P$ to $Q$ (the tangent if $P = Q$) intersects the curve at a third point $R$. The sum $P + Q$ is the inverse of $R$, in other words $P + Q + R = \mathcal{O}$. Useful introductions into the theory of elliptic curves are given in [Silverman 1987; Cohen 1996b; Menezes 1993]).

We obtain the following analogue of Pocklington's theorem:

Let $n \in \mathbb{Z}$ with $\text{GCD}(6, n) = 1$, and let $E_n$ be the point set of an elliptic curve over $\mathbb{Z}/n\mathbb{Z}$. Furthermore, let $s > (\sqrt[4]{n} + 1)^2$ a divisor of order $\#E_n$. If there exists a point $P \in E_n$ with $(\#E_n/q) \cdot P \neq \mathcal{O}$ for all divisors $q$ of $s$, then $n$ is prime.

The theorem leads to the the *Goldwasser-Kilian-Atkin (GKA) algorithm*: first, we choose a suitable group order, i.e., one which has a large factor $s$, where $s$ is supposed to be prime. Now, we determine an elliptic curve whose point set has, as a group over $\mathbb{Z}/n\mathbb{Z}$, the chosen order. Now, we are able to check the premises of the theorem, and to verify that $n$ is a prime number, always under the hypothesis that $s$ is in fact prime. Then, we apply the same procedure recursively to $s$. In the course of this process, the numbers which have to be tested become smaller and smaller, until we reach a range of integers which allows to decide primality by other means (e.g., by a precomputed table, or the *sieve of Eratosthenes*).

Finally, there remains only one question: how can we determine an elliptic curve $E_n$ whose point set has a given order over $\mathbb{Z}/n\mathbb{Z}$? To that end, we assume once more that $n$ is prime. Then, by a theorem of Hasse, the point set $E_n$ has order $n + 1 - c$, where $|c| \leq 2\sqrt{n}$. Furthermore, for a special class of elliptic curves – so-called ordinary elliptic curves – there exists an element $\pi$ in the maximal order of an imaginary quadratic number field with trace $c$ and norm $n$. Therefore, we randomly choose some imaginary quadratic number field, and check whether there exists such $\pi$ in the maximal order of the number field. If we can find a maximal order and such an element $\pi$, then we are able to construct an elliptic curve $E_n$ to our requirements. For that, we compute the so-called *Hilbert polynomial* with respect to the maximal order. If factorized modulo $n$, such a polynomial gives us the $j$-invariant of $E_n$, an invariant of isomorphy classes of elliptic curves. From the $j$-invariant we can easily derive the desired elliptic curve $E_n$.

Practical implementations can handle very large numbers. For example, the primality of $(2^{7331} - 1)/458072843161$ (2196 digits) has been successfully verified by this method (see [Morain 1998]) in one month on a DecAlpha 400 MHz.

An interesting feature of the elliptic curve primality proof is that is also yields a certificate for the primality which can be verified very quickly.

The largest known prime numbers however, are obtained in a different way. They all have a special form, they are so-called *Mersenne numbers* $M_n = 2^n - 1$. For Mersenne numbers, there exists a simple criterion for primality. First, we define the sequence $(e_n)_{n \in \mathbb{N}}$ by $e_1 = 4$, and $e_{k+1} = e_k^2 - 2, k \geq 1$. Then $M_n$ is prime if and only if $e_{n-1} \equiv 0 \bmod M_n$. This condition can be checked by basic calculations even though involving huge numbers. This method is often used to benchmark the capabilities of supercomputers. The largest prime number known to date, $M_{6972593}$ ($\approx 2 \cdot 10^6$ digits), was computed in that manner.

Johannes Buchmann (Darmstadt), Volker Müller (Yogyakarta, Indonesia) and Thomas Pfahler (Darmstadt)

### 2.4.2 Integer Factorization

The problem of decomposing a large number into its prime factors has been of special interest to number theorists for quite some time. During the past twenty years however, the problem also gained considerable practical importance. The presumption that factoring large integers is difficult became the basis of security in some modern crypto systems (see also 2.14).

The most important concurrent methods for factoring are the *elliptic curve method (ECM)* [Lenstra and Lenstra Jr. 1990], the *quadratic sieve (QS)* [Silverman 1987], and the *number field sieve (NFS)* [Lenstra and Lenstra Jr. 1993; Buchmann et al. 1994].

While the running time of QS and NFS depends on the size of the number $n$ to be factored, the running time of ECM depends on the size of the largest prime factor of $n$. Therefore, ECM can be used to factor very large numbers with moderate prime factors. The largest prime factor ECM has found so far has 54 decimal digits (see [Brent 2001]).

We explain the idea of ECM. Let $N \in \mathbb{N}$ be a composite number, $\mathcal{E} = E(\mathbb{Z}/N\mathbb{Z})$ an elliptic curve over $\mathbb{Z}/N\mathbb{Z}$, and $P = (x_p, y_p, z_p) \in \mathcal{E}$ a point on this curve. Let $p$ be the smallest prime factor of $n$. If the number of points on the curve defined modulo $p$ has only small prime factors, it is easy to compute a multiple $k$ of that order by multiplying together appropriate powers of small primes. By attempting to compute $k$ times $P$, one will eventually find a factor of $n$. This is the basis of the *elliptic curve method (ECM)*:

Repeat the following three steps until a factor is found:

1. randomly choose a curve $E$ and a point $P$,
2. choose suitable $B_1, B_2 \in \mathbb{N}$, and compute $k = \prod_{p \in \mathbb{P}, \ p < B_1} p^{max\{l : p^l < B_2\}}$,
3. compute $k \cdot P$ making use of the formulas for addition over fields, and fast techniques for exponentiation. The division required in addition of points can be reduced to computing inverses modulo $N$ via GCDs. If the inverse can not be determined then the GCD is greater than 1, i.e., we found a factor.

QS and NFS are both based on the construction of quadratic congruences. For any two numbers $x, y \in \mathbb{Z}$ with $x^2 \equiv y^2 \bmod N$ and $x \not\equiv \pm y \bmod N$, we have $N | (x^2 - y^2) = (x - y)(x + y)$, and $N \not| (x - y)$, as well as $N \not| (x + y)$. Therefore $\gcd(N, x - y)$ is a proper factor of $N$.

In QS, we choose a finite set $\mathcal{F}$ of prime numbers, a so-called *factor base*. Then, we compute sufficiently many arguments $y$ such that the quadratic polynomial $z(x) = (x + m)^2 - N$, where $m = \lfloor \sqrt{N} \rfloor$, has only prime factors in $\mathcal{F}$,

### 2.4.3 Algebraic Number Fields and Algebraic Function Fields

when evaluated at $y$. This leads to congruences $(y+m)^2 \equiv z \mod N$, where $z = \prod_{p \in \mathcal{F}} p^{e_{p,z}}$. If we can find sufficiently many congruences of that form then, by multiplying together suitable congruences, we can compute the squares we are looking for. The selection of congruences involves solving a linear system of equations over the field with two elements. The linear system is sparse, and typically has up to 1 000,000 equations and unknowns.

To obtain enough suitable arguments $y$, we test all $y$ with $|y| \leq M$ for some bound $M \in \mathbb{N}$. We accomplish that in three steps:

1. For all $p \in \mathcal{F}$ compute $y_p^{(1)}$, and $y_p^{(2)} \in \{0, \ldots, p-1\}$, such that $z(y_p^{(j)}) \equiv 0 \mod p$, $j \in \{1, 2\}$. If there is no such $y_p^{(j)}$, then remove $p$ from $\mathcal{F}$.
2. Select a bound $M \in \mathbb{N}$, and initialize all elements of an array $lz$ of length $2M+1$ to zero.
3. For all $p \in \mathcal{F}$, $j = 1, 2$, and all $y = y_p^{(j)} + l \cdot p$ with $l \in \mathbb{N}$ and $|y| \leq M$, replace $lz(y)$ by $lz(y) + \log p$.
4. At all evaluation points $y$ for which $lz(y)$ is sufficiently close to $\log z(y)$, try to construct the factorization of $z(y)$.

The efficiency of *QS* essentially depends on the speed of the sieving method used in the algorithm. Several variants and improvements have been developed in order to enhance efficiency. The largest number factored by *QS* to date has 129 decimal digits (see [Atkins et al. 1995]). To accomplish this task, massively parallel computers were employed. On a single computer, the running time would have been several hundred years. Several Computer Algebra systems have implemented *QS*; among those are Magma 4.1.5, Pari 4.2.30, and LiDIA 4.2.22.

The running time of *NFS* is asymptotically faster than the running time of QS. For that method, the quadratic congruences are generated via algebraic number fields. A famous record of NFS is the factorization of the ninth Fermat-Number $2^{2^9}+1$ which has 155 decimal digits (see [Lenstra et al. 1993]). Another, very recent one is the factorization of the 512-bit number RSA-155 (see [te Riele and et al. 2000]).

Additional reference: [LiDIA-Group 1998].

Johannes Buchmann (Darmstadt), Thomas Pfahler (Darmstadt) and Jörg Zayer (Saarbrücken)

### 2.4.3 Algebraic Number Fields and Algebraic Function Fields

In recent years algorithmic algebraic number theory has attracted rapidly increasing interest. There is now a regular meeting ANTS (algebraic number theory symposium) every two years, whose proceedings [Adleman and Huang 1994], [Cohen 1996a] give a good survey about ongoing research. Also there are several computer algebra packages concentrating on number theoretical calculations. At present, the most important ones, which are available for free, are KANT V4[Daberkow et al. 1997], PARI[Batut et al. 2000] and SIMATH[Zimmer 1997].

KANT V4comes with a data base for algebraic number fields, which already contains more than a million fields of small degree. We note that almost all of KANT V4 and parts of PARI are also contained in the MAGMA system [Bosma et al. 1997].

Hans Zassenhaus, one of the pioneers in this area, stated four principal computational tasks: the calculation of the Galois group, of an integral basis, of the unit group, and of the class group. Methods for determining these invariants (and also their complexity) are of considerable importance for the solution of diophantine equations, the factorization of large integers (using a *number field sieve*, see 2.4.2), and in cryptography (especially for *public key cryptosystems*, see 2.14). In the last 30 years, numerous algorithms for computing all invariants mentioned above have been developed, and are continuously improving. Important ingredients of these algorithms are methods from the geometry of numbers (see 2.4.6), as well as probabilistic methods. For the former group of methods, it is crucial that a number field $F$ of degree $n$ becomes a Euclidean vector space of dimension $n$ by choosing $\sum_{j=1}^{n} x^{(j)} \overline{y^{(j)}} = y^H x$ (where $^H$ denotes the conjugate transpose) as the scalar product of $x, y \in F$. The integral elements of $F$ then form a lattice. On the other hand, probabilistic methods are used for generating relations among ideals (power products which are principal ideals). Also sieving techniques are frequently applied. Current methods allow computations in number fields of small degree ($n \leq 4$), with discriminants in the order of up to $10^{80}$, as well as fields of degree up to 30 with small discriminants.

Since number fields of larger degree are usually not given as extensions of the rationals directly, but rather by a sequence of extensions or as a composite of fields, the subject of relative extensions has been studied in greater detail for more than five years. The relative point of view causes considerable new difficulties. We just mention the non existence of a relative integral basis in general, the different structure of a relative unit group and the problem of an adequate definition of the relative class group. Among the results obtained we should mention normal forms for modules over Dedekind domains. Several new algorithms require a detour via Kummer extensions, i.e. one has to adjoin appropriate roots of unity first, extend the larger field and then intersect to the subfield of that field in which one is interested. This led to the development of efficient algorithms for computing subfields. Variants of them can be used to determine (relative) automorphisms.

The results on arbitrary relative extensions were then used for computing Hilbert and ray class fields. A generation of class fields was possible only over imaginary quadratic fields by analytic functions beforehand. A generalization of that theory to arbitrary base fields (*Stark's conjectures*) could be recently used for calculating Hilbert class fields over small totally real cubic number fields. For arbitrary fields, a purely algebraic method is applied, which follows the lines of the proof of the existence theorem. Again a detour via Kummer extensions is necessary. Best results are now obtained by extensively using *Artin symbols*.

The potential of doing efficient calculations in number fields was used by several people for calculating integer solutions of diophantine equations and even parametric families of equations. We especially point out the results on Thue equations, unit equations and index form equations, for which considerable progress was made in the last three years. The study of such equations for relative extensions is just beginning.

Algebraic function fields in one variable over finite fields possess similar properties as algebraic number fields. Hence, it was only natural that people tried to transfer the algorithms for number fields to the function field case. This was easy for quadratic extensions, in which Artin's thesis showed the existence of a continued fraction expansion and therefore Shanks' infrastructure theory could be applied. For extensions of larger degree, difficulties occured because of a missing scalar product. They were overcome in a recent thesis by M. Schörnig, in which the author developed a new concept of reduced bases which can be calculated in an efficient way (see [Cohen 1996a]).

An introduction to the theory can be found in [Cohen 1996b] and [Pohst and Zassenhaus 1989], which also contains a survey on recent results and the corresponding references.

Michael E. Pohst (Berlin)

## 2.4.4 Galois Groups

The constructive Galois theory is divided into two subproblems. First there are algorithms to determine the Galois group of a given polynomial (direct problem), second there are methods to construct polynomials with given Galois group (inverse problem).

For the first problem deterministic exponential time algorithms were already used more than 100 years ago (see [Tschebotaröw and Schwerdtfeger 1950]). Nevertheless until today no general polynomial time algorithm is known. Since the factorization of polynomials over number fields is in polynomial time [Lenstra et al. 1982; Landau 1985], one can compute automorphisms of normal fields, compute a maximal subfield, and decide solvability in polynomial time [Landau and Miller 1985]. Moreover, polynomials with alternating or symmetric Galois group, which have asymptotic density 1, can be detected in polynomial time [Landau 1984]. Over number fields (and global function fields) there exists a fast probabilistic (deterministic using ERH) algorithm for this problem [Davenport and Smith 2000]. This is based on effective versions of the density theorem of Chebotarev (Tschebotaröw) [Lagarias and Odlyzko 1977].

All practical algorithms use the classification of transitive groups, which is completed up to degree 31 [Hulpke 1996a]. These are divided into the resolvent method [Soicher and McKay 1985; Mattman and McKay 1997] and the Stauduhar method [Stauduhar 1973]. The first one uses the factorization of resolvent polynomials which can be computed from the coefficients of the given polynomial [Casperson and McKay 1994]. There are no precision problems, but the degree of the resolvent polynomials can be very large. There is an implementation of this method in MAPLE up to degree 8.

In the Stauduhar method one has to decide if there are integer roots of a polynomial. There are implementations in PARI [Eichenlaub and Olivier 1995] (up to degree 11), and Kash (up to degree 15) which use complex approximations of the roots. In [Yokoyama 1997] the use of p-adic methods is suggested to avoid precision problems. There is an efficient implementation of this method in Kash (up to degree 15) [Geissler and Klüners 2000] which gives improvements for imprimitive groups using the computation of subfields [Klüners and Pohst 1997; Klüners 1998]. Another theoretically interesting variant based on symbolic evaluation of invariants is proposed in [Colin 1997b].

It is still not known whether every finite group is a Galois group over $\mathbb{Q}$. In particular, there does not exist a general method which would allow, starting from some finite group $G$, to construct an extension $N/\mathbb{Q}$ with Galois group $G$. First, one has to realize simple groups as Galois groups. As a second step, field extensions with composite groups can be obtained by solving embedding problems.

Using the rigidity method [Malle and Matzat 1999] one can realize all abelian groups and many simple groups using computational representation theory or the Katz algorithm [Dettweiler and Reiter 2000] from differential Galois theory, respectively. For the explicit construction of a polynomial one has to solve an algebraic system of equations. This can be done using a (modular) version of the algorithm of Buchberger. The algorithm of [Nauheim 1998] is very useful for problems of this kind since it allows bad reduction.

The book [Malle and Matzat 1999] gives an overview which embedding problems can be solved. Over a Hilbertian field every split embedding problem with abelian kernel can be solved. If a group $G$ is realized as a Galois group, one can easily realize each factor group of $G$ as Galois group. Combining these two methods one gets a class of groups called semi-abelian groups [Dentzer 1995]. For the solution of embedding problems with non-abelian kernel there exists a construction method based on so-called GAR-realizations of the composition factors of the kernel [Malle and Matzat 1999].

Complete lists of polynomials over $\mathbb{Q}$ for transitive permutation groups up to degree 11 are contained in [Eichenlaub 1996; Malle and Matzat 1999] and for the degrees 12 to 15 in [Klüners and Malle 2000]. Moreover in [Klüners and Malle 2000] it is proved that all transitive groups up to degree 15 have a regular realization over $\mathbb{Q}(t)$. Explicit polynomials for such regular Galois extensions are computed for nearly all transitive groups up to degree 11 and for most primitive non-solvable groups up to degree 30 [Malle and Matzat 1999].

Jürgen Klüners and B. Heinrich Matzat (Heidelberg)

### 2.4.5 Rational Points on Elliptic Curves

Elliptic curves and higher dimensional abelian varieties take center stage in algebraic number theory, and in arithmetic algebraic geometry. They also gain more and more importance in computer science, e.g., for factoring large integers [Lenstra Jr. 1987], and in primality tests [Atkin and Morain 1993; Silverman and

### 2.4.5 Rational Points on Elliptic Curves

Tate 1992]. Being the most basic abelian varieties, namely those of dimension one, elliptic curves are explicitly suited for computation. The curves are being investigated over global, local, and finite fields [Silverman 1986]. One of the main problems is the algorithmic computation of the group of rational points on elliptic curves over the fields mentioned above, the so-called *Mordell-Weil group*.

Over finite fields, we have a very efficient, combined algorithm of *"counting points"* by Shanks-Schoof at our disposal [Schoof 1985; Shanks 1971; Lehmann et al. 1994]. In this case, the order of the (finite) Mordell-Weil group is bounded by a constant depending on the cardinality of the underlying field, by the theorem of Hasse (an analogue of the Riemann hypothesis). The construction of cryptographically relevant elliptic curves over large finite fields is accomplished by means of class field theory (see [Lay and Zimmer 1994]).

However, over algebraic number fields, there is no general algorithm for computing the Mordell-Weil group. Although the finite part of that group, the so-called *torsion group*, can be computed relatively easily, the free part causes considerable difficulties. One way out was shown by Manin, through his *"conditional"* algorithm [Manin 1971]. It is based on the correctness of the famous conjectures by Birch and Swinnerton-Dyer, by Shimura, Taniyama, Weil and Hasse. Under this assumption, the Mordell-Weil group can be determined algorithmically, at least in principle. Manin's algorithm was implemented at Saarbrücken in the computer algebra system SIMATH[Gebel and Zimmer 1993]. Another Mordell-Weil algorithm, based on 2-descent and using the theory of modular forms is due to Cremona ([Cremona 1992], see also [Siksek 1995]). The basic ideas of this algorithm are already contained in the famous paper [Birch and Swinnerton-Dyer 1963] by Birch and Swinnerton-Dyer. Of course, 3-descent (see [Gebel 1996]) and 4-descent (see [Merriman et al. 1996]) can also be applied for computing the Mordell-Weil group. The 2-descent method was extended to elliptic curves over real quadratic number fields by P. Serf [Serf 1995].

The computation of the torsion groups of elliptic curves over number fields, and of their integral points, as well as the construction of elliptic curves of high rank over such fields, constitute further important goals, which came within reach by application of computer algebra methods [Zimmer 1989, 1994; Zagier 1987; Gebel 1996; Gebel et al. 1994; Stroeker and Tzanakis 1994; Mestre 1992; Nagao and Kouya 1994; Fermigier 1996]. Meanwhile, all $S$-integral points can also be determined, where $S$ denotes a finite set of places of $\mathbf{Q}$ [Gebel 1996; Gebel et al. 1996; Smart 1994].

Furthermore, computer algebra systems allow for extensive numerical experiments for testing fundamental conjectures in algebraic number theory and arithmetic algebraic geometry. In this context, we only mention the various versions of the Birch and Swinnerton-Dyer conjecture, and its generalizations, as well as Beilinson's conjecture (see, e.g., the books [Cohen 1996b; Hulsbergen 1992]).

Analog questions for elliptic curves over algebraic function fields (in lieu of number fields) can also be treated algorithmically. Some problems turn out to be more difficult than in the number field case, some, on the other hand, become easier to solve. Here, Drinfeld modules play a crucial role [Deligne and Husemöller 1987; Goss et al. 1992; Goss 1996; Gekeler 1985], because, according to Jacquet-Langlands and Weil, the classical theory of modular forms on the upper complex half-plane can be formulated for arbitrary global fields, i.e., for number fields as well as for function fields.

Increasingly, abelian varieties of higher dimension than elliptic curves are being investigated, e.g., Jacobian varieties of hyper-elliptic curves of genus two (see [Cassels and Flynn 1996]). Also in this case, one area of interest is the computation of the rational point group over number fields, and over function fields. Applications in computer science are again related to factoring large integers, and primality testing [Adleman and Huang 1992].

Horst G. Zimmer (Saarbrücken)

### 2.4.6 Geometry of Numbers

A considerable number of computer algebra problems can be translated into the language of arithmetic lattices, i.e. discrete subgroups $L$ of finite dimensional Euclidean spaces. Because of the existence of a scalar product the objects of $L$ are equiped with a length and there exists an angle between any two non-zero elements of $L$. Every lattice is a free $\mathbb{Z}$-module with a $\mathbb{Z}$-basis, say $\mathbf{b}_1, ..., \mathbf{b}_n$. The geometrical structure is then contained in the corresponding *Gram matrix* $A = (a_{ij}) \in \mathbb{R}^{n \times n}$, where the entry $a_{ij}$ is the scalar product of the basis vectors $\mathbf{b}_i$ and $\mathbf{b}_j$.

In recent years methods of the geometry of numbers were applied, for example, to problems like *the knapsack problem*, various problems in algebraic number fields (finding a good generating polynomial, calculating fundamental units), problems of diophantine approximation (prove that given real numbers are not zeros of integer polynomials, whose degree and coefficients are bounded), proving that factorizing polynomials over number fields is polynomial time, solving systems of integral linear equations.

When these and other problems are formulated in the language of the geometry of numbers they usually lead to one of the following tasks:

- the determination of suitable bases for a lattice as well as any of its sublattices (Hermite and Smith normal form),
- the computation of a lattice basis consisting of short (or shortest) vectors (LLL–reduction, Korkine-Zolotareff reduction, Minkowski reduction),
- the computation of a lattice basis from a system of generators (MLLL–method, Hermite normal form),
- the computation of short (or shortest) lattice vectors, or more general, the nearest lattice vector to a given vector,
- the computation of successive minima,
- the computation of automorphism groups of lattices.

In general, a lattice $L$ is given either by a system of generating vectors in Euclidean $n$–space or by the Gram matrix of the scalar products of a basis.

Hence, the usual presentations of finitely generated abelian groups apply, i.e. $L$ can be presented by a set of generators and a relation matrix. Normal forms of representations are then obtained by transforming the relation matrix into its Hermite normal form. Besides the existence of polynomial time algorithms for computing Hermite normal forms, for large matrices there are no known algorithms which solve this task efficiently. This is because of the occurrence of large intermediate entries. G. Havas developed several strategies which do the elimination procedure in a way to keep all entries at least fairly small, see [Havas and Majewski 1997]. If the matrix is quadratic and non singular, modular versions can be applied, for which the size of the entries is bounded by the absolute value of the discriminant.

For higher dimensions, only LLL–reduction can be performed in general. Since LLL–reduction was introduced by A.K. Lenstra, H.W. Lenstra, and L. Lovasz in [Lenstra et al. 1982] about fifteen years ago, this method has become very popular due to the great variety of potential applications, the favorable complexity, and its efficiency in actual calculations. The method was later generalized, so that it can also be used to compute a lattice basis from a system of generators, see e.g. [Pohst 1987a]. Though these methods work well for integral lattices (whose Gram matrices have rational integers as entries), their application to real lattices is usually not stable inasmuch as the LLL–condition (an inequality between the lengths of orthogonal projections of two consecutive basis elements) is hard to verify in that case.

Unfortunately, except for the original papers there is hardly any literature on this, especially a good survey.

Michael E. Pohst (Berlin)

## 2.5 Algorithms of Commutative Algebra and Algebraic Geometry

### 2.5.1 Algorithms for Polynomial Ideals and Their Varieties

A central problem of algorithmic commutative algebra and algebraic geometry is solving of systems of polynomial equations in several variables. In the following, we restrict ourselves to the solution over algebraically closed fields. For real solving compare 2.5.3.

Let $K$ be an arbitrary field, and $L$ an algebraically closed extension field of $K$; let $F$ be a finite set of polynomials in $R = K[X_1, \ldots, X_n]$, and let $I = \mathrm{Id}(F)$ be the ideal generated by $F$ in $R$; finally, let $V(F) = V(I)$ be the zero set (variety) of $F$ or $I$, respectively, in $L^n$.

First, it is important to decide whether $V(I)$ is the empty set, or non-empty and finite, or infinite. This is easily derived from the leading terms of an arbitrary Gröbner basis $G$ of $I$ (see 2.2.5). By Hilbert's Nullstellensatz $V(I) = \emptyset$ iff $G$ contains a non-zero constant; $V(I)$ is finite, iff for each $X_i$ a pure power of this indeterminate occur as highest term of a polynomial in $G$. The decision, whether $I = R$, requires only computation of a partial Gröbner basis of $I$, using the effective Nullstellensatz in [Fitchas and Galligo 1990].

Next, we consider the case where $V(I)$ is non-empty and finite, i.e., where the ideal $I$ is *zero-dimensional*. In this case the quotient ring $R/I$ is a $K$-vector space of finite dimension. Furthermore, its dimension $d$ is a bound for the number of zeros of $F$, i.e., for $|V(I)|$. If $I$ is even a radical ideal, then the bound is sharp. The set of reduced terms with respect to $G$ forms a linear basis for $R/I$ over $K$. By reduction of products of reduced terms with respect to $G$ one obtains the structural constants of $R/I$ as $K$-algebra [Becker et al. 1998] ( compare 2.2.5). An alternative approach to algorithmic computing in $R/I$ is described in [Auzinger and Stetter 1988].

The computation of the radical of $I$ in the zero-dimensional case can by done using $G$ by first computing e. g. from a Gröbner basis of $I$ all univariate polynomials in $K[X_i] \cap I$, and replacing them by their squarefree part.

The exact algebraic computation of the finitely many solutions as tuples over the algebraic closure $\overline{K}$ of $K$ requires the transformation of the system given by $F$ to some managable form:

The following approaches can accomplish this.

1. Computation of the Gröbner basis of $I$ with respect to lexicographical ordering (often very costly).
2. Conversion of the Gröbner basis of $I$ with respect to total degree ordering to one with respect to lexicographical ordering by [Faugère et al. 1993], or more efficiently by a Gröbner walk [Collart et al. 1997; Amrhein et al. 1997a; Amrhein and Gloor 1998].
3. Combination of a Gröbner basis computation with factorization of all intermediate polynomials. For many practical problems this heuristic has yielded surprisingly good results (compare [Gräbe and W. 1994; Gräbe 1997]).
4. Reduction to the computation of the solutions of a single univariate polynomial be computation of a shape basis [Becker et al. 1994] or rational univariate form [Rouiller 1999].
5. computation of a characteristic set of $F$ using the method of Ritt-Wu (see section 2.2.7),
6. computation of a triangular set in the sense of Lazard (e.g., from a Gröbner basis of $I$, see [Lazard 1992]),
7. iterative computation of resultant systems together with suitable transformations of the variables; that is the approach of classical elimination theory (see [van der Waerden 1940]). Alternative more efficient resultant concepts are discussed in [Kapur and Lakshman 1992; Kapur 1995; Emiris 1996]

Most of these methods above have been implemented in several computer algebra systems.

From a triangular form of the system of equations, the solutions can be derived explicitly as an $n$-tuple of elements in a tower of simple field extensions over $K$. However, this method entails GCD computations and factorization of univariate polynomials over such extension towers (see section 2.2.2). For base fields $K$ of characteristic zero, one can, if necessary, replace the towers by simple algebraic extensions of $K$, by computing primitive elements (e.g., using Gröbner bases).

### 2.5.1 Algorithms for Polynomial Ideals and Their Varieties

Alternatively, factorizations can be avoided by computing directly with solutions, using Lazard's triangular sets (see [Lazard 1992]). The computations are implemented as *lazy evaluation*, where comparisons and GCD calculations are only performed, if they are actually needed (see [Dicrescenzo and Duval 1989]).

Of course, varieties of *positive dimension* cannot be described anymore by listing all elements. Instead, one wants to determine algorithmically structural properties and invariants of those varieties.

In the first place, we should mention here, computing the dimension of $V(I)$, and determination of independent sets of parameters. Both tasks can easily be accomplished by insection of the leading terms of a Gröbner basis for $I$. The same applies for computing the Hilbert function and the Hilbert polynomial (see [Cox et al. 1992; Becker et al. 1998]). An alternative approach to the first task is again the method of Wu-Ritt in its complete form (see section 2.2.7). A refinement of this algorithm was developed by Lazard in [Lazard 1991], applying ideas of Chistov-Grigoriev.

Another fundamental task is decomposition of varieties into their irreducible components. Using Hilbert's Nullstellensatz, the problem can be reduced to primary decomposition of ideals in $R$.

For zero-dimensional ideals, the primary decomposition can be performed using Gröbner bases, transformation of the ideal into normal position, and finally factorization of univariate polynomials over the base field $K$. The higher dimensional case is far more difficult, and requires multiple expansion and contraction of ideals, then factorization of univariate polynomials over rational function fields, as well as computing intersections and localizations of ideals, which can again be accomplished by Gröbner basis methods (see [Becker et al. 1998]). A more direct method for the primary decomposition is given in [Eisenbud et al. 1992]. As a by-product, the prime ideals associated with $I$, and the radical of $I$ are computed in both methods.

The algorithms described so far can easily be modified for projective varieties.

Another important topic of algorithmic commutative algebra are quantifier elimination methods for algebraically closed fields. Here, the problem is (compare section 2.15.2) as follows: Given a system of polynomial equations and inequalities, depending on certain parameters, compute a Boolean combination of polynomial equations in those parameters that it is a necessary and sufficient condition for the solvability of the system in any algebraically closed field. In geometrical terms, an algorithm that solves this problem yields an explicit representation of the projection of a Boolean combination of varieties onto an affine subspace as Boolean combination of varieties in that subspace. In principle, the problem can be solved by elementary methods of classical elimination theory as observed by A. Tarski, about 1930. Those algorithms however, are not practical. Their *worst case* asymptotic complexity is doubly exponential in the number of variables. A more careful analysis shows that the problem is doubly exponential in the number of quantifier changes, singly exponential in the number of variables, and polynomial in the size of the input expression [Basu et al. 1994]. The asymptotic parallel complexity (w.r.t. arithmetic circuits) has also been analyzed (see [Caniglia et al. 1989], and the literature cited therein). Most of the asymptotically optimal algorithms have not been implemented so far. Exceptions are the quantifier elimination algorithm for linear and quadratic problems [Weispfenning 1988; Loos and Weispfenning 1993; Dolzmann and Sturm 1997b; Weispfenning 1997a] and the method of comprehensive Gröbner bases [Weispfenning 1992a; Dolzmann 1999a].

For the special case of solving an overdetermined system of $n + 1$ parametric polynomials in $n$ variables sparse elimination theory offers an implemented approach of practical significance (see [Emiris 1996]).

Multivariate polynomials given by straight–line programs were implicitly considered in elimination theory since the early eighties, but it is only in the early nineties that they have been shown to behave as an efficient data structure in order to improve existing complexity estimates for elimination problems. Indeed, problems such as elimination, consistency, dimension or even division procedures for special cases (Nullstellensatz or complete intersection varieties) can be performed within polynomial running time in the length of a dense input encoding (cf. [Giusti and Heintz 1991], [Giusti and Heintz 1993], [Fitchas (J. Heintz and J. Sabia L. M. Pardo and P. Solernò) et al. 1995], [Krick and Pardo 1994], [Krick and Pardo 1996]) or even more on a suitable defined intrinsic degree. Moreover these algorithms keep track of the possible good evaluation properties of the input, since they run polynomially in the length of the straight–line program input encoding. As a particular case, input systems in sparse encoding can be treated in time that depends polynomially on the toric degree of the system (cf. the series of papers [Giusti et al. 1995], [Pardo 1995], [Giusti et al. 1997a], [Bank et al. 1997], [Giusti et al. 1997b], [Giusti et al. 1998], [Hägele et al. 2000]). The description of a prototype based on these alternative ideas may be found in [Giusti et al. 2001]; some applications were described in [Giusti and Schost 1999].

Studies on suitable data structures for elimination problems have also been done from a lower complexity bounds point of view. Results obtained recently show that universal data structures (those that may answer any question concerning elimination objects) require exponential output length, whereas short data structures lose local information around elimination objects (cf. [Heintz et al. 1998], [Castro et al. 1999]).

Volker Weispfenning (Passau), M. Giusti (Palaiseau), L.M. Pardo (Palaiseau, Santander)

### 2.5.2 Singularities of Varieties

Many algebraic systems of equations feature singularities, i.e., points or higher dimensional sub-varieties, on which the functional matrix of the defining polynomials locally does not have constant rank. For this case, the implicit function theorem does not hold, and the solution set can therefore locally not be parametrized. Such singularities arise naturally, and often have a crucial influence on the behavior of the solution set in the presence of small perturbations of the parameters, which renders the systems, for numerical treatment in particular, practically intractable.

The structure of these singularities is encoded in the local ring (of convergent or formal power series) of the variety at the singular point. Many invariants of such a ring, like dimension, multiplicity, Hilbert function, and syzygies module, as well as structural information like, e.g., deformation properties, can be computed using standard bases (see 2.2.6). The fundamental algorithm for computing standard bases in local rings is Mora's variant of Buchberger's algorithm. Implementations of this algorithm (e.g., in SINGULAR, see 4.2.39) already contributed interesting results to the theory of singularities: e.g., regarding the exactness of the Poincaré complex at singularities of complete intersections or regarding the structure of moduli spaces of plane curves or in connection with the construction of many resp. high singularities on curves of small degree (see Section 3.2.2).

Almost everything that can be computed by Gröbner bases in affine or projective algebraic geometry, can be computed accordingly in local algebraic geometry, or within the theory of singularities, using standard bases (see section 2.2.6, and 2.5). However, to do this we need general standard bases, not only for local (i.e. tangent cone) orderings, but also for mixed local–global orderings (see [Greuel and Pfister 1996]). The computation then takes place in a localization of the polynomial ring with respect to the given ordering (see [Greuel and Pfister 1996], [Greuel and Pfister 1998]). Evidently, the input to the algorithm has to be polynomial. The general standard basis algorithm makes sure that the output will also consist of polynomials (as opposed to power series, like in Hironaka's standard bases method).

Gert-Martin Greuel (Kaiserslautern)

## 2.5.3 Real Algebraic Geometry

The zero sets $V(F_1, \ldots, F_r) \subseteq \mathbb{R}^n$ of polynomials $F_1, \ldots, F_r$ are called affine real algebraic sets. Projections, or more general polynomial images as well as connected components of real algebraic sets generally are not algebraic sets anymore; however, they are so-called semi-algebraic sets, i.e., they are finite unions of sets which themselves are defined by finitely many equations and inequalities:

$$F_1 = 0, \ldots, F_r = 0,$$
$$G_1 \geq 0, \ldots, G_s \geq 0,$$
$$H_1 > 0, \ldots, H_t > 0.$$

The class $\mathcal{S}(\mathbb{R}^n)$ of semi-algebraic subsets of $\mathbb{R}^n$ is very extensive; e.g., it contains "half-spaces" which are bounded by hyper- surfaces, the interior or exterior of cubes, ellipsoids and so on. It turns out that $\mathcal{S}(\mathbb{R}^n)$ is closed under the following operations: finite intersection, finite union, complement, projections, determination of connected components, and topological closure of the interior and the boundary. Semi-algebraic sets can be triangulized (c.f. [Bochnak et al. 1987]). They form the most basic objects of real algebraic geometry. The theory and analysis of algorithms are based on ideas and results from the following mathematical subject areas:

- real commutative algebra,
- the model theory of real-closed fields,
- classic algebraic geometry.

In real commutative algebra, basic concepts of classical commutative algebra are enriched by imports from the theory of ordered rings and fields. Model theory has a natural relationship to real algebraic geometry, e.g., the projection theorem for semi-algebraic sets is a geometric version of quantifier elimination in the theory of real algebraic fields (cf. 2.15.2).

Following this outline, algorithmic questions can be attributed to three areas, which maintain close interrelationships:

I) real commutative algebra,
II) model theory,
III) geometric problems.

In contrast to standard commutative algebra and classic algebraic geometry, implementations are still primordial, even though for many problems efficient algorithms have already been presented.

Ad I): Computations in ordered field extensions, especially in the real closure of an ordered field, are of cardinal importance. Currently, there exist implementations in ALDES-SAC2 and AXIOM, both based on interval arithmetic. Symbolic algorithms were proposed by Zassenhaus ([Kempfert 1968; Zassenhaus 1970]) and by Coste and Roy [Coste and Roy 1988], however without being implemented so far.

In the theory of real ideals, current focus is on real radicals

$$\sqrt[r]{\alpha} = \{f \in A \mid f^{2n} + g_1^2 + \ldots + g_s^2 \in \alpha \text{ for particular } n \in \mathbb{N}, g_1, \ldots, g_s \in A\}$$

of an ideal $\alpha$ of $A$ due to its occurrence in the real Nullstellensatz ($IV(\alpha) = \sqrt[r]{\alpha}, V(\alpha) \subset R^n$, $R$ real-closed). Lombardi's [Lombardi 1991] approach to solving the *membership-problem* has hyper-exponential complexity. In [Becker and Neuhaus 1993], an algorithm which computes a generating system of $\sqrt[r]{\alpha}$ for $\alpha \lhd R[X_1, \ldots, X_n]$ is given, its complexity is studied in [Neuhaus 1998]. Other approaches are discussed in [Galligo and Vorobjov 1995; Roy and Vorobjov 1995]. Implementations are still due to date.

Ad II), III): Decision procedures for the validity of (logic) formulas as well as algorithms for quantifier elimination offer solutions, from the geometric point of view, to many of the following algorithmic problems:

1) Decide whether a semi-algebraic set is empty or not.
2) Describe the connected components of a semi-algebraic set, generate a list of points, containing at least one point from each component.

3) Generate projections and polynomial images of semi-algebraic sets.
4) Construct semi-algebraic paths between points which can be be connected.
5) Quantifier elimination in general, decision procedures for the (formal) theory of ordered fields.
6) Description of semi-algebraic sets by special polynomial systems, e.g., by very few polynomials, or by polynomials of low degree.

Collins' CAD method from the seventies (including current optimizations) is the base of many of the implemented algorithms for problems 1) to 5), so far. An very interesting account of CAD and Quantifier Elimination can be found in [Caviness and Johnson 1998]. Several groups have attacked various of the problems above, cf. [Gonzalez-Vega et al. 1998b] for a survey. Also, Quantifier Elimination for problems expressible by polynomials of small degrees have been treated successfully by V.Weispfenning and his group [Weispfenning 1998, 1995; Dolzmann et al. 1998b; Weispfenning 1997b; Sturm and Weispfenning 1998b; Dolzmann et al. 1997, 1998c]. Substantial theoretical advances were made during the past years, especially by Grigor'ev-Vorobjev [Grigor'ev 1988; Grigor'ev and Vorobjov 1992], as well as by Heintz, Roy, and Solerno [Heintz et al. 1989, 1990, 1991b,c, 1995] and Renegar [Renegar 1992a,b,c]. The papers by Basu, Pollack and Roy are strongly recommended as well [Basu et al. 1997b,a, 1996b,c, 1998b,a].

See also [Galligo and Vorobjov 1995; Roy and Vorobjov 1995; Roy 1996].

Improved versions of Sturm's or Hermite's method for the counting of real zeros will certainly play a major role in implementations in the next 10 years [Gonzàlez et al. 1989, 1990; Pedersen 1991; Pedersen et al. 1993; Becker and Wörmann 1994; Gonzalez-Vega et al. 1998b]. Solutions to problems under 6) are still in the initial stages.

Additional references: [Basu et al. 1995; Pollack and Roy 1993; Basu et al. 1996a, 1994; Gonzalez-Vega et al. 1998a].

Eberhard Becker (Dortmund)

## 2.6 Algorithmic Aspects of the Theory of Algebras

Mathematical structures with addition $+$, a scalar multiplication $\cdot$ for scalars from a commutative ring or field $R$ and an algebra multiplication, denoted by various symbols as $\cdot, \times, \otimes, \wedge$ or $[\cdot, \cdot]$, satisfying the usual distributivity laws,* play an important role in many areas. Well-known examples of such *R-algebras* $A$ are the commutative polynomial algebras – see also sections 2.2.5 and 2.5.1, or the associative matrix (endomorphism) algebras or function algebras. However, the concept of an algebra is of a more general nature. As examples, we would like to mention here group algebras – compare section 2.7 –, path algebras – see 2.6.8, quantum groups, i.e. non-commutative algebras in quantum mechanics, – see 3.2.5, which are often given by generators and transformation relations,

---

* Also it is required that $(\lambda x)(\mu y) = \lambda\mu(xy)$ holds for all $\lambda, \mu \in R, x, y \in A$.

compare also 3.1 – *Lie*\* and *Jordan algebras*\*\*, as well as *genetic algebras* for modelling non-associative rules of heredity.

There are various ways to present algebras on a computer depending on the nature of available computational information of the algebra:

- finite-dimensional algebras of dimension $n$ with multiplication given by $n^3$ structure constants,
- algebras given by generators and relations,
- finite-dimensional associative algebras over fields are subalgebras of endomorphism algebras of a finite-dimensional vector space, and so can be given by means of a finite set of generating matrices
- nilpotent algebras\*\*\* can be given by a so-called *power commutator presentation*, just like for groups, cf. [Sims 1994], and
- monad algebras like group or path algebras.

### 2.6.1 Structure Constants

For a finite-dimensional $R$-free algebra $A$ over a ring (or a field) $R$ the most straightforward presentation is the abstract one, which assumes the knowledge of a basis $\{x_i \mid i \in I\}$, on which the multiplication is given by the the so-called *structure constants* $c_{ij}^{(k)}$ $(i, j, k \in I)$ which are defined by the relation (multiplication table)

$$x_i \cdot x_j = \sum_{k \in I} c_{ij}^{(k)} x_k.$$

Using the bilinearity of the multiplication, these are sufficient to calculate the product of two arbitrary elements of $A$. Implementations are straightforward and either are dense – each multiplication requires $n^3$ operations for $n := \#I$ – or sparse, where only the non-zero structure constants stored in a list are used.

### 2.6.2 Generators and Relations, Swapping and G-Algebras

Another way to represent an algebra is by generators and relations. In the case of free Lie algebras several $R$-bases are known, the most famous ones being *Hall bases* and *Shirshov bases*, cf. [Reutenauer 1993; Ufnarovski 1998], which play the role of the free monoid generated by the set of algebra generators in the associative case. If the algebra is associative, the algebra can directly be viewed as a quotient of the free associative algebra on the generators by the ideal generated by the relations. In order to present elements uniquely, the

---

\* The algebra multiplication, usually denoted by the *Lie bracket* $[x, y]$, is *anticommutative*, i.e. $[x, y] + [y, x] = 0$ and the *Jacoby identity* $[x, [y, z]] + [y, [z, x]] + [z, [x, y]] = 0$ is satisfied by all $x, y, z \in A$.

\*\* The characteristic of the ground ring $R$ is not 2 and the commutative algebra multiplication $\cdot$ satisfies the identity $(x \cdot y) \cdot x^2 - x \cdot (y \cdot x^2) = 0$ for all $x, y \in A$

\*\*\* There exists a natural number $n$ such that all possible products of $n$ elements of the algebra are 0.

strategy used most frequently is to work with a Gröbner basis for the ideal, see 2.6.5. *Swapping algebras*, also called *algebras of solvable type*, *skew polynomial rings* or considered as *Ore extensions* have the same underlying free $R$-module structure as polynomial algebras, but the (*twisted*) multiplication $*$ is defined by relations $y * x = a_{y,x} xy + p_{y,x}$ for $x < y$, where $x$ and $y$ are polynomial generators, $a_{y,x}$ is an invertible constant, and the element $p_{y,x}$ satisfies $p_{y,x} \prec xy$ with respect to n arbitrary fixed well-founded admissible term-ordering $\prec$, see 2.2.5. For Gröbner basis considerations see 2.6.5 and the references given there. They naturally generalize the Weyl algebra, generated by $\{x, Dx\}$ with the rule $Dx * x = x \cdot Dx + 1$ or the enveloping algebra of a Lie algebra, which has a filtration whose associated algebra is the polynomial algebra on the Lie algebra by the Birkhoff-Poincaré-Witt theorem, which can be used for efficient computations there by employing the relations $e_i * e_j = \begin{cases} e_j e_i - [e_j, e_i] & \text{for } i \geq j \\ e_i e_j & \text{for } i < j \end{cases}$ for elements $e_i$ and $e_j$ of a basis. Given a total ordering on the set of variables, the knowledge of the terms $a_{y,x}$ and $p_{y,x}$ for all polynomial generators $x$ and $y$ and some further conditions determine the full multiplication table for the basis of monomials. In the particular case that $R$ is a field see [Kandry-Rody and Weispfennning 1990] for these conditions and an alternative, more general construction as a factor algebra of the polynomial algebra in non-commuting variables, which is called *G-algebra*[*], was presented in [Apel 1988]. The case of arbitrary coefficient rings $R$ is considered in [Apel 1998a]. General implementations can be found – among other systems – in FELIX 4.2.15 and MAS 4.2.26 and are also reported in [Ekedahl et al. 1995].

Other computer algebra methods for Lie algebras opened up new areas of computer applications in Lie methods, especially in physics. Symbolic representations for envelopes of Lie algebras [Lassner 1991] allow, on the one hand, effective computations in such algebras (*enveloping algebras*), and in some of their localizations, on the other hand, they are suited to express the operator ordering problem for various quantizations in phase space methods of quantum theory, and Lie optics, in a way amenable to efficient computations on a computer algebra system.

Also, methods have been developed to produce a matrix representation directly from a presentation by generators and relations, at least for associative algebras, cf. [Labonte 1990; Linton 1993].

### 2.6.3 Monad Algebras, Path Algebras and Generalizations

A natural generalization of the concept of a group algebra $RG$ for a group $G$, where the group elements are used as an $R$-basis of $RG$ and the multiplication of the algebra is given by extending linearly the multiplication of the basis elements to arbitrary multiplicative structures $(M, \cdot)$ with $\cdot : M \times M \to M$ is a *monad*. Such *monad algebras* can easily be implemented, in particular if the monad is finite, see e.g. [Ekedahl et al. 1995].

---
[*] G is short for Gröbner.

Slight generalizations are *path algebras*, where in addition 0 is also allowed as the result of the multiplication of two basis elements. Let $\Gamma$ be a finite directed (oriented) graph with edges and arrows including loops and possibly multiplicities, also called *quiver*. Then the path algebra $R\Gamma$ is defined to be the free $R$-module generated by the basis of all paths in the graph. The algebra multiplication $pq$ is defined on this basis by concatenation of path $p$ and the path $q$, if the terminating edge of $p$ coincides with the starting edge of $q$, otherwise $pq := 0$. A graph consisting of one edge only and $n$ loops determines the polynomial algebra in $n$ non-commuting variables.

A further generalization is to allow that the result of multiplying two basis elements is a scalar times a basis element. Examples are the *exterior algebras*[*] or the *divided power algebras*[**].

### 2.6.4 Finite-Dimensional Lie Algebras

Every finite-dimensional Lie algebra can be presented using structure constants or by generators and relations. For performing computations in finite-dimensional Lie algebras the presentation by means of structure constants is often more suitable. For this and obvious other reasons, it is convenient to be able to pass from the presentation by generators and relations to a multiplication table and vice versa, but this way is not so hard. This problem generalizes the case of groups, where the Todd-Coxeter algorithm is known to provide a solution (due to Mendelsohn's theorem, cf. [Cohen et al. 1998]). Todd-Coxeter-like algorithms for very general kinds of algebra, still satisfying Mendelsohn's theorem, that is, they terminate if the quotient algebra is finite-dimensional, are known, see e.g. [Leeuwen and Roelofs 1997].

### 2.6.5 Non-commutative Gröbner Bases

In the last years, the method of Gröbner bases – see 2.2.5 – and its applications have been extended from commutative polynomial algebras over fields to various types of non-commutative algebras over fields and other rings. In general for such algebras arbitrary finitely generated ideals will not have finite Gröbner bases. Nevertheless, there are interesting classes for which every finitely generated (left or right) ideal has a finite Gröbner basis which can be computed by appropriate variants of Buchberger's algorithm.

Since the development of computer algebra systems for commutative algebras enabled to perform tedious calculations using computers, attempts to generalize such systems and especially Buchberger's ideas to non-commutative algebras

---

[*] All subsets of the set of $\{1, \ldots, n\}$ of natural numbers determine a basis. Multiplication of two non-disjoint subsets is 0, otherwise the result is the union of the two subsets decorated with the sign of the permutation to reorder the numbers after concatenation of the ordered lists of elements.

[**] As an $R$-module it is isomorphic to a multivariate polynomial ring, but additional scalar factors are given as binomial coefficients $\binom{i+j}{i}$ for the 'exponents' $i$ and $j$ of each variable.

followed. Originating from special problems in physics, Lassner in [Lassner 1985] suggested how to extend existing computer algebra systems in order to handle special classes of non-commutative algebras, e.g. *Weyl algebras*. He studied structures where the elements could be represented using the usual representation of polynomials in commutative variables and the non-commutative multiplication could be performed by a so-called *twisted product*, see 2.6.2, which required only procedures involving commutative algebra operations and differentiation. Later on together with Apel he extended Buchberger's algorithm to one-sided ideals of enveloping algebras of Lie algebras [Apel and Lassner 1985; Apel and Laßner 1988]. The motivation was to give an algorithmic method for the transformation of left fractions into right ones, which is the basic constructive problem in quotient skew fields of enveloping algebras of Lie algebras. Because these ideas use representations by commutative polynomials, Dickson's lemma can be carried over. The existence and construction of finite Gröbner bases for finitely generated left ideals is ensured. In [Galligo 1985] algorithmic questions on ideals of differential operators were studied.

On the other hand, Mora gave a concept of Gröbner bases for a class of non-commutative algebras by saving another property of the polynomial ring while losing the validity of Dickson's lemma. The usual polynomial algebra can be viewed as a monoid algebra where the monoid is a finitely generated free commutative monoid. Mora studied the class where the free commutative monoid is substituted by a free monoid – the class of finitely generated free monoid algebras, compare e.g. [Mora 1986, 1994]. The algebra operations are mainly performed in the coefficient domain while the terms are treated like words, i.e., the variables no longer commute with each other. The definitions of (one- and two-sided) ideals, reduction and Gröbner bases are carried over from the commutative case to establish a similar theory of Gröbner bases in *free non-commutative polynomial algebras over fields*. But these algebras are no longer Noetherian* if they are generated by more than one element. A terminating completion procedure exists for finitely generated one-sided ideals which allows to solve the membership problem. This cannot be achieved for the case of even finitely generated two-sided ideals where only an enumeration procedure exists, which in general is not terminating. Gröbner bases and Mora's Algorithm have been generalized to path algebras, see [Farkas et al. 1993; Keller 1998]. An introduction to non-commutative Gröbner bases theory can also be found in [Ufnarovski 1998].

Another class of non-commutative algebras where the elements can be represented by the usual polynomials and which allow the construction of finite Gröbner bases for arbitrary ideals are the algebras of solvable type – see 2.6.2, a class intermediate between commutative and general non-commutative polynomial algebras, studied in [Kandry-Rody and Weispfennning 1990; Kredel 1992; Pesch 1998]. Apel showed in [Apel 1988] that for the more general G-algebras finite Gröbner bases can be still constructed for arbitrary ideals. The more special case of twisted semi-group rings, has been studied in [Mora 1989b]. A general-

---

* The ascending chain condition for ideals does not longer hold.

ization to iterated *Ore extensions* defined by power substitution rules* can be found in [Pesch 1998].

In [Weispfenning 1992b] Weispfenning showed the existence of finite Gröbner bases for arbitrary finitely generated ideals in non-Noetherian skew polynomial algebras in two variables.

Most of the results cited so far assume admissible well-founded orderings on the set of terms so that in fact reduction can be defined by considering the head monomials only. This is essential to characterize Gröbner bases in the respective algebras with respect to the corresponding reduction in a finitary manner and to enable to decide whether a finite set is a Gröbner basis by checking whether the s-polynomials – see 2.2.5 – are reducible to zero**.

There are algebras combined with reduction where admissible well-founded orderings cannot be accomplished and, therefore, other concepts to characterize Gröbner bases have been developed. For example in case the algebra contains zero-divisors a well-founded ordering on the algebra is no longer compatible with the algebra multiplication***. This phenomenon has been studied for the case of zero-divisors in the coefficient domain and for the special case of regular rings in [Weispfenning 1989b]. Additional special critical pairs, or more precisely homogeneous left syzygies between the head terms, were used in [Apel 1988, 1992] in order to overcome the difficulties arising from non-scalar zero divisors in G-algebras. The basic idea is to utilize the theory of Gröbner bases in non-commutative graded structures, see [Mora 1988]. Further constructive instances of this widely non-constructive theory can be found in [Apel 1997].

Another possibility to attack the lack of an admissible term ordering is saturation, i.e. the generating sets are enlarged in order to ensure that enough head terms exist to do all necessary reductions. This idea was used by Madlener and Reinert when introducing reduction relations and Gröbner basis techniques to monoid and group algebras and they obtained positive results for several classes of monoids and groups, see e.g. [Madlener and Reinert 1993, 1998c] for the details. There exist the following explicit connections:

– between the word problem for monoids and the ideal membership problem for free monoid rings;
– between the word problem for groups and the ideal membership problem for free group rings;
– between the submonoid problem and the subalgebra problem for monoid rings;
– and between the subgroup problem and the one-sided ideal membership problem for group rings.

An overview can be found in [Madlener and Reinert 1998b]. Hence many results known from the field of string rewriting as well as from combinatorial group

---

  * A power substitution rule has the form $Yx = ax^iY + p$.
 ** Note that we always assume that the reduction in the algebra is effective.
*** When studying monoid algebras over reduction rings it is possible that the ordering on the algebra is not compatible with scalar multiplication as well as with multiplication with monomials or polynomials.

theory are strongly related to non-commutative Gröbner bases [Reinert et al. 1998].

Finally Gröbner basis techniques in non-commutative settings can be described in an axiomatic fashion using the general framework for reduction rings and algebra constructions as presented by Madlener and Reinert in [Madlener and Reinert 1998a]. This approach points out what to look for when studying reduction relations in algebras in order to define Gröbner bases and how such results are preserved under standard algebra constructions such as quotients and sums of reduction rings, and polynomial and monoid algebras over reduction rings.

### 2.6.6 Structural Issues and Classification

For an associative or Lie algebra given by structure constants, finding the *(nil) radical*\* is an important first step in the determination of its structure. In the case of characteristic 0, this is easy as the radical is the kernel of an associative bilinear form. For the general case, more intricate algorithms have been found, cf. [Cohen et al. 1997]. The quotient of the algebra with respect to the radical is semisimple. Algorithms for finding the simple components are related to the search for idempotents and can be viewed as generalizations of polynomial factorization, see [Cohen et al. 1998]. Algorithms for the structure determination of Lie algebras given by structure constants are implemented in MAGMA and GAP, cf. [Graaf 1997].

A natural question regards algorithms for solving the more general isomorphism problem for Lie algebras given by structure constants. A Gröbner basis method [Gerdt and Lassner 1993] is able to solve this problem for complex and real Lie algebras of small dimension. A different approach is taken by structural analysis, using structure constants of the particular algebra as a starting point, cf. [Graaf 1997]. The general implementations [Grabmeier and Wisbauer 1993] of non-associative categories and data structures in AXIOM allow in particular the determination of idempotents in algebras (also, e.g., in genetic algebras) from their structure constants by using Gröbner basis methods.

The same methods can be applied to the analysis of finite-dimensional Lie algebras, which arise as symmetry algebras of differential equations, see 2.11 and in particular 2.11.3.

Algorithms for the representation of Lie groups and their corresponding Weyl groups (root systems, maximal weights, etc.) are implemented in the package LiE, see 4.2.23.

### 2.6.7 Identities

Another topic of research in general algebras and rings is the study of identities. The most popular example is the *associative identity* $x(yz) - (xy)z = 0$, other examples are the *Jacobi-identity* $x(yz) + (yz)x + (zx)y = 0$ of Lie algebras or

---

\* For finite dimensional algebras the (nil) radical is the largest nilpotent ideal.

the alternative identities $(xx)y - x(xy) = 0$ and $y(xx) - (yx)x = 0$. The left-hand sides can be considered as objects (non-associative polyomials) in the free algebra generated by $x, y, z$, which e.g. can be constructed in a computer algebra system like AXIOM as the monad ring over the monad of all binary trees, see [Grabmeier and Wisbauer 1993]. One technique to check such identities in a finite dimensional algebra is to use linear algebra methods after linearization of the identities by substitutions like $x \mapsto (a+b)$ (polarization). D. Jacobs has built the special purpose system ALBERT – see 4.2.2 or [Jacobs 1994], which is devoted to identities exclusively. It verifies whether a given identity holds in an algebra by constructing the free algebra determined by the given set of defining identities of the algebra and testing whether the polynomial in question is zero, see [Hentzel and Jacobs 1991]. Other techniques use term rewriting systems directly, see e.g. [Widiger 1995].

### 2.6.8 Computational Aspects in the Representation Theory of Quivers and Path Algebras

The aim of the theory is to study the category of representations of a given associative unital* $R$-algebra $A$ of finite dimension over a field $R$. Familiar examples for algebras of this kind are the group algebras $RG$ for a finite group $G$ or the finite-dimensional factor algebras of polynomial algebras $R[X_1, \ldots, X_n]$ or the incidence algebras of finite partially ordered sets. Note, that these examples have specific properties. The first are Hopf algebras, the second are commutative, the third are of finite global dimension. Nevertheless, in all these cases $R$ is a splitting field for $A$ which is a useful assumption in order to avoid technical problems when describing $A$ by combinatorial data.

If one wants to approach finite-dimensional algebras $A$ in general, it has been found suitable to use the language of quivers as introduced in [Gabriel 1972]. It occurs as an unexpected advantage that a quiver can also be associated to the category of indecomposable finite-dimensional representations of $A$ in a natural way. Recall that *quiver* is a shorthand for *oriented graph with possibly multiple edges and loops*. The quiver describing the algebra is sometimes called the *Gabriel quiver* whereas for the quiver of the category of indecomposable finite-dimensional representations meanwhile *Auslander-Reiten quiver* is the standard name. The monographs [Auslander et al. 1995], [Gabriel and Roiter 1992] and [Ringel 1984] provide general introductions to this area of mathematics.

To describe a finite-dimensional algebra by a quiver $Q$ means to rewrite the algebra up to Morita equivalence** as a factor algebra of the path algebra $RQ$.

The concept of path algebras may be considered as a natural generalisation of the concept of free algebras in non-commuting variables. Consequently, a suitable adaption of the Gröbner basis technique was established for path algebras (see [Farkas et al. 1993]) and implemented by Blacksburg in a package of C programs called GROEBNER. This was used efficiently for the computation of Ext-algebras for finite groups, see [Carlson et al. 1997].

---

* It possesses a unit element.
** Two algebras are *Morita equivalent* if there module categories are equivalent.

In the Auslander-Reiten quiver an indecomposable representation shrinks to a mere vertex. In this way the category of representations of an algebra is transformed into a completely combinatorial data structure which is perfectly appropriate for being handled by a computer. Additional data are obtained by looking at the corresponding element of a representation in the Grothendieck group $K_0(A)$.* For general algebras the Auslander-Reiten quiver together with the Grothendieck group does not yet lead to satisfactory computational algorithms. The situation improves drastically if one considers algebras satisfying suitable simply connectedness conditions. Associated with these algebras and their representation theory other combinatorial structures like partially ordered sets, quadratic forms and translation quivers arise in a natural way.

Quivers, Grothendieck groups, partially ordered sets together with various actions on these structures form the combinatorial background for the system CREP – see 4.2.12 and [Draexler and Noerenberg 1996], which also contains classification lists of algebras. For a short introduction into the mathematical background we refer to [Draexler and Noerenberg 1997].

Additional reference: [Schöbel and Schöbel 1992].

Joachim Apel (Leipzig), Arjeh M. Cohen (Eindhoven), Peter Dräxler (Bielefeld), Johannes Grabmeier (Heidelberg), Wolfgang Laßner (Senftenberg), Klaus E. Madlener (Kaiserslautern), and Birgit Reinert (Kaiserslautern)

## 2.7 Computational Group Theory

Computational group theory has a history going back more than 80 years. It is a collaborative effort of researchers in a wide range of areas in both mathematics and computer science. The field has produced important algorithmic ideas and large software packages that have been used to obtain valuable results in mathematics and in other disciplines. The algorithmic achievements include the development of complicated algorithms designed to have the best possible asymptotic complexity and algorithms of exponential or perhaps unknown complexity that are nevertheless extremely useful in actual computations. There are also techniques for attempting to study instances of problems that are known not to have algorithmic solutions in general.

This article is intended as a survey of the field for the nonexpert. Thus it is written to provide a mathematician, physicist, or chemist who has encountered a particular group with guidance about the kinds of information one should expect to be able to determine about the group and in some cases to offer insight into the size of problems that can be attacked with current hardware. The article also is designed to show the computer scientist unfamiliar with the field the rich variety of problems concerning the complexity of group-theoretic algorithms.

---

* $K_0(A)$ is the factor group of the free abelian group generated by the finite-dimensional representations of $A$ by the subgroup generated by all alternating sums over representations occurring in short exact sequences.

The reader is introduced to several data structures that are unique to computational group theory. These include bases and strong generating sets for permutation groups and coset tables used to describe finitely generated subgroups of free groups. Other themes are the role of the classification of finite simple groups in the development of the subject and efforts to cope with the fact that many interesting problems do not have general algorithmic solutions.

A number of other surveys of computational group theory have appeared over the years. This article draws on many of them, including [Babai 1991], [Cannon 1990, 1991], [Neubüser 1995], [Seress 1997], and [Kantor 1998].

### 2.7.1 A Crash Course in Group Theory

The reader is assumed to have a knowledge of abstract algebra at roughly the level of an introductory undergraduate course. Thus the following terms will be used without definition: *group*, *commutative* or *abelian group*, *subgroup*, *normal subgroup*, *quotient group*, *generating set*, *cyclic group*, *homomorphism*, *isomorphism*, *permutation*, *even permutation*, *orbit*, *coset*, *Sylow subgroup*, *center*, *conjugacy class* of elements or of subgroups, *centralizer*, *normalizer*, and *simple group*.

We will also assume a basic familiarity with linear algebra, particularly linear algebra over finite fields. For each power $q$ of a prime, the field with $q$ elements will be denoted $GF(q)$.

Many properties of groups are defined using series of subgroups. Let $G$ be a group and let
$$G = G_1 \supseteq G_2 \supseteq \cdots \supseteq G_k \supseteq G_{k+1} = 1$$
be a series of subgroups of $G$. The series is said to be *subnormal* if $G_{i+1}$ is normal in $G_i$ for $1 \leq i \leq k$. A *composition series* is a subnormal series in which each quotient group $G_i/G_{i+1}$ is a (nontrivial) simple group. If $G$ has a composition series, the simple groups which occur are called the *composition factors* of $G$. They do not depend on the particular composition series. The group $G$ is *solvable* if it has a subnormal series in which the quotients $G_i/G_{i+1}$ are abelian. If $G$ has a subnormal series with cyclic quotients, then $G$ is said to be *polycyclic*.

The $G_i$ form a *normal series* if each is a normal subgroup of $G$. If in this case each quotient $G_i/G_{i+1}$ is in the center of $G/G_{i+1}$, then $G$ is *nilpotent*. All finitely generated nilpotent groups are polycyclic and all polycyclic groups are solvable. Every finite solvable group is polycyclic. If $p$ is a prime, then a *p-group* is a group in which all elements have orders which are powers of $p$. All finite $p$-groups are nilpotent.

Let $n$ be a positive integer. The *symmetric group*, which consists of all permutations of the set $\{1, 2, \ldots, n\}$, will be denoted $\Sigma_n$ and $A_n$ will denote the *alternating subgroup* of $\Sigma_n$, the set of all even permutations. If $K$ is a field, then the *general linear group* $GL(n, K)$ is the group of all $n$-by-$n$ invertible matrices with entries in $K$. If $K = GF(q)$, then we write $GL(n, q)$. The group $GL(n, \mathbb{Z})$ is the subgroup of $GL(n, \mathbb{Q})$ consisting of the invertible integer matrices whose inverses also have integer entries. These are the integer matrices with determinant $\pm 1$.

To every set $X$ there is associated the *free group* $F(X)$ on $X$. Intuitively, $F(X)$ is "the most general group generated by $X$". Formally we proceed as follows: Let $X^{-1}$ be a set disjoint from $X$ and having the same cardinality as $X$. Let $x \mapsto x^{-1}$ be a bijection of $X$ onto $X^{-1}$. Define $(x^{-1})^{-1}$ to be $x$. A word over $X$ is a finite sequence $u = u_1 \cdots u_r$, where the $u_i$ are in $X \cup X^{-1}$. The word $u^{-1}$ is defined to be $u_r^{-1} \cdots u_1^{-1}$. The word $u$ is reduced if it contains no subwords of the form $xx^{-1}$ or $x^{-1}x$. The product of two words is defined to be their concatenation. We may take $F(X)$ to be the set of reduced words. If $u$ and $v$ are reduced words, then there are unique words $a$, $b$, and $c$ such that $u = ac$, $v = c^{-1}b$ and $c$ has maximal length. The product of $u$ and $v$ in $F(X)$ is defined to be $ab$. The identity element of $F(X)$ is the empty word.

Here is an example with $X = \{x, y\}$: If $u = xyx^{-1}yxy^{-1}$ and $v = yx^{-1}y^{-1}x^{-1} \times y$, then $a = xyx^{-1}$, $b = x^{-1}y$, and $c = yxy^{-1}$. Thus the product of $u$ and $v$ in $F(X)$ is $ab = xyx^{-1}x^{-1}y$.

Let $R$ be a subset of $F(X)$. The subgroup of $F(X)$ generated by $R$ may not be normal in $F(X)$. However, if we form the subgroup $N$ generated by all conjugates of elements of $R$, then $N$ is normal in $F(X)$. The group $G = F(X)/N$ is denoted $\langle X \mid R \rangle$. The pair $(X, R)$ is called a *presentation* for $G$ by generators and relators. The presentation is *finite* if both $X$ and $R$ are finite. If $X = \{x_1, \ldots, x_r\}$ and $R = \{U_1, \ldots, U_s\}$, then we write $G = \langle x_1, \ldots, x_r \mid U_1 = \cdots = U_s = 1 \rangle$. For example, the group $\langle x, y \mid x^2 = y^3 = (xy)^7 = 1 \rangle$ is the most general group generated by an element of order 2 and an element of order 3 such that the product of these elements has order 7.

### 2.7.2 Describing Groups

Some computational questions in group theory refer to many groups at once. For example, in [O'Brien 1991] it is shown that the answer to the question "How many nonisomorphic groups of order 256 are there?" is 56092. However, more commonly a single group $G$ is specified and questions about the structure of $G$ are asked.

Computational group theory is divided into subfields based on the way the group $G$ is specified. The questions asked and the techniques used to answer them depend critically on the nature of the specification of $G$.

Here are four ways of specifying a group $G$.

1. One can define a finite subset $S$ of some previously specified group $M$ and say that $G$ is the subgroup of $M$ generated by $S$. Frequently $M$ is a symmetric group $\Sigma_n$, a general linear group $GL(n, K)$, or a free group $F(X)$ for some finite set $X$. In this context, the first problem one encounters is the *membership problem*: Given an element $g$ of $M$, is $g$ in $G$?
2. One can define $G$ by a finite presentation.
3. One can define $G$ to be the group of all automorphisms of some algebraic or combinatorial structure. Possible combinatorial structures are graphs and block designs. Galois groups are automorphism groups of field extensions.
4. One can give a *black box* description of $G$. Black box descriptions are an abstraction of the minimal information one would need to begin computing

in a finite group. They arise in algorithms for studying a finite group given by a set of generators. A black box description involves the following:

- A positive integer $N$, which is assumed to be $O(\log |G|)$.
- An encoding of elements of $G$ by bit strings of length $N$. By this is meant a bijection between $G$ and a subset of the bit strings of length $N$. The details of the encoding are "hidden". In particular, the set of bit strings which are encodings of elements of $G$ need not be known initially. All that one knows is that $|G| \leq 2^N$.
- Two oracles or "black boxes". Let $U$ and $V$ be the bit strings corresponding to two elements $g$ and $h$ of $G$. Given $U$, the first oracle returns the bit string for $g^{-1}$. Given $U$ and $V$, the second oracle returns the bit string for $gh$.
- The bit strings corresponding to a generating set for $G$. (Note that without these generators there would be no way to begin computing in $G$. Using the two oracles, one can start with the generators and construct more elements of $G$ as products and inverses of previously obtained elements.)

When the way $G$ is specified makes it obvious that $G$ is finite, then the existence of algorithms for studying properties of $G$ is usually not an issue. The question becomes how hard is it to determine the answer to a particular question. This is not literally true, however. For example, one might ask, of the homomorphisms of $G$ into some infinite group $H$, which ones are essentially different in some sense. Since there might be infinitely many such homomorphisms, it is not clear that there is a finite description of the answer.

Some specifications define infinite groups or groups for which it is not immediately clear whether they are finite or infinite. In this case, the first issue is: Does there exist an algorithm to answer a particular question? If the answer is no, then discussions of practicality or efficiency are irrelevant.

It is now known that there are natural computational questions about groups which can not be answered algorithmically. There may be instances of the questions which can be answered, but there is no algorithm which is guaranteed to provide an answer in every case. In fact, most computational questions about finitely presented groups turn out not to have algorithmic solutions.

Except in groups given by finite presentations, questions about individual elements of a group tend not to be particularly challenging. Frequently one asks questions involving large sets of elements, either in the statement of the question or in its answer. For example, elements $g$ and $h$ of $G$ may be given and we want to determine all the elements of $G$ which commute with $g$ and to describe the elements which conjugate $g$ to $h$. The answer to the first question is the centralizer $C$ of $g$ in $G$, which is a subgroup of $G$. The answer to the second question is either the empty set or a coset of $C$. For each method of describing a group it is useful to have data structures which permit the efficient representation of subgroups.

### 2.7.3 A Brief History

The definition of a group was not formalized until late in the 19th century. The first algorithmic questions about groups appeared almost immediately. The subject of computational group theory is generally considered to have originated with three questions posed in [Dehn 1911] by Dehn in 1911. All three concern groups defined by finite presentations $(X, R)$. Elements of such a group are described by words over $X$.

**The Word Problem.** Let $G = \langle X|R \rangle$ and let $U$ be a word over $X$. Does $U$ represent the identity element of $G$?

**The Conjugacy Problem.** Let $G = \langle X|R \rangle$ and let $U$ and $V$ be words over $X$. Do $U$ and $V$ represent conjugate elements of $G$?

**The Isomorphism Problem.** Let $G = \langle X|R \rangle$ and $H = \langle Y|S \rangle$. Are $G$ and $H$ isomorphic groups?

All three of Dehn's questions turned out not to have algorithmic solutions. The word problem was shown undecidable by Novikov in [Novikov 1955]. A finite presentation $(X, R)$ has been constructed for which the word problem is undecidable. The word problem is the special case of the conjugacy problem in which the word $V$ is taken to be empty. Thus the conjugacy problem is undecidable too. The isomorphism problem was shown to be undecidable in [Rabin 1958]. Even the case in which $H$ is a group of order 1 is undecidable.

Many other properties of finitely presented groups have been shown to be undecidable. For example, it is not possible to determine in general whether $G = \langle X|R \rangle$ is finite, nilpotent, or solvable.

Dehn showed the word problem does have a solution for a class of presentations which arise naturally in topology. The algorithm given by Dehn is now known to work for large classes of "small cancellation groups", groups in which distinct relators have no long subwords in common. With certain fairly natural probability measures on presentations, a random presentation turns out to define a small cancellation group and hence to have a solvable word problem. Unfortunately, the presentations which arise in many areas of research are not random.

One of the early positive contributions to computational group theory was [Todd and Coxeter 1936]. That paper describes a technique now known as coset enumeration. Coset enumeration is one of the most used tools for studying finitely presented groups. It is discussed below.

The first digital computers were used primarily for numerical computation. However, in the mid-1940's Alan Turing suggested that a proposed new computer might be used to answer questions in group theory.

Attempts to implement coset enumeration began in the early 1950's, making coset enumeration one of the first nonnumerical procedures to be programmed. Unfortunately, computer memories of the time were too small to permit machines to go beyond what human experts could do by hand.

Perhaps the first mathematician to make computational group theory the primary focus of his professional activity was Joachim Neubüser. His first paper with a computational content [Neubüser 1960] appeared in 1960.

The first conference to include a substantial portion of computational group theory was held in Oxford in 1967. Its proceedings [Leech 1970] is the first volume to contain a collection of papers in this field. The first conference to be devoted entirely to computational group theory was held in Durham, England, in 1982. The proceedings [Atkinson 1984a] of that conference is still a useful reference.

The most important result of the 20th century in finite group theory is the classification of finite simple groups, which was completed in the early 1980's. The classification has had an impact on computational group theory in two important ways.

Beginning in the mid 1960's, new simple groups were discovered at fairly regular intervals. The existence of these large groups was difficult to demonstrate. Efforts to provide computer proofs of the existence of new simple groups provided strong motivation to improve algorithmic techniques, particular techniques for working with permutation groups.

Early algorithms for studying finite groups used relatively elementary theoretical results. However, once the classification of finite simple groups was finished, a new generation of algorithms began to appear which required the classification in their proofs of correctness or in the analysis of their complexity. These include many of the algorithms for studying finite permutation groups which have the best asymptotic complexity.

As important as the classification of simple groups is, there are large parts of finite group theory which are unaffected by it. In a certain sense most finite groups are $p$-groups, in fact, groups of order a power of 2. Finite $p$-groups lend themselves to computation more readily than general finite groups. Around 1970 there began line of algorithmic development related to finite $p$-groups. These techniques now constitute another major tool.

Until the 1970's, most research in computational group theory was undertaken by individuals trained as group theorists. This changed dramatically with the appearance of [Luks 1982], which demonstrated the close connection between permutation group algorithms and efforts to solve the graph isomorphism problem, one of the most famous problems in theoretical computer science. Until computer scientists became actively involved, the primary focus of research in computational group theory consisted of efforts to solve specific problems with the then available hardware. Issues of complexity were not usually considered.

Computer scientists brought a new perspective to the subject. They insisted that researchers specify carefully what they meant by good algorithms. With their distinct point of view they constructed new data structures and algorithms with goal of obtaining better asymptotic complexity. It frequently turned out that these structures and algorithms provided practical improvements for people interested in actual machine computation. A health rivalry continues between the two "camps" within computational group theory. Two conferences have been held at DIMACS with the specific purpose of encouraging dialogues between group theorists and computer scientists about algorithmic questions. The proceedings of these conferences [Finkelstein and Kantor 1993, 1997] make interesting reading. Another conference, sponsored jointly by DIMACS and the Geometry Center, also produced a useful proceedings [Baumslag and et al. 1996].

Two other major collections of papers on computational group theory were published as special issues [Cannon 1990] and [Cannon 1991] of the Journal of Symbolic Computation.

### 2.7.4 Permutation Groups

The algorithmic theory of permutation groups is one of the most developed areas of computational group theory. Space does not permit giving here either a complete discussion of the available algorithms or a complete bibliography. An elementary introduction to the subject can be found in [Butler 1991]. The survey [Kantor 1998] provides a good overview of recent results. A more complete treatment [Seress 1999] is in preparation.

One of the first lessons learned about machine computation with permutation groups is that, although cycle notation is useful for hand computation, it is not a good representation for elements of $\Sigma_n$ in a computer. In most cases, an element $g$ of $\Sigma_n$ is represented by a vector of length $n$ whose $i$-th component is the image $i^g$ of $i$ under $g$. Thus the number of bits need to describe our element $g$ is $n \log n$. However, it is traditional to consider $n$ to be the size of a permutation.

The first question one normally asks about a permutation group $G$ is whether or not $G$ is transitive or, more generally, what the orbits of $G$ are. Finding the orbit $\Gamma$ of $G$ containing a point $\alpha$ is easy, but it is basic. One usually needs to construct not only $\Gamma$ but also, for each $\gamma$ in $\Gamma$, an element $v(\gamma)$ in $G$ such that the image of $\alpha$ under $v(\gamma)$ is $\gamma$. Suppose $G$ is given by a generating set $S$. Then we initialize the orbit $\Gamma$ to $\{\alpha\}$ and set $v(\alpha)$ equal to the identity. Then we consider the elements $\beta$ of $\Gamma$ in the order they are discovered and for each $x$ in $S \cup S^{-1}$ we compute $\gamma = \beta^x$. If $\gamma$ is not already known to be in $\Gamma$, we add $\gamma$ to $\Gamma$ and define $v(\gamma)$ to be $v(\beta)x$. An easy but fundamental fact is that $|\Gamma|$ is the index $|G : G_\alpha|$ of the stabilizer $G_\alpha$ of $\alpha$ in $G$.

If $G$ has only one orbit, then $G$ is *transitive*. In this case we can ask if $G$ is imprimitive. That is, we can ask if there is a nontrivial equivalence relation on $\Omega_n$ which is preserved by $G$. This question can be answered in time $O(n^2)$.

Subgroups of $\Sigma_n$ can have orders which are exponential in $n$. Thus we can not solve the membership problem in polynomial time for a subgroup $G$ of $\Sigma_n$ by listing the elements of $G$. A better data structure is needed.

A *base* for $G$ is a sequence $B = \beta_1, \ldots, \beta_r$ of elements of $\Omega_n$ such that the only element of $G$ fixing each $\beta_i$ is the identity. An element of $G$ is determined by what it does to $B$. If $G$ contains $A_n$, then any base for $G$ must have at least $n-2$ elements. However, many interesting groups have bases of length $O(\log^c n)$ for small values of $c$.

The group $\Sigma_n$ induces a permutation group $T(n)$ on the set $\Gamma$ of two-element subsets of $\Omega_n$. If $n = 3m$, where $m$ is an integer, then the following sequence of two-element sets is a base for $T(n)$:

$$C = \{1,2\},\ \{2,3\},\ \{4,5\},\ \{5,6\},\ \ldots,\ \{3m-2, 3m-1\},\ \{3m-1, 3m\}.$$

If an element $g$ of $\Sigma_n$ fixes the sets $\{1,2\}$ and $\{2,3\}$, then it fixes 1, 2, and 3. Thus if $g$ fixes all $2m$ of these sets, $g$ must be the identity. Small modifications of this sequence form bases in the case that $n$ is not congruent to 0 modulo 3.

Fix a base $B = \beta_1, \ldots, \beta_r$ for a permutation group $G$. For $1 \leq i \leq r+1$ let $G^{(i)}$ denote the subgroup of $G$ fixing $\beta_1, \ldots, \beta_{i-1}$. Then

$$G^{(1)} = G \quad \text{and} \quad G^{(r+1)} = 1.$$

For $1 \leq i \leq r$ let $\Delta_i$ be the orbit of $G^{(i)}$ containing $\beta_i$ and let $U_i$ be a set of right coset representatives for $G^{(i+1)}$ in $G^{(i)}$. Since $G^{(i+1)}$ is the stabilizer of $\beta_i$ in $G^{(i)}$, it follows that $|\Delta_i| = |G^{(i)} : G^{(i+1)}|$ and that every element of $G$ can be uniquely written in the form $u_r \cdots u_1$, where $u_i$ is in $U_i$. The $\Delta_i$ are referred to as the *basic orbits* relative to $B$. If we know the basic orbits and sets $U_i$, we can decide membership in $G$ using a straightforward algorithm which runs in time $O(n^2)$.

If we take $G$ to be the group $T(3m)$ on $\Gamma$ with the base $C$, then $r = 2m$ and $\Delta_1$ is all of $\Gamma$. The set $\Delta_2$ is the set of $2(n-2)$ two-element sets which intersect $\{1,2\}$ in one point.

Suppose that we are given a generating set $S$ for our group $G$ with base $B$. Then it is easy to find $\Delta_1$. However, finding the other $\Delta_i$ is more difficult, since we do not immediately know generating sets for the subgroups $G^{(i)}$ with $i > 1$. If $S$ contains generators for all the subgroups $G^{(i)}$, then we say $S$ is a *strong generating set* relative to $B$.

We can get generators for the stabilizers of points using the following theorem.

**Theorem 1.** *Let $\Delta$ be an orbit of a permutation group $G$, let $\delta$ be a point in $\Delta$ and for each $\gamma$ in $\Delta$ let $v(\gamma)$ be an element of $G$ taking $\delta$ to $\gamma$. Suppose that $G$ is generated by a set $S$. Then $G_\delta$ is generated by*

$$T = \{v(\gamma) s v(\gamma^s)^{-1} \mid \gamma \in \Delta, \ s \in S\}.$$

The generating set $T$ in Theorem 1 can be much larger than $S$. To use this theorem iteratively, we must show that a large generating set for a permutation group can be "boiled down" to one of manageable size. A permutation group on a set with $n$ elements can always be generated by fewer than $n$ elements. Here we exhibit an algorithm for producing a generating set of size $O(n^2)$. The algorithm uses variables $u_{ij}$ for $1 \leq i < j \leq n$. Either $u_{ij}$ is nil or it is a permutation taking $i$ to $j$.

    Procecure BOIL($S$)
      Initialize all $u_{ij}$ to nil.
      For $g$ in $S$ do
        h := g;
        For $i$ from 1 to $n-1$ do
          $j := i^h$;
          If $j > i$ then
            If $u_{ij}$ is nil then $u_{ij} := h$; break;

else $h := hu_{ij}^{-1}$; fi;
    fi;
  od; od;
  Return the set of $u_{ij}$ which are not nil;
end.

If $T$ is BOIL($S$), then $S$ and $T$ generate the same group and $|T|$ is $O(n^2)$. The running time for BOIL is $O(|S|n^2)$.

Suppose we are given a generating set $S$ for a permutation group $G$ on $\Omega_n$. Assume that $|S|$ is $O(n^2)$. The sequence $B = 1, 2, \ldots, n-1$ is a base for $G$. We can obtain the basic orbits $\Delta_i$ and generating sets $S_i$ for the subgroups $G^{(i)}$ by defining $S_1$ to be $S$, computing $\Delta_1$ as sketched above, and computing a generating set $T$ for $G^{(2)} = (G^{(1)})_1$ using Theorem 1. Then $S_2$ is taken to be BOIL($T$). Since $|S_2|$ is $O(n^2)$, we can proceed inductively.

The time of the $i$-th iteration of this procedure is dominated by the time needed to apply BOIL. The initial generating set $T$ for $G^{(i+1)}$ has size $O(n^3)$. Thus applying BOIL takes time $O(n^5)$. Since $n-1$ iterations may be needed, the time needed to compute the strong generating set is $O(n^6)$.

It is not hard to improve this approach to $O(n^5)$, but "breaking the exponent 5 barrier" has been possible only using the classification of finite simple groups. The best deterministic algorithm known runs in time $O(n^4 \log^c n)$ for some constant $c$. Details, along with historical information, can be found in [Seress 1999].

The Rubik Cube group is a permutation group on 48 points. Showing that the order of this group is 43252003274489856000 takes practically no time at all. For groups with relatively short bases, it is now possible to find strong generating sets even for degrees in the millions.

Let $G$ be a permutation group given by a generating set $S$. A *nearly linear time* algorithm answering a question about $G$ is one that runs in time $O(|S|n \log^c |G|)$ for some fixed constant $c$. If $G$ has a base of length $O(\log^d n)$, then the order of $G$ is $O(n^{\log^d n})$. Thus $\log |G|$ is $O(\log^{d+1} n)$. Therefore in this case a nearly linear time algorithm runs in time $O(|S|n \log^r n)$, where $r = c(d+1)$. This is not much larger than the time needed to read in the data for the problem.

Most known nearly linear time algorithms are randomized and thus have a small probability of error. There are randomized nearly linear time algorithms for finding orbits and blocks of imprimitivity of a group $G$, for computing the order of $G$, for deciding membership in $G$, and even for constructing a composition series for $G$.

There are still some algorithmic questions about permutation groups which are not known to have polynomial time solutions. For example, the problem of finding generators for the intersection of two permutation groups is equivalent to graph isomorphism and may not have a polynomial time solution.

Another class of difficult problems are those which ask about properties of specific generating sets of a permutation groups. As remarked above, it is easy to determine the order of the Rubik Cube group and thus to compute the number of possible configurations the cube can have. However, we do not yet know how to find out the minimum number of moves which is needed to return an arbitrary configuration to the starting configuration. That is, we do not know the minimum number of factors needed to express an arbitrary element of the Rubik Cube group as a product of the generators corresponding to the rotations of the six faces of the cube.

### 2.7.5 Matrix Groups

It is natural to consider a group $G$ generated by a set of invertible matrices.

Analogous to the problems of determining orbits and deciding primitivity for permutation groups, one has the following questions:

- Does $G$ act reducibly on the underlying vector space? That is, does there exist a proper nontrivial subspace which is invariant under the action of $G$?
- If $G$ acts irreducibly, does $G$ act imprimitively? That is, can the vector space be expressed as a nontrivial direct sum of subspaces in such a way that the elements of $G$ permute the direct summands among themselves?

As with any group defined as a subgroup, one has the membership problem for $G$. One would also like to be able to say something about the structure of $G$. Is $G$ finite? If so, what is its order and what are its composition factors? Is $G$ solvable or nilpotent?

What can be said about $G$ depends very much on the field over which $G$ is defined.

Let us first consider the case in which $G$ is a subgroup of $\mathrm{GL}(n,q)$ for some positive integer $n$ and some prime power $q$. The size of a matrix for analyzing complexity is $n^2 \log q$.

Computer scientists are quick to point out that there are some very difficult problems even when $n = 1$. For example, we do not know how to solve the membership problem in polynomial time in this case. This problem is usually called the discrete log problem. Given two nonzero elements $a$ and $b$ of $\mathrm{GF}(q)$, is $b$ a power of $a$? If so, find an integer $r$ such that $b = a^r$. The order of the multiplicative group of $\mathrm{GF}(q)$ is $q - 1$ and we do not know how to compute the prime factorization of $q - 1$ in polynomial time. Thus we can not determine in polynomial time the possible orders of subgroups of $\mathrm{GL}(1, q)$.

Group theorists tend not to worry about the discrete log problem as much as computer scientists. Group theorists are most often interested in the case in which $q$ is relatively small. They are prepared to assume that

$$|\mathrm{GL}(n,q)| = q^{\binom{n}{2}} \prod_{i=1}^{n}(q^i - 1)$$

can be factored into primes. This means that the prime factors of the numbers $q^i - 1$ can be determined.

A remarkably simple but extremely useful algorithm for deciding whether $G$ is reducible and if so, for finding an invariant subspace was found by R. Parker, who christened it the Meat-Axe. This algorithm has been improved by Holt and Rees.

So far it has not been possible to find a data structure analogous to a base and strong generating which permits a polynomial time solution to the membership problem even when the discrete log problem is not the obstacle. Researchers are currently focusing on what has been dubbed the "matrix group recognition project". This is a project to develop algorithms, possibly randomized, for providing information about the nonabelian composition factors of $G$ and about the maximal subgroups of $\mathrm{GL}(n,q)$ which contain $G$. An introduction to this project can be found in [Niemeyer and Praeger 1997].

Now suppose that $G$ is a finitely generated subgroup of $\mathrm{GL}(n,K)$, where $K$ is an infinite field. The main case which has been studied is the one in which $K$ is a finite extension of the rational numbers. If $m$ is the degree of $K$ over $\mathbb{Q}$, then any vector space of dimension $n$ over $K$ is a vector space of dimension $mn$ over $\mathbb{Q}$. This gives an embedding of $\mathrm{GL}(n,K)$ into $\mathrm{GL}(mn,\mathbb{Q})$. Thus it suffices to consider the case $K = \mathbb{Q}$.

If $n \geq 4$, then the membership problem for finitely generated subgroups of $\mathrm{GL}(n,\mathbb{Q})$ is undecideable. The group $\mathrm{GL}(2,\mathbb{Q})$ contains a subgroup isomorphic to the free group $F$ on two generators. Since $\mathrm{GL}(4,\mathbb{Q})$ contains a copy of $\mathrm{GL}(2,\mathbb{Q}) \times GL(2,\mathbb{Q})$, it follows that $\mathrm{GL}(4,\mathbb{Q})$ contains a copy of $F \times F$. The subgroup membership problem for finitely generated subgroups of $F \times F$ is equivalent to the word problem for finitely presented groups and hence is undecideable.

Somewhat surprisingly, it is possible to decide whether a finitely generated subgroup of $\mathrm{GL}(n,\mathbb{Q})$ is finite.

Any polycyclic group is isomorphic to a subgroup of $\mathrm{GL}(n,\mathbb{Z})$ for some $n$ and all solvable subgroups of $\mathrm{GL}(n,\mathbb{Z})$ are polycyclic. Thus there is a close connection between the algorithmic theories of polycyclic groups and solvable subgroups of $\mathrm{GL}(n,\mathbb{Z})$.

More information about algorithms referred to in this section can be found in [Beals 1997] and [Ostheimer 1997].

### 2.7.6 Black Box Groups

Black box groups are not just of theoretical interest. At times, when one is studying a permutation group $G$, one may have subgroups $H$ and $K$ of $G$ with $K$ normal in $H$. The quotient $H/K$ may not have a description as a permutation group and may have to be treated as a black box group. In fact it is sometimes useful to consider $G$ itself to be a black box group.

Most algorithms related to black box groups are randomized. An important class of such algorithms are algorithms for recognizing simple groups. The idea is to decide whether the black box group $G$ is isomorphic to a given simple group $H$ and, if it is, to exhibit an isomorphism. It is surprising that polynomial time randomized recognition algorithms exist. In some cases, additional information about the group $G$ is required, such as the ability to determine orders of elements quickly. See [Kantor 1998] for more details.

## 2.7.7 Abelian Groups

One class of groups in which computation is relatively easy is the class of finitely generated abelian groups. Any such group is isomorphic to a quotient group of a free abelian group $M = \mathbb{Z}^n$, the direct sum of copies of the additive group of integers.

A subgroup $H$ of $M$ is finitely generated and thus may be described as the group generated by the rows of an $m$-by-$n$ integer matrix $A$. Given $A$, the first question, as usual, is to decide membership in $H$. This is usually done by replacing $A$ by its (row) Hermite normal form by applying integer row operations to $A$. Sometimes one needs to know an element $P$ of $\text{GL}(m, \mathbb{Z})$ such that $PA$ is in Hermite normal form. It is important to keep the entries in $P$ as small as possible. Polynomial time algorithms for computing Hermite normal form are known.

Sometimes lattice reduction is used instead of computing Hermite normal forms. The LLL lattice reduction algorithm is slow, but it sometimes produces multipliers $P$ with quite small entries. See [Sims 1994]. Other approaches to computing nice bases of subgroups of $\mathbb{Z}^n$ are discussed in [Havas and Majewski 1997].

## 2.7.8 Polycyclic Groups

Although computation with infinite groups usually leads quickly to unsolvable problems, there is one interesting class which provides an exception. This is the class of polycyclic-by-finite groups, groups which possess a polycyclic subgroup with finite index. All such groups have finite presentations and most computational problems concerning these groups have algorithmic solutions. In some cases, however, it is not known whether there are algorithms which are practical with current hardware.

Since space is limited, the discussion here will be restricted to polycyclic groups. The finite subgroup at the top of a polycyclic-by-finite group adds only technical difficulties. A good general reference on polycyclic groups is [Segal 1983].

An alternative characterization of polycyclic groups gives some insight into the reason polycyclic groups are nice from a computational point of view. A group $G$ is polycyclic if and only if it is solvable and all subgroups are finitely generated. Finite generation of subgroups is equivalent to the ascending chain condition on subgroups. As noted previously, we are frequently looking for a subgroup $H$ of $G$. Suppose that we know a subgroup $K$ of $H$. If we have a procedure which either confirms that $K = H$ or produces an element $h$ in $H$ but not in $K$, then we are guaranteed to be able to find $H$. If $K \neq H$, then we replace $K$ by the subgroup generated by $K$ and the element $h$ produced by the procedure. The ascending chain condition implies that we will only be able to iterate this step a finite number of times.

Suppose that $G$ is a polycyclic group and let

$$G = G_1 \supseteq G_2 \supseteq \cdots \supseteq G_k \supseteq G_{k+1} = 1$$

be a polycyclic series for $G$. Since each group $G_i/G_{i+1}$ is cyclic, for $1 \le i \le k$ we can find an element $a_i$ such that the coset $a_i G_{i+1}$ is a generator for $G_i/G_{i+1}$. By an easy induction, an element of $G$ may be expressed in the form $a_1^{x_1} \cdots a_k^{x_k}$, where the $x_i$ are integers. If $G_i/G_{i+1}$ is finite of order $m_i$, then we can assume that $0 \le x_i < m_i$. With this condition, the exponents $x_i$ are uniquely determined by $g$. The $k$-tuple $(x_1, \ldots, x_k)$ is called the *exponent vector* of $g$ and $a_1^{x_1} \cdots a_k^{x_k}$ is called the *normal form* for $g$.

Let $I$ be the set of those indices $i$ for which $G_i/G_{i+1}$ is finite. For $1 \le i < j \le k$ and each choice of $\alpha$ and $\beta$ in $\{1, -1\}$ the element $a_i^{-\alpha} a_j^{\beta} a_i^{\alpha}$ is in $G_{i+1}$. Thus there are words $U(i,j,\alpha,\beta)$ over $\{a_{i+1}, \ldots, a_k\}$ such that

$$a_j^{\beta} a_i^{\alpha} = a_i^{\alpha} U(i,j,\alpha,\beta).$$

If $i$ is in $I$, then $a_i^{m_i}$ is in $G_{i+1}$. It follows that there are words $V_i$ and $W_i$ over $\{a_{i+1}, \ldots, a_k\}$ such that

$$a_i^{m_i} = V_i \qquad \text{and} \qquad a_i^{-1} = a_i^{m_i - 1} W_i.$$

To these relations it is useful to add the redundant relations

$$a_i a_i^{-1} = 1 \qquad \text{and} \qquad a_i^{-1} a_i = 1,$$

for $1 \le i \le k$.

Not only do these relations give a presentation for $G$, they permit the computation of normal forms. If $W$ is a word over the $a_i$ which is not in normal form, then $W$ contains as a subword the left side of one of these relations. If we replace that left side with the corresponding right side and iterate, we will eventually obtain a word in normal form representing the same element of the group as $W$. This process is an example of rewriting, which is discussed below.

The rewriting approach to computing with elements of polycyclic groups is usually called *collection* by analogy with commutator collection introduced by P. Hall.

Subgroups of our polycyclic group $G$ can be represented by integer matrices which are analogous to the matrices in Hermite normal form which represent subgroups of $\mathbb{Z}^n$. Let $H$ be a subgroup of $G$. For $1 \le i \le k$ let $H_i = H \cup G_i$. Let $J$ be the set of indices $i$ such that $H_i \ne H_{i+1}$, that is, $H_i$ is not contained in $G_{i+1}$. For $i$ in $J$ we can chose an element $h_i$ in $H_i$ whose image generates the image of $H_i$ in the cyclic group $G_i/G_{i+1}$. If $i$ is in $I$, then we can assume that $x_i$, the first nonzero exponent of $h_i$ is a proper divisor of $m_i$. The elements $h_i$ with $i$ in $J$ generate $H$ and the matrix whose rows are the exponent vectors for the $h_i$ gives a good description of $H$. With this matrix, or equivalently a knowledge of the $h_i$, we decide membership in $H$ and decide equality of cosets.

Finite solvable groups form an important subclass of polycyclic groups. If $G$ is a finite solvable group, then one normally chooses the polycyclic series to be a composition series. In this case the integers $m_i$ are all primes.

Algorithms for many constructions in finite solvable groups are available in both of the packages Magma and GAP discussed in Section 10. These include computing centralizers, normalizers, and intersections of subgroups and constructing Sylow subgroups. One can also find representatives for the conjugacy classes of elements. Which computations are feasible depends on the order of the group and the type of computation. Intersections can be found in very large groups, but complete information about conjugacy classes of elements can be difficult in groups with orders in the millions.

For infinite polycyclic groups algorithms for most constructions have been shown to exist, but many of these algorithms are not practical. See [Baumslag et al. 1991]. A practical algorithm for deciding conjugacy of elements in a polycyclic group would be quite interesting.

For nilpotent polycyclic groups the situation is better. In [Lo 1998a] practical algorithms for computing intersections and normalizers are described.

### 2.7.9 Finitely Presented Groups

Most computational questions about finitely presented groups are undecideable in general. One of the few exceptions are questions about certain quotient groups. This section describes some of the tools available to attempt to study a finitely presented group. Details can be found in [Sims 1994].

**2.7.9.1 Coset Enumeration** Computation of products of elements in a free group $F = F(X)$ is very easy. Most other calculations with elements of $F$ pose few difficulties. For example, the conjugacy problem for $F$ has an efficient solution. Given two reduced words $U$ and $V$ over $X$, one writes $U$ as $ABA^{-1}$ and $V$ as $CDC^{-1}$ with the words $A$ and $C$ of maximal length. Then $U$ and $V$ are conjugate in $F$ if and only the words $B$ and $D$ have the same length and are cyclic permutations of each other.

Subgroups of free groups are free. Most computational questions about a finitely generated subgroup $H$ of $F$ have algorithmic solutions. The solution to the membership problem for $H$ is particularly important. There are in fact two approaches to this problem. One uses Nielsen reduction of words and the other uses coset tables. The coset table approach will be sketched here. Nielsen reduction in described in [Lyndon and Schupp 1977].

A coset table relative to $X$ is an array $T$ whose rows are indexed by a finite set $\Omega$ of positive integers, whose columns are indexed by $X \cup X^{-1}$, and whose entries are elements of $\Omega$. Some entries may not be defined. The following conditions must hold:

1. The set $\Omega$ contains 1.
2. If $k$ is the entry in row $i$ and column $x$, then the entry in row $k$ and column $x^{-1}$ is $i$.
3. The graph $\mathcal{G}$ with vertex set $\Omega$ and labeled edges consisting of the triples $(i, x, j)$ such that $j$ is an entry in the $i$-th row and $x$-th column of $T$ is connected.

Here is an example of a coset table relative to $X = \{x, y\}$.

## 2.7 Computational Group Theory

|   | $x$ | $x^{-1}$ | $y$ | $y^{-1}$ |
|---|---|---|---|---|
| 1 | 2 |   | 2 | 3 |
| 2 | 4 | 1 |   | 1 |
| 3 |   |   | 1 | 5 |
| 4 |   | 2 | 5 |   |
| 5 | 5 | 5 | 3 | 4 |

Given an element $i$ of $\Omega$ and a word $u$ over $X$, there is at most one path in $T$ which starts at $i$ and has the property that the product of the labels on the edges of the path is $u$. If $j$ is the end point of this path, then we write $j = i^u$. In the above table, if $u = xyxy^2$, then $2^u = 1$.

The set of reduced words $u$ such that $1^u = 1$ is a finitely generated subgroup of $F$, which we denote $\mathcal{H}(T)$. Deciding membership in $\mathcal{H}(T)$ is linear in the length of the input word. There is a natural notion of isomorphism of coset tables and isomorphic tables define the same subgroup of $F$. Every finitely generated subgroup $H$ of $F$ is defined by a coset table and all coset tables $T$ such that $H = \mathcal{H}(T)$ and the cardinality of $\Omega$ is minimal are isomorphic. For any coset table $T$, the subgroup $\mathcal{H}(T)$ has finite index in $F$ if and only if every entry in $T$ is defined. In this case $|F : H| = |\Omega|$.

In its simplest form, *coset enumeration* refers to the construction of a coset table $T$ such that $\mathcal{H}(T) = H$, where $H$ is given as the subgroup of $F$ generated by a finite set $S$. The basic step in coset enumeration consists of the following: Given a coset table $T$ and an element $g$ of $F$, find a coset table $T_1$ such that $\mathcal{H}(T_1)$ is the subgroup generated by $\mathcal{H}(T)$ and $g$.

To find $T_1$, we write the word $g$ as a product of subwords $g = uvw$, where $i = 1^u$ and $j = 1^{w^{-1}}$ are defined, the length of $u$ is maximal, and subject to this the length of $w$ is maximal.

If the length of $v$ is 1, so $v$ is in $X \cup X^{-1}$, then we change $T$ by making $j$ the entry in row $i$ and column $v$ and making $i$ the entry in row $j$ and column $v^{-1}$.

If $v$ has length greater than 1, let $x$ be the first factor of $v$. We add a new row to $T$ indexed by a new index $k$ and add entries so that $i^x = k$ and $k^{x^{-1}} = i$. Then we recompute $u$, $v$, and $w$.

If $v$ is empty, then either $i = 1$ and $w$ is empty or $i \neq j$. In the first case, $g$ is in $H$ and no change to $T$ is needed. If $i \neq j$, then we must identify $i$ and $j$. More precisely, we find the finest equivalence relation $\sim$ on $\Omega$ such that $i \sim j$ and whenever $x$ is in $X \cup X^{-1}$, $r \sim s$ and both $r^x$ and $s^x$ are defined in $T$, then $r^x \sim s^x$. Let $\Omega_1$ be the set of elements in $\Omega$ which are first in their $\sim$-class. We can take $T_1$ to be the coset table with rows indexed by $\Omega_1$ such that $i^x = j$ in $T_1$ if and only if there are elements $r$ and $s$ in $\Omega$ such that $i \sim r$, $s \sim j$ and $r^x = s$ in $T$. This construction is known as the coincidence procedure. The fact that it can be carried out efficiently is what makes coset enumeration such a useful algorithm.

Now let $(X, R)$ be a finite presentation for a group $G$. There is a homomorphism of $F$ onto $G$ with kernel equal to N, the subgroup of $F$ generated by the conjugates of the elements of $R$. There is a one-to-one correspondence between subgroups of $G$ and subgroups of $F$ which contain $N$. If the subgroup $H$ of $G$ corresponds to the subgroup $K$ of $F$ then $|G:H| = |F:K|$.

Suppose that $X$ and two finite subsets $\mathcal{R}$ and $\mathcal{S}$ are given. Let $G = \langle X \mid R \rangle$ and let $H$ be the subgroup of $G$ generated by the image of $\mathcal{S}$. If $|G:H|$ is finite, then it is possible to verify this fact and to compute $|G:H|$. However, if $|G:H|$ is infinite, then there is no algorithm which is guaranteed to demonstrate that fact.

In its more general form, coset enumeration refers to a family of related procedures for attempting to verify that $|G:H|$ is finite. Suppose that $|G:H|$ is finite and let $K$ be the inverse image of $H$ in $F$. Then $|F:K|$ is finite and $K$ is generated by $\mathcal{S}$ and a finite set of conjugates of elements of $\mathcal{R}$. All variants of coset enumeration proceed by constructing the coset table for the subgroup $L$ of $F$ generated by $\mathcal{S}$ and larger and larger finite sets of conjugates of elements of $\mathcal{R}$ until $|F:L|$ is finite. As noted above, this occurs when every entry in the coset table $T$ for $L$ is defined. Let $\Omega$ be the set of integers indexing the rows of $T$. For each element $i$ of $\Omega$ we chose one word $U$ such that $1^U = i$ in $T$. We add all elements of the form $URU^{-1}$ to the generators of $L$, where $R$ ranges over the elements of $\mathcal{R}$. This may cause $L$ to get bigger. At this point $L = K$.

Efforts to program coset enumeration go back at least to 1953. Despite the intervening 45 years, we are still finding ways to carry out the procedure more quickly or with less memory. Currently it is possible to work with coset table having tens of millions of rows.

**2.7.9.2 The Knuth-Bendix Procedure for Strings** The Knuth-Bendix procedure for strings is a special case of the very general technique of universal algebra described in [Knuth and Bendix 1970]. It provides an alternative to coset enumeration for studying finitely presented groups. With coset enumeration the basic data structure is the coset table. With the Knuth-Bendix procedure for strings the basic data structure is the rewriting system.

Rewriting systems are defined with the aid of special orderings on words. Let $X$ be a finite set. A *reduction ordering* on the set $M$ of group words over $X$ is a well ordering $<$ of $M$ with the property that the empty word is less than any other word and, for all words $A$, $B$, $U$, and $V$, if $U < V$, then $AUB < AVB$. A rewriting system relative to a reduction ordering is a set $\mathcal{R}$ of ordered pairs of words such that for each element $(V, U)$ of $\mathcal{R}$ we have $V > U$. Elements of $\mathcal{R}$ are called rewriting rules.

Given a rewriting system $\mathcal{R}$, we shall be interested in the group $G$ generated by $X$ and defined by the relations $V = U$ for all $(V, U)$ in $\mathcal{R}$. It is useful to assume that $\mathcal{R}$ contains the rules $(xx^{-1}, 1)$ for all $x$ in $X \cup X^{-1}$. The corresponding relations are redundant in the context of groups but not in the context of monoids.

Once we have fixed a rewriting system $\mathcal{R}$, we can rewrite a group word $W$ over $X$ using the following procedure:

    Procecure REWRITE($W$)
        $P := W$;

## 2.7 Computational Group Theory

    While $P$ contains the left side of a rule in $\mathcal{R}$ do
      Let $P = AVB$, where $A$ and $B$ are words and $(V, U)$ is in $\mathcal{R}$;
      $P := AUB$;
    od;
    Return $P$;
END.

Because an occurrence of the left side of a rule $(V, U)$ is replaced by the right side, we often write the rule in the form $V \to U$.

Each time $P$ is changed in REWRITE, the new value is less than the old with respect to $<$. Since $<$ is a well ordering, $P$ can change only finitely often. Thus the procedure must terminate. The image of $P$ in $G$ does not change during the execution of REWRITE. Therefore the word returned by REWRITE defines the same element $g$ of $G$ as $W$ and is *reduced* with respect to $\mathcal{R}$ in the sense that it contains no left side of a rule as a subword. Executing REWRITE poses no problems if $\mathcal{R}$ is finite. If $\mathcal{R}$ is infinite, then it may still be possible to execute REWRITE.

It would be nice if the result of REWRITE depended only on $g$. Unfortunately this need not be the case. In the operation of REWRITE there may be many choices for the words $A$ and $B$ and the rule $(V, U)$. The final result may depend on those choices. Thus there may be many reduced words which define the same element of $G$.

The rewriting system $\mathcal{R}$ is called *confluent* if the word returned by the call REWRITE($W$) depends only on $W$ and not on the choices made during execution. When this is the case, each element of $G$ is defined by a unique reduced word. Thus, when $\mathcal{R}$ is confluent and we can execute REWRITE, then we can solve the word problem for $G$.

If $\mathcal{R}$ is finite, then there is a finite test for confluence. The test either confirms that $\mathcal{R}$ is confluent or produces a new rule $(V, U)$ which must be added to $\mathcal{R}$ if confluence is to hold. The Knuth-Bendix procedure for strings proceeds by iterating the test for confluence, adding any new rules produced. The procedure may not terminate, but if it does, the resulting rewriting system is confluent and gives a solution of the word problem for $G$.

The Knuth-Bendix procedure is most easily run which $<$ is a *lenlex* ordering. That is, words are first ordered by length and then lexicographically according to a specified ordering of the elements of $X \cup X^{-1}$. However, other orderings are frequently needed. For example, the polycyclic presentation for a polycyclic group defined above is a confluent rewriting system with respect to an ordering on words where it is possible to have $U < V$ even if $U$ is longer than $V$.

The following is a confluent rewriting system:

$$aA \to 1, \quad Aa \to 1, \quad bB \to 1, \quad Bb \to 1,$$
$$bA \to Abb, \quad bba \to ab, \quad Ba \to baB, \quad BA \to ABB.$$

Here the "case convention" is being used. That is, $A$ and $B$ represent $a^{-1}$ and $b^{-1}$, respectively. Again the ordering $<$ is not compatible with length. The group defined by this rewriting system is infinite since all words of the form $a^i$ with $i > 0$ are reduced.

**2.7.9.3 Quotient Groups** Let $G$ be a group given by a finite presentation $(X, R)$. Despite the unsolvability of many computational questions about $G$, there are algorithms which are guaranteed to construct certain nilpotent quotients of $G$. There are also procedures for attempting to determine certain other solvable quotients of $G$.

The most important case is the determination of the largest abelian quotient $G/G'$. Suppose that $|X| = n$ and $X = \{x_1, \ldots, x_n\}$. There is a homomorphism $f$ from $F = F(X)$ to $\mathbb{Z}^n$ which maps $x_i$ to the $i$-th standard basis vector of $\mathbb{Z}^n$. The map $f$ is clearly surjective and the kernel of $f$ is $F(X)'$. It is not hard to show that $G/G'$ is isomorphic to the quotient $\mathbb{Z}^n/M$, where $M$ is the subgroup of $Z^n$ generated by the image of $R$ under $f$. Using the methods of Section 7, we can determine the structure of $G/G'$.

More generally, for any $s > 0$ we can determine a polycyclic presentation for $P = G/\gamma_s(G)$ and thus compute effectively in this quotient of $G$. An algorithm for computing this presentation is described in [Nickel 1996].

A special case of nilpotent quotients are quotients which are $p$-groups. Very powerful algorithms have been developed to compute quotients of $G$ which are $p$-groups for some given prime $p$. Quotients of order $p^n$ with $n$ in the thousands are often within reach.

The quotients of $G$ by terms of the derived series other than the first need not be polycyclic and cannot in general be computed. There are two approaches for computing finite solvable quotients of $G$. They are described in [Niemeyer 1994a] and [Plesken 1987].

The only approach available for computing infinite nonnilpotent quotients of $G$ is given in [Lo 1998b]. This method is based on the fact that if $P$ is a polycyclic group, then right ideals in the group ring $\mathbb{Z}(P)$ are finitely generated. It is possible to generalize the Gröbner basis methods described in Section 2.2.5 to submodules of finitely generated free right modules over $\mathbb{Z}(P)$. Let $N$ be a normal subgroup of $G$ such that $G/N$ is isomorphic to $P$. Then $N/N'$ is a finitely generated right module over $\mathbb{Z}(P)$ and it is possible to give a finite presentation for this module and using the Gröbner basis algorithm one can decide whether $N/N'$ is finitely generated as an abelian group. In this way one can decide whether $G/N'$ is polycyclic. If it is, then one can find a consistent polycyclic presentation for it.

**2.7.9.4 The Reidemeister-Schreier Algorithm** Let $G$ be a finitely presented group and let $H$ be a subgroup of $G$. It is possible to describe a presentation for $H$. If the index of $H$ in $G$ is finite, then this presentation is finite. The Reidemeister-Schreier algorithm is one method for constructing presentations for subgroups of finitely presented groups.

One of the few ways available to prove that $G$ is infinite is to find a subgroup $H$ of finite index such that $H/H'$ is infinite, which can be confirmed by finding the Reidemeister-Schreier presentation for $H$ and then using the methods of the previous subsection to determine the structure of $H/H'$.

Even if the presentation for $G$ is quite manageable and the index of $H$ is only a few hundred, the presentations for $H$ obtained are very complicated. It is usually necessary to find some way of simplifying the presentation before it can be used further. Two approaches to simplification are the Knuth-Bendix method for strings and the systematic use of Tietze transformations.

## 2.7.10 Group-Theoretic Software

There are three software packages which offer general support for group theoretic computation. These are Magma, GAP, and Magnus. In addition, Maple provides some tools for group theory and there are many smaller packages which are either intended to be run by themselves or with one or more of the larger systems.

Magma with its predecessor system Cayley is the oldest of the existing comprehensive systems. It provides support for algebraic number theory and several other areas besides group theory. The system is described in Section 4.1.5. A license fee is involved.

GAP has been around for about 13 years. A major revision of the system is being tested as this article is written. The focus of GAP is more directly on group theory. The system is described in Section 4.2.18.

The system Magnus is much newer and is less developed than Magma and GAP. It is intended to support research on infinite groups. For example, while Magma and GAP only attempt to find presentations of subgroups of finite index in finitely presented groups, Magnus is able to produce partial presentations for subgroups of infinite index.

Among the more specialized packages are ANUQA (section 4.3.1), C-meataxe (section 4.2.10), MAT (section 4.3.11), MOC (section 4.2.28), and QUOTPIC (section 4.2.32).

## 2.7.11 Another Perspective

The point of view up to this point has been to look at a class of groups and ask how hard it is to compute with elements and subgroups in groups belonging to that class. There is a growing body of research which looks at computation in groups quite differently. One make assumptions about the ease or difficulty of computing in a group and asks what this implies about the structure of the group.

The most common assumption is that the group is automatic. An *automatic group* is a group $G$ in which multiplication and comparison of elements can be performed using finite automata. The standard reference on automatic groups is [Epstein et al. 1992].

The sets of elements of automatic groups are naturally described by families of words called regular languages. It is possible to consider classes of groups whose elements require more complicated types of languages for their description.

Additional references: [Beals 1993; Leon 1997].

Charles C. Sims (Rutgers U.)

## 2.8 Algorithms of Representation Theory

Soon after the first applications of computers in group theory—to coset enumeration—algorithmic methods have been introduced in representation theory, beginning with the character theory of finite groups.

### 2.8.1 Ordinary Representation Theory

**Characters.** Early implementations, by S. Comet, of algorithms for the computation of characters of symmetric groups, date back to the late fifties [Comet 1960]. Questions arising during the classification project of the finite simple groups lead to the demand for computing character tables of specific finite groups from incomplete knowledge of some of its characters. For this purpose interactive methods were implemented at various places (B. Fischer and T. Gabrysch (Bielefeld), D. Livingstone (Birmingham), J. H. Conway (Cambridge)). In the late seventies these methods were collected, extended and enhanced by an arithmetic for cyclotomic fields in the Aachen CAS system [Neubüser et al. 1984]. This included routines to compute tensor products, inner products, and symmetrizations of characters, induced characters and fusions of subgroups. The CAS system as well as its new implementation in GAP (see also [GAP 1997] and 4.2.18) have lead to the computation of numerous character tables, some of them included in the *Atlas of Finite Groups* [Conway et al. 1985]. This contains the character tables of the sporadic groups, their covering groups and automorphism groups (in compound form). All of these tables, and many more, are now also available in GAP. It contains, for example, the character tables of most of the maximal subgroups of the sporadic simple groups.

In the sixties J. D. Dixon suggested an algorithm for computing the character table of a finite group from its class multiplication coefficients. A substantially improved version of *Dixon's algorithm*, due to G. Schneider, is included in GAP and MAGMA [Bosma et al. 1997]. These implementations are capable of calculating the character tables of groups with up to several hundreds of conjugacy classes, provided the degrees of the irreducible characters are not too large.

More powerful algorithms exist for computing character tables of groups of special types, e.g., $p$-groups (Conlon [Conlon 1990]). For other classes of groups e.g., the symmetric groups, the rows and columns of the character table have natural labelings in terms of certain combinatorial objects, e.g., partitions. Moreover, there are algorithms, the *Murnaghan-Nakayama rules* and generalizations thereof, for computing the entries of the character tables in terms of the labels for the rows and columns. Such algorithms are known for all Weyl groups of classical types and have been implemented in GAP.

**Representations.** The irreducible matrix representations of a finite group over a field of characteristic 0 are considerably more difficult to construct than the corresponding characters. Nevertheless, some methods have emerged over the past few years.

Baum and Clausen [Baum and Clausen 1994] have described an algorithm to compute the irreducible matrix representations of a supersolvable group from a power commutator presentation.

Labonté [Labonte 1990] and Linton [Linton 1991] (independently) described a method—analogous to the Todd-Coxeter coset enumeration discussed in Sims' article 2.7—to construct representations for finitely presented algebras. While this *Vector Enumerator*, as it came to be called, works most efficiently over finite fields, it is in principle applicable to algebras over fields of characteristic 0. One major application has been the construction of representations of Iwahori-Hecke algebras.

Methods for constructing rational irreducible representations of finite groups have been introduced by Plesken. One such method, for soluble groups, is used in the *Soluble Quotient Algorithm* of Plesken [Plesken 1987], implemented by Brückner [Brückner 1998]. Another method, based on liftings of representations, has recently been applied by Plesken and Souvignier to prove the infiniteness of certain finitely presented groups. A survey of these ideas and their applications is given in [Plesken 1999, Section 2].

Richard Parker has suggested an integral version of his **Meat-Axe** (see also 2.8.2 below) [Parker 1998]. Rational and integral representations of finite groups play an essential role in the investigation of integral lattices and their automorphism groups. This information is used in the study of finite subgroups of the general linear groups over the integers or the rational numbers (see [Plesken 1998] for a survey). The monumental work of Plesken and Nebe [Nebe and Plesken 1995] has lead to the classification of all finite subgroups of the groups $GL_n(\mathbf{Q})$ for $n \leq 31$.

Since crystallographic groups are constructed from integral representations of finite groups, these are of great importance in crystallography. The **CARAT** package (see [Opgenorth et al. 1998] and 4.2.7) contains tables and implementations of various algorithms, including the *Zassenhaus algorithm* for computing extension groups, for handling enumeration and recognition problems for crystallographic groups.

The **GAP** share package **AREP** by Egner and Püschel (see [Egner and Püschel 1998] and 4.3.2) computes, symbolically, with structured representations of finite groups. Examples for structured representations are induced representations or tensor products of representations. Applications of **AREP** include the automatic construction of fast algorithms for discrete linear signal transforms.

### 2.8.2 Modular Representation Theory

The computation of Brauer character tables of finite groups was begun in the seventies with the work of Gordon James on the Mathieu groups. While James was still working by hand, Richard Parker soon applied his **Meat-Axe** [Parker 1984] (see also 4.2.10), originally designed to construct the largest Janko group, to this kind of problems.

Since the **Meat-Axe** is of fundamental importance in computational representation theory, I shall sketch its main ideas. Given $d \times d$-matrices $a_1, \ldots, a_n$

over a field $F$, let $A$ denote the (unitary) $F$-algebra generated by $a_1, \ldots, a_n$. The **Meat-Axe** aims to find a composition series of the natural left $A$-module $F^d := F^{d \times 1}$. Inductively, it suffices to find a non-trivial $A$-invariant subspace of $F^d$ or to prove that $A$ acts irreducibly on $F^d$. Let $v \in F^d$. By using a variation of the orbit algorithm for permutation groups (see 2.7), and the Gauß algorithm, the **Meat-Axe** computes the smallest $A$-invariant subspace $Av$ of $F^d$ containing $v$, and matrices for the actions of $a_1, \ldots, a_n$ on the subspace $Av$ and the quotient $F^d/Av$. The search for vectors lying in proper $A$-invariant subspaces of $F^d$ (if there are any) is guided by the following result of S. Norton. (We denote the transpose of a matrix $b$ by $b^t$ and by $A^t$ the $F$-algebra generated by $a_1^t, \ldots, a_n^t$.)

**Proposition.** (Norton's irreducibility criterion) *Let $b \in A$. Then at least one of the following occurs.*

(1) *$b$ is invertible.*
(2) *$Av$ is a proper subspace of $F^d$ for at least one non-zero $v$ in the nullspace of $b$.*
(3) *$A^t v$ is a proper subspace of $F^d$ for all non-zero $v$ in the nullspace of $b^t$.*
(4) *$A$ acts irreducibly on $F^d$.*

Thus one has to find a non-invertible element $b \in A$ with nullspace of small dimension (preferably 1). If $F$ is a (small) finite field a random choice of elements of $A$ is a reasonable strategy to find such a $b$. For larger fields more sophisticated methods, suggested by Holt and Rees [Holt and Rees 1994], have to be applied. If all non-zero vectors of the nullspace of $b$ fail to lie in a proper $A$-invariant subspace, chose a non-zero vector $v$ in the nullspace of $b^t$ and compute $A^t v$. If this is all of $F^d$, then $F^d$ is an irreducible $A$-module. On the other hand, if $F^d$ is a reducible $A$-module, one finds a proper invariant subspace this way.

A large number of Brauer character tables of sporadic groups have been computed by Parker and others with the help of the **Meat-Axe**. This method came to its limits with the degrees of the representations to be considered growing larger and larger.

The applicability of the **Meat-Axe** is greatly extended by *condensation* techniques, where the original algebra is replaced by a Morita equivalent one of considerably smaller dimension. Additional support for the **Meat-Axe** is provided by the **MOC** system, developed by Parker, Lux, Jansen and Hiss (see also 4.2.28). This system works with Brauer characters, rather than representations, so that there are no degree constraints. Computations with these systems lead to the construction of the Brauer character tables of all the Atlas groups of order at most $10^9$ which were published in the [Jansen et al. 1995]. Many more tables are now known (see http://www.math.rwth-aachen.de/~MOC/). The modular character tables for the symmetric groups $S_n$ can also be computed with **SPECHT**, a **GAP** share package written by Andrew Mathas. They are known completely for $n \leq 16$.

The **Meat-Axe** is also used, of course, in the investigation of module structures, e.g., submodule lattices or Loewy series, and also for computing endomorphism rings of modules, direct decompositions of modules and Green correspondents. The matrix group recognition project provides a further field of applications for the **Meat-Axe**. Moreover, it is still used according to its original design, namely to construct specific finite groups via some of their linear representations. A spectacular example is provided by the matrix representation of the Monster sporadic group, constructed by Linton, Parker, Walsh, and Wilson, [Linton et al. 1998]. Robert A. Wilson has initiated and maintains a data base of representations of finite groups (http://www.mat.bham.ac.uk/atlas/index.html).

Explicit modular representations of finite groups were used in a substantial way in the work of Holt and Plesken in the construction of perfect groups [Holt and Plesken 1989]. In addition, this work also used the algorithms and implementations by Derek Holt [Holt 1985] for computing first and second cohomology groups of finite groups. For example, there are programs to determine the Schur multiplier of a given finite group, or to find the extension classes of a finite group with a finite module.

New directions of algorithmic research aim to compute projective resolutions of modules and cohomology rings of finite groups (Adem, Milgram, Jon Carlson, David Green, Ed Green, Schneider, see, e.g., [Adem and Milgram 1994; Carlson et al. 1997]). A recent approach uses non-commutative Gröbner bases. These are also used in the Virginia Tech Hopf project for the algorithmic investigation of finite dimensional algebras, in particular Hopf algebras. Finite dimensional algebras and their representations are conveniently studied via a directed graph, a so-called quiver, which is a purely combinatorial object. The Bielefeld **CREP** system (see also 4.2.12) provides algorithms for using the quiver approach to finite dimensional algebras for research and teaching.

We close by pointing out the various recent applications of computer algebra in invariant theory, mainly due to Kemper (see [Kemper and Malle 1999] for a survey).

### 2.8.3 Generic Character Tables

It is often possible to encode the character tables of an infinite series of groups in a single table or in a program. Such a table or program is then called a *Generic Character Table*. The first example of a generic character table was computed by Frobenius in 1897: the generic table for the series of Chevalley groups $SL(2, 2^n)$, $n$ a positive integer. This example already has all the features common to a generic character table for a series of Chevalley groups. The conjugacy classes and the irreducible characters of the groups in the series are parameterized in a suitable way, and the character values of the generic table are given as functions of these parameters. Here, a series is a set of groups arising from one particular Dynkin diagram with a fixed symmetry, when the underlying field is allowed to vary. Usually, there are a finite number of tables for a fixed Dynkin diagram with symmetry. For example, there are two generic tables for the series $SL_2(q)$, one for even and one for odd $q$.

Many computations with characters can be performed symbolically on such a generic table. One can compute scalar products of character types, calculate tensor products of characters or compute class multiplication coefficients. Such computations are valid for all groups in the series. The **CHEVIE** system (see also [Geck et al. 1996] and 4.2.9) contains a library of generic tables for Chevalley groups and a collection of **MAPLE** routines to perform such computations. As an application the 6-dimensional symplectic groups were shown to be Galois groups over abelian number fields [Lübeck 1993].

Generic tables in form of programs have been implemented for Weyl groups of type $A_n$ (i.e., the symmetric groups), $B_n$ and $D_n$ and are available in **CHEVIE**. The tables for the symmetric groups are also available in **ACE**, a **MAPLE** share package written by Sebastian Veigneau and in **SYMMETRICA** [Kerber et al. 1992]. We are not aware of any implementation of generic character tables for the covering groups of the symmetric groups.

Related to the generic character tables of the Weyl groups are the generic character tables of the corresponding Iwahori-Hecke algebras, also available in **CHEVIE**, and, for type $A$, in **ACE**.

### 2.8.4 Summary of Systems

In this section I summarize the various systems presented above as well as some other packages not mentioned there. The most comprehensive systems for computational group and representation theory are **GAP** [GAP 1997] and **MAGMA** [Bosma et al. 1997].

Special purpose systems are **AREP** (Eindhoven, Pittsburgh) for computing symbolically with structured representations, **CARAT** (Aachen), for working with crystallographic groups, **CHEVIE** (Aachen, Kassel, Paris) for computing with generic character tables of Chevalley groups, Iwahori-Hecke algebras and Weyl groups, **CREP** (Bielefeld, see also 4.2.12) for the investigation of finite dimensional algebras, **LIE** (Eindhoven, see also 4.2.23) for computations with Lie algebras, Coxeter groups, and their representations, **QUOTPIC** (Warwick, see also 4.2.32), for the construction of quotients of finitely presented groups, **SISYPHOS** (Stuttgart), for computing in modular group algebras of finite $p$-groups, and **SYMMETRICA** (Bayreuth, see also 4.2.41), for combinatorics related to, and applications of the symmetric groups.

Combinatorics related to Lie algebras, Weyl groups, and symmetric functions are also contained in the two **MAPLE** share packages **ACE** by Sebastian Veigneau (see also 4.2.1) and **SF** by John Stembridge. Another **MAPLE** package, **INVAR** by Gregor Kemper, computes invariant rings of finite groups. The **GAP** share package **SPECHT** by Andrew Mathas contains algorithms for computing decomposition numbers of symmetric groups and Iwahori-Hecke algebras.

Finally, there is a large collection of stand alone programs related to the Meat-Axe, its *Condensation* enhancements, the *Vector-Enumerator* or MOC. Particular versions of these are available in **GAP** and **MAGMA**, as system commands as well as external packages.

Additional references: [Curtis and Wilson 1998; Atkinson 1984b; Matzat et al. 1999].

Gerhard Hiss (Aachen)

## 2.9 Algebraic Methods for Constructing Discrete Structures

Constructing and investigating discrete structures by algebraic methods has the following goals:

- lowering the complexity by using algebraic instead of combinatorial methods (symmetry approach, symmetry adapted bases etc.),
- identifying isomorphism classes of certain discrete structures (e.g., graphs, physical states, chemical isomers) with orbits of finite groups acting on finite sets,
- transforming combinatorial problems into equivalent algebraic ones, especially into problems of group theory, or problems of linear and multilinear algebra (e.g., t-designs expressed as 0-1 solutions of Diophantine equations, where the rows and columns belong to certain double cosets),
- enumerating such structures for given parameters (for example graphs with given numbers of vertices and edges) by applying the lemma of Cauchy-Frobenius, and its refined variants,
- generating a complete list (without redundancy) of all structures with prescribed properties for given parameters (the numbers obtained from enumeration steps can be used as stopping rules),
- generating such structures uniformly at random, using the Dixon-Wilf algorithm [Dixon and Wilf 1983] (the numbers mentioned above are used for the selection of appropriate probabilities), e.g., for checking a hypothesis before a mathematical proof is attempted.

The main algebraic ideas used in this context are (see [Kerber and Laue 1998]):

- the *fundamental lemma* of the theory of group actions, which associates orbits and double cosets,
- *coset enumeration,* for determining a permutation representation on the cosets of a subgroup of finite index in a finitely presented group. Coset enumeration is used in computations with amalgams for finding graphs, diagram geometries, polytopes, etc, with prescribed local properties.
- the *homomorphism principle,* which leads to efficient algorithms (it basically leads to the logarithm of the original complexity), and
- the *gluing lemma* that controls the identification of substructures along which structures can be glued together.
- *character theory* for finite groups, especially character tables, and permutation characters.
- *lattice basis reduction,* for solving systems of Diophantine equations.

They are mostly combined with a very helpful combinatorial (better say: order theoretic) method, the

- *orderly generation* according to R.C. Read, and I.A. Faradzev [Read 1978; Faradzev 1977; Ivanov 1985].

Coset enumeration is used in computations with amalgams for finding graphs, diagram geometries, polytopes, etc, with prescribed local properties.

Both, the homomorphism principle and orderly generation, break up the generation process into steps. So constraints may be used in order to bound the solution set early and to perform costly isomorphism testing only for the surviving candidates. This is an important achievement for all practical applications. Orderly generation is a special case of intelligent backtrack search, applications combine the use of symmetry with available combinatorial constraints at the bounds of feasible computation. Illustrative examples are the discovery of a new infinite family of finite projective planes [Penttila and Williams 1997] and of new line-transitive linear spaces (where the linear space restriction plus the line-transitivity allow searches to be performed at far larger values of the parameters than would be expected for general design construction methods), see [Nickel et al. 1992].

Lattice basis reduction methods have been applied to find $t$-designs with prescribed automorphism group, especially the first 7- and 8-designs with small parameters and projective automorphism group and new starting points for infinite series of $t$-designs with large $t$, see [Betten et al. 1997], [Betten et al. 1998b, 1999; Laue 1997].

A typical non-mathematical example arises with isomers in chemistry: for a given chemical formulas one has to construct all connected multigraphs where the degree of each vertex corresponds to the atom's valences. From this problem (as well as from the well-known solution to the Königsberg bridge problem, given by Euler, and the theory of electric circuits, introduced by Kirchhoff) originated graph theory as well as combinatorical enumeration theory.

Standard methods of computer science, like dynamic trees, union-find, and optimization algorithms of graph theory play an important role in solving such problems. Their application can often lead to a re-evaluation of algebraic methods.

Other aspects of this method, and interesting applications, can be found in section 3.5.1. Further information and literature are contained in [Kerber 1991] and [Laue 1993]. Recent breakthroughs based on these applications are concerned with counting as well as constructing representatives of isometry classes of linear codes ([Fripertinger and Kerber 1995][Betten et al. 1998a]).

Examples of geometrical studies using coset enumeration can be found in [Soicher 1992] and [Yoshiara 1991]. Neubüser gives a good general account of coset enumeration methods in [Neubüser 1982]. An interesting review of applications of cosets in sciences is [Ruch and Klein 1983].

Character tables and permutation characters have been used effectively for constructing low rank permutation representations, and small diameter arc-transitive and edge-transitive graphs, especially distance transitive graphs. Examples of their use may be found in [Praeger and Soicher 1997][Ivanov et al. 1995].

General group theoretic packages can be used in order to investigate the structure and properties of discrete structures, for example, the comprehensive computer algebra systems GAP and MAGMA (see above).

Both of these systems incorporate a variety of algorithms for investigating discrete structures, and some additional packages and other special purpose systems deserve separate mention; principally:

- GRAPE [Soicher 1993] is a GAP share library package for computing with G-graphs, that is, a graph together with a group G of automorphisms of that graph. Using G-graphs and special algorithms for G-graphs allows highly efficient structural computations in very large symmetric graphs which are of great interest to group theorists and finite geometers. GRAPE has been used in the classification of the primitive distance-transitive graphs for sporadic groups and in the discovery of new distance-regular graphs, as well as many investigations in discrete combinatorial geometries and the discovery of new designs for agricultural experiments.
- Co-Co [Faradžev and Klin 1991] was originally developed to investigate graphs and cellular rings associated with almost simple groups.
- DISCRETA is a package that covers at present the construction of various kinds of $t$-designs, see above.

Adalbert Kerber und Reinhard Laue (Bayreuth)

## 2.10 Summation and Integration

### 2.10.1 Definite Summation and Hypergeometric Identities

**2.10.1.1 The Definite Breakthrough** Until quite recently the algorithmic task of *summation*, compared to its continuous analogue *integration* (see 2.10.2) has received relatively little attention within the computer algebra community. The survey article [Lafon 1983] by J. Lafon characterizes the situation in the eighties: in textbooks and monographs the problem was hardly mentioned; only few methods had been implemented, at least in "commercial" systems. Indeed, down to date: B. Gosper's method [Gosper jr. 1978] for indefinite hypergeometric summation served as the work horse most of the time; sometimes variations of R. Moenck's approach [Moenck 1977] for rational summation were used, however the work of S. Abramov [Abramov 1971, 1975] and of M. Karr [Karr 1981, 1985] remained widely unnoticed.

The current intensified interest in summation was mainly initiated by two achievements of D. Zeilberger: his successful application ("creative telescoping") of Gosper's algorithm for *indefinite* hypergeometric summation to problems of *definite* hypergeometric summation [Zeilberger 1990a, 1991] and, based on I. Bernstein's theory of holonomic systems, his "holonomic systems approach to special function identities" [Zeilberger 1990b]. We want to emphasize that algorithmic proofs of definite hypergeometric summation identities had sometimes even been considered as unfeasible. After this breakthrough in the single sum case, Wilf and Zeilberger further extended this algorithmic proof theory to closed form evaluation ("WZ-method") and to multisums and integrals ranging over hypergeometric and hyperexponential terms [Wilf and Zeilberger 1992a,b]. The approach is based on the fact that these sums and integrals are holonomic and that annihilating linear difference and differential operators can be constructed effectively. This constitutes the realization of an old idea, known as *Sister Celine's technique* to researchers in the area of special functions for quite some time. One characteristic of the method is that, in addition to the operators, "proof certificates" in the form of multivariate rational functions will be constructed, whose verification by itself requires only exact rational function arithmetic.

A comprehensive presentation of this development and the algorithms around is now available in book form [Petkovšek et al. 1996; Koepf 1998a]. These monographs describe carefully the five basic algorithmic methods, their applications and extensions, in the field: Sister Celine's technique, Gosper's algorithm, creative telescoping, WZ-method, and Petkovšek's algorithm hyper. The latter [Petkovšek 1992] finds closed form solutions of linear higher order polynomial recurrences, and hence provides a decision procedure for closed form evaluation of definite hypergeometric summations.

For further references to original papers and other relevant literature we refer to the extensive bibliographies of [Petkovšek et al. 1996] and [Koepf 1998a], see also Section 3.2.6.

**2.10.1.2 Further Developments** A concise survey of some recent developments in symbolic summation is given in [Paule and Strehl 1995]. It is shown that normal form aspects such as the Gosper-Petkovšek form for rational functions or Paule's "greatest factorial factorization" for polynomials, a shift-analogue of square-free factorization, lead to efficient algorithms (e.g., the optimal solution by R. Pirastu und V. Strehl [Pirastu and Strehl 1995] in the case of rational summation) and to better understanding and extensions (e.g., Paule's algebraic approach to Gosper's algorithm [Paule 1995] and to its $q$- and multibasic analogues [Paule and Riese 1997; Riese 1996; Bauer and Petkovšek 1999]).

As to the holonomic aspects of the field, N. Takayama developed non-commutative Groebner bases methods for holonomic elimination problems (e.g., recurrence and contiguous relations for hypergeometric functions, zero-recognition, etc.); see [Takayama 1992, 1995]. F. Chyzak and B. Salvy formulated holonomic elimination in the generic setting of Ore algebras, which provides additional flexibility and tools in automatically proving special function identities. This context allows, e.g., to treat systems of mixed differential-difference type including ordinary as well as $q$-hypergeometric recurrences; see [Chyzak and Salvy 1998]. In [Chyzak 1998] F. Chyzak underlines the central role of multivariate $D$-finiteness and exploits it algorithmically.

**2.10.1.3 Applications** Proving binomial identities and finding recurrences satisfied by such sums, a task notoriously difficult for the human mathematician, has become a routine business for computers, even in the $q$-case and also

## 2.10.1 Definite Summation and Hypergeometric Identities

for ordinary hypergeometric multisums. The literature abounds with successful applications of this "technology"; for sake of brevity we only mention a few instances:
— The paper by I. Nemes et al. [Nemes et al. 199] describes how the typical Monthly problem (on summation) can be treated automatically with the computer.
— In [Gessel 1995], I. Gessel shows the effectiveness of the WZ-method if applied systematically.
— The celebrated Rogers-Ramanujan identities in additive number theory find "one-line" computer proofs as explained by P. Paule in [Paule 1994].
— Spectacular determinant evaluations with essential help of creative telescoping have been achieved by C. Krattenthaler et al. [Ciucu and Krattenthaler 1999; Krattenthaler and Zeilberger 1997] and Petkovšek/Wilf [Petkovšek and Wilf 1996].

**2.10.1.4 Implementations** There are now various implementations available of the work mentioned above. In particular:
— The algorithms discussed in "A=B" [Petkovšek et al. 1996] are contained on a diskette which comes with the book or can else be taken in Maple or Mathematica versions from the web site http://www.cis.upenn.edu/~wilf/AeqB.html (from which even the entire book can be downloaded).
— Mathematica implementations of creative telescoping in the ordinary and $q$-case have been written at RISC-Linz [Paule and Riese 1997; Paule and Schorn 1995]; efficient multisum algorithms are available through a package developed by K. Wegschaider [Wegschaider 1997]. Karr's algorithms and extensions have been implemented by C. Schneider [Schneider 1999]. All can be taken from http://www.risc.uni-linz.ac.at/research/combinat/risc/. See [Caruso 1999] for a Macsyma implementation of Zeilberger's method.
— Maple code (ordinary and $q$-case) is provided by T. Koornwinder [Koornwinder 1993] (http://turing.wins.uva.nl/~thk/recentpapers/zeilbalgo.html), by W. Koepf [Koepf 1998a; Böing and Koepf 1999], (http://www.imn.htwk-leipzig.de/~koepf/research.html), and by B. Gauthier [Gauthier 1999] (http://www-igm.univ-mlv.fr/~gauthier/HYPERG.html).
— The Maple library now contains the package sumtools, written by W. Koepf [Koepf 1995a]; a Reduce version is also available, see [Koepf 1995b; Böing and Koepf 1997].
— For implementations of N. Takayama's work consult http://www.math.s.kobe-u.ac.jp/HOME/taka/index.html
— The packages developed by F. Chyzak can be taken from http://www-rocq.inria.fr/algo/libraries/libraries.html
— Finally we mention C. Krattenthaler's Mathematica packages HYP and HYPQ [Krattenthaler 1995], though different in character, are invaluable tools for the investigation of hypergeometric transformation identities; they are to be found at http://radon.mat.univie.ac.at/People/kratt/hyp_hypq/hyp.html

Peter Paule (Linz) and Volker Strehl (Erlangen)

## 2.10.2 Symbolic Integration

An *elementary function* of a variable $x$ is any function that can be obtained from the rational functions in $x$ by repeatedly adjoining a finite number of nested logarithms, exponentials, and algebraic numbers or functions. Since $\sqrt{-1}$ is elementary, the trigonometric functions and their inverses are also elementary (when they are rewritten using complex exponentials and logarithms) as well as all the "usual" functions of calculus. The *problem of integration in finite terms* is, given an elementary function $f(x)$, to decide in a finite number of steps whether $\int f(x)dx$ is also an elementary function, and to compute it explicitly if it is. This problem has interested mathematicians since the invention of calculus, long before computer algebra existed. In fact, the partial fraction decomposition method for integrating rational functions can be traced back to Newton, Leibniz and Johann Bernoulli. Factoring denominators over the real numbers was however, and remains, computationally difficult, so in the nineteenth century, the russian and french mathematicians Ostrogradski and Hermite both presented improved algorithms, in which they were able to avoid this factorization step [Hermite 1872; Ostrogradski 1845]. Those algorithms are implemented within the rational function integrators of the mainstream commercial computer algebra systems, and interestingly remain the best methods for integrating rational functions to date. Since the nineteenth century, mathematicians have been looking for algorithms for larger classes of functions, for example Abel and Chebyshev worked on integrating special classes of algebraic functions. In the 1830's, Liouville began to develop a formal theory of elementary functions [Liouville 1833, 1835], which laid the foundations for modern differential algebra (see Sect. 2.11.6). Differential algebra was then formalized in the twentieth century by Ritt and Kolchin: the work of Ritt lead to algorithms for the symbolic integration of arbitrary elementary functions as well as to effective differential elimination theory (see Sect. 2.11.6), whereas the work of Kolchin led to algorithms for solving ordinary differential equations in closed form (see Sect. 2.11.2). The first general algorithm for integrating elementary functions was presented by Risch in a series of reports [Risch 1968, 1969a,b, 1970], unfortunately not all of them published. That algorithm was not ripe for a complete implementation however, as a number of computational problems remained open:

- The method used to deal with algebraic functions within an integrand was, and still is, too complicated, in that it relied extensively on computations of power series with fractional exponents and coefficients in the algebraic closure of the ground field. To date, this method has never been fully implemented.
- Even if there are no algebraic functions in the integrand, the Risch algorithm required factoring denominators, unlike the Ostrodgradski and Hermite algorithms. Those factorizations can exceed the current capabilities of computer algebra systems, in particular in the presence of algebraic numbers among the constants.
- Trigonometric functions are converted by that algorithm into complex exponentials and logarithms, making the output a complex elementary function,

which can be difficult to use in further computations, in particular when real integrals are needed.
- The Risch algorithm cannot handle special functions, neither as part of the integrand, nor in the integral.

Most of the above problems have been solved in the past 30 years, and their solutions implemented: Rothstein [Rothstein 1977] eliminated all the factorization steps in the pure transcendental case, *i.e.* when there are no algebraic functions in the integrand. This led to efficient commercial implementations for transcendental functions. A direct treatment of real trigononometric functions was first proposed through the Risch-Norman algorithm [Davenport 1982], which is currently a heuristic however, as it can fail to find some elementary integrals. More recently, Bronstein [Bronstein 1997] generalized Rothstein's algorithm to extensions given by tangent and arc-tangents, thereby eliminating the need to convert them to complex functions (the other trigonometric functions are converted to tangents and arc-tangents via the half-angle formulas), while the problem of completing the Risch-Norman integration method to an algorithm remains open. The case of algebraic functions remains more complicated, although improvements have been made: Davenport [Davenport 1981] gave a detailed version of the algorithm for pure algebraic functions, and was able to implement it for a restricted class of integrands (nested square roots) in REDUCE. A few years later, Trager [Trager 1984] developed an alternative method, which computes in the minimal field extension needed to express the integral, and which avoids unnecessary factorizations. His algorithm has been implemented in the AXIOM and MAPLE systems. More recently, Bertrand [Bertrand 1995] significantly improved Trager's algorithm for the case of hyperelliptic integrands (square roots of rational functions), and his improvements have been implemented in AXIOM. Trager's algorithm has also been extended by Bronstein [Bronstein 1990] to general elementary functions, allowing nested algebraic and transcendental terms in the integrand. That algorithm is only partially implemented in AXIOM, and the problem of devising a complete rational algorithm for the general case is still open, there are however indications that an elegant solution could come out of algorithms for computing solutions of linear ordinary differential equations in their coefficient fields [Singer 1991b]. The Risch algorithm has also been extended to some classes of special functions, first by Cherry [Cherry 1985, 1986], who allowed logarithmic integrals and error functions to appear in the integrals, then by Knowles [Knowles 1992, 1993], who also allowed them in the integrand. Those two algorithms are restricted to purely transcendental integrands, *i.e.* without algebraic functions. This restriction was removed by Weileder [Weileder 1990] and Baddoura [Baddoura 1994], who added dilogarithms to the class of functions handled.

Computing definite integrals, *i.e.* expressions of the form

$$I = \int_a^b f(x)dx$$

for particular values of $a$ and $b$, is an entirely different problem, one whose solution is not as advanced as indefinite integration. Even when $\int f(x)dx$ is an elementary function, the integral returned by the Risch algorithm might not be applicable on the interval $(a, b)$. This problem has been solved for rational functions by Rioboo [Rioboo 1991], who gave a method for returning an integral defined on the same domain than the integrand, and some progress has been made recently for some categories of trigonometric integrands [Jeffrey 1997], even for real piecewise continuous integrands [Jeffrey et al. 1997; Jeffrey and Rich 1998]. There are however examples where $I$ can be represented in closed form, whereas $f(x)dx$ does not have an elementary antiderivative, and there is currently no complete algorithm for computing such integrals.

Further information and references can be found in [Bronstein 1997; Singer 1990a].

Manuel Bronstein (Sophia Antipolis)

## 2.11 Symbolic Methods for Differential Equations

### 2.11.1 Introduction

Differential equations are of central importance in applied mathematics. Whenever a continuous process must be modeled in natural or engineering sciences (and increasingly often in social sciences), chances are high that differential equations are used. Thus it is not surprising that their study also plays an important rôle within computer algebra.

Traditionally, the main emphasis has been laid on solving differential equations in closed form. This can still be observed in *differential Galois theory* (Subsect. 2.11.2) and in *Lie symmetry theory* (Subsect. 2.11.3). However, many differential equations appearing in applications do not possess closed form solutions or one can determine only special solutions. Nevertheless, computer algebra can still support their analysis in manifold ways.

The *Painlevé theory* (Subsect. 2.11.4) studies the behavior of solutions in the neighborhood of singularities and provides an important test for complete integrability. For general systems of differential equations *completion* (Subsect. 2.11.5) is an important first step; it checks the consistency and determines the size of the solution space. In its algebraic variants this leads to *differential ideal theory* (Subsect. 2.11.6) where one tries to lift as much as possible of commutative algebra to differential equations.

*Dynamical systems theory* (Subsect. 2.11.7) is often considered as being dominated by numerical computations, but as more and more complicated systems are studied the need for symbolic manipulation is rapidly increasing. Last but not least more and more people notice that computer algebra is not in contrast with but rather complimentary to *numerical analysis* (Subsect. 2.11.8).

There exist several survey articles on the topic of computer algebra and differential equations. An overview with a structure similar to this chapter can be found in [Seiler 1997]. Singer [Singer 1990a] emphasizes more algebraic aspects;

MacCallum [MacCallum 1995b] concentrates on the integration of ordinary differential equations and Hereman [Hereman 1994, 1996] reviews mainly symmetry approaches and related fields. Furthermore there have been three workshops devoted to computer algebra and differential equations. Their proceedings [Singer 1990b; Tournier 1988, 1992] contain a number of useful introductory or review articles on more specialized topics.

W.M. Seiler (Mannheim)

### 2.11.2 Differential Galois Theory

In differential Galois theory [Magid 1994; van der Put 1999; Ramis and Martinet 1990; Singer 1990c] one starts with a field $k$ (e.g. $C(x)$, where $C$ is the field of complex numbers) and a derivation $\delta : k \mapsto k$ (e.g. $d/dx$) with constants $\mathcal{C} = \{a \in k \,|\, \delta(a) = 0\}$ and considers linear ordinary differential equations over $k$:

$$L(y) = \delta^n(y) + a_{n-1}\delta^{n-1}(y) + \cdots + a_1\delta(y) + a_0 y = 0 \qquad (a_i \in k).$$

Like in classical Galois theory, one looks for solutions of $L(y) = 0$ in *nice* differential field extensions $K$ of the coefficient field $k$:

1. *rational* solutions (i.e. solutions in $k$). For a recent algorithm see [Barkatou 1999; van Hoeij and Weil 1997].
2. *exponential* solutions (i.e. solutions whose logarithmic derivative is in $k$) [Barkatou 1997; Bronstein and Petkovsek 1996; Pflügel 1997a; van Hoeij 1997].
3. *algebraic* solutions (i.e. solutions which are algebraic over $k$) [Fakler 1997b; Singer 1979; Singer and Ulmer 1993b]
4. *Liouvillian* solutions (i.e. constructed from $k$ using combinations of $\int$, $e^x$ and algebraic extensions) [Duval and Loday-Richaud 1992; Ragot et al. 1999; Singer and Ulmer 1993b, 1997; Ulmer and Weil 1996]
5. Formal power series [Barkatou 1997; Pflügel 1997a; van Hoeij 1997].

If $\mathcal{C}$ is algebraically closed, there exists for each $L(y) = 0$ a differential splitting field $K$ (i.e. a Picard-Vessiot extension), in which there exist $n$ solutions which are linearly independent over $\mathcal{C}$. The differential Galois group $\mathcal{G}(L)$ is the group of field automorphisms of $K/k$ which commute with $\delta$. The elements of (the linear algebraic group) $\mathcal{G}(L)$ send solutions of $L(y) = 0$ into solutions, giving a faithful representation of $\mathcal{G}(L) \subseteq GL(n, \mathcal{C})$. Among the algorithmic questions currently under investigation are:

1. Factorization of (the differential operator of) $L(y)$ [Bronstein and Petkovsek 1996; Singer 1996; Tsarev 1996; van Hoeij 1997].
2. Computation of the differential Galois group of $L(y) = 0$. No general algorithm exists yet [Bertrand 1996; Beukers 2000; Compoint and Singer 1999; Duval and Loday-Richaud 1992; Hendriks and van der Put 1995; Ramis and Martinet 1990; Singer and Ulmer 1993a].

3. Construction of linear ordinary differential equations for a given algebraic group $\mathcal{G}(L) \subseteq GL(n, \mathcal{C})$ [Hendriks and van der Put 1995; van der Put and Ulmer 1998; Singer and Ulmer 1993a; Geiselmann and Ulmer 1997]. New results concerning the inverse problem can be found in [Mitschi and Singer 1996; Singer 1999].

Recently a differential Galois theory for difference equations has been developed [Singer and van der Put 1997]. For existing software see http://www-lmc.imag.fr/CATHODE2 or contact the authors of the packages: DIFFOP (MAPLE share library), $\Sigma^{IT}$ ([Bronstein 1996]), ISOLDE ([Pflügel 1997a]).

F. Ulmer (Rennes)

### 2.11.3 Lie Symmetries

Solving differential equations is one of the most important problems in applied mathematics since Leibniz and Newton introduced the concept of the derivative and the integral of a function about 300 years ago. Shortly after this discovery, numerous phenomena in the physical sciences were described by formulas involving differentiations and integrations, and the need arose to determine the functional dependencies between the variables involved satisfying these expressions. In other words, the problem of solving a differential equation was born. As a result, in the course of time there have been developed several *ad hoc* integration methods for special classes of ordinary differential equations which occurred in the description of physical phenomena, many of them based on the intuition originating from the problem at hand.

This was the state of affairs Sophus Lie arrived at in the second half of the last century when he created a theory for solving differential equations in closed form based on its transformation properties under various kinds of variable changes. His ideas were strongly influenced by Galois' theory for solving algebraic equations that had become widely known a few decades earlier. But despite the fact that Lie's theory is virtually the only systematic method for solving nonlinear ordinary differential equations, and almost all known closed form solutions may be obtained from it, it has never been used for practical problems. The collections of solved examples like e. g. the one by Kamke [Kamke 1961] that are the most important means for solving a given equation do not even mention Lie's name. This is mainly due to the enormous amount of calculations that are usually involved in applying his theory.

Due to the advent of computer algebra systems during the last two decades however this is not a serious hindrance any more for its application and the interest in Lie's theory has been revived. This process has been strengthened by the appearance of several good textbooks on the subject, e. g. the books by Olver [Olver 1986] and Bluman and Kumei [Bluman and Kumei 1989], see also Lie's original publication on this subject [Lie 1881] and the elementary introduction into the field given in [Schwarz März 1996]. If Lie's theory is to be utilized as the basis for developing solution algorithms, various parts have to be replaced by constructive methods and it is desirable to avoid unnecessary field extensions in intermediate steps whenever possible. Of particular importance turned out to be the theory of differential equations by Riquier and Janet and the canonical form for the determining system of the symmetries that follows from it [Schwarz 1996]. The term Janet base has been chosen because Janet recognized its importance and described an algorithm for obtaining this canonical form.

The answers obtained by applying the solution algorithms based on the symmetry analysis have a completely different quality in comparison with the traditional database like approach of using a collection of solved examples. Within the well defined context of Lie's theory, they provide a decision procedure for the existence of a solution of the equation at hand. If a solution is not returned, it is *guaranteed* that it cannot be obtained as a consequence of a symmetry. This is particularly significant because virtually all solutions described in the collection by Kamke are due to a symmetry. The computer algebra software based on these algorithms is on the verge of changing completely the work with differential equations. Very much like the advent of numerical calculators have rendered unnecessary slide-rules and tables of logarithms, collections like the one by Kamke will be replaced by working with a computer algebra system. A good survey on the available software may be obtained from the review article by Hereman [Hereman 1994, 1996].

There are various generalizations of the point symmetries discussed so far. Lie himself introduced the contact symmetries that transform explicitly the first derivatives. Later on transformations depending on higher-order derivatives, so called generalized or Lie-Bäcklund symmetries were introduced. They are discussed in the books by Bluman and Kumei [Bluman and Kumei 1989] and Olver [Olver 1986].

F. Schwarz (St. Augustin)

### 2.11.4 Painlevé Theory

The Painlevé test is a widely applied and quite successful technique to investigate the integrability [Flaschka et al. 1991] of nonlinear ordinary and partial differential equations by analyzing the singularity structure of the solutions. The test is named after the French mathematician Paul Painlevé (1863-1933) [Levi and Winternitz 1992], who classified second order differential equations that are solvable in terms of known elementary functions or new transcendental functions [Ince 1956]. The Painlevé test allows one to verify whether or not a differential equation (perhaps after a change of variables) satisfies the necessary conditions for having the Painlevé property. If so, the equation is a prime candidate for being completely integrable [Ablowitz and Clarkson 1991].

As originally formulated by Ablowitz *et al.* [Ablowitz et al. 1980], the Painlevé conjecture asserts that all similarity reductions of a completely integrable partial differential equation should have the Painlevé property (or be of Painlevé-type), i.e. their general solutions should have no movable singularities other than poles in the complex plane.

A later version of the Painlevé test due to Weiss *et al.* [Weiss et al. 1983] allows testing of partial differential equations directly, without recourse to the reduction(s) to ordinary differential equations. A partial differential equation is said to have the Painlevé property if its solutions in the complex plane are single-valued in the neighborhood of all its movable singularities. In other words, the equation must have a solution without any branching around the singular points whose positions depend on the initial conditions. The traditional Painlevé test does not test for essential singularities and therefore cannot determine whether or not branching occurs about these.

**2.11.4.1 The Algorithm.** The Painlevé test can be applied to nonlinear polynomial systems of ordinary or partial differential equations with (real) polynomial terms. For brevity, we give the three steps of the test for a single equation, $\mathcal{F}(x, t, u(x, t)) = 0$, in two independent variables $x$ and $t$. Following [Weiss et al. 1983], the Laurent expansion of the solution

$$u(x,t) = g^\alpha(x,t) \sum_{k=0}^{\infty} u_k(x,t)\, g^k(x,t)$$

should be single-valued in the neighborhood of a non-characteristic, movable singular manifold $g(x,t)$, which can be viewed as the surface of the movable poles in the complex plane. Here, $u_0(x,t) \neq 0$, $\alpha$ is a negative integer, and $u_k(x,t)$ are analytic functions in a neighborhood of $g(x,t)$.

Note that for ordinary differential equations the singular manifold is $g(x,t) = x - x_0$, where $x_0$ is the initial value for $x$. For partial differential equations, if $u(x,t)$ has simple zeros and $g_x(x,t) \neq 0$, one may apply the implicit function theorem near the singularity manifold and set $g(x,t) = x - h(t)$, for an arbitrary function $h(t)$ [Kruskal et al. 1990; Ramani et al. 1989]. This considerably simplifies the computations.

*Step 1: Leading order analysis.* Determine the (negative) integer $\alpha$ and $u_0$ by balancing the minimal power terms after substitution of $u = u_0 g^\alpha$ into the given equation. There may be several branches for $u_0$, and for each the next two steps must be performed.

*Step 2: Determination of the resonances.* For selected $\alpha$ and $u_0$, calculate the non-negative integers $r$, called the *resonances*, at which arbitrary functions $u_r$ enter the Laurent series. To do so, substitute $u = u_0 g^\alpha + u_r g^{\alpha+r}$ into the equation, only retaining its most singular terms. Require that the coefficient $u_r$ is arbitrary by equating its coefficient to zero. Compute the integer roots of the resulting polynomial. For the series to represent the general solution, the number of roots (including $r = -1$) must match the order of the given equation. The root $r = -1$ corresponds to the arbitrariness of the manifold $g(x,t)$.

*Step 3: Verification of the compatibility conditions.* Verify that such a Laurent series is indeed admissible as solution, and that it has the necessary number of free coefficients $u_r$. Substitute the series, truncated at the largest resonance, into the equation. Determine $u_k$ at non-resonance levels $k$. At resonance levels, $u_r$ should be arbitrary, and since we are dealing with a nonlinear equation, a *compatibility condition* must be unconditionally satisfied.

An equation for which these three steps can be carried out consistently and unambiguously passes the Painlevé test. In the case of systems, for every dependent variable $u_i$ one substitutes $u_i = g(x,t)^{\alpha_i} \sum_{k=0}^{\infty} u_k^{(i)} g(x,t)^k$, and carefully determines all branches of dominant behavior corresponding to various choices of $\alpha_i$ and/or $u_0^{(i)}$. For each branch, the single-valuedness of the corresponding Laurent expansion must be tested, i.e. the resonances must be computed and the compatibility conditions must be verified. Details and an abundance of worked examples can be found in [Ablowitz and Clarkson 1991; Conte 1993, 1998; Flaschka et al. 1991; Kruskal et al. 1990; Ramani et al. 1989; Steeb and Euler 1988].

**2.11.4.2 Symbolic Programs.** The Painlevé test, although algorithmic, is cumbersome when done by hand. Several computer implementations of the Painlevé test exist [Conte 1993; Hereman et al. 1998; Hereman and Zhuang 1995]. A brief review is given in [Scheen 1997]. These symbolic codes are particularly useful for the verification of the self-consistency (compatibility) conditions, and in exploring all possibilities of balancing singular terms. Applied to equations with parameters, the software can determine the conditions on the parameters so that the equations pass the Painlevé test (see [Hereman et al. 1998; Hereman and Zhuang 1995]).

**2.11.4.3 Further Reading.** There is a vast amount of literature about the test and its applications to specific differential equations. Several well-documented surveys [Cariello and Tabor 1989; Conte 1993; Flaschka et al. 1991; Gibbon et al. 1985; Kruskal and Clarkson 1992; Kruskal et al. 1997; Lakshmanan and Sahadevan 1993; Newell et al. 1987] and books [Clarkson 1993; Conte 1998; Steeb and Euler 1988] discuss the basics, as well as subtleties and pathological cases of the test. The survey papers also deal with the many interesting connections with other properties of partial differential equations and by-products of the Painlevé test. They show, for example, how truncated Laurent series expansions allow one to construct Lax pairs, Bäcklund and Darboux transformations, and closed-form particular solutions of partial differential equations. Some shortcomings of the traditional Painlevé test have been identified by Kruskal and others [Kruskal 1992; Kruskal and Clarkson 1992; Kruskal et al. 1997]. Improved versions of the Painlevé test have been proposed, such as the poly-Painlevé test [Kruskal and Clarkson 1992]. Besides, other variants of the test exist [Conte 1993, 1998; Kruskal 1992; Kruskal et al. 1997], e.g the weak Painlevé test [Ramani et al. 1989], and a perturbative Painlevé approach [Conte et al. 1993] which allows for a deeper analysis of equations with negative resonances.

W. Hereman (Boulder)

## 2.11.5 Completion

Among the properties of systems of analytical partial differential equations which may be investigated without their explicit integration there are compatibility and formulation of a well-posed initial-value problem providing existence and uniqueness of the solution. The classical Cauchy-Kowalevsky theorem establishes a certain class of quasi-linear partial differential equations admitting well-posed initial value problems. The main obstacle in investigating other classes of partial differential equation systems of some given order $q$ form their *integrability conditions*, that is, such relations for derivatives of order $\leq q$ which are differential but not purely algebraic consequences of equations in the system.

An *involutive* system of partial differential equations has all integrability conditions incorporated in it. This means that no prolongations of the system do yield integrability conditions. Extension of a system by its integrability conditions is called *completion*. The concept of involutivity was introduced by Cartan in his investigation of the Pfaff type equations in total differentials. For these purposes he used the exterior calculus developed by himself. Cartan's approach was generalized by Kähler to arbitrary systems of exterior differential equations, and the underlying completion procedure [Cartan 1945] was implemented in [Arais et al. 1974; Hartley and Tucker 1991].

In his study of the formal power series solutions of partial differential equations Riquier [Riquier 1910] introduced a class of relevant rankings for partial derivatives and considered systems of *orthonomic* equations which are solved with respect to their highest ranking derivatives called *principal*. Thereby, these derivatives, by the equations in the system, are defined in terms of the other derivatives called *parametric*. An integrability condition gives a constraint for parametric derivatives, and the one of highest ranking becomes a principal derivative. Recently, Riquier's class of rankings was generalized in [Rust and Reid 1997].

Janet made the important observation [Janet 1920] that integrability conditions may occur only from prolongations with respect to certain independent variables called *non-multiplicative*. Prolongations with respect to the rest of variables called *multiplicative* never lead to integrability conditions. Given a set of principal derivatives, Janet gave the recipe how to separate variables into multiplicative and non-multiplicative ones for every equation in the system. He formulated, on this basis, the involutivity conditions for orthonomic systems and designed an algorithm for their completion. This approach to completion is known as *Riquier-Janet theory* and was implemented in [Schwarz 1984, 1992; Topunov 1989].

A system satisfying Janet's involutivity conditions is often called *passive*. This involutivity is generally coordinate dependent. On the other hand, the modern formal theory of partial differential equations developed in 60s-70s by Spencer and others (see [Pommaret 1994]) allows to formulate involutivity intrinsically, in a coordinate independent way. The formal theory relies on another definition of multiplicative and non-multiplicative variables which was known to Janet as long ago as in 20s, but called nowadays after Pommaret because of its importance in the techniques presented in [Pommaret 1978]. An implementation in AXIOM of completion based on the formal theory was presented in [Schü et al. 1993; Seiler 1995].

Thomas [Thomas 1937] used another separation of independent variables into multiplicative and non-multiplicative ones and generalized Riquier-Janet theory to non-orthonomic algebraic partial differential equations. Given a system of partial differential equations, he showed that in a finite number of step one can: (i) check its compatibility; (ii) if the system is compatible, then split it into a finite number of *simple* systems involving generally both equations and inequalities and such that their equation parts are orthonomic and can be completed to involution. This splitting is similar to that generated by the Rosenfeld-Gröbner algorithm [Boulier et al. 1995].

In [Zharkov and Blinkov 1996] it was shown for the Pommaret separation of independent variables that an involutive (passive) basis of a non-differential polynomial ideal is a Gröbner basis. The implementation in REDUCE of the proposed completion algorithm for polynomial bases demonstrated a high computational efficiency of the involutive technique. However, Pommaret bases may not always exist for positive dimensional ideals unlike Janet and Thomas bases. Recently, it was shown [Seiler 2000] that Pommaret bases provide an algorithmic tool for determining combinatorial decompositions of polynomial modules and for computations in the syzygy modules.

The above classical separations of variables into multiplicative and non-multiplicative ones are particular cases of involutive monomial divisions, a concept invented and analyzed in [Gerdt and Blinkov 1998a] (see also [Apel 1995]). The underlying polynomial completion algorithms for a general involutive division [Gerdt and Blinkov 1998a,b] were implemented in REDUCE for the Pommaret division. Recently different involutive divisions and the completion of monomial sets based on them were implemented in MATHEMATICA [Gerdt et al. 1999].

By the well-known correspondence between algebraic equations and linear partial differential equations (see, for example, [Gerdt 1995; Pommaret 1994]) the polynomial completion algorithms admit a straightforward generalization to linear differential equations. This is very important, among other things, for the Lie symmetry analysis of nonlinear differential equations. It is because of the fact that completion is the most general and universal method of integrating the determining system of linear partial differential equations for infinitesimal Lie symmetry generators [Hereman 1994, 1996]. Moreover, an involutive form of determining equations allows one to construct the Lie symmetry algebra without their explicit integration [Reid 1991a]

Another method for the completion of linear partial differential equations to an involutive form called *standard* which is not based on the separation of variables was developed and implemented in MAPLE in [Reid 1991b]. The generalization of this method to nonlinear partial differential equations described in [Reid et al. 1996; Rust et al. 1999] contains a rather comprehensive bibliography of papers on the completion of partial differential equations and related topics.

V.P. Gerdt (Dubna)

### 2.11.6 Differential Ideal Theory

**2.11.6.1 Algebraic Systems and Differential Ideals.** The pioneering work of Janet ([Janet 1929]) gave a clear link between the theory of analytic partial differential equations and the one of algebraic ideals. But Ritt was the first to introduce *differential algebra* (*c.f.* [Ritt 1932, 1950]) as a new field of mathematics. He also introduced *difference algebra* (*c.f.* [Cohn 1966] and the references therein).

Given a differential or partial differential field, *e.g.* $\mathcal{F} := \mathbf{R}(x_1, \ldots, x_n)$ equipped with the partial derivations $\partial/\partial x_i$, a system of differential algebraic equations in $m$ differential indeterminates $f_1, \ldots f_m$ is a subset $\Sigma$ of the differential polynomial algebra $\mathcal{F}\{f_j\}$. One will search for zeros in some *universal extension of* $\mathcal{F}$. Such a set is the *differential algebraic variety* $V(\Sigma)$ defined by the system. We have $V(\Sigma) = V([\Sigma])$, where $[\Sigma]$ is the differential ideal generated by $\Sigma$, *i.e.* the set of all linear combinations of derivatives of elements of $\Sigma$. If $\mathcal{F}$ is of characteristic 0, then $V(\Sigma) = V(\sqrt{[\Sigma]})$.

A differential ideal $I$ such that $I = \sqrt{I}$ is said to be *perfect*. A perfect differential ideal is a *finite intersection* of *prime* differential ideals, meaning that a differential algebraic variety is a *finite union* of *irreducible components*. This is a corollary of the *Ritt-Raudenbush theorem* which asserts that a perfect differential ideal is the radical of a *finitely generated* differential ideal.

One may refer to [Kaplansky 1957] for an introduction, to [Kolchin 1973] for much more details and to [Buium 1994] for more recent theoretical developments (chapter 2 is also a quite readable introduction).

**2.11.6.2 Gröbner Bases.** It is known that the theory of *standard* or *Gröbner* bases allows to find a canonical set of generators for every algebraic ideal and to answer many basic problems such as ideal membership or the computation of the dimension. Unfortunately, if Gröbner bases do exist in the differential case, they are no longer finite (*c.f.* [Ferro 1987; Ollivier 1990a,b]). This is related to the fact that the Ritt-Raudenbush theorem only provides a weak analog of noetherianity. It may be proved that differential ideal membership is undecidable (*c.f.* [Gallo et al. 1991]). It is not known whether it is decidable for *finitely generated* differential ideals.

**2.11.6.3 Characteristic Sets.** The key theoretical and algorithmic tool for differential ideals is the notion of *characteristic sets*. One may notice that the work of Wu Wentsun on automatic theorem proving was inspired by Ritt's work (*c.f.* [Chou 1988]). Ritt gave an algorithm to decompose a perfect ideal into a finite intersection of prime ideals, defined by the datum of some characteristic set. This requires some hypothesis on the ground field in order to perform factorization. Seidenberg (*c.f.* [Seidenberg 1956]) gave an algorithm for elimination in differential algebra without factorization, but it does not produce characteristic sets.

The recent work of Boulier shows that we can decompose any perfect differential ideal into an intersection of *perfect* differential ideals defined by characteristic sets, *without factorization*. This leads to greater efficiency. The method has been implemented in a MAPLE package (*c.f.* [Boulier et al. 1995, 1997]). This package has been improved by Hubert, who also had theoretical contributions, allowing a full decoupling between the differential and the algebraic completion [Hubert 1999] .

A great number of papers on the computation of characteristic sets for regular ideals, without factorization, followed Boulier's pioneering work: see [Bouziane et al. 1998; Maârouf 1996; Maârouf et al. 1998; Sadik 2000b]. On may notice the bound obtained by Sadik (see [Sadik 2000a]) on the order of derivations necessary to compute a characteristic set of a generic ordinary differential ideal.

**2.11.6.4 Ritt's Problem.** A great unsolved problem is that of testing inclusion. Given two prime differential ideals $I_1, I_2$ defined by characteristic sets, can we decide whether $I_1 \subset I_2$? This is equivalent to finding an effective version of the Ritt-Raudenbush theorem, *i.e.* knowing a characteristic set of a prime differential ideal $I$ to find a finite set $\Sigma$ such that $I = \sqrt{[\Sigma]}$. See [Péladan-Germa 1995; Hubert 1996, 1999] for more details on the subject. This problem is related to that of testing equalities in differential rings defined by differential algebraic systems and *initial conditions*.

**2.11.6.5 Complexity.** Improving Seidenberg's method, Grigor'ev obtained a triple exponential upper bound for quantifier elimination of nonlinear ordinary differential equations (*c.f.* [Grigor'ev 1987]). Using the classical construction of Mayr and Meyer, Sadik (*c.f.* [Sadik 1995]) proved a double exponential lower bound for the complexity of the *differential Nullstellensatz* for system of linear partial differential equations.

F. Ollivier (Palaiseau)

## 2.11.7 Dynamical Systems

Applications to dynamical systems are not really in the main stream of computer algebra. Conversely, numerical computations play a much more prominent rôle within dynamical systems theory than symbolic ones. Nevertheless, there is increasing interest in the use of symbolic computation. Two fundamental techniques in dynamical systems theory are especially well suited for computer algebra: center manifolds and various types of normal forms. Other applications include bifurcation analysis, the Poincaré map and Hilbert's $16^{th}$ problem.

Defining *normal forms* and deriving algorithms to compute them is a classical topic in computer algebra. For dynamical systems normal forms have already been introduced by Poincaré, Birkhoff, Gustavson and many others, often in the context of celestial mechanics [Della Dora and Stolovitch 1994]. In order to analyze the stability of the origin or bifurcations it suffices to study equivalence classes of vector fields under coordinate transformations. It also suffices to study a truncation of the Taylor expansion of the vector field and a truncation of the coordinate transformation. A special representative of the equivalence class gains the most insight. Thus the computation of an approximation of the Birkhoff normal form means computation with polynomials which is appropriate for computer algebra systems.

Complications arise, if the linear part of the vector field is not semi-simple, i.e. if it contains a nilpotent part. In this case an additional normalization is necessary, as the classical algorithms consider only the semi-simple part. This requires further tools from invariant and representation theory of Lie algebras. For the special case of $sl(2, I\!R)$ this is discussed in some detail in [Cushman and Sanders 1987]. For planar systems with nilpotent linear part generic normal forms up to eightth order have been computed in [Freire et al. 1991] using REDUCE.

An algorithm for computing normal forms that is suitable for implementation in a computer algebra system was presented by Walcher [Walcher 1993]. It is closely related to *Lie transforms* [Meyer 1991]. This technique has its origin in Hamiltonian mechanics where it yields a canonical transformation. However, it can be extended to general dynamical systems.

In contrast to this Birkhoff normal form Gatermann and Lauterbach [Gatermann and Lauterbach 1998a] took normal forms from singularity theory in order to study bifurcation phenomena, see also Sect. 3.2.7. For equivariant systems (see below) they automatically classify them using Gröbner bases. Jaquemard and Teixeira [Jacquemard and Teixeira 1998] also use Gröbner bases and factorization algorithms in order to classify normal forms of reversible, discrete dynamical systems. Another use of singularity theory is done by Lunter [Lunter 1999]. In the context of Hamiltonian systems the algorithm by Kas and Schlesinger determines the coordinate transformation to a model problem. The actual computation requires a generalization and combination of standard bases and Sagbi bases within local rings.

Computer algebra is also much used to determine (approximations of) *center manifolds*, a special form of invariant manifolds. If a dynamical system possesses a center manifold, it often suffices to study its behavior on this manifold. If the zero solution of the reduced system is stable, solutions of the original system for initial data sufficiently close to the center manifold will approach this manifold exponentially fast. Thus the reduced system completely describes the asymptotic behavior of such solutions.

Center manifold theory has such a great importance, because it yields a reduction of the dimension and thus often a considerable simplification of the analysis. Sometimes it is even possible to reduce an infinite-dimensional problem to a finite-dimensional one. There are two main computational steps. First we need an approximation for the center manifold, then we must compute the reduced system. As in normal form theory, this is done with a power series ansatz [Freire et al. 1988]. Laure and Demay [Laure and Demay 1988] showed for the Couette-

Taylor problem how computer algebra and numerical analysis can interact to solve a complicated bifurcation problem for an infinite-dimensional problem using a reduction to a finite-dimensional center manifold.

But also some classical (computer) algebraic problems are of great importance in the study of dynamical systems. For example, before a fixed point can be analyzed it must be determined. This requires the solution of a nonlinear system of algebraic equations. If the defining vector field is rational, this can be done with Gröbner bases. Often the vector field depends on some parameters. At certain values of these parameters, the properties of the vector field may change, i.e. a *bifurcation* occurs. The determination of these values is a fundamental problem in dynamical systems theory.

Of special interest are here *equivariant systems* where the vector field is invariant under the action of a symmetry group. Here one can use linear representation theory and polynomial invariant theory (Sect. 2.2.8) for determining the fundamental invariants and equivariants [Gatermann 1996b; Worfolk 1994] which in turn are used for the exact solution of algebraic equations, see also Sect. 3.2.7. Using normal forms they enable the local bifurcation analysis, i.e. the typical bifurcation diagram in the neighborhood of a critical point can be derived. Also the stability is investigated with the help of Gröbner bases since it enables to study the signs of the eigenvalues of the Jacobian on the solutions branch depending on the arbitrary coefficients of the vector field [Callaham and Knobloch 1997]. Another point is the study of exceptional values of the parameters where essential phenomena change [Lari-Lavassani et al. 1994].

If bifurcations of periodic solutions are to be studied, the *Poincaré map* is often a very useful tool. However, in general it is not possible to obtain it analytically. Thus one computes again a power series approximation of it. The bifurcation depends then on the Taylor coefficients. They satisfy a system of differential equations which must be integrated numerically. In order to set up this system one needs higher order derivatives of the vector field which can often be determined only by computer algebra. A combined numerical-symbolical approach to the Poincaré map is described in [Kleczka et al. 1990].

A more theoretical application concerns *Hilbert's $16^{th}$ problem* of bounding the number of limit cycles in a planar polynomial system. In [Lloyd and Pearson 1990] an ansatz for the Lyapunov function is made. The conditions for multiple bifurcation lead to a polynomial system of equations in the coefficients of the Lyapunov function and the coefficients of the vector field as unknowns. Because of the nature of the problem an exact solution of the system is required. For quadratic systems a lot of results are known [Schlomiuk et al. 1990]; however already the cubic case becomes very complicated. An important subproblem is the *center problem*, namely to distinguish between a focus and a center. The derivation of sufficient and especially of necessary conditions for a center can be very involved and is sometimes hardly feasible without computer algebra [Pearson et al. 1996]. In a recent study of cubic systems [Edneral 1997] a Cray-J90 had to be used.

K. Gatermann (Berlin), W.M. Seiler (Mannheim)

### 2.11.8 Numerical Analysis

Despite all theoretical progress in solving differential equations in closed form, one can still only numerically integrate many systems appearing in applications. Thus the combined use of symbolic and numerical computations becomes more and more important and indeed most general purpose computer algebra systems provide some numerical facilities.

The oldest und simplest approach is *interfacing* a computer algebra system and a numerical library. Typically, the symbolic part consists of deriving differential equations, the interface generates code in the language of the numerical library (possibly with some optimizations), and finally the equations are solved with routines of the library. For almost any of the larger computer algebra systems such interfaces have already been developed.

In a more sophisticated approach the computer algebra systems supports the user in *selecting* an appropriate method from the library. This includes an analysis of the properties of the differential equation (e.g. its stiffness) but also suggestions of reasonable values for the many parameters which allow for the fine tuning of modern numerical routines. An example for this approach is the AXIOM-NAG link developed by Dupée [Dupée and Davenport 1996].

Computer algebra is also useful to *derive* and *analyze* numerical schemes. The Butcher theory of Runge-Kutta methods is here a typical example. For higher order methods the order conditions become rather large and complicated. Computer algebra packages have been developed that derive and solve them (using Gröbner bases) [Gruntz 1995; Sofroniou 1994]. In the case of partial differential equations the same holds for the construction of higher order discretizations or finite elements [Mund 1995].

Another application is the stability analysis of finite difference schemes for partial differential equations [Ganzha and Vorozhtsov 1996]. Finally, one should mention the determination of the so-called modified equation for a given numerical method [Ahmed and Corless 1997], i.e. of a differential equation whose solution is much closer to the numerically computed one than the solution of the original equation.

However, none of these applications represents what one would call a hybrid method. In contrast, several such algorithms have been developed for algebraic equations. A good example (motivated by equivariant dynamical systems) is the exploitation of symmetry in the numerical solutions of systems of equations as outlined in 3.2.8.

Recently, also some relations between the completion theory discussed in 2.11.5 and numerical analysis have emerged in the study of the so-called *differential algebraic equations (DAE)*. These are systems consisting of both ordinary differential equations and algebraic equations. More generally, one can speak of systems which are not in Cauchy-Kowalevsky form; this definition includes partial differential equations. Numerical analysts have developed a number of indices measuring the difficulties one must expect in numerically solving such a system. One can show that these indices count essentially the number of steps needed for the completion of the system [Seiler 1999].

W.M. Seiler (Mannheim)

## 2.12 Symbolic/Numeric Methods

### 2.12.1 Computer Analysis

Whereas "classical" areas of computer algebra are dominated by algorithms which are based on algebraic concepts, and which aim to find closed (symbolic) solutions, the main emphasis of computer analysis is on using computer algebra systems for computing approximate solutions in a controlled manner. Its goals are:

- to find a simple and transparent form of an expression, while maintaining an appropriate level of precision,
- to determine error bounds, which are computed entirely by a program, the same way as approximations are determined without requiring user interaction.

Prevalent algorithms in computer analysis combine the tried methods of numerical computing with symbolic procedures. Mixed analytic-numeric techniques pave the way for novel applications of mathematical methods. Therefore, powerful floating-point arithmetic of computer systems becomes an equal partner of exact arithmetic implemented in computer algebra systems. Consequently, investigation and solution of problems arising from this interaction is of prime importance.

This area of research started out about 15 years ago. N.J. Lehmann is internationally recognized as one of the co-founders, and made important contributions to the advancement of the subject. He and his team in Dresden and later in Rostock developed and implemented a number of algorithms [Hantzschmann 1984; Lehmann 1985a, 1989]. A summary of activities with references to articles can be found in [Lehmann 1985b] and [Hantzschmann 1990].

Finding analytic approximate solutions for differential equations is one of the main tasks of computer analysis. In accordance with the purpose of computer analysis preference is given to mathematical approximation methods allowing an automatic adaption to a given problem and giving a transparent short formula expression at a reasonable expense.

Good results in this direction could be achieved by a two-step concept. In the first step the given problem will be adapted by a suitable chosen "neighbour" problem whose exact solution can be determined by means of computer algebra algorithms. The fundamental solutions form the base for an adequate approximate ansatz in the second step. This two-step concept has the advantage of allowing a close problem adaption. It offers a wide field for the development of several algorithms depending first on the approximation criterions used in both steps and second on the selected family of "neighbouring" problems [Hantzschmann and Thinh 1991].

Current investigations focus on advancement of this concept: integrating the wide variety of projection methods and perfection of the theoretical foundation, adequate controlling the precision, including other classes of exactly solvable differential equations. [Becken 1995; Hantzschmann and Jung 1995].

All algorithms developed by now are completed by possibilities for implementable error estimations. A special procedure recommended in [Lehmann 1967] has proved its worth as a basic concept which was modified and extended in many ways. Necessary Lipschitz-estimations for non-linear differential equations are supplied by a specially developed Lipschitz-calculus.

Karl Hantzschmann (Rostock)

### 2.12.2 Algorithms for Computing Validated Results

Verification algorithms are somewhere on the boundary between computer algebra algorithms and numerical algorithms. With the former they share the strict validity of every result; like the latter they are very efficient, usually requiring a computing time only slower by a factor of five to ten compared to the best known numerical algorithm.

The basic principle of "verification algorithms" or "algorithms for computing validated results", also called "algorithms with automatic result verfication" is as follows. First, a pure floating point algorithm is used to compute an approximate solution for a given problem. This approximation is, hopefully, of good quality; however, no quality assumption is used at all. Second, a final verification step is appended. After this step, either error bounds are computed for the previously calculated approximation or, an error message signals that error bounds could not be computed. Any computed result is always perfectly correct in the sense that all possible conversion errors, rounding errors, approximation errors or others are rigorously estimated.

The calculation of the actual error bounds must use some estimation of the errors of the individual operations. This can be done by standard error analysis or, most convenient, by using interval operations. However, it is well known that extensive use of interval operations tends to weaken results. Therefore it is of utmost importance to diminish the overestimation due to successive use of interval operations.

This is the reason why the verification step is frequently based on a fixed point argument. Let the problem to find bounds for a solution of $f(x) = 0$ for some $f : \mathbb{R}^n \to \mathbb{R}^n$ be given. Then first the problem is transformed into a fixed point equation $g(x) = x$ with the property that the set of zeros of $f$ and fixed points of $g$ coincide.

### 2.12.2 Algorithms for Computing Validated Results

The major advantage of interval arithmetic is the ability to estimate the *range* of a function, for example, the range of $g$ over some n-dimensional interval vector $X$. This is done by simply replacing all operations by corresponding interval operations. The result, say $Y$, is definitive a superset of the true range $g(X)$. If the result $Y$ is a subset of $X$, Brouwers Fixed Point Theorem ensures existence of a fixed point of $g$ within $Y$, and henceforth existence of a zero of $f$ within $Y$.

In other words, algorithms with result verification do verify the assumptions of mathematical theorems on the computer, where the assertion states a validated error bound for the solution.

Another advantage of this ability to estimate the range of a function over a certain domain arises in global search algorithms for zeros, like for example finding all (real or complex) zeros of a nonlinear function. The main problem of such algorithms - computer algebra, validated algorithms or others - is to *exclude* zeros from a certain region. This can be done by the cited range estimation, and sometimes in a very efficient way.

In the following we mention some standard problems for which validated algorithms are available with hints to the literature. General background on interval analysis, arithmetic and algorithms are in standard books, among them [Alefeld and Herzberger 1983; Hansen 1992; Moore 1966, 1979; Neumaier 1990].

For general systems of linear and nonlinear equations standard algorithms may be found in [Krawczyk 1969; Rump 1994; Shary 1991], for systems of equations with sparse matrix in [Rump 1993] and in the literature cited over there. For global optimization algorithms and global search of all zeros of a function within a certain domain see [Csendes and Pintér 1993; Hansen 1992; Jansson 1994; Kearfott and Du 1993]. Ordinary systems of differential equations are treated in [Lohner 1988] and partial differential equations for example in [Nakao 1992; Plum 1991, 1992]. For a good overview on current algorithms and methods see [Herzberger 1994].

There are a number of public domain libraries for interval arithmetic and algorithms, where newer implementations include [Kearfott et al. 1994; Knüppel 1994]. Commercial libraries are available, among them [IBM 1986; Siemens AG 1986; Klatte et al. 1991; Lawo 1992]. A very fast and user-friendly library under Matlab [The MathWorks Inc. 1997] has recently been developed. It is called INTLAB [Rump 1999], it is completely written in Matlab and henceforth portable on almost any computer, and it comprises of all interval operations, including real, complex and sparse matrices, standard functions, automatic differentiation and much more. It is freely available from our homepage.

In summary, the objective of verification algorithms is to provide a certainty for pure numerical approximations by validated error bounds. They are designed to provide such error bounds whenever computation is possible within the limited floating point precision. The algorithms can be extended into never failing algorithms by increasing precision after temporary failure. In this case estimations of worst case computing times are possible using standard techniques from computer algebra. But, once again, this is not the main objective of validation algorithms.

However, the performance of computer algebra algorithms may be improved by simply applying a validation algorithm first. The additional computing time is usually small compared to the general algorithm, the results are still verified to be correct, and a verified result will be obtained if the problem is not too ill-conditioned with respect to the available floating point precision.

Another approach receives some attention recently. These are hybrid algorithms combining advantages of computer algebra and validated algorithms. A recent special issue of the Journal of Symbolic Computation (Number 6, December 1997) is devoted to this kind of algorithms.

Siegfried M. Rump (Hamburg)

## 2.12.3 Hybrid Methods

**2.12.3.1 Introduction.** We here discuss an active research area: algorithms for symbolic/numeric computation. The main goal of hybrid symbolic-numeric computation is to extend the domain of efficiently solvable problems by combining the methods of numerical and symbolic computation.

One can take a broad point of view, and call any method a hybrid symbolic-numeric method provided that it solves a mathematical problem, and involves some aspects of numerical computing and some aspects of symbolic computing. This is similar to what Knuth calls a *seminumerical algorithm*, one that lies "on the borderline between numeric and symbolic calculation" [Knuth 1981a, p. v]. This definition includes, then, such varied topics as:

- conversion of polynomial problems to eigenvalue problems,
- polynomial arithmetic (including GCD) with numerical coefficients,
- interval arithmetic (especially when correlations between intervals are tracked symbolically),
- symbolic pre-computation of expressions for later efficient and/or stable numerical evaluation (e.g. in C or Fortran compilers, not just computer algebra systems),
- code generation by computer algebra systems for later solution of numerical problems, for example PDE,
- automatic differentiation of formulas and programs,
- construction of special-purpose numerical methods for differential equations that automatically preserve invariants, and
- exact computation by intermediate use of floating-point arithmetic, for sake of speed.

The problem formulation can be, as above, entirely symbolic. An example of practical interest is the computation of the sign of the determinant of a rational matrix, and the result may be validated (see also section 2.12.2). An interesting extension of this idea is the inverse symbolic calculator (http://www.cecm.sfu.ca/projects/ISC/), where for example a high precision floating point approximation of an algebraic number is computed first, and then a defining polynomial with integer coefficients is found via lattice basis reduction (see section 2.4.6).

### 2.12.3 Hybrid Methods

Clearly we will be unable to survey all possible hybrid symbolic/numeric methods in this short article, under this inclusive definition. We focus on Symbolic-Numeric Algorithms for Polynomials (SNAP), partly because of recent progress but also because polynomial problems play a historical role in the exposition and explication of difficulties characteristic for more general nonlinearities. See also the recent survey [Emiris 1999], for a detailed discussion of SNAP and many references.

**2.12.3.1.1 History.** We choose to date the nascence of SNAP, as a field of study, from the 1996 SNAP conference at INRIA (Sophia-Antipolis). Several papers presented at that conference were later published in a special issue of the Journal of Symbolic Computation edited by Hans J. Stetter and Stephen M. Watt, published in 1998.

One can find papers relevant to the SNAP field that appeared earlier than this conference. Some examples are [Kharitononv 1979; Kaltofen 1985a; Schönhage 1985; Auzinger and Stetter 1988; Corless et al. 1995]. Many other papers have shown that certain algorithms for problems with exact data are unstable for problems with approximate data. Perhaps the best known of these is the Chauvenet-prize paper by Wilkinson [Wilkinson 1984], which shows very clearly that computing the characteristic polynomial and then finding its roots is not a stable method to find the eigenvalues of a matrix. This can be used to show the well-known result that trying to find the roots of a multivariate system by first using the Buchberger algorithm to compute a lexicographic-ordered Gröbner basis is, when implemented in floating-point arithmetic, unstable for approximate polynomials. To see this, simply consider the eigenproblem $Ax = \lambda x$ with the normalization $x_1^2 + x_2^2 + \cdots + x_s^2 = 1$ as an $s+1$-variate polynomial problem of total degree 2. Applying the Buchberger algorithm to this problem with the lexicographic ordering $x_1 > x_2 > \cdots > x_s > \lambda$, we see that the resulting Gröbner basis will contain the characteristic polynomial of $A$ as its first element. Therefore, by the result of Wilkinson described above, this approach is numerically unstable as a method of finding roots.

#### 2.12.3.2 Definitions and Techniques.

**2.12.3.2.1 Continuity.** One of the most important ideas in symbolic-numeric computation is the mathematical notion of continuity. Pure symbolic computation in algebraic domains does not concern itself with continuity or the lack thereof, and this is one of the main differences between pure symbolic computation and symbolic-numeric computation. Continuity is also important for pure symbolic computation with the elementary functions of analysis and for problems with parameters, but this has heretofore received less attention in the computer algebra community than the important algebraic aspects of such computation. But, in using numerical computation, we are forced to confront the issue. The solutions of many problems in this area are discontinuous with respect to changes in the data. It follows that applying exact methods to problems with approximate data does not always approximate the desired result. A key aspect of all symbolic-numeric methods and SNAP in particular is to define well-posed problems that may be solved by any mathematical method.

**2.12.3.2.2 Approximate Polynomials.** An important idea of this framework for computer algebraists working with polynomials, superficially similar to an idea from the field of interval arithmetic (see section 2.12.2), is the notion of an *approximate polynomial*. An approximate polynomial is a polynomial with inexactly-known coefficients. More formally, consider the space of polynomials $\mathbb{R}[x]$ (or $\mathbb{C}[x]$), together with a metric $d(f,g)$ giving a "distance" from the polynomial $f$ to the polynomial $g$. Usually we assume that the metric $d(f,g)$ is given by some norm $d(f,g) = \|f - g\|$. In Section 2.12.3.2.5 we take norms up further. Given some polynomial $f_0$ and scalar $\varepsilon > 0$, define a polynomial $\varepsilon$-neighbourhood $N_{f_0,\varepsilon} \subset \mathbb{R}[x]$ (or $\mathbb{C}[x]$) as the set $N_{f_0,\varepsilon} = \{f : d(f,f_0) \leq \varepsilon\}$. Later we generalize this to allow a vector of tolerances, some of which can be zero. Every member of an $\varepsilon$-neighbourhood is considered indistinguishable as an approximate polynomial. This membership is not an equivalence relation, in general. Therefore we see that the notion of approximate polynomial is tied to $\varepsilon$ and to the given polynomial $f_0$. The multivariate generalization is straightforward.

The reason this idea is important is that in some problem contexts, any exact solution to an approximate polynomial problem (relative to the originally stated problem $P(f_0)$) is precisely as useful as an exact solution to $P(f_0)$ itself; moreover, solution of the approximate problem may be more efficient.

One context where approximate polynomials are used is as an intermediate stage in the solution of exact problems, for example by embedding an exact problem into an approximate framework in order to take advantage of any existing fast (e.g. iterative) algorithms for approximate solution in the approximate domain; one then has to ensure that the final answer is then transferable back to the desired exact answer. One example of this kind of problem is the search for a certified sign of the determinant of a matrix [Clarkson 1992; Yvinec 2000]. Another context is the solution of a discontinuous problem known to be singular but given with imprecise inputs, say floating point coefficients. An example here is the problem of computing the rank of a matrix. Clearly, an infinitesimal perturbation can change the rank, because the rank of a matrix is a discontinuous function of the entries of a matrix, so one may need to consider the set of approximate matrices in an $\varepsilon$-neighbourhood.

**2.12.3.2.3 Metrics and Norms.** Another important aspect of this framework is that the choice of metric is crucial to the success (and applicability) of the algorithm. Most of the algorithms developed so far have concentrated on metrics derived from the simple norms, for instance, (weighted) 2-norms of the coefficients, so $\|f\|^2 = \sum_{k=0}^{n} w_k f_k^* f_k$ (where the $f_k$ are the coefficients of $f$ in some basis, usually the monomial basis and $w_k$ are non-negative weights); the 1-norm; or the max (infinity, component-wise) norm. In Chebychev's and Solotareff's approximation problems (their problems seek nearby polynomials of lower degrees), the distance between two univariate complex polynomials is measured by the maximum difference of values on the unit interval: $d(f,g) = \max\{|f(a) - g(a)| : -1 \leq a \leq 1\}$. This is the infinity norm, considered as a function norm, not a coefficient norm, on the interval $[-1, 1]$. We do not use the symbol $\|\cdot\|_\infty$ for this because of the possibility of confusion with the corresponding coefficient norm $\|p\|_\infty = \max\{|p_k|, 0 \leq k \leq \deg p\}$. Victor Ya. Pan (see below) uses the "spectral metric" $d(f, g) = \max_i \min_j |\rho_i - \sigma_j|$, where the $\rho_k$ and $\sigma_k$ are the roots of $f$ and $g$ respectively (the case where $\deg f$ is not the same as $\deg g$ needs more care).

Finally, the choice of representation for $f$ plays a role. We consider $f$ fundamentally as a function $f\colon \mathbb{R}^s \to \mathbb{R}$ (or $\mathbb{C}^s \to \mathbb{C}$). Many norms can be understood better in this fashion. For example, by Parseval's identity, the 2-norm of the vector of coefficients of a univariate polynomial $f$ is also the integral norm

$$\|f\|_2^2 = \frac{1}{2\pi} \int_C \overline{f(z)} f(z)\, dz$$

where $C$ is the unit circle in $\mathbb{C}$. Therefore the size of $f$ in the unit disk determines (and is determined by) the size of the 2-norm of the vector of coefficients of $f$. This shows clearly that the location of the origin and the scaling of the variable $z$ (sometimes, but not always, at liberty in applications) matters to the algorithms that we will discuss here. Note that translating $z$ to $z + a$ is ill-conditioned, amplifying errors in the coefficients by as much as $(1 + |a|)^n$, where $n$ is the degree of $f$. Further, relative errors may be infinitely amplified [Corless et al. 1999]. This corresponds to loss of sparsity in the symbolic context and is thus to be avoided both symbolically and numerically.

**2.12.3.2.4  Conditioning and Posedness.** We now define some terms, commonly used in numerical analysis, that we find useful in this context. We follow standard usage in numerical analysis and say that a problem is *well-posed* if it has a unique solution that depends continuously on the problem parameters; a problem is *ill-posed* otherwise. Typically, in the SNAP context, a problem will be ill-posed not because of lack of existence or uniqueness, but rather because an important characteristic (such as rank, or degree of the GCD) fails to be continuously dependent on the problem parameters. We call the norm of the difference between a computed solution $f$ and the true, exact solution $f_{\text{true}}$ the *forward error* $\|f - f_{\text{true}}\|$. We call the difference between the problem that the computed $f$ solves, and the problem that we originally wished to solve, the *residual*; we call the norm of the residual the *backward error*. For example, the backward error corresponding to the computed solution $x$ of a linear system $Ax = b$ is the norm of the residual[*] $\|r\| = \|b - Ax\|$. Similarly, a residual (backward error) in the problem of finding the GCD of two approximate polynomials $f$ and $g$ is a set $(\Delta f, \Delta g)$ ($\|(\Delta f, \Delta g)\|$) of perturbations to $f$ and $g$ that allows us to write the GCD as $h = s(f + \Delta f) + t(g + \Delta g)$. Such residuals may not be unique.

We call any linear measure of the asymptotic sensitivity of our stated problem to changes in its input arguments a *structured condition number* of the problem

---

[*] Note that already here the deformation has a structure: the matrix $A$ is taken exactly, while the vector $b$ is approximate, as in the problem of least squares. The structured backward error for an approximate matrix can also be analyzed [Oettli and Prager 1964]. The corresponding notion for curve fitting is *total* least squares.

[Turing 1948; Higham 1996]. We emphasize that this linear measure does not help us to understand the effects of large changes in the data. We have, for small enough changes, that the forward error is roughly equal to the structured condition number times the backward error; but more to the point, the structured condition number measures the sensitivity of the problem to changes in its data.

The definition of structure that we are using is as follows. A problem that depends on more parameters and for which there exists a substitution for the parameters by functions in fewer parameters is more structured. For example, an $n \times n$ matrix whose entries depend only on $O(n)$ independent parameters has more structure than a general $n \times n$ matrix. The condition number of the structured problem is by the chain rule for differentiation equal to the product of the condition number of the unstructured problem times the condition number for the substitution, which is often less.

We say that a problem with a large* structured condition number is *ill-conditioned*, and *well-conditioned* otherwise. An ill-posed problem is of course ill-conditioned (with infinite condition number); but the notion "well-conditioned" is a significant, quantitative, refinement of the notion "well-posed"; it is perfectly possible for a well-posed problem to be hopelessly ill-conditioned.

Note that the conditioning of a problem is important even if there are no rounding errors whatever in a computation. Fundamentally, this is because the condition number of a problem is the reciprocal of the distance to the nearest singular problem (a fact noted in [Demmel 1987] as holding in a very general problem context). In the symbolic context, one may seek this nearest singular problem, for instance, a minimal perturbation of the inputs so that a curve factors. The condition number measures sensitivity to errors in the data, which the model of "approximate polynomial" assumes *a priori*. An ill-conditioned problem will also, of course, amplify any rounding errors that happen during computation, but this is often a secondary consideration.

Finally, we say that an algorithm is *numerically stable* if it produces the exact answer to a problem near to the one posed: that is, the answer it produces has small backward error. Therefore, a well-conditioned problem solved by a numerically stable algorithm will have small *forward* error. This nomenclature splits the difficulty of explanation of numerical behaviour into two useful components: *algorithms* are stable or unstable; *mathematical problems* are well-conditioned or ill-conditioned.

A good numerical analysis of an algorithm for approximate polynomials will take both forward and backward error into account. The algorithm itself should provide an estimate of the structured condition number of the problem.

**2.12.3.2.5 Dual Norms and a Useful Inequality.** We follow the approach of the survey [Stetter 1999]. The key observation is that some recent SNAP results are straightforward implications of the Hölder inequality. In finite-

---

* "Large" means large relative to the problem context; in some cases a condition number of 5 is "large", whilst in others a condition number of $10^{10}$ might be acceptably small. A condition number should be provided; it is up to the user to do something intelligent with it.

dimensional linear spaces, we have that the equality sign is attained in the Hölder inequality [Hardy et al. 1951, pp. 24–26] at explicitly-known "witness" vectors. These witness vectors allow us to solve useful minimax problems explicitly.

Following the notation of [Stetter 1999], $v^T$ denotes the transposition of a column vector $v$ into a row vector, without complex conjugation. Suppose $\mathbb{C}^n$ is equipped with a norm $\|\cdot\|$. The space of linear functionals $v^T$ on $\mathbb{C}^n$ may then be equipped with the dual norm $\|\cdot\|^*$ defined by

$$\|v^T\|^* := \max_{\|u\|=1} |v^T u| . \tag{1}$$

For example, it is well-known that the $p$-norm is dual to the $q$-norm if $1/p + 1/q = 1$, for $1 \leq p, q \leq \infty$.

Proposition 1 in [Stetter 1999] is: For each $u \in \mathbb{C}^n$ with $\|u\| = 1$, there exist vectors $v \in \mathbb{C}^n$ with $\|v^T\|^* = 1$ such that $|v^T u| = 1$. That is, *the maximum value is attained.* Moreover, for the associated $p$ and $q$ norms, the vector $v$ in this proposition is given explicitly for $p < \infty$ ([Stetter 1999] also gives formulas for $p = \infty$) by

$$v_k = \gamma |u_k|^{p-2} \overline{u_k} , \text{ for } 1 \leq k \leq n . \tag{2}$$

where $\gamma \in \mathbb{C}$ is an arbitrary unimodular constant (that is, $|\gamma| = 1$). The proof is a straightforward application of the Hölder inequality.

This general result allows one to derive useful bounds and formulae for such problems as explicitly finding the nearest polynomial with a given zero. Corollary 4 of [Stetter 1999] states that if $p(x) = \sum_{k=0}^{n} a_k x^k = \mathbf{a}^T \mathbf{x}$, where $\mathbf{x} = [1, x, x^2, \ldots, x^n]^T$, then $(p + \Delta p)(z) = 0$ with $\Delta p(x) = \sum_{k=0}^{n} \Delta a_k x^k$ requires

$$\|\Delta a\|^* \geq \frac{|p(z)|}{\|\mathbf{z}\|} , \tag{3}$$

and moreover equality is attained at perturbations $\Delta a$ known explicitly from equation (2). This is a generalization of the work reported in [Manocha and Demmel 1995; Corless et al. 1995; Hitz and Kaltofen 1998; Zhi and Wu 1998; Hitz et al. 1999], observed by Hitz in 1999. In words, pseudozeros, which are values of $z$ for which $p(z)$ is small, are the roots of explicitly-known approximate polynomials.

**2.12.3.2.6 Restrictions on the Coefficients.** Restrictions of coefficients are also addressed in [Karmarkar and Lakshman Y. N. 1995; Hitz and Kaltofen 1998], for example keeping the perturbed polynomials monic. In [Stetter 1999] the following definition is given. If we want to restrict our approximate polynomials so that only certain coefficients are allowed to vary by $|\Delta a_k| \leq \varepsilon_k$, for $k \in \mathcal{K} \subset \{1, 2, \ldots, n\}$, we must use a weighted max-norm:

$$\|\Delta a^T\|_\varepsilon^* := \max_{k \in \mathcal{K}} \frac{|a_k|}{\varepsilon_k} . \tag{4}$$

The dual norm is then

$$\|\mathbf{z}\|_\varepsilon := \sum_{k \in \mathcal{K}} \varepsilon_k |z|^k . \tag{5}$$

Setting individual $\varepsilon_k = 0$ prevents variation in the $k$th coefficient of $p$. Theorem 8 from [Stetter 1999] states that for a given $z \in \mathbb{C}^s$, then the neighbourhood $N_\varepsilon(p) := \{\tilde{p} : \|p - \tilde{p}\|_\varepsilon^* \leq 1\}$ contains polynomials $\tilde{p}$ with $\tilde{p}(z) = 0$ iff $|p(z)| \leq \|\mathbf{z}\|_\varepsilon$.

**2.12.3.2.7 Local Versus Global Solutions.** Currently, SNAP algorithms can be divided in two categories: local solutions (such as quotient-divisor iteration or Newton iteration to find approximate GCD [Chin et al. 1998]) and global solutions (such as parameterizing exactly nearest polynomial pairs by an unknown common root (using equation (2)), setting up polynomial equations for the unknown, and globally solving them [Karmarkar and Lakshman Y. N. 1995]). Local algorithms are simpler to find than global algorithms, and correspond naturally to iterative solutions for local minima of optimization problems. Local optimization, because of its limited scope, can be more efficient than global optimization. For example, factoring bivariate approximate polynomials that are near to exactly factorable polynomials is possible in this fashion [Huang et al. 2000; Corless et al. 2001b]. Local solution methods do not find the absolute nearest factorable polynomial, however, if the given polynomial is far from a factorable one. Global methods are more mathematically satisfying, and, if efficient, more useful in practice because they come with a guarantee that they produce the best of all possible answers.

**2.12.3.2.8 A Note on Asymptotically Valid Algorithms.** It is very common in numerical analysis, and applied mathematics generally, to rely on the validity or utility of a formula or algorithm that can be proved to be valid only in an asymptotic limit. For example, one uses a condition number as an estimate of the sensitivity of a problem to small changes, when technically this is only valid for "infinitesimal" changes; that is, in modern pure analytical terminology, for all $\delta > 0$ there exists an $\varepsilon_0 > 0$ for which the difference between the forward error $E$ and the estimate $C\varepsilon$, namely $\|E - C\varepsilon\|$, is less than $\delta$ if the backward error $\varepsilon$ is less than $\varepsilon_0$. This is summarized by saying that the condition number $C$ is an asymptotically valid estimate as $\varepsilon \to 0$. Notice that no prescription for computing $\varepsilon_0$ from $\delta$ is given; *but experience with many practical problems shows* that the mere existence of asymptotic validity is often enough to ensure the practical utility of the estimate: in practice, we have $\|E - C\varepsilon\| = O(\varepsilon^2)$ and moreover that the constant of proportionality is of moderate size, unless we are near a singularity.

So, reasoning by analogy, one would expect that a *local algorithm* that is asymptotically valid as the size of the necessary regularization goes to zero would be useful in practice, giving the desired answers even when we cannot *a priori* guarantee that it will. Classical examples of this include Newton's method, which often converges even when we can't prove ahead of time that it will. A further point is that these local algorithms are self-validating: when they work, you can detect convergence and verify that the residual is small. When they fail, which is detectable, we cannot say whether the problem has a solution or not. We see that finding an efficient global algorithm for a SNAP problem is both harder and more valuable. An example from statistics is the method of least squares curve fitting that finds the global minimum. More generally, whenever a problem is convex, a local method will give a global minimum.

**2.12.3.3 Selecta.** We present here discussions relating to a number of approximate polynomial problems. Because the GCD is one of the most fundamental operations one encounters beyond ring arithmetic, it has received the most study. In the approximate setting, it gives us our first interesting algorithms. We therefore devote considerable attention to the GCD here. We then discuss algorithms for other fundamental operations, including matrix eigenproblems, approximate polynomial decomposition and factoring.

**2.12.3.3.1 Approximate GCD.** We call the GCD of approximate polynomials an "approximate GCD". The precise definition is as follows.

*Definition: Approximate GCD.* Given two approximate polynomials $f$ and $g$, a polynomial metric $d$, and a tolerance $\varepsilon > 0$, we say that $h$ is an *approximate* GCD of $f$ and $g$ if there exist polynomials $\Delta f$ and $\Delta g$ such that the exact GCD$(f + \Delta f, g + \Delta g) = h$ and $d(f, f + \Delta f) \leq \varepsilon$ and $d(g, g + \Delta g) \leq \varepsilon$. [Variations on this definition include, for example, the case where the last condition is replaced by $\|\Delta g\|_2^2 + \|\Delta f\|_2^2 \leq \varepsilon^2$.]

Call the GCD *nontrivial* if the degree of $h$ is *larger than* 1; moreover if the degree of $h$ is the maximum over all approximate GCDs of $f$ and $g$ in the $\varepsilon$-neighbourhood, then we call $h$ a *maximal degree approximate* GCD or just maximal approximate GCD. Approximate GCDs (even maximal approximate GCDs) are not necessarily unique. If there is no approximate GCD with degree $h$ larger than 1, we say by abuse of notation that there is no approximate GCD; this is really a convenient shorthand for saying that there is no *nontrivial* approximate GCD. Note that this is not a "quasi-GCD" in the sense of [Schönhage 1985]; that work assumes the input polynomials are not known to complete precision, but are exact, not approximate, and that more digits of the input can be obtained on demand. Furthermore, the paper [Corless et al. 2001a] uses an example from [Schönhage 1985] to show that a step of the Euclidean algorithm does not preserve approximate GCD, irrespective of the arithmetic system the step is carried out in. Further, the result of the Euclidean algorithm is not necessarily of a degree that is a lower bound on the degree of the approximate GCD: sometimes it can be spuriously too high. This fact contradicts statements by several authors.

The problem of computing an approximate GCD in polynomial-time, the minimax problem, is solvable in exponential time by methods from the existential theory of the reals (see section 2.5.3) and was finally solved in [Karmarkar and Lakshman Y. N. 1995]. We will give their solution below. Two attempts at the solution are noteworthy, although they do not provide a complete solution.

The first is based on the use of the Singular Value Decomposition or SVD [Corless et al. 1995; Gianni et al. 1998]. The smallest singular value of the Sylvester matrix for $f$ and $g$ gives a lower bound on the distance to the nearest singular Sylvester matrix. The SVD method can give overly weak lower bounds, because the smallest singular value is the 2-norm distance to the nearest singular matrix of the same dimension [Eckart and Young 1936], but not the 2-norm distance to the nearest singular Sylvester matrix, which must necessarily be at least as far away, and can sometimes be much farther away. The partial algorithm given in that paper for solving the optimization problem for approximate GCD relied on the SVD clearly separating singular values, and that separation is prone to fail for even moderately large problems. Nonetheless, the paper [Corless et al. 1995] gives a precise definition of approximate GCD, and phrases the problem of finding it as an optimization problem.

A second attempt is based on root-matching. Instead of working with the polynomial coefficients and trying to generalize the rational algorithms so successful in the exact computation case, Victor Ya. Pan proposed in [Pan 1998] to compute approximate roots of univariate approximate polynomials, efficiently sort to identify the roots that allow us to compute the spectral metric, and decide which nearby roots are to be considered equivalent. Once these decisions are taken, it is relatively easy to construct the GCD by recovering its coefficients from the root representation.

An example of why this approach is interesting is the "multiplicity problem" $f = z^n - a^n$ compared to $g = z^n$. It is clear that for $|a| < 1$ and large enough $n$, any coefficient-based metric will imply that these two polynomials have nontrivial approximate GCD. However, in the spectral metric, since the roots of $f$ are on the circle of radius $|a|$ and the roots of $g$ are zero, the spectral distance between these two polynomials is independent of $n$. This implies that for problems where the approximate GCD has roots with very high multiplicity, this approach may give more satisfactory results than coefficient-metric based methods. The strength of this approach is the possibility that the root locations can be very sensitive to small changes in the coefficients of a polynomial, as Wilkinson has observed. Therefore, constraining the roots instead of the coefficients may give better results.

**A Global Algorithm for Approximate GCD.** Using the Hölder inequality, one can derive explicit formulas for the nearest polynomials with a common root. These can be used to give a global algorithm to compute the nearest polynomials with a non-trivial GCD. This approach was introduced and shown to be polynomial time in [Karmarkar and Lakshman Y. N. 1995]. Using the 2-norm, the algorithm runs as follows.

If $f(x) + \Delta f(x)$ and $g(x) + \Delta g(x)$ have a common zero, say $x = \alpha$, then by the Hölder inequality formula we have

$$0 = f(\alpha) + \Delta f(\alpha)$$

and

$$\|\Delta f\| \geq |f(\alpha)|/\|[1, \alpha, \alpha^2, \ldots, \alpha^n]\| .$$

Equality is attained (i.e. the minimum $\|\Delta f\|$ is attained) at a vector of coefficients for $\Delta f$ that is explicitly known as a rational polynomial in $\alpha$. A similar formula holds for $\Delta g$. Therefore the minimum value of $P(\alpha) = \|\Delta f\|^2 + \|\Delta g\|^2$ occurs at one of the finitely many zeros of the derivative $P'(\alpha)$. The search for the zeros of the numerator of this rational polynomial and comparison of the resulting values of $P(\alpha)$ can be carried out in polynomial time.

### 2.12.3 Hybrid Methods

Note that the method determines the nearest pair of polynomials with a common root. That pair may share another root in common. It is possible to force the GCD to be of higher degree, but then the method becomes exponential in the number of common roots, see [Karmarkar and Lakshman Y. N. 1995] for details. The parametric optimization approach by Karmarkar and Lakshman has led to several other polynomial-time solutions for global optimization problems:

– Compute the nearest polynomial in Euclidean distance with a root on a piecewise rational parametric curve [Hitz and Kaltofen 1998].
– Compute the nearest polynomial in coefficient-wise distance with a real root [Hitz et al. 1999].
– Compute the nearest polynomial in Euclidean distance with a $k$-fold root, where $k$ is arbitrary [Zhi and Wu 1998].

The task of proving that two given approximate polynomials are relatively prime, i.e. that they have no nontrivial approximate GCD, is likely to be very common in practice. The algorithm of Beckermann and Labahn [Beckermann and Labahn 1998] is currently the fastest known. It uses fast matrix techniques, with look-ahead to skip ill-conditioned blocks, to compute a better bound than the unstructured condition number for the distance to the nearest singular Sylvester matrix. This is, at the time of writing, being implemented by Claude-Pierre Jennerod and George Labahn.

**Certified Approximate GCD.** André Galligo, and his co-workers David Rupprecht, Henri Lombardi, Ioannis Emiris and Laureano Gonzalez-Vega, have developed a refined point of view on resolution of approximate polynomial systems. Their work is based both on geometry (stratification of the parameter space of the coefficients of the inputs) and *a priori* inequalities obtained by a refined analysis of the SVD of Sylvester-like matrices [Galligo and Rupprecht 2001; Emiris et al. 1996, 1997; Rupprecht 1999; Galligo et al. 2001].

For the computation of approximate GCD of two univariate polynomials, here as above, the coefficients of these polynomials are known with a limited precision, governed by a tolerance $\varepsilon$. They proved so-called gap theorems and described an algorithm which produces a partition into intervals of tolerance, say $d = 0$ if $\varepsilon < 5 \cdot 10^{-6}$, $d = 3$ if $4 \cdot 10^{-5} < \varepsilon < 2 \cdot 10^{-3}$, and $d = 6$ if $9 \cdot 10^{-2} < \varepsilon < 1$, where $d$ is the maximum of the degrees of the GCD of two near-by polynomials, with respect to the given tolerance. Notice that there are gaps in these bounds where the maximal degree $d$ is not certified; nonetheless this information is useful and may allow deductions to be made about the underlying problem.

This means that in the corresponding gap of tolerances, the algorithm judges that the situation is too costly (in term of effective computations) to be rigorously solved. They try to keep these interval gaps as small as possible. Geometrically this situation corresponds in the parametric space to the case where the point (attached to the set of coefficients of the two polynomials) is almost equidistant to several strata. These strata are singular varieties in high dimensional spaces and the optimization process will be ill-conditioned.

At the time of writing, implementations of this algorithm are able to compute a certified approximate GCD of two polynomials of degree 100 in under one minute of cpu time.

**2.12.3.3.2 Interval Polynomials and the Kharitonov Theorem.** The Kharitonov theorem from control theory [Kharitononv 1979; Minnichelli et al. 1989] predates most SNAP work, and leads to a powerful stability criterion of an interval polynomial. Recall that an interval polynomial includes all approximate polynomials in a neighbourhood. Therefore, results about interval polynomials imply results about approximate polynomials. We state the simplest form of the Kharitonov theorem. Given are $2n$ rational numbers $\underline{a}_i, \bar{a}_i$. Let $P$ be the *interval* polynomial

$$P = \{x^n + a_{n-1}x^{n-1} + \cdots + a_0 \mid \underline{a}_i \leq a_i \leq \bar{a}_i \text{ for all } 0 \leq i < n\}.$$

Then every polynomial in $P$ is *Hurwitz* (all roots have negative real parts), if and only if the four "corner" polynomials

$$g_k(x) + h_l(x) \in P, \quad \text{where } k = 1, 2 \text{ and } l = 1, 2,$$

with

$$g_1(x) = \underline{a}_0 + \bar{a}_2 x^2 + \underline{a}_4 x^4 + \cdots, \quad h_1(x) = \underline{a}_1 x + \bar{a}_3 x^3 + \underline{a}_5 x^5 + \cdots,$$
$$g_2(x) = \bar{a}_0 + \underline{a}_2 x^2 + \bar{a}_4 x^4 + \cdots, \quad h_2(x) = \bar{a}_1 x + \underline{a}_3 x^3 + \bar{a}_5 x^5 + \cdots$$

are Hurwitz.

The corner polynomials are easily tested for the Hurwitz condition, for example by a variant of Sturm sequences, and the condition constitutes the stability criterion for the corresponding differential equations (see [Gantmacher 1960, Ch. XV]). There now exist many generalizations of Kharitonov's theorem.

**2.12.3.3.3 Polynomial Systems and Matrix Eigenvalues.** In the paper [Auzinger and Stetter 1988] it is shown how to reduce the solution of a system of multivariate polynomials to an eigenvalue problem for a commuting family of matrices. As pointed out in [Stetter 1996], this is a key step forward, because given an eigenproblem we may use efficient and stable numerical methods to solve it, while computing a triangular basis (see sections 2.2.5 and 2.2.6) may not be numerically stable. This approach has also been used for resultant formulations [Manocha and Demmel 1995; Emiris 1999].

The paper [Corless et al. 1997b] gives a theorem showing that a nearly-commuting family of matrices can be simultaneously placed in nearly-upper triangular form by a unitary transformation (and therefore this process is numerically stable). Using clustering heuristics [Manocha and Demmel 1995] this method can then be used to solve problems with multiple roots numerically. Using a generic linear combination of the multiplication matrices to remove apparent but not actual multiplicities is recommended.

*A Simple Example.* Consider the problem of finding the intersections of $x^2 + y^2 - 1$ and $x^2 - y^2 = 1/2$. A short computation shows that the normal set is $[1, y, x, xy]$

and the multiplication matrices corresponding to $x$ and $y$ are

$$M_x = \begin{bmatrix} 0 & 0 & 1 & 0 \\ 0 & 0 & 0 & 1 \\ \frac{3}{4} & 0 & 0 & 0 \\ 0 & \frac{3}{4} & 0 & 0 \end{bmatrix} \quad \text{and} \quad M_y = \begin{bmatrix} 0 & 1 & 0 & 0 \\ \frac{1}{4} & 0 & 0 & 0 \\ 0 & 0 & 0 & 1 \\ 0 & 0 & \frac{1}{4} & 0 \end{bmatrix}.$$

These matrices commute, as can be verified by direct computation. Both of these matrices have multiple eigenvalues, because of the symmetry in the problem. However, the roots (intersections) of the original problem are all simple. Taking $M = \alpha M_x + (1 - \alpha) M_y$ for some random $\alpha$ in $(0, 1)$, we find a matrix whose eigenvalues are all simple. The eigenvectors are necessarily (from the normal set) of the form $[1, y, x, xy]$ where $(x, y)$ are the roots of the original system. Computation of the (all simple) eigenvectors of $M$ gives us (to complete accuracy) the four roots $(\pm 1/2, \pm \sqrt{3}/2)$.

*A Larger Example.* In the paper [Li et al. 1989] we find the following problem, used to show the effectiveness of a particular trick for homotopy methods (the trick was called the 'cheater's homotopy' in that paper). The example is the following system of two polynomial equations in two variables, $x$ and $y$:

$$x^3 y^2 + c_1 x^3 y + y^2 + c_2 x + c_3 = 0$$
$$c_4 x^4 y^2 - x^2 y + y + c_5 = 0. \tag{6}$$

The symbols $c_k$, $k = 1, \ldots, 5$, represent parameters, which can take values in $\mathbb{C}$. Generically, as noted in [Li et al. 1989], there are 10 complex roots of this system—by "generically" one means for almost all sets of values of the parameters; for some exceptional values (which are of course interesting) the number of roots may be different.

A lexicographic order Gröbner basis computation fails for this example (the answer is too large to be useful). A total-degree order basis is computable in seconds, and is small enough that the normal set can be computed again in seconds, and 10 by 10 multiplication matrices for $x$ and $y$ can be found, with entries rational in the parameters, from which eigenvalues can be found easily once the parameters are specified.

One can compare the Gröbner basis approach with the resultant approach, see [Emiris 1999], or with the homotopy approach, see [Sommese et al. 2001]. At the time of writing there is no one "best" method for all problems.

**2.12.3.3.4 Functional Decomposition.** Writing a polynomial $f$ as a composition of two smaller polynomials, say $f(x) = g(h(x))$, can dramatically simplify working with that polynomial. Polynomial-time algorithms for computing the exact decomposition of exactly-known polynomials, when such decompositions exist, have been known for some time [Kozen and Landau 1989b; von zur Gathen 1990a; von zur Gathen and Weiss 1995]. Recently, these algorithms have been extended to the approximate polynomial case [Corless et al. 1999]. In the exact case, the algorithm is polynomial-time, and is guaranteed to succeed in finding a decomposition if such exists. For approximate polynomials, the algorithm is a local algorithm, and thus is guaranteed to find a decomposition only if the given approximate polynomial is sufficiently close to being a decomposable polynomial.

The algorithm of the paper is an iterative one. It takes as a starting point the $h(x)$ computed by the series reversion technique of the exact algorithm [von zur Gathen 1990a] (here normalizing so that $\|f\|_2 = 1$ and $g$ is monic) and then solves a linear least-squares problem to identify the corresponding $g$. Thereafter an alternating sequence of linear least-squares problems is solved to improve the $(g, h)$ pair. At every stage the residual $\Delta f = f - g \circ h$ is available so the convergence of the algorithm can be monitored. This technique of solving a nonlinear optimization problem by optimizing over alternating subsets of the problem parameters iteratively is well-known, and is linearly convergent at best. A Newton iteration can be used to speed up convergence, but as usual at the cost of more complicated steps at each iteration. According to the tests of the paper, the linearly-convergent method is expected to be faster in practice, as usually only a few iterations are sufficient to give approximate decompositions in the cases where the given polynomials are close to decomposable polynomials (which is likely to be the only case of practical interest).

**2.12.3.3.5  Bivariate Factoring.** Already in 1992, Kaltofen stated the problem of finding the nearest bivariate polynomial with nontrivial complex factors [Kaltofen 1992c]. Subsequent work includes [Galligo and Watt 1997; Hitz et al. 1999; Rupprecht 2001]. In fact, the results in [Hitz et al. 1999] give a polynomial-time algorithm that finds the globally nearest polynomial with a complex factor of a fixed given degree. The method is based on the parametric optimization of [Karmarkar and Lakshman Y. N. 1995]. If the degree of a factor can be arbitrary, no such polynomial-time algorithm is known. Therefore, several heuristic numerical algorithms have been proposed. Those algorithms suppose that the input polynomial is near a polynomial that factors.

The paper [Huang et al. 2000] reports a local algorithm, apparently numerically stable, to find factors of bivariate approximate polynomials. The algorithm is of complexity exponential in the degree of the input. The paper [Corless et al. 2001b] gives another local algorithm to factor bivariate approximate polynomials. This algorithm is also apparently numerically stable, and is of polynomial complexity in the degree of the input (modulo an unproven conjecture). The algorithm uses numerical path following in the Riemann surface of one factor only and numerical implicitization to recover the factor from the path.

The papers [Kaltofen 1985a; Sasaki et al. 1991, 1992; Sasaki 2001] discuss another interesting algorithm based on zero-sum identities of power-series solutions of $f(x, y) = 0$. In [Sasaki and Sasaki 1993] this algorithm is shown to be quite general, also being applicable to the problem of factoring over algebraic number fields and algebraic function fields. This algorithm is also apparently numerically stable and of polynomial complexity.

In all cases, when they succeed the algorithms produce answers which can, of course, be verified a posteriori to be factors of an explicitly constructible polynomial, which can be examined to see how near to the original it was. Moreover, the computed factors can be used as the starting point for a Newton refinement.

**2.12.3.3.6 Matrix Spectra under Perturbation** There is now a substantial theory of how matrix spectra and matrix canonical forms behave near a given matrix. For brevity, we only give three references [Edelman et al. 1997; Jeanerrod and Pflügel 1999; Hitz et al. 1999]

**2.12.3.4 Outlook.** Hybrid symbolic-numeric methods have proved their worth in applications such as Computer-Aided Geometric Design. Good implementations of many symbolic-numeric algorithms for polynomials are now publicly available. Nonetheless, much work remains to be done.

Computational complexity is an important consideration. Some of the problems in this area are computationally intractable (for example, the problem of computing the nearest singular matrix in entry-wise distance is NP-hard [Poljak and Rohn 1993]), or even recursively undecidable (for example, computing the GCD over the computable reals [Schönhage 1985]). Therefore, there is not much hope to attack these problems by numerical or any methods. Cook's hypothesis states that most instances of this problem will be computationally intractable. If a polynomial time algorithm becomes known that computes a singular matrix whose nearness to the input matrix is within a certain factor to the nearest one, the situation would change for that problem. Such is the case, for example, for lattice basis reduction problems (see section 2.4.6). The recent polynomial-time algorithms for computing the globally optimal approximate GCD or approximate factors show, however, that some of the problems are computationally tractable. The complexity class of others remains to be discovered. Full use of randomization has not yet been made; combinatorial methods exploiting sparsity are being developed. Other mathematical areas, such as optimization and perturbation theory, are being scrutinized for useful results. The challenge of hybrid symbolic numeric algorithms is to explore the effects of imprecision, discontinuity, and algorithmic complexity by applying mathematical optimization, perturbation theory, and inexact arithmetic and other tools in order to solve mathematical problems that today are not solvable by numerical or symbolic methods alone.

Robert M. Corless (U. Western Ontario),<br>
Erich Kaltofen (North Carolina State U.),<br>
Stephen M. Watt (U. Western Ontario)

## 2.13 Algebraic Complexity Theory

In *algebraic complexity theory*, algebraic problems (such as polynomial evaluation, interpolation, continued fractions, matrix multiplication, or factorization of polynomials) are studied within an algebraic model of computation (straight-line programs, computation trees, etc.). Such a model of computation specifies the possible inputs, the admissible operations which can be performed in each step, and the computational cost of a single step. That way, we limit ourselves to special algorithms. On the other hand, we may now ask for lower bounds for the *complexity* of a computational problem. That is, for lower bounds on the sum of the costs of all single steps of each admissible algorithm, which solves a given problem. The goal of algebraic complexity theory is to derive lower bounds for the complexity of algebraic problems. Ideally, the upper bound achieved by an algorithm coincides with a lower bound, which would show the *optimality* of that algorithm within the given model of computation.

Deriving lower bounds is a difficult task, since we have to make the case for any conceivable algorithm within the model which solves a given problem. Nevertheless, a deep theory has emerged, which we would like to present in extracts and in an exemplary fashion in the following paragraphs. More information and detailed references can be found in [Blum et al. 1998; Bürgisser 2000; Heintz et al. 1991a; Schönhage 1987; Strassen 1990; von zur Gathen 1988], and in the textbook [Bürgisser et al. 1996], which provides a thorough treatment of the subject.

Algebraic complexity theory is a comparatively recent area of research, witnessing increased activity since the early seventies. The subject originated already in 1954 with a short paper by Ostrowski, in which he asked for the optimality of Horner's rule. In the middle of the sixties Pan and Belaga proved its optimality with respect to the number of multiplications and additions, respectively.

Lower bounds which are nonlinear in the number of involved variables were first established by Strassen's *geometric degree bound*. The basic philosophy behind is that "geometric complexity" should imply "computational complexity." For instance for computing the coefficients of a univariate polynomial of degree $n$ from its roots, as well as for computing the interpolating polynomial, we obtain this way a lower bound of order $n \log n$ for the number of required multiplications and divisions. This is because the algebraic geometric object under consideration – the graph of the elementary symmetric functions – features a high degree. This lower bound is optimal up to order of magnitude if we do not count linear operations. Schnorr, as well as Baur and Strassen derived efficient extensions of the degree bound for dealing with single rational functions. As a consequence, just computing a "middle" elementary symmetric function requires about $n \log n$ multiplications and divisions.

By relying on a result of Khovanski, Grigoriev and Risler were able to confirm the above fundamental philosophy not only for multiplicative operations, but also for the number of required additions for computing real polynomials.

Now, we turn to two important problems of low degree: *matrix multiplication* and *Fourier transforms*. Matrix multiplication plays a key part in many computational problems of numerical linear algebra. Strassen's attempts to prove the optimality of the standard algorithm for matrix multiplication led to his astonishing discovery that just $O(n^{2.81})$ arithmetic operations are sufficient for multiplying two $n$ by $n$ matrices. In the meantime, an extensive theory regarding

## 2.13 Algebraic Complexity Theory

matrix multiplication has been developed (Schönhage's asymptotic sum inequality, Strassen's laser method and asymptotic spectrum), and asymptotically even faster algorithms with exponents below 2.38 have been found (Coppersmith-Winograd). Today, it is even conjectured that any exponent greater than 2 can be achieved! This is supported by an amazing result due to Cooperstmith, which states that $n$ by $n$ matrices can be multiplied with $n$ by $n^\alpha$ matrices using only $O(n^2 \log^2 n)$ operations, where $\alpha \approx 0.17$. However, with the possible exception of Strassen's algorithm, no algorithms are known to date, which are fast for practically relevant sizes of the input matrices. On the other hand, only linear lower bounds are known for the matrix multiplication problem.

There are well-known fast algorithms for performing the discrete Fourier transform of length $n$ working with $O(n \log n)$ arithmetic steps (Cooley-Tukey, Rader, Bluestein, Winograd). These algorithms are of fundamental importance to computer algebra: among others, fast algorithms for multiplying integers (Schönhage-Strassen) and polynomials rely on it (cf. section 2.1). Let us mention Valiant's remarkable graph theoretical attempt for showing the optimality of the above fast algorithms for the discrete Fourier transform. Although this attempt had failed, it has led to the discovery of special graphs called superconcentrators, which turned out to be a useful tool in the information and communication sciences. Today, the question whether a Fourier transform of length $n$ requires at least $n \log n$ operations is still open. However, Morgenstern proved that any algorithm, which may only use scalars of absolute value at most 2, needs at least $0.5 \, n \log n$ steps. After Beth's pioneering work, the theory of fast Fourier transforms was extended to arbitrary finite groups by Baum, Clausen, Diaconis, Maslen, and Rockmore. For instance, Baum showed that supersolvable groups of order $n$ possess fast Fourier transforms using only $O(n \log n)$ operations. Furthermore, Baum and Clausen designed an essentially optimal algorithm for constructing all ordinary irreducible representations of such groups from consistent power-commutator presentations.

The degree method can also be applied to study programs with branchings (computation trees). A fast variant of Euclid's algorithm for univariate polynomials of degree $n$ due to Knuth and Schönhage requires only $O(n \log^2 n)$ arithmetic operations, and only $O(n \log n)$ multiplications and divisions (compare section 2.1.3). Based on his degree bound, Strassen succeeded in proving the optimality of this algorithm with respect to the required multiplicative steps. Recently, using techniques from real algebraic geometry (cf. section 2.5.3), Lickteig and Roy have extended this to computations trees over the reals. These results provide a definitive answer for a fundamental problem of computer algebra (in the model of coefficients of fixed size).

Various problems in *computational geometry* can be formalized as semi-algebraic membership problems, which are to be solved by computation trees over the reals. Based on a theorem due to Milnor and Thom, Ben-Or derived a lower bound on the complexity of such problems in terms of the number of connected components of semi-algebraic sets. This lower bound was later generalized to higher Betti numbers in a series of papers authored by Björner, Lovász,

Montaña, Morais, Pardo, and Yao. These results reveal that a complicated topological structure leads to high computational complexity, thus confirming the general philosophy. It is conceivable that it is hard to test membership to a polyhedron with many faces. Recently, this intuition was confirmed by a result of Grigoriev, Karpinski, and Vorobjov, which relies on techniques from differential geometry. A different approach for obtaining lower bounds to semi-algebraic membership problems based on differential methods, the degree, and the real spectrum was pursued by Lickteig.

The theory of *NP-completeness*, originating in the work of Cook, Karp, and Levin, aims at distinguishing between computationally feasible and infeasible problems and centers around the notion of a complete problem. This theory relies on the fundamental $P \neq NP$ hypothesis, which is undoubtedly the most famous open problem in theoretical computer science. While NP-completeness had been originally defined in the Turing machine model, analogous concepts in algebraic frameworks of computation have been developed meanwhile. Valiant described such a theory in connection with his famous hardness result for the permanent. Another approach is the computational model introduced by Blum, Shub, and Smale designed for discussing complexity problems of numerical analysis and scientific computing. More information on these topics can be found in the new monographs [Blum et al. 1998; Bürgisser 2000].

Algebraic complexity theory has active research branching out into many disciplines of mathematics. It provides not only theoretical foundations for computer algebra, but also new ideas of considerable practical value. We mention here the idea of using straight-line programs as efficient data structures (von zur Gathen, Giusti, Heintz, and Kaltofen).

This survey of algebraic complexity theory is incomplete. Many topics had to be left out, such as quantifier elimination, real algebra, models of parallel computing, polynomial factorization, and boolean complexity. For information on those subjects, we refer to [Blum et al. 1998; Bürgisser 2000; Bürgisser et al. 1996; Heintz et al. 1991a; Schönhage 1987; Strassen 1990; von zur Gathen 1988; Wegener 1987] (see also the sections 2.15.2, 2.5.3, and 2.2.2).

Peter Bürgisser (Paderborn), Michael Clausen (Bonn), and M.A. Shokrollahi (Bell Labs)

## 2.14 Coding Theory and Cryptography

Coding theory and cryptography are branches of communication technology parts of which are close to algebraic theories. As a result, computer algebra tools and methods can be used for supporting research, testing conjectures, and sometimes deriving efficient algebraic algorithms.

### 2.14.1 Coding Theory

Coding theory deals with the problem of reliable transmission of information through unreliable channels. The theory was born with Shannon's revolutionary

### 2.14.1 Coding Theory

paper "A Mathematical Theory of Communication" [Shannon 1948] in which he established the limits of the gains possible with coding and proved the existence of codes that could effectively reach those limits. Since then the theory of error-correcting codes has established itself as an independent mathematical discipline with various connections to other areas such as combinatorics [Assmus and Key 1993; van Lint and Wilson 1992; MacWilliams and Sloane 1988], lattice theory [Conway and Sloane 1988], cryptography [van Tilborg 1988], algebraic geometry [Tsfasman and Vladut 1991], bilinear complexity [Bürgisser et al. 1996], computational complexity [Arora 1995], graph theory [Gallager 1963; Luby et al. 1997; Sipser and Spielman 1996; Spielman 1996], or computational algebraic number theory [Shokrollahi 1996]. In many of these fields tools and methods from computer algebra have had impacts on theoretical and practical discoveries.

Applications of computer algebra to coding theory can be divided into (at least) two categories: applications at an experimental level, and applications at a conceptual level. In the first category, researchers use computer algebra systems to test conjectures by designing examples that are larger than those that can be constructed by hand. The data so gathered may form the basis for the proof of a general result, or may lead to counterexamples. For example, the problem of distribution of Hamming distances of a linear code can be attacked by computer-algebra methods which exploit symmetries inherent to the problem [Kalouti et al. 1995]. In this respect the use of computer algebra in the coding theory community does not essentially differ from its use in other scientific communities.

At a conceptual level, computer algebra and its methods are used for solving problems from computational coding theory. As an example we briefly discuss applications to the theory of algebraic geometric codes. These codes are constructed by evaluating certain functions on an irreducible algebraic curve over $\mathbb{F}_q$ at the rational points of the curve. Algebraic geometric codes are arguably the most powerful class of codes known as they contain the asymptotically best codes known over non-binary alphabets [Tsfasman et al. 1982]. Moreover, when the underlying curve is the projective line, we obtain the important class of Reed-Solomon codes used in many communication situations like satellite and deep-space communication or CD-players.

Already the construction of algebraic-geometric codes calls for methods from computer algebra, as one has to construct bases for linear divisors on the underlying curve [Huang and Ierardi 1991]. On the decoding end, one can use methods from linear algebra combined with Sakata's generalization [Sakata 1988, 1990] of the Berlekamp-Massey algorithm for multidimensional sequences to decode a number of errors which is less than or equal to $(d-1)/2$, $d$ being the minimum distance of the code [Sakata et al. 1995b,a]. (See also the survey article by Høholdt and Pellikaan [Høholdt and Pellikaan 1995] and the corresponding section in the Handbook of coding theory [van Lint et al. 1998].) Alternatively, one can use Gröbner basis algorithms for decoding [Heegard et al. 1995; Saints and Heegard 1995].

In another direction, factorization of multivariate polynomials over finite fields is an essential tool for efficient decoding of algebraic-geometric codes even if the number of errors exceeds $(d-1)/2$. Such *list-decoding algorithms* which were first invented in [Sudan 1997] for the class of Reed-Solomon codes, and later generalized in [Shokrollahi and Wasserman 1998] to the class of algebraic geometry codes, have an unprecedented ability to reconstruct meaningful data in spite of very high amounts of noise. An interesting theoretic application of the high-noise decoding algorithms was recently given in [Arora and Sudan 1997] where a powerful "low-degree test" was invented with diverse impacts on in-approximability results. One of the tools in that paper is that the above high noise decoding algorithms run in *random polynomial time*; this result is directly related with the existence of polynomial time factorization algorithms for multivariate polynomials [von zur Gathen and Kaltofen 1985; Kaltofen 1985b]—a result which is clearly rooted in modern computer algebra.

As algebraic coding theory grows, so does its connections to various other fields of algebra. As a result, future developments will lead to even more applications of methods and tools from computer algebra to coding theory, both at an experimental and at a conceptual level.

### 2.14.2 Quantum Coding Theory

The prospect of quantum algorithms outperforming classical procedures gave rise to the question of realizability of quantum computers. The main problem with quantum state engineering was that of stability of quantum states and the information stored therein. Since unknown quantum states cannot be copied, which makes e.g. repetition codes impossible, most physicists were pessimistic about the existence of quantum codes. Shor [Shor 1995] showed that one can store the quantum information of one qubit in a subspace of the state space of nine qubits in such a way that the decoherence of one of the nine qubits can be detected and corrected. A measurement of the syndrome projects the disturbed quantum state onto a subspace corresponding to the one-qubit-error with this syndrome, without further disturbing the quantum information. Thus one can reconstruct the original quantum state by a quantum operation depending on the syndrome measured*. This process was later generalized [Calderbank and Shor 1996], [Beth and Grassl 1996] to obtain quantum codes from classical codes. In [Grassl et al. 1997] it was shown that no three qubit erasure code exists, after searching such a code via the nonlinear equations for its coefficients. In [Calderbank et al. 1997a,b] a close connection between quantum codes and orthogonal geometries — or equivalently additive codes over $GF(4)$ — was demonstrated.

Gottesmann [Gottesman 1997] derived equivalent results using common eigenspaces of abelian subgroups of the group generated by the one-qubit-errors. Calculation of the normalizer of this stabilizer groups yields nontrivial operations which map codewords to codewords and give rise to error tolerant operations.

---

* It is one of the paradoxes of quantum mechanics that even the measurement alone helps stabilizing the quantum state. "As long as you keep an eye on it, nothing happens" [Pellizzari et al. 1996].

Most interesting from the computer algebra point of view is the ongoing classification of quantum codes via polynomial invariants of the so-called group of local unitary transforms. Invariants of this group are constructed via matrices which are invariant under conjugation with the operation induced on the monomials [Rains 1997]. At the IAKS a large number of invariants was calculated. However, a calculation of the generating function (*Molien series*) using computer algebramethods and combinatorics revealed that the invariants found so far do not separate the orbits under this group action.

## 2.14.3 Cryptography

Modern cryptography employs a lot of techniques from computational mathematics. Most prominently techniques from algebra and number theory.

Apart from the well-known public key schemes which are based on the difficulty of factoring (RSA), calculating the discrete logarithm (El Gamal), and decoding a linear code (McEliece) there are schemes relying on computational problems like finding a closest vector in an integer lattice (Goldreich, Goldwasser, and Halevi) or solving algebraic equations (Patarin, Imai/Matsumoto).

Some of the proposed algebraic schemes for public key cryptography can be broken using algebraic techniques (see [Koblitz 1998; Kipnis and Shamir 1998]). Breaking ciphers, i.e. cryptoanalysis, is another area where algebraic techniques can be applied successfully. Methods of digital signal processing, correlation analysis, and linear algebra with added stochastic components provide the basics for investigating the asymptotic behavior in large scale experiments. The algorithm of Berlekamp and Massey for determining the linear complexity of a shift register sequence is an example of a typical computer algebra application.

The hash function of Tillich and Zémor is calculated as a matrix product in $SL(2, 2^n)$ and inverting this function is supposed difficult since the word problem in this matrix group is hard. However, embedding both the multiplicative group $GF(2^n)^\times$ and $SL(2, 2^n)$ into $GL(2, 2^n)$ reduced the problem to something like finding discrete logarithms in $GF(2^{2n})^\times$. Thus the hash function cannot be more secure than exponentiation in $GF(2^{2n})^\times$ (see [Geiselmann 1995]).

Gollmann proposed a stream cipher which uses a cascade of linear feedback shift registers each one stepping up the next. This stop and go generator was analyzed using eigenvalues over the rationals computed by the system Axiom [Geiselmann and Gollmann 1989]. The Gollmann generator proved resistent against attacks via correlation analysis.

Data structures of computer algebra are also significant in the area of *system security*, especially in authentication, protocol verification, and in *shared control* schemes. The major implementations of *shared control schemes* are based on finite geometry, where the intersection point of appropriately chosen hypersurfaces over finite fields is the shared secret.

In *multiparty protocols*, like electronic voting, one calculates with points on curves which are described by polynomials. After the computation a public interpolation reveals the result but not the data which was private to the participants.

Another cryptographic primitive becoming more important with internet technology is *secure outsourcing*. There one wants to use an untrusted resource for linear algebra computations (Atallah et al.) or for computations necessary for RSA based protocols (Matsumoto et al.).

Computer algebra can also help to derive algorithms and (chip) architectures for cryptographic purposes. This is treated in Subsection 2.20.

Thomas Beth, Sebastian Egner, Willi Geiselmann, Markus Grassl, Dejan Lazic, Jörn Müller-Quade (Karlsruhe), Frank Schaefer-Lorinser (Darmstadt), and M. Amin Shokrollahi (Bell Labs).

## 2.15 Algorithmic Methods in Universal Algebra and Logic

### 2.15.1 Term Rewriting Systems

Term rewriting is a discipline of replacing (or "rewriting") first order terms according to a given set of rules.

*Example 1.* A rule
$$+(x, x) \to \mathit{twice}(x)$$
(where $x$ is a variable) can be used to rewrite the term $*(+(-(5), -(5)), +(6, z))$ to $*(\mathit{twice}(-(5)), +(6, z))$. □

A set of such *rewrite rules* is called a *term rewriting system* or *TRS* for short.

Term rewriting systems are omnipresent in computer science. They can be thought of as a generalisation of grammars that allows to parametrize productions. Under certain restrictions term rewriting systems qualify as a simple functional programming language. The most popular term rewriting mechanism is probably the macro processing known from programming languages like C and C++ (#define directives). Some Computer Algebra systems like Reduce or Mathematica use term rewriting as a major vehicle for computation.

From a more theoretical point of view the importance of term rewriting systems for the area of computer algebra lies in their potential to describe a canonical simplifier for many congruence relations. That is, some term rewriting systems can be used to compute canonical (i.e., unique) normal forms for all congruence classes of certain algebraic structures. This is what makes term rewriting systems particularly interesting in universal algebra and logic.

Computing with term rewriting systems means to rewrite an (input) term repeatedly until no more rewrites are possible. The resulting irreducible term is called a *normal form*. It can be shown that this computation mechanism is Turing complete. Therefore showing the termination of all possible rewrite chains for a given term rewriting system is an important yet undecidable challenge. If a term rewriting system does not allow for infinite rewrite chains it is called *terminating*. A further issue is that in general computation with term rewriting systems is indeterministic. *Confluence* is an important property of term rewriting systems that implies deterministic results (if any) through indeterministic rewriting. A term rewriting system that is both terminating and confluent always computes canonical normal forms for its input terms. Such term rewriting systems are often called *complete* or *canonical*.

The symmetric, reflexive and transitive closure of a term rewriting relation describes a congruence relation: the congruence relation associated with the term rewriting system. Hence a complete term rewriting system computes canonical normal forms for its associated congruence relation. A pair of congruent terms is called an *equation*.

An important challenge is to find a complete term rewriting system for a given congruence relation. Equivalently: given an input term rewriting system $\mathcal{R}$ find a complete term rewriting system $\mathcal{R}^*$ that has the same associated congruence relation as $\mathcal{R}$. This is the task of the *Knuth-Bendix completion procedure* [Knuth and Bendix 1970]. The Knuth-Bendix procedure consists of two key steps:

1. deducing a new equation (called *critical pair*) from two rules and
2. transforming a critical pair into a rule such that the term rewriting system computed so far augmented by the new rule is terminating.

The second step may fail causing the whole Knuth-Bendix procedure to fail. Another difficulty with the Knuth-Bendix procedure is that the process of (1) producing and (2) consuming critical pairs may run for ever. Only if this process terminates a complete term rewriting system results. Otherwise the process may be employed as an congruence test for pairs or terms: two terms are congruent w. r. t. input term rewriting system if they have a common normal form w. r. t. a term rewriting system computed as an intermediate result of the completion procedure.

Rules describing the commutativity law (C) or both associativity and commutativity (AC) of certain operators always conflict with the termination requirement and therefore Knuth-Bendix completion always fails if such properties are present. To overcome this difficulty all basic operations of term rewriting were extended to deal with finite congruence classes of terms. This is called term rewriting *modulo* a finite congruence class (e. g., modulo C or modulo AC). According extensions of the Knuth-Bendix completion procedure have been proposed by Peterson and Stickel [Peterson and Stickel 1981] or Jouannaud and Kirchner [Jouannaud and Kirchner 1986].

From a logical point of view a rewrite rule $l \to r$ describes a universally quantified equation $\forall \vec{x} : l = r$ where $\vec{x}$ is the list of variables occurring in $l$ or $r$. An *equational specification* consists of a set of operators (the *signature*) and a set of equations describing congruences over terms constructed from the operators of the signature and a countable set of variables. Each equational specification specifies a class of algebras which are called the models of the specification. This class of all models of an equational specification is also called the *variety* of the specification. The semantic validity of term rewriting as an approach to prove equations w. r. t. an equational specification goes back to a key result in universal algebra. A theorem by Birkhoff [Birkhoff 1935] states that

$$\mathcal{E} \models a = b \text{ iff } a \leftrightarrow^*_{\mathcal{E}} b.$$

I. e., two terms $a$ and $b$ are equal in all models of a set of equations $\mathcal{E}$ if and only if $b$ can be derived from $a$ by repeatedly applying equations from $\mathcal{E}$ (back and forth). Hence if there is a complete term rewriting system $\mathcal{R}$ associated with the congruence defined by a set of equations $\mathcal{E}$ then $\mathcal{E} \models a = b$ iff $a$ and $b$ have the same normal form w. r. t. $\mathcal{R}$.

There are many interesting classes of algebras that can be characterised by equational specifications and for which canonical term rewriting systems are known: free semi-groups, free monoids, free (Abelian) groups, free (commutative) rings (with 1), Boolean rings, distributive lattices, modules, algebras and multivariate polynomials over commutative ring structures. Commutative structures need of course term rewriting modulo AC.

*Example 2.* A free group can be specified by the following implicitly universally quantified equations[*]:

$$1 \cdot x = x, \quad x \cdot (x)^{-1} = 1, \quad x \cdot (y \cdot z) = (x \cdot y) \cdot z$$

where the signature consists of the binary infix operation $\cdot$, a unary postfix operation $(\ )^{-1}$ and a constant $1$. $x$, $y$ and $z$ are first order variables.

The complete term rewriting system associated with the free group is:

$$
\begin{array}{llll}
1 \cdot x & \to x & x \cdot 1 & \to x \\
x \cdot (x)^{-1} & \to 1 & (x)^{-1} \cdot x & \to 1 \\
(1)^{-1} & \to 1 & ((x)^{-1})^{-1} & \to x \\
x \cdot ((x)^{-1} \cdot y) & \to y & (x)^{-1} \cdot (x \cdot y) & \to y \\
(x \cdot y) \cdot z & \to x \cdot (y \cdot z) & (x \cdot y)^{-1} & \to (y)^{-1} \cdot (x)^{-1}
\end{array}
$$

$\Box$

If an equational specification for one of the base structures mentioned above is augmented by a finite set of constants (0-ary operators, generators) and a finite set of equations that do not contain variables then this specification describes a finitely presented structure. Proving the validity of an arbitray (variable free) equation in a finitely presented structure is called the word problem. For many such finitely presented structures term rewriting and completion can be employed to (semi-)decide the word problem.

The investigation of the completion of finitely presented structures has lead to the detection of more powerful deduction rules in the completion process like *symmetrisation* [Bücken 1979; Chenadec 1986; Bündgen 1998a] and research on *normalised term rewriting* [Marché 1996]. For structures like semi-groups, monoids and groups the idea of term rewriting and completion can be projected to a more efficient data structure namely to strings. This leads to *string (or word) rewriting systems* and completion procedures for string rewriting systems.

*Example 3.* A complete string rewriting system for the finitely presented group $S_3 = \langle r, t; rr, ttt, rtt \rangle$ is

$$
\begin{array}{llll}
t\bar{t} \to \epsilon & \bar{t}t \to \epsilon & \bar{r} \to r & rr \to \epsilon \\
lt \to \bar{t} & rt \to \bar{t}r & r\bar{t} \to tr & \bar{t}\bar{t} \to r
\end{array}
$$

---

[*] The second equation can be interpreted as the Skolemnized form of $\exists y : x \cdot y = 1$

where $\bar{r}$ and $\bar{t}$ are inverted generators and $\epsilon$ represents the empty string. □

Due to their large relevance and the close relation to formal languages string rewriting systems have become an independent branch of research [Jantzen 1988]. Similarly Buchberger's algorithm to compute Gröbner bases [Buchberger 1965] can be viewed as a specialized completion procedure operating on polynomials rather than using Knuth-Bendix completion modulo AC for terms in order to complete a finitely presented polynomial ring. This analogy is described in detail in [Bündgen 1996].

Note that for the completion of finitely presented structures variables are only needed to specify the base structures (free monoids, groups, rings, etc). If these base structures are implemented by special data structures like strings, multisets or polynomials the variables vanish from the specifications[*]. This is in contrast to applications from first order logic which we will review in the next paragraph. In first order logic variables play a central rôle in generalizing propositions to the instances of a term or formula.

Term rewriting has become a very important technique for equational theorem proving in automated theorem proving in first order logic. It is used intensively in theorem provers like OTTER. Similar to the resolution inference [Robinson 1965] the central operation of the critical pair deduction is the unification algorithm. So far we discussed procedures to prove the general validity of an equation w. r. t.an equational specification. In contrast the task of a unification is to check whether an equation has a solution and in the positive case to find the solution. Standard unification algorithms decide this problem for free term algebras and there is at most one most general solution for each equation.

Hsiang [Hsiang 1985] proposed a proof procedure for first order theorem proving based on the completion of Boolean rings. The normal forms produced by this completion procedures are the so-called exclusive-or normal forms. In the propositional case Hsiang's procedure is a completion of a finitely presented boolean ring. Then the normal forms are also known as Stone polynomials [Stone 1936] or Reed-Muller forms (by hardware circuit designers). Hsiangs's completion procedure can be seen as an alternative to the resolution theorem proving method. The relation between resolution and Knuth-Bendix completion (modulo AC) has been investigated by Paul [Paul 1985].

In the area of algebraic specifications equational specifications are used to characterize abstract data types. In contrast to applications from free algebras or first order logic the intension of algebraic specifications is to investigate the features of one particular model (and not of all models). The model of interest is the largest term generated model. I. e., a model where each object can be associated with a term without variables and only congruences that are derivable from the equations hold. This model is isomorphic to the initial model of the equational specification. Proving properties in this particular model requires stronger mechanisms than pure equational reasoning that was sufficient to prove the validity

---

[*] Note that generators of groups and unknowns in polynomials correspond to constants rather than to first order variables.

of equations in a variety. The additional inference needed is (structural) induction. Knuth-Bendix completion has been extended to refute invalid inductive theorems [Musser 1980; Huet and Hullot 1980; Jouannaud and Kounalis 1989; Küchlin 1989]. In cases where the completion yields a term rewriting system that is confluent on all variable free terms even a prove of an inductively valid equation is established. The key idea of these inductive completion procedures is to show that adding the theorem to the original specification does not lead to inconsistencies. I. e., no two variable free terms that had distinct normal forms w. r. t. a complete base specification become equal after adding the potential inductive theorem to the base specification. The operation used to detect such inconsistencies is a ground reducibility test.

Other issues in term rewriting research include

- finding criteria to prove the termination of all rewrite chains w. r. t. a term rewriting system. Typically such a criterion is described as well-founded ordering on terms which includes the rewrite relation. For such an ordering it is sufficient to show that for all rules the left-hand sides are greater than their respective right-hand sides. See [Dershowitz 1987] for a survey on termination criteria of term rewriting.
- term rewriting systems with additional information like (hierarchical) sorts, conditions, or constraints. Sorts restrict the set of well-formed terms analogously to types in programming languages: each argument of an operator must be of a particular sort (type). Sorts may either be disjoint (many sorted term algebras) or they may allow for sort hierarchies (order sorted term algebras [Smolka et al. 1989]).

In conditional term rewriting a rewrite rule may be constrained by additional conditions that must hold for a rule to be applicable.

Another approach tries to break down the unification procedure during critical pair completion into smaller tasks to be resolved by need. Augmenting each critical pair with information of the unresolved parts of the unification leads to so called constrained completion procedures. Constrained completion can be very efficient if unification is extremely expensive like in term rewriting modulo associativity and commutativity [Kirchner et al. 1990].
- efficient (parallel) implementation of rewriting and completion. Research in term rewriting has always triggered efforts to implement efficient term rewriting machines and completion procedures. Rather recently parallel and distributed completion procedures have been developed [Avenhaus and Denzinger 1993; Bündgen et al. 1996b].
- term rewriting for higher order terms. The goal of research in higher order term rewriting is to extend the result of standard (i. e., first order) term rewriting to rewriting terms that allow for $\lambda$-abstraction as a constructor. See [Nipkow and Prehofer 1998] for a survey.

A standard survey article on term rewriting is [Dershowitz and Jouannaud 1990]. The book of Baader and Nipkow [Baader and Nipkow 1998] gives a broad introduction to term rewriting systems. German text books that cover the topic of term rewriting are [Avenhaus 1995] and [Bündgen 1998b]. The major forum for current research in term rewriting systems is the annual conference on Rewriting Techniques and Applications (RTA). A survey of many rewriting approaches to problems in Computer Algebra can be found in [Bronstein et al. 1998].

Reinhard Bündgen (Böblingen)

## 2.15.2 Decision Procedures and Quantifier Elimination Methods for Algebraic Theories

Let $K$ be an elementary class of algebraic structures (see previous section), defined by its constants, operations, and relations (i.e., by a formal language $L$), as well as by elementary axioms on the given objects. Axioms are syntactically composed of formulas consisting of equations and relations of terms, the operators $\wedge, \vee, \neg$ of propositional logic, and the quantifiers $\exists x$ and $\forall x$. Semantically, the quantifier range over the given structures. Variables in the range of a quantifier are called *bound* variables, if they occur outside they are called *free* variables. Formulas which contain no free variables are called *sentences*. Sentences correspond, in terms of semantics, to true or false statements on structures in the given language $L$. On the other hand, the truth value of a formula containing free variables is depending on the structure under consideration, as well as on the the evaluation point, i.e., on the elements which are substituted for the free variables.

A *decision procedure* for an elementary class $K$ of algebraic $L$ structures is an algorithm which takes a sentence $\varphi$ of the given language as input, and returns the decision whether $\varphi$ is true in all structures from $K$ or not. The decision problem for $K$ is the question whether a decision procedure exists for $K$. The problem described in 2.15.1 is a sub-problem of the decision problem for $K$. Therefore, the decision problem is unsolvable for many classes of algebraic structures. This applies to groups, rings, fields, but also to particular structures with addition and multiplication, such as $\mathbb{Z}$ and $\mathbb{Q}$. The more surprising is the fact that the ordered field of real numbers, and the field of complex numbers, as well as $p$-adic number fields allow for decision procedures, despite the fact that their elements lacking a finite representation. In particular these fields are not computable. Further examples for classes which permit decision procedures, are, e.g., abelian groups [Szmielew 1954], ordered abelian groups [Gurevich 1965, 1977] and boolean algebras [Tarski 1949].

Even more important for practical applications are *quantifier elimination methods*. Such a method assigns to an arbitrary formula $\varphi$ of the given language $L$ a quantifier-free formula $\varphi'$ having the same free variables, such that $\varphi$ and $\varphi'$ are equivalent in all structures of the given class $K$, and at all evaluation points. Because the validity of quantifier-free formula at a given point in a structure can usually be decided with ease, we obtain from a quantifier elimination method, in general, a decision procedure for the corresponding class. This is achieved by applying the quantifier elimination method to a given sentence $\varphi$ first, and then deciding the resulting quantifier-free sentence $\varphi'$. In contrast to a decision procedure, which can decide *sentences* only, quantifier elimination methods yield a *uniform* decision for arbitrary formulas which depend on their parameters.

The first decision procedures and quantifier elimination methods for algebraic theories were discovered by A. Tarski's Polish school of logicians and by A. Robinson and his collaborators: for integers with addition (Presburger arithmetic) [Presburger 1929], for abelian groups [Szmielew 1954], for boolean algebras [Tarski 1949], for real and algebraically closed fields [Tarski 1951], for algebrically closed valued fields [Robinson 1956], and for archimedean ordered abelian groups [Robinson and Zakon 1960]. Ever since, a large number of other methods for different classes have been described, among them most prominently, decision procedures for $p$-adic fields [Ax and Kochen 1965, 1966; Ershov 1965; Cohen 1969], for the ring of integral algebraic numbers in the algebraic closure $\overline{\mathbb{Q}}$ of $\mathbb{Q}$ (Rumely - v.d. Dries) [van den Dries 1988]), for ordered abelian groups with quantification over convex subgroups [Gurevich 1977], and for the monadic second order theory of two successors [Rabin 1977]. A survey, covering the methods and results until 1983, can be found in [Weispfenning 1984]. Additional survey articles on this topic are [Macintyre 1977; Rabin 1977]. The most decisive progress on decision procedures and quantifier elimination methods for new algebraic theories since then concern extensions of these theories by additional functions, in particular by exponential functions. The theory of natural numbers with addition and exponentiation $2^x$ (or a similar unary function) was shown decidable in [Semenov 1984]; corresponding quantifier elimination methods where given in [Cherlin and Francoise 1986; Göttsch 1988]. The theory of the ordered field real numbers with the exponential function was shown to be decidable in [Macintyre and Wilkie 1996] under the proviso of Schanuel's conjecture; for related results compare also [van den Dries 1982, 1986; Richardson 1991a,b, 1992, 1993, 1995, 1998]. The theory does not admit complete quantifier elimination, but only quantifier elimination up to a block of existential quantifiers [Wilkie 1996]. For this restricted quantifier elimination to be algorithmic Schanuel's conjecture is required again. A complete elimination of *bounded* quantifiers is possible if sufficiently many real analytic functions are included in the language [Denef and van den Dries 1988; van den Dries et al. 1994; van den Dries and Miller 1994]. An extension of Presburgers results to the mixed linear theory of reals and integers has been found in [Weispfenning 1999].

Given their generality, decision procedures and quantifier elimination methods are algorithmically very complex, as a rule at least exponential; in many cases doubly exponential, and for theories involving an exponential function even non-elementary recursive. The articles [Ferrante and Rackoff 1979; Fitchas et al. 1990; Renegar 1992a,b,c; Basu et al. 1994] give an overview of the *worst case* complexity. Therefore, only few of these methods have practical implementations so far. Among them are decision procedures and quantifier elimination methods for reals and integers: The first implemented quantifier elimination method and decision procedure for the real numbers is the decision procedure and quantifier elimination method of Collins for real fields by cylindrical algebraic decomposition (compare also [Collins 1975; Collins and Hong 1991; Collins 1998a,b]).

### 2.15.2 Decision Procedures and Quantifier Elimination

Alternative approaches for implementations are based on Sturm-Habicht sequences, or parametrized versions of multivariate real root counting with side conditions [Ben-Or et al. 1986; Becker and Wörmann 1994; Pedersen et al. 1993; González-Vega 1996; González-Vega 1998; Rouiller 1996; Roy 1996; Dolzmann 1994; Weispfenning 1998]. Further information on decision procedures and quantifier elimination methods for real and complex algebra can be found in the survey article [Dolzmann et al. 1998c] and in sections 2.5 and 2.5.3. Other decision procedures and quantifier elimination methods, e.g., for boolean algebras, real linear and quadratic problems, Presburger arithmetic, and Presburger arithmetic with exponentiation, were successfully implemented in REDUCE, at the university of Passau [Taubeneder 1991; Burhenne 1990; Dolzmann and Sturm 1996, 1997b; Köppl 1991; Mnafeg 1992].

In the last decade various implemented real quantifier elimination methods have reached a stage, where they can be applied to solve non-trivial problems in mathematics, natural science, engineering, and economics (compare [Dolzmann et al. 1998c]). These include:

- Real implicitization of parametric algebraic surfaces [Dolzmann 1999b].
- Automatic theorem proving and finding in real geometry [Dolzmann et al. 1998b].
- Geometric reasoning about three-dimensional objects, including parallel and central projections of objects, the reconstruction of objects from projections, lighting and shading, equi-distance surfaces, and collision problems [Sturm and Weispfenning 1998c].
- Real constraint solving in constraint logic programming [Jaffar and Maher 1994; Jaffar et al. 1992, 1994; Hong 1992a].
- Rounding and blending of solids [Sturm and Weispfenning 1997].
- The Birkhoff interpolation problem [González-Vega 1996].
- Sign behaviour of univariate polynomials [González-Vega 1998].
- Computing the Newtonian graph of a complex univariate polynomial [Kozen and Stefansson 1997].
- Implementation of guarded expressions for coping with degenerate cases in the evaluation of algebraic expressions [Corless and Jeffrey 1992; Dolzmann and Sturm 1997a].
- Stability analysis for ODE's, PDE's and difference schemes [Hong et al. 1997; Liska and Steinberg 1993; Liska and Wendroff 1999].
- Control theory [Abdallah et al. 1996; Jirstrand 1997].
- Robust multi-objective feedback design [Dorato et al. 1997].
- Simulation and error diagnosis of physical networks [Weispfenning 1997b].
- Non-convex parametric linear and quadratic optimization problems [Weispfenning 1994a], transportation problems [Loos and Weispfenning 1993].
- Parametric scheduling [Dolzmann 1998].

Volker Weispfenning (Passau)

## 2.16 Knowledge Representation and Abstract Data Types

### 2.16.1 Mathematical Knowledge Representation and Expert Systems

From a cognitive point of view, one goal of computer algebra is to represent algebraic knowledge in a form which can be processed and manipulated by computer. One conceivable approach is considering different types of knowledge: *theoretic knowledge, tabulated knowledge, specification knowledge, algorithmic knowledge, deduced knowledge,* and *strategic knowledge.*

*Theoretic knowledge* is often given in form of mathematical formulas and theorems. They affect the choice of suitable data structures and algorithms in computer algebra systems, especially in terms of specification and verification of the system. In general, this knowledge is neither a formal part of a computer algebra system nor is it recognizable from the outside. Exceptions are, with some restrictions, formalized mathematical propositions within a logic calculus, or formulas stored in data bases.

To date, such interconnection to data bases is realized in very few computer algebra systems, thus *tabulated knowledge* was hardly used in this context so far. Examples for application of tabulated knowledge are implementations of special functions, or orthogonal polynomials.

*Specification knowledge* comprises the formal description of abstract data types used in the system. In particular, it contains the definition of signatures for all operators and relations of the data structures under consideration. Furthermore, the specification of corresponding concrete computational structures has to be added. It includes declarations for certain mathematical properties of the structures; properties which are also used for automatic verification by type checking. The treatment of algebraic data types in computer algebra systems raises the following question: for correctly programming mathematical procedures, implementation of new data types is often unavoidable, because algebraic algorithms in generic form are usually applicable only to certain data types. Unfortunately, most current computer algebra systems offer only predefined data types.

Specification knowledge can be realized in the framework of algebraic specification. One such attempt can be found in AXIOM, where so-called *categories* represent abstract, and *domains* represent concrete data types respectively. Properties of operators, which should specify the semantics of the operators, can be defined, however they are not checked by the system later on. A conceivable approach for handling properties of operators is the application of knowledge representation formalism whose semantics is clearly defined.

*Algorithmic knowledge* is at the very core of computer algebra systems. An algorithm is realized by procedures and programs. It means the processing of predominantly theoretic knowledge, but also specification knowledge. The set of programs finally contains the algebraic knowledge for solving mathematical problems.

*Deduced knowledge* is used to obtain new propositions from a set of given propositions (formalized theoretic knowledge), and general inference rules. Most of the time it is confined to a special form of knowledge representation (e.g., logic

## 2.16.1 Mathematical Knowledge Representation and Expert Systems 141

calculus). One could argue that deduced knowledge combined with (formalized) theoretic and specification knowledge already constitutes an expert system. In such systems, the search for solutions can be represented as search in a decision tree. So far, this kind of strategy does not seem to be very promising, without the aid of additional heuristics.

*Constraint Logic Programming* (CLP) is an approach which combines algorithmic knowledge with deduced knowledge [Jaffar et al. 1992; Aiba et al. 1988; Hong 1992b; Colmerauer 1990]. In the case of computer algebra, the process of syntactic resolution (by unification), as used, e.g., in PROLOG, is amended by semantic resolution over algebraic structures, for which decision procedures or quantifier elimination methods are known. In CLP systems, the validity of, e.g., algebraic equations over the complex or real numbers (see section 2.15), is verified in a special computer algebra part of the program, and depending on the result, the resolution process continues. Evidently, this is just one example of a series of articles on this topic.

The heuristics, or rules of thumb, mentioned earlier, are part of *strategic knowledge*. That is where experience of users and experts (such as mathematicians) becomes integrated into the system. Strategic knowledge can be found in many programs. For example, a procedure for factoring integers would perform trial divisions, before it switches to more complicated methods. A procedure for symbolic integration tries to apply simple heuristics first, Risch's algorithm is used as a last resort. Additionally, tabulated knowledge in the form of integral tables can come into play again. One has to recognize that accessing such data bases is a non-trivial task, which requires application of involved techniques like unification and pattern matching.

In 1967, J. Moses [Moses 1967] already employed heuristics in his SOLDIER system. Along the same line, A.V. Bocharov's DELiA system (see [Bocharov 1990]) should be mentioned here. Another example for appropriate exploitation of heuristics is solving differential equations exactly, again because pure algorithmic methods are in general very time consuming.

How can the relationship between computer algebra systems and the theory and practical use of expert systems be described? The basic architecture of current expert systems can be sketched as follows: The two major parts which make up the system are on the one hand, a knowledge base, containing the knowledge about a subject, and on the other hand, an inference engine, containing deduced knowledge as well as strategic knowledge via application of heuristics, for solving problems. Part two in particular, has only rudimentary implementations in current computer algebra systems.

MACSYMA was often cited as an expert system from within the AI community (see for example the classic textbook by Hayes-Roth, Waterman, and Lenat [Hayes-Roth et al. 1983]). A reason for acquiring this reputation might be that MACSYMA used to be the most extensive computer algebra system at the time of its creation. In addition, the original draft of the system was motivated by results of AI at that time. However, MACSYMA's structure differs, with regard to its architecture, from what we would call an expert system nowadays. If we use the characterization of expert systems in the strict sense, as mentioned above (for more precise characterizations, see [Richter 1989; Schnupp and Leibrandt 1986; Hayes-Roth et al. 1983]), then none of the current computer algebra systems would qualify.

There are also projects to use computer algebra systems, and symbolic computing in general, as test platforms for AI methods. Examples to that end are the LEX and META-LEX system by T. Mitchell (see [Mitchell et al. 1983]), and the experiments of M. Vivet in [Vivet 1984]; he is using the integration package of REDUCE. An introduction into common research topics of AI and symbolic mathematical computing is given by the proceedings of the conferences on Artificial Intelligence and Symbolic Mathematical Computing [Calmet and Campbell 1993, 1994; Calmet et al. 1996]. Finally, we would like to make reference to special issue of the journal *Annals of Mathematics and AI*, which originated from that conference [Calmet and Campbell 1997].

Jacques Calmet (Karlsruhe)

### 2.16.2  Abstract Data Types

The concept of abstract data types arose more than twenty years ago, in response to a crisis in software development. It was intended to contribute mathematical methods to efforts targeted towards producing reliable (and safe) software both faster, and more cost-efficiently. To that end, the (mathematical) semantics of programming languages were introduced, as well as the consequent application of abstract data types during specification and implementation of programs. Both approaches developed into independent areas of research in the meantime.

An abstract data type consists of specification of its operators, and a set of axioms (i.e., equations) defining the properties of those operators. The theory of algebraic specification, which had its origin in universal algebra (see section 2.15), defines an abstract data type to be the initial model (see for example [Goguen et al. 1978]) within the equational theory given by the axioms; more succinctly, it is the isomorphism class of all initial algebras of that equational theory.

Using algebraic specifications techniques, problems such as specification, correctness, or error handling can be treated exactly. Furthermore, algebraic specification could make reusing software easier, a property which is of prime importance, not only for computer algebra. A good introduction to the theory of algebraic specification is given in the textbook [Ehrig and Mahr 1985].

Abstract data types are a powerful tool for developing algorithms; they realize the two major concepts in program development, namely separation and abstraction, in an ideal way. Algebraic structures, such as `Ring`, are naturally suited for representation by abstract data types. In practical applications however, data types which are typical for computer algebra do often not allow for efficient computing, or they cannot be handled constructively at all (cf. problems of the Knuth-Bendix completion in section 2.15.1).

In order to still be able to design efficient computer algebra systems based on abstract data types, we have to weaken the rigorous definition – specification of operators plus equations – somewhat: for example in AXIOM, properties such as associativity of functions are described by "attributes" in lieu of equations. These declarations have to be heeded by programmers, although they can neither be enforced nor verified by the system.

Via abstract data types, a type system including parametrized types, and types of higher order ("types of types") can be realized in a natural manner (AXIOM). This concept of types is able to express algebraic structures directly: the type `Polynomial` is of type (second order) `Ring`, and has one parameter, namely the type of its coefficients (also of type `Ring`).

Parametrized types enable us to write generic functions. For example: a function which operates on objects of type `Polynomial` has to be written only once, using the arithmetic operations of the underlying coefficient domain.

While strong typing is a considerable advantage (modularity, error avoidance, and easier debugging) during algorithm design, the situation for interactive user interfaces is totally different: users should not, as far as possible, have to deal with complicated type information (although it cannot be avoided in all cases). They should be able to enter formulas the way they are used to. The computer algebra system then has to employ some mechanism for type inference in order to invoke a suitable (and mathematically correct) function. The process of inference is based on *coercion* (see also [Fortenbacher 1990]).

To illustrate the required steps: given the basic expression 3/7 + x, a system with type inference is able to determine the type of 3/7 as rational number, using a *bottom-up* process. Next, it can *not* just add the symbol x to that number. The system will rather infer that both x and 3/7 can be converted (coerced) to type `Polynomial(Fraction(Integer))`. Finally, it will add the two polynomials.

Additional reference: [Wirsing 1990].

Albrecht Fortenbacher (Berlin)

## 2.17 On the Design of Computer Algebra Systems

If one follows the development of programming languages from a computer algebra point of view, one comes to realize that many issues modern programming languages try to address, have been encountered in the area of computer algebra first, and that those problems were—more or less successfully—resolved for that area.

Among these issues are memory management, program verification, abstract data types, modularization, parallelization of algorithms, and the extensibility of systems. In the following, we want to briefly discuss some of them.

### 2.17.1 Memory Management

Exact arithmetic for various areas of algebra requires data structures capable of a dynamic adaption to the size of the represented algebraic object. Therefore, memory management has always been a major issue for the implementation of computer algebra systems.

For practical algebraic computing, memory management is of prime importance when large computational problems cannot finish due to lack of sufficient memory. Maybe memory is, contrary to other theories, a more critical resource than processor speed. It takes much to expand physical memory, whereas longer running times can make up for less compute power.

The reference counter method for memory management was mentioned in the literature for the first time in connection with the implementation of a polynomial system. Reference counting nowadays is implemented in several object-oriented languages, most notably Java.

### 2.17.2 Program Verification and Abstract Data Types

Dynamic algebraic objects generally are implemented with pointers. However, explicit pointer manipulation is known to be rather error prone. computer algebra, therefore, introduced the feature of data types to detect errors on the level of data structures early on, without having to de-reference pointers.

The method of specifying data types algebraically proved useful for general programming languages: a well-defined set of values of objects and operations on them lays the foundation to address the verification problem successfully.

Abstract data types follow up on issues of program verification insofar they integrate an essential part of a type's specification, the set of legal operations on it. Along with encapsulation and the separation of interface and implementation, abstract data types support programming algorithms in an abstract form, a method in computer science usually referred to as object-based programming.

To date, only a few computer algebra systems, for example AXIOM, support abstract data types. On this subject, see also section 2.16.2.

### 2.17.3 The Concept of Types

Looking at the development of general purpose programming languages, we can see the proliferation of a general type and class concept in that area as well. Algebraic algorithms realize objects from structures that have been systematically developed and well analyzed in algebra. Carried over to the formalism of types in programming languages mathematical structures prove to go beyond the type concept introduced in programming languages.

To a certain degree, computer algebra pioneers in the development of programming languages, both in a positive and in a negative sense. On the positive side, we have the testing of new concepts, like the ones previously mentioned: recursive and dependent types, expansion and reduction, composition and homomorphy of structures. Negative issues are speed, user friendliness, and the safety of systems.

### 2.17.4 Genericity

Of the various aspects of genericity, overloading is the one that is directly inspired by algebra. Known in particular in the context of algebraic operators, overloading allows to denote algorithms of a related functionality by the same identifier. Since overloading helps to avoid artificial function identifiers almost every computer algebra system and every programming language today makes use of operator overloading. Declarative languages generalize the concept to overloading of functions, procedures, and constants. Another aspect of genericity, parametric polymorphism, is realized in Lisp based computer algebra systems. Parametric polymorphism refers to functions that work uniformly on different types; recursive data types can be handled in an elegant way. A third aspect of polymorphism applies to computer algebra systems based on object-oriented languages, which model relations between mathematical structures by an inheritance hierarchy. Here, subtyping as a form of inclusion polymorphism makes functions applicable to objects not just of one class, but of any of its subclasses. Recently developed computer algebra packages in C++, finally, exploit the C++ template feature to develop functions for parameterized types the instantiation of which can be statically checked.

### 2.17.5 Modularization

It is obvious that computer algebra systems should be modularized the same way algebraic structures are. However there are differences to other program modules in computer science: in algebraic algorithms, subroutines have more often their own purpose, functionality, and re-usability. Consequently, the application interface of algebraic modules is more extensive than one might expect and the proper design of name spaces becomes significantly more sophisticated.

Only few computer algebra systems explicitly support some notion of modules. This might have been the main reason why large, and temporarily successful systems were discontinued; the lack of structuring prevented further growth, the passing on to the next generation of programmers, and might have even moved it beyond a user's intellectual comprehension and skills.

### 2.17.6 Parallel Implementation

Computer algebra is a treasure trove of non-trivial and time consuming algorithms which are particularly suited for distributed and parallel computing. An almost classic example is the family of algorithms based on the Chinese remainder theorem, and its numerous modular applications. They are often used to benchmark novel parallel computer systems.

The theory of parallel computing was applied to algebraic algorithms with considerable success. Still open is the question, whether general purpose computer algebra systems will benefit from parallelization to the same extent special purpose systems did, where remarkable speed-ups have become common.

In this context, we also refer to section 2.18 on parallel computer algebra systems.

## 2.17.7 Continuing Development of Computer Algebra Systems

In academe, development of computer algebra in the area of algorithms—which features a strongly experimental component—can only be sustained if source code is openly accessible. We recognize the obvious commercial success of closed systems. However, it is desirable to have tools—in similar way to UNIX—freely available, to promote continuing the development and adaptation of computer algebra techniques for novel applications and research in special fields. Hereby, the increasingly pressing problem of embedding computer algebra components into other program systems (such as logic programming, data bases, expert systems, as well as numeric and graphic systems) would come somewhat closer to a satisfactory solution.

Sibylle Schupp (Troy) and Rüdiger Loos (Tübingen)

## 2.18 Parallel Computer Algebra Systems

On a practical point of view, many computer algebra applications require a large amount of resources, both in computation time and memory space. Although, various algorithms in computer algebra have been exhibited that present a high degree of parallelism [von zur Gathen 1984; Gibbons and Rytter 1988; Bini and Pan 1994]. Hence, since the 80's, numerous researches have focused on parallel computer algebra [Roch and Villard 1997] (see conferences [Della Dora and Fitch 1989; Zippel 1992; Hong 1994; Hitz and Kaltofen 1997] dedicated to this subject).

This section give an overview on the implementation of the fundamental components required in a parallel computer algebra system: the operating system interface to manage parallel architecture, the mapping and scheduling techniques to ensure efficient execution of programs and the application programming interface to express parallel algorithms.

### 2.18.1 Parallel Architectures and Operating Systems Supports

The main challenge of parallel computing is to efficiently exploit independent hardware units (loosely coupled processors) with limited resources (memory, network) that are hierarchaly organized. A general parallel architecture is composed of SMP* machines (each of them has several processors with cache memory and a high speed bus to access to memory banks) connected by a network that permits access to remote memory. Due to the hierarchy, the costs of memory access are not uniform and usually define the *grain of the architecture*. This is a key-point relevant to the difference between sequential and parallel computers: in the latter there are potential large overheads in memory accesses**.

---

\* Symmetric multi-processors.
\*\* The performance of the network ranges from Ethernet link at 10Mbit/s to high-performance dedicated network as found in SGI Origin 2000 super-computer (about $1Gbit/s$).

Operating system (Unix) tools and libraries give an interface to the underlying parallel architecture. They provide three mains features. First, an access to the network using point-to-point communication (*e.g.* BSD socket, PVM, MPI) or collective communication (*e.g.* PVM, MPI, BSP). Also, they permit to exploit the processors inside a SMP machine using *lightweight processes* or *threads* (*e.g.* Posix Thread, OpenMP) with functions for the synchronization (*e.g.* semaphore, mutex, monitor). At last, a shared memory (an address space inside a Unix process shared by threads or through a distributed shared memory library) or a remote memory access library (Cray SHMEM, Scalable Coherent Interface, Memory Channel interface).

In addition to hardware costs, conversion of algebraic data structure from a representation to another one can introduce over-costs. For instance, such a conversion appears during serialization through a network. An other classical source of inefficiency is related to the synchronization of threads, e.g. to ensure exclusive access to a global data structure. As a consequence, due to intensive memory allocation in computer algebra programs, a good implementation of a memory manager for many threads is reached using hierarchical data structures that allow, in the average case, *per*-thread memory allocation [Küchlin 1990] without synchronization.

### 2.18.2 Parallel Execution: Mapping and Scheduling

To ensure independence between the program and the architecture, the scheduling of threads on computational units and the mapping of data on memory modules is handled by the system. Due to the dynamic evolvement of the load on architectures that may be shared by several users, and also irregularity of most computer algebra applications, this schedule is computed on-line.

On-line scheduling algorithms are mainly based on the management of a task queue: when a processor becomes idle, it takes a ready thread if any in the queue. When overheads related to the computation of the schedule and communications are neglected, such a popular strategy ensures provable performances on architectures with identical processors: at most at a factor two from the optimal [Graham 1969] or even asymptotically close to the optimal for fine-grain computations [Blumofe and Leiserson 1998]. If a thread can execute a blocking synchronization operation, a migration mechanism is required to eventually move a task that was blocked and becomes ready to an idle processor. Relying on the closure of the context of a thread, most systems do not implement this feature: once started on a processor, a thread is executed on it until its completion.

Basically, the runtime of most parallel systems schedules threads following this general scheme: from operating systems on SMP to high-level parallel languages such as MultiLisp [Halstead 1986], Linda [Ahuja et al. 1986] or Java and most computer algebra systems such as DSC [Díaz et al. 1995], PARSAC-2 [Küchlin 1990], MAPLE/LINDA–SUGARBUSH [Char et al. 1994], PACLIB/PD [Hong et al. 1993], STURM [Hong et al. 1994] GIVARO/ATHAPASCAN-1 [GALILE ET AL. 1998] and MUPAD [Heckler et al. 1997]. However, distinctions appear in the management of the queue, which is, both on a theoretical and practical point of view, a critical point for efficiency in memory and time.

Various heuristics have been developed in order to avoid memory space exhaustion [Arvind and Culler 1988; Blelloch et al. 1997]. If a correct serial execution order is known, managing a priority queue according to this order enables to bound the memory space with respect to the one required by the serial execution. Such a serialization restricts the programming model: overheads in memory cannot be avoided for a general multithreading model [Blumofe and Leiserson 1998] where synchronization operators, such as semaphores, are unpredictable and forbid a serial execution. Among programming models that enables such a strategy are fully-strict computations in Cilk [Blumofe and Leiserson 1998], nested computations in Nesl [Narlikar and Blelloch 1997] planar graphs [Blelloch et al. 1997] and non-blocking tasks in Athapascan [Galile et al. 1998].

Serialization versus parallelization is also related to time efficiency concerning the tuning of the granularity [Gautier and Roch 1994; Char et al. 1994]: how to decide if a computation should be splitted into subsequent concurrent threads or not ? Assuming the correctness of a depth-first serial execution, Cilk implementation answers this question by compiling thread creation into a local function call with a very small overhead; effective creation of a closure occurs only when a processor receives a work request from another one. Since idle time is only related to the parallel time on an unbounded number of processors, only few steal operations are performed for program having a small critical path. Cilk implementation [Joerg 1996] has proven the practical performances of such a strategy for fine grain fully-strict programs [Joerg 1996] that encompasses the fork-join programming model.

Beyond granularity, reducing synchronization between processors and remote data access is also crucial for time efficiency. To decrease concurrency on the queue of tasks, it is often implemented in a distributed way following a *work-stealing scheme*. Each processor manages its own queue and attempts to steal work from other processors only when it becomes under-loaded [Rayward-Smith et al. 1990].

The processor to steal can be chosen according to cpu and memory load informations of the machine. In DSC, this feature appeared useful to efficiently solve sparse linear systems where the matrix require a huge amount of memory [Díaz et al. 1995].

Furthermore, due to the magnitude of the ratio between local and remote memory access costs on a distributed architecture, additional informations about relationships between data access and threads can also be used. In parallel numerical computation, scheduling threads according to this information has been successfully used in linear algebra. PYRROS [Gerasoulis and Yang 1992] performs a clustering of tasks and has been used for matrix factorization; METIS [Karypis and Kumar 1998], based on data dependency graph partitioning, is widely used for sparse matrix-vector product. Those informations can be synthesized at run-time from a macro-data-flow analysis [Feo et al. 1990; Rinard 1998; Galile et al. 1998].

The computation of fine schedules introduces large overheads which may appear suited to only a specific class of programs. Then, various approaches are distinguished. Cilk argues for a restricted model of parallelism that enables a low-overhead runtime schedule with provable performances. Most languages, like Java, provide only local scheduling; global scheduling has to be implemented by the user. Between both, some languages like OPEN MP provide code annotations to tune the schedule.

### 2.18.3 Parallelism Expression and Languages

Basic parallel programming paradigms in computer algebra are based on explicit synchronization: message passing, fork/join. Current trend in parallel programming is to abstract these low level details by relying on a high level language where parallelism is explicit but independent from the target architecture; synchronizations become implicit using compilation techniques (data flow analysis,...).

The common idea behind the parallel programming paradigms is to abstract the need of writing explicit synchronization. Using a message passing interface (*e.g.* PVM, MPI, BSP), the descriptions of synchronization are given at the level of remote data accesses. Within the *fork/join* paradigm, the computations are splitted into tasks following the dependencies* between data accesses, thus the synchronizations are expressed at the level of call of function rather than data access. The *functional parallelism* paradigm is based on the fact that, in a purely functional program, no side-effects are performed; then, the detection of parallelism and synchronizations can be achieved automatically

Nevertheless, experiments have shown that it is difficult to have good efficiency using automatic parallelization. Therefore, code annotations have been defined in order to specify a good *grain of parallelism* for both data and computations. For instance, high-order functions enable to abstract standard control structures such as *data parallel* evaluation over collection (*e.g.* parallel map, parallel reduction). Other categories of annotations concern the description of potential concurrent function calls (*e.g.* Cilk, pD), the specification of shared or private data (*e.g* Athapascan, OpenMP) or the template for distribution of arrays in the data parallel approach (HPF).

Current trends in computer algebra are to build parallel algorithms on top of an existing parallel language with implicit synchronization and explicit parallelism. Such a language is built on both the operating systems tools to exploit parallel architectures and include mapping and scheduling mechanisms. Since the runtime of the language manages synchronization, intermediate shared data are automatically garbaged (data flow analysis)[Arvind et al. 1989; Feo et al. 1990]: synchronization is defined from completion of access. Then, the implementation of the completion algorithms, especially for distributed architectures, is a key point for efficiency.

Let us notice to conclude, that the performances of a parallel program are not only related to the efficiency of the underlying system (communication of

---

* given by the user

algebraic objects, memory manager, dynamic load-balancing) but also on the choice of the algorithm that should enforce data locality and cache reuse.

Thierry Gautier (INRIA, LMC-IMAG)
Hoon Hong (NCSU)
Jean-Louis Roch (LMC-IMAG)
Wolfgang Schreiner (RISC-Linz)
Gilles Villard (CNRS, LMC-IMAG)

## 2.19 Interfaces and Standardization

### 2.19.1 Interfaces to Word Processors

Most, if not all, of the current systems are able to produce output in TeX or LaTeX format, which in turn can be inserted into text documents by the usual cut-and-paste techniques. In addition, the document-style interfaces of both MAPLE and MATHEMATICA [Soiffer 1995] are now able to generate documents approaching publication quality. Two programs have also embedded MAPLE in a typesetting system: SCIENTIFIC WORKPLACE is a technical word processor which generates TeX and MATHOFFICE is an interface between MAPLE and MICROSOFT WORD.

A major development has been the adoption of XML[Bray et al. 1998] by many software vendors, and the support for MathML[Ion and Miner 1998] being introduced into many mathematical editors. REDUCE supports MathML as both an input and output language [Alvarez-Sobreviela 1998], and MATHEMATICA version 4 can export objects as MathML. Although mainly designed for presentation of mathematics, MathML does have a small number of symbols which describe semantic information and so it is possible to do a round trip from a computer algebra package into an editor and back again without any loss of information.

### 2.19.2 Graphics

In recent years, we have witnessed a meteoric improvement in the quality of graphics generated by computer algebra systems. However, while the rendering quality has improved, these systems do not yet exploit fully the potential offered by their ability to do symbolic analysis of the expressions which they are plotting ([Fateman 1992] discusses some of the deficiencies of current systems while [Avitzur et al. 1995] shows how some of these deficiencies can be addressed in practice). Some packages are now exploiting standards for platform-independent graphical primitives such as Open-GL and VRML [Walton and Dewar 1997] which also allow geometries to be shared with more specialised visualisation packages or exported across the worldwide web.

### 2.19.3 Interfaces to Numerical Software

In current computer algebra systems, there exist three different types of interfaces for linking them to numerical packages. The first type allows a user to generate expressions in another programming language, such as FORTRAN or C. The second type connects object modules, originally created by the compiler of an arbitrary programming language, to computer algebra systems. With the third type, numerical and symbolic software are seemlessly integrated into a common environment.

The state-of-the-art in code generation for external programming languages is currently represented by the GENTRAN system of REDUCE [Gates 1985, 1976; Borst et al. 1994]. It provides for the generation and segmentation of large expressions (the size of expressions in FORTRAN-77 is limited by the standard [Institute 1978] to twenty lines of 72 characters each, and C compilers are notorious for being unable to handle the large expressions often produced by computer algebra systems), as well as for translating REDUCE expressions, and generating entire programs from built-in program skeletons and templates. At the moment, GENTRAN offers the target languages FORTRAN-77 and -90, C, RATFOR and Pascal. It can also be linked to the SCOPE system [van Hulzen et al. 1989], which is capable of optimising code by common subexpression elimination and other reductions. Computing symbolic derivatives, Jacobians and Hessians is a widespread application of computer algebra systems, a process which is tedious and error prone if done by hand. In this case, symbolic optimisations lead to spectacular improvements in efficiency. All systems include some method of generating FORTRAN (and increasingly also C) code but they are of varying quality for use in real applications which tend to involve very large expressions.

GENTRAN was originally written for MACSYMA [Gates and Wang 1984], and an updated version has been ported to Common Lisp [Keady 1991]. There exist similar packages for AXIOM [Dewar 1994], MAPLE [Gomez 1990] and MATHEMATICA [Kant et al. 1990]. All these systems can be used to generate expressions and program fragments, and some can be used to create complete, runnable programs, which can be executed on appropriate machines afterwards. In fact, several packages have been described, which take as input a user's problem description, and return a FORTRAN program for solving the given problem (see, e.g., [Wang 1986] and [Barbier et al. 1990, 1992]).

The second sort of approach to linking symbolic and numerical systems is through a direct link between two such systems at runtime. This is most easily achieved by starting two separate processes and allowing them to communicate via sockets, although with modern operating system technology it is increasingly the case that dynamic linking of shared objects is both practical and more efficient. Currently MATHEMATICA, MAPLE and MuPAD offer a general API for this kind of application, however these are often clumsy to use because of the need to convert data to and from the computer algebra system's native format. The OpenMath project [Abbott et al. 1996; Dalmas et al. 1997] seeks to address this through creating a standard format for mathematical software packages to use for input and output of objects. In future it is unlikely that computer algebra systems will have to develop their own strategies for inter-process communication since operating systems are increasingly providing standard mechanisms for this.

The most advanced approach usually builds on this second kind of facility, and consists of developing high level interfaces to numerical libraries, particularly the NAG library [NAG Ltd. 1999]. Implementations range from the most simple form of a FORTRAN-like interface (e.g., MATHEMATICA's *InterCall* system) to rather powerful interfaces, where the user does not know that a FORTRAN program is actually solving the problem (see the IRENA interface between REDUCE and the NAG library [Dewar 1991]). A comparison of several systems is given in [Broughan et al. 1991]. Since that article was written an interface to the NAG library has been incorporated into AXIOM. Another system which combines both symbolic and numerical packages (albeit from the opposite viewpoint of the other examples discussed here) is Matlab which, in addition to its own numerical and graphical functionality, offers a toolkit which provides access to a limited subset of MAPLE. The most recent version of MAPLE (version 6) includes over a hundred linear algebra routines from the NAG Library which have been specially adapted to run using both the machine arithmetic and arbitrary precision software floating point. MATHEMATICA 4 implements a number of numerical methods directly.

At an even higher level, some work has been done in REDUCE [Dewar 1992] and AXIOM [Dupée and Davenport 1996] which uses the power of the host algebra system to simplify the use of a numerical library by creating an "intelligent" interface which can analyse the problem in hand and select a good strategy for solving it.

### 2.19.4 User Interfaces

Traditionally, computer algebra systems used to have rather rudimentary user interfaces; an interpreter would process commands, and display the resulting mathematical formulas in an ASCII representation, which has the big advantage of being portable but can be ugly and hard to read. More and more systems now make use of bitmaps for improved graphical representation of the expressions which they produce, and some such as MATHEMATICA allow for the input expressions to be typeset as well. The widespread availability of components based on XML and MathML should make the production of high quality user interfaces much easier in future.

There is considerable interest in developing ergonomic input methods for computer algebra systems. For example, [Rimey 1990] describes an advanced formula editor which can be used as a front-end to an experimental system. The commercial package Theorist (now called MathView) provides a similar function for the Maple system. Additionally, experiments to recognise hand-writing as input data are underway (e.g., by the team led by Hotz in Saarbrücken).

### 2.19.5 General Problem-Solving Environments

[Kajler 1990, 1992] suggests an approach which integrates several computer algebra systems into a uniform environment including formula editors, graphics, and other utilities. This project tried to address issues which are generally neglected by other systems, such as representing large expressions in a user-friendly way. This is one of many projects which has attempted to develop some kind of integrated problem-solving environment encompassing symbolic, numeric and graphical tools. In fact there are many similar projects throughout the field of scientific computing; a proper treatment is beyond the scope of this article but examples can be found in [Ford and Chatelin 1987] and [Gaffney and Houstis 1992].

### 2.19.6 Standardisation

All the interfaces and environments described in this article have used ad-hoc techniques to pass objects and data between their constituent systems. In addition to those described, there are also a number of interfaces between a computer algebra system and *another* computer algebra system. For example GB [Faugère 1994b], which (with its successor FGB) is one of the fastest systems for computing Gröbner bases, uses AXIOM or MuPAD to provide a user interface and other kinds of general functionality. In many ways this is an example of how computer algebra technology could develop in the future: a few general systems using a large number of fast, highly-specialised servers to perform computations. For this to be practical we need to develop standard mechanisms for linking systems together.

Any standards will need to address two issues: the *transport mechanism* to be used to move data from one machine to another, and the *data representation* to be used. In the first case we should look to adopt the software component technology emerging in the latest generations of operating systems rather than inventing our own, so that the kind of "plug and play" paradigm already commonplace for office software can become widespread for scientific software.

Data representation, on the other hand, does require a solution designed specifically for computer algebra (and indeed for other kinds of mathematical software as well). There are many reasons for this: the objects manipulated by a computer algebra system are often quite complicated, notation can vary and the precise semantics associated with an object (such as an integral for instance) can differ quite significantly from one system to another. Thus there is a requirement for a framework in which we can describe both the abstract semantics and concrete representation of an object. One general mechanism that has been developed to address these issues is MP [Gray et al. 1994; Bachmann et al. 1997] which has been used to connect special purpose packages for polynomial manipulation to MATHEMATICA [Bachmann et al. 1995]. The biggest activity in this area is the OpenMath project [Abbott et al. 1996; Dalmas et al. 1997] which includes developers of AXIOM, MAPLE and REDUCE as well as representatives of the XML and electronic publishing communities. This project has already developed a framework to address the data representation issues outlined above to meet the requirements not just of computer algebra, but also of publishing in both its traditional and electronic forms.

Michael Dewar (NAG Ltd., UK)

### 2.19.7 MathML

MathML [Ausbrooks et al. 2001] is an XML representation for mathematical objects, allowing expressions to be stored in databases, transmitted between applications and operated upon by programs. MathML can be used to express mathematical content in web pages and digital libraries, and has become an accepted form for input and output of computer algebra systems.

With MathML available in a broad range of software tools, we can now share mathematical data in new ways. For example, with a judicious choice of software, one can now cut an expression from a web page, paste it into an E-mail message to a colleague, who can then use it in a computer algebra system, and paste the result into a patent application.

MathML provides a vocabulary for such things as identifiers, numbers, operators, grouping *etc.* There are two broad classes of constructions: The set of elements that describe the appearance, or notation, of an expression form what is called *presentation MathML*. The elements that describe the meaning, or semantics, of an expression are known as *content MathML*.

**Presentation MathML**

MathML has a quite complete set of elements to describe mathematical notation. There are primitives for various kinds of tokens, and others to describe relative position and grouping of subexpressions. Together, these notational primitives form a subset known as "presentation" MathML. A few examples of presentation MathML are given in Figure 1.

To illustrate the basic concepts, we examine the presentation MathML for the expression $x^2 + y^2$:

```
<mrow>
    <msup> <mi>x</mi> <mn>2</mn> </msup>
    <mo>&times;</mo>
    <msup> <mi>y</mi> <mn>2</mn> </msup>
</mrow>
```

Here the `<mi>` elements give math identifiers (variables or parameters), the `<mn>` denote numbers and the `<mo>` denotes an operator. The `<msup>` elements express superscripts and the `<mrow>` is used for a horizontal sequence. The form `&times;` is a named entity that expands to the Unicode character "×" (U+00D7). Note that there is no meaning ascribed to this expression. For example, $x^2$ is merely a superscripted quantity. There is no implication that it is a power—it could equally well denote the second component of a contravariant vector, for example.

$$\frac{a \pm \sqrt{b}}{c}$$

```
<mfrac>
  <mrow><mi>a</mi><mo>&PlusMinus;</mo><msqrt><mi>b</mi></msqrt></mrow>
  <mi>c</mi>
</mfrac>
```

$$\begin{bmatrix} a & b \\ c & d \end{bmatrix}$$

```
<mfenced open="[" close="]"> <mtable>
    <mtr><mtd><mi>a</mi></mtd> <mtd><mi>b</mi></mtd></mtr>
    <mtr><mtd><mi>c</mi></mtd> <mtd><mi>d</mi></mtd></mtr>
</mtable> </mfenced>
```

$$\sum_{i=1}^{n} e^i \omega_i$$

```
<mrow>
    <msubsup>
        <mo>&sum;</mo>
        <mrow> <mi>i</mi> <mo>=</mo> <mn>1</mn> </mrow>
        <mi>n</mi>
    </msubsup>
    <mrow>
        <msup><mi>e</mi><mi>i</mi></msup>
        <mo>&InvisibleTimes;</mo>
        <msub><mi>&omega;</mi> <mi>i</mi></msub>
    </mrow>
</mrow>
```

$$\lim_{h \to 0} \frac{f(t+h)}{h}$$

```
<mrow>
   <msub>
      <mo>lim</mo>
      <mrow><mi>h</mi> <mo>&rarr;</mo> <mn>0</mn></mrow>
   </msub>
   <mfrac>
      <mrow>
          <mi>f</mi>
          <mo>&ApplyFunction;</mo>
          <mfenced><mi>t</mi><mo>+</mo><mi>h</mi></mfenced>
      </mrow>
      <mi>h</mi>
   </mfrac>
</mrow>
```

**Fig. 1.** Presentation MathML examples

When writing, or generating, presentation MathML it is important to use sufficient markup used to capture the syntactic structure of the expression. In effect, all subexpressions should be explicitly grouped with <mrow>s. For example, to express $a = b + c$ one would use

```
<mrow>
   <mi>a</mi>
   <mo>=</mo>
   <mrow><mi>c</mi><mo>+</mo><mi>d</mi></mrow>
</mrow>
```

and not

```
<mrow><mi>a</mi> <mo>=</mo> <mi>b</mi> <mo>+</mo> <mi>c</mi></mrow>.
```

This way, line breaking and sub-expression selection can be handled correctly.

## Content MathML

MathML provides facilities to describe the meaning of mathematical expressions. The subset designed for this purpose is known as "content" MathML. Content MathML has a set of built-in tags to express the concepts which occur in elementary mathematics, up to the level corresponding approximately to the last year of secondary school or first year of university. More advanced concepts are expressed using markup with external references.

Content MathML expressions consist typically of expressions with operators applied to arguments, and are thus reminiscent of Lisp S-expressions. Supposing the presentation example above were intended as a sum of two squares, it could be represented in content MathML as

```
<apply>
   <plus/>
   <apply> <power/> <ci>x</ci> <cn>2</cn> </apply>
   <apply> <power/> <ci>y</ci> <cn>2</cn> </apply>
</apply>
```

Here the <ci> and <cn> give content markup for identifiers and numbers respectively. The <plus/> and <power/> elements denote operators. Expressions are formed from these leaves by giving function application elements with <apply>.

A few examples illustrating content MathML are given in Figure 2.

## Characters and Symbols

Mathematical expressions can make use of a broad range of special symbols. MathML uses the full Unicode character set [The Unicode Consortium 2000] to provide the numerous symbols and technical characters required.

Sometimes it is not convenient to work directly with the full range of Unicode characters. This is the case when using tools which do not support Unicode or when one must embed MathML data in a non-Unicode format. For this purpose MathML provides more than 2000 named entities for special characters. These

$$\sin^2\theta$$

```
<apply>
   <power/>
   <apply> <sin/> <ci>&theta;</ci></apply>
   <cn>2</cn>
</apply>
```

$$\log^{-1} = \exp$$

```
<apply>
   <eq/>
   <apply><inverse/><log/></apply>
   <exp/>
</apply>
```

$$\int_a^b f(t)\,dt$$

```
<apply>
   <int/>
   <bvar><ci>t</ci></bvar>
   <lowlimit><ci>a</ci></lowlimit>
   <uplimit><ci>b</ci></uplimit>
   <apply> <ci>f</ci> <ci>t</ci> </apply>
</apply>
```

$$\forall x, x \in \mathbb{R}, x > 1 : x^2 > x$$

```
<apply>
   <forall/>
   <bvar><ci>x</ci></bvar>
   <condition>
      <apply>
         <and/>
         <apply><in/> <ci>x</ci> <reals/></apply>
         <apply><gt/> <ci>x</ci> <cn>1</cn></apply>
      </apply>
   </condition>
   <apply>
      <gt/>
      <apply><power/><ci>x</ci><cn>2</cn></apply>
      <ci>x</ci>
   </apply>
</apply>
```

**Fig. 2.** Content MathML examples

allow Unicode characters to be referred to by mnemonic names, *e.g.* `&alpha;` for $\alpha$, and for any MathML expression to be written in ASCII. The MathML entities include alphabets used in mathematics (*e.g.* Greek, Cyrillic, Hebrew, fraktur, script and open-face), as well as numerous brackets, operators and other symbols.

The Unicode character set is organized in a set of *planes* of $2^{16}$ points. The *Basic Multilinugal Plane* (BMP) consists of the characters with values from 0 to $2^{16} - 1$. At the time of writing, many programs do not properly handle Unicode characters outside the BMP. While most characters for symbols used by MathML lie in the BMP, certain mathematical alphabets (*e.g.* open face, script and fraktur) lie in other planes. To give access to these alphabets in applications that handle only BMP values, the `mathvariant` attribute may be used. For example, one can give a script A identifier as `<mi mathvariant="script">A</mi>`. Unlike using a numeric or named entity (`<mi>&#x1D94C;</mi>` or `<mi>&Ascr;</mi>`), the `<mi>` with a `mathvariant` attribute is not automatically expanded to use U+1D94C.

### Annotations

It is often the case that MathML objects can have additional associated information. For example, if a computer algebra system generates MathML output, it may be desired to associate the system's original expression with the MathML. The `<semantics>` element is used for this purpose. The first child of this element is an expression to be annotated, and the second and any subsequent children are annotations either in textual or XML form. Figure 3 gives an expression with annotations providing meanings in Maple, TeX and OpenMath [Caprotti et al. 2000; Dalmas et al. 1997; Naylor and Watt 2001; Caprotti et al. 2000].

### Combining Presentation and Content

It is not uncommon to work with both presentation and content for the same mathematical expression. This can be done with a `<semantics>` element, giving either the presentation or the content as the first child and the other as the annotation, as shown in Figure 4. Joining a presentation expression and content expression in this way gives what is known as *top-level parallel markup*.

In many applications, it is desirable to be able to select subexpressions and to be able to find both their content and presentation markup. Top-level parallel markup is insufficient for this purpose. MathML provides `id` and `ref` attributes which may be used to cross-reference the subexpressions of content and presentation trees. This gives what is known a *fine-grained parallel markup*, as shown in Figure 5.

### Current Status

MathML has been an official Recommendation of the World Wide Web Consortium (W3C) since March 1998 as Version 1.0, and since February 2001 as Version 2.0.

```
<semantics>
   <mrow>  <mi>x</mi>  <mo>&times;</mo>  <mi>y</mi>  </mrow>

   <annotation encoding="Maple"> x * y </annotation>
   <annotation encoding="TeX"> x \times y </annotation>
   <annotation-xml encoding="OpenMath">
      <OMOBJ xmlns="http://www.openmath.org/OpenMath">
        <OMA>
           <OMS cd="arith1" name="times"/>
           <OMV name="x"/>
           <OMV name="y"/>
        </OMA>
      </OMOBJ>
   </annotation-xml>
</semantics>
```

**Fig. 3.** Example of annotations for alternative encodings

```
<semantics>
    <mrow>  <mi>a</mi>  <mo>+</mo>  <mi>b</mi>  </mrow>
    <annotation-xml encoding="MathML-Content">
        <apply> <plus/> <ci>a</ci> <ci>b</ci> </apply>
    </annotation-xml>
</semantics>
```

**Fig. 4.** Example of top-level parallel markup

```
<semantics>
    <mrow id="G1">
        <mi id="G2">a</mi>
        <mo id="G3">+</mo>
        <mi id="G4">b</mi>
    </mrow>
    <annotation-xml encoding="MathML-Content">
        <apply xref="G1">
            <plus xref="G3"/>
            <ci xref="G2">a</ci>
            <ci xref="G4">b</ci>
        </apply>
    </annotation-xml>
</semantics>
```

**Fig. 5.** Example of fine-grained parallel markup

At the time of writing, MathML can be imported and exported from major computer algebra systems, is supported natively or via plug-ins in the most popular web browsers, and handled by certain editors. Of the computer algebra systems, MathML may be imported or exported by both **Maple** (version 8 and higher) and **Mathematica** (version 4 and higher). **Maple** places greater emphasis on Content MathML while **Mathematica** emphasizes presentation MathML. The browsers **Amaya**, **Mozilla** 1.0 and **Netscape** 7.0 PR1 support MathML natively (though **Amaya** supports only Presentation MathML). IBM's **techExplorer** and Design Science's **MathPlayer** can both be used to display MathML in **InternetExplorer** 5.5 and later. Used naïvely, some of the browsers/extension combinations require browser-specific markup (`<object>`, `<applet>`, or `<embed>` tags) to view pages containing MathML. It is however possible to write *browser-independent pages* by making use of the MathML universal style sheet [W3C Math Working Group 2002] from the W3C Math Working Group. With this, an XHTML [Pemberton et al. 2000] page containing MathML would appear as shown in Figure 6. The two key items that must be included are the `<?xml-stylesheet...?>` processing instruction and the `xmlns` attribute on the `<math>` tag.

```
<?xml version="1.0"?>
<?xml-stylesheet type="text/xsl"
                 href="http://www.w3.org/Math/XSL/mathml.xsl"?>
<html xmlns="http://www.w3.org/1999/xhtml">
  <head>My Page</head>
  <body>
    <h1>Example</h1>
    <math xmlns="http://www.w3.org/1998/Math/MathML">
        <mrow> <mi>a</mi> <mo>+</mo> <mn>b</mn> </mrow>
    </math>
  </body>
</html>
```

**Fig. 6.** Use of W3C universal MathML style sheet

MathML is now supported by a wide range of software, and has found many applications outside of the original context of math for web pages. For example, it is used now in web-services for computer algebra systems and as the archival form for mathematics in all United States patents.

Stephen Watt (London Ontario)

## 2.20 Hardware Implementation of Computer Algebra Algorithms

The connection between computer algebra and hardware lies mostly in the use of computer algebra for the design or improvement of hardware with algebraic methods. It can even be found in the context of optical computing. Sometimes the so designed hardware can be used to speed up computer algebra calculations, but hardware specially built for computer algebra aplications is seldom found.

**Hardware for computer algebra applications**

In contrast to numerical analysis, there have been only sporadic attempts to implement computer algebra algorithms in hardware. Among the historically successful projects are D. H. Lehmer's machines for number theory, the *Analytik* constructed by Russian designers *(see Sigsam Bulletin, number 30, page 9)*, and machines for finite field arithmetic.

In a more general sense, the *Symbolics* computer which uses a second processor for on-line support of memory management, also belongs to this category.

Cryptography, coding theory, and digital signal processing are areas in which major efforts to realize basic algorithms of computer algebra on the chip level are underway. For example, arithmetic in finite fields of characteristic 2 has been intensively investigated by means of polynomial, as well as normal bases, and practically implemented [Geiselmann and Gollmann 1990]. Research into aspects of implementation in return, yielded several theoretic results, such as the complete classification of normal bases [Gao and Lenstra Jr. 1992]. Several implementations of long integer arithmetic, mostly based on the Karatsuba algorithm, originated as part of RSA hardware implementations. Also arithmetic of elliptic curves has been realized by CMOS technology. Various fast transforms of signals, like the fast Fourier transform in Fermat rings, have been laid out in parametric VLSI designs [Creutzburg et al. 1992].

**Computer Algebra as a Tool in Specification and Design of Algorithms and Architectures**

At the IAKS, a VLSI-Design tool was implemented, incorporating computer algebra systems and VLSI-Design tools into one homogeneous environment. For applications in signal processing, CMOS-VLSI-Implementations can be derived. In addition to classical design tools, the so called IDEAS environment (Intelligent Design Environment for Algorithms in Signal processing) is able to optimize algorithms and to synthesize architectures by incorporating algebraic optimization algorithms. These algorithms are used to determine good bases for finite field arithmetic or the inherent parallelism in signal transformations. A remarkable feature of the system is that the user is allowed to specify the problem in algebraic terms. Recent research is concerned with the extension of the design method to generate parallel algorithms for symmetric multiprocessor architectures.

Algebraic algorithms are used to analyze the problem i.e. factoring polynomials, computing minimal polynomials, investigation of conjugation properties, optimization of bases and more. Equipped with this structural knowledge, a process called *technology refinement* is started. The question to be answered is, which architecture is good for a given implementation technology. A good algorithm for a sequential machine may be a bad one for hardware implementation. To handle this question, the concept of parametrization is important. By inserting parameters into an equation, it is possible to generate generic implementations and to specialize the parameters on a quite low level of abstraction. Specialization on a low level of abstraction is advantageous, because information concerning the implementation technology can be used comparably easily. To give a simple example, consider the polynomials $x^3 + x + 1$ and $x^3 + x^2 + 1$ over $GF(2)$. Both are irreducible, thus each of them can be used to construct a three dimensional extension of $GF(2)$. A reformulation of the problem may read like this: Let $t \in GF(2)$ a parameter and $f(x;t) := x^3 + (1-t)x^2 + tx + 1$, a polynomial, then any value for $t$, a specialization, results in an irreducible polynomial. Symbolic computations can be used to characterize e.g. multiplication matrices without specializing the parameter $t$. As a consequence, the resulting generic matrix describes a set of multiplications. According to technological parameters like gate sizes, specialization algorithms are used to optimize the gate count or chip size.

**Computer Algebra for the Design of Optical Setups**

At the IAKS research is done concerning multivariate rational decomposition. The question which linear transforms are realizable with three arbitrary diffractive elements led to a large polynomial equation not solvable with standard methods. A functional decomposition solved the problem by reducing it to a system of linear equations and a system of binomial equations both of which can be solved efficiently. This solution of a design problem for diffractive systems motivated a more general approach to functional decomposition using intermediate fields.

Thomas Beth, Andreas Klappenecker, Jörn Müller-Quade, Armin Nückel
(Karlsruhe), Rüdiger Loos (Tübingen), and Frank Schaefer-Lorinser
(Darmstadt)

# 3 Applications of Computer Algebra

Applications of computer algebra range over the entire spectrum of research, development, production, and education. Computer algebra problems arise in industry, commerce, software engineering, and also in banking and insurance applications, although sometimes hidden. We compiled several interesting applications which are exemplary for the most important areas.

Classical areas of computer algebra applications are physics — see page 163 – and mathematics — see page 195. However, more and more methods and systems of computer algebra are also utilized in computer science, — see page 217 – in engineering, — see page 221 – and in other natural sciences — see page 242.

A very important application area of computer algebra is education. More and more computer algebras are used for teaching — both in schools and universities — see page 244.

## 3.1 Physics

Physics is one of the classical and traditional areas of computer algebra application, most likely being the reason why physicists — given the need to deal with mathematical formulas and relations of highest complexity — always participated at the forefront in the practical development of computer algebra. Nowadays computer algebra is used in essentially all areas of physics. Therefore below we didn't try to cover the whole of physics, rather we picked some typical applications in physics in order to give the reader an idea of how to select the suitable computer algebra system which may solve his or her own problem in physics. Moreover, the merits and the limitation of computer algebra systems may also come into focus.

- Elementary Particle Physics in 3.1.1.1: In this field, computer algebra tools are indispensable for generation, evaluation, and summing up of Feynman integrals. J. Fleischer gives a general overview, whereas M. Steinhauser concentrates on the strong interaction, i.e., on quantum chromodynamics (QCD), and G. Weiglein on the electroweak interaction. J. Vermaseren sketches the use of his Mincer Form-package for the evaluation of certain loop diagrams.
  - Fleischer
  - Steinhauser
  - Weiglein
  - Vermaseren
- Gravity: In gravity and general relativity, computer algebra was used right from its beginnings. C. Heinicke and F.W. Hehl survey computer algebra applications in the whole field, whereas I. Kotsireas turns his attention to one important problem: The N-body problem of Newtonian mechanics with gravitational interaction.
  - Heinicke & Hehl

– Kotsireas
- computer algebra-Systems for Differential Geometry and Applications: Differential geometry plays a fundamental role in physics such as in mechanics, in field and relativity theory, and, in particular, in gauge field theory, in string theory etc.. Therefore in this section computer algebra systems are discussed which support differential geometric applications in physics (and also in mathematics, of course). E. Schrüfer describes how Excalc can directly handle exterior differential forms. Yu. Obukhov and S. Tertychniy explain the merits of the GRG and the $GRG_{EC}$ computer algebra packages, respectively, which can be used in general relativistic field theories. Tools for the manipulation of spinors are available and many special functions.
  – Schrüfer
  – Obukhov
  – Tertychniy
- Differential Equations in Physics: As two typical examples we selected T. Wolf's work on the determination of Killing tensors and conserved quantities for partial differential equations by means of his packages Crack and ConLaw and G. Baumann's use of his Mathematica based MathLie package for solving differential equations via the method of Lie symmetries.
  – Wolf
  – Baumann

Friedrich W. Hehl (Cologne)

### 3.1.1 Elementary Particle Physics

**3.1.1.1 Computer Algebra in Elementary Particle Physics** Applications of computer algebra in particle physics are mainly concerned with calculations in the framework of perturbation theory for Quantum Field Theories (QFT). The $SU(3) \times SU(2) \times U(1)$ gauge field theory is the so called 'Standard Model' (SM) and unifies the known interactions apart from gravity: The $SU(3)$ group generates the 8 massless 'gluons' which mediate the interactions among the quarks (i.e. the 'strong interaction', quantum cromodynamics, QCD), the $SU(2) \times U(1)$ 'electroweak' part unifies the electric (QED) and 'weak' interaction. The corresponding gauge bosons are the photon and the heavy gauge bosons $W^\pm$ and $Z$, mediating the weak interaction (like $\beta$-decay).

The perturbation theory is expressed in terms of 'Feynman diagrams' which are built from 'vertices', describing 'point-interactions' of the various particles, and their connections in terms of particle 'propagators'. Their representation is usually done in 'momentum' space, i.e. the space of four-dimensional momentum vectors. The 'order' of the perturbation theory can be identified with the number of closed loops in the diagrams, each 'loop' corresponding to a (fourdimensional) integration in momentum space. The lowest order (Born diagrams) has no loops.

The precision in high energy accellerator experiments, with the Large Electron Positron Collider (LEP) at CERN in Geneva and others, e.g., is so high (of the order of 1 part per thousand for many observables) that on the side of the theory a corresponding accuracy in precision is needed in order to verify the SM or to find 'new physics' in case deviations would show up. This is the reason why higher order calculations are needed where quite often thousands of diagrams contribute and even one-loop calculations for multiple particle production have to be performed. This requires from the very beginning a large amount of automation, and several groups have provided packages for such calculations based on computer algebra. For an extensive review see [Harlander and Steinhauser 1999].

On of the first implementations of computer algebra programs came from particle physics. In 1967, Veltman wrote his once widely used program SCHOON-SCHIP [Veltman 1967; Strubbe 1974] of which subsequently many ideas were taken over in FORM[Vermaseren 1991]. This latter one is nowadays the most widely used program for Feynman diagram evaluation in particle physics since it is tailored for this particular purpose. Nevertheless, also programs like MACSYMA — see 4.1.4 and [Macsyma 1977], REDUCE — see 4.1.9 and [Hearn 1993] — and MATHEMATICA — see 4.1.7 and [Wolfram 1998] — have partly their roots in particle physics.

Technically the calculations are to be performed in the following steps:

**1. Generation of the diagrams.** Here two aspects are important: If thousands of diagrams are to be calculated (for the two-loop anomalous magnetic moment of the moon, 1832 diagrams contribute in the SM), one will not even be able to investigate each of them separately, i.e. one would produce 'input' for computer algebra programs like FORM, e.g.. These programs provide the momentum representation of each diagram in terms of the Feynman rules.

Nevertheless, to investigate questions like what type of topologies occur, how to adjust the integration momenta in an optimal way, to see which diagrams might be infrared divergent etc., also a graphical representation of the diagrams is needed. This also guides the physical intuition.

**2. Simplifications.** Usually one projects out certain scalar amplitudes. This mostly results in scalar products of external and/or integration momenta, which one tries to cancel against scalar propagators in the denominator. Such a cancellation, however, is not always possible and there remain 'irreducible' numerators. For each process one puts the external particles on their mass shell by introducing corresponding conditions for their external momenta.

**3. Integration.** The integration of multi-loop integrals is very complicated in general and many techniques have been developed for this purpose. First of all the integration in most cases is performed in $d = 4 - 2\varepsilon$ dimensions ('dimensional regularization'), the infinite parts of the integrals manifesting themselves in poles in $\varepsilon$. After 'renormalization' all these poles must cancel.

For on-shell selfenergy diagrams with one non-zero mass, special packages have been developed, one [Fleischer and Tarasov 1992] especially for QED and QCD diagrams and another one [Fleischer and Kalmykov 2000] for diagrams occurring in electroweak interactions, where one has a larger variety of masses. The latter package also provides a basis for an expansion in mass differences.

In general tensor integrals and integrals with irreducible numerators can be represented as a combination of scalar ones with a higher space-time dimension. They are reduced to basic integrals by recurrence relations in the dimension (reducing the dimension in steps of 2) and in indices (powers of scalar propagators) [Tarasov 1996a,b].

Finally scalar 'master integrals' are evaluated by various expansion techniques (w.r.t. small and large masses and momenta and differences of masses, e.g.) [Tkachov 1984, 1993; Chetyrkin 1988; Smirnov 1988, 1990]. The results of expansions are quite often numerically improved by the application of Padé approximants [Fleischer and Tarasov 1994].

All these methods require large computer algebrapackages. Several packages have been developed with different areas of applicability. For example, FEYNARTS/FEYNCALC — see 4.3.8 and [Küblbeck et al. 1990; Mertig et al. 1991a] — are MATHEMATICA packages convenient for various aspects of the calculation of radiative corrections in the SM. In particular FEYNARTS also offers the possibility to produce graphics of the diagrams up to two loops for self-energies and vertices. There are several FORM packages for evaluating multiloop diagrams, like MINCER — see 3.1.1.4 and [Larin et al. 1991] —, and a package [Avdeev 1996] for the calculation of 3-loop bubble integrals with one non-zero mass. Other packages for automation are GRACE [Ishikawa 1993] and COMPHEP [Boos and et al. 1994], which partially perform full calculations, from the process definition to the cross-section values.

A somewhat different approach is pursued by XLOOPS [Brücher and et al. 1997; Frink et al. 1998]. A graphical user interface makes XLOOPS an 'easy-to-handle' program package, but it is mainly aimed at the evaluation of single diagrams. To deal with thousands of diagrams, it is necessary to use special techniques like databases and special controlling programs. In [van Ritbergen and et al. 1995], for evaluating more than 11 000 diagrams, the special database-like program MINOS was developed. It calls the relevant FORM programs, waits until they finished, picks up their results and repeats the process without any human interference.

A further program DIANA (DIagram ANAlyser) [Tentyukov and Fleischer 2000], written in C, for the automatic Feynman diagram evaluation is available and is presently extended in its graphical part.

The evaluation of the QED radiative corrections to the anomalous magnetic moment of the electron ($g_e - 2$) is one of the greatest triumphs of theoretical physics. This has always been one of the most accurate tests of QED since its very beginning in the late 40's. Now one is reaching the level of 1 part per billion (ppb) in precision, the experimental error being at present of order 4 ppb. In [Laporta and Remiddi 1996] a closed analytic form for the contribution of the three-loop non-planar 'triple-cross' diagrams has been obtained and thus the QED calculation of the anomalous magnetic moment of the electron has been completed in this order. This result has also been obtained by intensive use of computer algebra, i.e. by FORM and the nowdays outdated ASHMEDAI.

In the next 4-loop order, there contribute 495 diagrams, which are evaluated numerically (for details see [Kinoshita 1990]).

Jochem Fleischer (Bielefeld)

**3.1.1.2 Computer Algebra and its Applications in QCD** The main task of modern particle physics is the exploration of the fundamental laws of nature at very small distances. The experience of the last two decades has shown that a tight connection between experiment and theory is important in order to reach this aim. A recent example is the discovery of the top quark — the heaviest particle known so far. Its direct production at the Large-Electron-Positron-Collider (LEP) at CERN is not possible due to its large mass. Nevertheless, it was possible to use precision measurements in combination with calculations involving quantum corrections in order to predict the mass $M_t$ of the top quark. Finally the top quark was discovered in the expected mass range at TEVATRON (Chicago), a proton-anti-proton collider. This example makes clear that both the experimental methods and the theoretical tools have to be further developed.

At present the interaction of the fundamental particles are contained in the so-called Standard Model of elementary particle physics. They are described by the gauge group $SU(3) \times SU(2) \times U(1)$. Here, $SU(2) \times U(1)$ constitutes the electroweak sector of the model with three massive gauge bosons ($W^\pm, Z$) and a massless one, namely the photon which is already known from Quantum Electrodynamics (QED). The group $SU(3)$ generates the strong interaction among the quarks and eight massless gauge bosons — the gluons. This part is in general referred to as Quantum Chromodynamics (QCD) with the coupling constant $\alpha_s$.

At the moment the majority of the calculations are based on perturbation theory in the coupling constant. There is a very intuitive approach representing the perturbative expansion by so-called Feynman diagrams which have a one-to-one translation to mathematical expressions. The evaluation of quantum corrections introduces closed loops which themselves manifest in integrations over the momenta running in them. Nowadays it is almost impossible to evaluate these expressions by hand as they are in general rather large and also very complicated. Thus modern particle physicists rely to an increasing extend on the use of computers accompanied with powerful computer algebra programs. This article considers as an example the computation of higher order quantum effects induced by the strong interaction between quarks and gluons.

Almost 20 Million hadronic events have been collected at the $Z$ boson resonance at LEP which lead to an impressive precision of quite a lot of observables especially of the properties of the $Z$ boson itself. An important quantity is the decay of the $Z$ boson into bottom quarks mainly because there the top quark appears as a virtual particle in the loop diagrams giving rise to corrections which grow quadratically with its mass. Actually those kind of corrections have been very important in the indirect determination of $M_t$ mentioned above.

Quite a lot of observables are significantly affected by quantum effects arising from QCD. Unfortunately we will not be able to mention all of them in this article. In the following we will describe by means of the example of the partial width $\Gamma(Z \to b\bar{b})$ how it was possible to evaluate higher order effects exploiting state-of-the-art computer algebra programs [Harlander et al. 1998].

The first problem one encounters in the evaluation of higher order quantum correction is the large number of diagrams. It may easily happen that it is of the order of a few thousand. In the case of $\Gamma(Z \to b\bar{b})$ they amount to 69 which is still a moderate number. Nevertheless it is saver to use a generator in order not to forget a diagram or some symmetry factors.

Once the amplitudes of the diagrams is available containing all the relevant information about the particles involved and the distribution of the loop and external momenta, programs like AWK or PERL can be used in order to bring the output in the desired form. In our case the output is transformed into a format which is readable by MATHEMATICA— see 4.1.7 and [Wolfram 1988]. The amplitudes are read one-by-one and the given information is used in order to determine the topology of the diagrams which becomes essential in the evaluation of the integrals. Also the Feynman rules — the translation to the mathematical expressions — are inserted by a MATHEMATICA program and administrative files are generated which rule the calculation and take care that all diagrams are computed.

In general it is not possible to solve the momentum integrals exactly. Thus one has to rely on approximations. One possibility it to compute the integrals (at least partly) numerically. Another promising attempt is the use of asymptotic expansions which is applicable as soon as a certain hierarchy exist between the mass scales involved in the process. Well-defined prescriptions provide rules which specify the actions on the individual diagrams (for a review see e.g. [Smirnov 1995]). In general each diagram generates several subgraphs which have to be expanded in their small quantities. This leads to a factorization of the original diagram into integrals of lower loop level which are, of course, simpler to evaluate. However, the price one has to pay is a further increase the complexity of the calculation. In the case of $\Gamma(Z \to b\bar{b})$ the original 69 diagrams blow up to 234 subgraphs. Thus it is desirable to automatize the asymptotic expansion procedures. This has been done in the programs EXP [Seidensticker 1998] (written in Fortran 90) and LMP [Harlander 1998] (written in PERL). They work on a diagram-by-diagram basis and apply the rules for the asymptotic expansion procedures in turn generating the subgraphs, the corresponding administrative files and also the files which rule the expansion in the small quantities. Here special care has to be taken in order not to generate too many terms in the intermediate steps which would significantly slow down the performance.

Of course, also the very computation of the single (sub-)diagrams needs to be done by the computer as the size of intermediate expression become rather large. Although the application of the asymptotic expansion technique leads to significant simplifications, there are still many complicated integrals which have to be solved. The method used for the type of diagrams contributing to $\Gamma(Z \to b\bar{b})$ takes advantage of recurrence relations which express one complicated integral in terms of several simpler ones [Chetyrkin and Tkachov 1981; Broadhurst 1992]. The clever application of such relations leads at the end to only a few complicated integrals which need a hard calculation and a lot of simple ones. Both the input and the output of such kind of calculations are quite handy. In intermediate steps, however, it might be that up to several Giga bytes of disk space are needed, The implementation is realized in the programs MINCER— see 3.1.1.4 and [Larin et al. 1991] — and MATAD [Steinhauser 1996]. They are written in FORM [Vermaseren 1991], which was written for the purpose to deal with a large amount of data.

The package GEFICOM was written by K. G. Chetyrkin and M. Steinhauser in order to automatize the single steps of the calculation discussed above. The user has to provide a few small files specifying the process under consideration and the particle content. If the asymptotic expansion technique should be applied, the hierarchy of scales has to be provided. Then the generation procedure is initiated by specifying the number of loops. It results in a huge database containing all relevant files needed for the very computation and the supervision of it. The very calculation of the diagrams again has to be started by the user. Each diagram is treated separately and the result is stored on disk. At the end they are summed taking into account the proper prefactors.

Up to now GEFICOM, has successfully been applied to Higgs and $Z$ decays and to the computation of higher order corrections to renormalization group functions. However, there are still plenty of interesting applications in particle physics which will be considered in future projects.

Matthias Steinhauser (Bern)

### 3.1.1.3 Evaluation of Feynman Diagrams in Electroweak Interactions

The electroweak and strong interactions of elementary particles are very successfully described by quantized gauge field theories. The quantized nature of these theories manifests itself via corrections beyond the lowest order in the perturbative expansion, which is based on Feynman diagrams. The evaluation of higher-order Feynman diagrams (which are called loop diagrams) is a very tedious but on the other hand algorithmic procedure. Some of the first computer algebra programs were in fact developed in order to facilitate this kind of calculations [Veltman 1967; Hearn 1985], and computer algebra has been applied in this field now for several decades.

Powerful computer-algebraic tools are needed in particular for the evaluation of higher-order corrections in the electroweak Standard Model (SM) and its extensions, most notably the Minimal Supersymmetric Standard Model (MSSM), since the large number of different fields in these models gives rise to a large number of contributing Feynman diagrams (at the one-loop level typically at the order of 100, at the two-loop level at the order of 1000) and the massiveness of the fields makes the evaluation of the diagrams very complicated in general. The precision tests of the SM allow in particular to set constraints on the mass of the Higgs boson, which is the last missing ingredient of the SM and plays a crucial role for a consistent description of massive particles. By comparing the SM predictions for the precision observables with those of extended models, it can be investigated whether the data allow a distinction between different kinds of possible models. Furthermore, the supersymmetric models provide a very stringent direct test since they predict the existence of a relatively light Higgs boson, whose mass can be calculated from the other parameters of the model.

In recent years many applications of computer algebra in the theory of electroweak interactions have been based on the collection of *Mathematica* packages *FeynArts* [Küblbeck et al. 1990] (see 4.3.8), *FeynCalc* [Mertig et al. 1991b] (see 4.3.9), *FormCalc* [Hahn and Pérez-Victoria 1999], (see 4.3.8) and *TwoCalc* [Weiglein et al. 1992] (*FormCalc* is partially written in FORM [Vermaseren 1991]), which use a common syntax and can be linked together. A further program for calculations in the SM is the *Maple* package *xloops* [Brücher et al. 1998].

*FeynArts* [Küblbeck et al. 1990] is a program for generating all Feynman amplitudes contributing to a certain process to a given order in *Mathematica* format and for drawing the corresponding Feynman diagrams. This is done by inserting the propagators and vertices of the model considered (which has to be specified in a model file) into the topologically different Feynman diagrams in all possible ways. As a feature of particular importance for higher-order calculations in the electroweak theory, *FeynArts* generates not only the unrenormalized diagrams at a given order but also the counterterm contributions at this order and the counterterm diagrams needed for the subloop renormalization. The model file for the electroweak SM is predefined in *FeynArts*. The program has been applied in the SM for calculations up to three-loop order (some examples are given below). In applications to other models or extensions of the SM, like chiral perturbation theory [Bürgi 1996], the SM in the background-field formulation [Denner et al. 1995b], the two Higgs-doublet model [Beenakker et al. 1993], and the MSSM [Arhrib and Moultaka 1999; Djouadi et al. 1997; Heinemeyer et al. 1998b; Krauss and Soff 1998], the relevant model file was implemented by the user.

At one-loop order, general algorithms exist for the different steps of the calculation of the Feynman amplitudes, i.e. algebraic simplifications of the Lorentz and the Dirac structure of the amplitude, reduction of the tensor integrals to a set of standard scalar integrals, and evaluation of the latter integrals in terms of known analytic functions. The algebraic reduction of the amplitudes generated by *FeynArts* can be performed at one-loop order in an automatic way using the programs *FeynCalc* [Mertig et al. 1991b] or *FormCalc* [Hahn and Pérez-Victoria 1999]. These programs can then be linked to routines performing the numerical evaluation of the standard integrals. They have been applied for evaluating the complete one-loop contributions for many processes in the SM, as e.g. $e^+e^- \to ZH$ [Denner et al. 1992], $\gamma\gamma \to W^+W^-$ [Denner et al. 1995a], and $W^+W^- \to W^+W^-$ [Denner and Hahn 1998]. Examples for applications of *FeynArts* and *FeynCalc* to one-loop processes in the two Higgs-doublet model are $e^+e^- \to t\bar{t}$ [Beenakker et al. 1993] and associated hadroproduction of $W^\pm H^\pm$ [Bendezu and Kniehl 1999]. Within the MSSM, the programs have been used at one-loop order for example for the process $e^+e^- \to H^+H^-$ [Arhrib and Moultaka 1999].

In contrast to the one-loop case, no general algorithm exists so far for the evaluation of two-loop corrections in the electroweak theory. Besides the large number of contributing diagrams, the main obstacles in two-loop calculations in the electroweak theory are the complicated tensor structure of the Feynman amplitudes, the fact that the two-loop scalar integrals in general are not expressible in terms of polylogarithmic functions but can only be solved numerically, and the need for a renormalization at the two-loop order, which has not yet been worked out in detail. The program *TwoCalc* [Weiglein et al. 1992, 1994] is based on an algorithm for the tensor reduction of general two-loop 2-point functions. It can be used for an automatic reduction of the Feynman amplitudes for two-loop self-energies with arbitrary masses, external momenta, and gauge parameters to a set of standard scalar integrals. These scalar integrals can be evaluated by using one-dimensional integral representations in terms of elementary functions, which allow a fast numerical evaluation with high precision. The corresponding routines can directly be linked to *TwoCalc* [Bauberger and Weiglein 1997]. The programs *FeynArts* and *TwoCalc* have been applied within the SM, e.g., for deriving exact two-loop results for the Higgs-mass dependence of the fermionic contributions to the electroweak precision observables [Bauberger and Weiglein 1998; Weiglein 1998]. In the MSSM, the leading two-loop contributions to the electroweak precision observables [Djouadi et al. 1997, 1998] and to the masses of the neutral $\mathcal{CP}$-even Higgs bosons [Heinemeyer et al. 1998b,a] have been evaluated using *FeynArts* and *TwoCalc*. The algebraic results of these calculations have been converted into *Fortran* code which has been implemented into the program *FeynHiggs* [Heinemeyer et al. 2000].

Georg Weiglein (Geneva)

### 3.1.1.4 MINCER for Feynman Diagrams

MINCER is a collection of FORM routines for the evaluation of three loop propagator-type diagrams in which all particles are massless. In that case the answer will contain just numbers, $\zeta$-constants and powers of $\epsilon = -(Dim-4)/2$. It has been used for a number of three loop calculations in QCD (compare pg. 167) by a variety of authors. Additionally it has been used in cases in which four loop calculations could be reduced to three loop calculations, because only divergencies had to be computed. It has also been incorporated inside other packages like MATAD and BUBBLES.

Documentation about the algorithms can be obtained from the homepage of the system FORM (http://www.nikhef.nl/~form) and the program, with some examples, is available on request. Currently there are two versions: one for version 2 of FORM and one for version 3 of FORM. The difference is just in some of the syntax being used.

Jos Vermaseren (Amsterdam)

172     Chapter 3   Applications of Computer Algebra

### 3.1.2   Gravity

#### 3.1.2.1   Computer Algebra and Relativity

**3.1.2.1.1   Introduction** A. Einstein's gravitational theory, *general relativity* (GR), is the valid theory for describing gravitational effects. In the search for making GR compatible with quantum theory and/or unifying it with the other interactions of nature (strong, electro-weak, superweak,...), different schemes have been developed, like the gauge approach to gravity, including supergravity and metric-affine gravity, higher-dimensional Kaluza-Klein type models, string models, but also more conventional Hamiltonian (canonical) or Feynman quantization schemes or, more far-fetched, models based, e.g., on noncommutative spacetime geometries.

The computer algebra programs applied in GR can be and partially have been extended to these more general frameworks. Still, it is probably true that most computer algebra programs in gravity are applied in the context of GR followed by those for evaluating gravity-based Feynman integrals and for executing computations in the framework of gauge models encompassing non-Riemannian spacetimes. In this note we mainly concentrate on GR and will give a couple of examples.

Detailed overviews of computer algebra in GR are given by Brans [Brans 1995], Hartley [Hartley 1996b], Lake [Lake 1998], and MacCallum [MacCallum et al. 1994], [MacCallum 1996], e.g.

**3.1.2.1.2   Riemannian Curvature in Tensor and Exterior Calculus** In GR and gravity, computer algebra was used as soon as it became available. The reason for this is that for solving standard problems it is required to manipulate a large number of terms and equations. We will clarify this by an example. A generic problem in gravity is to calculate the Ricci tensor from a given metric. A general form of a spacetime metric $g$ in four dimensions is given by 10 independent functions $g_{ij} = g_{ji}$ of the coordinates $(x^0, x^1, x^2, x^3)$:

$$g = \sum_{i,j=0}^{3} g_{ij}(x^0, x^1, x^2, x^3)\, \mathrm{d}x^i \otimes \mathrm{d}x^j \ . \tag{7}$$

The so-called Christoffel symbols are determined from the functions $g_{ij}$ by means of the following equations:

$$\Gamma_{ij}{}^k = \frac{1}{2} \sum_{m=0}^{3} g^{km} \left( \frac{\partial g_{jm}}{\partial x^i} + \frac{\partial g_{mi}}{\partial x^j} - \frac{\partial g_{ij}}{\partial x^m} \right) = \Gamma_{ji}{}^k \ . \tag{8}$$

Here the $g^{km}$ denote the matrix reciprocal to $g_{ij}$. Since $i, j, k$ are running from 0 to 3, the $\Gamma_{ij}{}^k$ represent 64 functions. Because of the symmetry in $i, j$, only 40 are independent. In our conventions, the Riemannian curvature tensor is derived from the Christoffel symbols in the following way:

$$R_{ijk}{}^l = \frac{\partial \Gamma_{jk}{}^l}{\partial x^i} - \frac{\partial \Gamma_{ik}{}^l}{\partial x^j} + \sum_{m=0}^{3} \left( \Gamma_{im}{}^l \Gamma_{jk}{}^m - \Gamma_{jm}{}^l \Gamma_{ik}{}^m \right) \ . \tag{9}$$

Eventually we find the Ricci tensor as

$$R_{jk} = \sum_{i=0}^{3} R_{ijk}{}^{i}. \tag{10}$$

Now one can easily estimate that the number of terms in each of the components $R_{ij}$ may be very large. In [Lake and et al. 2001] it is shown that in the general case this number is in the order of 10 000 for each of the components. Thus, only in simple cases these calculations can conveniently be done by hand.

It would be most desirable to have a computer algebra system which allows to enter mathematical expressions in an analogous way as one would write them down on paper. That is, defining objects with abstract properties, doing calculations, and assigning explicit values to these objects should be possible in a natural way. Such kind of systems already exist. We illustrate this by an example.

In terms of Cartan's calculus of exterior differential forms, Eq.(9) can be very compactly displayed as

$$R_\alpha{}^\beta = d\Gamma_\alpha{}^\beta - \Gamma_\alpha{}^\gamma \wedge \Gamma_\gamma{}^\beta . \tag{11}$$

Here, $R_\alpha{}^\beta$ is the curvature 2-form and $\Gamma_\alpha{}^\beta$ the connection 1-form. The summation convention is assumed, i.e. summation is understood over $\gamma$. Let us check the *Bianchi identity*

$$dR_\alpha{}^\beta - \Gamma_\alpha{}^\gamma \wedge R_\gamma{}^\beta + \Gamma_\gamma{}^\beta \wedge R_\alpha{}^\gamma = 0 \tag{12}$$

on the computer. This can be deduced from Eq.(9) and the properties of the exterior product and the exterior derivative.

In Excalc, a Reduce package for exterior calculus, — see section 3.1.4.1 — we first have to declare that all indices run from 0 to 3. Then we declare $\Gamma_\alpha{}^\beta$ to be a 1-form and $R_\alpha{}^\beta$ a 2-form.

```
indexrange 0,1,2,3;
pform gamma(a,b) = 1 , curv(a,b) = 2;
```

Eq.(11) and the left hand side of Eq.(12) can almost literally be translated,

```
curv(-a,b) := d gamma(-a,b) - gamma(-a,c) ^ gamma(-c,b);
```

and

```
d curv(-a,b) - gamma(-a,c) ^ curv(-c,b) + gamma(-c,b) ^ curv(-a,c);
```

The negative sign in front of an index indicates that it is subscript whereas a superscript is denoted by a positive (or no) sign. The last command yields zero, i.e. the Bianchi identity is confirmed. In [Parker and Christensen 1994, p.234] it is presented how to handle this in the MATHEMATICA package MathTensor.

**3.1.2.1.3 Abstract and Component computer algebraSystems** Packages (or systems) capable of performing symbolic calculations as with the Bianchi identity are called *abstract* or *indicial* calculus systems [Hartley 1996b]. They are necessary if one wants to investigate general properties of objects. So called *component* calculus systems are designed to calculate the components of unknown quantities from known ones. In our example, it was not necessary to introduce a basis or a metric in order to define $\Gamma_\alpha{}^\beta$ and $R_\alpha{}^\beta$. In a typical component system, one first would have to enter the components of a metric. Then, by means of build–in routines, the components of the connection and curvature could be calculated. Both, abstract as well as component systems, allow to define new objects by means of various mathematical operations like products or derivatives, e.g.. The difference here is that in the case of a component system it is always necessary to assume a specific basis and/or metric. In turn there are abstract calculus systems which do not support computations of explicit values of components. Some packages, like MathTensor or Excalc, for instance, allow both, abstract and component calculations (see Table 1). In our Excalc example, we could assign explicit expressions to the components of $\Gamma_\alpha{}^\beta$. Then we would find the components of $R_\alpha{}^\beta$ by calling `curv(-a,b)`.

**3.1.2.1.4 General Versus Special Purpose Systems** Today it seems that most people use relativity packages of *general purpose systems* like MACSYMA, MAPLE, MATHEMATICA, DERIVE, and/or REDUCE— see section 4.1. These programs offer very user-friendly front-ends and, moreover, a wealth of useful facilities like simplification routines, programs for solving algebraic or differential equations exactly or numerically, TEX and FORTRAN interfaces, etc..

*Special purpose systems* are, as the name suggests, specialized to handle only a specific class of problems — see section 4.2. Therefore the set of instructions is usually very limited and programming these systems normally requires much more effort than is the case of general purpose systems. However, these systems are rather compact, very fast, and may sometimes be the only available facility to solve a problem. In the case of calculating Feynman diagrams, e.g., special purpose systems like Schoonship or FORM are often used, but there are also packages available for MATHEMATICA and REDUCE. A fairly widely used special purpose system for tensor calculus and general relativity is Sheep.

**3.1.2.1.5 Applications** Today many authors use computer algebra in order to obtain or confirm their results *without* mentioning it explicitly. In the following we present some articles which explicitly illustrate the applications of computer algebra in gravity.

It was already stated that quantities like the *Ricci tensor* can reach an enormous size. Thus, for most applications (classification, numerics), these objects have to be put on the computer. The use of computer algebra has the advantage that there is no need to enter very large expressions for the curvature, e.g., but a comparably small input program which calculates these. Moreover, one can use the possibility to transmit programs or results by electronic means, cf. the closing remark in [Ernst et al. 1987].

| System | Component | Abstract |
|---|---|---|
| Macsyma [Macsyma 1998], Chap. 4.1.4 | CTensor | ATENSOR ITENSOR CARTAN |
| Maple [Maple 2001], Chap. 4.1.6 | tensor cartan NPspinor debever oframe GRTensorII [GRTensor 2001] Riemann [Riemann 1998] | difforms forms |
| Mathematica [Mathematica 2001], Chap. 4.1.7 | Cartan [Soleng 1996] TTC [Balfagón et al. 2001], Chap. 4.3.15 MathTensor [Parker and Christensen 1994], [Mathtensor 2001] | EinS [Klioner 1999], Chap. 4.3.7 Ricci [Lee 2000] MathTensor DifferentialForms |
| REDUCE [Reduce 1999], Chap. 4.1.9 | EXCALC Chap. 3.1.4.1 REDTEN [Harper and Dyer 1994] GRG Chap. 3.1.4.2 $GRG_{EC}$, Chap. 3.1.4.3 | EXCALC RICCIR [Kadlecsik 1996] GRGlib |
| Sheep [MacCallum 1995a], [MacCallum et al. 1994], [Skea 1994] | CORD FRAME | STENSOR |

**Table 1.** Relativity systems/packages

A standard application of computer algebra in GR is the *classification of exact solutions*. These are necessarily found in special coordinates. However, for various reasons it is necessary to characterize the corresponding spacetimes in a coordinate independent way. This includes determination of Petrov and Segre types, calculations of curvature invariants (the trace of the Ricci tensor of Eq.(10), for instance), the maximal isotropy group, etc.. The appropriate algorithms involve an enormous amount of work. Programs for the widely used Petrov classification are available for most computer algebra systems. By means of the Sheep-package Classi [MacCallum et al. 1994], it was possible to create a searchable online-databank which includes nearly 200 exact solutions and is still growing [Skea and et al. 1997].

To find out whether two solutions which look different are not just the same solution in different coordinates, one has to solve the so-called *equivalence problem*. This involves differentiation of the curvature tensor up to the seventh order. In [MacCallum et al. 1994], appropriate programs are presented for Sheep.

Computer algebra is also very useful for finding *new solutions* of the field equations. The Einstein vacuum equation reads $R_{ij} = 0$. As we can recognize from Eq.(4) together with Eqs. (2) and (3), it represents a system of ten second order partial differential equations for the $g_{ij}$ which are obviously very difficult to solve. A simple example is given in [Vulcanov and Ghergu 1998] where the Schwarzschild solution is derived from a spherical symmetric line element by using the Reduce package Excalc. In [Tertychniy 1998] it is illustrated how to use the Reduce-based system $GRG_{EC}$ for searching for solutions of the Einstein-Maxwell equations. In [Wolf 1996] the Reduce package Classym was used to derive the Killing vector and Killing tensor equations and their integrability conditions from a general form of a metric. Subsequently the Reduce package Crack, see Chap. 4.3.5, has been used for solving these equations.

In [Socorro et al. 1998] it is shown how to construct solutions of *metric-affine* gravity from solutions of the Einstein-Maxwell equations under heavy use of the computer algebra system Reduce.

A further application is the derivation of a field equation from an action principle by means of *variational calculations*. A simple example is given in [Parker and Christensen 1994, p. 303] where the Einstein vacuum equation is rederived. In [Tsantilis et al. 1996], the gravitational field equation of a unified field theory of Pawłowski and Rączka has be derived (and corrected) from the corresponding Lagrangian by means of MathTensor.

In [van de Ven 1996] an example is given of how to use the special purpose system Form to calculate *Feynman diagrams* in the context of quantum gravity.

As mentioned previously, in the case of general purpose systems one can benefit from many additional features. In [Vulcanov and Ghergu 1998] the *visualization* of Schwarzschild type solutions by means of Mathematica is presented.

Another feature of many computer algebra systems is a *Fortran interface* which converts equations into Fortran readable form. This helps to develop programs for numeric calculations, as illustrated in [Brandt and Seidel 1995].

In [Davies 1998] it is outlined of how to apply the computer algebra system GRTensorII to second-order Black Hole perturbations.

Christian Heinicke and Friedrich W. Hehl (Cologne)

### 3.1.3 'Central Configurations' in the Newtonian N-Body Problem of Celestial Mechanics

**3.1.3.1 Introduction** The Newtonian $N$-body problem is one of the important issues in Celestial Mechanics. (See also section 2.11.7.) It is concerned with describing the motion of $N$ bodies (particles or point masses) under their mutual gravitational attraction. The differential equations of motion, as given by Newton's laws, are not integrable in general. However, there is an important special class of solutions which can be computed analytically. These particular solutions are called 'central configurations' and geometrically they describe motions in which the configuration of the bodies remain self-similar in time. When

### 3.1.3 'Central Configurations' in the N-Body Problem

the center of mass is shifted at the origin, central configurations are described by the fact that the acceleration vector of each body is a common scalar multiple of its position vector.

In what follows, we concentrate our attention to the case of the $N$-body problem in a euclidean space of dimension $N-2$. We will be exclusively involved with the cases $N = 4$ (planar 4-body problem) and $N = 5$ (spatial 5-body problem). In [Dziobek 1900], there appeared first the innovative idea to used the mutual distances of the bodies as unknowns in order to formulate the equations of motion. In [Albouy and Chenciner 1998] the authors use this idea and a construction in linear algebra to provide a reformulation of the equations of motion in the general case of homographic motions. Their work reduces the problem of searching for central configurations to the study of systems of non-linear polynomial equations that exhibit many symmetries.

Central configurations are of fundamental importance in the study of changes in the topology of the integral manifolds of the N-body problem (see [Smale 1970], [Cabral 1973], [Albouy 1993]) as well as in the analysis of expanding gravitational systems, since they are the only configurations that can be formed asymptotically by the bodies in the neighbourhood of a multiple collision.

**3.1.3.2 Equal Masses** The equations of central configurations are homogeneous w.r.t. the masses. In the case of equal masses, this fact allows us to simplify the equations considerably.

**3.1.3.2.1 Planar 4-Body Problem** A symmetry theorem proved by Albouy [Albouy 1995] asserts that a central configuration of the newtonian planar 4-body problem has at least one axis of symmetry. The complete classification of central configurations in the newtonian planar 4-body problem is given in [Albouy 1996]. This classification is obtained again in [Kotsireas 1998a] by using various computer algebra techniques. More specifically, there are three kinds of such configurations: the square, the equilateral triangle with its barycenter and the pseudo-centered isosceles triangle. There is only one configuration (the third one) with exactly one symmetry axis. This configuration is defined by a real root of an irreducible univariate polynomial of degree 37, see [Albouy 1996] and [Kotsireas 1998a].

**3.1.3.2.2 Spatial 5-Body Problem** The original proof of the symmetry theorem for the 4-body problem (see [Albouy 1995]) cannot be generalized to the 5-body problem because it uses geometric ideas and facts. Nevertheless, many parts of the original proof can be carried out symbolically by using Gröbner bases (cf. 2.2.5). This approach is used in [Faugère and Kotsireas 1999] to establish a symmetry theorem for the 5-body problem in the convex case. The lack of a complete symmetry theorem renders the computations more difficult. A systematic process of eliminating the variables representing the oriented tetrahedra volumes of the five tetrahedra formed by the five bodies has been introduced in [Kotsireas and Lazard 1999]. It uses elementary linear algebra and is most conveniently presented by using the graphical interface of MAPLE (see section 4.1.6). For a detailed presentation of the method and applications to many different cases, the reader is referred to [Kotsireas 1998a]. The resulting equations show clearly the underlying symmetry of the problem and are tractable, in the simple cases, by many computer algebra methods such as direct Gröbner bases computations, triangular sets methods (see [Lazard 1991], [Aubry et al. 1999]) and a method based on invariant theory of finite groups (see [Colin 1997a]). This time there are four types of central configurations: the regular square pyramid, the regular tetrahedron and its barycenter, and two more configurations, each with one symmetry, defined by univariate polynomials of degrees 12 and 43.

In [Kotsireas and Lazard 1999] it is proved that there is no central configuration with exactly two planes of symmetry. The Gröbner bases computations involved require the use of the FGb program (see [Faugère 1994a] and [Faugère 1999]). The final system consists of 11 polynomial equations in 8 variables. The Gröbner basis for a lexicographical ordering of this fairly small system is computed with FGb in approximately 4 minutes. It is interesting to note that this computation is not feasible in Gb. The resulting univariate polynomial is of degree 216 with big integer coefficients. This polynomial has been treated with RS (see [Rouillier 1996]) in order to find isolating intervals with rational extrema for its positive real roots. The approximate values for the real roots have been found with RS and MAGMA(see 4.1.5) independently. In view of this fact, it seems probable that there is no central configuration with one (or none) plane of symmetry. This would mean that the list presented here is complete, just as in the case of the 4-body problem.

**3.1.3.2.3  The C.C. Program** The results on central configurations obtained with Symbolic Computation methods, agree with the pictures produced by the program C.C. written by R. Moeckel at the University of Minnesota. In C.C. the user can specify the number of bodies and their masses and visualize planar or spatial central configurations. The program works by applying repeatedly a combination steepest descent Newton's method iteration with randomly chosen initial conditions. The center of mass is taken to be the origin of the axes. For each picture the program displays the distances of the masses form the origin, the potential energy and the index and nullity of the Hessian of the potential at that particular central configuration. For a more detailed presentation of the C.C. program as well as the corresponding pictures for the planar newtonian 4-body problem and the spatial newtonian 5-body problem see [Kotsireas 1998a].

**3.1.3.2.4  Polynomial Systems Used As Benchmarks** Some polynomial systems arising in the study of central configurations in the planar 4-body problem and the spatial 5-body problem have been used as benchmarks (see [Faugère 1994a], [Aubry and Maza 1999]) and have also been included in the FRISCO polynomial test suite maintained at INRIA by D. Bini and B. Mourrain and in the CABRI initiative maintained at the University of Mannheim by H. Kredel. We mention for example the following system, which arises in the study of the

### 3.1.3 'Central Configurations' in the N-Body Problem

planar 4-body problem:

$$(b-d)(B-D) - 2F + 2 = 0$$
$$(b-d)(B+D-2F) + 2B - 2D = 0$$
$$(b-d)^2 - 2b - 2d + f + 1 = 0$$
$$B^2 b^3 - 1 = 0$$
$$D^2 d^3 - 1 = 0$$
$$F^2 f^3 - 1 = 0$$

This system has been studied by K. Gatermann in [Gatermann 2000] using algorithms for equivariant systems in Invariant Theory (see 2.2.8) and by J. Verschelde in [Verschelde 1999] using polyhedral homotopies.

**3.1.3.2.5 Generalized Central Configurations** The study of the algebraic system that describes central configurations of the newtonian planar 4-body problem with equal masses, for various potential energy functions, is fundamental in understanding the bifurcations of the solutions. By changing the potential energy function we get a system of equations which is similar to the system in the case of the 4-body problem considered above. Practically this change amounts to replacing the exponent $-3/2$, which corresponds to the newtonian potential, by another exponent. It is conjectured (see [Albouy 1996]) that leaving aside two easily identified degenerate cases, the result proved in the case of the 4-body problem remains true for every negative exponent. That is, there is only one generalized central configuration with one symmetry axis. This conjecture is verified up to $n = 10$ for the classes of exponents $1/n$ and $n/(n+1)$ in [Kotsireas 1998b] by using Gröbner bases computations in Gb. The conjecture has also been verified for other exponents by Faugère, using FGb, since the corresponding systems were no longer tractable in Gb.

**3.1.3.3 Unequal masses** The situation is more complicated when the masses are different. As a first example one can consider a newtonian 3+1-body problem in the plane. Namely we consider three equal masses and another mass $m$. It can be shown by using techniques based on the theory of triangular sets (see [Lazard 1991], [Aubry et al. 1999]) or the theory of comprehensive Gröbner bases (see [Weispfenning 1992a], section 2.2.5). that for every positive value of $m$ the number of solutions of the system remains finite. A proof of this result using results related to the shape of a Gröbner basis of a zero-dimensional ideal, can be found in [Kotsireas and Schicho 2000]. In view of this result it is logical to assume that further results can be obtained in this direction using these and/or other computer algebra methods.

**3.1.3.4 Conclusion** The search for central configurations in the N-body problem of celestial mechanics offers great computational challenges and computer algebra methods are well suited for attacking them. This interaction of two seemingly unrelated research areas is particularly fruitful for both of them. A better understanding of symmetry is the key point that can lead to easier proofs and thus allow us to treat more difficult problems in this area, with the help of computer algebra techniques.

Ilias Kotsireas (Paris)

### 3.1.4 CA-Systems for Differential Geometry and Applications

**3.1.4.1 Differential Geometry, Computer Algebra** Modern differential geometry has established itself as a fundamental mathematical framework for theoretical and mathematical physics. It permits an intrinsic formulation of a wide variety of theories and provides at the same time an efficient calculus for solving problems in these theories. In mathematics the application of methods of modern differential geometry to the study of differential equations has been especially fruitful.

Therefore it comes as no surprise that substantial fragments of modern differential geometry can be found nowadays in all major computer algebra systems. For REDUCE, we implemented one important part, the so-called EXterior CALCulus and coined the resulting software package EXCALC [Schrüfer 1987]. A major design objective for EXCALC was to build a *general purpose* system for differential geometry which is fully integrated into the underlying computer algebra system and which provides a syntax that is as close as possible to the notations used in standard textbooks. The first goal was chosen to give the user the full power of a computer algebra system without needing to play tricks in order to manipulate the data of EXCALC with the operations provided by the underlying REDUCE system. With the realization of the second goal it is hoped that even the novice user can write programs immediately since it is not necessary to learn a complicated programming language (the formulas essentially represent the program). This feature also makes EXCALC an ideal tool for use in teaching exterior calculus.

Applications of the exterior calculus in physics, which can be implemented with EXCALC, are numerous. They range, for example, from computing the component-wise representation of equations modeling the growth of crystals on a time-dependent, boundary layer adapted coordinate system, to computing the variation of super-symmetric Lagrange densities and determining their Noether currents. EXCALC was put to extensive use in analyzing dynamic systems, especially in determining generalized symmetries (G. Prince, Australia), in Poincaré gauge-invariant gravitation theory (F.W. Hehl (Köln), J.D. McCrea (Dublin) [Schrüfer et al. 1987]), in higher-dimensional cosmological theories (J. Demaret, H. Caprasse (Liège) [Caprasse et al. 1989]), in ADM formalism (D. Vulcanov [Vulcanov 1995]) and for Clifford Algebra valued forms (J. Vaz [Vaz 1996]).

The computer programs verify lengthy calculations done by hand, or they produce results which could not be obtained by pencil and paper within a reasonable amount of time. Consequently, computing times can vary from seconds up to several days. The following example taken from cosmology should give an impression of the ease of programming in EXCALC. Here, we want to compute

### 3.1.4 CA-Systems for Differential Geometry and Applications

the curvature and Weyl tensor in a 6-dimensional Riemannian space. First, we define a (non-holonomic) reference system (coframe) of 1-forms. The curvature and Weyl forms can then be computed by virtue of well-known formulas found in any standard textbook.

```
pform {R,S}=0;
fdomain R=R(t),S=S(t);

coframe theta(t) = d t,
        theta(1) = R*d u/(1 + k*(u**2)/4),
        theta(2) = R*u*d th/(1 + k*(u**2)/4),
        theta(3) = R*u*sin(th)*d phi/(1 + k*(u**2)/4),
        theta(4) = S*d v1,
        theta(5) = S*sin(v1)*d v2
  with metric g =-theta(t)*theta(t)+theta(1)*theta(1)+theta(2)*theta(2)
                 +theta(3)*theta(3)+theta(4)*theta(4)+theta(5)*theta(5);

frame e;

riemannconx om;

pform {curv(k,l),weyl(k,l)}=2,rie(a)=1,{Riemann(a,b,c,d),rie}=0;

index_symmetries curv(a,b)    : antisymmetric in {a,b};

curv(k,l) := d om(k,l) + om(k,-m) ^ om(m,l);

Riemann(a,b,c,d) := e(d) _| (e (c) _| curv(a,b));

rie(-a) := e(b) _| curv(-b,-a);

rie := e(a) _| rie(-a);

weyl(a,b) := curv(a,b) + 1/20*rie*theta(a)^theta(b)
                      - 1/4*(theta(a) ^ rie(b) - theta(b) ^ rie(a));
```

The legibility of output can be greatly enhanced by using the X-interface of REDUCE. Once MATHML-enabled browsers are available, interactive dialogue in real textbook notation will become a reality.

The study of ordinary or partial differential equations within the framework of exterior calculus is facilitated by their conversion into a system of equivalent differential forms. All manipulations known for differential equations have their counterpart in the calculus of exterior systems. For example, the determination of the generating equations for Lie symmetries is given by the condition that the Lie derivative of the exterior system vanishes (modulo the exterior system). In EXCALC, we can express this condition simply by a few lines of program statements (see e.g. [Satir 1998]). Bäcklund transformations within the Estabrook-

Wahlquist formalism can also be handled fairly easily. Tools for constructing generalized characteristics of PDEs as well as for R. Gardner's formulation of Cartan's equivalence problem are a pending project. Algorithms for the quite demanding Cartan-Kähler theory of exterior differential systems have been implemented by D. Hartley [Hartley 1996a] in the EDS software package which coexists with EXCALC. Also here, programming is very easy. For example, calculating the involutive prolongation of the Janet system could be done in the following way. First we construct the contact system on the second order jet bundle. Then we calculate the pullback under the map defining the system of second order partial differential equations of Janet (yielding the corresponding exterior system) and finally, we compute its involutive prolongation.

```
pform {x,y,z,u,v,w}=0,{j2,janet}=1;

j2 := contact(2,{x,y,z},{u,v,w});

janet := pullback(j2,{u(-y,-y)=w,u(-z,-z)=y*u(-x,-x)+v});

involution janet;
```

More information about the algorithms used in EDS can be found in [Hartley 1997].

EXCALC as a software package is still under development, even though no new major release has appeared for some time. A new version which will implement a considerably larger part of modern differential geometry is planned for the next major release of REDUCE. As a long-term goal, we want to provide the functionality to the extent given, for example, in the lecture notes "Differential Geometry for Physicists" by A. Trautman [Trautman 1984].

Eberhard Schrüfer (St. Augustin)

### 3.1.4.2 Computations with GRG in Gravity Theory

The computer algebra system GRG [Zhytnikov 1997] has been developed for calculations in theories of gravity, classical field theory and modern differential geometry. GRG grew from the same root as the system $GRG_{EC}$[Tertychniy and Obukhova 1997; Tertychniy 1998], see 3.1.4.3, although at present the two systems are essentially different from the point of view of their internal structure and input/output interface.

The system requirements of GRG are as follows. Being based on the general purpose computer algebra system REDUCE [Tertychniy 1998], GRG needs a preliminary installation of REDUCE, see 4.1.9, versions 3.3, 3.4, 3.4.1, 3.5, or 3.6, in the professional version (with the compiler included). The system is distributed in the form of a source code written in LISP which is available via anonymous ftp, e.g., from ftp.maths.qmw.ac.uk in the directory /pub/grg3.2 (the total size of the code is about 500 Kb). In order to install the program, it

### 3.1.4 CA-Systems for Differential Geometry and Applications

should be compiled with the help of the REDUCE compiler. Both the PSL and CSL dialects of REDUCE are supported, and thus GRG can run practically on all DOS/Windows, Unix, and VAX/VMS platforms. The completely compiled system requires about 0.7 Mb disk space. At present the distribution includes a detailed user's manual together with the short reference guide.

Although being based on REDUCE, GRG cannot be considered on the same foot with the numerous packages distributed together with REDUCE. The system GRG has its own input language with an extremely simple structure, which actually makes it unnecessary for a user to know either LISP or REDUCE. The user can run a GRG-session either in the interactive or in the batch mode by writing the instructions in the form very close to the common English phrases using the standard mathematical notation. The development of the maximally user-friendly interface was one of the main priorities in the GRG (and $GRG_{EC}$) project which was aimed at a typical user who has only knowledge of the differential-geometric and/or gravitational and field-theoretic subjects and is not an expert in programming.

The GRG system performs analytic calculations with all kinds of geometrical objects: spinors, vectors, tensors, exterior differential forms, connection and related structures defined on a smooth manifold of arbitary dimension. More than 150 built-in geometrical and field-theoretic quantities are known to the system, together with their covariance properties and formulas for their computations. This enables the user to obtain immediate answers to practically all standard computational problems in the gravity theory, such as derivation of the curvature, making its irreducible decomposition, finding the field equations, verifying symmetry properties, calculating covariant and Lie derivatives, etc.. GRG can represent the output of the results in the form of a Latex file as well as in the format of Maple (see 4.1.6), Mathematica (see 4.1.7), Macsyma (see 4.1.4) or REDUCE (see 4.1.9). Such an export feature allows to combine conveniently the advantages of several major computer algebra systems for a better processing of the computational data. It is worthwhile to mention also that GRG supports the REDUCE graphical shells.

At present, there is a number of systems and packages which are solving more or less the same range of problems in the gravitational theory, see an overview in 3.1.2.1. In this respect, GRG can be compared with the REDUCE based systems $GRG_{EC}$ [Tertychniy and Obukhova 1997; Tertychniy 1998] (see 3.1.4.3) and EXCALC [Schrüfer et al. 1987; Schrüfer 1994] (see 3.1.4.1) and with the Mathematica (see 4.1.7) based packages MathTensor [Parker and Christensen 1994] and Cartan [Soleng 1993]. The two latter systems require large computer resources, and usually the run times are very big. On the contrary, all the three REDUCE systems (GRG, $GRG_{EC}$, and EXCALC) generally demonstrate higher performance rates. In comparison to $GRG_{EC}$ which is confined to the dimension 4 of spacetime, GRG provides a convenient tool for the gravitational computations in arbitrary dimension.

Perhaps one of the main advantages of GRG over EXCALC and other systems is that the GRG system operates with all the physical and geometrical objects in an explicitly covariant matter. In particular, a unique possibility of making arbitrary coordinate, spinor and frame transformations is provided. As soon as the form of transformation is specified, GRG automatically recalculates the transformed values for all the quantities under consideration. The list of covariant objects includes, besides the tensors and exterior forms, also the pseudo-tensorial quantities (i.e., densities). Further, as compared to EXCALC which demands from a user a good deal of programming experience in REDUCE, GRG does not require a deep knowledge of programming languages such as LISP and REDUCE.

A great number of built-in geometrical and physical quantities and formulas provide an efficient support of analytic computations both in Einstein's general relativity and in the generalized gravity models (such as gauge gravity, supergravity, string theory, etc.). For example, starting with a metric or a (co-)frame, a user can obtain the curvature and the Einstein equations, the geodesic equations, optical scalars for null congruences, the kinematics for time-like congruences, or he can use the Newman-Penrose formalism. The list of matter sources known to the system includes the scalar field with minimal and non-minimal interaction, the electromagnetic field, the Yang-Mills field with an arbitrary gauge group, the Dirac spinor field, the ideal fluid and the spin fluid. The most general manipulations are available for post-Riemannian geometries with an arbitrary linear connection, for which torsion and nonmetricity do not vanish. Among other computations, irreducible decomposition of the curvature, torsion, and nonmetricity can be done in any dimension, and the generalized gravitational equations can be derived for a theory with arbitrary gravitational Lagrangian in Riemannian and post-Riemannian spaces.

A very convenient feature of the system is its flexibility, in the sense that GRG usually provides several alternative built-in computational schemes for complicated objects and equations. Moreover, the user can override the built-in formulas when he/she needs to compute something specific by writing an own program in GRG. However, such a programming is very different from a programming in REDUCE or other system: One rather writes equations in the form which is very close to the traditional mathematical notation. This flexibility, combined with the possibility to export the output in the format of other systems for further processing in Maple, Mathematica, Macsyma, or REDUCE, makes GRG an effective calculational tool in gravity and geometry.

The number of researches and groups using GRG on the regular or quasi-regular basis is constantly growing. Most frequently, this system is applied to computations in string (effective string) models, supergravity, Kaluza-Klein and gauge gravity models (e.g., in Einstein-Cartan, Poincaré gauge and metric-affine theories).

Yuri N. Obukhov (Moscow)

### 3.1.4.3 The Computer Algebra System GRG$_{EC}$ for Gravity and Differential Geometry
The specialized computer algebra system GRG$_{EC}$ [Terty-

### 3.1.4 CA-Systems for Differential Geometry and Applications

chniy and Obukhova 1997; Tertychniy 1997] is intended for applications to the theory of gravity and related problems of geometry and classical field theory.

In particular, $GRG_{EC}$ is 'aware' of the majority of basic characteristics of the geometry of a curved space-time utilized in Einstein's gravitational theory. These are such notions as the coframe forms, the bases in higher order foliations of exterior forms, the connection (including its Newman-Penrose representation), the curvature with its irreducible decomposition and algebraic invariants, the general relationships connecting the objects above (Cartan's structural equations, Bianchi's identities, various algebraic identities), and the field equations of gravity theory (Einstein equations for vacuum and various matter contents). This list can be continued with the basic elements of the Rainich theory, the theory of the Lanczos potential, and methods for describing symmetries.

The $GRG_{EC}$ system operates with major characteristics of many classical fields. These are the electromagnetic field, the massless spinor field, the massive spinor field including the cases with and without interaction with the electromagnetic field, the massless and massive scalar field, the conformally invariant scalar field, the massive vector field, pressure-free dust matter, both massive and lightlike, *etc.*. Resorting to numeric characteristics, the $GRG_{EC}$ system currently enables one to handle more than two hundred of the so called *data objects* modeling the basic notions and relations (equations) originating from the geometry and field theory in curved space-time.

At the same time, $GRG_{EC}$ is not accommodated for the purpose of abstract index manipulating. The handling of data objects endowed with extrinsic indices (tensors, spinors, *etc.*) is carried out utilizing, essentially, the explicit sets of their components in a definite gauge. Partially at the price of lack of capability to handle the 'general' expressions with 'abstract' indices, $GRG_{EC}$ gains advantage of the optimizing of the low level representations of the data processed. As a result, its performance rate is comparable — or superior — to the one of computer algebra systems of similar capabilities and an analogous application field.

The mathematical tool applied by $GRG_{EC}$ is, in the case of four-dimensional Lorentz geometry, a version of the formalism of null tetrads based on the method of exterior forms and Cartan's structural equations incorporated with spinor theory. Accordingly, the system includes the full-featured package realizing the calculus of exterior forms. The classical tensor formalism is also implemented.

One of the goals pursued during development of $GRG_{EC}$ was to liberate the user from the tedious duty to 'develop a program' — at least in the sense usually associated with the latter pursuit. To that end, $GRG_{EC}$ maintains the input language which is maximally close to the one used for the representation of the relevant notions and the relationships taking place in the application field itself. Dealing with the $GRG_{EC}$ system, a researcher has to merely describe, employing mostly conventional words arranged in a sequence of natural 'phrases' and operating with more or less standard mathematical notations what initial data are given and indicate what a result has to be generated. This allows one to focus mostly on the essence of the problem considered.

As to the topic of the system realization, GRG$_{EC}$ is, intrinsically, a 'superstructure' over the well known general purposes computer algebra system REDUCE developed by A. Hearn, [Hearn 1996]. Accordingly, GRG$_{EC}$ potentially possesses the same degree of portability characteristic of REDUCE, the list of appropriate hardware/OS families comprising the majority of the widespread computer platforms, from personal to super-computers. Concerning the relation of GRG$_{EC}$ and background REDUCE, GRG$_{EC}$ utilizes the following constituents of the latter:

- the algebraic processor, i.e. the routine which carries out simplifications (transformations) of general mathematical expressions;
- the corresponding facilities realizing the control over the algebraic processor such as the handler of flags (the dynamic REDUCE options) and substitution support (REDUCE's pattern matching facility), *etc.*;
- some additional applied packages such as polynomial factorizer, the simple fraction routine, the package COMPACT, *etc.*;
- the routine realizing the output of mathematical expressions and the corresponding controlling facilities.

The programming language utilized for the GRG$_{EC}$ realization is the Lisp dialect known as STANDARD LISP [Marti et al. 1979], supported in frames of the *PSL* package, which constitutes the bottom level of the REDUCE infrastructure hierarchy. It has to be emphasized that, at the same time, working with GRG$_{EC}$, a user does not need to be familiar with either Lisp or with REDUCE itself. The point is that GRG$_{EC}$ creates an 'opaque casing' which conceals the lower level data structures and the inner ways of their handling. Although the background REDUCE system imposes certain inevitable signs on GRG$_{EC}$, the latter should not be regarded as one of the elements of the subpackage library of REDUCE. In fact, it constitutes a fully structured programming system.

The structure of GRG$_{EC}$ can be roughly divided into the following more or less separated constituents:

- the input interface realized in the form of an interpreter of the *language of problem specification*;
- the package realizing a number of mathematical methods of differential geometry including exterior form calculus and spinor theory, the methods of classical differential geometry and some other algorithms augmenting the capabilities of the algebraic processor of REDUCE;
- an extensive library of applied routines realizing mathematical relations taking place in general relativity, classical field theory on curved background, and in Riemannian geometry.

The source code of GRG$_{EC}$ currently amounts to approximately 2 Mb. The compiled (32-bit) binary code occupies about 1.2 Mb of disk space.

The system can also be considered as a promising base for the further development of efficient tools for doing computer analysis of a wide scope of problems of theoretical physics.

Sergey I. Tertychniy (Moscow)

## 3.1.5 Differential Equations in Physics

### 3.1.5.1 Conservation Laws for Geodesic Motion in Curved Spaces and PDEs
This report describes current work on problems related to the computation of conservation laws. Although the computation of conservation laws of geodesic motion in curved space and the computation of conservation laws of arbitrary systems of partial differential equations (PDEs) are rather different, in both cases overdetermined PDE-systems have to be solved which is accomplished with the same computer algebra package CRACK (included on the book's CD). For more details about CRACK and other packages used for investigating differential equations also refer to section 4.3.5.

#### 3.1.5.1.1 Computation of Killing Vectors and Killing Tensors
If $u^i$ is the velocity of a test particle moving in a curved space, then "force free" (geodesic) motion is described by $u^i{}_{;j} u^j = 0$ where ';' denotes covariant differentiation and summation over equal indices is performed. Any constants of this motion that are polynomial in $u^i$ and that are independent of the path parameter have the form
$$\text{const} = K_{i_1 \ldots i_r} u^{i_1} \ldots u^{i_r}.$$

The conditions on the tensor $K_{i_1 \ldots i_r}$ (called a Killing vector if the rank is one, otherwise Killing tensor) which is completely symmetric in its $r$ indices are
$$K_{(i_1 \ldots i_r; i_{(r+1)})} = 0 \tag{13}$$
where () stands for symmetrization.

The need to compute Killing tensors is not only restricted to problems in General Relativity or Differential Geometry. Let a dynamical system be described by a Hamiltonian
$$H = \frac{1}{2} g^{ij}(x^k) p_i p_j + V(x^k),$$
with Jacobi metric $g^{ij}$, momentum $p_i$ and potential $V(x^k)$. Then the question of the existence of first integrals
$$F = K_s(p)^s + K_{s-1}(p)^{(s-1)} + \ldots + K_0 \quad \text{(indices suppressed)},$$
polynomial in the momentum $p$ with completely symmetric coefficient tensors of rank $s, s-1, \ldots$, requires the solution of the Killing tensor condition for $K_s$ and $K_{s-1}$. The other $K_{s-i}$ are calculated from them subsequently.

Another application is the question of separability of the Hamilton-Jacobi equation
$$g^{ij} \partial_{x^i} W \, \partial_{x^j} W = E, \quad g^{ij} = g^{ji}, \ 1 \le i, j \le n,$$

for an $n$-dimensional Hamiltonian system, which leads to the question of the existence of enough Killing tensors in a space with metric $g^{ij}$.

One way to solve Killing tensor equations (13) is to use the program CLASSYM [Wolf and Gebot 1994] in order to formulate the equations for a given metric including additional integrability conditions and to use the package CRACK to solve them all. Such overdetermined PDE-systems can easily have thousands of terms.

To give an example, recent improvements in lowering the memory consumption during the formulation and application of integrability conditions within CRACK (computation of the differential Gröbner basis) have made it possible to compute the 4 Killing vectors (KV) and 13 Killing tensors (KT) of rank 2 of the Kimura metric [Kimura 1976]

$$g_{tt} = -r^2/b, \; g_{rr} = 1/(rb)^2, \; g_{\theta\theta} = r^2, \; g_{\phi\phi} = r^2 \sin^2\theta,$$

where all other $g_{ij} = 0$ and $i,j = t, r, \theta, \phi$. The 4 KV come from its static spherical symmetry and among the 13 KT are 10 reducible KT which are all symmetrized products of 2 of the 4 KV, one trivial KT which is the metric $g_{ij}$ but also 2 non-trivial KT:

$$K_{tt} = 2tr^4/b, \; K_{tr} = r/b^2, \; K_{rr} = -2t/(rb)^4, \; K_{\phi\phi} = -2tr^4 \sin^2\theta \quad \text{and}$$

$$K_{tt} = t^2 r^4/b + r^2/b^2, \; K_{tr} = tr/b^2, \; K_{\theta\theta} = -t^2 r^4, \; K_{\phi\phi} = -t^2 r^4 \sin^2\theta.$$

In a different approach an algorithm was developed [Wolf 1998d] to write conditions (13) for KT of rank $r$ and the integrability conditions of order up to $r$ as a system of structural equations

$$(F_A)_{;k} = \sum_B \Gamma_{kAB} F_B \tag{14}$$

for unknowns $F_A$ which are the components of a Killing tensor $K_{i_1...i_r}$ and its symmetrized covariant derivatives of order up to $r$. The $\Gamma_{kAB}$ are expressions in the Riemann tensor for the given metric and its first $r$ derivatives. One advantage of this formulation is that all integrability conditions are linear algebraic conditions for the $F_A$. After completing

- lengthy tensor algebra computations with an especially written computer tensor program, to formulate the large tensorial system (14) (which has to be computed only once),
- programs that compute the relevant tensors in (14) for a given metric and in this way formulate (14) for a given metric,
- a module to compute integrability conditions for systems of the structure of (14),

the current efforts concentrate on computer algebra algorithms to solve large linear algebraic systems (the integrability conditions) which are too big for currently available algorithms and software.

**3.1.5.1.2 Computation of Conservation Laws of ODEs and PDEs** For a given system of differential equations $\Delta_\mu = 0$ for unknown functions $u = (u^1, u^2, \ldots, u^\nu)$ of independent variables $x = (x^1, x^2, \ldots, x^p)$, conservation laws can be formulated as a condition

$$\text{Div}\, P = 0 \quad (\text{mod}\ \Delta = 0,\ D_i\Delta = 0, \ldots) \tag{15}$$

for a conserved current $P = (P^1, \ldots, P^p)$. Another possible formulation is the condition

$$\text{Div}\, P = \sum_\nu Q^\nu \Delta_\nu \tag{16}$$

for a conserved current $P$ and characteristic functions $Q = (Q^1, \ldots, Q^\mu)$. Eliminating Div$P$, and hence $P$ with the Euler-Lagrange operator, we obtain conditions for the $Q_\nu$

$$\forall \nu: \quad 0 = E_\nu \left( \sum_\mu Q_\mu \Delta_\mu \right) = \sum_J (-D)_J \left( \frac{\partial}{\partial u_J^\nu} \sum_\mu Q_\mu \Delta_\mu \right) \tag{17}$$

which on the space of solutions of $\Delta_\mu = 0$ give the necessary conditions

$$0 = \sum_{\mu, J} (-D)_J \left( Q_\mu \frac{\partial \Delta_\mu}{\partial u_J^\nu} \right) \bigg|_{\Delta_\mu = 0} \quad \forall \nu. \tag{18}$$

Equations (18) are the conditions for adjoint symmetries of $\Delta = 0$.

Each of the approaches (15)-(18) has different advantages, therefore four programs CONLAW1/2/3/4 have been written to be able to use the most appropriate approach for any given problem (see [Wolf et al. 1999; Wolf 1998a]). Compared with other programs, e.g. [Göktaş and Hereman 1997], CONLAW can be used to find conservation laws that are non-polynomial (next example) and have an explicit variable dependence. The number of equations and variables are only restricted by complexity issues.

*Examples:* The Liouville equation for a function $u = u(x, y)$ reads $\Delta = u_{xy} - e^u$. Conservation laws with $Q$ of order zero found by CONLAW3 are

$$(f_x + f u_x)\Delta = D_x(-e^u f) + D_y(f_x u_x + f u_x^2/2), \quad f = f(x)\ \text{arbitrary},$$

with equivalent laws for $x \leftrightarrow y$.

To give another example, with CONLAW2 it was possible for the Kadomtsev-Petviashvili equation in its potential form

$$0 = [v_t + v_{3x} + v_x^2]_x - v_{2y}, \quad v = v(t, x, y),$$

to find non-trivial conservation laws, like the one given by

$$Q = -c_{4t}y^4 - 12c_{3t}xy^2 + 24c_{2t}y^2 v_x - 12c_{2t}x^2$$
$$+ 48c_t x v_x + 96 c_t y v_y + 48 c_t v + 144 c v_t,$$

with $c = c(t)$ being an arbitrary function. (The conserved current $P$ has been omitted due to its length.)

Allowing the computation of non-polynomial conservation laws, even involving arbitrary functions enabled the linearization of non-linear evolutionary PDE-systems of the Schrödinger type as described in [Sokolov and Wolf 1999].

Demonstration versions of the programs CRACK and CONLAW are accessible online under http://cathode.maths.qmw.ac.uk and the code including manuals on ftp.maths.qmw.ac.uk in the directory pub/tw/crack.

Thomas Wolf (London)

**3.1.5.2 Solving Differential Equations with** *MathLie* We discuss the symmetry analysis of differential equations by using *MathLie*. *MathLie* is a package written in MATHEMATICA 3.0 supporting the calculation of classical, non-classical, potential, approximative, and generalized symmetries. Two examples demonstrate the application to specific problems in thermodynamics and cosmology.

**3.1.5.2.1 Introduction** The package is able to solve ordinary differential equations of orders greater than two by quadratures. We show how Lie's theory becomes vital again by using computer algebra calculations with *MathLie*[Baumann 1998].

The theory behind *MathLie* is the symmetry analysis of Lie [Lie 1874] — see also section 2.11.3. His theory is useful in solving any kind of differential equations in an algorithmic way. The question raised by Lie at the end of the 19th century was of how differential equations can be solved systematically. Up to the present day this question is of topical interest for physicists, engineers, and mathematicians alike.

**3.1.5.2.2 Theory of Lie's Method** In his work Lie pointed out [Lie 1899] that a symmetry of any differential equation is defined as follows: *A Lie (point) symmetry is characterized by an infinitesimal transformation which leaves the given differential equation invariant under the transformation of all independent and dependent variables.*

Let us consider the general case of a nonlinear system of differential equations for an arbitrary number $q$ of unknown functions $u^\alpha$ which may depend on $p$ independent variables $x^i$. We denote these sets of variables simply by $u = (u^1, u^2, \ldots, u^q)$ and $x = (x^1, x^2, \ldots, x^p)$, respectively. The general case of a PDE is given by a system of $m$ nonlinear differential equations

$$\Delta^i(x, u_{(k)}) = 0, \quad i = 1, 2, \ldots, m, \tag{19}$$

of order $k$. The term $u_{(k)}$ is understood as $k$th derivative of $u$ with respect to $x$. We note that $m$, $k$, $p$ and $q$ are arbitrary, positive integers. Consider further a one-parameter $\epsilon$-Lie group of transformations

$$x^* = \Xi(x, u, \epsilon), \tag{20}$$
$$u^* = \Phi(x, u, \epsilon), \tag{21}$$

### 3.1.5 Differential Equations in Physics

under which (1) must be invariant. The asterisk * on the variables $x$ and $u$ denote the new variables. Invariance of (1) under the action of (2) and (3) means that any solution $u = \Theta(x)$ of (1) maps into some other solution $v = \Psi(x;\epsilon)$ of (1). Let $u = \Theta(x)$ be a solution of (1). If we replace the dependent and independent variables $u$ and $x$ by $v$ and $x^* = \Xi$, equations (1) become

$$\Delta^i(x^*, v_{(k)}) = 0 \quad i = 1, 2, \ldots, m. \tag{22}$$

The central observation made by Lie was that (2) and (3) can be represented by an infinitesimal transformation. Expanding equations (2) and (3) around the identity $\epsilon = 0$, we can generate the infinitesimal transformations

$$x^{*i} = x^i + \epsilon \xi^i(x, u) + O(\epsilon^2), \quad i = 1, 2, \ldots, p, \tag{23}$$
$$u^{*\alpha} = u^\alpha + \epsilon \phi^\alpha(x, u) + O(\epsilon^2), \quad \alpha = 1, 2, \ldots, q, \tag{24}$$

where the functions $\xi^i$ and $\phi^\alpha$ are the infinitesimals of the transformations for the independent and dependent variables, respectively. The aim of Lie's theory is to find the unknown infinitesimals for a specific differential equation. In order to find them, we have to extend or to prolong the transformation group in order to include the properties of the derivatives. This is an infinitesimal approach which considers the Lie algebra $L$ corresponding to the Lie group $G$. The infinitesimal transformation (5,6) can be put into the form of

$$\vec{v} = \sum_{i=1}^{p} \xi^i(x, u) \frac{\partial}{\partial x^i} + \sum_{\alpha=1}^{q} \phi^\alpha(x, u) \frac{\partial}{\partial u^\alpha}, \tag{25}$$

where $\vec{v}$ represents a linear combination of the vector fields generating $L$ which in turn is based on the characteristic quantities $\xi^i$ and $\phi^\alpha$ of the transformation (5,6). The algorithm used in *MathLie* for finding the infinitesimals $\xi^i$ and $\phi^\alpha$ is described below. We emphasize that the infinitesimals in this simple form only depend on independent and dependent variables. Transformations (5,6), together with the transformations for the first, second, ... derivatives of the $u^\alpha$'s, are called first, second, ... prolongations. Using these various extensions, the infinitesimal criterion for the invariance of (1) under the group (2,3) is derivable by

$$\mathrm{pr}^{(k)}\vec{v}\Delta|_{\Delta=0} = 0, \tag{26}$$

where $\mathrm{pr}^{(k)}\vec{v}$ is the $k$th prolongation of the vector field $\vec{v}$ [Olver 1993].

From an algorithmic point of view the steps to calculate the invariance condition (8) and thus the infinitesimals are:

1) calculate the prolongation of the system of differential equations up to $k$th order by equation (11) $\mathrm{pr}^{(k)}\vec{v}\Delta = 0$,
2) use the equations themselves to eliminate redundant information of the prolongation $\mathrm{pr}^{(k)}\vec{v}\Delta|_{\Delta=0} = 0$,
3) extract the determining equations from the prolongation by setting equal to zero the coefficients of the derivatives in the dependent variables,

**4)** solve the resulting determining equations.

Steps one to three are standard in any computer algebra program on Lie symmetries. The fourth step in *MathLie* deals with the solution of the determining equations using the theory by Janet [Janet 1920] and Riquier [Riquier 1910] and differential Gröbner bases [Baumann 1998] — cf. sections 2.11.5 and 2.11.6.

So far we have briefly discussed Lie's theory to introduce the basic notation on point symmetries. Other kinds of symmetries are discussed in [Baumann 1998]. The following section will demonstrate how the theoretical concepts are realized in *MathLie*.

**3.1.5.2.3 Joining Lie With *MathLie*** The program is connected with Lie's theory by special functions. These functions are located around the central parts in Lie's theory. The main mathematical notions in Lie's theory are prolongations, determining equations, infinitesimals, reductions etc. Each of these mathematical expressions is represented by a single function in *MathLie*; i.e. Prolongation[], DeterminingEquations[], Infinitesimals[], and LieReduction[]. In order to demonstrate how these functions can be used to calculate the specific expression in Lie's theory, let us consider two examples.

**3.1.5.2.4 Example: The Cattaneo Equation [Cattaneo 1948]** An equation closely related to the Cattaneo equation is the heat equation, a favorite example of Lie [Lie and Scheffers 1891]. The Cattaneo equation is given by a second-order partial differential equation for a field $u$ by

$CattaneoEquation= \partial_t u[x,t] + \tau \partial_{t,t} u[x,t] - \delta \partial_{x,x} u[x,t] == 0;$
**CattaneoEquation//LTF**

$$u_t + \tau u_{t,t} - \delta u_{x,x} == 0$$

The function LTF[] of *MathLie* changes the standard MATHEMATICA notation to a more readable index notation. The Cattaneo equation is discussed in connection with a finite propagation velocity in *Extended Irreversible Thermodynamics* by Nonnenmacher [Nonnenmacher 1980] and Jagher [de Jagher 1980]. The hyperbolic equation by Cattaneo allows the infinitesimals

$infCattaneoEquation =$
**Infinitesimals**$[CattaneoEquation, \{u\}, \{x,t\}, \{\tau, \delta\}];$
**infCattaneoEquation//LTF**

$$-\frac{(F_1)_t + \tau(F_1)_{t,t} - \delta(F_1)_{x,x}}{\delta} == 0$$
$$\xi_1 == k1 + \frac{k4 t \delta}{\tau}$$
$$\xi_2 == k3 + k4 x$$
$$\phi_1 == u\left(k2 - \frac{k4 x}{2\tau}\right) + F_1$$

consisting of an infinite-dimensional symmetry group, given by $F_1$, containing a four-dimensional finite subgroup denoted by $k1, \ldots, k4$. The infinitesimals are

### 3.1.5 Differential Equations in Physics

automatically determined by *MathLie*'s function Infinitesimals[]. For its calculation this function needs the equation of motion, the dependent and independent variables, and an optional list of parameters occurring in the equation. The derived symmetries are useful to calculate a similarity reduction and thus to solve the equation.

One of two possible reductions follows, if we extract from the main group a subgroup by choosing $k1 = k2 = k3 = 0$, $k4 = 1$, and $F_1 = 0$.

$$subgroup2 = \{\{\xi[1][x,t,u], \xi[2][x,t,u]\},$$
$$\{\phi[1][x,t,u]\}\}/.infCattaneoEquation[[1]]/.$$
$$\{k1 \to 0, k2 \to 0, k3 \to 0, k4 \to 1, free[\_][\_\_\_] \to 0\}$$

$$\{\{\frac{t\delta}{\tau}, x\}, \{-\frac{ux}{2\tau}\}\}$$

The related similarity reduction is calculated by the function LieReduction[] as follows:

$reduction2 = LieReduction[$
 $CattaneoEquation, \{u\}, \{x,t\}, subgroup2[[1]], subgroup2[[2]]];$
$reduction2/.zeta1 \to \zeta//$**Flatten**$//$**LTF**

$InverseFunction :: "ifun" : "$Warning: Inverse functions are being used. Values may be lost for multivalued inverses."

$-\zeta + \frac{1}{2}\left(-x^2 + \frac{t^2\delta}{\tau}\right) == 0$
$e^{\frac{t}{2\tau}}u - F_1 == 0$
$-F_1 + 8\delta\tau(F_1)_\zeta + 8\delta\zeta\tau(F_1)_{\zeta,\zeta} == 0$

The similarity solution of the Cattaneo equation follows by solving the second-order ordinary differential equation of the similarity function $F_1$:

$SimilaritySolution2 = DSolve[reduction2[[3]], F1, zeta1]$

$\{\{F1 \to \left(\text{BesselI}\left[0, \frac{\sqrt{\frac{1}{\delta\tau}}\sqrt{\#1}}{\sqrt{2}}\right]C[1] + \right.$

$\left. \text{BesselK}\left[0, \frac{\sqrt{\frac{1}{\delta\tau}}\sqrt{\#1}}{\sqrt{2}}\right]C[2]\&\right)\}\}$

The solution derived contains Bessel functions $I_0$ and $K_0$. The representation of the solution in original coordinates follows from the inversion of the similarity transformations to be

$solution2 =$
**Solve**$[reduction2[[2]]/.SimilaritySolution2, u]//$**Flatten**

$$\{u \to \exp\left[-\tfrac{t}{2\tau}\right]\left(\text{BesselI}[0, \tfrac{1}{2}\sqrt{\tfrac{t^2\delta}{\tau} - x^2}\sqrt{\tfrac{1}{\delta\tau}}]C[1] + \right.$$
$$\left. \text{BesselK}[0, \tfrac{1}{2}\sqrt{\tfrac{t^2\delta}{\tau} - x^2}\sqrt{\tfrac{1}{\delta\tau}}]C[2]\right)\}$$

Since the solution follows from a second order ODE, it depends on two constants of integration $C[1] = c_1$ and $C[2] = c_2$.

**3.1.5.2.5 Example: Early Epochs of the Universe** The second example is concerned with the solution of a second-order ordinary differential equation used to describe the early epoch of the universe. The model of the early universe is discussed by D'Inverno [D'Inverno 1992] by means of the equation

$$EarlyEpoch = \frac{\partial_{t,t}R[t]}{R[t]} + \frac{(\partial_t R[t])^2}{R[t]^2} + \frac{k}{R[t]^2}; \textbf{EarlyEpoch}//\textbf{LTF}$$

$$\frac{k}{R^2} + \frac{R_t^2}{R^2} + \frac{R_{t,t}}{R} == 0$$

where $R$ is the scale function and $k$ a constant. The point symmetries of this second order ordinary differential equation (ODE) are determined by

$InfiEarlyEpoch = \textbf{Infinitesimals}[EarlyEpoch, R, t, \{k\}];$
**InfiEarlyEpoch//LTF**

$$\phi_1 == -\frac{2k8R^4 + kR^2(k2+2k1t) - 2k^2\left(k4+t\left(k5+k6t+k7t^2+k8t^3\right)\right)}{2k^2R}$$

$$\xi_1 == \frac{R^2(k1-k7-4k8t) - k\left(k3+k5+(k2+2k6)t+(k1+3k7)t^2+4k8t^3\right)}{2k^2}$$

The resulting symmetry group consists of an eight-dimensional group. The dimension of the group is sufficiently large to find a solution. In fact, for a second order ODE, we need at least a two-dimensional symmetry group to apply Lie's method of quadrature [Lie and Scheffers 1891; Baumann 1998]. The solution for the above model follows by applying the *MathLie* function SecondOrderIntegrate[] to the equation

$$solution = \textbf{SecondOrderIntegrate}[EarlyEpoch, R, t, \{k\}]$$
$$\{R \to$$
$$\text{Function}\left[t, -\frac{\sqrt{k}\sqrt{4k^2 - 4kt\textbf{CI}[1] + t^2\textbf{CI}[1]^2 - 8kt^2\textbf{CI}[2]}}{\sqrt{-\textbf{CI}[1]^2 + 8k\textbf{CI}[2]}}\right]\}$$

The symbols CI[1] and CI[2] denote constants of integration for the second order ODE. Inserting the solution found into the original equation of motion, we can verify the result:

$$EarlyEpoch/.solution//\text{Simplify}$$
$$0$$

We note that the solution derived was calculated without any use of the function DSolve[] of MATHEMATICA. The solution is purely based on the theory of Lie to integrate ordinary differential equations.

**3.1.5.2.6 Conclusions** We presented a small part of *MathLie*'s capabilities to determine point symmetries and solutions. The point symmetries and solutions are derivable by the functions Infinitesimals[], LieReduction[] and SecondOrderIntegrate[]. These functions combine, in a nutshell, the essential tools for a symmetry analysis. The functions are designed in such a way that all standard applications of a symmetry analysis are manageable. However, *MathLie* contains additional functions broadening its application to non-classical, potential, generalized, and approximate symmetry analysis. The functions of *MathLie* are designed in such a way that each mathematical notion in Lie's theory has its counterpart in *MathLie* [Baumann 1998].

Gerd Baumann (Ulm)

## 3.2 Mathematics

In Mathematics, the development of algorithms for computer algebra computer algebra systems and applications often are intertwined. Hence, separating the topics seems to be somewhat superficial. For that reason, many sections of chapter 2 directly can be interpreted as application of computer algebra to mathematics, too.

Nevertheless, we would like to list a few selected mathematical subjects which require the use of sufficiently powerful computer algebra systems and algorithmus:

- classification of finite groups, and their presentations (including the Atlas project for ordinary and modular representations),
- systematic investigation of algebraic number and function fields, and determination of their invariants,
- study of systems of nonlinear algebraic equations with regard to problems from commutative (and also non-commutative) algebra, and algebraic geometry,
- experimental and theoretical investigation of special classes of diophantine equations,
- phenomenological and structural investigation of dynamic systems,
- coupling of symbolic and numeric methods for solving numerical problems effectively, while reducing round-off errors at the same time,

The contributions to this chapter give examples for some of the topics in more detail.

Johannes Grabmeier (Deggendorf)

## 3.2.1 Computer Algebra in Group Theory

Among the reasons that *Computational Group Theory* was much more systematically and constantly developed than computational methods in other parts of mathematics, were problems arising in the classification of the finite simple groups. A typical task was to construct a new sporadic simple group whose existence had been conjectured before.

For example, C. Sims constructed the Baby Monster, a sporadic simple group of order 4 154 781 481 226 426 191 177 580 544 000 000, as a permutation group on 13 571 955 000 points [Sims 1980]. To achieve this remarkable result he developed special methods for the computation in large permutation groups. A further example is provided by R. Parker's **Meat-Axe** [Parker 1984] for splitting modular representations of finite groups. The **Meat-Axe** was used to prove the existence of the group $J_4$, the largest of Janko's sporadic groups.

Some time later, using similar methods, R. Wilson constructed a 4 370-dimensional matrix representation over the field with two elements for the Baby Monster [Wilson 1993]. As an application, this representation was used to prove the existence of a rationally rigid generating system and thus to realize the Baby Monster as a Galois group over the rationals. (So far no existence proof for such a Galois realization of the Baby Monster is known which does not use Wilson's explicit matrix representation). The most spectacular recent application of the **Meat-Axe** is the explicit construction of the largest sporadic group, the Monster, as a group of $(196\,882 \times 196\,882)$-matrices over the field with two elements [Linton et al. 1998], see also 4.2.10. Many more explicit matrix representations for specific finite groups are now available in Wilson's world-wide-web *Atlas of Finite Group Representations* [Wilson 1998], see http://www.mat.bham.ac.uk/atlas.

After the classification of the finite simple groups an important task is the systematic investigation of the quasi-simple groups. The character tables and information about the subgroup lattices of all sporadic groups as well as of the first few members of the infinite series are contained in the *Atlas of Finite Groups* [Conway et al. 1985]. These data are also available through **GAP** [GAP 1997] (see also 4.2.18). Supplementary information is provided by collections of modular character tables [Hiss and Lux 1989; Jansen et al. 1995], also available through **GAP**, as well as generic character tables of groups of Lie type (of reasonable small rank) through **CHEVIE** [Geck et al. 1996] (see also 4.2.9). Such generic character tables are applied, e.g., to prove the existence of rigid generating systems providing Galois realizations for the groups in question [Malle 1991].

Another typical area of applications of computer algebra in group theory—in fact the one with the longest history—is the analysis of finitely presented groups (see [Sims 1994] and 2.7.9). One of the problems is to decide whether such a group is finite or infinite. New ideas and techniques have led to solutions of various longstanding problems in this field, in particular to proofs of infiniteness of certain groups given by specific presentations [Holt and Hurt 1998]. It is remarkable that some of the techniques involve the construction of explicit matrix representations of the groups in question [Plesken 1999].

A further objective of finite group theory is the classification of $p$-groups, or, more generally, of solvable groups. To compute with such groups, special presentations, such as *power commutator presentations* (respectively *polycyclic presentations*) and special algorithms such as the *p-quotient method* and *collection* have been developed (see 2.7.8 for more details). These approaches led to impressive new results, e.g. the classification and construction of all 56 092 groups of order 256 [O'Brien 1991] and the enumeration of the 10 494 213 groups of order 512 [Eick and O'Brien 1999].

Other ambitious projects making substantial use of computer algebra are concerned with the classification of groups of small order [Besche and Eick 1998], primitive and transitive permutation groups [Hulpke 1996b], perfect groups [Holt and Plesken 1989], matrix groups [Nebe and Plesken 1995], or crystallographic groups [Plesken et al. 1998], see 4.2.7.

Computer algebra has also been successfully applied to the computation of cohomology groups and cohomology rings of finite groups [Holt 1985; Carlson 1996; Green 1997; Adem and Milgram 1994], compare also 3.2.3.5. Finally, recent algorithmic developments [Kemper 1996] have lead to new and unexpected results [Kemper and Malle 1997] in invariant theory of finite groups, see also 2.2.8.

Additional references: [Durham II 1980; Curtis and Wilson 1998; Atkinson 1984b; Matzat et al. 1999; Michler and Ringel 1991].

Gerhard Hiss (Aachen)[*]

---

[*] This is a translated, updated and extended version of Section 3.2.1 of [Fachgruppe Computeralgebra der GI and DMV and GAMM 1993], written by B. Heinrich Matzat (Heidelberg) and Joachim Neubüser (Aachen)

### 3.2.2 The Tangent Cone Algorithm and Applications in the Theory of Singularities

In this section, we should like to present some interesting conjectures and problems in local algebraic geometry. Two of these conjectures were decided with the help of computer algebra systems. One is still open. However, experiments conducted on the system SINGULAR (see 4.2.39) already yielded a be it small but positive partial result. Moreover, standard bases for global and local orderings (i.e., well–orderings and tangent cone orderings) were both used in order to construct polynomials of small degree with surprisingly high $A_k$–singularities. In all cases, the results could be proved after the fact without computer support, nevertheless they would not have been found without application of these systems.

**3.2.2.1  A Generalization of a Theorem by K. Saito** The first conjecture is based on a beautiful and often used theorem by K. Saito [Saito 1971]: let $f \in \mathbb{C}[[x_1, \ldots, x_n]]$ define an isolated hypersurface singularity $(X, 0) \subset (\mathbb{C}^n, 0)$,

i.e.,
$$\tau(f) := \dim_{\mathbb{C}} \mathbb{C}[[x_1,\ldots,x_n]] / \left(f, \frac{\partial f}{\partial x_1},\ldots, \frac{\partial f}{\partial x_n}\right) < \infty.$$

Then the following statements are equivalent:

(i) $f \in \left(\frac{\partial f}{\partial x_1},\ldots, \frac{\partial f}{\partial x_n}\right)$, i.e.,
$$\tau(f) = \mu(f) := \dim_{\mathbb{C}} \mathbb{C}[[x_1,\ldots,x_n]] / \left(\frac{\partial f}{\partial x_1},\ldots, \frac{\partial f}{\partial x_n}\right).$$

(ii) There exists an automorphism $\varphi : \mathbb{C}[[x_1,\ldots,x_n]] \to \mathbb{C}[[x_1,\ldots,x_n]]$, such that $f \circ \varphi = p$ is a weighted homogenous polynomial (i.e., there exist positive integers $w_1,\ldots,w_n,d$, such that $p(t^{w_1}x_1,\ldots,t^{w_n}x_n) = t^d p(x_1,\ldots,x_n)$).

(iii) The holomorphic Poincaré complex
$$0 \to \mathbb{C} \to \mathcal{O}_{X,0} \to \Omega^1_{X,0} \xrightarrow{d} \ldots \xrightarrow{d} \Omega^n_{X,0} \to 0$$
is exact.

Property (ii) states that the germ $(X,0)$ has a holomorphic contraction onto the origin. It immediately implies the exactness of the holomorphic Poincaré complex by the usual proof of Poincaré's lemma. The inverse is interesting, in that for hypersurface singularities the validity of the holomorphic Poincaré lemma already implies holomorphic contractibility.

Now, all statements are meaningful if we consider complete intersections $(X,0)$ in lieu of hypersurfaces resp. general reduced curve singularities: $\mu = \mu(X,0)$ is the Milnor number (i.e., the number of vanishing cycles when being smoothed), and $\tau = \tau(X,0)$ is the Tjurina number (i.e., $\dim_{\mathbb{C}} T^1_{X,0}$, which is the number of linearly independent, infinitesimal deformations of $(X,0)$). In [Greuel et al. 1985] has been shown that for Gorenstein curve singularities (especially for complete intersections) (i) and (ii) are equivalent, and that this is equivalent to exactness of the modified Poincaré complex $\Omega^\bullet_{X,0}/\text{torsion}$. However, we were unable to prove (iii) $\Rightarrow$ (ii).

With the help of an implementation of the tangent cone algorithm (in a preliminary version of SINGULAR), Pfister and Schönemann were able to find first examples in, [Pfister and Schönemann 1989], for which the Poincaré complex is exact, yet they are not quasi-homogeneous. Among them are, e.g., the unimodular complete intersections of

$$f = xy + z^{\ell-1}, \quad g = xz + y^{k-1} + yz^2, \quad 4 \leq \ell \leq k, \quad k \geq 5 \text{ with}$$
$$\mu = k + \ell + 2, \quad \dim_{\mathbb{C}} \Omega^2_{X,0} = k + \ell + 3, \quad \dim_{\mathbb{C}} \Omega^3_{X,0} = 1, \quad \tau = k + \ell + 1.$$

**3.2.2.2 A Conjecture by Mumford** The second conjecture is by Mumford, and is concerned with structure of the moduli space of curve singularities.

Let $f \in \mathbb{C}[[x,y]]$ such that $f = 0$ defines an irreducible, isolated, plane curve singularity $(X,0) \subset (\mathbb{C}^2,0)$. We are looking for a moduli space of curve singularities which are of fixed topological type (which have fixed Puiseux pairs), i.e., an algebraic variety whose points uniquely correspond to the isomorphy classes of such singularities.

One way of constructing such moduli spaces is the following [Laudal and Pfister 1988; Greuel and Pfister 1996]: first, we create a family $F(x, y, t)$, $t \in \mathbb{C}^N$ containing all curve singularities we are interested in (the $\mu$-constant deformation of the "most special" object). In general, this may be difficult, but it is easy for one Puiseux pair $(p, q)$, with $\text{GCD}(p, q) = 1$, where

$$F = f + \sum_{\substack{qi+pj>pq \\ i \leq p-2, j \leq q-2}} t_{ij} x^i y^j, \qquad f = x^p + y^q, t = (t_{ij}) \in \mathbb{C}^N$$

is such a family. However, that family contains trivial sub-families, i.e., there exist sub-varieties in the parameter space $\mathbb{C}^N$ along which the analytical type of the curve singularity defined by $F_t(x, y) = 0$ remains constant under variations of $t$.

One can show that for a family of isolated hypersurface singularities $F = f + \Sigma t_\alpha x^\alpha$ with constant Milnor number these trivial sub-families are exactly the integral manifolds of a Lie algebra $L$, where

$$L = \text{Ker}(\text{Der}_{\mathbb{C}} \mathbb{C}[t] \to \mathbb{C}[t][[x, y]]/(F, \tfrac{\partial F}{\partial x}, \tfrac{\partial F}{\partial y})), \qquad \delta \mapsto [\delta(F)]$$

is the kernel of the Kodaira-Spencer map.

Although the orbit space $\mathcal{M} = \mathbb{C}^N / L$ parametrizes the isomorphism classes of singularities, in general it does not carry any useful algebraic structure, but is only a topological space. If the orbit dimension (the dimension of the integral manifolds) is not a constant then $\mathcal{M}$ is not even Hausdorff.

Therefore, it is necessary to first decompose $\mathbb{C}^N$ into invariant strata $S_i$ over which the quotient $S_i/L$ exists as an algebraic variety.

Now, the claim of the conjecture was that stratification by the orbit dimension is not only necessary but also sufficient. For strata of maximal dimension, this was in fact proved in [Laudal and Pfister 1988]. At the same time, using the tangent cone algorithm, it was possible to construct a counterexample for strata of low dimension (for generalizations to higher dimensions see [Greuel and Pfister 1993]).

**3.2.2.3 Zariski's Multiplicity Conjecture** The third, and the most interesting conjecture is still unsolved. It is Zariski's multiplicity conjecture: Let $f \in \mathbb{C}[[x]] = \mathbb{C}[[x_1, \ldots, x_n]]$, $f = \Sigma c_\alpha x^\alpha$ be a non-constant, convergent power series, and let $\text{mult}(f) = \min\{|\alpha| \mid c_\alpha \neq 0\}$ be the multiplicity of $f$. Furthermore, let $B \subset \mathbb{C}^n$ be a sufficiently small ball centered at 0, and let $X = f^{-1}(0) \cap B$ be the hypersurface singularity defined by $f$. If $g \in \mathbb{C}[[x]]$ is another power series, and $Y = g^{-1}(0) \cap B$ is the singularity defined by $g$, then $f$ and $g$ are called topologically equivalent if and only if there exists a homeomorphism $h : B \xrightarrow{\sim} B$ such that $h(X) = Y$. In 1971, Zariski posed the question [Zariski 1971] whether topological equivalence implied identical multiplicity. The general problem is open to date, although a positive answer could be found in the following special cases:

- $n = 2$, i.e., plane curve singularities (Zariski, Lê Dũng Tráng);
- $f$ weighted homogeneous, and $g$ a small deformation of $f$ (Greuel [Greuel 1986], O'Shea [O'Shea 1987]).

However, homeomorphisms are highly transcendental objects. Without making use of suitable invariants, the question of topological equivalence cannot be decided in general. If $f_t(x) = f(x) + t f_1(x) + t^2 f_2(x) + \ldots$ is a small deformation of $f$, such that $f_t$ is analytically isomorphic to $g$ (for $t \neq 0$, and sufficiently small) then $\mu(f_0) = \mu(f_t)$ for small $t$ if $f$ is topologically equivalent to $g$. Here $\mu(f) = \dim_{\mathbb{C}} \mathbb{C}[[x]]/(\frac{\partial f}{\partial x_1}, \ldots, \frac{\partial f}{\partial x_n})$ is the Milnor number of $f$. If $f$ has an isolated singularity then $\mu(f) < \infty$, and consequently $\mu$ can be computed by the tangent cone algorithm.

Given the fact that Zariski's conjecture withstood decision attempts for more than 20 years, it is worthwhile looking for counter–examples. However, due to the cited partial results, the range for possible counter–examples is within very high Milnor numbers. Computing the Milnor number by hand is practically impossible for such examples. So far, no example violating the multiplicity conjecture has been discovered. On the other hand, investigating a series of examples revealed the proof of Zariski's conjecture for deformations of an isolated singularity of type $f_t(x, y, z) = g_t(x, y) + z^2 h_t(x, y, z)$ with $\text{mult}(g_t) < \text{mult}(f_0)$ for $t \neq 0$ (cf. [Greuel and Pfister 1996]).

#### 3.2.2.4 Curves of Minimal Degree With High Singularities
Another way to use computer algebra is to construct interesting, explicit examples. A recent application of this kind is due to C. Lossen in connection with the classical problem concerning the existence of irreducible projective curves $C$ in $\mathbb{P}^2_{\mathbb{C}}$ of small degree $d$ with many resp. complicated singularities.

Let $z$ be a point of $C$ and $\mu(C, z) = \dim_{\mathbb{C}} \mathbb{C}[[x, y]]/(f_x, f_y)$ the Milnor number of $C$ at $z$, where $f$ is a local equation of $C$ at $z$. Note that $\mu(C, z) \neq 0$ if and only if $C$ is singular at $z$ and that, e.g., $\mu(A_k) = k$ where an $A_k$–singularity is given by $x^2 - y^{k+1} = 0$. In particular, for a node $(A_1)$ $\mu = 1$ and for an ordinary cusp $(A_2)$ $\mu = 2$. The problem, which can be traced back to Euler, Plücker, Segre, Severi, is to determine the maximal possible number of singularities, each counted with its Milnor number, on a curve of degree $d$.

The only complete answer so far is known for nodal curves and is due to Severi (1921): sufficient and necessary for the existence of an irreducible nodal curve is the condition

$$\#(\text{nodes}) \leq \frac{(d-1)(d-2)}{2}.$$

For arbitrary (topological) types $S_1, \ldots, S_r$ a classical, necessary condition for the existence of an irreducible curve $C$ of degree $d$ with exactly $r$ singularities, of types $S_1, \ldots, S_r$, is

$$\sum_{i=1}^{r} \mu(S_i) \leq (d-1)^2,$$

i.e., the number of singularities, counted with Milnor number, is bounded by a quadratic function in $d$.

Until recently, only linear functions in $d$ were known as a general, sufficient bound for the existence. In [Greuel et al. 1998], the following sufficient bound for existence was found,

$$\sum \mu(S_i) \leq \frac{d^2 + 4d}{81},$$

which is quadratic in $d$ and, hence, asymtotically optimal. The proof of this result is, however, not constructive and the question of explicitly given equations for such curves with high singularities remained open.

Using the system SINGULAR for checking several ideas to construct such singularities, Lossen [Lossen 1999] was able to construct a series of curves o f small degree and high $A_k-$ resp. $D_k$-singularities (even higher than the above bound). For example, the irreducible curve with affine equation

$$y^2 - 2y\bigl(x^{14} + \tfrac{1}{2}x^{13}y^2 - \tfrac{1}{8}x^{12}y^4 + \tfrac{1}{16}x^{11}y^6 - \tfrac{5}{128}x^{10}y^8 + \tfrac{7}{256}x^9 y^{10}$$
$$-\tfrac{21}{1024}x^8 y^{12} + \tfrac{33}{2048}x^7 y^{14} - \tfrac{429}{32768}x^6 y^{16} + \tfrac{715}{65536}x^5 y^{18} - \tfrac{2431}{262144}x^4 y^{20}$$
$$+\tfrac{4199}{524288}x^3 y^{22} - \tfrac{29393}{4194304}x^2 y^{24} + \tfrac{52003}{8388608}xy^{26} - \tfrac{185725}{33554432}y^{28}\bigr)$$
$$+ x^{28} + x^{27}y^2$$

has degree 29 and an $A_{432}$–singularity as its only singularity. Indeed, computing a standard basis of $(f, f_x, f_y)$ with respect to a local ordering we get $\dim_\mathbb{C} \mathbb{C}[[x,y]]/(f, f_x, f_y) = 432$, hence the curve $f = 0$ has an $A_{432}$–singularity at 0. Then, computing a standard basis with respect to a global ordering, we get $\dim_\mathbb{C} \mathbb{C}[x,y]/(f, f_x, f_y) = 432$, hence, $f = 0$ has only one singularitity in the affine plane. Now it is easy to see that $f = 0$ is irreducible and has no singularity at infinity. In this case, checking the correctness of the example with the help of computer algebra is much safer than a proof "by hand", if at all possible.

Gert-Martin Greuel (Kaiserslautern)

### 3.2.3 Automatic Theorem Proving in Geometry

**3.2.3.1 Problems and Approaches** The automatic proving of geometric theorems is a particularly successful application area of computer algebra within mathematics. Extensive and detailed studies [Chou 1988; Dolzmann et al. 1998b; Kapur 1986b; Kutzler 1988; Kutzler and Stifter 1986; Wang 1995; Wu 1984a, 1986] have produced automatic proofs for a large number of classical and more recent geometric theorems. They have even supported the discovering of new such theorems. Though the emphasis is on plane geometry, the approaches are valid quite generally for finite-dimensional real geometry, i.e. for $\mathbb{R}^n$ as a Euclidean space. In the following we shall restrict ourselves to the plane geometry of points, lines, triangles, quadrangles, and circles.

The general approach follows Descartes' idea of algebraization of geometry: By introducing a Cartesian coordinate system the considered geometric configurations are described via polynomial equations. The position and the scaling of the coordinate system can be chosen arbitrarily. In practice they will both be chosen in such a way that the resulting algebraic description of the geometric configuration becomes as simple as possible. Angles are coded by introducing extra variables for their real-valued trigonometric functions. Altogether, the configurations consists only of polynomial equations and inequalities.

Many geometric theorems in the plane are *closure theorems*. Their conclusion is of the type "three points lie on a straight line or on a circle" or "three straight lines meet in one point." Such a conclusion translates algebraically via the chosen coordinate system into a polynomial equation. The resulting algebraic translation of the closure theorem is then of the following form:

$$\bigwedge_{i=1}^{l} f_i(x_1, \ldots, x_k) = 0 \longrightarrow f_0(x_1, \ldots, x_k) = 0. \tag{27}$$

The $f_i(x_1, \ldots, x_k)$ are polynomials with integer coefficients. As a rule, they are linear or quadratic according to the geometric figures they describe. The following simple example illustrates the approach.

*Example 4.* Let $ABCD$ be a parallelogram, and let $E$ be the intersection point of its diagonals. We claim that $E$ is the midpoint of both diagonals. We choose the Cartesian coordinate system and the scaling such that $A = (0,0)$, $B = (1,0)$, $C = (1+u, v)$, $D = (u, v)$, and $E = (x, y)$. The situation is pictured in Figure 7.

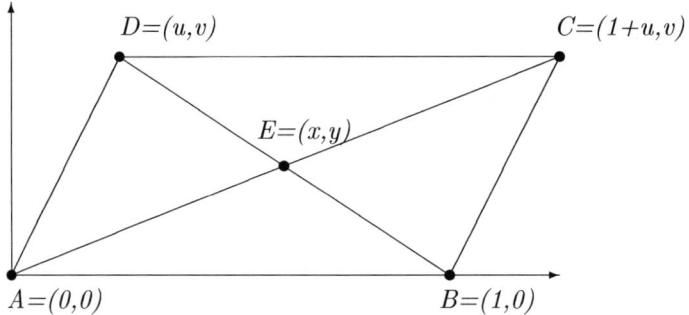

**Fig. 7.** Parallelogram diagonals intersect in their midpoint

The condition that $E$ is the intersection point of the diagonals $\overline{AC}$ and $\overline{BD}$ is expressed by the equations

$$\frac{y}{x} = \frac{v}{1+u} \quad \text{and} \quad \frac{y}{1-x} = \frac{v}{1-u}.$$

After clearing denominators a first algebraic translation of our theorem reads as follows:

$$\big(y(1+u) = vx \wedge y(1-u) = v(1-x)\big) \longrightarrow x^2 + y^2 = (1+u-x)^2 + (v-y)^2.$$

### 3.2.3 Automatic Theorem Proving in Geometry

An obvious approach is now to try and prove the geometric statement by verifying its algebraic translation for all real values of the variables $x_1, \ldots, x_k$. In general this will fail. The reason is that in the algebraic translation of the geometric configuration one forgets to enter *non-degeneracy* conditions in the hypothesis. Such conditions would, e.g., guarantee that certain points do not coincide or that a certain triangle does not collapse to a line. Non-degeneracy conditions are reflected in the algebraic description as additional polynomial *disequations*, i.e. negated equations in contrast to inequalities involving orders, $g_i \neq 0$ in the hypothesis. In the example above such a condition could be $v \neq 0$. Instead of conjunctively adding disequations to the hypothesis, the product $g = \prod_i g_i$ of the corresponding left hand side polynomials can be multiplicatively added to the equation in the conclusion. This yields again a formula of the form (27):

$$\bigwedge_{i=1}^{l} f_i(x_1, \ldots, x_k) = 0 \longrightarrow f_0(x_1, \ldots, x_k) g(x_1, \ldots, x_k) = 0. \tag{28}$$

Adding non-degeneracy conditions by hand is extremely tedious. In all the approaches described below such conditions can therefore be generated automatically. The user is only required to split the set of variables into two disjoint parts: the *parameters* of the configuration, which can be chosen arbitrarily, and the remaining *dependent variables*, whose values are determined (not always uniquely) by the values of the parameters and the configuration. Based on this information the system will generate polynomial disequations in the parameters that are sufficient for the validity of the geometric statement in its algebraic form. These side conditions can frequently be interpreted as geometric non-degeneracy conditions. In the example above $u$, $v$ can be chosen as parameters, and $x$, $y$ can be chosen as dependent variables.

The existing methods for geometric theorem proving can be roughly divided into two classes, namely *complex* and *real* methods. The following approaches belong to the first class:

- The Wu–Ritt method using characteristic sets, cf. [Chou 1988; Wu 1984a, 1986] and 2.2.7,
- Gröbner Basis techniques, cf. [Kapur 1986b; Kutzler 1988; Kutzler and Stifter 1986] and 2.2.5, complex elimination methods based on ideas by A. Seidenberg, cf. [Carrá-Ferro and Gallo 1987; Wang 1993, 1995].

Here the decisive idea is to prove the desired implication (27) or (28) for all complex values of the variables instead of considering only real values of these variables. If this goal succeeds, the desired geometric theorem is proved *a fortiori*. If it fails, however, there can be no conclusion made about the validity of the geometric theorem.

It is an amazing fact without sufficient theoretical explanation today, that the overwhelming majority of geometric theorems in their "proper" algebraic formulation hold over the complex numbers as well. Trivial exceptions are theorems that claim the existence of points that exist only in the complex, but not in the real plane. Some non-trivial exceptions are listed in [Kutzler 1988].

The second class comprises *genuinely real methods*. This includes decision methods and quantifier elimination methods for the elementary theory of the real numbers as an ordered field. They admit as input an arbitrary first-order formula in the language of ordered fields. Formulas of this kind are obtained from polynomial equations and polynomial inequalities via boolean combinations and quantification of some of the variables. A *decision method* answers the question of universal validity of such formula in the real numbers with either "yes" or "no." A *quantifier elimination method* assigns to an arbitrary formula with quantified variables an equivalent formula without quantified variables. Such quantifier-free formulas can then be easily evaluated for all real values of the variables. The existence of decision methods and quantifier elimination methods for the elementary theory of the real numbers was proved by A. Tarski around 1935. Since then the asymptotic complexity of such methods has been improved considerably, cf. [Davenport and Heintz 1988; Weispfenning 1988; Renegar 1992a,b,c; Basu et al. 1994] and 2.15.2.

For geometric theorem proving these methods have the advantage of being able to prove also theorems that hold only over the reals but not over the complex numbers. Moreover the inputs can be much more general than closure theorems. In particular they can include polynomial inequalities and both existential and universal quantification of variables in arbitrary combinations. Finally, a real quantifier elimination method can also be used to *find* new geometric theorems: On input of a geometric conjecture with some of the real parameters as free variables and the remaining variables universally quantified, the method outputs necessary and sufficient conditions for the validity of the conjecture. In our example the input formula to such a method could be $\varphi$ or $\forall u(\varphi)$ or $\forall v(\varphi)$ or $\forall u \forall v(\varphi)$, where $\varphi$ is the following formula:

$$\forall y \forall x \big( (y(1+u) = vx \wedge y(1-u) = v(1-x)) \longrightarrow \\ x^2 + y^2 = (1+u-x)^2 + (v-y)^2 \big). \quad (29)$$

A prominent example for an implemented real decision and quantifier elimination method is the CAD method due to Collins and Hong, cf. [Collins and Hong 1991] implemented in the QEPCAD package by Hong et al.

There is an alternative real decision and quantifier elimination method based on the *virtual substitution of parametric test points*, cf. [Weispfenning 1988; Loos and Weispfenning 1993; Weispfenning 1997a]. Here the inputs are (with some exceptions) restricted to formulas, where the variables to be eliminated occur only linearly or quadratically. For most geometric problems this restriction is satisfied by the input formula. During successive variable elimination, the degrees may, however, increase in such a way as to preclude complete quantifier elimination. Again, geometric theorem proving problems are surprisingly well-behaved in this concern. This method is implemented in the REDUCE-package REDLOG* [Dolz-

---

* http://www.fmi.uni-passau.de/~redlog/

mann and Sturm 1997b, 1999]. A variant that automatically computes a set of non-degeneracy conditions [Dolzmann et al. 1998b] has proved to be very adequate for generating non-degeneracy conditions in geometric automatic theorem proving.

We finally sketch the available methods, and illustrate them by application to our example. All the example computations have been performed on a SUN workstation in less than a second.

**3.2.3.2 Characteristic Sets** This method was originally developed by Ritt for differential algebra; later it was adapted by Wu and his school to algebraic and geometric problems [Wu 1984a, 1986]. In essence the method extends a given system of polynomial equations to an equivalent disjunction of systems of polynomial equations and disequations by iterated pseudo-division with remainder. These resulting systems are called *characteristic sets*. Together with each of the characteristic sets, we obtain the *initial*, i.e. the product of all polynomial factors that have been introduced for pseudo-division. If the input system constitutes the hypothesis of a geometric closure theorem, then the equation describing the conclusion can be reduced via pseudo-division with respect to each of the characteristic sets. If all the reductions lead to a trivial equation, then the conclusion follows from the hypothesis over the complex numbers [Mishra 1993; Becker et al. 1998], and thus over the reals. The Wu–Ritt method has been applied with great success to a large collection of geometric problems [Chou 1988]. In our example we choose $x$, $y$ as main variables and $u$, $v$ as parameters. We obtain on input of the $\{y(1+u) - vx, y(1-u) - v(1-x)\}$ one characteristic set

$$C = \{v - 2y, v(u - 2x + 1)\}$$

with initial $v$. The conclusion $g = -v^2 + 2vy - u^2 + 2ux - 2u + 2x - 1 = 0$ reduces to $0 = 0$ by pseudo-division with the two polynomials in $C$. From the initial $v$ we derive the non-degeneracy condition $v \neq 0$.

**3.2.3.3 Gröbner Bases** Gröbner Bases have been introduced by Buchberger [Buchberger 1965] for computing in linear spaces of polynomial rings modulo zero dimensional ideals. They have meanwhile turned out as an extremely general and powerful tool in algorithmic algebra [Becker et al. 1998]. Let $S = \{\,f = 0 \mid f \in F\,\}$ be a system of polynomial equations. Let $g = 0$ be another polynomial equation. Then by Hilbert's Nullstellensatz the equation $g = 0$ is a consequence of the system $S$ over the complex numbers if and only if the number 1 is a polynomial linear combination of the polynomials in $F$ and of the polynomial $1 - gz$, where $z$ is a new variable. This condition can be tested be computing a monic Gröbner basis $G$ of the system $F \cup \{1 - gz\}$ and then checking, whether $1 \in G$, cf. [Becker et al. 1998]. In comparison to characteristic sets the Gröbner basis method is frequently somewhat slower [Chou 1988; Kutzler 1988; Kutzler and Stifter 1986]. In our example, we have

$$F = \{y(1+u) - vx, y(1-u) - v(1-x)\}, \quad g = x^2 + y^2 - ((1+u-x)^2 + (v-y)^2).$$

The following system $G_1$ is a monic Gröbner basis of $F \cup \{1 - gz\}$:

$$G_1 = \{u^2z - 2uxz + 2uz - 2xz + z + 1, v, y\}.$$

Since $1 \notin G_1$, we have disproved the theorem over the complex numbers. Next, we add the condition $v \neq 0$ by multiplying $g$ with $v$ as discussed earlier. As a Gröbner basis of $F \cup \{1 - gvz\}$ we obtain $G_2 = \{1\}$, which proves the theorem under the additional assumption $v \neq 0$ over the complex numbers and thus over the reals. Kutzler et al. have improved this method to automatically detect non-degeneracy conditions. See [Kutzler 1988; Kutzler and Stifter 1986] for details.

**3.2.3.4 Cylindrical Algebraic Cell Decomposition** This method was developed by Collins [Collins 1975] and refined by Hong and others [Collins and Hong 1991]. It constructs—by successive variable elimination—from a given real polynomial system $F$ a cylindrical algebraic decomposition of the real space of all variables into cells in such a way that all polynomials in $F$ are sign invariant over these cells, and the sign of each polynomial over each cell can be determined. One furthermore obtains a description of projections of these cells by polynomial equations and inequalities in the parameters. All the information can be combined to a quantifier elimination method over the reals for formulas constructed from polynomials in $F$. If there are no parameters, the quantifier elimination amounts to a decision procedure. For automatic theorem proving in geometry this method is as a rule limited to problems with few variables. In our example the method computes for the formula (29) the quantifier-free equivalent $v \neq 0 \lor u^2 + 2u + v^2 + 1 \leq 0$. This is easily shown to be equivalent to $u + 1 = 0 \lor v \neq 0$, cf. [Dolzmann and Sturm 1997c]. Next, we add $v \neq 0$ to the premises and quantify all the variables:

$$\forall u \forall v \forall x \forall y \big( (y(1+u) = vx \land y(1-u) = v(1-x) \land v \neq 0 \big) \longrightarrow$$
$$x^2 + y^2 = (1+u-x)^2 + (v-y)^2 \big). \quad (30)$$

Notice that the extra condition $v \neq 0$ can be added straightforwardly here. The elimination result for (30) is $0 = 0$, which means "true."

**3.2.3.5 Virtual Substitution of Test Points** Based on ideas by Ferrante and Rackoff [Ferrante and Rackoff 1979] for decision problems, this real quantifier elimination method has been introduced for linear formulas by the third author [Weispfenning 1988]. It has later been extended to higher degrees [Weispfenning 1997a, 1994b]. The method proceeds by successive elimination of real variables. These variables are substituted by a finite number of test points that depend on the remaining variables. The substitutions are, however, not carried out explicitly but in a virtual fashion so that the resulting formula can be processed further [Loos and Weispfenning 1993; Weispfenning 1997a]. The method has been successful in many application areas [Dolzmann 1999b; Dolzmann and Sturm 1997a; Sturm 1997; Sturm and Weispfenning 1997, 1998a,c]. As a real quantifier elimination procedure, it can be used in the style of CAD as described above. In fact, our example would not violate the degree restrictions.

For the purpose of geometric theorem proving, however, the method has been extended to construct non-degeneracy conditions in the style of Wu–Ritt ignoring the possibility that these conditions do not hold [Dolzmann et al. 1998b]. This extended procedure provides automatic proofs for many of the problems treated by complex methods, and moreover for a significant number of genuinely real problems [Dolzmann et al. 1998b]. In our example, it will yield "true" on input of the formula (29) with automatically generated non-degeneracy condition $v \neq 0$.

Andreas Dolzmann, Thomas Sturm, and Volker Weispfenning (Passau)

### 3.2.4 Homological Algebra

**3.2.4.1 Introduction** Homological algebra [Lane 1995], [Cartan and Eilenberg 1956] is an important area of mathematics which deals with derived functors (cf. 3.2.4.2) and related concepts. The cohomology of groups and Lie algebras, which play a role in theoretical physics, are examples of derived functors. Homological algebra also plays a role in the theory of differential equations [Bryant et al. 1991] and is used in algebraic geometry and topology.

Throughout this article, $R$ will denote a commutative ring with unit. All algebras over $R$ will be graded and of the form $A = \oplus_{i=0}^{\infty} A_i$ where the operation of multiplication satisfies $A_i \otimes A_j \longrightarrow A_{i+j}$ (and all of the usual algebra axioms are satisfied). Modules over $A$ will be of the form $M = \oplus_{i=0}^{\infty} M_i$ where each $M_i$ is an $A$-module. Note that this includes the ungraded case when $A_i = 0$ and $M_i = 0$ for $i > 0$. We write $|a| = i$ if $a \in A_i$ and we say that $A$ is a commutative algebra if $A$ is an algebra over $R$ and $ab = (-1)^{|a||b|} ba$ for all $a, b \in A$.

We will need the following basic notions. A chain complex (over $A$) is an $A$-module $M$ which is equipped with a sequence of maps

$$\ldots M_{i+1} \xrightarrow{d_{i+1}} M_i \xrightarrow{d_i} M_{i-1} \xrightarrow{d_{i-1}} \ldots \xrightarrow{d_2} M_1 \xrightarrow{d_1} M_0 \qquad (31)$$

which satisfy $d_i d_{i+1} = 0$ for all $i \geq 0$. In particular, note that $\mathrm{im}(d_{i+1}) \subseteq \ker(d_i)$. We often think of the sequence of maps above as a single map $M \xrightarrow{d} M$ where $d|_{M_i} = d_i$ and write $(M, d)$ (or simply $M$ if the context is clear) to denote (31). The map $d$ is called the *differential* of the complex $(M, d)$. Note then that $\mathrm{im}(d) \subseteq \ker(d)$ and the quotient, which we call the *homology* of $M$,

$$H(M, d) = \ker(d)/\mathrm{im}(d)$$

can be formed. We denote this quotient by $H(M)$ when the context is clear. Note that $H(M)$ is graded by $H(M)_i = H_i(M)$ where $H_i(M) = \ker(d_i)/\mathrm{im}(d_{i+1})$.

The sequence (31) is said to be *exact at $i$* if $H_i(M) = 0$. It is an *exact sequence* if $H(M) = 0$, i.e. $H_i(M) = 0$ for all $i$.

**3.2.4.2 Derived Functors and Computer Algebra** Let $_A\mathcal{M}$ be the category of all left $A$-modules and let $_A\mathcal{M} \xrightarrow{F} {_A\mathcal{M}}$ be a (covariant) functor [MacLane 1988]. Assume that $F$ is additive, i.e. $F(0) = 0$ and $F(f + g) =$

$F(f) + F(g)$ for all module maps $f$ and $g$. We can apply $F$ to the sequence (31) to obtain

$$\ldots F(M_{i+1}) \xrightarrow{F(d_{i+1})} F(M_i) \xrightarrow{F(d_i)} F(M_{i-1}) \xrightarrow{F(d_{i-1})} \ldots \xrightarrow{F(d_1)} F(M_0).$$

Since $d^2 = 0$ and $F$ is a functor, we have that $F(d)^2 = 0$ so we obtain a new complex $(F(M), F(d))$, however if (31) is exact, it does not follow that this new complex is also exact. For example, let $\mathbb{Z}$ denote the integers and consider the functor on $Z$-modules (abelian groups) given by $F(X) = X \otimes_{\mathbb{Z}} \mathbb{Z}_2$, where $\mathbb{Z}_2$ is the group with two elements. The sequence $0 \longrightarrow \mathbb{Z} \xrightarrow{\alpha} \mathbb{Z} \xrightarrow{\beta} \mathbb{Z}_2 \longrightarrow 0$ where $\alpha(n) = 2n$ and $\beta$ is the quotient map, is exact, but it is not hard to see that upon applying $F$, we obtain a sequence that is not exact. The *derived functors* of $F$, to be defined shortly, are a measure of how much deviation from exactness occurs when $F$ is applied to an exact sequence. More precisely, for a given $A$-module $N$, there is a class of exact sequences

$$\ldots M_{i+1} \xrightarrow{d_{i+1}} M_i \xrightarrow{d_i} M_{i-1} \xrightarrow{d_{i-1}} \ldots \xrightarrow{d_2} M_1 \xrightarrow{d_1} M_0 \xrightarrow{d_0} N \longrightarrow 0 \quad (32)$$

with the property that $H(F(M))$ is the same (up to isomorphism) for any member of the family. This is the class of *projective resolutions*. A projective (free) resolution of $N$ over $A$ is just an exact sequence (32) in which each $M_i$ is a projective (free) $A$-module. A fundamental theorem in homological algebra (the *comparison theorem*) ensures this invariance property [Lane 1995].

Thus, if (32) is an exact sequence, we can form the well defined module $\mathcal{L}(F)(N) = H(F(N))$ and it is not hard to see that $\mathcal{L}(F)$ is itself functorial. $\mathcal{L}(F)(N)$ is graded by $\mathcal{L}(F)(N)_i = \mathcal{L}_i(N)$ where $\mathcal{L}_i(N) = H_i(F(M))$ and these components are called the left derived functors of $F$. There is a dual notion of *right derived* functors $\mathcal{R}_i(G)$ for a contravariant functor $G$. For example, one has the functor $G_N(X) = \hom_A(X, N)$ of $A$-modules (contravariant) and $F_N(X) = N \otimes_A X$ (covariant). Their derived functors have special names. $\text{Tor}^A(N, X) = \mathcal{L}(F_N)(X)$ and $\text{Ext}_A(X, N) = \mathcal{R}(G_N)(X)$.

Two special cases of Ext and Tor should be pointed out. If $L$ is a Lie algebra over a field $k$ and $A = \mathcal{U}(L)$ is its universal enveloping algebra [Cartan and Eilenberg 1956] then if $A \xrightarrow{\epsilon} k$ is the map which vanishes on elements of degree bigger than zero and is the identity on elements of degree zero and $k$ is given the $A$-module structure using this map, by definition, $H_*(L; k) = \text{Tor}^A(k, k)$ is the Lie algebra homology of $L$ and $H^*(L; k) = \text{Ext}_A(k, k)$ is the Lie algebra cohomology of $L$. If $G$ is a group and $A = k(G)$ is the group ring over $k$, $A \xrightarrow{\epsilon} k$ is the map $\epsilon(\sum r_g g) = \sum r_g$ and $k$ is an $A$-module via this map, $H_*(G, k) = \text{Tor}^A(k, k)$ is the group homology of $G$ and $H^*(G; k) = \text{Ext}_A(k, k)$ is the group cohomology of $G$ by definition.

Note that derived functors may be defined for non-additive functors using simplicial methods [Dold and Puppe 1958] and homological algebra may be developed in categories other than module categories [Lane 1995].

There are two main areas in which computer algebra has been employed in investigating derived functors. One is in computing the ranks of the objects and the other is in using symbolic manipulation to actually computing resolutions. Usually, the goal in the latter is to get complexes small enough so that computer methods in linear algebra may used to compute the corresponding homology groups. These methods will now be described.

**3.2.4.3 Computing Ranks of Derived Functors** One set of tools for dealing with Ext and Tor on a computer involves the manipulation of formal power series. The reason for this is the well known identity ([Lemaire 1974, Appendix])

$$\frac{1}{P_A(-1,y)} = H_A(y). \tag{33}$$

In Equation (33), $A$ is an algebra over a field $k$,

$$H_A(y) = \sum_{i=0}^{\infty} \dim_k(A_i) y^i$$

is the *Hilbert Series* of $A$, and

$$P_A(-1,y) = \sum_{j=0}^{\infty} \sum_{i=0}^{\infty} (-1)^i \dim_k(\text{Tor}_{i,j}^A(k,k)) y^i$$

is the the *Poincaré-Betti* series of $A$.

**Example 31** *Let $k$ be any field, $A = k[x_1, \ldots, x_n]$ be the polynomial algebra over $k$ in the variables $x_i$ (assigned even degrees), $1 \leq i \leq n$. Let $A \xrightarrow{\epsilon} k$ be the map such that $\epsilon(p) = 0$ if $\deg(p) > 0$ and $\epsilon(p) = p$ otherwise.*

*As is well known ([Cartan and Eilenberg 1956]), one has the Koszul resolution $(X, d)$ of $k$ over $A$ where $X = A \otimes_k E$ where $E = E[u_1, \ldots, u_n]$ is the exterior algebra (or Grassman algebra) of skew-symmetric polynomials over $k$ in the variables $u_i$ (assigned odd degrees), $1 \leq i \leq n$, and $d$ is defined by extending the $A$-linear map $d(x_i) = u_i$ as a derivation of the tensor product algebra structure on $X$. I.e. $d(ae) = ad(e)$, for $a \in A$ and $d(e_1 e_2) = d(e_1)e_2 + (-1)^{|e_1||e_2|} e_1 d(e_2)$, for $e, e_1, e_2 \in E$.*

*Its easy to see that $\bar{X} = k \otimes_A X \cong E$ and that the corresponding differential $\bar{d} = id_A \otimes d = 0$. This gives the classical result that $\text{Tor}^A(k,k) = E[u_1, \ldots, u_n]$. Now consider the case $n = 1$. Then clearly, $\dim_k(k[x]_i) = i$ and $H_{E[u]}(y) = 1+y$ so that (33) is just a restatement of the well known formula $\frac{1}{1+t} = \sum_{i=0}^{\infty} (-1)^i t^i$. The case for a general $n$ can be obtained by observing that $A$ is the tensor product of $n$ copies of $k[x]$, $E$ is the tensor product of $n$ copies of $E[u]$, and the series of a tensor product is the product of the series. Note that there is no inner summation in using the formula from 33 since, in this case, $Tor_{i,j}$ is non-zero only for $i = j$.*

Because of Equation (33), it is important to be able to compute the Hilbert series of a graded algebra $A$. It turns out that this can be done using Gröbner bases 2.2.5. The interested reader should see [Bayer and Stillman 1992a] for the case of commutative algebras. In fact, there are built-in commands in MACAULAY [Bayer and Stillman 1992b; Grayson and Stillman 1996] (see also 4.2.25) in case $A$ is *commutative* (also see SINGULAR [Greuel et al. 1995], 4.2.39 and CoCoA [Capani et al. 1999] 4.2.11 for this case). One has to include to non-commutative Gröbner bases in order to compute Hilbert series in general. The program BERGMAN [Backelin 1998], 4.2.5 was designed to do this, among other things. A very good reference to study for interesting applications of Hilbert series calculations in both MACAULAY and BERGMAN is [Roos 1996] where extensions of (33) are given and used for special classes of (not necessarily commutative) algebras (also consult the references listed there).

Finally, it is important to point out that these series methods can be used to derive deeper information about (co)homology than just the ranks of individual modules. An interesting example of this is given in [Lambe and Löfwall 1995] where computer algebra is used in connection with series calculations to derive previously unknown properties of cyclic homology and the cohomology of free loop spaces.

**3.2.4.4  Computing Resolutions**  Computer algebra has been used to compute resolutions. In the case when $A$ is a (commutative) algebra, MACAULAY [Bayer and Stillman 1992b; Grayson and Stillman 1996] has some built-in commands for computing resolutions over various algebras. At present, there are analogous facilities (and extensions) for non-commutative algebras in joint work of J. Backelin (Stockholm), V. Ufnarovski (Lund), and A. Podoplelov (Linz) [Backelin and Ufnarovski 1998]. This work involves a construction from [Anick 1986] which recently has been shown to follow from a very general theorem in homological algebra (the *Perturbation Lemma* – see [Barnes and Lambe 1991] and the references therein) along with the use of non-commutitive Gröbner bases [Sköldberg 1997]. For more on the use of Gröbner bases in computing resolutions in the commutative case, see [Möller and Mora 1986] and the MACAULAY documentation. The use of the Perturbation Lemma is less well known but gaining in usage, so it will be described here in a little more detail.

The Perturbation Lemma has a long history in mathematics [Brown 1965], [Gugenheim 1972] and has been used to derive an algorithm for small resolutions of the integers $R = \mathbb{Z}$ over the group ring $A = R(G)$ of any finitely generated torsion-free nilpotent group $G$ [Lambe 1991] (also see the algorithm in [Lambe and Stasheff 1987] which is related). Computer algebra has been used to implement this algorithm [Lambe 1994] and this method of calculation was extended to a wide class of algebras (not necessairly commutative) in [Lambe 1992] and [Lambe 1993] where computer algebra was also used.

Intuitive terms will be used here to describe the method of homological perturbation. Precise terms can be found in [Barnes and Lambe 1991] and the other references just given (as well as the references they contain). The idea is to start with an equivalence between one chain complex $X$ and a larger one $Y$ (they are required to have the same homology and this must be via some well defined and well controlled maps). One then has a change in the differential in $Y$ and one seeks a change in the differential in $X$ so that the new complexes have the same homology *and* there are new well controlled maps that realize this. The

### 3.2.4 Homological Algebra

Perturbation Lemma gives an explicit formula for the new differential and the new maps. It is this explicit nature of the formulas that makes the Perturbation Lemma convenient in computer calculations.

In terms of resolutions, the intuitive picture is this (rather explicit examples are given in the references). One has an algebra $A$ which is a perturbation of another algebra $A_0$ and one is given a module $M$ over $A$ which can be viewed as a perturbation of a module $M_0$ over $A_0$. An example of this is the class of groups $G$ whose group laws may be expressed in the form $\rho : R^n \times R^n \longrightarrow R^n$ for a ring $R$, where $\rho$ is a polynomial function. This includes finitely generated torsion free nilpotent groups ($R = \mathbb{Z}$) and finite $p$-groups ($R = \mathbb{Z}_p$). In these cases, the group ring $R(G)$ can easily be seen to be a perturbation of the ring of Laurent series in $n$ variables over $R$ [Lambe 1992]. The problem is to

a) find a resolution $(X_0, d)$ of $M_0$ over $A_0$ for which a change in differential gives a resolution $(X, d + t)$ of $M$ over $A$ (often this is given by some universal construction, but rather large in size)

such that

b) there is a strong comparison of another – but "smaller" and more computationally accessible – resolution $(L_0, d_0)$ of $M_0$ over $A_0$ to $(X_0, d)$ in the sense above.

This sets up what is called a *transference problem* in [Barnes and Lambe 1991] and the Perturbation Lemma may be used to transfer the perturbed differential $d + t$ "down to" $L_0$ (when it converges). The idea here is to set this up in a way that $L_0$ is effectively managable on a computer. The interested reader can find examples of this – including the use of AXIOM [Jenks and Sutor 1992] for these purposes – in [Lambe 1991], [Lambe 1994], [Lambe 1992], [Lambe 1993] and [Grabmeier and Lambe 2001]. It should be noted that these examples do not exhaust the possibilities and there is still a lot of ground to be covered in the use of these methods and those in [Sköldberg 1997].

It should also be pointed out that there are programs such as GAP [GAP 1997], 4.2.18 and MAGMA [Cannon and Playoust 1996], 4.1.5 which can be used to compute the ranks of the first and second (co)homology of finite groups given a suitable presentation of the group, but currently, they are restricted to only finite groups and to only these dimensions.

Finally, it is noted that the methods discussed here can be used to unify and understand various constructions from the literature in an organized fashion. For example, the "twisted tensor product" resolutions of Wall [Wall 1961] can be seen to be related to the methods described here because of the *uniqueness* theorem from [Barnes and Lambe 1991] which states that under mild assumptions on the underlying objects, any two solutions to the transference problem are essentially the same. Thus, for example, the twisted tensor product resolutions for the semi-direct products given explicitly in [Lambe 1993] are essentially Wall resolution complexes. Also, this kind of observation could be used to get explicit perturbation formulas for the resolutions in [Gugenheim and Milgram 1970] which could be useful in machine calculations.

Larry A. Lambe (Rutgers and Bangor)

The author is greatful to Johannes Grabmeier and Werner M. Seiler for pointing out typos and suggesting changes in the first few drafts of this exposition which improved its readability.

### 3.2.5 Study of Differential Structures on Quantum Groups

**3.2.5.1 Abstract** Non-commutative differential calculi are important tools in studying non-commutative differential geometry on quantum spaces. There is no constructive method to define canonical calculi as in case of classical differential geometry. But by imposing natural conditions, such as covariance with respect to a quantum group action and dimensionality of the bimodule of first-order forms, one or a few calculi can be selected. There are multiple ways to continue with differential forms of higher order since various exterior algebras resulting from different definitions of symmetric forms can be considered. This article shows that the study of differential calculi and exterior algebras related to quantum groups opens a wide field of interaction between theoretical mathematics and computer algebra.

**3.2.5.2 Mathematical Background** A *quantum group* $\mathcal{A}$ is a Hopf algebra, i.e. an associative unital algebra together with algebra homomorphisms $\Delta : \mathcal{A} \to \mathcal{A} \otimes \mathcal{A}$ and $\varepsilon : \mathcal{A} \to \mathbb{C}$ fulfilling the coassociativity and counit axioms with an antipode $S : \mathcal{A} \to \mathcal{A}$. By a *right quantum space* for $\mathcal{A}$, we mean a pair $(\mathcal{X}, \varphi)$ of a unital algebra $\mathcal{X}$ and an algebra homomorphism $\varphi : \mathcal{X} \to \mathcal{X} \otimes \mathcal{A}$, called the *coaction* of $\mathcal{A}$ on $\mathcal{X}$. Thus the map $\varphi$ satisfies $(\varphi \otimes \mathrm{id}) \circ \varphi = (\mathrm{id} \otimes \Delta) \circ \varphi$ and $(\mathrm{id} \otimes \varepsilon) \circ \varphi = \mathrm{id}$. Classical examples of quantum spaces are quantum vector spaces (see [Wess and Zumino 1991]) and Podleś' quantum spheres $S_{qc}^2$ introduced in [Podleś 1987]. A *first-order differential calculus* (abbreviated by FODC) over $\mathcal{X}$ is a pair $(\Gamma, \mathrm{d})$, where $\Gamma$ is a bimodule over $\mathcal{X}$ which as linear space is spanned by the elements $x \cdot \mathrm{d}y$ and $\mathrm{d} : \mathcal{X} \to \Gamma$ is a linear mapping satisfying the Leibniz rule $\mathrm{d}(xy) = \mathrm{d}x \cdot y + x \cdot \mathrm{d}y$. A FODC $(\Gamma, \mathrm{d})$ on a quantum space $(\mathcal{X}, \varphi)$ is called *covariant* if there exists a linear mapping $\Phi : \Gamma \to \Gamma \otimes \mathcal{A}$ such that $(\Gamma, \Phi)$ is a right comodule, $\Phi(x \omega y) = \varphi(x)\Phi(\omega)\varphi(y)$ and $\Phi(\mathrm{d}x) = (\mathrm{d} \otimes \mathrm{id})(\varphi(x))$ for all $x, y \in \mathcal{X}$ and $\omega \in \Gamma$. In [Woronowicz 1989] Woronowicz invented the theory of non-commutative differential geometry on quantum groups. Further studies in this direction can be found (e.g.) in [Podleś 1989, 1992; Wess and Zumino 1991]. In order to introduce curvature, connections, and other differential geometrical notions one has to study also differential forms of higher order. For this purpose one has to pass to the $k$-fold algebraic tensor product $\Gamma^{\otimes k} = \Gamma \otimes_{\mathcal{X}} \cdots \otimes_{\mathcal{X}} \Gamma$. The direct sum $\Gamma^{\otimes} = \sum_{k=0}^{\infty} \Gamma^{\otimes k}$ is an associative $\mathbb{Z}$-graded algebra. Introducing a space of symmetric forms $\mathcal{S}$, which in particular has to be a two-sided ideal and a bicovariant $\mathbb{Z}$-graded subbimodule of $\Gamma^{\otimes}$, there can be defined an *exterior algebra* $\Gamma^{\wedge} = \Gamma^{\otimes}/\mathcal{S}$ of $\Gamma$. For a comprehensive overview on the theory of differential

### 3.2.5 Study of Differential Structures on Quantum Groups

calculi on quantum groups we refer to the textbook [Klimyk and Schmüdgen 1997].

**3.2.5.3 Classification of FODC's** Let $\mathcal{X}$ be a complex quantum space for the quantum group $\mathcal{A}$. For any algebra generating set $\{e_1,\ldots,e_l\}$ of $\mathcal{X}$ the set $T = \{e_{k_1}\cdots e_{k_r}\ ; r \in \mathbb{N}, 1 \leq k_1,\ldots,k_r \leq l\}$ generates $\mathcal{X}$ as $\mathbb{C}$-vector space. Suppose that the set $\{de_i\ ;\ 1 \leq i \leq l\}$ generates $\Gamma$ as a left $\mathcal{X}$-module. Obviously, any such FODC $\Gamma$ of $\mathcal{X}$ contains elements

$$de_j \cdot e_i = \sum_{k=1}^{l} p_k^{i,j} \cdot de_k\ ,\ \text{where } p_k^{i,j} \in \mathcal{X}\ (1 \leq i,j,k \leq l). \tag{34}$$

The $p_k^{i,j}$ are finite linear combinations of the elements of $T$. Leibniz rule and covariance condition can be applied in order to deduce further information about the structure of the $p_k^{i,j}$. Let $\prec$ be a well-founded order on the set $\{u \cdot de_i \cdot v\ ;\ u,v \in T, 1 \leq i \leq l\}$ which is admissible for the computation of Gröbner bases in the bimodule $(\mathcal{X} \otimes_\mathbb{C} \mathcal{X})^l = \bigoplus_{i=1}^{l} \mathcal{X} \otimes_\mathbb{C} \mathcal{X}$. For the theory of Gröbner bases in noncommutative algebras and modules we refer to Section 2.6.5 of this volume and the references therein. If $u \cdot de_k \prec de_j \cdot e_i$ for all $u \in T$ appearing with nonzero coefficient in $p_k^{i,j}$, $1 \leq i,j,k \leq l$, then the covariant FODC $\Gamma$ of $\mathcal{X}$ is freely generated as a left $\mathcal{X}$-module by the differentials $de_1,\ldots,de_l$ if and only if its system (34) is a Gröbner basis with respect to $\prec$. $\mathcal{X}$ is a quotient of the free unital complex algebra with generators $e_1,\ldots,e_l$ modulo a two-sided ideal $I$. If a finite Gröbner basis of $I$ with respect to a term order which is compatible with $\prec$ can be computed then the generating set $T$ can be replaced by a basis $B$ of $\mathcal{X}$. Though the computation of Gröbner bases in the bimodule $(\mathcal{X} \otimes_\mathbb{C} \mathcal{X})^l$ is not algorithmic the Gröbner basis property of a given system (34) can be checked in an algorithmic way. If for almost all triple $(i,j,k)$ the equation $p_k^{i,j} = 0$ can be proved a priori then there remain only finitely many complex parameters $p_k^{i,j}$. Moreover, instead of a single quantum space $\mathcal{X}$ we can consider simultaneously a family $(\mathcal{X}_{c_1,\ldots,c_n})_{c_1,\ldots,c_n \in \mathbb{C}}$ of quantum spaces with finitely many complex parameters $c_1,\ldots,c_n$. We are lead to the computation of Gröbner bases of parametric modules. As a byproduct we obtain systems of non-linear algebraic equations and inequations whose solutions correspond to the covariant FODC of the quantum spaces $\mathcal{X}_{c_1,\ldots,c_n}$ during the Gröbner basis calculations. Possible superfluous solutions have to be excluded by verification of Leibniz rule and covariance condition. The above approach involves various types of difficulties. First of all, a Gröbner basis for $\mathcal{X}$ has to be computed. Then applying theoretical arguments the structure of the $p_k^{i,j}$ has to be specified in a sufficiently precise way. Third, comprehensive Gröbner bases of bimodules have to be computed. Finally, systems of non-linear algebraic equations must be solved.

Podleś classified two-dimensional differential calculi on the quantum spheres $S_{qc}^2$ for the quantum group $SL_q(2)$ using hand calculations in [Podleś 1992]. However, the case of three-dimensional covariant FODC requires the use of a computer. A complete classification was given in [Apel and Schmüdgen 1994].

**3.2.5.4 Exterior Algebras** For a right comodule $(\mathcal{Y}, \varphi)$ we denote the subspace of right invariant elements by $\mathcal{Y}_R = \{y \in \mathcal{Y}\,;\, \varphi(y) = y \otimes_{\mathbb{C}} 1\}$. In the special case $\mathcal{X} = \mathcal{A}$ we have $\Gamma = \Gamma_R \otimes_{\mathbb{C}} \mathcal{A}$. Moreover, $\Gamma^{\wedge} = \Gamma_R^{\wedge} \otimes_{\mathbb{C}} \mathcal{A}$ and $\mathcal{S} = \mathcal{S}_R \otimes_{\mathbb{C}} \mathcal{A}$.

Since $\mathcal{S}_R$ is graded the dimension of the subspace $\Gamma_R^{\wedge k}$ of right invariant differential $k$-forms can be computed for any given $k \in \mathbb{N}$. This is done by computing a suitable truncation of the Gröbner basis of $\mathcal{S}_R$ and counting the irreducible (non-commutative) monomials of degree $k$. Furthermore, if the right invariant exterior algebra $\Gamma_R^{\wedge}$ is a finite dimensional vector space then the Gröbner basis computation for $\mathcal{S}_R$ will eventually terminate and the dimension can be read off from the result.

In [Heckenberger and Schüler 1998] this method was applied to the analysis of various exterior algebras related to the quantum group $\mathrm{O}_q(3)$.

**3.2.5.5 Computer Support** The described methods involve Gröbner basis computations in very general non-commutative situations and the size of the problems to be solved is remarkable. Though more and more computer algebra systems contain implementations of non-commutative Gröbner bases the non-commutative algebras and modules appearing here are still not supported in most systems. In [Apel and Schmüdgen 1994] and [Heckenberger and Schüler 1998] the special computer algebra system FELIX (see 4.2.15) was applied to handle the calculations.

Joachim Apel, István Heckenberger, and Axel Schüler (Leipzig)

### 3.2.6 Orthogonal Polynomials and Computer Algebra

Orthogonal polynomials have a long history, and are still important objects of consideration in mathematical research as well as in applications in Mathematical Physics, Chemistry, and Engineering. Quite a lot is known about them. Particularly well-known are differential equations, recurrence equations, Rodrigues formulas, generating functions and hypergeometric representations for the classical systems of Jacobi, Laguerre and Hermite which can be found in mathematical dictionaries (see e.g. [Abramowitz and Stegun 1964], [Erdélyi et al. 1953–1955], [Magnus et al. 1966]). Less well-known are addition theorems and connection relations between different systems and other identities for these families, their discrete counterparts [Nikiforov et al. 1991] and other systems of orthogonal polynomials ([Magnus et al. 1966], [Askey and Gasper 1971], [Ronveaux et al. 1995]). The ongoing research in this still very active subject of mathematics expands the knowledge database about orthogonal polynomials continuously. In the last few decades the classical families have been extended to a rather large collection of polynomial systems, the so-called Askey-Wilson scheme and its $q$-analogue ([Askey and Wilson 1985], [Koekoek and Swarttouw 1998]), and they have been generalized in other ways as well.

In the last decade major steps towards an algorithmic treatment of orthogonal polynomials and special functions have been made, notably Zeilberger's brilliant extension of Gosper's algorithm on algorithmic definite hypergeometric summation (considered in Section 2.10.1) ([Gosper jr. 1978], [Zeilberger 1991], [Koornwinder 1993], [Koepf 1995a]). By implementations of these and other algorithms symbolic computation has the potential to change the daily work of everybody who uses orthogonal polynomials or special functions in research or applications. Zeilberger's algorithmic approach enables one to compute differential, recurrence and similar equations from sum or integral representations ([Petkovšek et al. 1996], [Koepf 1998a], [Koepf 1997]). These methods turn out to be quite useful to prove or detect identities for orthogonal polynomial systems (see [Koepf 1998a]).

Further algorithms to detect connection coefficients or to identify polynomial systems from given recurrence equations have been developed ([Ronveaux et al. 1995], [Koepf and Schmersau 1998]). Although some algorithmic methods had been known already in the last century (see e.g. [Beke 1894a]–[Beke 1894b]), their use was rather limited due to the immense amount of calculations. Only the existence and wide distribution of computer algebra systems makes their use simple and useful for everybody.

It can be expected that symbolic computation will also play an important role in major revisions of existing formula books in the area of orthogonal polynomials and special functions. By an algorithmic approach Roach [Roach 1997] found that about every tenth representation formula for hypergeometric functions of the database [Prudnikov et al. 1990] contains a misprint. But proof-reading is only a rather trivial application of symbolic computation, and one can expect that algorithmic approaches have much stronger impact as Swarttouw's on-line version CAOP of the Askey-Wilson scheme shows [Swarttouw 1996].

Many articles on the algorithmic treatment of orthogonal polynomials and special functions and on applications of such algorithms have been published. Such articles are distributed widely in the literature. To collect articles about the interaction between computer algebra and the field of orthogonal polynomials and special functions Dick Askey, Wolfram Koepf and Tom Koornwinder recently were co-editing a special issue of the *Journal of Symbolic Computation* [Askey et al. 1999] on the topic *Orthogonal Polynomials and Computer Algebra*.

Wolfram Koepf (Kassel)

### 3.2.7 Computer Algebra in Symmetric Bifurcation Theory

Theoretical investigation of bifurcation problems with symmetry has been a very active area of research in the last decade. The investigation of the symmetry leads to questions related to algorithmic invariant theory as presented in the book by Sturmfels [Sturmfels 1993] (see also Sect. 2.2.8). The investigation of symmetric bifurcation problems typical start with a general vector field having the symmetry of a certain group action being relevant in applications. The algorithmic determination of such a generic equivariant vector field involves symbolic computations, especially Gröbner basis computations. The papers [Callaham and

Knobloch 1997; Gatermann and Werner 1996; Lari-Lavassani et al. 1999; Worfolk 1994] are typical examples of this approach. The Hilbert basis of an invariant ring and the generators of the module of equivariants are computed. The Packages INVAR [Kemper 1993] and SYMMETRY [Gatermann and Guyard 1996] and others provide software for these tasks; the first package concentrating on invariants (also over finite fields); the second package including as well computations for equivariants and question like complete generation, see the description in [Gatermann 1996b] and [Gatermann and Guyard 1999]. In order to perform these computations efficiently an improved MAPLE Package for computations of Gröbner bases has been necessary [Gatermann 1996a].

The first task is to find the stationary solutions of an equivariant vector field. The exact methods exploit the knowledge of fundamental invariants and fundamental equivariants, see [Gatermann 1996b; Worfolk 1994].

Secondly, the theoretical classification of bifurcation points of higher codimension by singularity theory involve objects with algebraic structure such as the tangent space whose systematic treatment necessitates Gröbner bases. For example the codimension and thus the minimal number of unfolding parameters are computed with Gröbner bases. The idea in [Gatermann and Lauterbach 1998b] is the classification of bifurcation phenomena by classification of possible Gröbner bases of the tangent space.

The exploitation of symmetry also leads to a reduced system which is defined on a variety which naturally leads to questions concerning the structure of the variety, appropriate coordinates etc. This method called *orbit space reduction* leads to algorithmic determination of the dimension, algorithmic representation of a polynomial in a ring generated by polynomials and choosing a symmetry adapted coordinate system [Gatermann 2000].

Karin Gatermann (Berlin)

### 3.2.8 Symbolic-Numeric Treatment of Equivariant Systems of Equations

In certain applications, we encounter a class of nonlinear systems of equations which depend on an additional parameter, and which feature symmetries originating from geometric properties. Examples of such *equivariant systems* are discrete versions of reaction-diffusion equations, problems in structural engineering [Gatermann and Hohmann 1991a], and neural nets.

SYMCON is a hybrid symbolic-numeric algorithm which exploits symmetries in a symbolic REDUCE part generating C source code, and finally performs the actual computation of solutions in a numeric part (written in C). Basically, the numeric part uses path-following, computes special points such as bifurcation points and Hopf points, and performs stability calculations. It requires the problem to be input in pre-processed form.

The symbolic part takes care of many derivations which otherwise would have to be done by paper and pencil. For example, it determines the bifurcation groups and isotropy groups (see also Sect. 3.2.1). Furthermore, it computes the systems of equations resulting from the reductions by symmetry and the block structure of the Jacobian, and finally generates C source code via the GENTRAN module. The mathematical tool which facilitates the exploitation of symmetries is the theory of linear representations, which is implemented in [Gatermann 1991c].

The combination of symbolics and numerics has considerable advantages: the equivariance condition is checked, therefore application instances are implemented reliably and error-free. Symbolic differentiation used to compute bifurcation points is by far superior compared to numeric differentiation. Special structure of the Jacobian matrices is exploited by generating the functions which are evaluated numerically in a highly sophisticated way.

The major advantage however, has to be seen in the way the method makes use of the particular symmetry property. The symmetry exploitation in the numerical part is automated by the computer algebra part. A REDUCE program takes over the task of mathematicians who write papers on this very subject.

The numeric part provides an interactive graphical user interfaces based on the *X-Windows* system. Application of computer algebra methods combined with an interactive interface make the system both user-friendly and reliable, facilitating the study of many examples as well as the illustration of results from calculus [Gatermann and Werner 1994].

Additional references: [Gatermann and Hohmann 1991b; Gatermann 1991a,b, 1997, 1993].

Karin Gatermann (Berlin)

## 3.3 Computer Science

### 3.3.1 Computer Algebra in Computer Science

A connection between computer algebra and computer science is not only given by computer algebra systems being computer programs, but also by formal and algebraic methods used in computer science. Below we exemplify this for the areas of signal processing, wavelets, and algebraic specification.

Another link between computer algebra and computer science is the use of inference on mathematical databases for *knowledge based systems in mathematics* (see 2.17, 2.16). Also algebra provides the foundations for *coding theory* and *cryptography* their connection to computer algebra is described in 2.14. Computer algebra yields a systematic approach to *efficient algorithms*, exemplified below by the items signal processing and wavelets. Algebraic techniques are also used for the design of hardware architectures and *VLSI design* (see 2.20).

Further examples for applications are decision problem solving within algebraic structures by *term rewrite and reduction systems* (see 2.15.1) and *automatic theorem proving* for which Gröbner bases and characteristic set methods (see 2.2.5) are important tools.

Algorithms from computer algebra are successfully applied to the theory of *lattices and ordered structures*, with applications to data analysis and knowledge representation (see [Wille 1992]).

Furthermore the correspondence of the theory of monoids and *automata theory* allows the application of computer algebra in theoretical computer science (see 4.2.4).

Three applications in more detail:

**Signal Processing**

Signal transforms as the Fourier transform, the Hartley transform, or the cosine transform often exhibit a symmetric structure, i.e., for a signal transform matrix $M$ there are representations $\rho, \rho'$ of a group $G$ such that $\forall g \in G : M = \rho(g) M \rho'(g)$. If one can find sparse base transform matrices decomposing these representations the matrix $M$ can be factored into a product of sparse matrices which yields a fast algorithm for the matrix multiplication with $M$. At the IAKS computer algebra is used to derive these sparse base transform matrices with methods of representation theory which are described in [Egner 1997]. For this application of representation theory many theorems had to be refined to statements involving *equality* rather than similarity. For literature and more mathematical details see [Püschel 1998].

**Wavelets**

The discrete wavelet transform is a recent technique in computational harmonic analysis. There is a close connection between the fast algorithms for these transforms and perfect reconstruction filter banks, which are studied in signal processing. A typical application of fast wavelet transforms is image compression, where the transform is used to reduce the correlation between adjacent image pixels. Computer algebra systems are used* to reduce the computational complexity of fast wavelet transform algorithms. While most wavelet based compression methods are lossy (i.e. degrade the image quality), a lossless compression method was derived using wavelets over finite rings [Klappenecker et al. 1997].

**Algebraic Specification**

Formal methods, i.e. the systematic use of mathematics in software (or hardware) design, have become standard in the development of high-integrity systems for safety-critical applications. For software systems, algebraic specification (see for example [Astesiano et al. 1999]) allows for a formal design process in terms of abstract datatypes, starting from "loose" requirement specifications and ending up with executable specifications close to program code. An open collaborative effort in this area is the CoFI initiative (Common Framework Initiative) [CoFI 2001], which not only provides a specification language CASL (Common Algebraic Specification Language) [CoFI 2001] with a formal semantics, but also supports this language by methodologies, tools, standard libraries, etc.

For high-integrity system design, "correctness" is of course *the* crucial point. In algebraic specification, correctness is achieved by proving the consistency of specifications (i.e. there exists a model which has the properties described in the specification), validating requirement specifications (i.e. the specification

---

* e.g. at the IAKS, the institute of the authors, see http://iaks-www.ira.uka.de.

describes the class of models which one has in mind), and proving that a specification refines another one (i.e. that a development step towards a computer program is correct). Tool support is needed to deal with these items within a large system's design.

"Classical" tools for algebraic specification are theorem provers, e.g. Isabelle, Inka, KIV, or term rewriting systems, e.g. OBJ, ELAN. Usually, the development of a complex system requires a great variety of specialised tools as no single tool is able to deal with all its aspects. At this point, computer algebra systems come into scope as a welcome supplement to the established tools. The reason is that computer algebra systems are able to deal effectively and efficiently with certain datatypes. From an algebraic specification point of view, datatypes of special interest are not only the classical algebraic structures like groups, rings, and fields, but also the more "practical" datatypes, for example numbers (naturals, integers, rationals, even reals with the concepts of differentiation and integration [Roggenbach et al. 2000b]) or structured datatypes like lists and bags, which all exhibit a large amount of algebraic structure. The CASL library of standard datatypes [Roggenbach et al. 2000a], for example, includes specifications of groups, rings, and fields, explicitly states the algebraic properties of datatypes, and makes even use of algebraic properties to specify standard datatypes. This should open up a direct connection to computer algebra systems.

Currently, such an integration of computer algebra systems into the algebraic specification development process lacks a semantically sound basis, but a lively research community is going to deal with this problem. For example, first solutions to this problem have been stated and partially realised by the OpenMath initiative [The OpenMath Society 2001] (see www.openmath.org).

Thomas Beth (Karlsruhe), Karsten Homann (Karlsruhe),<br>
Andreas Klappenecker (College Station), Jörn Müller-Quade (Tokyo),<br>
Armin Nückel (Karlsruhe), and Markus Roggenbach (Bremen)

### 3.3.2 Decomposable Structures, Generating Functions and Average-Case of Algorithms

Although the use of generating functions has a long tradition in enumerative combinatorics, a systematic investigation and exploitation of this tool with its "mechanization" in mind has been made only quite recently. It has become evident that a considerable portion of enumeration problems can be dealt with in a routine and highly efficient way. Using a setup which is (not accidentally) reminiscent of context-free grammars and languages, one may specify many interesting classes of combinatorial structures from atomic building blocks by using a few standard constructors, such as *union, product, sequence, (multi-)set, cycle*. As a simple specimen, the class FD of *functional digraphs* may be specified as

$$\text{FD} = set(\text{CFD}) \quad \text{CFD} = cycle(\text{RT}) \quad \text{RT} = product(\text{Z}, set(\text{RT}))$$

which expresses the fact that a functional digraph is a set of connected components (CFD), each of which is a cycle of rooted trees (RT), where rooted trees of nodes Z are defined recursively in an obvious way. Such a specification for *decomposable* structures can be compiled into a system of equations for the corresponding generating functions, which in our example reads (in a universe with distinguishable atoms)

$$fd(z) = \exp(cfd(z)) \quad cfd(z) = -\texttt{log}(1 - rt(z)) \quad rt(z) = z \cdot \exp(rt(z))$$

where in the exponential generating function $fd(z) = \sum_{n \geq 0} fd_n z^n/n!$ the coefficient $fd_n$ denotes the number of functional digraphs on $n$ points.

In fortunate cases such a system of equations can be solved explicitly, but even if this is not possible, lots of useful information can be obtained: initial segments of the counting sequence (such as $(fd_n)_{n \geq 0}$ in the example) can be computed, efficient algorithms for random and exhaustive generation of the structures under consideration can be constructed automatically, and detailed information about the asymptotic behaviour of the counting sequence can be extracted using methods such as *singularity analysis* or *saddle point methods*.

One can go farther: using the technique of *tagging* and bivariate generating functions, parameters recording the number of occurences of substructures, such as leaves in a tree or the number of connected components of a graph, can be analyzed within the same setup, giving precise and/or asymptotic information about the average, the variance etc. of the corresponding distribution over structures of fixed size.

It is well known since the work of D. E. Knuth that for many algorithms the task of analyzing the quantitative behaviour can be reduced to combinatorial counting problems, hence generating functions and recurrence relations play an important rôle in the analysis of average-case complexity ; see [Flajolet and Sedgewick 1996] for a textbook exposition of this approach. Viewing the definitions of decomposable structures as specifications of datatypes, one may analyze algorithms which systematically traverse these structures and operate relative to the substructures encountered. Think of tree searching methods, rewriting algorithms, unification, pattern matching etc. as typical examples. See [Flajolet et al. 1991] for a systematic exposition and [Flajolet et al. 1989] for many examples.

The programme outlined above has been carried out systematically by researchers at INRIA (under the direction of Ph. Flajolet) over the last decade, see [Flajolet and Salvy 1995] for an overview of these activities. The MAPLE package combstruct represents much of the current status of these efforts, integrating many (not yet all) features of its predecessors such as gaia [Zimmermann 1997] and luo [Flajolet et al. 1989]. The package gdev provides the function equivalent responsible for asymptotic analysis. In this context the highly useful package gfun [Salvy and Zimmermann 1994] must be mentioned: it deals with *holonomic* generating functions , i.e. generating functions defined by linear differential equations with polynomial coefficients, or (equivalently) linear recurrent sequences with polynomial coefficients, quite often encountered in combinatorial situations (as described above) and otherwise. gfun implements the closure properties of this class of functions and has strong guessing capabilities which lets you find plausible candidates for a differential equation (recurrence relation) satisfied by a generating function (its sequence of coefficients) once you know a sufficiently long initial segment of the sequence. Finding such equations (recurrences) can serve for various purposes: proving identities (in the spirit of Zeilberger, cf. Sec. 2.10.1), fast computation of coefficients, search for closed form solutions, and asymptotics. For a considerable extension of the gfun-approach see the work on Mgfun by F. Chyzak [Chyzak 1998], which is also referred to in Sec. 2.10.1.

For the programs mentioned above we refer the home page http://www-rocq.inria.fr/algo/libraries/ of the Algorithms Project at INRIA.

Volker Strehl (Erlangen)

### 3.3.3 Telecommunication Management Networks

A telecommunication management network (TMN) is a data network for administering and maintaining network nodes such as switches, cross-connects, and large telecommunication networks, like the synchronous digital hierarchy (SDH), from a central operations center.

For a commercial telecommunication operator, real-time response and performance are of prime importance. Malfunctioning network nodes and lines have to be detected in a timely manner in order to keep down-time at a minimum.

Massive data transfer within short time intervals is required, e.g., to collect billing information, or to update hundreds of network nodes with new software.

The protocols used in TMNs have been standardized by international organizations like the ITU-T. For manufacturers and operators of networks, there still remains the task to determine a large number of design parameters, to achieve smooth operation and performance.

At Philips Kommunikations-Industrie AG in Nürnberg, we were using computer algebra systems to improve performance of nodes and networks. Commercial "general" purpose computer algebra packages with their libraries and their flexible programming language allowed to generate quickly e.g. simulations, which included generation of statistical input data, Fourier transform, filtering and discrete mathematics (finite automata=protocol state machines).

Kurt Behnke (Düsseldorf)

## 3.4 Engineering

### 3.4.1 Computer Algebra, a Modern Research Tool for Engineering

Looking back more than two decades of development and application of computer algebra in science and engineering, a search in physical journals results in a list of more than thousand entries, indicating the dominant role of computer algebra in nearly all research areas, especially in the last ten years. Besides the well developed multimedia methods and the numerical simulation of physical problems, the computer algebra has opened a totally new field of scientific investigations in engineering sciences which is based on powerful methods for the analysis of multiparametric solutions and the characterisation of solutions of equations which model physical or industrial processes. Nearly in each branch of scientific research the computer algebra has led to an enormous impact on the development of new algebraic methods or the application of known algebraic algoritms to even complex problems. Symbolic data processing has steadily become a powerful tool for mathematicians, physicists, and engineers. Without symbolic utilities, effective mathematical problem solving in a theoretical as well applied environment is almost unthinkable of, nowadays.

Historically, the first computer algebra system available for engineers was implemented on IBM-machines in the midseventies, based on PL/1. A pioneering work on the treatment of singular perturbation problems in general, and especially for fluid dynamic problems which are connected to boundary layer approximations according to Prandtl's boundary layer theory, is due to Feuillebois and Lasek [Feuillebois and Lasek 1977]. One of the early publications on computer algebra application is due to Wirth [Wirth 1979] who demonstrated the usefulness of MACSYMA for the symbolic vector and dyadic analysis. Another application of computer algebra dates back to 1987, when Beyer et al. [Beyer et al. 1987] published a paper on the derivation of a formula for the calculation of the volume of the set of points which are common to the interiors of two congruent cones the axes of which intersect at one point equidistant from the vertices of the cones. Such a complicated formula has a universal meaning to physical and engineering problems of optics and irradiation of any kind. A striking example for the fact that the computer algebra is sometimes superior to a numerical analysis is given by Zahalak et al. [Zahalak et al. 1987] where the large deformations of a cylindrical liquid-filled membrane are analysed under the action of a viscous shear flow. Using REDUCE for a fifth order series solution of the free-boundary-value problem led to results which described much better the physical behavior of this system than the numerical analysis could do. For example, modern materials research relies on simulation of physical processes which can only be described by extremely complicated models in order to achieve the required accuracy. The complexity of the model equations makes the use of suitable utilities a necessity, before we can even start with unavoidable numeric computations. Symbolic programs enable us to derive, swiftly and without making errors, systems of differential equations for particular materials models, and convert them to difference form.

But also such practical questions as e.g. the lubrication of a four-parameter Oldroyd fluid in the slider bearing was solved successfully by Cheng et al. influence of the elasticity on the loading capacity and the friction. Another example for the combination of an algebraic and numerical procedure to solve a stability problem for the Taylor-Couette system is given by Roesner and Zikanov [Roesner and Zikanov 1997], where the time-independent exact solution in parametric form was found for a polymer solution by computer algebra solving an ordinary

### 3.4.1 Computer Algebra, a Modern Research Tool for Engineering

differential equation of first order, but fifths degree. This last example links to another area of computer algebra application in engineering science: Transforming equations with respect to different coordinate systems chosen for particular problems. Any attempt to perform this mechanical procedure by hand would be anachronous. Famous mathematicians, like Sophus Lie, already complained about how time consuming and error-prone such activities would be if carried out by humans. In this sense the doctoral dissertations of Kilgenstein [Kilgenstein 1984] and Viehl [Viehl 1989] illustrate how the partial differential equations originating from boundary layer theory and the Navier-Stokes equations can be exactly solved by series expansions in complicated three-dimensional coordinate systems. It means up to the solution of ordinary differential equations of Euler type. In the following, we use non-trivial examples from different research areas of engineering to foster the claim that computer algebra has become a useful and essential tool. In mechanics, we often have to analyze structures composed of a great number of individual parts which in turn are supply connected among each other, and are agitated by external forces. For example, this is the case when studying a sequence of robotic motions, where we have to derive and solve systems of differential equations whose number can easily be in the order of several thousand. Nowadays, computer programs generate those gigantic systems of equations, making extensive use of symbolic manipulation of data. However, not only for deriving, but also for analyzing systems of equations, software packages such as AXIOM, MACSYMA, MAPLE, MATHEMATICA, or REDUCE are utilized, also to avoid tedious, purely numerical investigations – so-called parameter studies – for most of the time.

For practical purposes, it is usually sufficient to reduce large systems of equations to a smaller number of equations, and then to analyze the *qualitative* properties of the solutions of the reduced system. A typical example for this procedure is given by the methods in perturbation theory of ordinary differential equations. This special field of applied mathematics marks a major area of computer algebra application. The monograph [Rand and Armbruster 1987] by Rand and Armbruster gives convincing testimony for the power of symbolic data processing, presenting the application of the theory of central manifolds and Lie transforms by sample programs. For the theory of nonlinear dynamics in technical sciences and physics, computer algebra based methods of investigation became indispensable. Polyanin et al. [Polyanin and Zhurov 1994] have described a method to find the general solution of a class of ordinary differential equations of polynomial type, which means that the coefficients are polynomials of the dependent and independent variables. A survey among researchers from engineering and physics revealed that there are basically no limits for applications of computer algebra to many areas of research. A typical example is given by the Ph.D. thesis of Ikl'e [Ikl'e 1993] where exact solutions to a discrete velocity model are derived for the coagulation-fragmentation problem in particle physics. The mathematical background is the Boltzmann equation which represents normally the most difficult basic equation for the calculation of elementary processes in particle physics. Although the employment of computer algebra is still not as wide-spread as it would deserve, given its versatility and potential for applications, it is currently successfully used in several areas, such as:

- robotics,
- nonlinear dynamics,
- elastomechanics,
- fluid dynamics,
- visco-elastic materials,
- rarefied gas dynamics,
- aerodynamics,
- computational fluid dynamics,
- control theory, and
- real time systems.

These applications prove that computer algebra has become an integral part of analytical research methods which are essential in exploring complex interrelations. Computer algebra can be integrated into algorithms for solving problems from areas of engineering in many different ways:

- On the lowest level, it is used to perform *basic operations* of real and complex calculus, e.g., symbolic differentiation, integration, summation, or computing limits.
- On a higher level, computer algebra assists investigating *linear systems of differential equations* which arise, e.g., in modeling nonlinear, mechanical systems by linearization.
- Another important area of application is given by *initial* and *boundary value* problems of partial differential equations whose analytic solutions provide more insights for parameter studies than numeric computations could do. This technique is widely used in elastomechanics.
- If we look at ordinary differential equations then the cumbersome task of thumbing through Kamke's classic monograph on ordinary differential equations [Kamke 1961] will soon be a matter of the past; nowadays, computer algebra programs for notebook computers, which can solve certain classes of differential equations, are already at our disposal.
- For a numerical analysis of the basic equations of fluid dynamics the investigation of the numerical stability of an algorithm is essential. When the numerical schemes become complicated such an investigation can be carried out by using a computer algebra system like REDUCE.

For a project engineer, e.g., dealing with combustion, problems from flow dynamics and thermodynamics, which can be treated by nonlinear field theory, are of special interest. For analyzing and solving practical problems, we probably will not be able to do without numeric computations as a last resort. Yee [Yee 1993] has discussed a solution of the reaction-convection-diffusion equation applying the decomposition method. This paper shows how powerful the use of computer algebra can be solving practical engineering problems. However, particularly in computational flow dynamics, computer algebra is used

### 3.4.1 Computer Algebra, a Modern Research Tool for Engineering   225

with increasing tendency, especially for deriving difference approximations to given partial differential equations. In order to be able to check the relevance of numerical data, we have to know the properties of the solutions for the discrete equations first. To that end, our engineer makes intensive use of computer algebra in the following way: First, differential equations, given by physical models, are analyzed for such point transforms which leave the equations invariant. This method is feasible, e.g., for equations from gas dynamics. From a group-theoretic point of view, one determines the generating system of the Lie group defined by the differential equations. In a next step, the *first differential approximation*, a differential relation which mirrors the effects of discretization, is derived from the corresponding difference equations. Again, the algorithm for determining the generators of the Lie group is applied to the new system. If the two groups do not differ, i.e., if the discrete equations admit the same invariant transforms as the original differential equations then the system of difference equations is called an *invariant scheme*, and one is assured that the symmetry properties of the given equations carry over to solutions of the approximation. Graphically speaking, a numeric solution remains spherically symmetric if the initial-value distribution had this property. It will only loose its symmetric characteristics for physical reasons if instability forces the medium to break the symmetry. As a practical application, where one has to conclude the stability or instability of the result from numeric computations, deriving detonation and flame fronts in combustion chambers of piston engines is of prime importance. For an engineer, this requires a priori knowledge of the validity of the results, i.e., he has to be certain that no artificial instabilities are introduced by numerics.

Computer algebra systems are also valuable tools for numerical analysts who are developing software. They allow them to convert formulas into source code of programming languages such as FORTRAN in a convenient and error-free way.

However, not only engineers working in theoretical areas can take advantage of computer algebra also in the experimental field, computer algebra programs provide assistance to swiftly obtain correct results: we are talking about *dimensional analysis* involving many independent variables during the design phase of an experiment. Determining all dimension-less variables is at the beginning of any lab-experiment. Nowadays, a computer with a good computer algebra system can accomplish within seconds what, in the past, had to be done by analytically solving numerous linear equations.

Finally, we should mention that computer algebra systems combined with word-processors allow to transfer mathematical formulas, error-free and ready for printing, therefore helping in cutting down the time one needs to compose scientific papers.

We can conclude that computer algebra has spread into many areas which, about twenty years ago, were only accessible to mathematically skilled experts. In many technical fields, we enjoy the benefits of expert systems as a result of software development in computer algebraCombined with readily available fast work-stations, computer algebra provides researchers in engineering with relief in analytical processing of problems. In the last years several monographies on computer algebra were published which covered even industrial aspects of the application of algebraic methods to very practical problems. The Proceedings of the 1991 SCAFI Seminar at CWI, Amsterdam, [Cohen 1993], contain many applications of computer algebra in industry. Another seminar, devoted to computer algebra in science and engineering [Fleischer et al. 1995], gives an overview of the many fields of application of computer algebra in science and engineering. Especially problems of Mechanics are treated in detail in the book of Klimov and Rudenko [Klimov and Rudenko 1989b] which reflects the activities in Russia at an early stage of this subject.

Summarizing all the results which were gained by the application of computer algebra one can easily conclude, that like the sliderule some decades ago became the hallmark of engineering, computer algebra will become an integral part of the software environment in engineering offices of the future.

Karl G. Roesner (Darmstadt)

### 3.4.2 Critical Load Computations for Jet Engines

In manufacturing of aircraft turbines, there is a considerable difference in temperature between the exterior and the interior during the hardening process. If we look at a nozzle as two cylinders, joined into one another, then tempering exerts forces onto the connection of the parts. In the most basic, one-dimensional case, it is just the classic critical load problem. For that particular problem, we have good numerical methods for computing the critical load for any given model – for example, for a linear combination of several polynomial basis functions – of a joint at our disposal.

Of special interest would be to conduct critical load computations which could be applied directly during the design of the joints, such that the critical load could be maximized with respect to a given class of functions modeling the joints. Typically, such a class is defined by a set of basis functions, where a general function is represented by a linear combination with appropriately many parameters.

If, in contrast to numerically computing which only can yield one numerical value for the critical load for a particular situation, we are able to derive a formula with respect to the given design parameters, then we can use this formula as a function of its parameters to determine the maximal values.

In a joint project of IBM Heidelberg and MTU Munich, we have used AXIOM and the classic numerical methods (Raleigh-Ritz) symbolically. Typically, the resulting expressions and intermediate results are quite large, and often require special treatment. For example, in during the computations it turned out that solving a certain quadratic equation for the critical load by the explicit formula would not be advantageous. Instead, it is carried along as implicitly defined function with respect to the parameters mentioned above, its partial derivatives are computed, and finally a system of equations involving three polynomials is solved for determining the maxima. Furthermore, we realized that in this case, a resultant approach (see section 2.2.1) is superior than computing with Gröbner bases (2.2.5). Details can be found on [Grabmeier 1995].

Additional reference: [Fleischer et al. 1995].

Johannes Grabmeier (Deggendorf)

### 3.4.3 Audio Signal Processing

Audio signal processing (ASP) arose as one of the oldest research fields within the engineering society. Meanwhile ASP has become an interdisciplinary topic with deeper connections to mathematics, computer science, psychoacoustics, among others. Delicate open research problems emerged from application areas like digital audio broadcasting, multimedia, the Internet, audio-on-demand, as well as broad-band communication. Computer algebra is an important tool for solving those problems. In the sequel we illustrate the use of computer algebra in ASP by several examples. For a detailed account on general ASP the reader is referred to the books [Jayant and Noll 1984; Kahrs and Brandenburg 1998; Zölzer 1996] and the relevant journals, the IEEE Transactions on Signal Processing, the IEEE Transactions on Speech and Audio Processing, and Journal of the Audio Engineering Society (AES).

Fundamental constructs in audio signal processing are described properly using algebraic structures. Among those structures are polynomials, (formal) power series and matrices over certain fields or rings modeling signals, filters, and linear systems. Many systems are described by highly structured linear transformations which facilitates the design of efficient algorithms. Among those transforms are Fourier, cosine, and (in connection with wavelets) block Toeplitz transforms.

Discrete signals may be modeled by sequences $(x_j)_{j \in \mathbb{Z}}$ belonging to certain sequence spaces, like the Hilbert space $\ell^2(\mathbb{Z})$ of square summable sequences. In practice one is mostly concerned with finite support sequences. Another form of signal representation which is especially suited to simulate (hardware) implementations represents the time-flow by delay elements. One delay element represents one time step of the discrete signal. In a formal representation delays are denoted by a complex variable $z$. Sequences are mapped to the delay representation, their *z-transform*, by $(x_j)_{j \in \mathbb{Z}} \mapsto \sum_j x_j z^{-j} =: X(z)$. Negative exponents denote the past and positive ones the future behaviour of the signal. In implementations one has to consider the limitation that future events are inaccessible, i.e., circuits with "positive" delays cannot be realized physically. Of particular importance is the Fourier series representation of the signal, $\omega \mapsto X(e^{i\omega})$ for $\omega \in [0, 2\pi)$. Here, the signals sampling rate, i.e., the number of samples or delay elements per second, implicitly determines the *frequency range* that is analyzed by the Fourier transform. This way, frequency behaviour of signals as well as filters may be determined or specified.

Transformations of the sequences are frequently written as certain operations on their $z$-transforms. In the case of filters, i.e., convolutions in the sequence space, the $z$-domain operation is a multiplication, e.g., $Y(z) = T(z)X(z)$ for a filter $(t_j)_{j\in\mathbb{Z}}$ and a signal $(x_j)_{j\in\mathbb{Z}}$. Important classes of filters or filter banks (combination of filters composed with translation, down- or upsampling operators) are obtained by factorizations and decompositions of $T(z)$ in the case that $|T(z)| = 1$ for $|z| = 1$. Examples are filters of the form $z^{-\ell} = T(z) = \sum_{j=0}^{n} S_j(z)A_j(z)$, $\ell \geq 0$, with $j$-th *analysis* filter $(a_{j,k})_{k\in\mathbb{Z}}$ and *synthesis* filter $(s_{j,k})_{k\in\mathbb{Z}}$. Because $Y(z) = z^{-\ell}X(z)$ describes just a delayed version of the original signal, those filter banks are called *perfect reconstruction (PR)* filter banks. Theory and design of filters and filter banks hence relies on methods for the treatment of polynomials and formal power series, like fast multiplication or factorization. For *multirate* PR filter banks, there is a natural correspondence between possible filter factorizations and the linear representation of the component filters gcd [Vetterli and Herley 1992]. In the multidimensional case, Gröbner basis techniques are used [Park et al. 1997]. Since the zeros of the $z$-transform play a crucial role in the frequency behaviour of filters, the ability to construct PR systems under constraints of certain predetermined zeros is very important. As a prominent example we mention the wavelet filters of Daubechies [Daubechies 1988].

Digital audio of CD quality (44.1 kHz sampling rate PCM, 16 bit linear quantization) amounts to a data rate of about 1.5 MBit/sec. Various multimedia and Internet applications call for data compression and at the same time for coding techniques to protect data against bursty and transient packet losses. The Moving Picture Expert Group (MPEG) within the International Organisation of Standardization (ISO) has developed a series of audio-visual standards known as MPEG-1 and MPEG-2 [Noll 1997]. Compression rates up to 1 : 12 while retaining CD quality are obtained combining psychoacoustic models with tools from computer algebra like FFT, fast modified DCT, and filter banks. It is suspected that improved filter banks and psychoacoustic models will yield even higher compression rates.

The FFT is not only an analysis tool but has also many other applications. The iterative procedure for modeling equiripple FIR filters using FFTs [Çetin et al. 1997] is only one example from filter design. Efficient Fourier transform algorithms facilitate fast convolution and hence fast filtering for real-time applications and implementation on digital signal processors (DSPs). For an overview on FFT algorithms we refer to [Blahut 1987; Burrus 1995; Clausen and Baum 1993]. Filter design algorithms as well as other signal processing applications frequently require high precision outputs. Especially when using fixed-point DSPs, the accumulation of round-off errors typically results in unacceptable computational noise. A new FFT-algorithm [Buhler et al. 1997], based on fast approximation of complex numbers by cyclotomic integers and Chinese remaindering, circumvents this problem.

Many software packages well-suited for ASP are now available. We only mention the widely used Matlab package [The Mathworks Inc. 1996] for numerical signal processing and visualization. A plug-in to this package, the symbolic toolbox, contains the kernel software of the Maple computer algebra package (see 4.1.6), allowing symbolic calculations within Matlab. Several toolboxes supporting the design of wavelets and filter banks, for example the Wavelab package [Buckheit et al. 1996] and the FBT-tools [Kurth and Clausen 1999], have been implemented. Additional toolboxes complementing computer algebra tools are important for many other ASP applications. We mention the design of near perfect reconstruction multirate filter banks, e.g., the filters used in the MPEG audio coders, which amounts to solving a least-squares optimization problem with quadratic constraints [Nguyen 1995].

Michael Clausen and Frank Kurth (Bonn)

### 3.4.4 Robotics

Robotics integrates a very large spectrum of technical sciences ranging from electrical and mechanical engineering to Mathematics and Computer Science. The primary area of robot-specific applications of algorithms from computer algebra is robot kinematics.

Kinematics investigates the relationship between the joint(variable)s of arbitrary mechanisms and the pose (position and orientation) of its links (bodies) in space. For a given robot, the so-called *inverse kinematics problem* ("IKP") requires to determine all sets of joint values that take the robot hand to a given pose in space. A similar problem is the *direct kinematics problem* ("DKP") of so-called (generalized) *Stewart Platforms*, i. e. flight-simulator-type mechanisms consisting of a stationary and a mobile platform which are joined via passive balljoints and actuated telescope legs. The DKP asks for all mobile platform poses that can be attained for a given set of leg lengths. Both problems require to solve non-linear systems of sine-cosine equations (some of the variables being trigonometric functions) which relate joint-variables, pose-parameters and (constant) mechanism-dimensions.

Industry demands all solutions of the IKP for any given hand pose within less than 5 milliseconds because otherwise the robot hand cannot be driven fast and smoothly along arbitrary desired trajectories. This speed requirement prohibits the use of iterative algorithms. Thus, kinematic systems of equations must be solved *symbolically*, i. e. in terms of implicit or even explicit "solution formulae" that reduce computations at run-time to a minimum - see below for an example of a symbolic solution. However, even simple manipulator geometries are far too complex for a direct application of state-of-the-art symbolic solution techniques like the Buchberger-Algorithm; see 2.2.5 or [Buchberger 1985b] for an introduction as well as applications to polynomial system solving and beyond. Specific kinematical strategies for solving the IKP symbolically were developed since the 1970s. They work far better than elaborate computer algebra techniques for elementary manipulator classes but fail for difficult classes.

The widespread Puma 560 is used to give an example of a kinematic system of equations and its symbolic solution. Fig. 1 shows the manipulator and its six revolute joint axes which are labeled traditionally $Z1$ to $Z6$. Without going into details, the kinematic system of equations is obtained via elementary geometric considerations by calculating the pose of the hand as a function of the joint

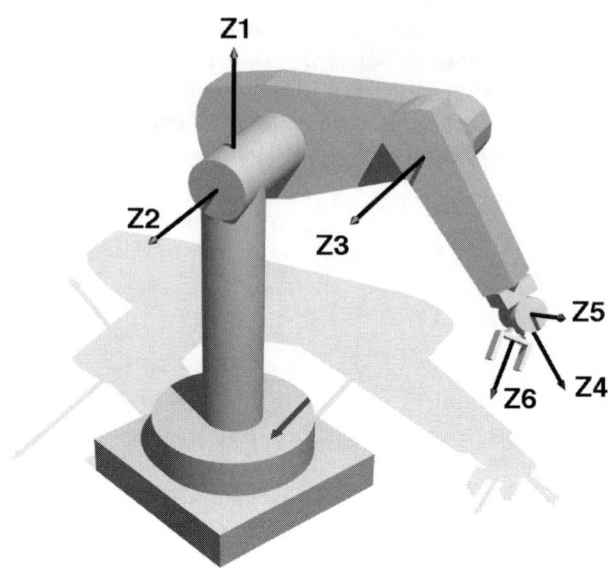

**Fig. 8.** The Puma 560 manipulator with joint axes Z1 to Z6

variables $\theta_1, \theta_2, ..., \theta_6$, which specify the rotation angle of each of the six revolute joints. The resulting system is

$$t_{11} = ((c_1 \, c_{23} \, c_4 - s_1 \, s_4)c_5 - c_1 \, s_{23} \, s_5)c_6 - (c_1 \, c_{23} \, s_4 + s_1 \, c_4)s_6$$
$$t_{12} = ((-c_1 \, c_{23} \, c_4 + s_1 \, s_4)c_5 + c_1 \, s_{23} \, s_5)s_6 - (c_1 \, c_{23} \, s_4 + s_1 \, c_4)c_6$$
$$t_{13} = (c_1 \, c_{23} \, c_4 - s_1 \, s_4)s_5 + c_1 \, s_{23} \, c_5$$
$$t_{14} = c_1(s_{23} \, d_4 + c_{23} \, a_3 + c_2 \, a_2) - s_1 \, d_2$$
$$t_{21} = ((s_1 \, c_{23} \, c_4 + c_1 \, s_4)c_5 - s_1 \, s_{23} \, s_5)c_6 + (c_1 \, c_4 - s_1 \, c_{23} \, s_4)s_6$$
$$t_{22} = ((-s_1 \, c_{23} \, c_4 - c_1 \, s_4)c_5 + s_1 \, s_{23} \, s_5)s_6 + (c_1 \, c_4 - s_1 \, c_{23} \, s_4)c_6$$
$$t_{23} = (s_1 \, c_{23} \, c_4 + c_1 \, s_4)s_5 + s_1 \, s_{23} \, c_5$$
$$t_{24} = s_1(s_{23} \, d_4 + c_{23} \, a_3 + c_2 \, a_2) + c_1 \, d_2$$
$$t_{31} = (-s_{23} \, c_4 \, c_5 - c_{23} \, s_5)c_6 + s_{23} \, s_4 \, s_6$$
$$t_{32} = (s_{23} \, c_4 \, c_5 + c_{23} \, s_5)s_6 + s_{23} \, s_4 \, c_6$$
$$t_{33} = -s_{23} \, c_4 \, s_5 + c_{23} \, c_5$$
$$t_{34} = c_{23} \, d_4 - s_{23} \, a_3 - s_2 \, a_2$$

where the $s_i$, $c_i$, $s_{ij}$, $c_{ij}$ stand for $\sin(\theta_i)$, $\cos(\theta_i)$, $\sin(\theta_i + \theta_j)$, $\cos(\theta_i + \theta_j)$ respectively and the remaining entities $d_i$, $a_j$ on the right hand side of the equations specify the robot geometry, i. e. they give the length of arm segments, etc. The $t_{ij}$ on the left hand side represent the position and orientation of the manipulator hand in a redundant way (being elements of a rotation matrix and a position vector). The redundancy of the pose representation induces a (helpful) algebraic dependency on this system of 12 equations in 6 variables $\theta_1, \theta_2, ..., \theta_6$.

The system below is equivalent to the above system. It is called a symbolic solution because the first equation contains only one variable (here: $\theta_1$, i. e. $s_1$ and $c_1$) and each subsequent equation contains at most one variable more than the set of its predecessors. Consequently, for given values of the pose parameters $t_{ij}$ and given robot geometry parameters all variables $\theta_i$ can easily be determined by consecutively solving the equations for the corresponding unknown variable, thus solving the IKP for the Puma 560. Due to the particular type of equations, variables with a unique solution like $\theta_4$, $\theta_6$ or $\theta_{23}$ in the example must be determined from the explicit simultaneous solution of a pair of independent equations. This can be achieved through simple textbook formulae. For each given hand pose the system below yields eight different solution sets, i. e. sets of joint values taking the hand to the desired pose. Solution sets contain complex joint values iff the desired pose is not reachable by the Puma 560.

$$0 = -s_1 t_{14} + c_1 t_{24} - d_2$$

$$a_3^2 + d_4^2 = d_2^2 + a_2^2 - 2 a_2 c_2 ( c_1 t_{14} + s_1 t_{24}) + 2 s_2 a_2 t_{34} + 2 d_2(s_1 t_{14} - c_1 t_{24}) + t_{14}^2 + t_{24}^2 + t_{34}^2$$

$$c_{23} d_4 - s_{23} a_3 - s_2 a_2 = t_{34}$$
$$s_{23} d_4 + c_{23} a_3 + c_2 a_2 = c_1 t_{14} + s_1 t_{24}$$

$$c_5 = s_{23}(c_1 t_{13} + s_1 t_{23}) + c_{23} t_{33}$$

$$s_4 s_5 = -s_1 t_{13} + c_1 t_{23}$$
$$c_4 s_5 = c_{23}(c_1 t_{13} + s_1 t_{23}) - s_{23} t_{33}$$

$$c_6 = -s_4 c_{23}(c_1 t_{12} - s_1 t_{22}) + s_4 s_{23} t_{32} - c_4(s_1 t_{12} + c_1 t_{22})$$
$$s_6 = -s_4 c_{23}(c_1 t_{11} - s_1 t_{21}) + s_4 s_{23} t_{31} - c_4(s_1 t_{11} + c_1 t_{21})$$

The Buchberger Algorithm and other elaborate, universal algebraic techniques fail to yield this symbolic solution in practice due to the large number of formal parameters appearing in the original kinematic system of equations. In contrast, the above solution for the Puma 560 and for some limited other classes of robot geometries can be found by relatively simple geometric considerations. For example, the first equation above, which contains only $\theta_1$, is derived from the observation that joints 2 to 6 move the common intersection point of axes 4 to 6 in a plane which is permanently perpendicular to axis Z2. See [Craig 1986] for an introduction to robot kinematics.

Effective elementary applications of computer algebra to the IKP specialize all formal parameters of kinematic systems of equations, investigate these systems by the Buchberger-Algorithm and use results to optimize the solution of corresponding unspecialized systems. Based on a combination of traditional and new kinematic techniques large numbers of simple unspecialized elimination ideals are compiled and searched by elementary artificial intelligence methods and the optimal one is solved symbolically by specific adaptions of (multivariate) resultant algorithms to systems of sine-cosine polynomials with large numbers of formal parameters.

After 45 years of intense research a universal symbolic solution of the IKP for general robots with six revolute joints was presented 1988 in a chinese master's thesis by H. Y. Lee [Lee and Liang 1988]. It is based on the discovery of (some relevant relationship yielding three low degree elimination ideal members and sufficiently simple elimination steps leading to) a low degree basis of a third elimination ideal and a subsequent application of multivariate resultants. This pioneering work was significantly improved and extended to completely general robots with revolute or prismatic joints by Raghavan and Roth [Raghavan and Roth 1992]. The univariate equation of $16^{th}$ degree with coefficents consisting of polynomials in the pose- and structure parameters is estimated to contain 15000 lengthy terms in expanded form. A posteriori, it becomes apparent that the Buchberger-Algorithm could have supported the detection of this handcrafted solution significantly.

D. Manocha and J. Canny accelerated the universal solution of the IKP considerably by extending an approach by Auzinger and Stetter who compute solutions of a resultant system very efficiently via a matrix-eigenvalue approach [Manocha and Canny 1992]. The former algorithm comes close to the performance required by industry but up to now, industry still circumvents the problem by restricting commercial production to those manipulator types that can be solved with elementary, traditional methods.

The universal solution of the IKP and its improvements cannot yield parameterized symbolic solutions in practice. Consequently a number of prominent problems remain to be solved effectively. One of them, which was recently promoted by T. Levelt [Levelt 1997], is the closed form solution of the kinematics of the cycloheptane molecule. A symbolic solution of a slightly more general ring-molecule will be given in [Kovács et al. 2000].

Today, the DKP of generalized stewart platforms is considered the primary unsolved problem of kinematics. For nearly a decade, it was an open problem to prove a sharp upper bound for the number of solutions of the DKP. A surprising proof was given by D. Lazard [Lazard 1993] who represented rigid body motion with more redundancy than usual instead of less. After splitting up the resulting (completely general, highly complex) system of equations into two parts it became possible to calculate some total degree Gröbner Basis, corresponding Hilbert Functions and the Bezout number of the first part due to the particular

### 3.4.4 Robotics

representation. The Bezout number of the other part could be determined directly, proving immediately that generalized Stewart platforms possess at most 40 real or complex "assembly configurations".

The most significant progress so far towards a solution of the DKP is an approach by M. Husty which combines traditional methods with modern computer algebra techniques [Husty 1996]. Husty uses a classical transformation of kinematic problems ("Studys kinematic mapping") to transform the original system into a 7-dimensional quasi-elliptic space and is able to derive a basis of some third elimination ideal in the corresponding residue class ring with limited effort. The final equation of $40^{th}$ degree is obtained via repeated resultant and GCD-calculations.

An active area of research are essential simplifications of symbolic solutions of kinematic problems since it may frequently be impossible to satisfy tight industrial real-time demands without additional simplifications. An initial negative result was shown by M.-J. Gonzalez-Lopez and T. Recio who prove that kinematic ideals of open kinematic chains and Stewart-platforms are irreducible [Gonzalez-Lopez and Recio 1994].

The primary remaining alternative is to solve equations through efficient identification of radical decompositions. The simplest variant of solvability by radicals is functional decomposition i. e. to express a given function $f(x)$ as a composition $f(x) = g(h(x))$, see 2.2.4. The identification of decomposition factors $g$ and $h$ is well known for polynomials, but a direct application of the original algorithms to sine-cosine polynomials proved to be impossible. The problem was solved initially by an adaption of an algorithm by Kozen and Landau and subsequently by a significantly faster and qualitatively new algorithm in [Kovács and Hommel 1993]. Kinematical problems induced a deeper investigation of radical decomposition methods. Weiss and v. z. Gathen presented an algorithm for computing homogeneous bivariate decompositions ("HBD") of polynomials [von zur Gathen and Weiss 1995]. HBD is a significant generalization of the concept of functional decomposition and is in a certain sense a conceptual closure, integrating the functional decomposition of ordinary polynomials and sine-cosine polynomials. The algorithm is based on an investigation of imprimitivity domains (or block decompositions, resp.) and factorizations of polynomials over algebraic number fields. A significant acceleration of the HBD-algorithm can be obtained by a modular approach.

Currently, M. Pohst and the author are trying to transfer effective concepts for radical decomposition of single (sine-cosine-)polynomials to systems of (sine-cosine-)polynomials.

Methods of computer algebra also play an important role in the so-called "piano movers" or collision avoidance problem where a robot of certain geometry is supposed to move a payload with given shape through an environment containing a number of known obstacles without collision. All involved objects are usually specified by semi-algebraic sets. Schwarz and Sharir [Schwarz and Sharir 1987] used cylindrical algebraic decomposition and Sylvester resultants to obtain a general algorithm of double exponential complexity in the number of points, lines and faces of the involved objects.

A completely new approach was presented in J. Cannys thesis which won the ACM Dissertation Award 1987 [Canny 1987]. His "Roadmap-Algorithm" solves the general problem in single exponential time and optimality of this bound is proven. Most prior algorithms for restricted special cases turned out to be significantly slower in practice than the new, general algorithm. If a collision-free path exists, the algorithm yields a certain set of "critical points" in the space of robot joint-configurations which are joined by one-dimensional, collision-free "roads". The mathematical core of the algorithm computes "silhouettes" of forbidden regions and reduces complexity through "stratification" based on results on real algebraic varieties by H. Whitney and finally uses multivariate resultants to identify the critical points in projections of the silhouettes onto the 2-dimensional strata.

An improved roadmap algorithm was developed by Basu, Pollack and Roy in [Basu et al. 1996a]. It significantly reduces the computational effort by avoiding multivariate resultants through a well controlled application of the Buchberger Algorithm plus an intelligent new way of managing infinitesimals appearing in the computations. The algorithm is very sensitive to the geometry of the considered problem, i. e. geometric parameters appear in complexity estimates.

A field which is closely related to kinematics is mechanism synthesis. In the classical form of the problem a set of desired "precision points" and the kinematical structure of a mechanism are given and dimensions of the mechanism have to be calculated such that the mechanism tool is driven through the precision points by movement of some actuated "input-joint". Potentially this field poses a similar scope of computer algebra problems as kinematics, however, the industrial relevance of the problem is decreasing. An interesting example of computer algebra applications to mechanism-synthesis is the design of so-called cam-follower mechanisms for peeling clams [Cohen and Heck 1995].

Peter Kovács (Berlin)

### 3.4.5 Computer Aided Design and Modelling

The usefulness of CAD/CAM (Computer Aided Design/Computer Aided Modelling) systems as a means of increasing the efficiency of the design process is nowadays uncontested. Advantages such as

- reduction of lead times,
- quality improvements, and
- cost reduction by saving time spent implementing engineering changes in the design process

are often cited as the major benefits resulting from the introduction of specialized software for CAD/CAM. From a mathematical point of view almost all the CAD/CAM problems are related to the manipulation of geometric objects into the two or three dimensional space, mainly curves and surfaces and combinations of both. Since these geometric entities are usually presented through polynomials via their implicit of parametric representation, it is clear that the intersection between Computer Algebra and Computer Aided Geometric Design must be non-empty. There are four different problem classes to be considered here where Computer Algebra and Computer Aided Geometric Design meet: the implicitization (or variable–elimination) questions, the intersection of parametric curves and surfaces, Computer Aided Geometric Design with exact arithmetic and the algebraically guided tracing of algebraic curves and surfaces implicitly defined.

**3.4.5.1 Implicitization and Variable Elimination Questions.** One of the main problems arising in the manipulation of parametric curves and surfaces in computer aided geometric design is the finding of efficient algorithms for computing the implicit equation of curves and surfaces parametrized by rational functions (see for example [Mandache 1993], [Cox et al. 1992] or the chapters 5 and 7 in [Hoffmann 1989b]). This is due, for example, to the fact that, if for tracing the considered curve and surface the parametric representation is the most convenient, to decide in an efficient way the position of a point with respect to the curve or surface considered, the implicit equation is desired.

The implicitization problem for hypersurfaces (in real applications, curves in the plane or surfaces in the three dimensional space) parametrized in a rational way can be stated in the following terms: let $\mathcal{V}$ be a hypersurface in $\mathbb{R}^n$ (in real applications $n = 2$ or $n = 3$) parametrized by ($i \in \{1, \ldots, n\}$):

$$x_i = \frac{f_i(t_1, \ldots, t_{n-1})}{g_i(t_1, \ldots, t_{n-1})}$$

where $f_i$ and $g_i$ belong to $\mathbb{Z}[t_1, \ldots, t_{n-1}]$ with $\gcd(f_i, g_i) = 1$. The implicitization problem for $\mathcal{V}$ is the finding of a non zero element $\mathcal{R}_\mathcal{V}(x_1, \ldots, x_n)$ in $\mathbb{Z}[x_1, \ldots, x_n]$ with the smallest posible total degree and such that:

$$\mathcal{R}_\mathcal{V}\left(\frac{f_1(\underline{t})}{g_1(\underline{t})}, \ldots, \frac{f_n(\underline{t})}{g_n(\underline{t})}\right) = 0.$$

More general formulations of the implicitization problem for arbitrary parametric varieties can be found in [Alonso et al. 1995a], [Canny and Manocha 1992], [Chionh 1990], [Kalkbrener 1991], [Gonzalez-Vega 1997] or [Gao and Chou 1992].

Another problems with a similar formulation than the implicitization problem described before and where the solution is obtained by eliminating from the initial equations some variables, are (see [Hoffmann 1989b]):

- Computation of offset curves and surfaces.
- Computation of constant–radius blending surfaces.
- Computation of the convolution of two plane curves or surfaces.
- Computation of the convolution of two plane curves.
- Computation of the common tangent of two plane curves.
- Computation of the inversion formula for parametric surfaces.

These geometrical operations are oftenly used when generating the boundary of a configuration space obstacles, in order to construct collision free motion paths for translating objects.

Two main difficulties are encountered when trying to use the usual elimination technics offered by Computer Algebra (resultants, Gröbner bases, etc.) to deal with the variable elimination problems mentioned before. For example the implicitization of a rational surface defined by

$$x = \frac{X(s,t)}{W(s,t)}, \; y = \frac{Y(s,t)}{W(s,t)}, \; z = \frac{Z(s,t)}{W(s,t)}$$

appearing into a real–world problem is difficult to achieve by applying directly resultants or Gröbner bases because, first, it is usually a very costly algebraic operation and, second, the coefficients of the polynomials in the parametrization are usually floating–point real numbers.

These difficulties can be currently overcome in two different ways:

- By using multivariate resultants (see [Canny and Manocha 1992]), the implicit equation is described as a non evaluated determinant. Then any question about the considered surface requiring the implicit equation is reduced to a Numerical Linear Algebra question over such matrix (usually an eigenvalue problem).
- By taking into account that, in general, a concrete object to model is made by several hundreds (or thousands) of small patches, all of them sharing the same algebraic structure: for such an object a database is constructed containing the implicit equation of every class of patch appearing in its definition. This database must also contains the inversion formulae (giving the parameters in terms of the cartesian coordinates) and must be pruned to avoid specialization problems. Moreover the database for a specific objetc is kept into a bigger and general database for a further use see [Espinola et al. 1999].

The main drawback of the first approach is due to the existence of base points, ie solutions of the polynomial system

$$X(s,t) = 0, \; Y(s,t) = 0, \; Z(s,t) = 0, \; W(s,t) = 0,$$

since their existence implies the vanishing of the determinant defining the implicit equation. This problem is solving by looking for an appropiated submatrix of full rank as shown in [Manocha 1992].

The main drawback provided by the second approach is due to the fact that some algebraic structures arising in the database construction are very complicated and the implicit equation can not be generated or even difficult to use due to its huge size. Namely:

$$x = \frac{X(s,t)}{W(s,t)}, \; y = \frac{Y(s,t)}{W(s,t)}, \; z = \frac{Z(s,t)}{W(s,t)}$$

### 3.4.5 Computer Aided Design and Modelling

with
$$W = \sum_{j=0}^{3}(A_j t^2 + B_j t + C_j)s^j$$

and $(U \in \{X, Y, Z\}, i \in \{x, y, z\})$

$$U = \sum_{j=0}^{3}(\alpha_j^{(i)} t^2 + \beta_j^{(i)} t + \gamma_j^{(i)})s^j.$$

It is not also easy to deal in advance with specialization problems: upto this moment these are detected by substituting several points in the surface, uniformly generated by the parametrization, into the candidate to be the implicit equation.

Recently, a new method for the implicitization of rational curves and surfaces has appeared: the using of the moving line ideal basis for planar rational curves (see [Cox et al. 1998b]) and the using of moving surfaces for rational surfaces (see [Sederberg 1998]). In principle they provide easier respresentations for the implicit equation but they have the same drawbacks than the using of resultants. In fact this method is still not completely understood for the case of rational surfaces.

The inverse problem to the one considered here: given a curve or surface by its implicit equation, to determine if it could be presented in a rational way is, from the algebraic point of view, very well understood at least for the case of planar curves (the curve admits a rational parametrization if and only if its genus is zero, see [Sederberg 1998], for example). For surfaces is more complicated but also a genus computation gives the answer, at least for several kind of surfaces (see [Schijo 1998]). Anyway let us indicate that all these algorithms require the exact knowledge of the coefficients appearing in the implicit equation.

**3.4.5.2 Algebraically Guided Tracing of Algebraic Curves and Surfaces Implicitely Defined.** Many important problems in Computer Aided Geometric Design are reduced to the computation of the graph of a planar algebraic curve presented implicitely. For example if we want to section the surface

$$x = \frac{X(s,t)}{W(s,t)}, \quad y = \frac{Y(s,t)}{W(s,t)}, \quad z = \frac{Z(s,t)}{W(s,t)}; \quad s, t \in [0,1]$$

with respect to the plane $X = x_0$ then we have two possibilities: either we draw "into the square unit" $[0, 1] \times [0, 1]$ the planar algebraic curve defined by

$$x_0 = \frac{X(s,t)}{W(s,t)}$$

and then this picture is lifted to the considered surface or, if the implicit equation $H(x, y, z)$ of the considered surface is available, the lifting procedure can be avoided by merely computing the graph of the planar curve $H(x_0, y, z) = 0$.

The problem of computing the graph (even topologically) of a planar algebraic curve defined implicitely has received a special attention from Computer

Algebra since it has been responsible of many advances regarding subresultants, real root counting, infinitesimal computations, etc. From the seminal papers [Gianni and Traverso 1983], [Arnon and McCallum 1988] and [Roy 1991], the interested reader can see in [Cucker et al. 1991], [Feng 1992], [Cellini et al. 1991], [Hong 1996], [Gonzalez-Vega and Kahoui 1996] and [Gonzalez-Vega and Necula 1999], how the theoretical and practical complexities of the algorithms dealing with this problem have been dramatically improved.

The usual strategy to compute the graph (even topologically) of a planar algebraic curve defined implicitly by a polynomial $f(x,y) \in \mathbb{R}[x,y]$ proceeds in the following way:

- Step I: Computation of the discriminant of $f$ with respect to $y$, $R(x)$, and characterization of the real roots of $R(x)$, $\alpha_1 < \ldots < \alpha_r$.
- Step II: For every $\alpha_i$, computation of the real roots of $f(\alpha_i, y)$, $\beta_{i,1} < \ldots \beta_{i,s_i}$.
- Step III: For every $\alpha_i$ and $\beta_{i,j}$ computation of the number of half–branches to the right and to the left of the point $(\alpha_i, \beta_{i,j})$.

Following [Gonzalez-Vega and Necula 1999] and in order to avoid the numerical problems arising from the computation of the roots of $R(x)$ and of every $f(\alpha_i, y)$ which has always multiple roots, before starting the computations, a generic linear change of variables is performed in order to have the following condition for every $\alpha \in \mathbb{R}$:

$$\#\{\beta \in \mathbb{R} : f(\alpha, \beta) = 0, \frac{\partial f}{\partial y}(\alpha, \beta) = 0\} \leq 1.$$

This assures that for every $\alpha_i$ real root of $R(x)$, there is only one critical point of the curve in the vertical line $x = \alpha_i$ whose $y$-coordinate can be rationally described in terms of $\alpha_i$. Moreover this allows to symbolically construct, from every $f(\alpha_i, y)$, a squarefree polynomial $g_i(\alpha_i, y)$ whose real roots are needed to compute in order to finish with the so called Step II. Step III is thus accomplished by merely computing the number of real roots of the squarefree polynomials $f(\gamma_i, y)$ ($i \in \{0, 1, \ldots, r\}$) with $\gamma_0 = -\infty$, $\gamma_r = \infty$ and $\gamma_i$ any real number in the open interval $(\alpha_i, \alpha_{i+1})$. These computations provide a graph of the considered curve which is very helpful when this is going to be traced numerically since we know exactly how to proceed when close to a complicated point: for the curve defined by the polynomial

$$P = 2y^3 - (3x - 3)y^2 - (3x^2 - 3x)y - x^3$$

the following graph is obtained

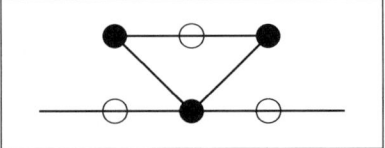

where the black points represent the critical points which include those singular.

When dealing with problems as the sectioning without the implicit equation of the considered surface, the tracing of the considered algebraic curve $f(s,t) = 0$ must be done into the "unit square" $[0,1] \times [0,1]$ and, in many cases, for physical and/or design reasons it is known that no singular points will appear. This information is very useful since it implies that all the information needed can be easily derived from the computations of the real roots in $[0,1]$ of the polynomials

$$f(0,t), \ f(1,t), \ f(s,0), f(s,1), \ R(s)$$

where $R(s)$ is the discriminant of $f(s,t)$ with respect to $t$.

The problem of tracing surfaces implicitely defined is considerably more complicated since, first, their topology structure is no so easy as the one for curves and, moreover, the degrees of the univariate polynomials to deal with are extremely big (see for example [Cellini et al. 1991]).

### 3.4.5.3 Intersection of Parametric Curves and Surfaces in Computer Aided Geometric Design by Using Computer Algebra.

The problem of computing the intersection between curves and surfaces is very important in Computer Aided Geometric Design and is usually reduced to the resolution of a polynomial system of equations, one of the most studied problems in Computer Algebra.

The intersection of two planar curves without common components given parametrically can be determined by merely computing the implicit equation of the first curve and then substituting the parametrization of the second one into this implicit equation: the problem is solved by solving an univariate polynomial equation. The intersection when the two curves are given by their implicit equations $f(x,y)$ and $g(x,y)$ can be determined by using a similar argument that in the previous paragraph: by performing a generic change of coordinates it is verified the following condition for every $\alpha \in \mathbb{R}$:

$$\#\{\beta \in \mathbb{R} : f(\alpha, \beta) = 0, g(\alpha, \beta) = 0\} \leq 1$$

This allows to represent every $\beta$ rationally in terms of $\alpha$ for every $\alpha$ real root of the resultant of $f$ and $g$ with respect to $y$ which are determined via an eigenvalue problem.

The intersection of a rational surface (or a surface given by its implicit equation) and a rational curve into the 3–space, when the implicit equation of the surface, is available is reduced to the computation of the real roots of the univariate polynomial obtained after the susbstitution of the parametrization of the curve into the implicit equation of the surface. If the implicit equation of the considered rational surface is not available then, by equating both parametrizations we have to solve a polynomial system of 3 equations and 3 unknowns (with floating point numbers as coefficients). If using Computer Algebra, the only used technique currently is that based on multivariate resultants (see [Manocha 1992]) and mainly in Dixon's formulation (see [Dixon 1908]) for eliminating two of the unknowns from the system.

The intersection of two rational surfaces is reduced to the tracing of a implicitely defined plane algebraic curve (a problem already considered into the previous paragraph) plus a lifting process if the implicit equation of one of the surfaces is available. If no implicit equation is available then we have to solve a polynomial system of 3 equations and 4 unknowns (with floating point numbers as coefficients) and, again as before, if using Computer Algebra, the only used technique currently is that based on multivariate resultants (see [Manocha 1992]) and mainly in Dixon's formulation (see [Dixon 1908]) for eliminating two of the unknowns from the system. If the two surfaces are presented by their implicit equation then the system to solve has only 2 equations and 3 unknowns. Note that in both cases the intersection curve is, in general, no longer rational.

### 3.4.5.4 Computer Aided Geometric Design With Exact Arithmetic.
This is the probably best CAGD topic where Computer Algebra techniques can be perfectly applied. It is assumed that all the input objects are known exactly and with their presentation involving only rational numbers or algebraic numbers exactly presented. In this case all the problems are reduced to the resolution of a polynomial system of equations or to the elimination of one or several variables and for these problems the using of Gröbner bases (see for example 2.2.5, [Hoffmann 1989b], [Hoffmann 1989a] and [Hoffmann and Vermeer 1991]) or resultants (see [Manocha 1992] and [Sederberg 1998]) are specially well suited.

The only implemented experience of this strategy appears in [Keyser et al. 1997] where purely algebraic techniques such as multivariate Sturm sequences ([Milne 1992] and [Gianni and Traverso 1983]), exact algebraic numbers computations and multivariate resultants are used to deal with low degree sculptured solids by using boundary representations.

### 3.4.5.5 Future Applications.
Computer Aided Geometric Design is sometimes considered as the intersection of Geometry and Computer Science. Since Computer Algebra has a very big intersection with these two domains, it is expected in the near future to find new and significant applications of Computer Algebra algorithms to the efficient manipulation of curves ans surfaces in CAGD. In this last paragraph four different problems in CAGD are described where Computer Algebra may be applied.

The first problem to be described is motivated by the fact that *Computer Aided Geometric Design is interested only in real solutions*. Polynomial System Solving in CAGD deals only with real solutions since all the involved geometric objects live into a 2 or 3 dimensional real space. Nevertheless most of the techniques already mentioned are purely algebraic techniques designed usually to deal with the manipulation of complex solutions. Moreover it would be very natural the inclusion, into the polynomial systems to solve, of constraints by adding also polynomial inequalities: for example to be inside or outside of a given region with non linear boundary or to be to a distance less than or equal of a given point or geometric entity. For example, the implicitization of the parametric curve given by ($t \in \mathbb{R}$)

$$x = 1 + t^2, \; y = t(1 + t^2)$$

produces the equation
$$y^2 + x^2 - x^3 = 0.$$
But the curve defined by this implicit equation has one more (and singular) real point than the initial parametric curve.

In the case of exact arithmetic there are algorithms for dealing with these problems (see [Rouillier 1996]) when the initial system is known to have a finite number of complex (yes, complex !) solutions but when we enter into the general case, the only available techniques are those known as Quantifier Elimination Algorithms over the reals which is a very lively area of research (see 2.5.3 and compare 2.15.2). In [Sturm and Weispfenning 1998c, 1997], several examples of using Quantifier Elimination to CAGD problems can be found. For the previous example, and by using Quantifier Elimination terminology, the implicitization problem is no more than the elimination of the existential quantifier into the formula:
$$\exists t \in \mathbb{R} \quad t^2 + 1 - x = 0, \ t(1+t^2) - y = 0$$
whose solution is
$$y^2 + x^2 - x^3 = 0, \ x \geq 0.$$
For the non–exact arithmetic case the situations is considerably more complicated.

The second problem to be described deals with *geometric formats conversion: from rational to polynomial parametrizations*. Two are the most used formats in the CAD/CAM systems to represent and deal with the data required to describe and communicate the essential engineering characteristics of physical objects such as manufactured products: the VDA and IGES formats. The VDA (Verband der Automobilindustrie) format was started at 1982 by the VDA Committee founded mainly by several German automobile and automotive supply industry. The IGES (Initial Graphics Exchange Specification) created by the IGES/PDES Organization (USA).

The main difference between these two formats is found in the fact that the VDA format only accepts surfaces defined by polynomial parametrizations while the IGES format accepts both, rational and polynomial. This is a very important problem in many companies in the automobile industry since they get the information in the IGES format but the specific CAD/CAM software they use only allows the use of the VDA format. The only way of solving this problem upto this moment is by means of a method based upon an uniform subdivision of the parameter domain plus a polynomial interpolation procedure whose efficiency depends strongly on the degree of the polynomials to appear into the searched polynomial parametrization (see [Bardis and Patrikalakis 1989]). The Computer Algebra behind this problem is still not completely well understood and new algorithms for solving this problem could appear by using new Hermite–Birkhoff interpolation schemes for multivariate polynomials.

The third problem tackles the using of "*hybrid (Symbolic/Numeric) methods for Polynomial System Solving*" to Computer Aided Geometric Design. Recently several proposals for mixing symbolic and numerical methods for solving polynomial systems of equations have been proposed. In [Corless et al. 1995, 1997a] and

[Mourrain and Pan 1998] this is achieved by applying well stablished methods in Numerical Analysis as the Singular Value Decomposition to several matrices very close to those defining multivariate resultants. In [Bonini et al. 1998] and [Stetter 1997; Thallinger and Stetter 1998] this mixture is achieved by providing several adaptations of Buchberger algorithm to compute Gröbner bases of sets of polynomials are only known upto a limited accuracy (see also 2.12).

The application of these methods to Computer Aided Geometric Design has not been still pursued but their impact in CAGD is estimated to be very significant since they bring together the accuracy of an almost exact method with the high speed and stability of standard and well–known numerical methods.

Last problem introduces *Parametric Computer Aided Geometric Design*. A very usual request from any user of CAD/CAM software is the possibility of manipulating parametrically geometric entities: for example a cylinder whose radius is a function whose values are provided at the screen by a mouse movement and whose intersection with a fixed sphere is to be represented in the screen. Again this is no more that a polynomial system solving problem but now involving a parameter and if, at this moment and theoretically, Computer Algebra provides several algorithms dealing with this kind of problems their real applicability is far from being achieved.

Laureano Gonzalez–Vega (Cantabria)

## 3.5 Chemistry

### 3.5.1 Computer Algebra in Chemistry and Crystallography

One of the basic tasks in chemistry consists in exploring the relationship between reaction kinetics of molecular structures and the spatial arrangement of their atoms or molecular sub-structures respectively. This leads to a number of problems with computer algebra appeal.

One of these problems was mentioned in section 2.9 *algebraic methods for constructing discrete structures* by A. Kerber and R. Laue: The design of reasonable *branch and bound* methods for constructing all connected multigraphs for a given sum formula, where the degree of a vertex corresponds to the number of valences of an atom (cf. http://www.mathe2.uni-bayreuth.de/molgen4). Closely related to this *isomerism problem*, yet much more intricate, is the *stereoisomerism problem* pertaining to the area of algorithmic real algebraic geometry. A paradigm is the determination of all configurations of cyclo-hexane and cyclo-heptane (see, e.g., [Jenks and Sutor 1992], pp. 377–378). Here, restrictions with respect to bonding distances and angles are known for certain atomic bonds, and we are looking for ways of computing the space of all configurations satisfying these constraints – or, at least, the number of its connected components. By means of Cayley-Menger determinants, the problem can be transformed into that of finding all real solutions of a system of polynomial equations and inequalities, or the number of its connected components, respectively. This relates to the theory of semi-algebraic sets. Further information can be found in [Dress et al. 1982],

[Crippen and Havel 1988], or in the literature cited therein. Currently, the most important field of application is probably the determination of protein structure from 2D-NMR data.

Other interesting problems arise when symmetry properties of the molecular structures under investigation (including crystals) are either known or can be presumed to exist. In this case, many questions address combinatorial group theory. It starts with the classification of all three-dimensional crystallographic groups, – a task which was completed more than 100 years ago without the help of computers, see [Conway et al. 2000] for a modern approach not using computers. However, one should note that the classification [Brown et al. 1978] of all four-dimensional crystallographic groups (important for the study of incommensurable crystals, and possibly also for quasi-crystals), by Brown, Bülow, Neubüser, Wondratschek, and Zassenhaus, and the more recent classification of all 5- and 6-dimensional crystallographic groups by W. Plesken [Opgenorth et al. 1998] could not have been accomplished without development and implementation of methods from *computational group theory*, cf. 2.7. Other problems relate to the theory of tilings (see, e.g., 4.2.34, resp. [Dress et al. 1993b], for a computer-oriented discussion of chemically relevant tilings), and to *orbifold* theory as well as the theory of so-called *Delaney symbols* (see [Dress 1987]). Note that Delaney symbols "present" orbifolds by appropriately combining the way simplicial complexes present topological spaces with the way Coxeter matrices present a reflection group; in particular, they allow a computational analysis and manipulation of orbifolds. Correspondingly, their study requires algorithmic methods from both combinatorial group theory – the construction of ramified coverings with given branching properties corresponds for instance to the computation of subgroups with particular properties – and algorithmic topology, such as developed by Wolfgang Haken in recent years. Remarkably, the efficiency of his approach depends almost exclusively on the development of efficient methods for finding all those non-negative and non-zero, integral solutions of large linear systems of equations that are not themselves already the sum of such solutions.

Advances in this area should have considerable potential for the classification of the geometry of crystalline structures (see [Delgado Friedrichs et al. 1999; Dress et al. 1993a] as well as 4.2.34), e.g., the derivation of criteria for distinguishing geometrically (if possible) high temperature supra-conductors from other crystals.

Examples for application of computer algebra to chemical equilibrium computations are given in the next article by H. Melenk, as well as in section 3.2.8 by K. Gatermann.

Andreas W.M. Dress (Bielefeld)

### 3.5.2 Chemical Reaction Systems

The dynamics of *chemical reaction systems* is often described by an explicit system of first order ordinary differential equations for the concentrations of the given species. The right hand side is pure polynomial, typically of degree

2, and most of the time consists of one or two terms in each row. Of special interest are *stationary states* $\frac{dc_i}{dt} = 0$ of such systems, which lead to a polynomial system of equations. For small or medium systems (e.g., about 40 species), the solutions can be algebraically determined by a combination of combinatorial methods, factorization, and *Gröbner bases techniques*. In contrast to numerical approximation methods, algebraic solutions provide a summary of all possible stationary states. For small systems, the solutions can be computed as functions of the reaction constants (which will be symbolic in this case).

Further information can be found in [Melenk et al. 1989].

Herbert Melenk (Berlin)

## 3.6 Computer Algebra in Education

Twentyfive years ago when I studied Physics, only one of the students who participated in a laboratory course that I took was possessing one of the first *calculators* to process the data we were obtaining. Everybody else, including me, used a *slide-rule* for this purpose. Nowadays, the calculator is used by everybody, by far not only for academic purposes. Hence it is the responsibility of school education, and here in particular of Mathematics education, to take this situation into account, and to teach our children the (intelligent) use of a calculator.

In my opinion, there is no doubt that sooner or later computer algebra systems or hand-held computer algebra tools will be used by everybody in the same way as (numeric) calculators are used today. Obviously this gives us a new responsibility to integrate computer algebra in the Math curriculum and to teach the students the use of them. When I realized this, I began to use DERIVE in my calculus courses at the Free University Berlin, in particular for the Math teacher education [Koepf et al. 1993; Koepf 1994, 1996] as a didactical tool.

Whereas calculators brought more numeric computation into the classroom, computer algebra systems enable the use of more symbolic computation. In particular, the interaction between numeric and symbolic computation can enhance the Math education substantially [Koepf 1998b].

The usage of computer algebra in the classroom is increasing worldwide, and slowly these steps are institutionalized and Math curricula are adapted accordingly.

In the articles below, several of the main contributors to education describe their activities in this direction. These activities are spread from the use of hand-held devices like the TI-89 in undergraduate education to the use of special purpose computer algebra systems in graduate studies.

Wolfram Koepf (Kassel)

## 3.6.1 New Hand-Held Computer Symbolic Algebra Tools in Mathematics Education

An unparalleled opportunity exists today to deliver better mathematics education than we ever thought possible. And it can be delivered to *all* students because of the rapid expansion of inexpensive powerful hand-held computer technology with built-in computer symbolic algebra software. These amazing products are now available from Casio and Texas Instruments (TI-89, TI-92, CASIO CFX-9970G and CASIO Algebra FX2.0). We fear, however, that our community is not ready to deal with the implications of their use due to misunderstanding, fear, and inexperience.

It is a fact that hand-held scientific calculators have *significantly* changed the high school and university mathematics curriculum around the world in the past 25 years. For example, many topics that dealt with paper and pencil "computation" involving transcendental functions have been deleted. Many sections and even some chapters in textbooks dealing with paper-and-pencil computation methods became obsolete and disappeared from the curriculum. Why? Because hand-held scientific calculators provided better ways to "compute" than paper-and-pencil methods. *The same thing (obsolescence) will soon happen with paper-and-pencil symbolic algebraic manipulations common today because of student use of inexpensive hand-held computer algebra systems that now exist and soon will proliferate.*

It is important to note that *less time is now spent* on certain topics (ones made obsolete by scientific calculators) but we still "do" the same things. For example, we still "compute" the sine of 14.25 degrees but not by the time consuming method of paper-and-pencil linear interpolation. What changed was *not* the "to do's" but the "how to" do the "to do's." It is also equally important to note that many educators found pedagogical ways to use scientific calculators that enhanced the teaching and learning of mathematics.

We should continue to teach the same *content* topics, but we should expect the methods we will use "to do" or "to apply" the topics will change (and likely be much faster) because of advancing technology. For example, some reformers have said it is no longer necessary or desirable to teach factoring. We believe they are wrong. The mathematical topic of factoring *is a major and important topic*. It *must* remain in the curriculum. However, in the past factoring was a mental or tedious paper-and-pencil exercise that often hid the really beautiful underlying mathematics. Recall using the "rational zeros theorem" to factor $2x^3 - 5x^2 - 9x + 18$? What a painful experience for students—and it took a good deal of time to do just one example! With computer algebra this polynomial can be factored instantly. What is important and was often lost in the fog of tedious computations was recognizing what the factors can tell us about the behavior of the expression. The *concept* topic of factoring *is* important! Integrating computer algebra into the curriculum means the same topics can be taught in less time so more time can be devoted to new mathematics, better mathematics, understanding, proof, problem solving and so forth.

Consider the "exercise" of evaluating the definite integral given in the example below. We use the TI-89 "*integrate*" command to do the computation.

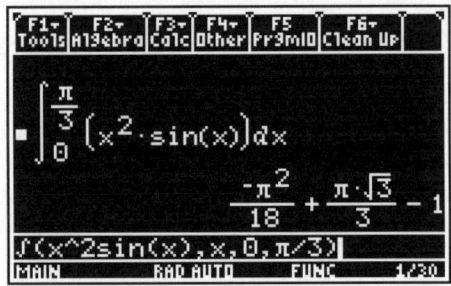

The answer shown is "exact" but what "*is*" $\frac{-\pi^2}{18} + \frac{\pi\sqrt{3}}{3} - 1$ really? How do we know it is correct? Rather than asking students to do the tedious "paper-and-pencil" manipulations, with no real understanding of the integral concept necessary to find the "exact" answer, a much better series of questions can now be asked.

a. How do you know this definite integral exists?
b. Describe a "problem" for which the integral is the "answer."
c. *Estimate* the answer without using computer algebra and compare your estimate with the computer algebra solution.

This solution is easy and involves a computer or graphing calculator graph in the interval $[0, \pi/3]$ by $[0, 1]$ together with the observation that the area under the curve is "about" the same as that under the line shown $(1/2) \cdot$ base $\cdot$ height $\approx 0.5 \cdot 0.5 \cdot 1 = 0.25$.

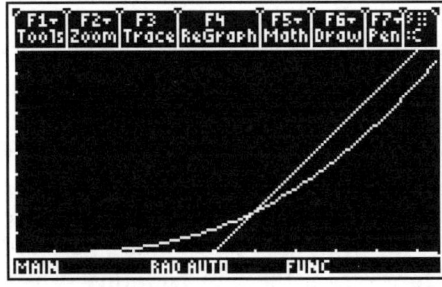 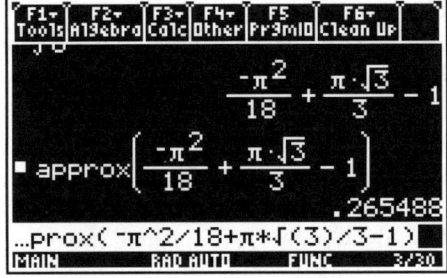

To conclude that the computer algebra answer *must* be near 0.25 requires real understanding of calculus concepts (not low order manipulative skills).

Students will demand the use of computer algebra because it provides a "better" tool to do the tedious algebraic manipulations common in "mathematics" today. To do otherwise is a waste of valuable teaching time and learning opportunities. We have wonderful examples of innovative curricula from Austria "better mathematics better" using computer algebra [Heugl et al. 1996]. Use of hand-held computer algebra *together with a recognition that some of what we once did is now obsolete* can provide the time to spend in the classroom on more worthwhile topics!

What is needed today and in the future is a school and university mathematics curriculum that takes advantage of computer algebra technology to assist students in gaining mathematical understanding, in becoming powerful and thoughtful "thinkers," communicators, and problem solvers. There should be a *balanced* approach to the use of computer algebra technology in mathematics teaching and learning. Some of us in mathematics education have an appreciation of the deeper and richer understanding of mathematics that is possible when technology is used effectively. The great challenge for mathematics educators in the future is to make clear to the "public" that such good mathematics is both possible and desirable [Demana and Waits 1997].

Bert K. Waits (Columbus)

### 3.6.2 The Dutch Perspective

Computer algebra has been an issue in mathematics education in the Netherlands for some years already. This does not imply that the discussion on this phenomenon has resulted in an agreement; no consensus on the role of computer algebra in the mathematics classroom has emerged so far.

Below, I describe some recent developments in my country from a personal perspective. I confine myself to mathematics education at upper secondary, pre-university level. As far as the university level is concerned, I refer to the Internet site of Computer Algebra Netherlands (www.can.nl), that provides information on the use of computer algebra in academic education.

Firstly, I describe the Dutch situation concerning curriculum and assessment. Secondly, I briefly consider the first educational experiments with computer algebra. Thirdly, the rise of the graphics calculator is discussed. In the end, computer algebra comes into the picture again, but now in a hand-held format. I conclude with an imaginary jump into future.

Understanding the developments in my country requires some knowledge of the organisation of the curriculum and assessment.

As far as the curriculum is concerned, it is important to notice that there is no detailed curriculum that prescribes which topic should be taught when. The curriculum is defined by a description of skills and concepts that will be assessed by the end of secondary school; the schools are free to choose how they get there. They can also decide on the textbooks they want their students to use. It is because of this relative freedom that the final assessment is so important.

The final assessment at upper secondary level consists of two parts: a school set assessment and a final national examination that is externally set and internally graded. This national examination is very important for the implementation of technology; if a certain technology device is not allowed at the final examination, it will not easily become popular in the classroom. The current regulation is that the graphics calculator is required at the final examination, whereas computer algebra is excluded. Two arguments guided this legislation. Firstly, it would be hard to organize a national examination throughout the country with computer access for all the candidates; hand-held computer algebra was not yet available. Secondly, the financial aspect was important. If computer access was required at the final examination, schools would need money to buy them, whereas hand-held technology devices are supplied by the students themselves.

For the school assessment, the authorities recommend the use or partial use of a computer, but again, the schools are free to decide. I have the impression that the number of schools that use a computer in their examination is increasing. The computer is often used in combination with investigation tasks where a written or oral report forms the assessment.

Obviously, the Dutch policy on technology is a careful one. Information about the different strategies concerning technology use and assessment in other countries can be found in [Drijvers 1998].

The first project on computer algebra at upper secondary level started in 1990. The idea of this two-year project was to develop short instructional units that were tested in pilot schools. Although the production of these materials was useful, the project as a whole was not very successful. This was caused by the lack of computer facilities at schools and by the difficulties students had with the user friendliness: they had little 'computer literacy' and a windows interface was not available. Obviously, the time was not yet ripe for the implementation of computer algebra at this level.

By the end of this project, a group of volunteering teachers decided to continue the work. This group, called CAVO, existed until 1998 and was a lively and important platform for further development and discussion (see [Drijvers et al. 1997]). In the mean time, however, the graphics calculator came on the market, and attracted much attention.

The development of the graphics calculator elicited discussion on which technology platform should be used in secondary education (see [Drijvers 1994]). The Dutch authorities decided that the implementation of the graphics calculator would be the first step to take. Therefore, they supported a research project on this issue in 1992. This project was carried out by the Freudenthal Institute, a research group on mathematics education. Later it became an integrated part of a larger curriculum development project called Profi. Results of the Profi-project included student textbooks that integrated the use of the graphics calculator, and experimental examinations that required the availability of a hand-held graphing device. The role of the graphics calculator in this project is summarized in [Drijvers and Doorman 1997].

Some educators and teachers were, however, opposed to the implementation of the graphics calculator. Their arguments were that computer algebra is a much more sophisticated mathematical tool, and that a graphics calculator is only a temporary step backwards compared to the possibilities that PC's offer. In 1996, however, a questionnaire revealed that PC's were hardly ever used during mathematics lessons, although they were available in schools. This supports the idea that real implementation of technology requires that the student has direct access to the device. The limited mathematical power of the graphics calculator is not an important disadvantage: it allows teachers, textbook authors and examination boards to have sufficient time to carefully integrate a graphical and numerical tool without having to cope with computer algebra in the mean time.

The choice of the graphics calculator may be a temporary preference indeed: the symbolic calculator raises the issue again. Nowadays, computer algebra is also available in a hand-held format. A first pilot experiment using the TI-92 revealed that the students appreciated this machine as an 'algebraic calculator', but not so much as a dynamic geometry tool [Drijvers 1997b].

When the Dutch Association of Mathematics Teachers became aware of the possible impact of symbolic calculators on secondary mathematics education, an Advisory Board on Computer Algebra and Symbolic Calculator was formed. In May, 1998, this Board concluded that:

- computer algebra should be implemented in upper secondary education;
- research was needed in order to find answers to the pedagogical and curriculum issues that will be raised by this;
- as a computer algebra platform, the PC would be preferred to the symbolic calculator, at least in the long term.

For the full report of the Board (in Dutch!) I refer to the Dutch Association of Mathematics Teachers site: http://www.nvvw.nl.

In the fall of 1998, the Freudenthal Institute conducted an explorative case study using the symbolic calculator. This machine turned out to be quite useful in investigation tasks. The sophisticated use of variables and parameters, however, was not always clear to the students. Furthermore, some students were reluctant to use computer algebra for the application of techniques that they had not yet mastered manually.

At present, many research questions concerning the role of computer algebra in secondary education are still unanswered (see [Drijvers 1997a]). No decisions on its implementation in the Netherlands have been made so far. In the next few years, I expect three developments to take place.

Firstly, teachers, examination boards and school book authors will get used to the graphics calculator and will take advantage of the pedagogical possibilities that these devices offer.

Secondly, research will be carried out concerning the role of computer algebra in the learning of mathematics and, more specifically, in the learning of algebraic concepts. Such a study was started recently by the Freudenthal Institute.

Thirdly, research will be carried out on the possibilities of computer algebra as a wide-range technology tool. A project that focuses on the use of a computer algebra environment in combination with a text editor (to write mathematical reports) and an Internet browser has been started at the Algemeen Pedagogisch Studiecentrum, an institute for improvement of (mathematics) education.

It is my hope that these developments will lead to a carefully considered implementation of computer algebra in secondary education.

Paul Drijvers (Utrecht)

### 3.6.3 Computer Algebra in Teaching and Learning Mathematics: Experiences at the University of Plymouth, England

There are many ways of using computer algebra systems in the teaching and learning of mathematics. Research in the literature shows that student learning can be enhanced when a computer algebra system is used as an add-on through 'laboratory activities' (see for example the work of Mayes [Mayes 1998] and Heid et al. [Heid et al. 1998]) and when more fully integrated into the curriculum (see for example the report of the Austrian Experiment [Austrian Experiment 1996]).

Our experience of using the computer algebra system DERIVE in some of our courses in Plymouth began in the late 1980s with some experimental research work with engineering students. Encouraged by the outcomes of this research we have increased the use of computer algebra systems to the mathematics degree programmes. Inevitably we have experienced resistance along the way from many of our colleagues. Their concern is that students may become too dependent on a computer algebra system and will not develop appropriate mental mathematics skills. These concerns have encouraged the team of computer algebra systems enthusiasts at Plymouth to develop ways of using computer algebra systems to enhance learning through investigational activities that develop concepts and understanding.

Recruitment of students to engineering degrees in has been falling steadily over the past decade. Among the many reasons for this phenomenon is the perception among school students that mathematics and science, physics in particular, are hard and unglamorous subjects of study, and that to be an engineer is a low status occupation. Attempts to redress this perception through the National Curriculum in secondary schools have led to a widening gap between a student's knowledge of mathematics and physics at the age of eighteen and the traditional starting points of degrees in engineering subjects.

At the same time access to higher education in Great Britain has widened considerably to admit far more mature students (aged 21 or over) than hitherto. This widening of access has enriched the undergraduate population in many ways but has usually meant that the mathematical and scientific background of such students is weak.

An alternative approach is to broaden the curriculum by developing undergraduate engineering degree courses that are focused more on Design, Communications or Technology Management rather than the more traditional and mathematically more demanding areas of Mechanical or Electrical Engineering. As a consequence the customary entry requirements of Advanced level Mathematics (or equivalent) (Advanced level is a school leaving public examination in England and Wales taken at 18 years of age) has been removed and about one third of the intake to these new courses has not studied any mathematics beyond the age of 16 years. Consequently they begin their undergraduate studies lacking much of the mathematical ability, thinking and confidence which earlier cohorts have displayed.

How then can such students be taught an appropriate mathematics course given their weak knowledge base? Fortunately the great majority of today's new undergraduates possess significant IT skills. It was therefore decided to exploit the fact that they are comfortable with IT and use it in a central role to support the teaching and learning of their mathematics.

We encourage the use of graphics calculators, particularly the TI-83, and the computer algebra package DERIVE. Once their prices have become competitive, we intend to use the TI-92 and TI-89 as well. We have a dedicated computer laboratory with 16 PCs and students will usually spend two hours a week in the laboratory for the mathematical methods modules. In this time they normally follow guided investigations which either follow up some work introduced in a lecture or prepare for work which will then be followed up in a subsequent lecture. We regard this approach as a modification of Buchberger's white-box/black-box principle [Buchberger 1989]. Our work with DERIVE has been ongoing for some years now and may be read about elsewhere ([Watkins 1993], [Berry et al. 1994], [Watkins 1994]). In lectures we may often use a laptop computer and overhead viewscreen for demonstrations with DERIVE and likewise with the graphics calculators.

The various topics of the module syllabus in the Design, Communications and Technology Programme are introduced via integrated case studies whose origins are based either in the student's previous experience or engineering knowledge. The case studies are integrated in the sense that a scenario is presented, requiring a model and/or solution, and the students identify mathematics that they feel appropriate. The solution is then progressed until further support is required. Given the academic background described above this often means some sort of support with the implementation of a piece of mathematics (e.g. solution of an equation, differentiation of an expression). This is where the IT comes in—at this stage the students make use of a computer algebra system (currently DERIVE) to provide the support and help them to progress their solution. The strategy of 'identify the relevant mathematics, progress the solution, seek assistance of a computer algebra system, progress the solution,...' is continued until a satisfactory solution is obtained. En route the tutor notes any pieces of mathematics which will need to be revisited for expansion in order that the overall learning experience is mathematically coherent and not just a jumble of techniques.

The use of DERIVE with a group of students enrolled on building courses at the University of Plymouth is discussed now. As with the students discussed earlier these students have also rarely studied mathematics beyond the age of 16 and often have weak mathematical backgrounds. The mathematics module that they take covers mathematics and some statistics and is designed to prepare them for the mathematics that they need in their science and surveying modules, as well as introducing them to the statistics that they need for subjects like economics.

DERIVE is employed as a tool to help them develop a graphical understanding of some of the mathematics that they meet in the module. The ideas of transformations of graphs provide a theme that the sessions use in the context of different functions. The students have six computer lab sessions (once every two weeks), of which four make use of DERIVE and the other two a statistics package.

Computer Algebra has been less readily accepted into the mathematics degree programme because of the need to develop a different view of mathematics and the associated knowledge and skills and the more traditional curriculum. We would also suggest that for engineers, mathematics is seen as more of a tool for solving problems than in a mathematics programme in which understanding the concepts is a more important outcome. However we are encouraged that the situation is changing. Our experiences of using DERIVE and the move to a more student centred learning culture is encouraging colleagues in the Department to investigate the use of MAPLE across the programmes.

The first year calculus course is designed to explore the fundamental concepts of calculus and to introduce some of the applications. Students on the course are in the first semester of their first year at university and so have not usually met any type of computer algebra system before. DERIVE is used extensively as a problem solving tool and in investigations to introduce new mathematical concepts to the students.

The topic on boundary layers is part of an introductory course on non-linear systems. This course covers three broad topics: the use of geometric methods to study the solution of first and second order differential equations; the study of discrete systems (recurrence relations) including ideas of bifurcations and chaos; and asymptotic methods of solving differential equations. Computer algebra in the form of DERIVE or the TI-92 is used to draw direction fields for the geometric approach and to explore iterations on the discrete systems. The asymptotic methods part of the course is more algebraic in approach.

In describing these examples there has been an implicit assumption that the students still need to learn to do the same mathematics that was required in a pre-computer algebra age. In the short term computer algebra has revolutionised the way that we teach mathematics but not what is taught, with the assumption that computer algebra is a desirable but not an essential ingredient. In the longer term there is the need to change the mathematics that is taught as well as the way that we teach it. If this does happen then computer algebra will be an essential ingredient of any new style course.

To return to the present situation, where it is possible to enhance our teaching with computer algebra, students who have followed such courses soon realise what an asset computer algebra can be to their learning. They also expect to see it forming part of their other mathematical studies and are clearly disappointed if more traditional approaches are taken. Students once exposed to MAPLE, DERIVE or the TI-92 will place pressure on their teachers, now and in the future, to make full use of such technology in their teaching.

For further details of these examples visit
www.tech.plym.ac.uk/maths/ctmhome/ctm.html

John Berry, Ted Graham, Jenny Sharp, Stewart Townend, Anthony Watkins
(Plymouth)

### 3.6.4 The Educational Use of Computer Algebra Systems at the University of Illinois

This is a report on the development of computer algebra courses at the University of Illinois at Urbana-Champaign (UIUC). It describes the changes in the curriculum that have been taking place and the reasons for these.

In the fall of 1991, I offered a course on computational group theory to graduate students (see 2.7). This was a hands-on course, with students sitting at Sun workstations attempting problems and me roaming around giving them mathematical and computational suggestions as the need arose. The students were highly motivated to learn the material, often stayed on long after class finished, and ultimately solved one of the unsolved problems resulting in [Boston et al. 1993] being published. Further details on this course are given in [Boston 1997].

No further computer algebra course was offered until the fall of 1994, when I gave a course on elliptic curves by computer for graduate students (cf. 2.4.5). This did not lead as before to a publication by a large proportion of the class. Instead, some of the students came out with individual papers. It was in fact becoming increasingly standard for graduate students in algebra at UIUC to use software packages in their learning and research. Computers allowed a much wider class of examples to be investigated, giving students a more solid grounding in their area and sometimes allowing them to make discoveries that earlier researchers had overlooked. I had to be careful to avoid them using these systems as crutches. The right attitude to cultivate seemed to be one of partnership between human and computer.

The next step was to institutionalize these sporadic topics courses. We applied for a course number and Math 420, Computer Algebra Systems, was created. Together with Math 321 (on Groebner bases) and four other courses this formed the mathematics department's part of the campus-wide Computational Science and Engineering Option. We set up a web page (http://www.math.uiuc.edu/~boston/math420.html) with links to documentation on many software packages and the students learned how to pick the right system for the problem at hand, obtain on-line help or web help, and avoid various pitfalls. It was a very practical hands-on training with students learning how to use many advanced computer algebra systems, such as MAGMA, PARI-GP, KASH, Macaulay2, ...

Math 420 has been offered twice so far, in the spring of 1997 and of 1998. It was given again in spring 2000. Enrolment suggests that it should be offered once every two years. Computer algebra seems to be very popular with graduate students, who use it routinely in their work. Many other graduate courses now include short visits to the computer lab to supplement with examples the more theoretical approach usually used in lecture/discussion. Math 321, mentioned above, is an undergraduate course, intended for bright math majors. All indications are that the trends described above will continue and that in future computer algebra courses will be required parts of both the graduate and undergraduate curriculum.

Nigel Boston (Urbana-Champaign)

### 3.6.5 Mathematics Education from a Mathematica Perspective

In this section I would like to speak about changes in educational practice resulting from the availability of computer algebra systems, using MATHEMATICA as an exemplar of such systems. Clearly, the practice of mathematics itself has changed radically during the past twenty years, as we exploit our increasing ability to shift the burden of algorithmic processing from humans to machines. The expanding capacity of computer systems for graphically representing complex mathematical concepts and processes, and the possibilities offered by electronic communication have significantly altered traditional mathematical discourse. In the words of one mathematician active in the "calculus reform" movement, "The most visible force for change in the mathematics curriculum is the computer, a mathematics-speaking device that has totally transformed science and society" [Steen 1991].

All of the above capabilities are embodied in MATHEMATICA, a fully integrated environment for technical computing. Definitive quantitative studies showing the effects of using MATHEMATICA in education are relatively sparse. Qualitative information, on the other hand, is abundant, and one can infer that there is considerable impact on educational practice resulting from the availability of MATHEMATICA and other computer algebra applications.

An early and reportedly successful application of computer algebra systems to calculus instruction is the Calculus&MATHEMATICA project [Uhl et al. 2001], developed at the University of Illinois at Urbana-Champaign and the Ohio State University and tested at thirty other sites for about six years. The project has received high praise in a US National Research Council report [Kirwan et al. 1991]. The mathematics department at the University of Missouri-Columbia has adopted a variation of Calculus&MATHEMATICA, in which all undergraduate calculus is taught through MATHEMATICA-based instruction, and they report equally successful results [Saab 1997].

The journal MATHEMATICA in Education and Research, begun as a newsletter in 1991, just three years after the introduction of MATHEMATICA, provides a useful guide to the evolving uses of computer algebra in education. In an editorial introduction to Volume 1, Number 1, Wellin [Wellin 1991] described four components of MATHEMATICA's potential for changing education:

1. Active involvement of students in learning;
2. Experimentation as a means of understanding mathematical concepts;

3. Visualization of mathematical processes;
4. Access of students to real-world problems.

These themes have continued as the principle set of arguments for using computer algebra systems, particularly in the teaching of calculus. Wellin also raised the salient issues regarding the uses of computer algebra-based instruction; namely, the establishment of labs, the role of an instructor in a lab rather than a lecture setting, and the articulation of MATHEMATICA-taught courses with conventionally-taught courses in the university curriculum. Again, these issues have continued to be raised in any analysis of computer algebra-based instruction. In subsequent issues of the journal, authors describe specific calculus concepts that can be effectively taught with MATHEMATICA. In some cases, these authors temper their enthusiasm for computer algebra with cautionary advice. Cohen [Cohen 1994] provides suggestions for effectively using MATHEMATICA in a "new calculus" course, using arclength as a prototypical example. Among Cohen's suggestions for successful instruction is the need to provide coding templates, since students have problems with accurately entering correct MATHEMATICA syntax. A related difficulty is described by DeJong [DeJong 1994] in an article on symbolic algebra computer laboratories. DeJong describes an "empowerment problem": Students do not easily acquire confidence and ability to use MATHEMATICA.

Other authors concentrate on the advantages of using MATHEMATICA without mention of lab-based difficulties. For example, Prevost [Prevost 1994] extols the virtues of using MATHEMATICA graphics to reinforce the limit concept. From this and other articles in later volumes of the journal, one might infer that the pedagogical questions related to using MATHEMATICA in calculus instruction have been solved. However, Holdener [Holdener 1997], in 1997, reported that "continuing controversies" still existed concerning the use of Calculus&MATHEMATICA. She cited "Gadgetry over Intellect," "Proof-abuse," and "Lack of Necessary Hand Skills" as concerns raised by opponents of computer-based calculus courses in general and Calculus&MATHEMATICA in particular. Although there were rebuttals to each of these concerns by proponents of computer-based calculus, one can see that in the last third of the 1990s, computer algebra systems are still far from being in universal use for calculus instruction.

The range of educational articles in MATHEMATICAin Education and Research extends far beyond calculus instruction. In physics, for example, Gilfoyle [Gilfoyle 1995] describes the use of the transfer matrix method to present an approach to quantum tunneling for undergraduate physics students. In engineering, Sipcic [Sipcic 1995] argues for a large-scale revision of the traditional approach to teaching mechanics, and offers numerous MATHEMATICA examples for instruction. Benninga and Wiener [Benninga and Wiener 1997–1998] present a series of six articles from a graduate course in financial engineering. Akritas and Bavel [Akritas and Bavel 1998] describe the use of MATHEMATICA to teach historical topics in a college liberal arts course. Although MATHEMATICA was designed primarily for advanced technical computing, innovative teachers have explored its use in pre-college mathematics instruction. A comprehensive collection of visualizations for classroom use on CD-ROM is described by Gloor in the next subsection 3.6.6. Mathews and McCallister [Mathews and McCallister 1995] present a study that found a statistically significant difference in the problem-solving performance between groups of algebra students that did not use MATHEMATICA and those that did. Peckman [Peckman 1998] suggests that MATHEMATICA could be used to give high school students in the USA a deeper understanding of the concept of function and provides numerous examples. Holzinger [Holzinger 1997] discusses research to test the motivation of students using MATHEMATICA at Handelsakademie, a high school in Graz-Austria, using 35 different MATHEMATICA 3.0 notebooks. Holzinger found higher motivation and greater interest in problem-solving among students who used MATHEMATICA. He also found that the solution of mathematical problems did not become easier for the student after exposure to computer-aided math instruction.

Computer algebra systems are beginning to be used for introducing students to a new set of mathematical concepts. For example, a set of notebooks by this author includes novel visualization of very large and very small numbers, fractal dimension, cellular automata-generated music and other topics formerly outside the mainstream pre-college curriculum [Fowler under development].

During the first decade of use, MATHEMATICA has had a steadily growing influence on education. Some university instructors continue to express doubts about using symbolic algebra systems to teach basic subjects such as calculus and linear algebra. Consider, however, that an emerging group of students who learned these subjects using computers and advanced symbolic calculators are now becoming university professors. These instructors will organize their courses with the computer as an implicit tool, rather than as an add-on device for displaying an isolated animation or performing a few specialized computations. Simultaneously, computing machinery that can easily handle MATHEMATICA continues to become more widely available, so that students need not be limited to labs for their instruction.

Finally, the emergence of a web-based language, most probably MathML, will further blur the division between journal, textbook and universal electronic communication [Fowler 1998]. It is highly likely that symbolic algebra will provide computational power, and eventually inferential power—on-line proof-generating applications—that will be a standard resource for mathematical cognition.

David Fowler (Lincoln)

### 3.6.6 Visualization: Courseware for Mathematics Education

We report on the projects *Analysis Alive* and *Illustrated Mathematics*, which both deal with the application of computer algebra systems on mathematics education. In both projects, computer algebra is not the object to be taught but serves as an aid in the process of teaching and learning mathematics. Thus, the focus entirely lies on mathematics, and the computer algebra systems, on which our software relies, are mainly a tool.

## Visualization in Mathematics

The importance of visualization can hardly be overestimated in general cognitive skill acquisition and problem solving processes (see [Anderson 1995; Denis 1989]). Pictures activate mental processes such as the perception of spatial relationships, intuitive comprehension of complex processes, or the observation of patterns and, therefore, aid the process of understanding. Looking at a picture, we use it as a vehicle of thinking, but intend to understand processes and behaviors of the real world.

Learning can be achieved through the translation between representations at different levels of abstraction. Visualization can be seen as providing the relevant representations to assist the learner in carrying out this cognitive process. The useful aspects of visualization are the translation from representations which are more abstract to those which are less abstract. Therefore, current techniques of scientific visualization can bring invaluable insight to students.

In particular in mathematics we deal with abstract structures, which visual representation helps to enlighten. This is important, particularly for those students who have difficulty understanding abstract mathematical objects. The objects we have in mind are not primarily geometric figures, but arbitrary mathematical objects such as infinite sequences, complex functions, or conformal mappings. For beginners, these terms are most difficult to grasp. Therefore, their visualization is a key to understanding these complex topics.

## Computer Algebra Systems in Mathematics Education

Computer algebra systems provide the necessary algorithms needed to compute mathematics visualizations. computer algebra systems give teachers and students also another and more direct approach to using the computer. Applying a computer algebra system, much less effort to treat a simple practical problem with the computer is needed than is with the classical approach, learning a full programming language first. So the focus moves from computer handling to the application. This enables the possibility of applying the computer in education not as a teaching object but as a tool to solve problems in other disciplines. Therefore, an introduction to computer algebra systems belongs to a modern curriculum in the education of scientists and engineers.

These two reasons—using computer algebra systems for visualizations and introducing computer algebra systems in education—make it natural to choose visualizations for the first contact of students with computer algebra systems.

Therefore, it is not surprising that many mathematicians have already combined the teaching of mathematics with a course on a computer algebra system (see e.g. [Gray and Glynn 1991; Finch and Lehmann 1992; Braden et al. 1992; Devitt 1993; Koepf et al. 1993; Davis et al. 1994; Braun and Meise 1995; Boggess et al. 1995; Stroeker and Kaashoek 1999]). However, this approach leads to additional difficulties for the students as they have to acquire not only one but two skills: the understanding of mathematics together with the understanding of the special computer algebra system which has a priori nothing to do with mathematics itself. Thus, this approach might even produce negative interferences.

## The Grey Box Approach

Both in *Illustrated Mathematics* and in *Analysis Alive*, we tried to avoid overloading students (and teachers) by the issues to be taught (mathematics) and the technology (computer algebra system). For that, we shielded the user as much as possible from the intricacies of both operating system and computer algebra system, by providing a grey box* consisting of the following parts.

- Electronic documents containing the ready-made visualizations (graphics and animations) of mathematical objects.
- Programs providing commands for the creation of new visualizations according to user-specified parameters.

As the documents already contain the commands to compute the graphics contained therein, the user only has to change the parameters and to process the command in order to obtain his own visualizations. In particular, users do not need to learn the subtleties of the input syntax let alone to write programs in a computer algebra system. They merely need to handle the basic functionalities of its user interface such as opening and browsing documents and evaluating a command. (For a more detailed discussion of the advantages of the use of a computer algebra system for such a grey box, see [Amrhein et al. 1997b].)

Numerous authors chose similar approaches and provide programs which allow the user to perform experiments. The resulting packages usually treat only a few issues and do not cover an entire course. A notable exception is [CalculusLive 1999], which consists of a black box built on a limited version of MATHEMATICA.

## Illustrated Mathematics

The goal of this project (cf. [Gloor et al. 1994, 1995]) was to provide a comprehensive collection of graphics and animations for topics in mathematics at the high-school and undergraduate and graduate college level. The visualizations are inteded for classroom use and can be used for demonstration during class, printed as hardcopy, or included in other documents.

The collection (provided as MATHEMATICA notebooks) is organized by mathematical topic and is not intended to replace textbooks. Teachers can select the examples that fit their syllabus and incorporate them into their lectures and class notes.

The programs (written in MATHEMATICA's own programming language) allow users to experiment by seeing the effects of changing parameters on the objects they are studying.

The topics range from basics such as sequences and series and end up with complex functions and minimal surfaces.

For further information on *Illustrated Mathematics* please consult the world wide web at http://www.amrhein.ch.

---

* We use the term *grey box* like *black box* for indicating that it is not necessary to know the inside process but merely the functionality. However, the entire software is user readable, but unlike a *white box* its internals are not discussed.

## Analysis Alive

This project (cf. [Wolff et al. 1998]) addresses students of mathematics as well as of sciences and engineers. It offers a new complete interactive approach to "Analysis" as it is taught at German universities.* It connects the concept of a modern textbook on this field and the opportunities of a modern computer algebra system in such a way that the user can profit from the advantages of both systems for knowledge acquisition without being obliged to master secondary skills.

*Analysis Alive* comprises a textbook and a CD–ROM jointly forming a complex unit. First of all, the book itself can be used as a common modern textbook combining the representation of the material, hundreds of illustrated examples and exercises presented directly within the current text. The text, however, is tightly linked to the electronic documents on the CD–ROM. For almost all relevant issues, the user can find visualizations in electronic form. These graphics and animations are presented in a similar way as in *Illustrated Mathematics*. They are presented in the form of MAPLE worksheets, and the software providing the commands for the creation of user-chosen examples relies only on MAPLE.

The text, however, is tightly linked to the electronic documents on the CD–ROM. Icons and background shading show which parts of the book are represented as visualizations on the CD–ROM. This direct relation offers now easy creation of visualization examples for the hundreds of examples listed in the text. Moreover, and much more important, it provides a great opportunity for experiments.

The following graphics show the effect of the transformation from cartesian to polar coordinates for the integral approximation.

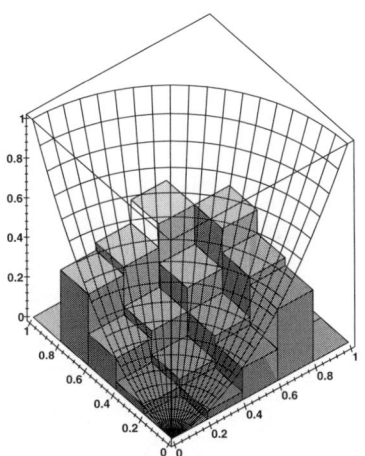 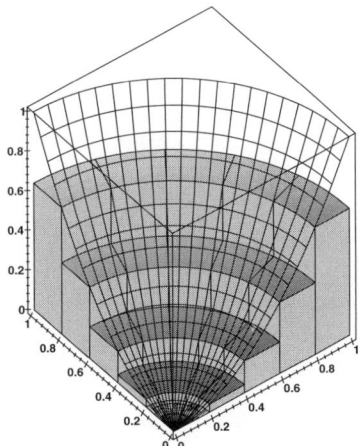

---

* This course roughly corresponds to a course in higher calculus.

For further information on *Analysis Alive* please consult the world wide web at http://WWW.amrhein.ch/AA.

Oliver Gloor (Bern)

# 4 Computer Algebra Systems

Short descriptions of computer algebra systems are presented in three sections: major systems, special purpose systems, and packages. However, the separation between special purpose systems and packages is not to be taken too literally. An older survey is the paper by Calmet and van Hulzen in [Buchberger et al. 1982]. There is now an excellent new book by Wester [Wester 1999] that covers computer algebra systems.

Additional systems are described in Chapter 3. For instance, section 3.1.1.1 refers to the SCHOONSCHIP program. The Nobel Prize in physics was awarded in 1999 to its author, and as the citation of the Nobel Foundation reads [www.nobel.se/announcement-99/physics99.html] "At the end of the 1960s ... [Martinus J. G.] Veltman had developed the Schoonschip computer program which, using symbols, performed algebraic simplifications of the complicated expressions that all quantum field theories result in when quantitative calculations are performed. ... With the help of Veltman's computer program [Gerardus] 't Hooft's partial results were now verified and together they worked out a calculation method in detail."

The editors

## 4.1 General Purpose Systems

### 4.1.1 AXIOM

AXIOM, a Computer Algebra System with Abstract Data Types

**4.1.1.1 Overview** AXIOM is a modern computer algebra system with a modular design which takes advantage of the intrinsic relationships between mathematical objects to allow them to be viewed and described in terms of their mathematical properties via a system of abstract data types. AXIOM is a fully interactive system and includes a command-line interpreter, a hypertext-based help system, graphics and a compiler which allows efficient extensions to the system to be produced by users. The system is based on Cambridge Common Lisp (CCL). There is also an interactive link to the NAG Fortran library for numerical processing.

AXIOM is a successor to the *Scratchpad* system developed by Richard Jenks' group at the IBM T.J. Watson Research Center in New York. In 1991 the technology was licensed to NAG Ltd. and AXIOM release 1.0 appeared. The current version is 2.2 and is available on all Unix platforms. In addition there is a version for Windows-95 and Windows NT which has a different user interface and graphics system.

**4.1.1.2 The AXIOM Type System** AXIOM recognises the fundamental relationships between mathematical objects and is designed to take advantage of them. Its object-centered design provides efficiency, consistency and robustness through a natural type system which categorises objects according to their algebraic properties. Users can add their own extensions to the system with the AXIOM Library Compiler (known as ALDOR).

At the top level of the AXIOM type system are *Categories*, which can be viewed as specifications of algebraic structures in terms of their abstract properties. Some of these will be familiar, such as Set or Ring, while others are less obvious, for example PolynomialCategory. At the next level are *Domains*, which are concrete instances of categories. For example the domain Integer is a Ring (and also a Set amongst other things). A major benefit of this type system is that it allows us to parameterise domains and categories in a mathematically rational way. For example the Polynomial domain is parameterised by a Ring allowing for multiple *instantiations*: Polynomial(Integer), Polynomial(PrimeField 3) and Polynomial(SquareMatrix(2, Polynomial Integer)) for example. This is just another way of saying that to create and manipulate polynomials we need to be able to add and multiply their coefficients according to certain axioms, which means that the coefficient domain is a member of the category Ring. However, because we have given the parameter a type, we cannot instantiate bogus objects such as Polynomial String. Armed with such data type constructor the user can build up hierarchies of data types. In each case all the appropriate operations on the objects are immediately available.

Finally, AXIOM has *Packages* which are collections of functions implemented outside a domain or category. Often these packages are parameterised by the types which they work over, so for example the code for computing the GCD of two polynomials is implemented in a package which is parameterised by any element of PolynomialCategory.

This type system provides **consistency**. In the AXIOM system you can build structures such as polynomials and matrices over any constituent objects where it makes sense. That includes new objects which a user has added to the system. Collections of algorithms can be parameterised as well, so that they work in any context which makes mathematical sense which means, for example, that if a user adds a new kind of polynomial to AXIOM he or she will not need to write any extra code to factorise it.

The type system thus allows **specialisation**. Algorithms can work over different representations of the same object (e.g. dense and sparse) because both representations belong to the same category. So the user can choose the most suitable representation for his or her problem. The type system provides quite a fine level of control, so for example a user can use it to determine the ordering of the monomials in a particular kind of polynomial, or how the exponents are to be implemented.

All this complexity might seem daunting at first, but the AXIOM interpreter allows a user to interact with the system without worrying about these details. However all AXIOM's sophistication is available on demand, whenever it is required. In addition, the source code for all the algorithms in the system is distributed with the software, allowing users to check its fidelity or adapt it to their own uses.

**4.1.1.3 The AXIOM Library** The AXIOM library contains 205 categories, 393 domains, and 468 packages. There are 6309 signatures representing functions, many of the are overloaded, e.g. the function name `factor` represents 39 different functions. An introduction to many of the data structures and their mathematical capabilities is given in the AXIOM book [Jenks and Sutor 1992]. The source code of all the data types is delivered with the system.

The AXIOM library provides the usual range of symbolic capabilities, and is particularly strong in the areas of symbolic integration and computing Gröbner bases. However its greatest strength is as a framework in which users can implement their own mathematical structures with a minimum of effort.

**4.1.1.4 The AXIOM Interpreter** The interpreter is the principle mechanism by which a user can interact with the AXIOM system. It uses sophisticated type inference techniques to remove the need for the user to specify the types of all objects, and so simplifies most straightforward interactions by allowing the system to be used in a similar way to Reduce or Maple. However it is possible to use the full power of the type system and sometimes this is more efficient, as well as allowing a user to specify exactly what he or she wants the system to do. In general this type interference problem is not easy to solve, a possible algorithmic solution is discussed in [Fortenbacher 1990].

On Unix platforms, the interpreter runs in a shell and uses the ASCII character set for input and output. Althoug, for consistency, such an interface is also provided under Windows, the default interface there uses a document-style interface which provides high-quality bitmapped output of mathematical expressions.

**4.1.1.5 The NAG Link** AXIOM users can have interactive access to NAG Fortran Library routines for numerical analysis via the NAG Link [Dewar 1994; Keady and Nolan 1994]. On Unix platforms this is achieved via a distributed client-server system, while under Windows it works via dynamic link libraries (DLLs). The exact mixture of functionality varies depending on which platform is being used, but includes algorithms for root finding, quadrature, curve and surface fitting, solving differential equations, linear algebra, optimisation and evaluating special functions. The combination of NAG numerical libraries and AXIOM's symbolic capabilities has been used to develop an expert system for numerical analysis [Dupée and Davenport 1996].

**4.1.1.6 Graphics and Visualisation** On the Unix platforms, AXIOM provides 2-D and 3-D graphics based around a special package which uses X-Windows. Graphs in AXIOM are objects and can be manipulated from the command line or from a special "control panel". They can be exported as PostScript objects or saved for future use.

Under Windows, AXIOM uses a special system built on top of the Open-Inventor toolkit to represent and manipulate 2- and 3-dimensional objects. Images can be saved in the VRML format to allow them to be exported to state-of-the-art visualisation packages such as Iris Explorer or to the latest generation of VRML-compliant web viewers [Walton and Dewar 1997].

**4.1.1.7 Availability & Contact Information** AXIOM is available from NAG Ltd. and its subsiduaries and distributors. More details are available on the NAG web site http://www.nag.co.uk or from:

NAG Ltd
Wilkinson House
Jordan Hill Rd
Oxford
OX2 8DR
United Kingdom

Telephone: +44 (0)1865 511245
Telefax: +44 (0)1865 310139
Email: infodesk@nag.co.uk

Additional references: [Bronstein et al. 1989; Davenport 1992a,b; Davenport et al. 1991; Davenport 1992c; Davenport and Trager 1990; Gollan and Grabmeier 1990; Grabmeier et al. 1991; Grabmeier and Scheerhorn 1992; Grabmeier and Wisbauer 1993; Jenks et al. 1988a; Burge 1991; Bronstein 1991, 1992b; Gil 1992; Rioboo 1992; Bronstein 1994; Broadbery et al. 1995; Corless et al. 1995; Brown and Tonks 1994; Duval 1994; Andrews 1995; Grabmeier 1995; Davenport and Faure 1994; Robidoux 1997; Jacobs 1997; Lambe 1991; Wang 1991; Duval and Jung 1992; Sit 1992; Petitot 1993; Seiler 1994a,b; Boulanger 1995; Roesner 1995; Bronstein 1992a].

Michael Dewar (Oxford) and Johannes Grabmeier (Heidelberg)

## 4.1.2 Aldor

*Aldor* is a programming language originally intended to develop compiled libraries for computer algebra. The design of the language was influenced by several factors: It had to be *expressive* enough to capture naturally the high-level objects and relationships which arise in modern mathematics. An implementation had to be *efficient* enough for resource-intensive symbolic and numeric computing needs. Finally, the language had to be *modular* enough to allow large libraries of independently developed facilities to be used together in any combination.

The resulting formulation of the programming language has attempted to balance the mathematical desire for generality and uniformity, on one hand, with the practical requirements of demanding symbolic and numeric computation, on the other. This has required certain trade-offs. As an example, types and programs are first class values in *Aldor*. This means that they may be created and used dynamically, and provides a natural realization of mathematical sets and functions. The language does require, however, that certain information about the types and functions be known statically (before program execution) for efficient execution.

An optimizing compiler has been developed for *Aldor*. Programs are first compiled to a low-level intermediate language, this undergoes a number of optimizing transformations, and from this C, Lisp or native object code is generated. Interlanguage interfaces support natural linking with programs written in C, C++ and Fortran. The *Aldor* compiler was originally used to produce libraries for *Axiom* and its predecessor, while more recent emphasis has been on the production of stand-alone programs or modules for use in Maple.

The main reference for *Aldor* is [Watt et al. 1994a].

### 4.1.2.1 Programming Language Characterization

*Aldor* may be characterized as a strongly typed functional programming language with a higher order type system and strict evaluation. All values are treated uniformly and memory is managed automatically.

The type system has two levels: Each value belongs to some unique type, known as its *domain*, and the domains of expressions can be inferred statically. Each domain is itself a value belonging to the domain **Type**. Domains may additionally belong to some number of subtypes (of **Type**), known as *categories*. Categories can specify properties of domains such as which operations they export, and are used to specify interfaces and inheritance hierarchies.

The biggest difference between the two-level domain/category model and the single-level subclass/class model is that a domain *is an element of* a category, whereas a subclass *is a subset of* a class. This difference eliminates a number of problems in the definition of functions with multiple related arguments.

Ex post facto extension allows existing domains to belong to new categories. This supports a programming style which has recently come to be known as "aspect oriented programming."

Dependent products and mapping types are fully supported. Dependent products are tuples where the *type* of some component depends on the *value* of another, e.g. (n: Integer, m: IntegerMod(n)). Dependent mappings are functions where the *type* of the result depends on the *value* of an argument, e.g. mod: (Integer, n: Integer) -> IntegerMod(n).

Generic programming is achieved through explicit parametric polymorphism, using functions which take types as parameters and which operate on values of those types, e.g. f(R: Ring, a: R, b: R): R == a * b - b * a. Object oriented programming can be achieved naturally using dependent product values, e.g. (R: Ring, r: R).

Control flow is explicit and uses a fixed set of language-defined primitives, including the usual conditionals, loops, function call and return, as well as stack-oriented exception handling (try/catch). Control abstraction is provided by suspendable/resumable generators which yield values over the course of a computation. Continuations are *not* supported in order to retain interoperability with languages such as C and Fortran.

In order that application-defined types have equal privilege and power as built-in types, language properties are defined independently of type. For example, overloading of names is not restricted to function valued quantities. Similarly, new types may support literals which appear in source programs. The language itself has very little in the way of predefined types, and in practice the basic types, e.g. Integer, DoubleFloat, Array(T) are all defined in standard libraries.

**4.1.2.2 History** The design of *Aldor* spanned a the period 1985-1995, and successive implementations, led by Stephen Watt. The original use of the language was as a compiled extension language for the Scratchpad II system [Jenks et al. 1988b]) at IBM Research. During its development, *Aldor* was known internally to IBM as $A^\sharp$ ("A sharp") [Watt et al. 1994b].

In 1990/91 IBM partnered with the Numerical Algorithms Group Ltd to release Scratchpad II, and did so under the under the trademark *Axiom*. The second release of Axiom included the $A^\sharp$ compiler. The interim name "Axiom XL" (for *Axiom* extension language) was used for a short period by NAG before the legal trade name *Aldor* was established. From 1996 to 2000, NAG continued to extend *Aldor*, partially through the support the ESPRIT Frisco project, e.g. adding a Fortran foreign function interface.

The language has drawn ideas from a number of programming languages, including Ada, C++, CPL, Clu, Fortran, Lisp, ML, Pebble and Russel. It has borrowed heavily from an earlier language designed by Richard Jenks and Barry Trager [Jenks and Trager 1981] at IBM Research. The most notable differences between *Aldor* and this predecessor are that *Aldor* is more heavily functional and dependent types are completely integrated so all expressions have types expressible in the language. The language of [Jenks and Trager 1981] required programs to be very highly stylized, with type constructions occurring in particular specialized settings and at top-level only. The ideas *Aldor* has adopted from this language have been reworked to be strongly orthogonal.

In 2001 Aldor.org was formed to distribute *Aldor* more widely, and independently of *Axiom*.

## Examples

### Function Definition

```
miniSqrt(x: DoubleFloat): DoubleFloat == {
        r := x;
        r := (r*r + x)/(2.0*r);
        r := (r*r + x)/(2.0*r);
        r := (r*r + x)/(2.0*r);
        r := (r*r + x)/(2.0*r);
        r := (r*r + x)/(2.0*r);
        r := (r*r + x)/(2.0*r);
        r
}
```

– This function computes a square root by six steps of Newton's method. The value returned by the function, is the value of the expression of its body.

### Category Definition

```
define Logic: Category == BasicType with {
        ~:         % -> %;          ++ Logical complement.
        /\:        (%, %) -> %;     ++ Logical 'meet', e.g. 'and'.
        \/:        (%, %) -> %;     ++ Logical 'join', e.g. 'or'.
        xor:       (%, %) -> %;     ++ Exclusive or.

        default (x: %) \/ (y: %): % == ~(~x /\ ~y);
        default xor(x: %, y: %): % == (x /\ ~y) \/ (~x /\ y);
}
```

– A with expression forms a category.
– BasicType is a previously defined category which this extends.
– The comments beginning with ++ are retained as documentation.
– The symbols "%" are ultimately replaced by the domain which belongs to the category. E.g. importing Boolean: Logic will give the operations such as xor: (Boolean, Boolean) -> Boolean.
– The keyword define makes the value of Logic public information. This is necessary in order to know what operations are exported when Logic is used. Normally only the name and type are public information.

### Category Producing Function

```
define FiniteLinearAggregate(S: Type): Category ==
        Join(FiniteAggregate S, LinearAggregate S);
```

– This definition gives a function which computes a category.
– The Join primitive constructs a category which is a subtype of all its arguments.

## Domain Producing Function

```
MiniList(S: BasicType): LinearAggregate(S) == add {
        Rep == Union(nil: Pointer, rec: Record(first: S, rest: %));

        import from Rep, SingleInteger;

        cons (s:S, l:%):%      == per(union [s, l]);
        first(l: %): S         == rep(l).rec.first;
        rest (l: %): %         == rep(l).rec.rest;
        empty (): %            == per(union nil);
        empty?(l: %):Boolean   == rep(l) case nil;
        sample: %              == empty();

        [t: Tuple S]: % == {
                l := empty();
                for i in length t..1 by -1 repeat
                        l := cons(element(t, i), l);
                l
        }
        [g: Generator S]: % == {
                r := empty(); for s in g repeat r := cons(s, r);
                l := empty(); for s in r repeat l := cons(s, l);
                l
        }
        apply(l: %, i: SingleInteger): S == {
                while not empty? l and i > 1 repeat
                        (l, i) := (rest l, i-1);
                empty? l or i ~= 1 => error "No such element";
                first l
        }
        (out: TextWriter) << (l: %): TextWriter == {
                empty? l => out << "[]";
                out << "[" << first l;
                for s in rest l repeat out << ", " << s;
                out << "]"
        }
}
```

- An add expression constructs a domain object. The domain defines a number of exports, including empty?, generator, << and [_].
- Within the domain the types Rep and % are distinct, begin the representation and abstract types for values in the domain. The operations per: Rep -> % and rep: %-> Rep allow values to be viewed in either way.
- The constructor [_] taking the generator as an argument first extracts the elements from the generator (for s in g repeat ...) secondly reverses the list to place them in the correct order.
- The expression apply(a,b) may be written simply as a b or a.b so defining apply allows the selection notations 1.3 . (Juxtaposition and dot associate as (f (g h)) and (f.g).h .)

## Conditional Category Definition

```
define SimplifiedComplex(R: Ring): Category == Ring with {

        if R has Field then Field;

        if R has DenseStorageCategory then DenseStorageCategory;

        complex: (R, R) -> %;
        *:      (R, %) -> %;
        real: % -> R;
        imag: % -> R;
        %i: %;
}
```

- The if expressions in the category expression assert additional categories under particular conditions.

## Post Facto Extension of a Domain

```
extend Symbol: Ordered == add {
    import from String;

    (u:%) <  (v:%): Boolean == printName(u) <  printName(v);
    (u:%) >  (v:%): Boolean == printName(u) >  printName(v);
    (u:%) <= (v:%): Boolean == printName(u) <= printName(v);
    (u:%) >= (v:%): Boolean == printName(u) >= printName(v);
}
import from List Symbol;

sort [+"a",+"x",+"i",+"o",+"m"];
```

- Add ordering operations to an existing domain for symbols.
- The syntax +"xxx" forms symbols.

## A Higher Order Program

```
Ag ==> (S: BasicType) -> LinearAggregate S;

swap(X:Ag,Y:Ag)(S:BasicType)(x:X Y S):Y X S==[[s for s in y] for y in x];

al: Array List Integer := array(list(i+j-1 for i in 1..3) for j in 1..3);
print << "This is an array of lists: " << al << newline;

la: List Array Integer := swap(Array,List)(Integer)(al);
print << "This is a list of arrays:  " << la << newline;
```

- The swap function takes two aggregate type constructors as arguments and produces a new function to swap aggregate data structure layers.

## An Algebraic Domain Constructor

```
PolynomialCategory(R: Ring, E: AbelianMonoid): Category == Ring with {
        *:                 (R, %) -> %;
        *:                 (%, R) -> %;
        degree:            % -> E;
        monomial:          (R, E) -> %;
        reductum:          % -> %;
        leadingCoefficient: % -> R;
        coerce:            R -> %;
}

Polynomial(R: EuclideanDomain, Expon: OrderedAbelianGroup):
    PolynomialCategory(R, Expon)

== add {
    Rep == List Term(R,Expon);

    -- Some details omitted ...

    0: % == per nil;

    degree(x: %): Expon == {
        empty? rep x => 0;
        first(rep x).expon
    }
    leadingCoefficient(x: %): R == {
        empty?(l := rep x) => 0;
        first(l).coef
    }
    (xx: %) + (yy: %): % == {
        (x, y) := (rep xx, rep yy);
        not x => yy;
        not y => xx;

        (x0,y0):= (first x,first y);
        y0.expon > x0.expon =>
                  per cons(first y,rep(xx + per rest y));
        x0.expon > y0.expon =>
                  per cons(first x,rep(per rest x + yy));

        r: R:= x0.coef + y0.coef;
        r = 0 => per rest x + per rest y;
        per cons([x0.expon,r],rep(per rest x + per rest y));
    }
}
```

Stephen Watt (London Ontario)

## 4.1.3 Derive and the TI-92

DERIVE is a computer algebra system developed for personal computers. It is available in two versions: DERIVE FOR DOS runs on any PC-compatible computer with 512 KB of memory and MS-DOS 3.0 and is available in English, French, and German. DERIVE FOR WINDOWS runs on any PC-compatible computer with 8 MB of memory and Windows95, NT, or 3.11 (or higher) and is available in English, German ,Italian and Spanish. The authors of DERIVE are Albert Rich, Theresa Shelby, and David Stoutemyer from Soft Warehouse, Inc. (Honolulu, Hawaii).

DERIVE offers a wide range of algebraic, numeric, and graphic algorithms. It comprises less built-in functions than other computer algebra systems and the programmability is restricted to function definitions using recursion, conditionals, and iterations. The scope of the built-in functions, its easy-to-use interface, and the low price make DERIVE particularly well suited for teaching and learning mathematics. According to the available literature DERIVE is used for teaching grade levels 7 and higher at schools (all abilities) and introductory courses at universities. Countries and regions that systematically equipped their schools with DERIVE include Austria, Slovenia, the United Arab Emirates, South Tyrol, and Hamburg. Standard school textbooks are available already that include references to DERIVE or are fully based on it.

In the sequel we present DERIVE with some examples. The major goal of this article is to demonstrate *how* to use DERIVE.

One works with DERIVE by issuing commands. Commands can be selected from pull down menus.

o  Open the **A**uthor menu by clicking on the menu name.

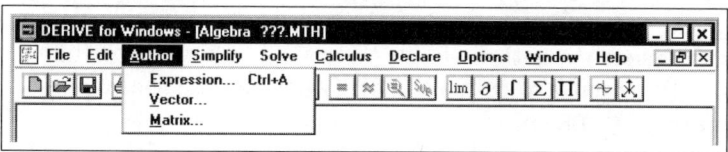

The first example is the expansion of the polynomial $(2x + \frac{3}{4})^5$. We enter the polynomial using the **A**uthor menu's **E**xpression... command:

o  **E**xpression... (2x+3/4)^5

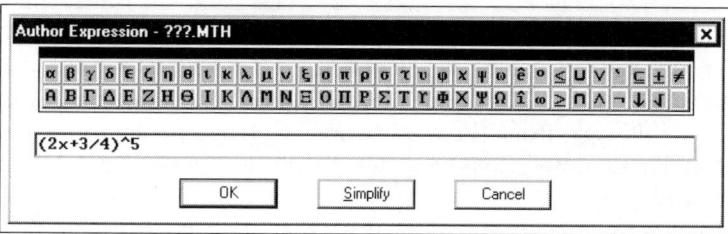

The input is done via the above dialog box. Note the Greek toolbar that allows convenient access to many special characters and symbols.

o  We conclude the input with  OK .

#1: $\left(2 \cdot x + \dfrac{3}{4}\right)^5$

The computer displays the expression with raised exponents and built-up fractions, then writes a label number (#1) in front of it. We can later refer to this expression by this label. The expansion of the polynomial is performed with the command

o  Simplify:Expand...  [ OK ]

#2: $32 \cdot x^5 + 60 \cdot x^4 + 45 \cdot x^3 + \dfrac{135 \cdot x^2}{8} + \dfrac{405 \cdot x}{128} + \dfrac{243}{1024}$

The result appears as the second expression on the screen, with the label number #2. It is marked and is the default expression for the next operation.

As our next example we will calculate the second derivative of the expression $x^3 \sin(x) \cos(x)$.

o  Author:Expression...  x^3sinxcosx  [ OK ]

#3: $x^3 \cdot \text{SIN}(x) \cdot \text{COS}(x)$

You see from this example how easy it is to input expressions into the system. Often we can simply omit the multiplication signs and the parentheses for the arguments of a function. To differentiate, we must invoke the Differentiate... command from the Calculus menu:

o  Calculus:Differentiate...  In the Order field change the 1 to a 2, then conclude with  [ OK ].

#4: $\left(\dfrac{d}{dx}\right)^2 (x^3 \cdot \text{SIN}(x) \cdot \text{COS}(x))$

The resulting expression #4 is then simply expression #3 with the differential operator applied. To actually calculate the derivative, we use the command

o  Simplify:Basic...  [ OK ]

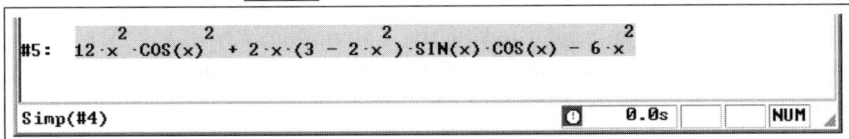

#5: $12 \cdot x^2 \cdot \text{COS}(x)^2 + 2 \cdot x \cdot (3 - 2 \cdot x^2) \cdot \text{SIN}(x) \cdot \text{COS}(x) - 6 \cdot x^2$

Simp(#4)                                          0.0s            NUM

Expression #5 is the second derivative of $x^3 \sin(x) \cos(x)$. The status bar (bottom of the DERIVE screen) shows the computation time (0.0 seconds, which means less than 0.1 seconds). The annotation Simp(#4) indicates the origin of the expression, in this case from the application of Simplify:Expression... to expression #4. These annotations exist for all expressions and can be freely modified by the user.

For the most frequently used commands such as Author:Expression... and Simplify:Basic... there are toolbar buttons for fast access. Clicking on one of these buttons is the same as issuing the respective commands. For easier reading we will continue referring to the commands by their names:

### 4.1.3 DERIVE and the TI-92

 for Author:Expression...

 for Simplify:Basic...

So far, so good; a first impression of algebraic computation and interaction with the computer. Let's go on to some other aspects of DERIVE.

DERIVE can also carry out numerical calculations, doing much more than a traditional pocket calculator. For instance the only limit to the size of the numbers used is the memory available, a restriction that means almost nothing today.

o   Author:Expression... 23^45

   Simplify:Basic...

```
#8:  23^45
#9:  1895625843011620279131971571327722762615928949974529023566354
```

This number has 62 digits, but that is by no means the limit of what we can do. How many digits does the factorial of 567 have? Let's have a look.

o   Author:Expression...   567!

   Simplify:Basic...

```
#10: 567!
#11: 649483984028395659661733295685901544934978750889527945304800057
```

In the standard font this result is about 1.8 meters long and reaches far over the right edge of the screen. To calculate the exact number of digits we use the Simplify:Approximate... command. While Simplify:Basic... gives exact results, Simplify:Approximate... calculates a numeric approximation and prints it as a decimal.

o   Simplify:Approximate...

```
#12: 6.49483·10^1316
```

The use of scientific notation for numerical values is a major bonus for us here. We can simply read from the screen that the number 567! has 1316+1=1317 digits. The prime factorisation of this number could be calculated with the Simplify:Factor... command.

With DERIVE it is very simple to plot functions. To plot the graph of $sin(x)$ we

o   enter the expression, then switch into the 2D-plot window, then issue the Tile-Vertically command, then use the Plot! command.

## 274   Chapter 4   Computer Algebra Systems

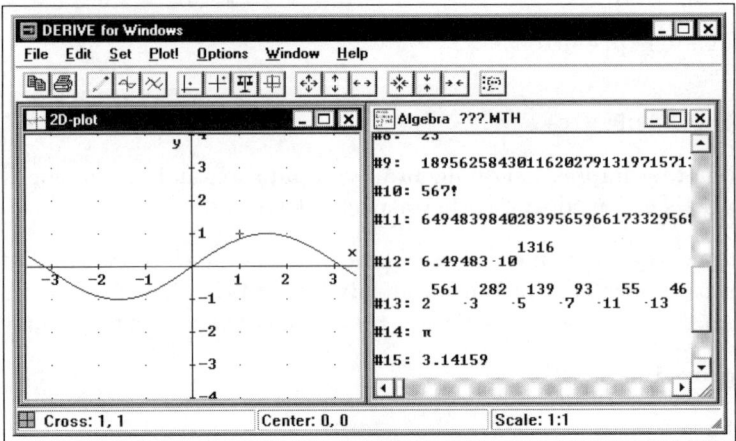

Each window is labeled with the window type in its upper left corner ('2D-plot' and 'Algebra'). The active window's title bar is dark. The inactive window's title bar is dimmed. Since the plot window is active, the status bar displays graphics information.

o   Now we add a plot of the corresponding Taylor polynomial.

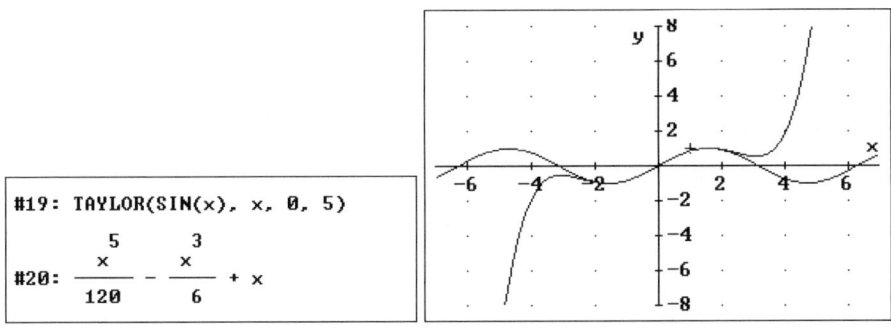

One can work interactively in the graphics window, in much the same way as one uses a graphic calculator. The graphic pointer, a cross, currently at position (1,1), can be moved with the arrow keys. The current coordinates of the cross are displayed in the bottom left of the screen.

The trace mode, switchable with the Options:Trace-Mode command or the $\boxed{\text{F3}}$ key, is very useful. When on, the graphics pointer moves up or down at the same x-coordinate onto the most recently plotted graph. The cross becomes a box and moves only along the graph.

o   Options:Trace-Mode

   Pressing $\boxed{\rightarrow}$ moves the graphics box to the right.

### 4.1.3 DERIVE and the TI-92

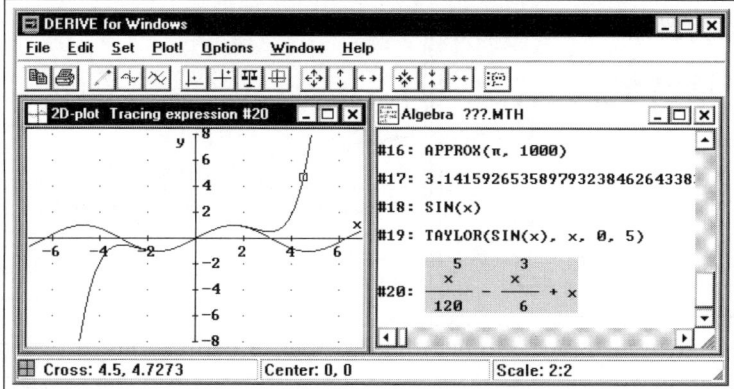

The coordinates of the box are displayed in the bottom left corner. The number of the expression that generated the inspected graph is displayed in the title bar as the text: `Tracing expression #20`.

To investigate the area around the cross (or box, if we are in trace mode), we can change the range of the plot. By using the Set:Range... command we can specify new numeric values for the minimal and maximal values of x and y. However, in general it is more convenient to use the mouse to mark the rectangular area that should be enlarged.

o  We click on the button ⊞ . Then we click and hold the left mouse button at the top left corner of the desired crop area. Drag the mouse down and to the right until the box encloses the desired plot range.

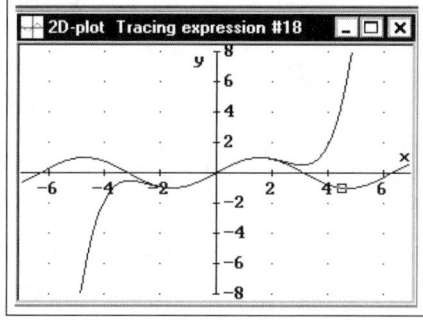 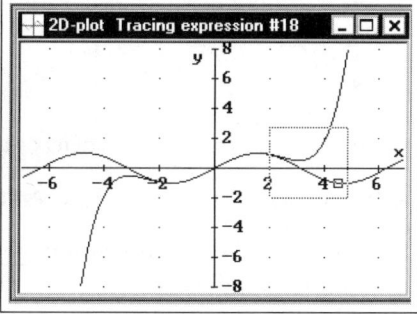

o  Release the mouse button and confirm the numerical equivalents of the graphical choices with  OK .

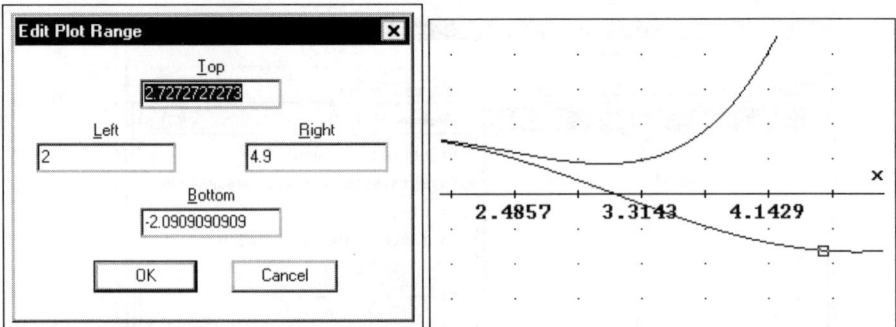

With DERIVE we can also plot 3-dimensional graphs of expressions in two variables.

o Enter x^4-y^4, switch into a 3D-plot window, then issue the Plot! command

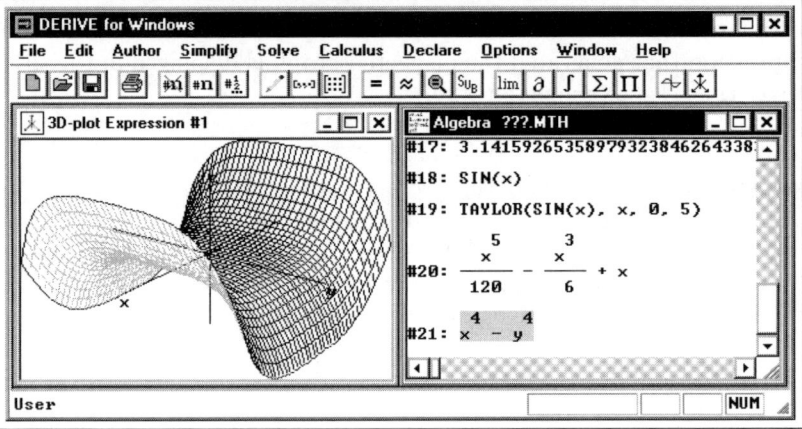

To finish, we look at an example of programming in DERIVE. The built-in factorial function has already been demonstrated. Let's define our own version named FACT. We use the IF function, which takes as first argument the condition to be tested, as second argument the result to return if the condition is fulfilled, and as third argument the value in case the condition is not fulfilled. We use the definition below:

$$fact(n) := \begin{cases} 1 & \text{if } n = 0 \\ n \cdot fact(n-1) & \text{otherwise} \end{cases}$$

o Author:Expression... fact(n):=if(n=0,1,n fact(n-1))

```
#22: FACT(n) := IF(n = 0, 1, n·FACT(n - 1))
```

We test function FACT on the number 7, using a trailing equals sign to force an immediate simplification (i.e. direct application of the Simplify:Basic... command).

o Author:Expression... fact(7)=

```
#23: FACT(7) = 5040
```

### 4.1.3 DERIVE and the TI-92

The preceding examples from the fields of algebra, arithmetic, graphics and programming give an introduction to what is possible with DERIVE. Those who want to learn more about the system should consult the manual that comes with DERIVE FOR WINDOWS: *"B Kutzler: Introduction to* DERIVE FOR WINDOWS*" [Kutzler 1997b]*. A book that describes how to use DERIVE (or other computer algebra systems) for teaching mathematics is book [Kutzler 1997a] available from Chartwell-Yorke, (info@ChartwellYorke.com).

The TI-92 is the first handheld calculator with true computer algebra capabilities. It is the result of a cooperation between Texas Instruments (Dallas, Texas), Soft Warehouse (Honolulu, Hawaii, the developers of DERIVE) and the University of Joseph Fourier (Grenoble, France, the developers of CABRI GEOMETRE). The TI-92 offers the following features:

- computer algebra (similar to DERIVE)
- 2D and 3D graphics
- interactive geometry (a subset of CABRI GEOMETRE II)
- data/matrix editor with column logic
- text editor
- program editor

In the sequel we give a brief overview of the TI-92. The following text describes all inputs and shows all screens. You could easily follow the examples with your own TI-92. We start with a simple symbolic operation:

o   x/2+x/3

As one types this, the characters appear in the entry line at the bottom of the screen:

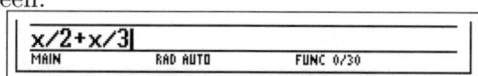

We press  ENTER  to conclude the input and perform a simplification.

o   ENTER

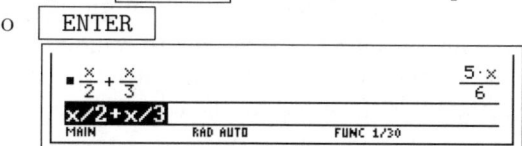

The first *history pair* is now displayed in the *history area* (this is the area between the toolbar on the top and the entry line). A history pair consists of the *entry* (left) and the *answer* (right). As a next example we enter the expression $\sqrt{\frac{x+1}{x-1}}$ in two steps:

o   (x+1)/(x-1)  ENTER

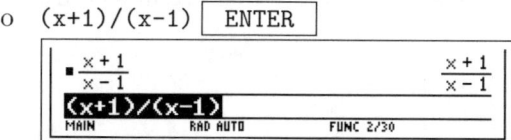

Pressing the square root key inserts a square root symbol and an opening parenthesis into the entry line:

o   [√]  (=  2nd   ×  )

278     Chapter 4  Computer Algebra Systems

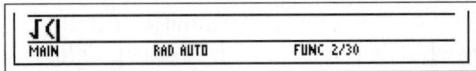

To auto-paste the last answer into the entry line (as the argument for the square root), we highlight the expression with the up arrow key, then press ENTER.

o  ▲ ENTER

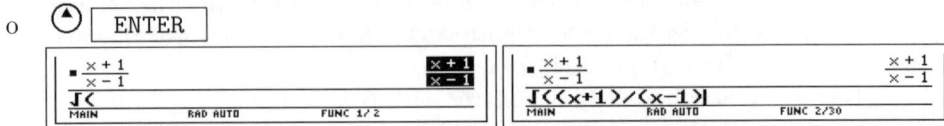

We type the closing parenthesis, then conclude the input:

o  ) ENTER

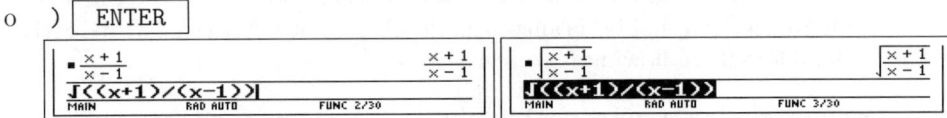

The next example shows how the TI-92 performs calculus operations. The differential operator can be entered, for example, by using the [ d ] key (below left). In order to differentiate the last answer, we first highlight it (below right) ...

o  [ d ] (= 2nd 8)       ▲

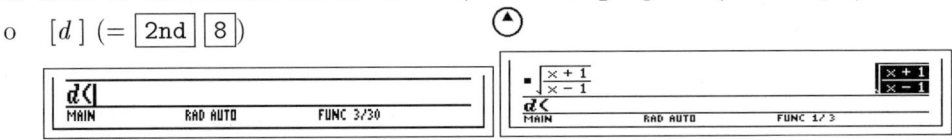

... then we auto-paste it into the entry line (below left). Next we insert a comma, the differentiation variable x, and a closing parenthesis. Then we conclude with ENTER (below right).

o  ENTER                  ,x) ENTER

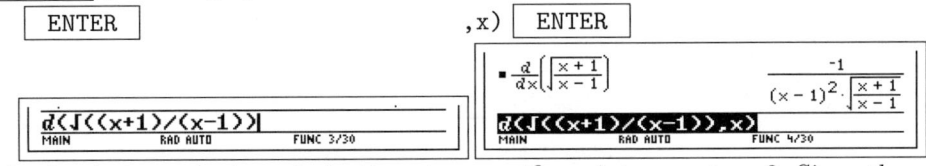

It is simple to evaluate a derivative at a specific point, say at x=2. Since the last entry still is displayed in the entry line, we use the right key to remove the highlighting and position the cursor at the end of the expression. Then we enter the 'With' operator (denoted by a vertical bar) followed by the constraint x=2:

o  ▶                      [ | ] x=2 ENTER

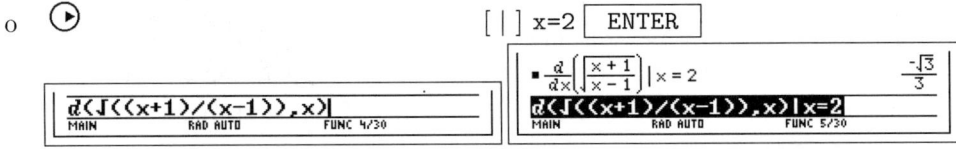

To obtain a numeric approximation instead of the above exact result, we use:

o  [♦] ENTER

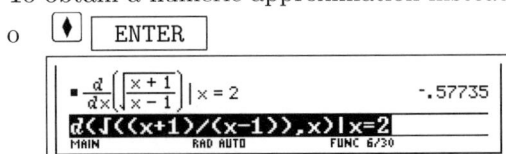

How about computing $\int \frac{1}{1+x^3} dx$? We first enter the expression $\frac{1}{1+x^3}$:

o  1/(1+x^3) ENTER

### 4.1.3 DERIVE and the TI-92

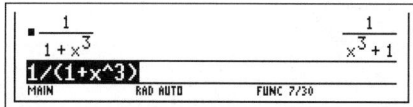

We enter the integral operator, then we highlight the last answer:

o  [ ∫ ]   ⊙

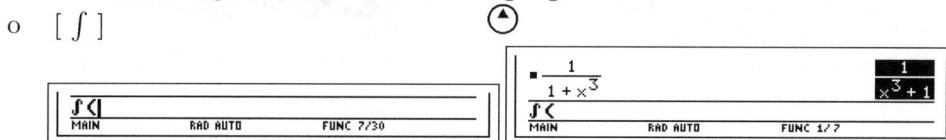

ENTER auto-pastes the expression into the entry line. We insert a comma, the integration variable x, and a closing parenthesis, then conclude the input.

o  ENTER    ,x) ENTER

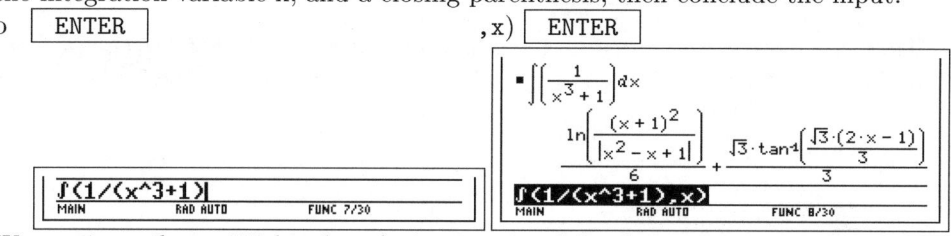

We want to pile up 150 bottles of wine in trianglular form:
$n$ bottles lie in the bottom row, then $n$-$1$ bottles on top of them, then $n$-$2$, etc. How many bottles should be in the bottom row? Employing the formula $\sum_{j-1}^{n} i = \frac{(n+1)n}{2}$, this leads to the problem of solving a quadratic equation. We will use the TI-92 to find $n$:

o  (n+1)n/2=150 ENTER

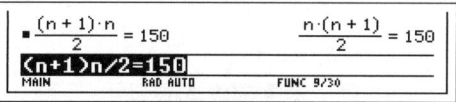

There is no 'Solve' key on the keyboard, but we find a respective function in the Algebra menu:

o  F2 :Algebra

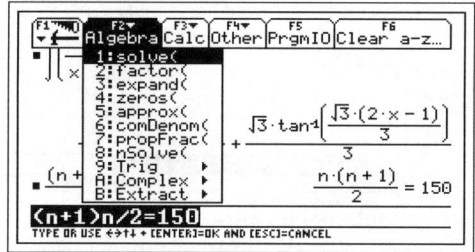

We use 1 to auto-paste the function solve into the entry line:

o  1 :solve

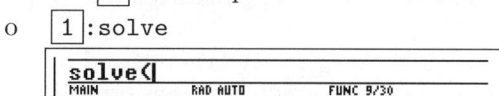

We auto-paste the last answer, then we add the variable with respect to which the equation should be solved.

280     Chapter 4  Computer Algebra Systems

o  Ⓐ  | ENTER |                                ,n)  | ENTER |

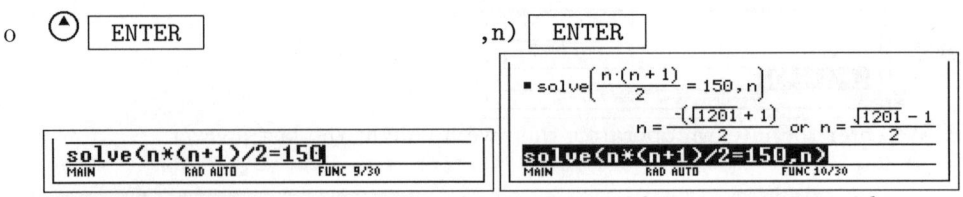

To obtain a numerical approximation also, we repeat the computation with

o  |◆|  | ENTER |

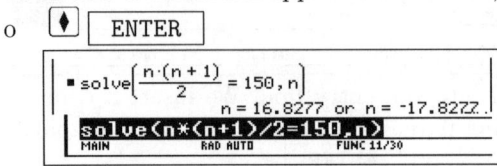

There are more methods of entering a function: there is a CATALOG to choose from and there is the MATH menu, both of which can be accessed via the keyboard. Or one simply types the function name:

o   solve(x^4-1=0,x)  | ENTER |

$$\blacksquare \text{solve}(x^4 - 1 = 0, x) \quad x = 1 \text{ or } x = -1$$
solve(x^4-1=0,x)

The solve function produces the real solutions only (properly separated by the logical or operator), which is what most teachers prefer anyway. For obtaining the complex solutions also, we need to change the function name to cSolve, which can be done, for example, by inserting the letter c in front of the expression. The left key removes the highlighting and puts the cursor at the left end of the expression (below left).

o  Ⓐ                                           c  | ENTER |

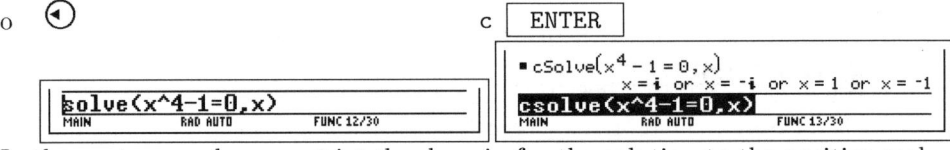

In the next example we restrict the domain for the solution to the positive real numbers. This is done by highlighting, then auto-pasting the penultimate entry into the entry line, then adding the 'With' operator and the constraint x > 0.

o  Ⓐ Ⓐ Ⓐ Ⓐ                                | ENTER |  | x > 0  | ENTER |

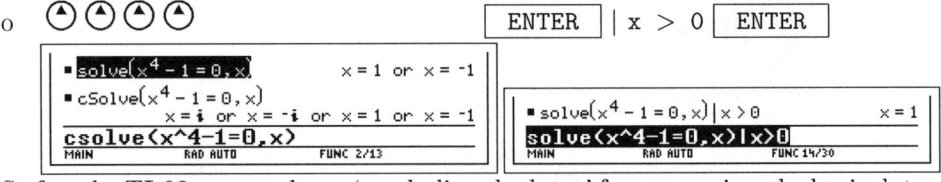

So far, the TI-92 was used as a 'symbolic calculator' for processing algebraic data (as opposed to the 'scientific calculator' which processes only numbers). However, the TI-92 also is a very useful educational tool for *learning* mathematical skills such as solving equations.

We start with the simple equation $5x - 6 = 2x + 15$. We choose and apply equivalence transformations step by step, aiming at gradually transforming the given equation into the form .

o   5x-6=2x+15  | ENTER |

### 4.1.3 DERIVE and the TI-92

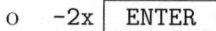

The subtraction of $2x$ is a reasonable first step. Here we make use of the feature that, if we start the input with the binary subtraction minus, a reference to the previous answer (ans(1)) is automatically inserted in front of the minus. Pressing ENTER at this point results in the application of the equivalence transformation $-2x$ to both sides of the equation.

o   -2x  ENTER

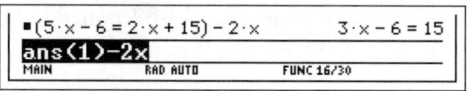

A useful next step is the addition of 6.

o   +6  ENTER

Many beginners make a fatal (but typical) mistake here by arguing that *"there is a 3 in front of the variable x. To get rid of the 3 I need to subtract 3"*. In a hand calculation they would end up with the wrong solution $x=18$, then would have difficulties in detecting this error. If, however, they use the TI-92 as shown here, ...

o   -3  ENTER

... they receive an *immediate* feedback about the quality of their choice: Since -3 did not *simplify* the equation further (it made it more complicated), this equivalence transformation was not a good choice. So we better delete the last history pair ...

o   ▲                                                                  CLEAR

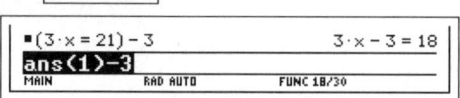

... then try something different:

o   ▼ /3  ENTER

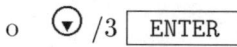

The command for plotting a graph can be found in the Other menu.

o   F4 :Other                                    2 :Graph 5sin(x)  ENTER

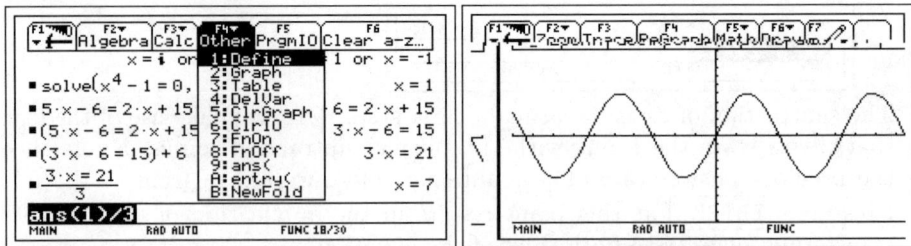

This command automatically switches to the graphics window, which is overlaid the Home screen we have been using so far. The graphing features are similar to that of DERIVE. This was only a glimpse of what the TI-92 can do. For a thorough introduction to working with the TI-92 see the book [Kutzler 1997c] available from EAI, info@eaiusa.com.

Bernhard Kutzler (Linz, Austria)

## 4.1.4 Macsyma

**4.1.4.1 Introduction.** MACSYMA is a comprehensive, powerful software system for doing *symbolic and numeric scientific computing*. MACSYMA runs on the following platforms: Windows 95, 98 and NT PCs, LINUX, most UNIX platforms, Symbolics Lisp Machines, and Alpha machines with Symbolics GENERA.

**4.1.4.2 Applications of Macsyma.** MACSYMA has been used in a wider variety of applications since 1968, a longer period of time than any other symbolic-numerical mathematics software package.

More information and demos are available from http://www.macsyma.com.

This report has three major sections: a brief synopsis of MACSYMA's capabilties, a short history of MACSYMA, and some selcted examples.

A brief bibliography of MACSYMA related publications and resources includes [Watanabe 1984; Kamke 1961; Wester 1994], [Macsyma Inc. 1997], [Macsyma Inc. 1998a], [Macsyma Inc. 1996b], [Braun and Hauser 1995], [Fell 1998], [Redfern et al. 1998], [Ben-Israel and Gilbert 1999], [Macsyma Inc. 1996a], [Macsyma Inc. and Three's Company 1996], [Macsyma Inc. 1998b], [Macsyma Inc. 1998d], [Macsyma Inc. 1998c], [Backstrom 1996], [Bialkowski 1995], [Mei 1994], [Beltzer 1990], [Rand 1994], [Heller 1991], [Rand and Armbruster 1987], [Pavelle 1985] and [Rand 1984].

**4.1.4.3 Brief Synopsis of Capabilitites of Macsyma.**

- **On-Line Help.**   Plain English queries entered on the command line, returning executable "tips" to complete the given task
- MathTips$^{TM}$ Advisor with solutions to 1,300 common computational tasks
- Math topic browser menus with 900 topics and commands
- Hypertext descriptions of 2,900 topics
- 1,350 executable examples / demos accessible from menus
- 100 template buttons and 900 function templates
- Interactive primer with six lesson scripts.

- **Arithmetic.**   Single, double, arbitrary precision floating point numbers
- Rational, algebraic, transcendental, complex numbers.

- **Algebra.**   Simplify, factor and expand expressions
- Trig simplifications and trig functions at special angles
- Summations: simplify, express in closed form
- Polynomials and special functions (many families)
- Exact symbolic, series and numerical solutions of systems of multivariate equations, inequalities, recurrence equations
- Numerical solutions of equations and minima of functions.

- **Linear Algebra.**   360+ commands support all popular (and some not-so-popular) symbolic and numerical linear algebra operations
- All major normal forms and decompositions, eigenanalysis
- Sparse matrix facilities

- Faster floatnum matrix operations with NumKit™ add-on
- Translator for Matlab commands into MACSYMA language.

- **Calculus.**   Differential Calculus
    - Differentiation and Limits
    - Analytic optimization, calculus of variations
    - Taylor and Laurent series, power series, Pade approximants
- Integral Calculus
    - Exact indefinite and definite integration
    - Numerical integration by six methods
    - Laplace and Fourier transforms and inverses, FFT
- Differential and Integral Equations
    - Exact, series and numerical solutions of first and second order O.D.E.s by 3 dozen methods; stiff numerical solver; systems solver
    - Symbolic systems of linear O.D.E.s, linear control problems
    - P.D.E.s: Lie symmetries and solutions of nonlinear systems
    - Generates input to PDEase(r) for finite element analysis
    - Integral equations of the 1st & 2nd kinds
- Vector and Tensor Calculus
    - Vector and tensor calculus in coordinate-invariant form and in specific coordinates. Includes variational tensor calculus.
    - Exterior forms, products, derivatives, homotopy operations.

- **Statistics & Data Analysis.**   DataViewer™ to import/export, view, edit graph large data sets.
- Univariate and multivariate descriptive statistics
- (Non)linear multivariate least squares fit of data
- Polynomial, rational, spline interpolation of tabulated data
- Many probability densities and cumulative distributions
- Dimensional analysis and units conversion
- Stochastic algebra and calculus.

- **Scientific Graphics.**   2D & 3D plots of data points, functions, implicit relations, parametric curves and surfaces, contours, vector fields
- Camera animation and data animation of plots
- Rotation, clipping, truck, zoom, roll with mouse controls
- Interactive mouse query for coordinates
- Publication-quality color rendering and lighting
- Labels: control of position, orientation, font, color, axes
- Graphics styles
- Export .bmp, .gif, .mfe, .pcx, .rle formats.

- **Notebook Interface.**   Re-executable notebooks combine textbook-quality math display, formatted text, editable graphics. (This and some selected other features are available in 1998 only in MACSYMA 2.3.1.)
- Text processing with formatting of characters and paragraphs with user-defined text styles

- Hypertext links within/between notebooks, and to other documents and URL addresses
- Collapsible outlines
- Textbook quality math output display
- Echo of input commands with textbook quality math display
- Editing of plots in notebooks
- Export/import of math, graphics as Windows metafiles.

- **Utilities.** Pattern matching with automatic and user-applied rules
- Mathematical properties assigned to variables and functions and used in computations
- FORTRAN and C code generation: translates math expressions and program control statements (DO, IF-THEN)
- Segmentation and optimization of expressions (optional)
- $\TeX^{TM}$ or $\LaTeX$ output
- Programming tools: compiler, trace, break, metering.

**4.1.4.4 A Short History of Macsyma.** The history of MACSYMA is divided into four distinct stages.

**4.1.4.5 Origins at M.I.T.: 1968 - 1982.** MACSYMA was born as a research project in 1968 at the Massachusetts Institute of Technology. Predecessor research projects reached back a decade. By the early 1970s MACSYMA was being widely used for symbolic computation in research projects at M.I.T and internationally. MACSYMA was one of the early applications that helped justify building the ARPANET. UNIX was first licensed outside Bell Labs to enable MACSYMA to be ported to DEC VAXs, thereby spawning the proliferation of UNIX. By 1980 MACSYMA was easily the most advanced symbolic-numerical-graphical mathematics software in the world.

**4.1.4.6 Commercialization by Symbolics, Inc.: 1982 - 1992.** In 1982 M.I.T. licensed MACSYMA to Symbolics, Inc., a workstation spin-off company from M.I.T. By 1986 over 600 research papers and other publications had been counted that referenced MACSYMA.

1987 - 1989: MACSYMA added a Gröbner basis package,[*] GENTRAN,[**] a $\TeX$ generator,[***] ODEFI,[†] a new Poisson series package,[‡] Taylor_solve, pattern matcher extensions, database extensions, special function libraries, and new O.D.E. perturbation and series solution methods. As Symbolics' main business began to decline in 1987, the MACSYMA business suffered constraints and declined with the rest of Symbolics between 1988 and 1992.

---

[*] Contributed by Gail Zacharias.
[**] Contributed by Barbara Gates and Paul Wang. GENTRAN translates entire Macsyma programs to FORTRAN and C, including program flow control statements.
[***] Based on code contributed by Richard Fateman.
[†] Contributed by Roman Shtokhamer and B. F. Caviness. ODEFI strengthened Macsyma's lead in solving first order ordinary differential equations symbolically.
[‡] Contributed by Richard Fateman.

**4.1.4.7 Revitalization by Macsyma Inc.: 1992 - 1994.** Macsyma Inc. acquired MACSYMA from Symbolics, Inc. On the 1992 benchmark tests that appeared in *The Notices of the A.M.S.* and in *PC Magazine*, MACSYMA solved more benchmark problems than any other product and gave more reliable symbolic results, even on some basic problems. In 1994 two independent studies* concluded that MACSYMA was the most capable symbolic-numerical mathematics software system.**

**4.1.4.8 Macsyma 2.X: 1995 - Present.** New features include vast improvements in user interface, graphics, linear algebra, numerical analysis, differential equations and speed. Independent reviewers consistently praised the new release.

"MACSYMA is remarkably easy to use. It will load automatically any math packages it needs; competing products require users to pitch in manually. MACSYMA has extensive on-line help ... Users with heavy math needs should insist on MACSYMA ." - *IEEE Spectrum*, December 1994 (USA)

1995: Dr. Michael Wester [Wester 1994] of the University of New Mexico assembled the most comprehensive test ever devised for symbolic mathematics software. He selected 131 mathematics problems and attempted to solve them with six mathematics software products. Results are summarized: a score of +1 for each correct solution, 1/2 for partial credit, 0 for no solution and -1 for an incorrect solution. With this scoring scheme, Wester's results yield the following scores. (Some software scores have changed. In particular MuPad which is a new system, and Macsyma 421, have increased their scores significantly since 1995.)

| Axiom 1.2 | Derive 2.0 | MACSYMA 419 | Maple V.3 | Mathematica 2.2 | Reduce 3.5 | MuPad |
|---|---|---|---|---|---|---|
| 59.5 | 73.5 | 108 | 90.5 | 88 | 50 | 24.5 |

Table 2. Results of Scoring the Wester Test (1995)

**4.1.4.9 Selected Capabilities of Macsyma.** This section illustrates some capabilities that enhance MACSYMA's reliability and/or enable it to solve more problems than other systems. Examples focus on MACSYMA's treatment of some elementary problems, rather than advanced problems [Macsyma Inc. 1996a].

**4.1.4.10 On-Line Help Systems and User Interface.** According to *Personal Computer World* (1993), "[MACSYMA's] enormity never compromises its ease of use." MACSYMA's user interface and on-line help systems play a critical role in making this possible. Many reviewers say that MACSYMA has the finest help system in mathematics software.

---

* Stefan Braun, Munich Germany was to carefully compared the mathematical capabilities of a range of mathematics software systems.
** Michael Wester (wester@amber.unm.edu) compared symbolic math capabilities of six math software systems on 123 carefully selected problems. This study is available by ftp at math.unm.edu . Change to directory `pub/cas`.

- **On-Line Help.**   MACSYMA includes a complete array of on-line help facilities. See also [Macsyma Inc. 1997], [Macsyma Inc. 1996b], [Macsyma Inc. 1996a], [Macsyma Inc. 1998b] and [Macsyma Inc. 1998d].

  - MATHTIPS$^{TM}$ Advisor: over 1,200 worked example problems with executable code, organized into over 100 hierarchical topic areas
  - Natural Language Query: response to help requests in plain English, provides a ranked list of relevant MATHTIPS
  - The Math Topic Browser menus with a plain English taxonomy of mathematical capabilities, leading to a small group of commands that address the area of immediate interest
  - 2,600 hypertext topic descriptions with cross-references, plus about 100 additional in-depth descriptions of external packages and special topics
  - Over 1,000 executable examples of individual commands and 340 executable demonstrations of applications and external packages which are included with MACSYMA
  - 900 function templates for individual MACSYMA commands
  - 100 of the most common templates available from icon buttons at the top of the MACSYMA Front End Window.

- **Document Processing:**   Creating Windows metafiles to export mathematical expressions and graphics to other applications. Combining fancy math expressions, text and graphics. Combining math input and output, graphics, and text sections in one re-executable document. Editing and rearranging sections. Editing graphs in place. Saving, re-opening or re-executing notebooks. Including hypertext links or a "navigator" diaglog to summarize each notebook section [Macsyma Inc. 1998d].

- **Graphics.**   Full spectrum of two-dimensional and three-dimensional scientific graphics, including plotting functions, parametric curves and surfaces, vector fields, and surface contours.

  The Macsyma Front End stores and replays animated motion Over 200 editable notebook attributes. Five graphics control dialogs (camera view, bounding box and axes, surfaces lines and points, titles, and decorations) with custom controls [Macsyma Inc. 1998d,b].

- **Fancy Display of Mathematical Expressions.**   In 1992 - 1993, MACSYMA pioneered formatted display of mathematical expressions using graphical images of mathematical symbols, Greek letters, and sizing and positioning of exponents. The following example shows some formatted mathematical expressions:

  **(c1)**   (p1: sqrt(ratexpand ((sqrt(alpha) + beta)^6)),
  p2: ratexpand ((c + d)^11),   p3: 'integrate(ratexpand((a + x)^12), x),
  p4: ratexpand((g + h)^13),   expr: p1 / p2 + p3 / p4)

(d1)
$$\int \frac{\begin{pmatrix} x^{12} & +12\,a\,x^{11} & +66\,a^2\,x^{10} & +220\,a^3\,x^9 & +495\,a^4\,x^8 \\ +792\,a^5\,x^7 & +924\,a^6\,x^6 & +792\,a^7\,x^5 & +495\,a^8\,x^4 & +220\,a^9\,x^3 \\ +66\,a^{10}\,x^2 & +12\,a^{11}\,x & +a^{12} & & \end{pmatrix}}{\begin{pmatrix} h^{13} & +13\,g\,h^{12} & +78\,g^2\,h^{11} & +286\,g^3\,h^{10} & +715\,g^4\,h^9 \\ +1287\,g^5\,h^8 & +1716\,g^6\,h^7 & +1287\,g^8\,h^5 & +715\,g^9\,h^4 & +286\,g^{10}\,h^3 \\ +78\,g^{11}\,h^2 & +13g^{12}\,h & +h & & \end{pmatrix}} dx$$

$$+ \frac{\sqrt{\begin{pmatrix} \beta^6 & +6\,\sqrt{\alpha}\,\beta^5 & +15\,\alpha\,\beta^4 & +20\,\alpha^{\frac{3}{2}}\,\beta^3 \\ +15\,\alpha^2\,\beta^2 & +6\,\alpha^{\frac{5}{2}}\,\beta & +\alpha & \end{pmatrix}}}{\begin{pmatrix} d^{11} & +11\,c\,d^{10} & +55\,c^2\,d^9 & +165\,c^3\,d^8 & +330\,c^4\,d^7 \\ +463\,c^5\,d^6 & +462\,c^6\,d^5 & +330\,c^7\,d^4 & +165\,c^8\,d^3 & +55\,c^9\,d^2 \\ +11\,c^{10}\,d & +c^{11} & & & \end{pmatrix}}$$

The main problem with "ugly display" is that you cannot easily identify which of the plus signs is the primary operator of the expression, and that the two division operators are the secondary operators. Formatted display, such as that on line (d1), makes identification of primary and secondary operators very simple.

### 4.1.4.11 Basic Arithmetic and Algebra.

- Arithmetic and Symbolic Procedures
    - Three types of floating point numbers (single floats, double floats, bigfloats)
    - two types of integers (fixnums and bignums), rational numbers, and transcendental numbers
    - Extensive use of inequality information and integer properties of exact arithmetic expressions. MACSYMA extracts inequality information from exact arithmetic expressions.

(c1) [ block( [heuristic_precision_limit:false],
csign (1 + cos (355 /113) ) ),   csign (1 + cos (355 / 113)) ]$
(d1)
$$[pz, \quad pos]$$
(c2)   [ sqrt ( (%e-%pi)^2), nummod (16, %pi)]
(d2)
$$[\pi - e, \quad 16 - 5\,\pi]$$

Compute in exact form trig functions of many special angles. The algebraic flag `ratsimp` enables MACSYMA to clear radicals from the denominator.
(c1)    1 / cot (%pi / 24)
(d1)
$$\frac{1}{(\sqrt{2}+1)\,(\sqrt{3}+\sqrt{2})}$$

**(c2)** ratsimp (%), algebraic : true
**(d2)**
$$\left(\sqrt{2}-1\right)\sqrt{3}+\sqrt{2}-2$$

- **Symbolic Algebra** MACSYMA's database stores information about mathematical expressions.

**(c1)** (-1)^n,
**(d1)**
$$(-1)^n$$

**(c2)** (declare (n, odd), (-1)^n)
**(d2)**
$$-1$$

**(c1)** /* Another example */ sqrt (x^2)
**(d1)**
$$|x|$$

**(c2)** (assume (x < 0), sqrt (x^2))
**(d2)**
$$-x$$

**(c3)** (declare (z, complex), sqrt (z^2))
**(d3)**
$$\sqrt{z^2}$$

**(c1)** integrate (x^n, x)
Is n + 1 zero or nonzero?

zero
**(d1)**
$$\log x$$

**(c2)** (assume (not equal(n, -1)), integrate (x^n, x))
**(d3)**
$$\frac{x^{n+1}}{n+1}$$

**(c1)** sin ( (n*(n + 1) / 2)*%pi)
**(d1)**
$$\sin\left(\frac{\pi n\,(n+1)}{2}\right)$$

**(c2)** (declare (n, integer), sin ( (n*(n + 1 )/ 2)*%pi))
**(d2)**
$$0$$

- **Complex Algebra.** Example: MACSYMA computes the rectangular and polar form of a basic complex exponential expression. A somewhat more intriguing problem in complex simplification is to find the arcsin of 2.

  (c1)     asin(2)
  (d1)
  $$\arcsin 2$$

  (c2)     rectform(asin(2))
  (d2)
  $$\frac{\pi}{2} - i \log\left(\sqrt{3}+2\right)$$

  (c3)     polarform(asin(2))
  (d3)
  $$\sqrt{\log^2\left(\sqrt{3}+2\right) + \frac{\pi^2}{4}}\; e^{-i \arctan\left(\frac{2\log(\sqrt{3}+2)}{\pi}\right)}$$

- **Algebraic Equations.** Solvers include a triangular solver for systems of algebraic equations and a `root_of` capability to express results as roots of polynomials without explicitly presenting the roots in terms of radicals.

### 4.1.4.12 Basic Calculus Operations.

- **Differential Calculus.** Uses user-declared variable dependencies and values of derivatives that do not appear explicitly in a symbolic expression.

  - Example: the user applies the chain rule to a function $u(x, y, t)$, where both $x$ and $y$ depend on $t$, though these dependencies do not appear explicitly in the differentiation command line.
  - New differentiation capabilities distinguish partial derivatives and total derivatives.

  (c1)     (depends(u,[x,y,t]), diff(u,t))
  (d1)
  $$\frac{du}{dt}$$

  (c2)     (gradef(x,t,alpha), gradef(y,t,beta), diff(u,t))
  (d2)
  $$\beta\frac{du}{dy} + \alpha\frac{du}{dx} + \frac{du}{dt}$$

- **Integral Calculus.** MACSYMA guards against making invalid assumptions that introduce subtle errors into results.

  - Example: MACSYMA explicitly asks about the exceptional value of n.
  - Example: in an inverse Laplace transform, MACSYMA employs the normal assumption for Laplace transforms without asking and points out that it made this assumption to get the result.

**(c1)**   integrate(x^n,x)
Is n + 1 zero or nonzero?
zero
**(d1)**
$$\log x$$

**(c2)**   assume(not equal(n,-1))$
**(c3)**   integrate(x^n,x)
**(d3)**
$$\frac{x^{n+1}}{n+1}$$

**(c1)/**   /* Inverse Laplace transform */   ilt(1,s,t)
Proviso: Assuming s > 0.
**(d1)**
$$\delta(t)$$

MACSYMA evaluates a double integral from Newtonian gravitational theory and applies series expansions in the presence of absolute values.
**(c1)**   expr: -r^2*sin(theta)/sqrt(x^2+r^2-2*x*r*cos(theta))
**(d1)**
$$-\frac{r^2 \sin\theta}{\sqrt{x^2 - 2r\,\cos\theta\,x + r^2}}$$

**(c2)**   integrate(integrate(expr, theta, 0, %pi), r, 0, 1)
Is x positive, negative, or zero?
pos
Is x - 1 positive, negative, or zero?
pos
**(d2)**
$$-\frac{2}{3x}$$

If we tell MACSYMA $x - 1$ is negative, we get: $\frac{x^2}{6} + \frac{x^2-6}{6}$.
MACSYMA asks to determine whether $x < 0$ or $0 < x < 1$ or $1 < x$. If MACSYMA can be told this information beforehand, it will not ask the questions.

**4.1.4.13 Differential Equations.** Extensive capabilities for exact symbolic solutions, perturbation solutions, series solutions and numerical solutions of ordinary differential equations (O.D.E.s). Finds symmetries and symbolic solutions of partial differential equations.

– Exact Solutions, First Order O.D.E.s.
  Symbols such as %c, %k1, %k2 represent arbitrary integration constants. Initial conditions are inserted into a general solution.
  **(c1)**   diff(y,x)*x=(x*log(x^2/y)+2)*y
  **(d1)**
  $$x\frac{dy}{dx} = \left(x\left(\log\left(\frac{1}{y}\right) + \log x^2\right) + 2\right)y$$

(c2)   ode(%,y,x)
(d2)
$$\log(2\log x - \log y) + x - \log 2 = \%c$$

(c3)   solve(ic1(d2, x=1, y=a), y)
(d3)
$$\left\{ y = e^{\log a\, e^{1-x}}\, x^2 \right\}$$

- **Exact Solutions, Second Order O.D.E.s**
  We start with a simple damped linear harmonic oscillator.
  (c1)   (depends(y, x), m*diff(y, x, 2) + b*diff(y, x) + k*y = 0)
  (d1)

$$m\frac{d^2y}{dx^2} + b\frac{dy}{dx} + ky = 0$$

(c2)   ode(m*diff(y, x, 2) + b*diff(y, x) + k*y = 0, y, x)
Is m^2 * (4 * k * m - b^2)
positive, negative, or zero?
pos
(d2)

$$y = e^{-\frac{bx}{2m}} \left( \%k1 \sin\left( \frac{\sqrt{\frac{4k}{m} - \frac{b^2}{m^2}}\, x}{2} \right) + \%k2 \cos\left( \frac{\sqrt{\frac{4k}{m} - \frac{b^2}{m^2}}\, x}{2} \right) \right)$$

We can isolate the x-dependence in this solution to obtain a more compact form.
(c4)   /* Isolate the x-dependency */   isolate(d2, x)
(e4)
$$\frac{\sqrt{\left(\frac{4k}{m}\right) - \left(\frac{b^2}{m^2}\right)}}{2}$$

(d4)
$$y = e^{-\frac{bx}{2m}} \left( \%k1 \sin(e4\, x) + \%k2 \cos(e4\, x) \right)$$

(c5)   /* Specify initial amplitude a and initial velocity 0. */
ic2(d4, x = 0, y = a, diff(y, x) = 0)

(d5)
$$y = e^{-\frac{bx}{2m}} \left( \frac{ab \sin(e4\, x)}{2\, e4\, m} + a \cos(e4\, x) \right)$$

- **Approximate Symbolic Solutions to O.D.E.s.**   Provides many approximate symbolic solutions to algebraic and differential equations, plus symbolic utilities to support numerical analysis (such as finite difference generators, exceptionally strong FORTRAN and C code generators, and a companion finite element product for solving P.D.E.s numerically).

- Series solutions give very accurate representation of exact solutions over a small interval around the center of expansion of the series. They are often poor approximations to the exact solutions for longer times.
- Perturbation solutions accurately represent the asymptotic long-time behavior of the exact solutions. Generally, they do not capture the transient behavior of the solution.

Example: Taylor series solution to fourth order for a difficult equation with a derivative in an exponent. We specify initial conditions $y(0) = a$ and $dy/dx(0) = b$.

(c1)  'diff(x, t, 2) = x + %e^(c*t*'diff(x, t))
(d1)
$$\frac{d^2 x}{dt^2} = e^{c\,t\,\frac{dx}{dt}} + x$$

(c2)  taylor_ode('diff(x, t, 2) = x + %e^(c*t*'diff(x, t)),x,t,4,[0,a,b])
(d2)
$$\left\{\left\{ x = a + b\,t + \frac{(a+1)\,t^2}{2} + \frac{(b\,c+b)\,t^3}{6} + \frac{\left(b^2\,c^2+(2\,a+2)\,c+a+1\right)t^4}{24} + \cdots \right\}\right\}$$

Example: a perturbative solution of Duffing's equation for an almost-harmonic oscillator by Lindstedt's method. We specify initial conditions x(0)=a and dx/dt(0)=b. The initial conditions affect the form of the asymptotic solution, even though the perturbation method is not computing transient responses.

(c1)  duffing: 'diff(x, t, 2) + x + e*x^3 = 0
(d1)
$$\frac{d^2 x}{dt^2} + e\,x^3 + x = 0$$

(c2)  lindstedt (duffing, e, 1, [a, b])
(d2)

$$\{ a\sin(\%\text{tau}) + b\cos(\%\text{tau}) \quad - \left(\tfrac{e}{32}\right)\left(b^3\left(9\sin\left(\%\text{tau}\right)\right)\right.$$
$$+a\,b^2\left(3\cos\left(3\%\text{tau}\right) - 3\cos\left(\%\text{tau}\right)\right) + a^2\,b\left(21\sin\left(\%\text{tau}\right) - 3\sin\left(3\%\text{tau}\right)\right)$$
$$+a^3\left(\cos\left(\%\text{tau}\right) - \cos\left(3\%\text{tau}\right)\right), \quad \%\text{tau} = t\left(1 + \frac{e\left(3\,a^2 + 3\,b^2\right)}{8}\right)\}$$

– Partial Differential Equations
  MACSYMA contains vector and tensor calculus packages that are useful for stating many physically important systems of partial differential equations (P.D.E.s) in a wide range of coordinate systems.
  - MACSYMA contains the PDELIE©* package for finding Lie symmetries and symbolic solutions of (systems of nonlinear) P.D.E.s.

---

* PDELIE copyright ©Trinity University and Peter Vafeades.

- In 1994, MACSYMA Inc. began shipping a companion to MACSYMA, PDEase2D[TM] *, for solving systems of P.D.E.s by finite element analysis. PDEase2D accepts a wide range of static and dynamic systems of P.D.E.s in two dimensions, including systems of mixed elliptic, parabolic and hyperbolic type. MACSYMA can produce symbolic equations for PDEase in string form. In 1996 MACSYMA Inc. introduced PDEase2D 3.0 which integrates PDEase2D into the MACSYMA Front End interface.
- MACSYMA automates the writing of systems of P.D.E.s in about 20 standard coordinate systems using vector calculus or tensor calculus.

**4.1.4.14 Linear Algebra.** MACSYMA has a broad portfolio of linear algebra capabilities, including inversion, solving matrix equations, matrix normal forms (Jordan, QR, LU, LDU), matrix exponentiation, and multilinear algebra.** MACSYMA has nearly all the functionality of MATLAB*** release 3.0, as well as symbolic versions of most of the operations. MACSYMA also has a dozen MATLAB-inspired language features for compact expression of linear algebra. The NUMKIT[TM] package greatly increases the speed of numerical linear algebra by seamlessly incorporating Lapack.

**4.1.4.15 Numerical Analysis.** Extensive capabilities in six types of basic algorithms.

- Numerical integration and integral transforms. (Six methods of numerical quadrature, including extrapolated Gaussian quadrature)
- Numerical solution of equations and critical point conditions
- Numerical solution of differential equations including for stiff systems (MACSYMA's finite element companion PDEase2D solves partial differential equations by finite element analysis.)
- Numerical linear algebra, including numerical eigen-analysis, with NUMKIT facility improve the speed of numerical linear algebra
- Numerical evaluation of special functions
- Interpolation, extrapolation, fitting and statistical description of data.

**4.1.4.16 Acknowledgements.** We would like to acknowledge the invaluable assistance of the Documentation Staff of Macsyma Inc. for reviewing and improving many aspects of this article. Any errors or omissions remain solely the responsibility of the author.

Macsyma Technical Staff
Macsyma Inc. 20 Academy Street, Arlington, MA 02476-6436
email: info@macsyma.com
Web: http://www.macsyma.com

---

* PDEase®️ is a registered trademark of Macsyma Inc. PDEase2D[TM] is a trademark of Macsyma Inc.
** MACSYMA's ATENSOR package performs computations in associative algebras, including universal tensor, Grassmann, Clifford, and symplectic algebras.
*** MATLAB is a trademark of The Math Works, Inc.

## 4.1.5 Magma

### 4.1.5.1 Introduction

MAGMA is a general-purpose computer system for algebra that is based on a computational model that closely reflects the way in which algebraists view their subject. More precisely, rather than designing around a particular data structure or programming paradigm, MAGMA has been designed to provide a computing environment that is constructed out of the fundamental concepts of algebra (set, mapping, structure, morphism). Its primary application areas include algebra, number theory, algebraic geometry, algebraic topology, algebraic combinatorics and all areas of mathematics which are algebraic in nature. The specific goals may be summarised as follows:

- ALGEBRAIC DESIGN PHILOSOPHY: The language attempts to approximate as closely as possible, the usual mathematical modes of thought and notation. In particular, the principal constructs in the user language are set, algebraic structure and morphism.
- UNIVERSALITY: In-depth coverage of all the major branches of algebra, number theory, algebraic geometry and finite incidence geometry.
- INTEGRATION: The facilities for each area are designed in a similar manner using generic constructors wherever possible.
- PERFORMANCE: MAGMA provides outstanding performance both in terms of the algorithms used and in their implementation. The major algorithms installed in the MAGMA kernel are state-of-the-art and give performance similar to, or better than, specialised stand-alone programs.

A major motivation behind the design and implementation of MAGMA has been a desire to open up advanced areas of algebra and geometry to effective computation. We believe that the system goes a long way towards achieving these aims. Examples of major MAGMA packages written for such abstract areas include algebraic geometry via schemes (Gavin Brown), polynomial equation solving (Gregoire Lecerf), hyperelliptic curves (Michael Stoll), modular symbols (William Stein) and homological algebra (Jon Carlson).

The core of the MAGMA system has been developed by the Computational Algebra Group at the University of Sydney. The basic design for MAGMA emerged during the late 1980's [Butler and Cannon 1989][Butler and Cannon 1990][Cannon 1981], building on extensive experience obtained through development and application of the Cayley system [Cannon 1982] [Cannon 1984]. Since the launch of MAGMA in London in 1993, the system has expanded enormously and its development continues at a rapid rate. MAGMA includes algorithms and code generously contributed by many mathematicians from around the world. In many ways the MAGMA project represents a major cooperative effort by mathematicians to develop a system properly tuned to their needs.

In following sections we discuss the design of MAGMA, give some examples of its use, and present some information on its applications and performance.

For a more complete description, the reader should consult [Bosma et al. 1997] and [Bosma et al. 1994] for a formal overview, [Bosma and Cannon 1995–2000], [Cannon and Playoust 2001a] and [Cannon and Playoust 2001b] for full user documentation, and [Bosma and Cannon 2000] for applications of MAGMA to the solution of diverse mathematical problems.

### 4.1.5.2 Design Features of Magma

#### 4.1.5.2.1 Computational Model

The computational model at the heart of MAGMA is based on a modern structural view of mathematics, in which the central notions are structure and morphism. Indeed, the name of the system reflects the fact that the fundamental concept in MAGMA is a set with structure; the term *magma* is borrowed from Bourbaki [Bourbaki 1970] who defines it to be a set with a law of composition.

Each object definable in MAGMA is considered to belong to a unique mathematical structure (*magma*), called its *parent*, and this parent must be present at the time the object is created. For instance, MAGMA does not have the concept of a matrix as a primitive object; rather, a matrix may only exist as an element of some particular ring or bimodule. Moreover, MAGMA employs *evaluation semantics*, so that an expression may only be evaluated if every identifier appearing in it has a value. Thus, notions such as "indeterminates" or "unknowns" must be defined as elements of the appropriate magma, such as a univariate polynomial ring over a field, before any use is made of them. This philosophy makes programming a great deal easier, since there is never any ambiguity as to the current interpretation of an object. In programming language parlance, MAGMA is a strongly typed language.

Magmas are organised hierarchically. At the top level are *varieties*, which are families of magmas sharing the same basic operations and axioms. Typical varieties include groups, commutative rings and modules. A variety has attached to it functions (or methods in OO parlance) that do not depend upon properties of a specific representation of a magma. When computing with a structure $A$ it is necessary that $A$ be represented in some concrete form; here we come to the level of *categories*. For example, within the variety of groups there are categories for finitely-presented groups, permutation groups and matrix groups (among others). The category to which a magma belongs determines not only how the magma and its elements are represented, but also the permitted operations. Note that operations independent of the representation are inherited from the parent variety while representation-dependent operations are attached to the category. Categories of magma that are specialised subcategories of an existing category $C$ may, in turn, inherit operations from $C$, in addition to having further operations that are specific to the subcategory.

The general model for constructing a magma is to start with a "free" magma and then successively form submagmas, quotients and extensions until the desired magma is obtained. This approach, which has its foundations in Universal

Algebra, provides a powerful and general mechanism for constructing a vast range of algebraic and geometric structures. Since most commonly occurring structures in computational algebra are finitely generated, in general, a magma $M$ is given in terms of a finite generating set. Special handling applies to the rare cases of infinitely generated magmas, such as $\mathbb{R}$, $\mathbb{C}$, and Laurent series rings.

#### 4.1.5.2.2 The User Language and Environment

The MAGMA user language is an interpreted high-level language with dynamic typing. The data structures are designed around the ideas of set, sequence, mapping and structure. While the language formally belongs to the class of imperative languages there is also a strong functional flavour. Powerful constructors allow complex sets and magmas to be constructed in a single statement. Typically, a magma is created from another magma $M$ by means of *constructor* expressions, such as the submagma and quotient constructors:

```
sub< M | generators >
quo< M | ideal or normal subgroup >
```

These constructors not only return the new submagma $S$ or quotient $Q$, but also return a morphism relating the new magma to the existing magma $M$ given in the left part of the constructor; here the morphisms are the inclusion monomorphism $S \to M$ and the natural epimorphism $M \to Q$. Other mappings from magma $M_1$ to $M_2$ may be directly created by means of the mapping or homomorphism constructors:

```
map< M₁ -> M₂ | x :-> expression in x >
hom< M₁ -> M₂ | images of generators of M₁ >
```

MAGMA has a powerful collection of set and sequence constructors whose syntax is based on traditional mathematical set notation and was influenced by SETL [Schwartz et al. 1986]. For example, $\{1..10\}$ is the set of integers $\{x \in \mathbb{Z} | 1 \leq x \leq 10\}$ and
```
> { i^2 : i in {1..10} | not IsPrime(i) };
{ 1, 16, 36, 64, 81, 100 }
```
is the set of squares of the non-primes in this set.

The language has a functional subset, providing functions and procedures as first-class objects. Functions are allowed multiple return values. Value arguments and parameters are supported both for functions and procedures while reference arguments are supported for procedures only. As first-class objects, functions and procedures are created as the values of expressions, rather than by means of special statements.

The run-time environment supports packages containing user-defined intrinsics with automatic compilation; command completion and interactive line editing; a history system with recall and editing of previous lines; a hierarchical online help facility; logging of output; redirection of I/O; the ability to save and restore user workspaces; environment variables for configuring the style of output, etc.; verbose options; a special file type for fully-featured file I/O; the ability to execute system commands from within Magma; input/output pipes for communication with external programs; UNIX commands and functions such as process ID, alarm setting and much else.

#### 4.1.5.2.3 Mathematical Facilities

A great deal of effort has been invested in the implementation of very efficient algorithms for working with most of the fundamental algebraic and geometrical structures in which computation is currently practical. It is estimated that the system currently contains around 10, 000 algorithms. A summary of the facilities may be found in [Cannon 2000]. The mathematical structures implemented in the current version of MAGMA (V2.8) may be loosely grouped into the following nine areas:

- *Semigroups*: Semigroups, monoids, rewriting systems.
- *Groups*: Permutation groups, matrix groups, Coxeter groups, finite groups of Lie type, finitely-presented groups, Abelian groups, soluble groups, (infinite) polycyclic groups, automatic groups.
- *Commutative rings*: Univariate and multivariate polynomial rings, residue class rings of polynomial rings, affine algebras, rings of group invariants.
- *Fields and their orders*: Ring of integers, galois fields, galois rings, number fields and their orders, algebraic function fields, $p$-adic and local fields, power series and Laurent series rings, real and complex fields, Puiseux series fields.
- *Modules*: Vector spaces, modules over Euclidean rings, modules over orders, modules over affine algebras, modules over associative algebras, modules Hom$(M, N)$, lattices, homological algebra.
- *Algebras*: General finite-dimensional (f.d.) algebras, f.d. associative algebras, quaternion algebras, matrix algebras, group algebras, Hecke algebras, (split) basic algebras, finitely-presented associative algebras, f.d. Lie algebras.
- *Algebraic geometry*: Projective and affine curves, surfaces, schemes.
- *Arithmetic geometry*: Modular forms, modular symbols, congruence subgroups of PSL$(2, \mathbb{Z})$, Dirichlet characters, Hecke modules, elliptic curves, hyperelliptic curves and their Jacobians, modular curves, binary quadratic forms.
- *Combinatorics*: Counting theory, directed and undirected graphs, incidence structures, incidence geometries, designs, finite planes, codes.

The MAGMA package includes a growing number of databases many of which are compressed versions of standard databases compiled by various mathematicians. The current databases include elliptic curves of conductor up to 5300 (J. Cremona), the maximal subgroups and automorphism groups of all simple groups of order less than $10^7$, all groups of order up to 2000 (excluding 1024) (U. Besche and B. Eick) and transitive groups of degree up to 23 (A. Hulpke). Some of these databases are used by internal algorithms. For example, an algorithm needing to factorise the order of the multiplicative group of the finite field $GF(p^n)$ would first consult a database containing 200, 000 factorizations of integers of the form $p^n - 1$ for primes $p$ up to 1000 and various $n$, before deploying an integer factorization algorithm.

### 4.1.5.3 Some MAGMA Examples

#### 4.1.5.3.1 Groups, Designs, Codes and Lattices

The advantage of an integrated system, such as MAGMA, over stand alone programs, is the ability to easily move between many different algebraic structures during a computation. Here we illustrate this by starting with the Mathieu group $M_{24}$ and constructing from it a 5-design, and from that a linear code (the extended binary Golay code) and finally a lattice. The blocks of the design are the images of b under all group elements.

```
> load m24; /* defines G to be the Mathieu group */
> orbs := Orbits(Stabilizer(G, [1,2,3,4,5]));
> b := &join [ o : o in orbs | #o le 3]; b;
{ 1, 2, 3, 4, 5, 8, 11, 13 }
> D := Design<5, 24 | b^G>; D; /* b^G is orbit of b under G */
5-(24, 8, 1) Design with 759 blocks
> C := LinearCode(D, GF(2)); C : Minimal;
[24, 12, 8] Linear Code over GF(2)
```

We now check that C is the binary Golay code and that its automorphism group is the permutation group as we started with. We apply rational function arithmetic to compute the weight enumerator of the dual code via the MacWilliams identity. This should be the same as the weight enumerator of C, as C is self-dual.

```
> IsIsomorphic(C, GolayCode(GF(2), true));
true
> A := AutomorphismGroup(C); A eq G;
true
> Q<a, b> := RationalFunctionField(Integers(), 2);
> MacWilliams := func< r, n |
>     Evaluate(r,[1,(a-b)/(a+b)])*(a+b)^n/Evaluate(r,[1,1])>;
> Q ! WeightEnumerator(C);
a^24 + 759*a^16*b^8 + 2576*a^12*b^12 + 759*a^8*b^16 + b^24
> MacWilliams(WeightEnumerator(C), Length(C));
a^24 + 759*a^16*b^8 + 2576*a^12*b^12 + 759*a^8*b^16 + b^24
```

We next construct a lattice such that every lattice element, when reduced mod 2, gives a code word. The automorphism group of this lattice is calculated as a matrix group over the integers. It has $M_{24}$ as a quotient group.

```
> L := Lattice(C, "A"); L : Minimal;
Lattice of rank 24 and degree 24
> A := AutomorphismGroup(L); A : Minimal;
MatrixGroup(24, Z) of order 2^34 * 3^3 * 5 * 7 * 11 * 23
> H := OrbitImage(A, Orbit(A, A'Base[1])); #H eq #A;
true
> K := PrimitiveQuotient(H);
> IsConjugate(Sym(24), G, K);
true
```

### 4.1.5.3.2 Function Fields and Geometric Goppa Codes

We consider the Klein quartic $X$ defined by $x^3y + y^3z + xz^3$ over $GF(8)$ and apply Goppa's construction to obtain a $[24, 20, 4]$ error-correcting code. We then apply a standard coding theory operation to expand this to a $[96, 60, 8]$ code. The details of the construction may be found in [van Lint and van der Geer 1988]. We actually work in the function field $K$ of $X$ using machinery provided by the KANT system [Daberkow et al. 1997], see 4.2.21.

```
> kx<x> := FunctionField(GF(8));
> kxy<y> := PolynomialRing(kx);
> K<u> := FunctionField(x^3*y + y^3 + x); Genus(K);
3
```

We need to find the 24 points of $X$ and we use the fact that these are simply the places of $K$ having degree 1. We also need the divisor $E$ of degree 2 over $GF(2)$ that corresponds to the bitangent $x + y + z = 0$. We omit the calculations which show that $E$ lies in the support of the divisor $D$ corresponding to the rational function $u + v + 1$.

```
> P := Places(K, 1); #P;
24
> D := Divisor(u + x + 1);
> l, e := Support(D);
> E := l[3]; E;
(x^2 + x + 1, u + x + 1)
```

Taking the divisor $G = 3E$, we construct a basis for the Riemann-Roch space $R$ of $G$. The generator matrix $M$ for the dual of the code we want is found by evaluating the 4 basis vectors of $R$ at the 24 places of $P$.

```
> B := Basis(3*E); #B;
4
> M :=  Matrix( #P, [ Evaluate(f, p) : p in P, f in B ]);
> C := Dual(LinearCode(M));
> d := MinimumDistance(C); C : Minimal;
[24, 20, 4] Linear Code over GF(2^3)
```

```
> F := ConcatenatedCode(C, EvenWeightCode(4));
> time d := MinimumDistance(F); F : Minimal;
Time: 73.229
[96, 60, 8] Linear Code over GF(2)
```

The code $C$ produced by the Goppa construction is a $[24, 20, 4]$ linear code over $GF(8)$. We then concatenated this code with the $[4, 3, 2]$ parity check code to obtain a $[96, 60]$ linear code $F$ over $GF(2)$. Using the built-in MAGMA function, MinimumDistance, in 74 seconds we found the minimum distance of $F$ to be 8. So we have a length 96 binary code of rate approximately 2/3 that is 3-error-correcting.

#### 4.1.5.3.3 Algebraic Curves

In the following example, prepared by Gavin Brown, we create the plane projective curve $C$ defined by $x^5 + y^4z + y^2z^3$ and compute its canonical embedding in projective 3-space. We then notice that the curve is actually trigonal and compute, by hand, its Maroni invariant. We begin by creating the projective plane with coordinates $x$, $y$ and $z$.

```
> k := Rationals();
> P<x,y,z> := ProjectiveSpace(k,2);
> f := x^5 + y^4*z + y^2*z^3;
> C := Curve(P,f);
> C;
Curve over Rational Field defined by x^5 + y^4*z + y^2*z^3
> SingularPoints(C);
{ ( 0 : 0 : 1 ) }
> Genus(C);
4
```

The genus is the dimension of the space of canonical divisors. We can compute a basis of this space inside the function field of the curve $C$.

```
> w := CanonicalDivisor(C);
> V := RiemannRochSpace(w); Dimension(V);
4
```

We use these functions to map $C$ into projective 3-space. At the time of writing, MAGMA doesn't have machinery to compute the image of this map automatically, so we study the image using other functions. We know that the image curve will span the space (since it's an mapping by a complete linear system) so we look to see which conics and cubics contain the image.

```
> phi := DivisorMap(w);
```

302    Chapter 4  Computer Algebra Systems

```
> P3<a,b,c,d> := Codomain(phi);
> Image(phi,C,2);
Scheme over Rational Field defined by a*d - c^2
> Image(phi,C,3);
Scheme over Rational Field defined by a^2*b + b^3 + c*d^2,
                                      a^2*d - a*c^2,  ...
> J := Ideal($1) + Ideal($2);
> MinimalBasis(J);
[
    a*d - c^2,
    a^2*b + b^3 + c*d^2
]
> D := Scheme(P3,J); Dimension(D);
1
> IsNonSingular(D);
true
```

We see that the conics containing $D = \phi(C)$ cut out the quadric cone

$$\overline{F} : ad - c^2 = 0.$$

which is the birational image of the plane $P$ by $\phi$. We could consider this to be a ruled surface, often called a rational surface scroll, mapped into space in a way that contracts its negative section. See [Reid 1997] Chapter 2 for details. The image curve $D$ itself is then cut out by a single cubic equation on this scroll. In general it could be fiddly to figure out how to pull back the curve $D$ to the scroll but in this case we can easily work it out. We have to permute the coordinates since the given order of equations defining $\psi$ would yield image $ac - b^2 = 0$ rather than $\overline{F}$.

```
> F<r,s,u,v> := RuledSurface(k,2,0);
> psi := RationalMap(F,P3,Sections(LinearSystem(F,[0,1])));
> psi;
Mapping from: PrjScrl: F to Prj: P3
with equations :
r^2*u
r*s*u
s^2*u
v
> g := PermutationAutomorphism(P3,Sym(4) ! (4,3,2));
> f := psi*g;
> [ pf : p in Equations(D) | pf ne 0 where pf is p@@f ];
[
    r^4*u^2*v + r*s^5*u^3 + v^3
]
> E := Curve(F,$1[1]);
```

```
> E;
Curve over Rational Field defined by r^4*u^2*v + r*s^5*u^3 + v^3
```

The ruled surface $F$ has a map to the projective line with fibres that are also projective lines: the functions $r, s$ are coordinates on the image line, while the functions $u, v$ are coordinates on the fibres. The resulting curve $E$ is a cubic in $u, v$ so really is a triple cover of the line, a so-called *trigonal curve*. The fibres cut out the $g_3^1$ on $E$. Trigonal curves have an invariant called the *Maroni invariant*, see [Reid 1997] Section 2.10. This can be computed from the basic invariants of the scroll in which their canonical embedding lies. In this case, the Maroni invariant is 2, being the difference of the two twisting weights of the scroll $F$. Note that we are in the genus equals 4 case here which is the only case in which the ambient scroll can be a cone.

The curve $E$ is certainly birational to the original curve $C$, although it is reassuring in this case to note the equivalence directly on one of the affine patches.

```
> E4<X,Y> := AffinePatch(E,4);
> E4;
Curve over Rational Field defined by X^5 + Y^3 + Y
```

### 4.1.5.3.4 Elliptic Curves and Cryptography

MAGMA includes algorithms for computing the group of rational points of elliptic curves over $\mathbb{Q}$ and over arbitrary finite fields. Of particular interest to cryptographers is the finite field case where the Schoof-Elkies-Atkin point counting algorithm with extensions due to Lercier is implemented.

With the following code, we produce curves suitable for digital signatures. The function selects a random $n$-bit prime $p$, creates the field $\mathrm{GF}(p)$, and then finds a random elliptic curve over this field such that the group order has a prime divisor of at least $m$ bits.

```
> function CryptographicRandomCurve(n, m)
>     p := RandomPrime(n);
>     K := FiniteField(p);
>     while true do
>         E := EllipticCurve([ Random(K) : i in [1,2] ]);
>         t := TraceOfFrobenius(E);
>         if t ne 0 then // avoid supersingularity
>             N := p-t+1;
>             fac := TrialDivision(N,2^(n-m));
>             q := #fac eq 0 select 1 else fac[#fac][1];
>             if Ilog2(q) ge m then
>                 break;
>             end if;
```

```
>                 N := p+t+1;
>                 fac := TrialDivision(N,2^(n-m));
>                 q := #fac eq 0 select 1 else fac[#fac][1];
>                 if Ilog2(q) ge m then
>                     E := QuadraticTwist(E);
>                     break;
>                 end if;
>             end if;
>         end while;
>         repeat
>             P := (N div q)*Random(E);
>         until P ne E!0;
>         return E, P, q;
> end function;
> E, P, q := CryptographicRandomCurve(160, 150);
> E;

Elliptic Curve defined by y^2 = x^3 +
946714064155810302070181252411777685876460224620*x +
907334624253194016909420788750120649950794338696
over
GF(1226120467300804884526644301229884426615264608677)

> Order(E);
1226120467300804884526644654153031419833946407634

> P;
(1096688647020852987921168349143083221198284694413 :
2006936496844468270323255745249918864000186437 : 1)
```

This function takes an average of 6 minutes to find a suitable curve over a 160-bit prime field and the determination of the number of points on a curve accounts for almost all of this time.

MAGMA includes a range of tools that have application to research in cryptography. For integer-based cryptosystems there is fast integer and modular arithmetic, integer factorization (many algorithms including ECM, MPQS) and primality proving. Finite fields come equipped with fast algorithms for arithmetic, polynomial factorization and linear algebra. In addition MAGMA contains a range of tools for attacking the discrete logarithm problem: Shanks, Pollard rho, Pohlig-Hellman and index calculus. Elliptic curves over finite fields are supported with fast arithmetic, point counting (as noted above) and tools for the discrete logarithm problem: the Pohlig-Hellman algorithm and the van Oorschot-Wiener parallel collision search algorithm. The study of stream ciphers is supported with tools for working with LFSR, BBS and RSA pseudo-random sequences. For working with lattices, MAGMA has a number of high-performance LLL reduction algorithms.

#### 4.1.5.4 Applications of MAGMA

MAGMA has been applied to a huge range of problems in many different areas. A few recent noteworthy applications are listed below.

– The discovery of the first infinite family of codes for quantum error correction by Peter Shor.
– The application of MAGMA's Groebner basis techniques to prove that a particular open quantum system cannot be simulated by any closed quantum system of dimension two by Terhal et al.
– The development of new algorithms for invariant theory and their application to solve numerous outstanding problems in the invariant theory of finite groups by Gregor Kemper.
– The construction of cohomology rings of $K[G]$-modules by Jon Carlson.
– Classification of low-rank geometries associated with simple groups by Francis Buekenhout and associates.
– The discovery of numerous theorems relating the properties of finite planes and codes by Ed Assmus and Jenny Key.
– The discovery of an infinite family of optimal packings of $m^2 + m - 2$ $m/2$-dimensional subspaces of $m$-dimensional space, whenever $m$ is a power of 2 by Peter Shor and Neil Sloane.

#### 4.1.5.5 Performance

A great deal of effort has gone into optimizing the algorithms used in MAGMA to provide outstanding performance. For numerous classes of problems, the MAGMA implementation outperforms all other software known to us. Some sample timings are presented below. Unless otherwise stated, the times were obtained on a 400 MHz Sun UltraSPARC.

*Integer Arithmetic:* MAGMA includes very high performance integer arithmetic including Karatsuba and FFT algorithms. MAGMA multiplies integers 25-40% faster than GNU gmp 2.0.2 running on a Sun UltraSPARC computer. The product of two arbitrary integers, each having one million *decimal* digits, may be computed in 3.1 seconds. The calculation of the GCD of the two Fibonacci numbers $F(10^8)$ and its predecessor takes 1.85 hours, and the extended GCD, 3.25 hours.

*Polynomial Arithmetic:* Factorizing the von zur Gathen challenge polynomial of degree 2000 over a prime field of 2000 bits takes 18.8 hours using a 64-bit 200 MHz SGI Origin 2000. The 3000 bit polynomial takes 75.8 hours. Magma includes the exciting new van Hoeij algorithm for factorization in $\mathbb{Z}[x]$, and factorizes the 7-th Swinnerton-Dyer polynomial (degree 128, with at least 64 modular factors modulo any prime) in 11 seconds. The Zimmermann challenge polynomials are all easily factored in a minute in the worst case.

*Linear Algebra:* Given a 301 × 300 matrix over $\mathbb{Z}$ with random entries between 0 and 10, Magma can compute its nullspace taking an average of 4.2 seconds (using a fast $p$-adic algorithm). The nullity is practically always 1, and the integer entries of the non-zero null vector typically have about 450 digits each.

*Commutative Algebra:* A Groebner basis (in lex order) for the "Cyclic 6" ideal over $\mathbb{Q}$ is found in 1.2 seconds. For the "Cyclic 7" ideal, MAGMA takes 6.4 minutes to find a grevlex basis and a further 2.6 minutes to convert this to the lex basis using a $p$-adic version of the FGLM algorithm.

*Number Fields:* Consider the number field obtained by extending $\mathbb{Q}$ by a root of $x^9 - 57x^6 + 165x^3 - 6859$. Using the KANT [Daberkow et al. 1997] package as incorporated into MAGMA the maximal order is found in 0.249 seconds, the class group ($\mathbb{Z}_3$) is found unconditionally in 41.750 seconds and the unit group ($\mathbb{Z}_2 + \mathbb{Z} + \mathbb{Z} + \mathbb{Z} + \mathbb{Z}$) is found in 3.760 seconds. The Galois group of order 18 is found in 0.269 seconds.

*Lattices:* The 98 280 shortest vectors in the Leech lattice are found in 5.8 seconds while the automorphism group of the lattice is found in 85 seconds. MAGMA takes 150 seconds to solve a 100 element knapsack problem using LLL reduction.

*Elliptic curves:* The points on a random elliptic curve over a 160 bit prime field may be enumerated in around 45 seconds. MAGMA finds that the elliptic curve $y^2 + y = x^3 - 180817x + 29664426$ over $\mathbb{Q}$ has rank 8 in 130 seconds. To determine that the curve $y^2 + xy = x^3 - 215x + 1192$ over $\mathbb{Q}$ has group $\mathbb{Z}_2 + \mathbb{Z} + \mathbb{Z}$ takes 280 seconds.

*Permutation Groups:* Starting with six permutations of degree 8 835 156, it takes MAGMA 5.0 hours to determine the order of the group generated, and a further 2.7 hours to determine that the group is the simple group of Lyons. The 29 normal subgroups of the wreath product of $PGL_4(7)$ by the dihedral group of order 8 (degree 1600, order 3692812761030005840485044311727673849990348800000000) takes 24 seconds. Finding the 20 maximal subgroups of the group of Rubik's cube (degree 48, order 43252003274489856000) takes 4.76 secs.

In an independent comparison of integer and polynomial arithmetic in a number of Computer Algebra systems (including Pari and Maple), carried out by Lewis and Wester [Lewis and Wester 1999], Magma outperformed all other systems for 27 of the 34 test problems (about 80% of the tests), usually by a factor of at least 2.

### 4.1.5.6 Documentation

MAGMA has an extensive online help system. It includes specially-written introductory material, suitable for the first-time user, and complete access to the Magma Handbook which provides a full description of all the facilities. The help system may be consulted from within MAGMA or as a web document. In addition to the online help, there are five main components to the printed documentation:

- J. Cannon and C. Playoust: **First Steps in Magma**, 16 pages.
- J. Cannon and C. Playoust: **An Introduction to Algebraic Programming with Magma: The Language**, 350 pages (Springer, to appear).
- W. Bosma, J. Cannon and C. Playoust: **An Introduction to Algebraic Programming with Magma: The Categories**, 500 pages (Springer, to appear).
- W. Bosma and J. Cannon (eds.): **The Magma Handbook**, (Five volumes), 2000 pages.
- W. Bosma, J. Cannon et al.: **Solving Problems with Magma**, 190 pages.

In addition, a book "A first course in MAGMA" by Lancelot Pecquet [Pecquet 2000] will shortly be published by Springer.

**4.1.5.7 Availability**
MAGMA is available on all standard workstations and PCs. These include Sun (SunOS and Solaris), DEC Alpha (OSF and Linux), SGI (IRIX), IBM RS 6000 (AIX), HP-Apollo 9000/700 (HP-UX) and IBM PC compatible machine (Linux and Windows). Details on acquiring a copy of the system may be obtained by sending an email to magma@maths.usyd.edu.au or by consulting the MAGMA home page at http://www.maths.usyd.edu.au:8000/u/magma.

John Cannon and Bill Unger (Sydney)

## 4.1.6 Maple

### 4.1.6.1 Introduction

Maple is a powerful, interactive and easy-to-use general-purpose computer algebra system for solving mathematical problems quickly and to any desired accuracy. It provides scientists, engineers, educators, and students with a complete mathematical environment for manipulating algebraic expressions; calculating numerical results; visualizing mathematics using two-dimensional, three-dimensional, and animated graphics. Maple's high level mathematical programming language permits users to extend the power of its comprehensive library by adding new scripts and procedures. Maple is available on Windows, Macintosh and Unix platforms. Specific information relating to compatibility is available on the Waterloo Maple Inc. Web site (http://www.maplesoft.com).

### 4.1.6.2 Background

The Maple system was created at the University of Waterloo, Canada in 1980 in response to the need for a compact, flexible, and efficient computer algebra system. In 1988, Waterloo Maple Inc. was founded to create a high quality document-oriented user interface for Maple, to augment the reseach innovations from the University of Waterloo, and to provide services and support for the expanding base of users. Currently, Maple development is carried out, not only by strong teams of mathematicians and computer scientists at Waterloo Maple Inc., but also by leading world-class research teams at the University of Waterloo, ETH Zurich in Switzerland, INRIA in France, Simon Fraser University in Canada, Moscow State University, and other research sites worldwide.

Maple is used worldwide and across a wide range of technical disciplines including mathematics, physical sciences, business management, economics, and engineering. Specific applications of Maple include modelling telecom systems, cryptography, turbojet design, control systems analysis, and brain research. Further examples of the use of Maple to solve problems in all disciplines are available from the Waterloo Maple Inc. Web site (http://www.maplesoft.com).

In education, the use of Maple as a learning tool encourages an exploratory approach to problem solving and enables students to concentrate on the concepts of a topic rather than simply the calculation methods. As a result, in recent years, within universities throughout the world Maple has become a standard tool in student laboratories for mathematics, the sciences, and engineering.

### 4.1.6.3 Mathematics

Maple offers huge flexibility in problem solving with its library of over 3,000 functions, which can be used for symbolic and numeric calculations. These include standard mathematical operations such as integration; differentiation; summation; products; solving many forms of equations; series expansions and limits; elementary functions such as logarithms, trigonometric, hypergeometric, and elliptic functions; special functions; matrix methods; as well as more specialized functions for specific fields of mathematics. Maple handles many different types of mathematical objects such as integers; rationals; irrationals; complex numbers; algebraic numbers; floating point numbers; sets; lists; arrays; matrices; vectors; polynomials; boolean, relational or logical expressions; symmetric functions; constants; as well as structured object types.The Maple library contains over 30 specialized packages. These include packages of routines for solving ODEs and classes of PDEs, teaching calculus, linear algebra, two-dimensional and three-dimensional Euclidean geometry, statistics, linear optimization, group theory, power series, combinatorics, tensors, differential algebra, Groebner Bases problems, Lie symmetries, and financial applications.

### 4.1.6.4 Interface

Maple combines powerful mathematics with an advanced, yet straightforward interface. Calculations, results, plots, and textual notes are displayed in a document-style worksheet in which you can record and annotate your session steps and then edit, save, and rerun your Maple commands. Maple uses standard mathematical notation, so your mathematics appears as you would write it. This makes expressions easy to interpret, check, and communicate to others. Maple V also includes extensive word-processing functionality for professional-looking documents as well as presentation tools such as hyperlinks, collapsible sections, and animation.

The worksheet interface is ideal for multi-stage analysis, solution creation, and also for teaching. Several worksheets can be opened at the same time and, under multi-tasking operating systems, you can choose whether these worksheets share mathematical information. Worksheets can be printed, emailed to colleagues, or converted to RTF, LaTeX or HTML (to automatically create WWW pages).

Maple allows several ways to enter and evaluate expressions using your mouse. Input can be displayed either in standard mathematical notation or in the Maple language. New users are able to start using Maple without any manuals or guides and without learning any command syntax. For new users, palettes containing common mathematical symbols and forms are available to help the users create mathematical formulae or expressions. Context-sensitive menus let the users work with mathematical objects produced by Maple, again without learning any programming language syntax and without memorizing command names. For each mathematical object, Maple will analyze the object and then produce individual context-sensitive menus of common operations for that object. For example, clicking on a polynomial expression brings up a menu that includes options for integrating, differentiating, and plotting the expression, using the variables of the polynomial. Alternatively, if the output is a matrix, then various matrix operations appear in the context-sensitive menu. These menus are also customizable by users.

Maple also enables users to carry out symbolic or numeric calculations in a spreadsheet format. The cells of a Maple spreadsheet can contain both symbolic and numeric data or formulae. They offer many other familiar spreadsheet features such as the filling of rows and columns, and relative and absolute referencing between cells. Or, with the Maple Add-In to Microsoft Excel 2000, users can access the Maple functionality from within a familiar speadsheet environment, Excel.

Recently Waterloo Maple Inc. partnered with the Numerical Algorithms Group (NAG) from the UK to incorporate the NAG libraries with the Maple technology. As a result, Maple now tightly integrates powerful computational linear algebra routines with the core Maple product. The Maple environment can handle complex numeric computations with ease. Users with MATLAB installed on their machines can access MATLAB via the Maple interface. The MATLAB link and the integration of the NAG librar results in greater flexibility and enables more mathematical and modelling applications to be solved completely within the Maple environment.

### 4.1.6.5 Plotting

Maple has extensive plotting capabilities for creating two-dimensional, three-dimensional, and animated graphics of mathematical functions and data. Among the plotting functions are contour plots, complex plots, implicit plots, parametric plots, differential equation plots, and statistical plots. Animations can be created in two or three dimensions. Animations using a time variable allows scientists to study models of a changing process.

Maple also includes "smartplots" - automatic plotting which makes it even easier to visualize mathematics. To plot a mathematical expression, users can either select the "plot" menu item automatically generated for that expression or they can simply drag the expression onto any existing graph. To examine 3D surfaces from different viewpoints, the user simply grabs the surface with the mouse and rotates it in real-time.

### 4.1.6.6 Programming

Maple contains a powerful high-level programming language, which is similar in form to Pascal or C. This language can be used to extend Maple by adding customized procedures. The open architecture of Maple allows users to examine the source code for the entire Maple library, where over 95maintained. New procedures and methods created by the user can be used to extend the power of Maple's existing code to provide automatic solutions for new problems. Problems can be modeled very quickly using Maple as a rapid, high level prototyping system. The results can be converted automatically by Maple into optimized C or FORTRAN code for solutions that require highly iterative numeric calculations. Maple's C and FORTRAN code generation facilities also allow scientists and engineers to merge their solutions from Maple into their existing numeric programs. Or, users can call their C or FORTRAN libraries into the Maple environment and incorporate their external libraries with Maple commands.

### 4.1.6.7 Help Available

A comprehensive online help system is provided with Maple, including details and examples of all commands, a help browser, hyperlinks to related help pages, balloon help, and topic and keyword searches. In addition, Waterloo Maple Inc. provides an excellent technical support service.

There are over 300 books in 11 different languages about Maple and its applications. Numerous Web sites offer solutions and additional resources for Maple. The interested reader is referred to http://www.maplesoft.com/cybermath/sh_resources.html.

The Maple User Group is a worldwide email community of highly qualified professionals who use Maple (see www.maplesoft.com/support/discussion.html) and are often helpful in supporting other users through guidance and practical advice. Or you can participate in a Maple newsgroup (comp.soft-sys.math.maple).

### 4.1.6.8 Example 1: An Eigenvector Problem

```
> restart;
> with(LinearAlgebra):
> B:=<<0|1|0|0>,<-u^2|2*u|0|0>,<-u*s|s|u|0>,<-u*t|t|0|u>>;
```

$$B := \begin{bmatrix} 0 & 1 & 0 & 0 \\ -u^2 & 2u & 0 & 0 \\ -us & s & u & 0 \\ -ut & t & 0 & u \end{bmatrix}$$

```
> Eigenvectors(B);
```

$$\begin{bmatrix} u \\ u \\ u \\ u \end{bmatrix}, \begin{bmatrix} 1 & 0 & 0 & 0 \\ u & 0 & 0 & 0 \\ 0 & 1 & 0 & 0 \\ 0 & 0 & 1 & 0 \end{bmatrix}$$

### 4.1.6.9 Example 2: A Differential Equations Problem

```
> restart;
> with(DEtools):
> eqn:=diff(y(x),x)=-y(x)-x^2;
```

$$eqn := \frac{\partial}{\partial x} y(x) = -y(x) - x^2$$

A General Solution

```
> dsolve(eqn,y(x));
```

$$y(x) = -x^2 + 2x - 2 + e^{(-x)}\_C1$$

A Particular Solution for the initial condition of y(0)=0

```
> dsolve({eqn, y(0)=0\}, y(x) );
```

$$y(x) = -x^2 + 2x - 2 + 2e^{(-x)}$$

A plot of the phaseportrait and nullclines for 3 different initial conditions: y(0)=-1, y(0)=0, y(0)=1

```
> phaseportrait( eqn, y(x),x=-1..2.5,
> [[y(0)=0],[y(0)=1],[y(0)=-1]],
> title='Asymptotic solution',colour=magenta,
> linecolor=[red,blue,green]);
```

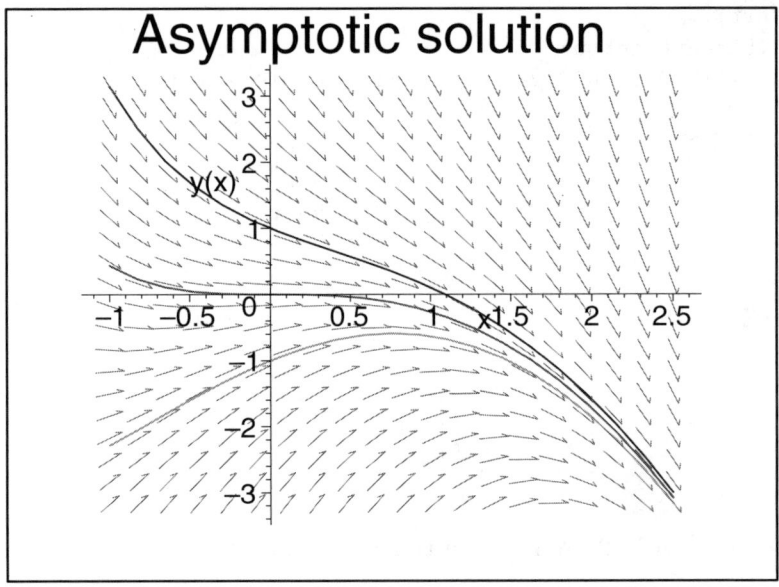

### 4.1.6.10 Example 3: A Programming Problem

```
> restart;
```

Write a procedure to use the Euclidean Algorithm to compute the GCD of 2 polynomials

```
> GCD := proc(A::polynom,B::polynom,x::name,p::prime)
>     a := Expand(A) mod p;
>     b := Expand(B) mod p;
>     if a = 0 and b = 0 then RETURN( 0 ) fi;
>     while b <> 0 do (a,b) := (b,Rem(a,b,x) mod p) od;
>     a/lcoeff(a,x) mod p; # return a monic GCD
> end:
Warning, 'a' is implicitly declared local
Warning, 'b' is implicitly declared local

> GCD(x^4+x+1,x^4+x^3+1,x,3);
```

$$x + 2$$

### 4.1.6.11 Conclusion

Maple is the result of over a decade of world-class research, development, and customer service. It provides the powerful combination of a huge library of mathematical methods and an advanced user interface enabling users to solve complex technical problems successfully and efficiently. In addition, it provides interactive graphics, standard mathematical notation, and a powerful programming language. Over a million researchers, educators, and technical professionals worldwide use Maple, or derivative products using Maple technology, as a mathematical assistant for exploring and solving problems in mathematics, engineering, science, and other disciplines.

Waterloo Maple Inc.<br>
Email: info.maplesoft.com<br>
URL: www.maplesoft.com

### 4.1.7 *Mathematica*

**4.1.7.1 Introduction** *Mathematica* is a fully integrated environment for technical computing, covering aspects such as numerical calculations, symbolic manipulations on various kinds of objects, graphical visualization, and programming. *Mathematica* is a commercial system designed, developed, distributed, and marketed by Wolfram Research Inc. (www.wolfram.com) since its first release in 1988. At first, *Mathematica*'s impact was felt mainly in the physical sciences, engineering and mathematics. But over the years, *Mathematica* has become important in a remarkably wide range of fields, such as biology, medicine, finance, social sciences, and other. At the same time, the system plays a major role in math education, where the applications range from high-school level up to advanced university courses supported by *Mathematica*.

Apart from its applications, *Mathematica*'s elegant language, the powerful user front end, and the marketing power of a professional software development company had substantial impact on the development of the whole computer algebra community. It is hard to estimate, how far the development of currently available mathematical software systems would have grown without the pressure put on all systems existing at the time of *Mathematica*'s first release in the late eighties.

In the subsequent sections, we try to give an impression of the potential of *Mathematica*, structured along the three main constituents of the system: a vast collection of mathematical algorithms, a powerful high-level programming language, and a comfortable, highly configurable user interface.

**4.1.7.2 The Algorithm Library of *Mathematica*** The algorithm library of *Mathematica* has its origin in a classical computer algebra system for doing symbolic and numerical computations. As a basis for all computations in such a system, there must be support for *arithmetic* in various domains available, such as integers of arbitrary length, rational numbers, floating point numbers, complex numbers, algebraic numbers, polynomials of arbitrary degree containing any number of variables over various coefficient domains, rational functions, transcendental expressions, matrices, and finite fields. Standard algorithms for arithmetic in these domains known from computer algebra are applied, however, for floating point arithmetic special techniques for preserving accuracy have been developed and, invented in Version 4, there is a specially tailored internal data structure called *packed arrays* that optimizes speed and memory consumption of numerical operations.

As for symbolic computation, *Mathematica* provides manipulation of symbolic expressions in various kinds, notably simplification to canonic forms, replacement of subexpressions, expansion of terms applying special rules for known functions occurring in the expression, putting terms over common denominator, canceling of common factors in rational expressions, and the like. Traditionally, polynomial-related computations play a key role in computer algebra systems such as polynomial expansion, factorization of polynomial terms, decomposition of polynomials, greatest common divisors, and Gröbner bases.

Based on its capabilities in symbolic manipulations, *Mathematica* offers algorithms for *solving equations* in many forms, most prominently systems of linear equations with particular support for sparse systems, systems of algebraic (polynomial) equations based on the above mentioned Gröbner bases method, and systems of ordinary or partial differential equations. For all of these types of equations there are algorithms for obtaining both symbolic (exact) and numeric (approximate) results.

In the area of calculus, *Mathematica* provides support for symbolic and numerical differentiation and integration, computation of limits, expansion of functions as Taylor- or Fourier-series, function transformations such as the Fourier-, Laplace-, or Z-transformation, or vector analysis. In addition, methods for optimization of linear and non-linear functions, interpolation and approximation of data sets, or eigenvalue and eigenvector computations are available.

An important role in the day-to-day use of *Mathematica* are its capabilities in graphical representation of functions in 2D and 3D. The spectrum of available plotting procedures ranges from ordinary function plots of univariate and bivariate functions, over plots of parametrized curves or surfaces, plots of curves implicitly defined by an equation, contour- and density plots, to list plots, visualizations of vector fields, charts, etc. See Figure 9 for an example of a 3D graphics generated by *Mathematica*. A particularly nice feature is the visualization of a sequence of images in an animated movie, which nowadays is a standard feature in virtually all computer algebra systems but has appeared for the first time already in *Mathematica*'s version 1 in the late eighties.

In addition to the *Mathematica* built-in functions there are several layers of extendibility of the system:

**Standard Packages** contain special purpose functions that are by default not loaded when the *Mathematica* system starts up. However, every version of *Mathematica* includes these packages for free, they just need to be loaded explicitly by the user to make the functions available. Among the over 100 available standard packages there are packages for solving polynomial inequalities, special partial differential equations, solving recurrence relations, spline fitting, and linear regression.

**Add-on Packages** are separate commercial products containing application-oriented functions for particular application domains that are developed outside of Wolfram Research. Add-on packages are available e.g. fro electrical engineering, fuzzy logic, finance, optical design, or digital image processing. These packages are distributed by Wolfram Research and, thus, need to fulfill certain quality standards defined by Wolfram.

**Math Source** is a common forum, where *Mathematica* users provide their own developments based on *Mathematica* for the community of *Mathematica*

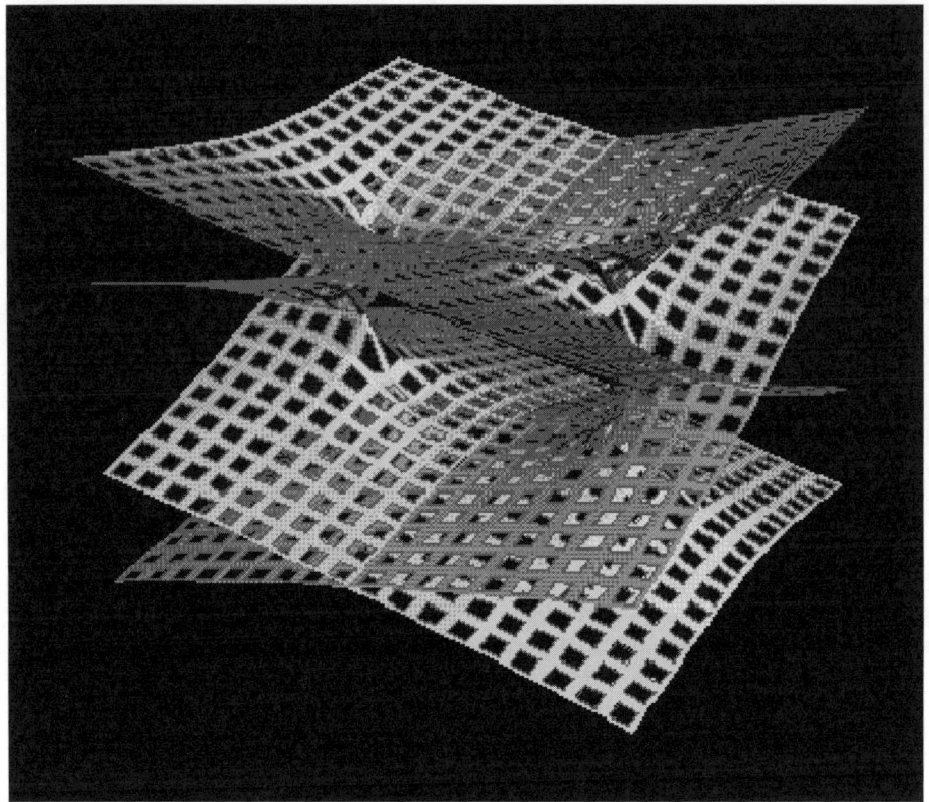

**Fig. 9.** *Mathematica* plots the Riemann surface of the function $(z^2 - 1)^{\frac{1}{4}}$

users worldwide. No quality standard is maintained for these packages but they are freely available on the net at www.mathsource.com.

**4.1.7.3 The Mathematica Language** One of the reasons of *Mathematica*'s great success is definitely its high-level programming language, which allows the user to extend the system by adding own functions to the system. Some of the striking features of the *Mathematica* programming language are:

**Multi-paradigm programming** A programmer developing software need not stick strictly to one programming paradigm, because the programming language supports procedural programming (assignments, loops, conditionals, and jumps), functional programming (recursion, application of unnamed function on lists, etc.), rule-based programming (applying transformation rules for application patterns), or even certain aspects of object-oriented programming (polymorphism).

**Higher order pattern matching** From the point of view of logic, the *Mathematica* language is an untyped higher-order language. In definitions of functions, the arguments may be described using a highly advanced pattern specification that allows a very elegant style of programming. During evaluation of functions, the parameter values are matched with the argument patterns by an efficient algorithm for higher order matching.

**Front end programming** Since Version 3, the *Mathematica* Front End (see Section 4.1.7.4) is accessible from within the programming environment, i.e. it is since then possible to manipulate the user interface from within *Mathematica* programs. In principle, a GUI for a mathematical software package written in *Mathematica* can be developed in the same programming environment, having interaction elements available such as windows, buttons, and hyperlinks.

**Name spaces** Keeping apart name spaces for symbols is of utmost importance in serious software development. In the *Mathematica* programming environment, this can be achieved by a mechanism of symbol contexts and packages, which is the basis for building up software exchange on the basis of *Mathematica*.

**MathLink** As for import and export capabilities, the MathLink communication standard gives the opportunity to set up communication channels to external programs, with which structured data can be exchanged in both directions. Toolkits for connecting to C/C++ and Java executables are available by now. Of course, reading and writing textual data from or to files is supported, mathematical expressions can be exported in HTML or MathML format, and graphics can be exported in various formats such as GIF, JPG, TIFF, or PostScript.

**Compilation** A subset of the language containing only machine-near data types such as small integers, small floats, and strings may even be compiled for gaining efficiency. However, for the majority of elegant pattern constructs compilation is unfortunately not supported yet.

### 4.1.7.4 The Mathematica Front End

*Mathematica* introduced the so-called *notebook interface* already at a time, when all other systems only had text-based command-line interfaces. It was certainly due to the emergence of *Mathematica*, that now most of the mathematical scientific software systems have a reasonable user interface, but it seems that *Mathematica* is always one step ahead. A *Mathematica* notebook is primarily *the user interface* for entering commands and mathematical expressions – even in a two-dimensional way like in a WYSIWYG formula editor – and displaying computation results. At the same time, the notebook is an *electronic document* that can store all data entered into it during a *Mathematica* session, be it entered by the user or generated by *Mathematica* computations. As a consequence, a notebook may contain only a "session log" consisting of input, output, and graphics, but it can also be enriched by textual comments, formatted in different fonts, styles, sizes, and colors, if desired structured into sections and subsections, converted to PostScript or HTML so that it ultimately serves as a complete scientific publication tool.

The user front end also contains a Help Browser, in which information on all available functions and standard packages can be found. The functions are grouped by topics, a search utility helps to find particular items in the help. The help system also contains examples, which can be evaluated immediately from within the help system. Moreover, the help browser can be extended, so that information on program packages developed by the user can be integrated smoothly into a homogeneous help system.

As for input, there are various ways to a accomplish two-dimensional mathematical expression input.

**Palettes** *Mathematica* provides standard palettes for supporting most common symbols and expression templates. Moving the mouse cursor over a palette template and clicking the mouse pastes the template into the notebook document, where pre-defined slots in the template can be filled out. A user can provide own templates for expression patterns that do not appear in the standard palettes provided by default.

**Menu** Some movements in two-dimensional input, e.g. moving to subscript or superscript position, entering matrices, entering radicals or fractions, can be supported in addition by menu commands.

**Keyboard Shortcuts** Most mathematical symbols and cursor movements can be typed in using keyboard shortcuts, which again are customizable.

An interesting feature is the possibility to customize the input parser, i.e. the translation from the optical appearance of an expression to its internal representation in the *Mathematica* system. Using this possibility, it is possible to define arbitrary syntax to be processable by *Mathematica* programs, e.g. $T^\dagger$ can be defined to be interpreted as `MoorePenroseInverse[T]` and then this function can be implemented in *Mathematica* providing for it the nice input syntax $T^\dagger$. In Figures 10 and 11 we show a screenshot of a *Mathematica* notebook containing formulae, input, text, output, graphics, and a user-defined input palette.

The structure of a notebook is determined by *cells*, whose style determine the behavior and the appearance of its contents. Input cells contain *Mathematica* input, which, by evaluation, is sent to the *Mathematica* computation engine that sends the result back to the notebook front end to be displayed in an output cell. Available cell styles are defined in *style sheets* so that formatting information can be strictly separated from document contents. An interesting aspect of the notebook document standard is that its entire cell structure is stored as plain text in form of a *Mathematica* expression in the sense of the *Mathematica* programming language. By this, as already mentioned in Section 4.1.7.3, the front end interface becomes accessible for *Mathematica* programs and *Mathematica* notebooks serve as programmable documents.

Wolfgang Windsteiger (RISC Institute, University of Linz, Austria)

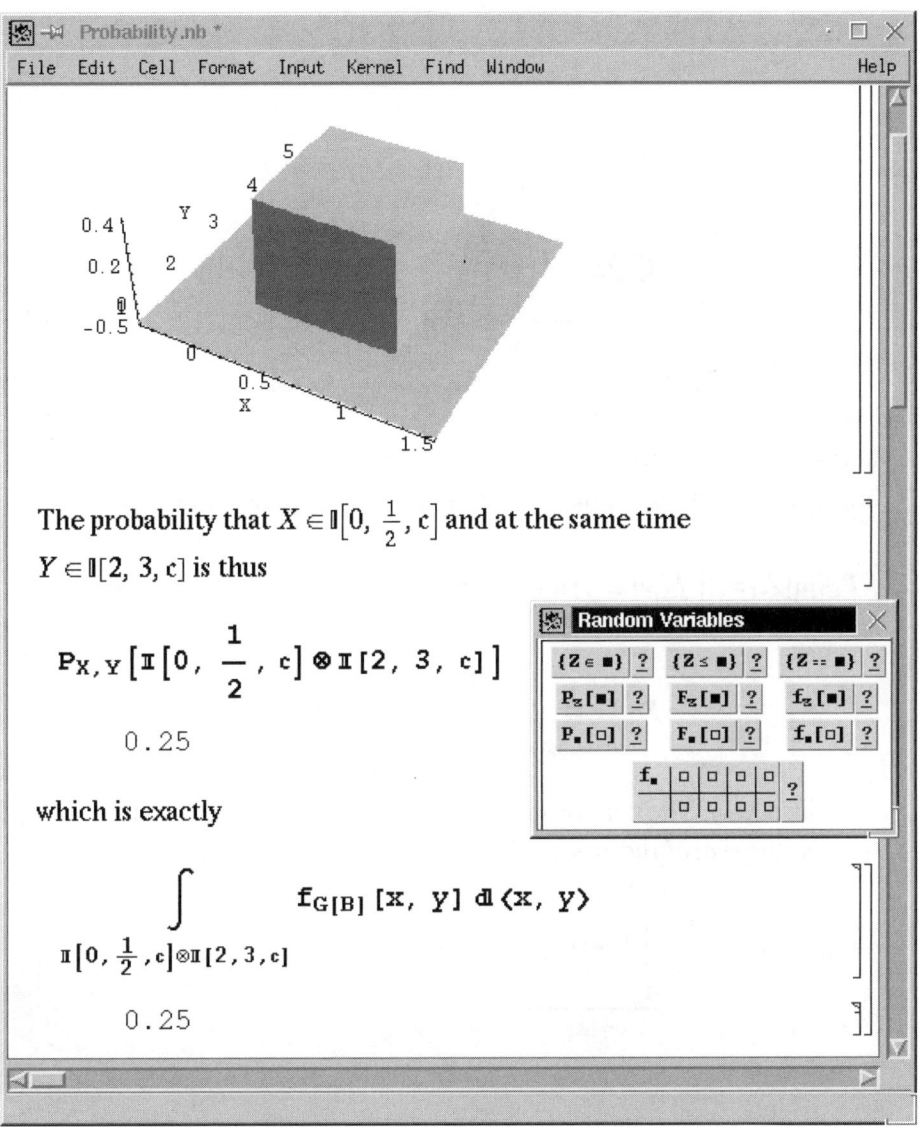

Fig. 10. Screenshot of a *Mathematica* notebook document ...

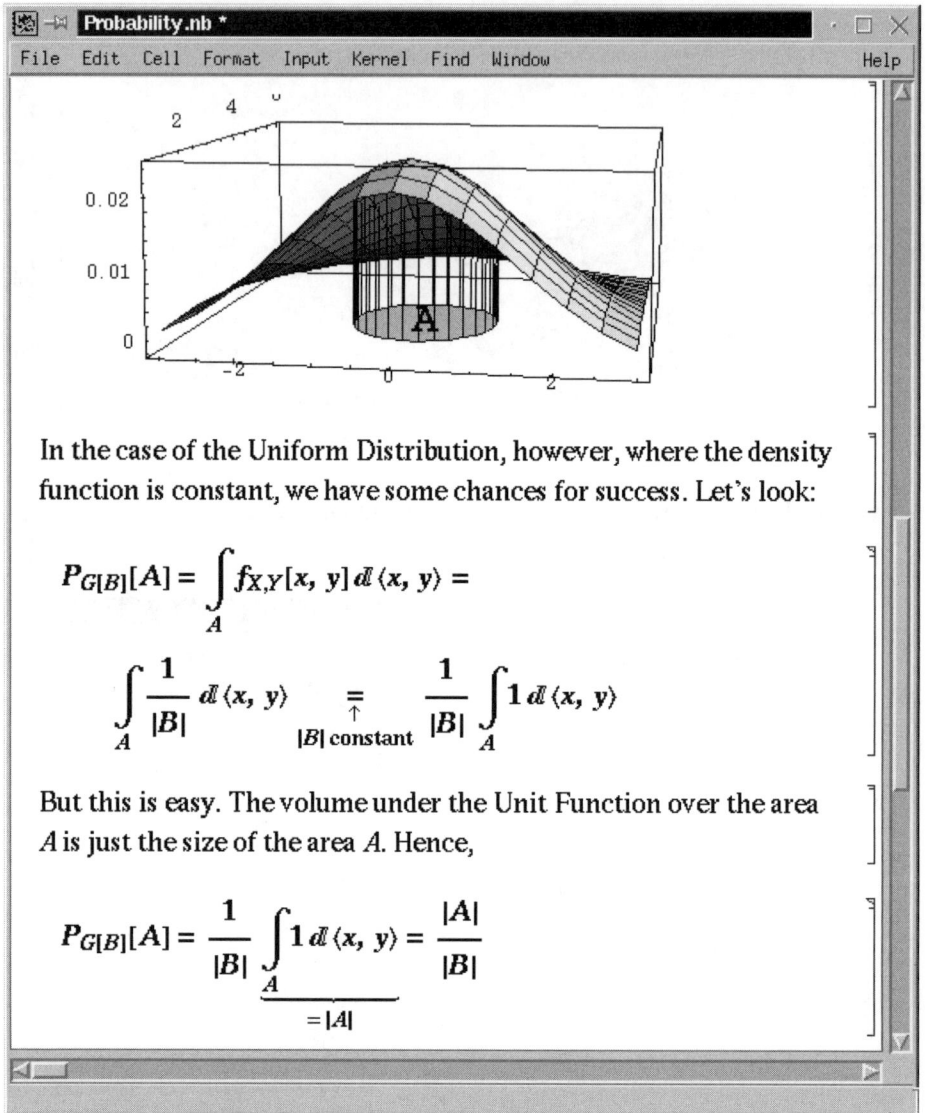

In the case of the Uniform Distribution, however, where the density function is constant, we have some chances for success. Let's look:

$$P_{G[B]}[A] = \int_A f_{X,Y}[x, y] \, d\langle x, y\rangle =$$

$$\int_A \frac{1}{|B|} \, d\langle x, y\rangle \underset{|B|\text{ constant}}{=} \frac{1}{|B|} \int_A 1 \, d\langle x, y\rangle$$

But this is easy. The volume under the Unit Function over the area $A$ is just the size of the area $A$. Hence,

$$P_{G[B]}[A] = \frac{1}{|B|} \underbrace{\int_A 1 \, d\langle x, y\rangle}_{=|A|} = \frac{|A|}{|B|}$$

**Fig. 11.** Continued: Screenshot of a *Mathematica* notebook document

### 4.1.8 MuPAD

**The Open Computer Algebra System**

*MuPAD* is a general purpose computer algebra system, which has been developed since 1989 mainly at the University of Paderborn, Germany. It features both symbolic and numerical algorithms for computing with mathematical objects such as numbers, polynomials, matrices, or functions. The system can be used as a problem solving environment in mathematics, engineering, natural sciences, and economy, but also as a tutorial system in science and education. It comprises a high-level programming language, a collection of library packages containing the mathematical knowledge, a window-based graphical user interface, an online hypertext help system, and a graphical component for rendering two- and three-dimensional mathematical data. Besides these typical aspects of a general purpose system, *MuPAD* offers the following features:

- a concept for *object oriented programming* and dynamical generation of new data types (*domains*),
- the possibility to link object code written in other programming languages, such as C++ or FORTRAN, to the system dynamically at run time (*dynamic modules*),
- a *source level debugger* for debugging code written in *MuPAD*'s programming language, with a graphical user interface, and
- the possibility to assume certain mathematical *properties* about identifiers, such as being prime or positive.

*MuPAD* is an open computer algebra system. Its programming language and the concept of dynamic modules provide a comfortable way of extending the system's functionality and linking *MuPAD* with other software packages. *MuPAD* is used as a back end in scientific and tutorial componentware. A version of *MuPAD* that can be linked to other programs as a C library is in preparation.

*MuPAD* is available for the following platforms: MS Windows, MacOS, Solaris, SGI, and Linux.

**4.1.8.1 System Architecture.** The main components of *MuPAD* are its *kernel*, the *libraries*, and the *user interfaces*, as illustrated in Figure 12.

The kernel provides the basic functionality of the system and is written in C and C++ for efficiency reasons. It contains the *parser* and the *evaluator*

322    Chapter 4  Computer Algebra Systems

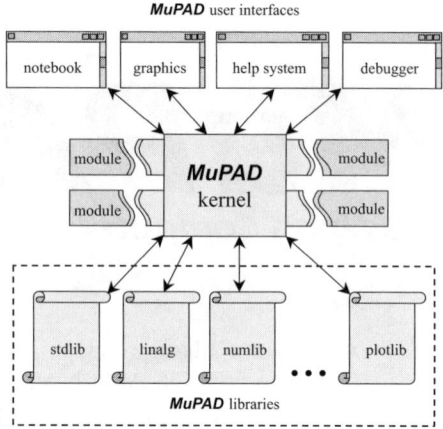

**Fig. 12.** *MuPAD*'s main components.

for *MuPAD*'s programming language, a *multiprecision arithmetic* package for integers and floating point numbers*, the *memory management*, and a collection of *built-in functions* for basic tasks.

**4.1.8.2   Libraries.** Most of *MuPAD*'s mathematical knowledge resides in the libraries. Each library is a collection of procedures written in *MuPAD*'s programming language. Users can access the source code of a library procedure interactively, and also from within the debugger. Besides the *standard library*, which contains the most commonly used routines, there are library packages for code generation, combinatorics, differential equations, factorization and gcd computation, functional programming, Gröbner bases, input and output, integral transforms, linear algebra, linear programming, mathematical properties of expressions, number theory, numerical algorithms, plotting mathematical data, polynomials, recurrence equations, series expansions, solving equations and inequalities, special functions, statistics, string manipulation, symbolic integration, type checking, and for defining abstract mathematical data types.

**4.1.8.3   Graphical User Interface.** Users interact with the system through a window-based graphical user interface. There is a separate window for each component: the main window containing a *MuPAD* session, the help window, the graphics window, and the debugger window.

On Windows platforms, the main window is organized in the form of a *notebook*, with full OLE support. A notebook records the user's inputs and the system's responses, but it is also possible to integrate text (comments) or embedded graphics. The user can edit all input fields and have them executed again. A notebook may be saved for later use, which is convenient for creating

---

* *MuPAD* uses parts of the PARI package, see Section 4.2.30.

presentations or tutorials. Figure 13 shows a screen shot. The main window also

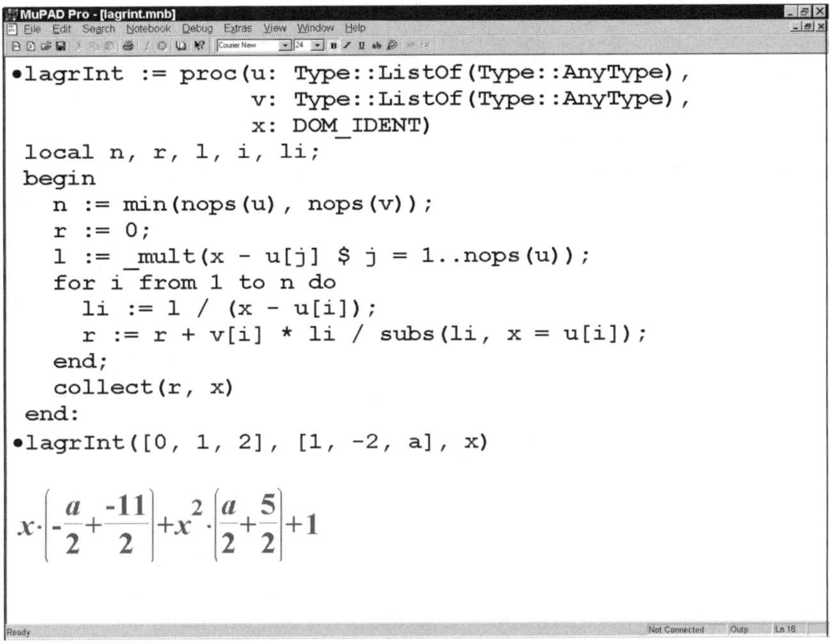
MuPAD Pro - [lagrint.mnb]
File Edit Search Notebook Debug Extras View Window Help
•lagrInt := proc(u: Type::ListOf(Type::AnyType),
v: Type::ListOf(Type::AnyType),
x: DOM_IDENT)
local n, r, l, i, li;
begin
n := min(nops(u), nops(v));
r := 0;
l := _mult(x - u[j] $ j = 1..nops(u));
for i from 1 to n do
li := l / (x - u[i]);
r := r + v[i] * li / subs(li, x = u[i]);
end;
collect(r, x)
end:
•lagrInt([0, 1, 2], [1, -2, a], x)
x·(-a/2 + -11/2) + x^2·(a/2 + 5/2) + 1
Ready
Not Connected
Outp
Ln 16

**Fig. 13.** A screen shot of a *MuPAD* notebook on a Windows platform.

features a *MuPAD* source code editor with syntax highlighting.

The *MuPAD* documentation, including some books and technical reports, is available online. Help for a specific *MuPAD* command can be obtained by entering `?command`. This opens the help window displaying the help page for the corresponding command. The help system has hypertext functionality: the user can click with the mouse on phrases that are highlighted and the system follows the corresponding link. The help pages contain examples illustrating the use of the system functions. By a mouse click, such an example can be copied to the main window for execution by *MuPAD*. The help system and many other online documents about *MuPAD* are generated by the LaTeX typesetting system with a special *MuPAD* style file. Users can employ this to write their own online documents; an example is the interactive book [Schwarz 1998].

The graphics window is used for displaying two- or three-dimensional plots of mathematical objects such as curves or surfaces. It is opened automatically by the corresponding plotting commands. The user can adjust the layout and the perspective of the graphics from within the graphics window or save the graphics in various formats.

The *MuPAD* source level debugger is opened from within the main window. After the user enters a *MuPAD* command to be debugged, the debugger window opens, showing the relevant part of the current command's source code. Many actions that are familiar from C-code debuggers are available: single stepping, inspecting variables, setting (conditional) breakpoints etc. Profiling of functions is possible from within *MuPAD*'s main window.

**4.1.8.4  Data Types.** Besides the standard data types, such as integers, rational numbers, floating point numbers, complex numbers, Boolean values, strings, and identifiers, *MuPAD* provides container types, such as sequences, lists, sets, tables, or arrays, and more complex mathematical types, such as (multivariate) polynomials, series expansions, functions, matrices, differential operators, rings, and fields. *Identifiers* can act both as unknowns in the mathematical sense and as variables that can be assigned values in the programming sense. A variety of elementary *operators*, such as +,-,*,/,<,>,=,div,mod,union,intersect, and system functions are available for handling all kinds of data types. *Expressions*, such as sin(PI+x)+y^3, are more complex symbolic objects composed of the basic data types by means of operators.

**4.1.8.5  Programming Language.** *MuPAD* features a high-level functional programming language with a C- and Pascal-like syntax and a concept for object oriented programming. Basic constructs are the branching statements if and case, for, while, and repeat loops, and the proc command for user-defined procedures.

Figure 13 shows a sample procedure. It takes as input two lists $u = (u_i)$, $v = (v_i)$, and an indeterminate $x$, and computes the interpolating polynomial $r$ in the indeterminate $x$ satisfying $r(u_i) = v_i$ for all $i$, using Lagrange's formula.

**4.1.8.6  Domains and Categories.** *MuPAD* provides a concept for object oriented programming that is tailored to mathematical objects such as polynomials, matrices, rings, or fields: the domains concept. A *domain* is an abstract data type (see also Section 2.16.2), comparable to a class in C++, that represents a certain structured mathematical set, such as the ring $\mathbb{Z}$ of integers, the field $\mathbb{F}_7$ of integers modulo 7, or the ring $\mathbb{C}[x,y]^{2\times 2}$ of all $2 \times 2$ matrices whose entries are polynomials in the indeterminates $x, y$ with complex coefficients. A domain usually has several *slots* attached to it. These are *MuPAD* procedures for handling objects belonging to that domain, e.g., for arithmetic, normalization, element generation, or screen output. Elements of a domain are created by calling domainname(..). In Figure 14, the domain R representing $\mathbb{C}[x,y]^{2\times 2}$ is created and two elements A and B of that domain are generated and multiplied. The domain R has—among others—the slots R::row and R::col for row and column extraction, respectively, and the slots R::_plus and R::_mult for arithmetic. The latter slot *overloads* the system function _mult, which is equivalent to the multiplication operator *. Thus the familiar mathematical notation may be used for arithmetic in R, as in Figure 14.

- **export(Dom):**
  ```
  R := SquareMatrix(2, DistributedPolynomial([x, y], Complex)):
  A := R([[1/2, x], [PI, x*y]]):
  B := R([[3.0 - 2*x^2, 0], [y^2, 1 + I*y^2]]):
  A*B
  ```

$$\begin{pmatrix} -x^2 + x \cdot y^2 + 1.5 & i \cdot x \cdot y^2 + x \\ -2 \cdot \pi \cdot x^2 + x \cdot y^3 + 3.0 \cdot \pi & i \cdot x \cdot y^3 + x \cdot y \end{pmatrix}$$

Fig. 14. An example with domains.

Domains may be *derived* from other domains and either *inherit* or *overload* their slots. A domain may also belong to several *categories* and inherits their slots. A category represents a more abstract mathematical notion, such as a group, a ring, or a field. This provides a concept for generic algorithms. For example, the parametrized domain SquareMatrix is derived from the domain Matrix, which in turn belongs to the category Matrix. In fact, the domain R inherits the slots row, col, _plus, and _mult from this category. Domains and categories may be associated with *axioms* asserting certain properties, such as the existence of a normal form or a unit element. Slot procedures may query such axioms and react accordingly.

The domains package of the *MuPAD* distribution contains a variety of predefined categories for sets, groups, rings, fields, modules, vector spaces, algebras, differential rings, polynomials, and matrices. It also features domains for the most commonly used mathematical structures, such as residue class rings, algebraic extensions, ideals, finite fields, quaternions, or tensor algebras. In addition, users can define their own domains and categories by means of the special keywords category and domain.

**4.1.8.7 Properties of Identifiers and Expressions.** The user can specify via the system function assume that a *MuPAD* identifier has certain algebraic properties. These properties are taken into account by system functions, such as the equation solver, the symbolic integrator, or the simplifier. For example, the system automatically simplifies $(ab)^n$ to $a^n b^n$ when $n$ is assumed to be an integer. The following basic properties are available: membership in

- a simple number domain, such as the real numbers, the rational numbers, the positive real numbers, the (positive) prime numbers, or the odd integers,
- an interval of real numbers, rational numbers, integers, etc., with arbitrary *MuPAD* expressions as bounds,
- a residue class of integers.

Boolean combinations of these basic properties are also possible. For a given Boolean expression containing identifiers with properties, the function is tries to deduce from the identifiers' properties one of the truth values TRUE or FALSE for the Boolean expression. If this is not possible, then is returns UNKNOWN. The function getprop returns the properties of a *MuPAD* expression. *MuPAD* also offers features to add new user-defined properties and deduction rules. In Figure 15, the limit of the function $x \longmapsto x^n$ for $x \to \infty$ is considered. The system can only determine this limit with additional information about $n$.

- limit(x^n, x = infinity)
  Warning: can not determine sign of n [limit]

  $$\lim_{x \to \infty} x^n$$

- assume(n < 0)

  $< 0$

- limit(x^n, x = infinity)

  $0$

- is(n^2, Type::Rational)

  **UNKNOWN**

- is(n^2 > 0)

  **TRUE**

- getprop(n^2 + 1)

  $> 1$

Fig. 15. Properties in *MuPAD*.

**4.1.8.8 Dynamic Modules.** *MuPAD* provides an efficient and flexible concept for plugging C/C++ code or other compiled machine code into *MuPAD* at runtime: dynamic modules [Sorgatz 1998]. After a module is loaded, it can be used like any *MuPAD* library package. A module consists of module functions implemented in C/C++, using the *MuPAD* Application Programming Interface (MAPI). This interface provides routines for creating, manipulating, and converting *MuPAD* data, accessing the *MuPAD* memory management and the multi-precision arithmetic, etc. With the *MuPAD* module generator, the module source code is compiled to a dynamically linked library. This binary file can be loaded into *MuPAD* at runtime and can be unloaded by the user or by automatic dis- and replacing strategies of the module manager. The fact that modules are represented as domains enables the user to implement module functions and new data types using all features provided by *MuPAD*'s concept of domains and categories. Examples for well known software components that have been linked to *MuPAD* via dynamic modules are the *NAG Library* by the Numerical Algorithms Group*, the GNU library for multiprecision arithmetic *GMP*\*\*, Victor Shoup's library for number theory *NTL*\*\*\*, the Parallel Virtual Machine *PVM*[†], Jean-Charles Faugère's Gröbner basis package *GB*[‡], and Fabrice Rouillier's *RealSolving* for computing real roots of univariate polynomials[§]. Users can create dynamic modules from other software packages by writing the corresponding module interface functions in C/C++ using MAPI.

**4.1.8.9 *MuPAD* in Education.** At the University of Paderborn, *MuPAD* has been used for several years not only in introductory courses for students of physics, engineering, computer science, and economy, but also in advanced courses for mathematics students, such as numerical analysis and computer algebra. The *MuPAD*-based interactive book [Schwarz 1998] on elementary number theory is a spin-off of one of these courses. The electronic version is organized as a *MuPAD* help document, with hypertext links and examples that can be executed in a *MuPAD* window by a mouse click. Another interactive book about cryptography is used in undergraduate courses of the Fernuniversität Hagen.

The *MuPAD* distribution comprises some libraries that are specially designed for use in school. Currently available are three packages for analytical geometry, plane geometry, and calculus. The former two libraries provide a variety of basic geometric data types, such as vectors, lines, planes, or circles, and functions for manipulating and plotting them. Example notebooks illustrating how to use these libraries are included as well.

*MuPAD* is used as a back end in mathematical tutoring software packages; this is described in the following paragraph. See also Section 3.6.

**4.1.8.10  Projects and Cooperations.**

- *Multimedia tutoring software and componentware.* *MuPAD* is used as a back end in *Mathlantis Algebra 1*, an algebra tutoring package disguised as a computer adventure game, by Cornelsen Software[¶]. In the math parts of the software, a pupil can learn and practice the handling of terms and their simplifications and of linear equations and inequalities. The system checks

---

\* http://www.nag.co.uk
\*\* http://www.gnu.org/software/gmp/gmp.html
\*\*\* http://www.shoup.net/ntl
[†] http://www.epm.ornl.gov/pvm/pvm_home.html
[‡] http://posso.lip6.fr/~jcf/GB.html
[§] http://www.loria.fr/~rouillie
[¶] http://www.cornelsen.de

the intermediate steps of the pupil's solution and gives assistance if desired. The pupil has to solve a certain number of mathematical tasks to complete the game. Pupils can also enter their own exercises, e.g., from school assignments, and solve them with the assistance of the system.

In a project funded by the BMBF*, graphical user interfaces are developed that make the power of a computer algebra system such as *MuPAD* and the possibility to render and to manipulate three-dimensional mathematical objects accessible from within standard publishing or presentation software, such as, for example, web browsers [Hillebrand 1999]. Teachers or publishers can use these components to create electronic tutorial documents with mathematical contents for interactive learning or for presentation. The components are based on the ActiveX technology for Windows platforms. In cooperation with the Institut für Mediendidaktik at the University of Koblenz, these tools are used to develop tutorial documents for linear algebra in schools and universities.

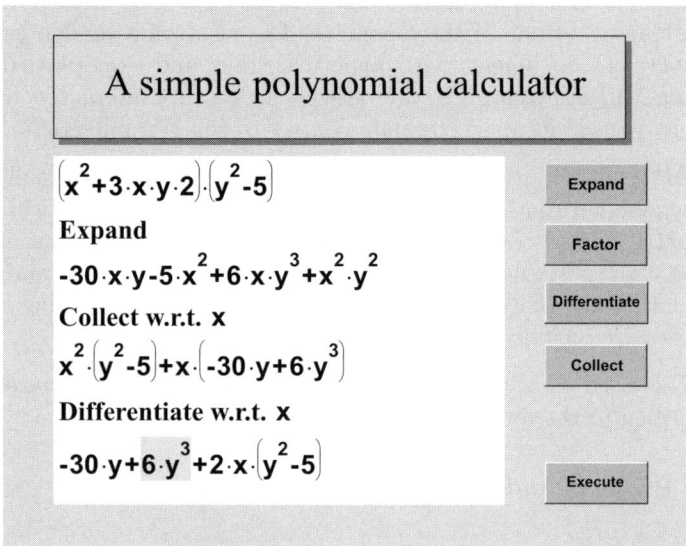

**Fig. 16.** The Calculator Control embedded in a Microsoft PowerPoint presentation.

One of the components, the *Calculator Control* (Figure 16), is essentially a simplified graphical interface to *MuPAD*. For example, the author of a tutorial about polynomials can create a "polynomial calculator" with buttons for expanding, factoring, and differentiating polynomials, and embed this component into a HTML document or any other ActiveX compliant software.

---

* Bundesministerium für Bildung und Forschung (German Department for Education and Research)

The user viewing this document can interact with the calculator by entering an arbitrary polynomial. After pressing a button, the Calculator Control sends the input to the *MuPAD* kernel, which executes the corresponding command, and then renders the result in the browser window. The user can manipulate this result again. By providing appropriate buttons, the author of a document decides which parts of *MuPAD*'s functionality are accessible; the user never gets in direct contact with the computer algebra system. The development of such documents is supported by *MuPAD* libraries providing special commands suitable for use in school.

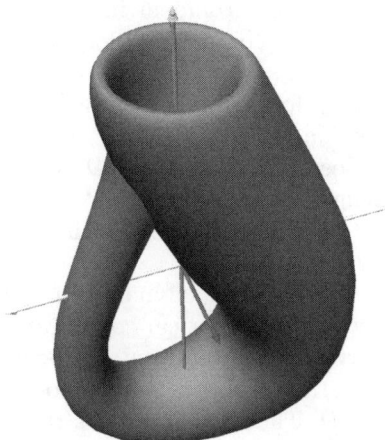

**Fig. 17.** A Klein bottle rendered by the Graph Control.

The *Graph Control* is a pixel-oriented graphics component for rendering and manipulating three-dimensional mathematical data, based on OpenGL. Like the Calculator Control, it can be embedded in any ActiveX compliant software. The Graph Control provides graphical primitives for points, vectors, planes, spheres, cones, function plots etc., and includes methods for rotating, zooming, or animating the graphics online. Figure 17 shows a sample plot. It is possible to pass a mathematical expression from the Calculator Control to the Graph Control for plotting. The Graph Control will be available as an add-on to *MuPAD* in the future.

- *MuPAD as a server.* In a project for the German "Schulen ans Netz e.V.", a *MuPAD* computing server for use in internet and intranet applications, in particular in schools, is under development. The idea is that remote clients, such as plugins for web browsers or Java applets, can connect to the server via sockets, send requests like "please compute an indefinite integral of $x^2 \sin(x)$ with respect to $x$", receive the answer "$2x \sin(x) - \cos(x)(x^2 - 2)$" from the server, and render it on the screen. Various communication protocols, such as

*OpenMath* or *MathML*, will be supported in the future. Prototypical clients are implemented as well.
- *Macro parallelism.* Based on *PVM*, a parallel *MuPAD* version for distributed computing in heterogeneous networks has been developed at the University of Paderborn\*. Following the master-worker paradigm, it features local and remote work queues, work groups, shared variables, and semaphores for synchronizing tasks in the network. The prototype is implemented as a dynamic module [Metzner et al. 1999].
Besides the concept of dynamic modules for flexible and efficient software integration and the domains concept providing object oriented programming features, macro parallelism is intended as the third pillar for using *MuPAD* as an open problem solving environment in mathematical applications and distributed high-performance computing [Sorgatz and Wehmeier 1999]. See also Section 2.18.
- *SINGULAR.* This package for commutative algebra, algebraic geometry, and singularity theory (Section 4.2.39), which is developed at the University of Kaiserslautern\*\*, has been linked to *MuPAD* via a dynamic module [Bachmann et al. 1999], using the protocol *MP*\*\*\* for exchanging mathematical data. As side effect of this project, a lot of improvements and extensions have been added to *MP*.
- *Scilab.* In cooperation with INRIA, a dynamic module is developed that provides the functionality of the numerical computation package *Scilab*[†] from within *MuPAD*. This is useful for efficient symbolic-numeric computations. For example, an engineering problem is first modeled in terms of symbolic equations, using *MuPAD*. Then the user calls the appropriate *Scilab* functions from within *MuPAD* to compute numerical solutions quickly. This is particularly useful for simulations or for generating graphical output, where large amounts of numerical data are needed and it would be too time-consuming to perform the numerical computations in *MuPAD* itself. The communication between *MuPAD* and *Scilab* is based on the *UDX*[‡] protocol, which has been developed by Jean-Charles Faugére and Fabrice Rouillier at INRIA.
- *Parallel robotics.* Another project together with INRIA is the development of software for modeling, analysis, and simulation of parallel robots (see also Section 3.4.4). *MuPAD* shall be used in the symbolic modeling stage, for generating systems of polynomial equations, and as a control environment. As a first step, the two packages *GB* for Gröbner basis computation and *RealSolving* for computing real roots of polynomials have been linked to *MuPAD* via dynamic modules. In the intended application, huge systems of

---
\* This work was supported by the Sonderforschungsbereich 376 *Massive Parallelität: Algorithmen, Entwurfsmethoden, Anwendungen* of the Deutsche Forschungsgemeinschaft.
\*\* http://www.singular.uni-kl.de
\*\*\* http://snake.mcs.kent.edu/areas/protocols/mp.html
† http://www-rocq.inria.fr/scilab
‡ Universal Data eXchange

polynomial equations comprising several megabytes of data are exchanged in heterogeneous networks. In order to handle these large amounts of data, the communication between *MuPAD* and *GB/RealSolving* uses the very efficient *UDX* protocol.
- *MuPAD in CAE.* MuPAD is used as a back end in the software package *CAMeL-View* for computer-aided engineering by iXtronics*. The package is a tool that supports modeling, analysis, synthesis, and realization of mechatronic systems, such as cars or robots, via "virtual prototyping". *MuPAD* is used in the modeling stage to generate systems of ordinary differential equations, which are then transformed into optimized C code for efficient numerical solving.
- *Symbolic solutions of ordinary and partial differential equations.* At the time of writing, *MuPAD* provides two library packages for computing exact solutions of ordinary differential equations (ODE) and for arithmetic of linear ordinary differential operators (see Section 2.11 for the theory). *MuPAD* has implementations of the most common methods for ODE's. Some new algorithms for Liouvillian solutions of second and third order linear ODE's are currently implemented [Fakler 1999]. A link is in preparation that makes the functionality of BERNINA**, an interface to Manuel Bronstein's $\Sigma^{IT}$ package for symbolic manipulation of differential and difference equations [Bronstein 1996], available from within *MuPAD*, using dynamic modules. In cooperation with Werner M. Seiler from the University of Mannheim, a library for symbolic solutions of partial differential equations is currently developed.

**4.1.8.11 More About** *MuPAD*. The *MuPAD* Tutorial [Oevel et al. 2000] gives an introduction to using *MuPAD*; for a quick reference see [Oevel 1999]. Demonstrations are available at http://www.sciface.com/support/papers/demo.shtml. Technical documentation is published in the *MuPAD Reports* series [Fakler 1999; Kluge 1996; Naundorf 1996, 1997; Sorgatz 1996] and at the web address http://www.sciface.com/support/papers/index.shtml. The mathPAD *journal* is a publication of the *MuPAD* group at the University of Paderborn. Recent mathPAD issues are available electronically at http://www.mupad.de/mathpad.shtml. Further information and documentation is available at the *MuPAD* web site:

http://www.mupad.de

For ordering information, please contact:

SciFace Software GmbH & Co. KG
Technologiepark 11
D-33100 Paderborn, Germany
WWW: http://www.sciface.com
e-mail: info@sciface.com

---

* http://www.ixtronics.de
** http://www-sop.inria.fr/cafe/Manuel.Bronstein/bernina.html

Additional references: [Drescher 1997; Fuchssteiner and the MuPAD Group 1996; Oevel 1998; Postel 1999; Postel and Zimmermann 1999].

Jürgen Gerhard and Andreas Sorgatz (Paderborn)

## 4.1.9 REDUCE

**4.1.9.1 Introduction** The first version of REDUCE was developed and published by Anthony C. Hearn about 30 years ago. The starting point were formal computations for problems in high energy physics (Feynman diagrams, cross sections etc.), which are hard and time consuming if done by hand. Although the facilities of the today REDUCE are no longer comparable with those of the early versions, the direction towards big formal computations in applied mathematics, physics and engineering has been stable over the years, however leading to a much broader spectrum of application.

Like symbolic computation in general, REDUCE has profited by the increasing power of computer architectures and by the facilitated information exchange based on recent network development. Spearheaded by A.C. Hearn now several groups in different countries take part in the REDUCE development, and contributions of users have significantly widened the field of application.

Today REDUCE can be used with a variety of hardware platforms from a Personal Computer up to a massively parallel supercomputer. However, the primary vehicle is the class of advanced UNIX workstations or powerful PC systems.

Although REDUCE is a mature program system, it is extended and updated in a contiguous process. Since the establishment of the REDUCE Network Library in 1990 users take part in the development such that the compatibility step caused by new releases is reduced. In April 1999 version 3.7 of REDUCE was released. Informations about all issues regarding REDUCE and this version in particular are available through the World Wide Web server at URL: **http://www.rrz.uni-koeln.de/REDUCE** . In addition to general information about REDUCE, this server has pointers to the REDUCE distributors home pages, the network library, the demonstration versions, examples of REDUCE programming, access to a test server, a set of manuals and the complete packages documentation, and the REDUCE online help system.

**4.1.9.2 Problem Solving** The primary scope of REDUCE is the solution of large scale formal problems in mathematics, science and engineering. REDUCE offers a number of powerful operators which often give an immediate answer to a given problem, e.g. solving a linear equation system or computing a determinant (with symbolic entries, of course). More typical however are more complicated applications where only the combination of several evaluation steps leads to the desired result. Consequently the development of REDUCE primarily is oriented towards a collection of powerful tools, which enable problem solving by combination.

In some cases even complete new algorithmic bases will be required for problem solving. REDUCE supports this by various interfaces to all levels of symbolic evaluation, and the modules of REDUCE and of the REDUCE Network Library demonstrate by example how this technique is to be used.

### 4.1.9.3 Data Types, Structures

**4.1.9.3.1 Elementary Expressions** The central object of REDUCE is the formal expression, which is built with respect to the common mathematical rules. Elementary items are

- numbers (integers, rationals, rounded fractionals, real or complex, modular and algebraic numbers); the domain can be selected dynamically,
- symbols (names with or without indexes)
- functional expressions(names followed by a parameter list)
- operator symbols $+,-,*,/,**,$
- parentheses for precedence control.

A symbol here can play the role of an unknown in the mathematical sense as well as the role of a placeholder for a value. An expression can be assigned to a symbol as a value such that later all references of the symbol are replaced by the assigned value.

Examples of elementary expressions are:

```
3.1415928       % fraction
a               % simple variable
(x+y)**2 / 2    % quadratic expression
log(u)+log(v)   % function
```

**4.1.9.3.2 Aggregates** There are data structures which collect a number of formal expressions: An equation is an object dominated by the operator = with two slots for expressions, the lhs and rhs.

```
p=u**2
```

A list is a linear sequence of expressions enclosed in curly brackets, where each member is elementary or an aggregate. There are operations for construction, join, decomposition and reordering of lists.

```
{2,3,5,7,11,13,17,19}.
```

An array is a rectangular multidimensional structure; the elements are identified by integer indices; its elements can be used as symbols. Elements always have a value which defaults to zero.

A matrix is a named structure of rows and columns, where elements are identified by two positive integers. For matrices with compatible dimensions and for matrices and scalars there are operations corresponding to the laws of linear algebra.

Example: using the derivative operator df to construct a Jacobian:

```
matrix jac(n,n);
for i:=1:n do for j:=1:n do
    jac(i,j):=df(f(i),x(j));
```

**4.1.9.4 Programming Paradigms** For specifying symbolic tasks and algorithms REDUCE offers a set of different programming paradigms.

**4.1.9.5 Algebraic Desk Calculator** Using REDUCE as a desk calculator for symbolic and numeric expressions is the simplest approach. Formulae can be entered, combined, stored and processed by a set of powerful operators like differentiation, integration, polynomial GCD, factorization etc. Any formula will be processed immediately with the objective of finding its most radical simplification, and the result will be presented on the screen as soon as it is available.

Example: Taylor polynomial for $x * sin(x)$

```
for i:=0:5 sum
   sub(x=0,df(x*sin(x),x,i)) * x**i / factorial(i);

       1   4    2
   - ---*x  + x
       6
```

**4.1.9.6 Imperative Algebraic Programming** Evaluation of a single formula with the immediate output of the result is a special case for a statement in the REDUCE programming language. It allows you to code complicated evaluation sequences like conditionals, groups, blocks, iterations (controlled by counters or list structures) and the definition of complete parameterized procedures with local variables.

Example: definition of a procedure for expanding a function to a Taylor polynomial:

```
procedure tay(u,x,n);
  begin scalar ser,fac;
    ser:=sub(x=0,u);fac:=1;
    for i:=1:n do
    << u:=df(u,x); fac:=fac*i;
       ser:=ser+sub(x=0,u)*x**i/fac >>;
    return(ser);
  end;
```

A call for this procedure: `tay(x*sin(x),x,5);` yields

```
    1   4    2
 - ---*x  + x
    6
```

Example: a recursive program for collecting a basis of Legendre polynomials from the recurrence relation:

$$P_{n+1}(x) = ((2n+1)xP_n(x) - nP_{n-1}(x))/(n+1)$$

The infix operator "." adds a new element to the head of a list.

```
procedure Legendre_basis(m,x);
  % start with basis of the order 1
  Legendre_basis_aux(m,x,1,{x,1});

procedure Legendre_basis_aux(m,x,n,ls);
     % ls contains polynomials for n, n-1, n-2 ...
  if n>=m then ls     % ready
  else Legendre_basis_aux(m,x,n+1,
  (((2n+1)*x*first ls - n*second ls)/(n+1)) . ls);
```

A call for this procedure:

`Legendre_basis(3,z);`

```
   5   3    3          3   2   1
 {---*z  - ---*z ,   ---*z - ---, z, 1}
   2        2          2        2
```

**4.1.9.7 Rule Oriented Programming** In REDUCE global algebraic relations can be formulated with rules. A rule links an algebraic search pattern to a replacement pattern, sometimes controlled by additional conditions. Rules can be activated (and deactivated) globally, or they can be invoked with a limited scope for single evaluations. So the user has an arbitrarily precise control over the algebraic simplification.

Example: Expanding trigonometric functions for combined arguments; the symbol ~ represents an implicit for all quantifier.

```
Sin_Cos_rules:={sin(~x+~y) => sin(x)*cos(y) + cos(x)*sin(y),
cos(~x+~y) => cos(x)*cos(y) - sin(x)*sin(y)};
```

Global activation is achieved by: `let Sin_Cos_rules;` and global deactivation by: `clearrules Sin_Cos_rules;`. An example for local activation:

```
sin (h + g) where Sin_Cos_rules;
```

Note: REDUCE has no predefined "knowledge" about these relations for trigonometric functions, as they can be used as production rules in both directions, in an expanding or collecting style; only the user can decide which mode is adequate for his problem.

Based on rules a complete calculus can be implemented; the rule syntax here is very close to the mathematical notation with multiple cases.

Example: Definition of Hermite polynomials:

```
operator Hermite;
Hermite_rules:= {Hermite(0,~x) => 1, Hermite(1,~x) => 2*x,
     Hermite(~n,~x) => 2*x*Hermite(n-1,x) -2*(n-1)*Hermite(n-2,x)
     when n>1};
let Hermite_rules;
```

Generation of a Hermite polynomial:

```
Hermite(4,z);

      4        2
 16*z   - 48*z   + 12
```

**4.1.9.8 Symbolic Imperative Programming** The paradigms described so far give access to the REDUCE facilities on the top level. They allow you a compact programming close to the application problem. No knowledge about the internal data structures is necessary as REDUCE converts data automatically for each evaluation step. On the other hand the frequent conversions are time consuming and for very large problems it might be desirable to keep intermediate results in the internal form in order to avoid the conversion overhead. Here the "symbolic" mode of REDUCE can be used, which gives you access to internal data structures and procedures directly with the same syntax as in top level programming. Of course, this level of programming requires some knowledge about LISP and about internal REDUCE structures. However, it enables the implementation of algorithms with the highest possible efficiency. For an introduction see 'Symbolic Mode Primer', Version 1, by Herbert Melenk.

**4.1.9.9 Algebraic Evaluation** The evaluation of expressions is the heart of REDUCE. Because of the great complexity it can be touched here only briefly.

One central problem in automatic formula manipulation is the detection of identity between objects, e.g. the confirmation $a + b = b + a$ under the assumption of a commutative addition.

It is well known, that this problem is equivalent to the problem of recognizing that an expression is zero, in other words to the existence of an algorithm for the transformation of a formula into an equivalent canonical normal form. Unfortunately there is no universal canonical form; only for subcases, for example polynomials, rationals or ideals are canonical forms known. Therefore, REDUCE evaluation is based on a canonical form for rational functions (= quotients of multivariate polynomials), where symbols or function expressions play the role of variables (REDUCE: kernels). REDUCE attempts to transform as many functions as possible into the canonical form by applying additional heuristic rules. A coarse sketch of evaluation:

- a symbol with an assigned value is replaced by that value,
- a call for a known procedure is replaced by the value produced by the procedure invocation,
- rules are applied as long as one of them matches,
- polynomials are expanded recursively due to a lexicographic order of variables(kernels): a multivariate polynomial is a polynomial in its highest variable with decreasing exponents, where the coefficients are polynomials in the remaining variables,
- a rational function is converted to a form with common denominator (= quotient of 2 polynomials); common factors of numerator and dominator are cancelled.

Of course, this is a highly recursive process, which is applied until no more transformations are possible.

**4.1.9.10 Approximations** In the context of symbolic computation mostly exact arithmetic is used, especially with algorithms from the classical computer algebra. That aspect is supported by REDUCE with arbitrarily long integer arithmetic and, built on top of that, rational numbers, modular numbers (p-adic) etc.

The values of transcendental functions with general numeric arguments do not fall into these domains, even if symbols like $\pi, e, i$ are attached. Nevertheless, symbolic computation can be used for fields beyond classical algebra, for example in the context of analytic approximations in numerical mathematics.

**4.1.9.11 Power Series** Power series are a valuable tool for the formal approximation of functions, e.g. in the context of differential equations. REDUCE supports several types of power series, among them univariate Taylor series with variable order and multivariate Taylor series with fixed order.

**4.1.9.12 Rounded Numbers** For several decades floating point numbers have been established as useful tool for numerical computations, although they miss most of the algebraic properties of number domains. In REDUCE they are incorporated as "rounded numbers" and if compared to classical floating point numbers (e.g. in the IEEE view) they offer interesting additional properties:

- the mantissa length can be selected arbitrarily (specified as number of decimal digits),
- there is no limit for the exponent and so no upper or lower limit for the magnitude of a number.

Technically, this arithmetic has been implemented by an embedding of the standard (hardware) floating point operations in a software package, which tries to execute as much as possible in (fast) hardware, stepping to software emulation as soon as the hardware limits are reached. Based on this number domain attractive algorithms can be implemented, which start with coarse approximations and refine the overall precision more and more in an adaptive style when approaching the desired solution.

**4.1.9.13 Interface for Numerical Programs** A field of growing importance for the symbolic computation is the use of the algorithms of mixed symbolic-numeric type, when e.g. a symbolic part evaluates formal properties of an equation system for control or conditioning of a numerical solver. Examples are e.g. the automatic programming of Jacobians for ODE solvers or the reduction of the order of a system by exploiting formal symmetries. For the cooperation of symbolic and numeric components REDUCE offers several facilities for the generation of partial or complete programs in languages like FORTRAN or C. As automatically generated programs tend to flood the target compilers, REDUCE is able to optimize the generated numeric code.

**4.1.9.14 Graphical User Interfaces** REDUCE uses standard GUI libraries like WINDOWS or X11 to display formulae in a decent form on the user device. An interface to the GNUPLOT system has been developed which uses this system as a graphical postprocessor. The evaluation of the pointset for a plot is done by the algebraic processor, though. The GNUPLOT interface is common to all distributed versions of REDUCE whereas some additional interfaces to other graphical postprocessors are available for certain versions.

**4.1.9.15 Interfaces to the Operating System** REDUCE has got interfaces to its hosting operating systems which allow the user to issue system commands like navigation in the file tree, start of command shells or pipes, reading environment variables, etc. This enables the system to connect to various (mostly special purpose) processors or libraries. In this way one is able to enlarge the class of solvable problems.

**4.1.9.16 Input/Output** Normally in interactive mode REDUCE prints results in a two dimensional "mathematical" form, where exponents are raised and quotients are printed with denominator below numerator and matrices are represented as rectangular blocks. The output can be influenced by a variety of switches, e.g. for reordering or collecting of terms.

For special purposes additional output forms are available:

- **linear form**: the data can be re-used for later input in REDUCE or another system,
- **foreign syntax**: the expressions are printed in syntax of FORTRAN, C or another programming language for the direct insertion in numeric codes,
- **TeX**: indirect formatting as input for the TeX layout program to be inserted into a publication.

Examples for $q := (x+y)^3$;

natural (default) output:

```
          3        2         2       3
q  :=  x    + 3*x  *y  +  3*x*y   +  y
```

for later re-use:

```
q := x**3 + 3*x**2*y + 3*x*y**2 + y**3
```

as contribution to a FORTRAN source:

```
q=x**3+3.*x**2*y+3.*x*y**2+y**3
```

```
\begin{displaymath}
q=x^{3}+3 x^{2} y+3 x
y^{2}+y^{3}
\end{displaymath}
```

Additionally to direct terminal access I/O can be redirected to or from files.

**4.1.9.17 Open System** In contrast to most other symbolic math systems REDUCE traditionally is completely open:

- REDUCE is written in the language **RLISP**, which incorporates the functionality of LISP in a user friendly syntax. At the same time RLISP is the language of application.

- From the first version on REDUCE is delivered with all sources. So the algorithmic basis is visible for any user. Even the REDUCE translator (compiling RLISP to LISP) is delivered as source code.
- Any internal REDUCE function and data structure can be accessed by the user directly (in "symbolic" style programming). Most of the REDUCE implementations contain a LISP compiler, such that the user can produce very efficient modules. REDUCE can be integrated into other (LISP-) packages as algebraic engine.
- REDUCE inherits automatically from LISP the ability of dynamic loading of modules, of incremental compilation and dynamic function redefinition. Even the kernel of REDUCE is open for local modification. A remarkable effect is that user code can achieve the same high speed operation as core routines, because the user code undergoes the same procedures of compilation.

One effect of the liberality of REDUCE is the growing number of application packages written by users and available to the whole community through the REDUCE Network Library.

**4.1.9.18 REDUCE Packages** (State: autumn 1999)

This list contains all packages which do not belong to the system core like the RLISP to LISP translator and basic algebraic support modules, even though these are organized as packages as well.

The complete documentation for these packages can be found on the Web, e.g. at *http://www.zib.de/Optimization/Software/Reduce/moredocs* .

- **ALGINT** integration for functions involving roots (James H. Davenport)
- **APPLYSYM** Applying infinitesimal symmetries of differential equations (Thomas Wolf)
- **ARNUM** algebraic real numbers (Eberhard Schrüfer)
- **ASSIST** general utilities (Hubert Caprasse)
- **AVECTOR** vector algebra (David Harper)
- **BOOLEAN** A Package for boolean algebra (Herbert Melenk)
- **CALI** commutative algebra (Hans-Gert Gräbe)
- **CAMAL** celestial mechanics (John Fitch)
- **CGB** Comprehensive Groebner Bases (Thomas Sturm, Andreas Dolzmann)
- **CHANGEVR** transformation of variables in differential equations (G. Uecoluk)
- **COMPACT** condensing of expression with polynomial side relations (Anthony C. Hearn)
- **CONLAW** Conservation Laws (Thomas Wolf)
- **CRACK** differential equations (Andreas Brand, Thomas Wolf)
- **CVIT** Dirac gamma matrices (V.Ilyin, A.Kryukov, A.Rodionov, A.Taranov)
- **DEFINT** Definite Integration (Kerry Gaskell, Stanley L. Kameny)

- **DFPART** Derivatives of generic functions (Herbert Melenk)
- **DESIR** differential equations and singularities (C. Dicrescenzo, F. Richard-Jung, E. Tournier)
- **DUMMY** Canonical form of expressions with dummy variables (Alain Dresse)
- **EDS** Exterior Differential Systems (David Hartley)
- **EXCALC** calculus for differential geometry (Eberhard Schrüfer)
- **FIDE** code generation for finite difference schemes (Richard Liska)
- **FPS** Automatic calculation of formal power series (Wolfram Koepf)
- **GENTRAN** code generation in FORTRAN, RATFOR, C (Barbara Gates)
- **GEOMETRY** mechanized (plane) geometry manipulations (Hans-Gert Gräbe)
- **GNUPLOT** Display of functions and surfaces (Herbert Melenk)
- **GROEBNER** computation in multivariate polynomial ideals (Herbert Melenk, H.Michael Möller, Winfried Neun)
- **GUARDIAN** (to be released) Guarded Expressions (Thomas Sturm, Andreas Dolzmann)
- **HEPHYS** high energy physics (Anthony C. Hearn)
- **IDEALS** polynomial ideal algebra (Herbert Melenk)
- **INEQ** Support for solving inequalities (Herbert Melenk)
- **INT** indefinite integration (A. C. Norman, P. M. A. Moore)
- **INVBASE** A package for computing involutive bases (A. Yu. Zharkov, Yu. A. Blinkov)
- **LAPLACE** Laplace and inverse Laplace transform (C. Kazasov et al.)
- **LIE** classification of Lie algebras (Carsten and Franziska Schoebel)
- **LIEPDE** symmetries of differential equations (Thomas Wolf)
- **LIMITS** A package for finding limits (Stanley L. Kameny)
- **LINALG** Linear algebra package (Matt Rebbeck)
- **MATRIX** basic Matrix operations (Anthony C. Hearn)
- **MATHML** MathML interface (Luis Alvarez Sobreviela)
- **MODSR** Modular solve and roots (Herbert Melenk)
- **MRVLIMIT** Exp-log Limit Package (Neil Langmead)
- **NCPOLY** Non-commutative polynomial ideals (Herbert Melenk)
- **NORMFORM** Computation of matrix normal forms (Matt Rebbeck)
- **NUMERIC** Solving numerical problems (Herbert Melenk)
- **ODESOLVE** ordinary differential equations (Malcolm MacCallum et al.)
- **ORTHOVEC** calculus for scalar and vector quantities (J.W. Eastwood)
- **PHYSOP** additional support for non commuting quantities (Mathias Warns)
- **PM** A REDUCE pattern matcher (Kevin McIsaac)
- **QSUM** Q-hypergeometric Summation (Wolfram Koepf, Harald Böing)
- **RANDPOLY** A random polynomial generator (Francis J. Wright)
- **RATAPRX** Rational Approximations Package (Lisa Temme, Wolfram Koepf)
- **RATINT** Rational Function Integration (Neil Langmead)
- **REACTEQN** manipulation of chemical reaction systems (Herbert Melenk)
- **REDLOG** Computer Logic Package (Thomas Sturm, Andreas Dolzmann)
- **RESET** Code to reset REDUCE to its initial state (John P. Fitch)
- **RESIDUE** A residue package ( Wolfram Koepf)

- **RLFI** REDUCE LaTeX formula interface (Richard Liska, Ladislav Drska)
- **ROOTS** roots of polynomials (Stanley L. Kameny)
- **RSOLVE** Rational/integer polynomial solvers (Francis J. Wright)
- **SCOPE** optimization of numerical programs (J. A. van Hulzen)
- **SETS** A basic set theory package (Francis J. Wright)
- **SINGULAR** Finding Singularities (David Waugh)
- **SPARSE** Sparse Matrices (Stephen Scowcroft)
- **SPDE** symmetry analysis for partial differential equations (Fritz Schwarz)
- **SPECFN** special functions (Chris Cannam, Winfried Neun, et. al.)
- **SPECFN2** Package for special special functions (Victor S. Adamchik)
- **SUM** sum and product of series (Fuji Kako)
- **SUSY2** Supersymmetric functions and operators (Ziemowit Popowicz)
- **SYMMETRY** bases and forms for symmetric matrices (Karin Gatermann)
- **TAYLOR** multivariate Taylor series (Rainer Schöpf)
- **TPS** univariate Taylor series with indefinite order (Alan Barnes, Julian Padget)
- **TRI** TEX REDUCE interface (Werner Antweiler, Andreas Strotmann, Volker Winkelmann)
- **TRIGINT** Weierstrass substitution (Neil Langmead)
- **TRIGSIMP** Simplification and factorization of trigonometric and hyperbolic functions (Wolfram Koepf, Andreas Bernig, Herbert Melenk)
- **WU** Wu algorithm for polynomial systems (Russel Bradford)
- **XCOLOR** Calculation of the color factor in non-abelian gauge field theories (Alexander Kryukov)
- **XIDEAL** Groebner Bases for exterior algebra (David Hartley)
- **ZEILBERG** A package for indefinite and definite summation (Wolfram Koepf, Gregor Stölting)
- **ZTRANS** Z-transform package (Wolfram Koepf, Lisa Temme)

There are some more packages available from other sources. Please refer the the REDUCE WWW server for information on these packages.

### 4.1.9.19 Books About REDUCE

F. Brackx, D. Constales: Computer Algebra with LISP and REDUCE, Kluwer, 1991

J.H. Davenport, Y. Siret, E. Tournier: Computer Algebra, second printing, Academic Press, London, 1989

A. C. Hearn: REDUCE User's and Contributed Packages Manual, Version 3.7, Santa Monica (CA) and Codemist Ltd, 1999

F. W. Hehl, V. Winkelmann, H. Meyer: Computer-Algebra, ein Kompaktkurs ueber die Anwendung von REDUCE (in German), Springer 1992

A. G. Grozin: Using REDUCE in High Energy Physics, Cambridge University Press, 1997

M. MacCallum, Francis Wright: Algebraic Computing with REDUCE, Oxford University Press, 1991

H. Melenk: REDUCE Symbolic Mode Primer, Version 1, Konrad-Zuse-Zentrum Berlin, 1995

G. Rayna, REDUCE, Software for Algebraic Computation, Springer, New York, 1987

D. Stauffer, F.W. Hehl, N. Ito, V. Winkelmann, J.G. Zabolitzky: Computer Simulation and Computer Algebra, Springer, 1993

W.-H. Steeb. D. Lewien: Algorithms and Computation with REDUCE, BI Wissenschaftsverlag, 1992

J. Ueberberg: Einfuehrung in die Computeralgebra mit REDUCE (in German), BI Wissenschaftsverlag, 1992

REDUCE Network Library, Bibliography, available through the REDUCE WWW server, permanently updated

Herbert Melenk (Berlin)

## 4.2 Special Purpose Systems

### 4.2.1 Algebraic Combinatorics Environment (ACE)

The Algebraic Combinatorics Environment *ACE* [Veigneau 1998] is a *Maple* library which includes various packages providing combinatorial tools (about 400 functions) useful in algebraic combinatorics. These functions are mostly related to the symmetric group and handle such objects as partitions, compositions, permutations, words, Young tableaux, divided differences, symmetric functions, non-commutative symmetric functions and Schubert polynomials.

Moreover, *ACE* gives several algebras related to the symmetric group together with their actions on the ring of multivariate polynomials: not only the group algebra of the symmetric group but also its Hecke algebra and several of its specializations (to compute with divided differences for example). The analog algebras for the hyperoctahedral groups are also included as well as some functions related to groups of type B, C and D.

One of the main feature of *ACE* is to furnish a toolbox to handle multivariate polynomials [Kohnert and Veigneau 1997]: linear bases (Schubert polynomials), symmetrizing operators, action of symmetric polynomials on multivariate polynomials ($\lambda$-ring structure), multivariate interpolation (generalization of Newton's interpolation) as well as a discrete analog of differential calculus.

Apart from classical symmetric functions, one also finds several non-commutative extensions of them: non-commutative Schur functions in the free algebra, and non-commutative functions derived from the quasi-determinants of Gelfand and Retakh.

This environment is freely available for any Unix, Linux, Ms-Dos, Windows and Macintosh systems. A Cd-Rom is available on request and the system can be obtained on Internet (http://phalanstere.univ-mlv.fr/~ace). More information is provided by email at ACE@univ-mlv.fr.

Several persons have been involved in this 1994–1998 project and have written some of the libraries: Bun-Chan-Vorac Ung [Ung 1996], Renaud Eppstein and Vincent Prosper. The main contributor has been Sébastien Veigneau [Veigneau 1996]. All developments have been supervised by Alain Lascoux and Jean-Yves Thibon, who have provided the necessary mathematical knowledge.

The development of *ACE* has been motivated by problems in algebraic combinatorics and algebraic geometry. It uses many new mathematical results published during the last decade in specialized journals. However, many of the tools contained in *ACE* have more general applications than the original problems, and can be described in elementary terms, leading to interesting and efficient algorithms.

For instance, symmetric functions and the symmetric group appear in different fields of mathematics, statistical mechanics and theoretical computer science:

characters of classical groups, cohomology of flag varieties, Lie idempotents, Yang-Baxter equation and sorting procedures. Now, if one wants to compute the number of $k$-th root of a permutation $w$ (*i.e.* the number of solutions of the equation $v^k = w$ in the symmetric group), one has in fact to decode this problem into the determination of certain coefficients in a symmetric function. It would be totally impossible to enumerate permutations and test their $k$-th power. *ACE* gives instantly that there are 36 roots of order 6 of the permutation $(12, 2, 16, 19, 10, 17, 7, 14, 13, 11, 9, 4, 5, 6, 15, 1, 18, 8, 20, 3)$.

Another example, in geometry, this time. Schubert varieties are in bijective correspondance with permutations, and to determine intersection numbers, one has to multiply two Schubert polynomials and take the scalar product with a third one. Thus, there are 2 components indexed by $(9, 10, 11, 2, 7, 6, 5, 4, 3, 1, 8)$ in the intersection of the varieties respectively indexed by $(2, 1, 5, 3, 6, 4)$ and $(1, 10, 9, 8, 7, 6, 5, 4, 3, 2)$. The polynomial representing the product contains 97240 monomials from which it would be difficult to extract the required information without using the combinatorics of Schubert polynomials.

We stress that most of the algebraic computations on polynomials in *ACE* have been implemented as manipulations of combinatorial structures: change of bases, products, specializations, etc.

Another project, $\mu$-EC [Prosper and Lascoux 1998], is developped by Vincent Prosper and Alain Lascoux, as a *MuPAD* library. It uses the same combinatorial objects as *ACE*, but allows the user to include $C$ procedures to accelerate certain algorithms.

Several libraries in *ACE* include functions already existing in the system *SYMMETRICA* [Kerber et al. 1992] developped at Bayreuth. However, this system is more devoted to representation theory and is written in $C$ (which makes it quite efficient for large scale computations). *ACE*, since it is included into a computer algebra system, offers greater flexibility and simplicity for experimentation.

Additional reference: [Ung 1998].

Sebastien Veigneau (Noisy-le-Grand)

### 4.2.2 Building Nonassociative Algebras With Albert

Albert is an interactive program to assist in the study of nonassociative algebras. The problem addressed by Albert is the recognition of *nonassociative polynomial identities*. That is, does an algebra satisfy a certain identity, assuming that it satisfies another given set of identities?

Experience has shown that recognizing nonassociative identities is difficult. A successful method by Hentzel makes use of group representation theory (see

[Beck and Kolman 1977], p. 13–40). Another approach, however, is to build the *free algebra* determined by the given set of defining identities. If the polynomial in question is zero in the free algebra then it must be an identity. This is the approach taken by `Albert`, using a dynamic programming algorithm described in [Hentzel and Jacobs 1991].

`Albert` has a simple user interface which will be illustrated in the following example. Suppose the researcher wishes to study algebras satisfying

$$xy - yx = 0 \qquad (35)$$
$$x^3 = 0. \qquad (36)$$

Such algebras are commutative with nilindex 3. Assume also that the researcher wishes to know if identities (35) and (36) imply

$$(((a^2b^2)c^2)d^2)e = 0. \qquad (37)$$

Note that we are dealing with identities (universally quantified equations) and not relations on generators. In terms of the free algebra, we are asking if the target polynomial (37) is a member of the so-called *T-ideal* generated by the defining polynomials ([Zhevlakov et al. 1982], p.4). We are not assuming associativity of multiplication. If we were, then the theorem of Nagata-Higman (which states that in associative algebras satisfying $x^n = 0$ and having no elements of additive order $n$, the product of any $2^n - 1$ elements is zero) would imply that this expression is zero [Zhevlakov et al. 1982].

To solve this using `Albert`, the user first enters identities (35) and (36):

```
identity xy - yx
identity (xx)x
```

Then the user inputs the degrees of the letters in the target polynomial:

```
generators 2a2b2c2d1e
```

To solve the problem, `Albert` constructs a sufficiently large homomorphic image of the free algebra. The above command means that the constructed algebra will include all products whose degrees in $a$, $b$, $c$, $d$, $e$ are at most 2,2,2,2,1 respectively. `Albert` does all work over a Galois field $Z_p$, where $2 \leq p < 256$, selected by the user. The construction is initiated with

```
build
```

Often, in practice, the construction fails due to the lack of sufficient memory or time. However, if the construction is successful, the user can ask whether the polynomial is an identity by typing:

```
polynomial ( ( (a^2 b^2) c^2) d^2) e
```

Here, **Albert** responds with:

```
Polynomial is an identity.
```

Interestingly, this identity is critical to understanding the structure of Bernstein algebras [Hentzel et al. 1994].

More information on **Albert** can be found in [Jacobs 1994], and other papers describing results obtained with its assistance are [Bremner 1999; Hentzel et al. 1993a,b]. The program is written in C, and its source code is freely available at

http://www.cs.clemson.edu/~dpj/dpj.html

It has been installed on several Unix-like platforms.

David P. Jacobs (Clemson)

### 4.2.3 ALGEB

ALGEB is a programming language (as opposed to a symbolic algebra system) designed specifically for working in algebra and number theory.

It was designed and first implemented in 1977 (DEC PDP-11) for the purpose of testing the Zassenhaus "Round Four" algorithm for computing integral bases. Since then it has been implemented on DEC VAX/VMS and on the IBM-PC.

Compared to symbolic systems (such as MAPLE), ALGEB is generally faster and easier to program. On the other hand, ALGEB lacks the large library of pre-programmed functions available in a good symbolic system.

Among the major projects in which ALGEB has been used are: Computing tables of quartic and sextic number fields; computing Galois groups; computing the Rational Normal (Frobenius) Form of a matrix.

Direct inquiries to: Prof. David Ford, Concordia University, Dept. of Computer Science, 1455, de maisonneuve boulevard west, Montreal, Quebec, H3G 1M8, Canada. E-mail: KBKFE24@vax2.concordia.ca.

David Ford (Montreal)<br>(Original 1993 contribution [Eds.])

### 4.2.4 AMORE

AMoRE (Automata, Monoids, and Regular Expressions)

AMoRE is a program written in C for synthesizing, analyzing and transforming various representations of regular (formal) languages, in particular representations by finite automata, finite monoids, and regular expressions. Among the available functions, which are called by menues, are the following:

- transformation of regular expressions into minimal deterministic finite automata
- transformation of regular expressions into small nondeterministic automata (found by a heuristic search procedure, not necessarily yielding an automaton of minimal number of states),
- computation of the syntactic monoid of a regular language (as presented by a regular expression or an automaton), as well as the decomposition of this monoid by Green's relations,
- the test on language properties such as "star-free", "locally testable", "piecewise testable",
- standard operations on regular languages: Boolean operations, concatenation, Kleene star, shuffle product, left and right quotient,
- the graphical display of small transitions graphs of automata (of up to around 25 states) on the screen, determined by the heuristic of the Sugiyama algorithm.

For the theoretical background see e.g. [Pin 1986]. A documentation of the program, including user's manual, detailed description of data structures and memory management, as well as an explanation of the implemented algorithms, is available as a technical report [Matz et al. 1995].

AMoRE was developed during eight years at the universities of Aachen and Kiel, partly supported by Deutsche Forschungsgemeinschaft and the ESPRIT Working Group ASMICS ("Algebraic and Syntactic Methods in Computer Science").

The program as well as the report can be obtained per anonymous ftp: ftp.informatik.uni-kiel.de:pub/kiel/amore. Any queries and suggestions should be mailed to amore@informatik.uni-kiel.de.

Wolfgang Thomas (Kiel)

### 4.2.5 BERGMAN

BERGMAN is a special-purpose system for computations in commutative and purely non-commutative graded algebra*. It is mainly developed by Jörgen Backelin (Stockholm University). Some additional facilities are implemented by

---
* In 1999: release 1.0

participants of the joint project "Non-commutative computer algebra" (Stockholm University, Institute of Mathematics and Computer Science of Moldavian Academy of Sciences).

BERGMAN is public domain software available by anonymous ftp from ftp.matematik.su.se. It is written in Standard Lisp, the Lisp dialect underlying Reduce implementation. There is also available an experimentative Common Lisp version.

BERGMAN presently runs on the following platforms with pre–installed Reduce or PSL:

- Sun–Sparc station under SunOS, Solaris and Linux;
- Alpha station under Dec OSF and Linux;
- PC under MS DOS.

The Common Lisp version runs under CLISP on:

- Sun–Sparc station under SunOS, Solaris and Linux;
- PC under Windows 95, Windows 98, Windows NT.

The main idea behind BERGMAN is to give the user a possibility to use Bergman's Diamond Lemma in efficient calculations, combined with a maximum flexibility. BERGMAN is mainly intended to be a powerful instrument for calculating Gröbner basis in several situations: commutative and non-commutative algebras, modules over them.

Besides Gröbner bases it provides some facilities to calculate appropriate invariants of the algebras and modules, such as the Hilbert series, and (in the non-commutative case only), the Poincaré series, Anick's resolution and the Betti numbers.

The essential tool for combining flexibility and time efficiency is to employ the inbuilt function pointer feature in Lisp, and to change a set of crucial procedures (by changing the function pointer values) when the context set-up is changed. Thus, in the parts which are time crucial, different modes of operation are not distinguished by repeated look-ups of flag or variable values.

The space handling is not as efficient.

Among the context set-up alternatives are: commutativity or not, various strategies of Gröbner basis computation, some different monomial orders, various coefficient fields. The set-up may be changed interactively during the session; but one then should take some care with partial results on set-up specific internal form. Most calculations can be done both for ideals and modules.

In the Reduce version it is possible to include indeterminates (with or without declared reduction rules) as coefficients for the commutative computations;

otherwise the coefficient field should be a prime field (i.e., either the field $\mathbb{Q}$ of rationals, or a field $\mathbb{F}_p$ of prime order).

Special efforts were done to customise input and output. Since the objects internally always are representated as homogeneous ones, and all Gröbner basis calculations are done degree by degree, it is possible to get degree-wise output of partial results.

BERGMAN works under Reduce, PSL or Common Lisp (at least one of them should be pre-installed) and the most part of commands should be written in respect to Reduce or Lisp syntax.

BERGMAN is not a full computer algebra system. However, it is an open system: one can add his own procedures, written in Lisp or R-Lisp. It is also possible to use BERGMAN as a REDUCE package, if some care is taken not to set up REDUCE and BERGMAN with contradicting coefficient domains.

Reference: [Backelin and Fröberg 1991].

Jörgen Backelin (Stockholm), Svetlana Cojocaru (Chisinau), Victor Ufnarovski
(Lund)

### 4.2.6 Cannes / Parcan

CANNES and its parallel counterpart PARCAN (Parallel computer algebra nucleus, cf. [Gloor and Müller 1998]) are computer algebra core systems that mainly serve two purposes:

1. They supply a fast and reliable basis for a number of implementations of sequential and parallel algorithms for algebraic and symbolic computations.
2. They provide a platform for the investigation of core issues of sequential and in particular parallel computer algebra systems. The latter includes shared memory as well as distributed memory parallelism.

Both CANNES and PARCAN are written in an object oriented way in C/C++. In particular, they are built up from modules with well-defined interfaces. The following modules comprise CANNES:

− The low level mpn-functions of GNU MP [Granlund 1996] for the arithmetic of big natural numbers ($\geq 2^{30}$).
− The integer arithmetic module that handles short integers without accessing the GNU MP module.
− The basic list cell management.
− The memory management module for the arrays storing integers and list cells. This includes a Mark-Sweep based garbage collection.*

---

* Due to a thorough evaluation of different garbage collection methods (cf. [Gloor and Müller 1998]), we decided for Mark-Sweep for parallel polynomial computations.

PARCAN relies on the modules of CANNES. In addition, it contains

- a modified S-thread system [Küchlin 1992] that provides thread functionalities for user applications, and
- a multi-thread safe version of the memory management module of CANNES, including a parallelized version of the garbage collection.

The development of CANNES and PARCAN started with a complete redesign of the system components of SACLIB (cf. [Buchberger et al. 1993]) and PARSAC (cf. Section 4.2.31), respectively. This involves the complete memory management mechanisms, the basic list cell administration, and the integer arithmetic (see also Section 2.1.1).

The motivation for the reinvestigation of these system issues was based on the following observations:

- The Mark-Sweep garbage collection method (cf. [Jones and Lins 1996, Chapter 4]), in the manner it is used in SACLIB, is not compatible with optimizing compilers. Optimizing compilers produce code that often stores intermediate results instead of proper references on the stack. That is, when a garbage collection takes place, pieces of active data can be undetected in the "Mark" phase and therefore get collected and hence lost in the "Sweep" phase.
  As the SACLIB system components provide support for sequential computations only, garbage collection can only take place if it is directly triggered or when trying to fetch a piece of data (a list cell).
  In PARSAC, which uses the same garbage collection principles, we run several threads of control concurrently. From the viewpoint of one of the operating threads, garbage collection can take place at any time as it can be triggered by any of the other threads. This results in much more opportunities for garbage collection failures, and even in unoptimized code.
- The integer arithmetic of SACLIB is list-based. From the theoretical point of view, this creates only a (small) constant factor of overhead with respect to an array-based integer arithmetic such as GNU MP (cf. [Granlund 1996]). However, the storage model itself slows the execution down as the list cells are usually distributed over the allocated memory and not in the same cache line. This is in contrast to array entries that are neighbored in memory. That is, SACLIB is hardly taking advantage of the latest development of different levels of on-processor cache, and an increase of processor speed does not lead to the corresponding decrease of overall computation time as the time needed to fetch and store the pieces of data in the memory remains the same. In addition, a lot of space (roughly 50%) is wasted by the list-based storage model.
- The allocation of single list cells during a computation in PARSAC produces a large overhead (compared with the allocation in SACLIB) due to the need for multi-threading safety. For example in the Gröbner basis application [Amrhein et al. 1996a], this amounts to up to 40% in the overall time. Thus,

memory management (allocation and collection) tends to dominate the computation time.
- To support Mark-Sweep garbage collection, the list cell management of SACLIB uses so-called SAC-pointers instead of C-pointers. A SAC-pointer is an index (with an offset) to the global array of list cells. That is, whenever a list cell hast to be fetched, the proper address has to be computed from the SAC-pointer. This produces a significant overhead over a pure list administration in C.

Speed and memory requirements are an issue in almost all computer algebra systems as the problem size often needs to be increased only very little to create unfeasible computations.

The redesign faced the following boundary conditions:

- The existing applications built on top of SACLIB and PARSAC needed to be ported to the new basis systems. This included our implementation of the parallel Buchberger completion algorithm for the computation of Gröbner bases (cf. Section 2.2.5) as described in [Amrhein et al. 1996a], the Gröbner bases conversions [Amrhein et al. 1996b, 1997a; Amrhein and Gloor 1998], and the parallel quantifier elimination [Dolzmann et al. 1998a].
Timings of sequential runs on the parallel Gröbner bases computation system should lie in the same order of magnitude as those on a state-of-the-art sequential system (cf. e.g. Section 4.2.11).
- The new systems needed to be implemented in an object-oriented manner. In particular, well-defined interfaces between the different modules need to provide the application code with a transparent access to the core functionalities of the system. In addition, every module (e.g., the integer arithmetic) should be easily exchangeable.

The results of the transition from SACLIB/PARSAC to CANNES/PARCAN can be summarized as follows:

- We obtained a speedup of 3–5 for the applications.
Computations that involve relatively few integer arithmetic operations (or involving mostly short integers) such as the Fractal Walk [Amrhein and Gloor 1998] are about three times faster. In computations where integer arithmetic is highly involved (e.g. Buchberger algorithm), we obtained a factor of up to 5.
- The overhead of the parallel system is now smaller and often even negligible.
- The garbage collection has become reliable even with highly optimized code.
- The application (source) code from SACLIB/PARSAC can identically be used for CANNES/PARCAN, provided it uses the appropriate integer and list operations (and hence does not make any assumptions about the storage model). That is, the transition is straightforward.

Oliver Gloor (Bern, Switzerland)

## 4.2.7 CARAT

**Crystallographic Groups**

A crystallographic space group is a discrete cocompact subgroup of the group of all motions of Euclidean n-space and describes the symmetries of n-fold periodic structures. For $n \leq 3$ such groups are used to describe conventional crystal symmetries, for bigger n they are used for quasicrystals and modulated crystal structures. Torsionfree space groups are the fundamental groups of flat Euclidean space forms (the torus and the Klein bottle being the best known examples). Beyond geometry and topology space groups turn up in group theory and the theory of lattices in Euclidean space.

The Bieberbach Theorems describe the algebraic structure of n-dimensional space groups as extensions of a free abelian group of rank n of translations by a finite unimodular group called the point group of the space group. Isomorphism and conjugacy in the group of all affine motions are the same and lead to a finite classification of space groups for each dimension n with 2, 17, 219, 4.783, 222.018, 28.927.922 *, affine classes in dimensions 1,2,3,4,5, resp. 6. The big number of these groups as well as their geometric understanding call for coarser equivalence relations than affine equivalence. The most important ones are arithmetic and geometric equivalence for the conjugacy of the point groups in the full unimodular group resp. in the general linear group over the rational numbers and Bravais types, which are associated with the range of realizations of the abstract space group as Euclidean group. Finally crystal families consist of full geometric equivalence classes and Bravais types.

**Problems To Be Attacked With CARAT**

CARAT handles enumeration and construction problems, as well as recognition and comparison problems for crystallographic groups up to dimension 6. The enumeration part is already beyond the the scope of a book because the big number of groups.

Enumeration problems include

(a) splitting arithmetic crystal classes into affine classes of space groups,
(b) splitting geometric crystal classes into arithmetic crystal classes,
(c) splitting Bravais types into arithmetic crystal classes,
(d) splitting crystal families into Bravais types,
(e) splitting crystal families into geometric classes,
(f) enumerating inclusions between Bravais groups.

Recognition problems include deciding equivalence with respect to any of the listed equivalence relations above.

---

* The last two figures were computed for the first time by the means of CARAT

## Methods of Solution

CARAT contains tables and implementations of various algorithms. The tables include representatives of the arithmetic classes of the Bravais groups up to degree 6 together with their normalizers and inclusions and tables of the geometric classes also up to degree 6. The main algorithms implemented include a centering (≡sublattice) algorithm, a lattice automorphism and isometry algorithm, and the Zassenhaus algorithm to compute vector systems. It is also possible to calculate generators for normalizers of point groups via perfect forms in the Bravais manifold.

## Documentation and Availability

There is a short LaTeX-file giving the basic details for handling the package. Each program comes with a short online help. A skeleton of the mathematical background is described in [Opgenorth et al. 1998], where further references are given.

CARAT is available via http://samuel.math.rwth-aachen.de/~LBFM/carat/.

## Software Environment

CARAT is a compilation of about 60 programs, all of which are written in C. It uses a normal UNIX environment on user defined files, which are basically of two types. Together with the system comes a library of C-functions which makes it possible to develop programs for special purposes. Great care is taken that the enumeration problems (a)-(f) and the corresponding recognition problems can easily be performed even by inexperienced users.

It is running successfully on various UNIX-machines, including those running Linux, HP-UX, and Solaris. It should be easily portable to any machine running UNIX.

Addresses:
Dr. Jürgen Opgenorth (juergen@momo.math.rwth-aachen.de),
Prof. Dr. Wilhelm Plesken (plesken@momo.math.rwth-aachen.de),
T. Schulz (tilman@momo.math.rwth-aachen.de),
all at Lehrstuhl B für Mathematik, RWTH Aachen, Templergraben 64, 52064 Aachen, Germany.

Jürgen Opgenorth, Wilhelm Plesken and Tilman Schulz (Aachen)

### 4.2.8 CASA

CASA is a special-purpose system for computational algebra and constructive algebraic geometry. The system has been developed since 1990. CASA is the ongoing product of the Computer Algebra Group at the Research Institute for Symbolic Computation (RISC-Linz), University of Linz, Austria, under the direction of the third author. The system is built on the kernel of the computer algebra system MAPLE (4.1.6).

**4.2.8.1 System Requirements** CASA is built on the kernel of MAPLE and is fully independent of the operating system. Hence, it can be used on every hardware where MAPLE is running. In order to run the current version of CASA (V2.3), one needs a computing environment that provides MAPLE.

**4.2.8.2 Availability** The system can be obtained by anonymous ftp at ftp.risc.uni-linz.ac.at. Major releases will be located in the directory /pub/CASA. Bug correction and minor updates will be put in /pub/CASA/update. We have also a home page on the World Wide Web at http://www.risc.uni-linz.ac.at/software/casa.

The system is freely distributed under the following conditions: any research activity which uses CASA should cite the authors and the system explicitly. The system can be freely distributed to other users. New users are encouraged to notify the CASA coordinator so that they can be included in a user list where they will be kept up to date about the progress of the system.

Bug reports, questions and suggestions should be sent to the e-mail address casa@risc.uni-linz.ac.at.

**4.2.8.3 Target Topics** CASA is a special-purpose system for computational algebra and constructive algebraic geometry.

Constructive methods in algebra and algebraic geometry have been gaining more and more importance with the availability of computers and computer algebra software. Since its first version, CASA has been designed to perform computations and utilize reasoning about algebraic and geometrical objects in classical affine and projective spaces over algebraically closed fields of characteristic zero. More precisely, the field has to be a computable field in the sense of the underlying computer algebra system MAPLE, i.e. all the arithmetic operations have to be available in the system. Usually, the field of computation is the rational numbers $\mathbb{Q}$ or a finite algebraic extension thereof.

The system has been developed since 1990. With the latest version (version 2.3) CASA is the ongoing product of the Computer Algebra Group at the Research Institute for Symbolic Computation (RISC-Linz), University of Linz, Austria, under the direction of Franz Winkler. The system is built on the kernel of the widely used computer algebra system MAPLE.

Further references on CASA can be found in [Gebauer et al. 1991; Mňuk and Winkler 1996; Tran and Winkler 1997a,b].

In the system, an algebraic set—a central notion in algebraic geometry—can be represented in four different ways:

- *Implicit representation:* An algebraic set is the set of common zeros of a system of polynomial equations. To give an algebraic set in implicit form means to give finitely many polynomials.
- *Projected representation:* As a consequence of the primitive element theorem every irreducible $d$-dimensional algebraic set in $n$-dimensional space is, after a suitable linear transformation of coordinates, birationally projectable onto an irreducible $d$-dimensional algebraic set in $(d+1)$-dimensional space, which can be specified by a single polynomial in $d+1$ variables. This can be generalized to unmixed-dimensional algebraic sets. An algebraic set in projected form is given by a polynomial and a tuple of rational functions (specifying the birational mapping).
- *Parametric representation:* Some irreducible algebraic sets can be parametrized by rational functions. An algebraic set in parametric form is given by a tuple of rational functions that parametrizes the algebraic set.
- *Representation by places:* All algebraic curves can be locally parametrized by a set of power series that are convergent around a point of the curve. An algebraic set is given by places if for each branch passing through a certain point on the algebraic set a tuple of power series that parametrizes the algebraic set around the point is specified.

The system provides a variety of operations on algebraic sets. As the efficiency of these operations is tightly bound to the way algebraic sets are represented, conversion routines have been provided to support various views on one object, to deepen the understanding of its principles, and to speed up algorithms working on algebraic sets. CASA also works with the polynomial ideals corresponding to these geometric objects.

The basic operations available in CASA include:

- ideal theoretic operations (sum, product, intersection, union)
- creating algebraic sets in different representations,
- generating curves of fixed multiplicities at given points,
- intersection, union, and difference of algebraic sets,
- computing tangent cones and tangent spaces,
- computation of the dimension of an algebraic set,
- decomposition into irreducible components,
- transformations of algebraic sets to hypersurfaces,
- computation of the singularities, genus, neighborhood graphs and adjoint curves of an algebraic curve.

Besides these basic operations, the following more advanced operations are available:

- rational parametrization of rational curves over an optimal extension field of coefficients,
- implicitization of parametrically given algebraic sets,
- Puiseux series expansions,
- multivariate resultants and Dixon's resultants
- Gröbner bases of ideals and modules,
- Gröbner walk,
- hybrid methods for finding solutions of an arbitrary system of equations,
- computation of rational points on conics,
- offset curves,
- plotting both explicitly and implicitly given curves and surfaces,
- computation of syzygy-bases.

**4.2.8.4 Distinction from other Systems** The major goal of CASA is to provide a comfortable, easy to use, efficient, flexible and mathematically exact working environment for computational algebra and constructive algebraic geometry where all basic theoretical concepts map easily to available data structures. There are software systems which partially cover some of the fields of CASA. However, no other system seems to provide all the functionalities of CASA.

**4.2.8.5 Example of Application: Scientific Visualization** A powerful feature of CASA is the capability to visualize algebraic sets. For the majority of tasks, only singular or other distinguished points of algebraic sets yield some interesting information. These objects have to be treated with care, requiring special analysis. Most commonly used computer algebra systems have problems with plotting sets in implicit representation. CASA makes a thorough effort to obtain correct information about the local topology in the neighborhood of singularities and other critical points while still keeping the algorithm efficient by using hybrid symbolic-numerical methods.

Figure 18 shows the graphs produced by CASA for the algebraic curves defined as zeros of the following equations:

1. $\frac{93392896}{15625}x^6 + (\frac{94359552}{625}y^2 + \frac{(91521024)}{625}y - \frac{249088}{125}x^4 + (\frac{1032192}{25}y^4 - 36864y^3 - \frac{7732224}{25}y^2 - 207360y + \frac{770048}{25}x^2 + (65536y^6 + 49152y^5 - 135168y^4 - 72704y^3 + 101376y^2 + 27648y - 27648$
2. $2x^4 - 3x^2y + y^2 - 2y^3 + y^4$
3. $y^4 - 96a^2y^2 + 100a^2x^2 - x^4$
4. $8(y^2 - x^2) - 8(y^3 + x^3) + 2y^4 + x^5$

Note that CASA also finds isolated singularities such as in curve 1.

Ralf Hemmecke, Erik Hillgarter, and Franz Winkler (Linz)

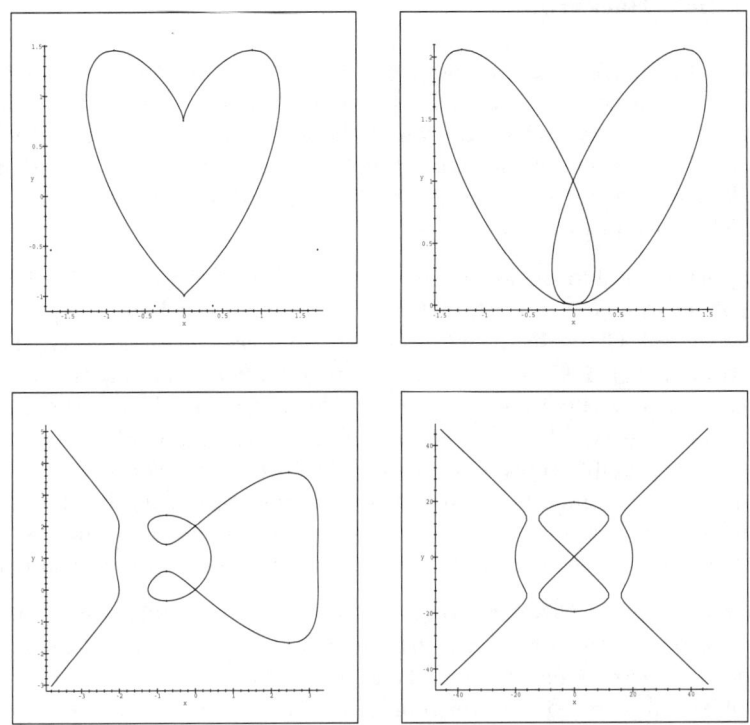

**Fig. 18.** Plotting

## 4.2.9 CHEVIE

**CHEVIE** is a computer algebra package for symbolic calculations with generic character tables of groups of Lie type (including Green functions), finite Coxeter groups and Iwahori-Hecke algebras. It consists of programs and library files in the formats of the computer algebra systems **GAP** [GAP 1997] and MAPLE [Char et al. 1992].

The name of this project is a combination of the names of two famous mathematicians: Claude Chevalley and Sophus Lie.

**CHEVIE** is a joint project of Meinolf Geck (Lyon), Gerhard Hiß (Aachen), Frank Lübeck (Aachen), Gunter Malle (Kassel), Jean Michel (Paris) and Götz Pfeiffer (Galway).

More information on the mathematical background of the package (compared to this short note) as well as some explicit examples of applications can be found in [Geck et al. 1996]. The current software and additional information can be found in:
http://www.math.rwth-aachen.de/~CHEVIE/

## Mathematical Background

Finite groups of Lie type are analogues over finite fields of the real or complex Lie groups associated with the various families of complex simple Lie algebras, whose classification was achieved around the end of the last century by Cartan and Killing. The aspects of the structure of such a group which are independent of the underlying field are to a large extent controlled by the corresponding Weyl group which is a finite Coxeter group.

An important tool for investigating any given finite group $G$ is the study of its homomorphisms $\rho : G \to GL_n(\mathbb{C})$. Here $\rho$ is called an irreducible representation if no subspace of $\mathbb{C}^n$ is invariant under all matrices $\rho(g)$, $g \in G$. The map $\chi : g \mapsto \mathrm{trace}(\rho(g)) \in \mathbb{C}$ is called the *character* of $\rho$. Characters are constant on *conjugacy classes* (i.e., $\chi(g) = \chi(xgx^{-1})$ for all $g, x \in G$) and the number of conjugacy classes is equal to the number of different irreducible characters (i.e., characters of irreducible representations). The (ordinary) *character table* of $G$ is a matrix giving the values $\chi(g)$ where the columns are labeled by the conjugacy classes and the rows are labeled by the irreducible characters. Many structural properties of a group are encoded in its character table in a very compact way.

In the case of finite Coxeter groups and groups of Lie type, one can naturally associate various invariants to the irreducible characters, which lead to some natural parameterizations of them. Iwahori–Hecke algebras arise naturally as endomorphism algebras of certain representations of groups of Lie type. There are also character tables for them, which are "generic" analogues of the character tables of finite Coxeter groups.

The aim of **CHEVIE** is to provide a computational framework in which one can work with these groups and algebras, as well as with their representations and characters. Moreover, the associated libraries contain the information about natural labelings of classes and characters, and the character tables themselves.

## The Maple Part of CHEVIE

This part of **CHEVIE** deals with *generic character tables* of series of groups of Lie type.

A series of groups of Lie type is an infinite set of groups associated to a certain combinatorial structure, called a complete root datum. Examples of such series are the groups $\{GL_3(q) \mid q \text{ a prime power}\}$ of invertible $3 \times 3$-matrices over a finite field with $q$ elements or the Suzuki groups $\{Sz(q^2) \mid q^2 = 2^{2m+1}, m \in \mathbb{N}\}$. These groups are closely related to a large part of the groups appearing in the classification of finite simple groups.

The number of conjugacy classes, and so of irreducible characters, in such a series grows with $q$ (it can essentially be expressed as a polynomial in $q$). But the classes and characters can be collected into subsets, called class types respectively character types, such that the classes or characters in a fixed type can be parameterized in terms of $q$ and some extra parameters. With these extra parameters and $q$ the character values for all characters in a fixed type on all classes in a fixed type can be described by a single expression.

While the smallest examples of generic character tables were already known to Schur and Frobenius and several others for groups of small rank can be found in the literature, a general concept for such tables could only be found with the help of the Deligne-Lusztig theory. There, the finite groups of Lie type are considered as finite subgroups of connected reductive algebraic groups and this allows to get information on the finite groups using deep results from algebraic geometry.

Working with printed generic character tables turned out to be quite cumbersome. Sometimes there are inaccuracies in the tables which are very difficult to spot. And the computation with generic character values by hand is already complicated for small tables.

In the MAPLE part of CHEVIE we have both, a library of known generic character tables (an electronic Atlas), hopefully with only very few inaccuracies left, and also programs which allow to compute with these tables.

Here is a short outline; there are:

- a library of known ordinary generic character tables of series of finite groups of Lie type
- a library of Green functions of such groups
- programs for printing (parts of) the tables on screen or in LaTeX format
- programs for printing information on conjugacy classes, characters or references to the literature
- programs for computing with the tables: orthogonality relations, structure constants, tensor products (without specializing the $q$)

Currently we distribute generic character tables for the following series of groups:

- $SL_2(q)$, $PGL_2(q)$, $GL_2(q)$, $PSL_2(q)$, $GU_2(q)$,
- $SL_3(q)$, $PGL_3(q)$, $GL_3(q)$, $PSL_3(q)$ and corresponding unitary groups,
- $Sp_4(q)$ ($q$ even), $CSp_4(q)$ ($q$ odd),
- $G_2(q)$,
- Suzuki groups $^2B_2(q^2)$ ($q^2 = 2^{2k+1}$),
- Ree groups $^2G_2(q^2)$ ($q^2 = 3^{2k+1}$),
- $Sp_6(q)$ ($q$ even), $CSp_6(q)$ ($q$ odd),
- Steinberg groups $^3D_4(q)$,
- Ree groups $^2F_4(q^2)$ ($q^2 = 2^{2k+1}$),
- tables of unipotent characters: $SO_7(q)$, $D_4(q)$, $^2D_4(q)$, $GL_n(q)$ ($n \leq 8$)

There were successful applications of this part of **CHEVIE** in different areas like constructive Galois theory, modular representation theory of finite groups of Lie type, and the subgroup structure and generation properties of these groups.

This part of **CHEVIE** runs with MAPLE -V3, it includes a manual and an online help for all commands.

### The GAP Part of CHEVIE

This part of **CHEVIE** deals with finite Coxeter groups and their characters, and the associated Iwahori-Hecke algebras and braid groups. Similar to the MAPLE part, there are both libraries containing the basic information for all irreducible types of finite Coxeter groups and programs for working with general finite Coxeter groups. For example, the character table of a general finite Coxeter group is automatically determined by first decomposing it into irreducible types and then building together the information on irreducible types from the library files. Here is a more complete outline of the range of this part; one can

- create Coxeter groups by type, Cartan matrix or root datum, as well as reflection subgroups
- compute with their elements as permutations on a root system, matrices or words in the Coxeter generators
- get character tables of the Coxeter groups together with the natural labelings of classes and characters
- compute induce/restrict matrices for ordinary induction, Macdonald-Lusztig-Spaltenstein $j$-induction and Lusztig's $J$-induction
- create and compute with "Coxeter cosets" and get their character tables
- create braid groups and compute with their elements, via the normal forms defined by Deligne and Brieskorn–Saito
- compute Bruhat order, Kazhdan-Lusztig polynomials and left cells
- create Iwahori-Hecke algebras and compute with their elements in the standard basis $T_w$ and the various Kazhdan–Lusztig bases $C_w$, $C'_w$, $D_w$, $D'_w$
- get character tables of the Iwahori-Hecke algebras, reflection and left cell representations, Poincaré polynomials, Schur elements, generic degrees
- have preliminary support for complex reflection groups, cyclotomic algebras and Hecke cosets

This part of **CHEVIE** is distributed as a share package with **GAP**-Release 3.4.4. The manual is a part of the **GAP** manual and can also be read in the online help. We have used this part of **CHEVIE** to study, for example, properties of the Bruhat–Chevalley order and of representatives of minimal length in the conjugacy classes of finite Coxeter groups. The various types of induce/restrict matrices for characters are an important ingredient in Lusztig's theory of characters of finite groups of Lie type.

Addresses:

Prof. Dr. Meinolf Geck, Institut Girard Desargues, Bâtiment 101, Université Claude Bernard - Lyon I, 43 Boulevard du 11 Novembre 1918, 69622 Villeurbanne Cedex, France
e-mail: geck@desargues.univ-lyon1.fr.

Dr. Frank Lübeck, Lehrstuhl D für Mathematik, RWTH Aachen, Templergraben 64, 52062 Aachen, Germany
e-mail: Frank.Luebeck@Math.RWTH-Aachen.De.

Meinolf Geck (Lyon) and Frank Lübeck (Aachen)

### 4.2.10   C-Meataxe

The following problems in the representation theory of finite groups and finite dimensional algebras have an efficient algorithmic solution over finite fields, see for example [Parker 1984], [Holt and Rees 1994], and 2.8. Given a matrix representation of an algebra $A$ in terms of matrices for a generating set we can determine

- the composition factors and their multiplicities,
- the socle and the radical series,
- the complete lattice of $A$-submodules, and
- the endomorphism ring and a decomposition of the representation into a direct sum of indecomposable representations.

The C-MEATAXE developed by M. Ringe, contains implementations of the corresponding algorithms, which rely up on the fundamental paper by R.A. Parker, [Parker 1984]. It is a collection of standalone programs written in ANSI-C. A typical C-MEATAXE program reads the (binary) input data from files, solves a specific problem and produces an answer in form of (binary) output files. These tasks can range from basic problems in linear algebra such as the determination of the rank, the nullspace and the order of a given matrix, to more sophisticated problems such as the ones mentioned above. The C-MEATAXE differs from other implementations of R.A. Parker's ideas in the sense that it has been mainly used for solving large research problems and the program design has therefore been driven by memory and speed considerations and not necessarily user friendliness. One of its main features is a fast vector arithmetic and a compact data type for vectors over finite fields up to order 256. The C-MEATAXE also supplies the programmer with a library of C-functions that implement the basic algorithms for matrices. This library can be used to develop new algorithms.

The C-MEATAXE version 2.3 is available from Michael Ringe, mringe@math.rwth-aachen.de. It runs under various UNIX flavours, for example HP-UX, DEC ULTRIX, Linux, under MS-DOS and WINDOWS NT. The version 2.2 is also available as a share package in the GAP-system, see 4.2.18. A large database of matrix representations for finite simple groups that can be used with the C-MEATAXE has been put on to the web by R. A. Wilson, under the URL http://for.mat.bham.ac.uk/atlas/.

Klaus Lux (Tucson)

### 4.2.11  CoCoA

**Introduction**

**What Is CoCoA?** CoCoA is a *special-purpose* system for doing **Co**mputations in **Co**mmutative **A**lgebra (see 2.5). It runs on the following platforms: SUN running Solaris, HP, SGI running IRIX, DEC Alpha, PC running Linux or DOS or Windows, Macintosh PPC running MacOS or LinuxPPC.

CoCoA is one of the products of a research team in Computer Algebra, whose members are: Lorenzo Robbiano, Gianfranco Niesi, John Abbott, Anna Bigatti, Massimo Caboara, Martin Kreuzer, David Perkinson, Alessandro Polverini, Antonio Capani and occasionally other researchers and students.

The system includes *complete on-line help* (also available as an *html-manual*). CoCoA is *freely available* software for research and educational purposes: the latest version (CoCoA *4*, June 2000) and more information about the system can be found at:
http://cocoa.dima.unige.it
as well as from the mirror sites
http://www.physik.uni-regensburg.de/~krm03530/cocoa
http://www.reed.edu/mirrors/cocoa

**The Main Features of** CoCoA  CoCoA's principal area of expertise is that of operations over commutative rings of polynomials. For example, it can readily compute *Gröbner bases* (see 2.2.5), *syzygies and minimal free resolutions* ([Capani et al. 1997]), *intersections, divisions* ([Caboara and Traverso 1998]), *the radical of an ideal* ([Caboara et al. 1997]), *the ideals of 0-dimensional schemes* ([Abbott et al. 2000a], [Abbott et al. 2000b]), *Poincaré series and Hilbert functions* ([Bigatti 1997]), *factorization of polynomials* ([Abbott 1998]), *toric ideals* ([Bigatti et al. 1999]).

The capabilities of CoCoA and the flexibility of its use are further enhanced by the *dedicated high-level programming language* CoCoAL. For convenience, the system offers a graphical user interface common to all platforms.

**The Users of** CoCoA Currently CoCoA is used by researchers in *several countries*. Most of them are Commutative Algebraists and Algebraic Geometers, but also people working in different areas such as Analysis and Statistics (see [Robbiano 1998]) have already benefitted from our system.

CoCoA is also used as the main system for *teaching* advanced courses in several Universities. Besides Italy, the most intensive use is by Tomas Recio at the University of Santander (Spain), by Anthony Geramita at Queen's University (Canada), by Martin Kreuzer at the University of Regensburg (Germany), by Dave Perkinson at Reed College (USA), and by Marie Vitulli at the University of Oregon (USA).

CoCoA was one of the few systems to have been invited to participate in the Special Session on Mathematical Software at ICM'98 and **3ecm**. It is also mentioned in some of the *most widely used text books* in Computational Algebra (see for instance [Adams and Loustaunau 1994a] pp. 275–276, and [Cox et al. 1992] pp. 493–494), and plays a major role in the forthcoming book [Kreuzer and Robbiano 2000].

**The Future of** CoCoA Aside from the normal continual development, a number of more specific plans are afoot to improve and extend CoCoA: the choice of coefficients is to be widened to handle parameters and finite algebraic extensions; and the mathematical core will be made available as a software library facilitating integration into other systems.

## Some examples

The CoCoA commands can be roughly divided into two classes: *simple commands* and *structured commands*. Examples of simple commands are: evaluation of an expression, assignment, and printing. Examples of structured commands are: if-then-else, for-loop, while-loop, foreach-loop, and function definition.

The simplest way to use CoCoA after starting it, is to set a ring (or use the default ring, i.e. $\mathbb{Q}[t, x, y, z]$), type a command, and ask the system to execute it.

In CoCoA a *ring* is a polynomial ring over $\mathbb{Q}$, $\mathbb{Z}$ or $\mathbb{Z}/n\mathbb{Z}$ with some annotations (about ordering or weights or other things).

*Example 5.* This example shows how to define and use a ring different from the default ring.

```
A ::= Z/(5)[xy];
Use A;
(2x^2 - 1/2y)^4;
```

The first command uses the operator "::=" to define $A$ to be the polynomial ring in the indeterminates $x$ and $y$ with coefficients in the field $\mathbb{Z}/(5)$ of the integers modulo 5. The second command chooses $A$ as the current ring. Then the expression $(2x^2 - 1/2y)^4$ is evaluated in $A$ and the result is displayed.

```
x^8 - x^6y + x^4y^2 - x^2y^3 + y^4
--------------------------------
```

*Example 6.* Now we use the structured command For to compute $(x+y)^7$ in the ring $\mathbb{Z}/n\mathbb{Z}$ for $n \in \{3, 5, 7, 9\}$. Note that any text from a -- to the end of the line is a comment.

```
For N := 3 To 9 Step 2 Do
  S ::= Z/(N)[xy];      --> define the ring S
  PrintLn Ring(S);      --> display the ring S
  S :: (x+y)^7;         --> evaluate the power in S and print it
  PrintLn;              --> start a new line
  PrintLn
End;
```

**Special Orderings** On this section we discuss two problems whose solutions require using a particular term-ordering. The first problem is how to find the minimal polynomial of an element of an algebraic extension of $\mathbb{Q}$. The second one is how to find a cartesian representation of a space curve given parametrically.

*Example 7.* The minimal polynomial of $\frac{4\alpha-1}{\alpha^3}$ over $\mathbb{Q}$, where $\alpha$ is a root of $x^7 - x - 1$ can be found by computing the reduced Gröbner basis of the ideal $(x^7 - x - 1, x^3y - 4x + 1)$ of the ring $\mathbb{Q}[x, y]$ with respect to the lexicographic term-ordering with $x > y$. In CoCoA this is achieved by using the following commands:

```
Use R ::= Q[xy],Lex;
Set Indentation;
GBasis(Ideal(x^7-x-1, x^3y-4x+1));
```

The output is:

```
[ x - 10022553737/89893683351809y^6 + 49925279149/89893683351809y^5
+ 17282591991/89893683351809y^4 - 2197476813566/89893683351809y^3
- 3212847751937/89893683351809y^2 + 1627614492145/89893683351809y
- 52555866039552/89893683351809,
  y^7 - 5y^6 + 147y^4 + 640y^3 - 31y^2 + 2176y - 20479 ]
--------------------------------
```

*Example 8.* Given the space curve $(t^{31}+t^6, t^8, t^{10})$, its cartesian equations can be found by eliminating the indeterminate $t$ in the ideal $(t^{31}+t^6-x, t^8-y, t^{10}-z)$. In CoCoA this is achieved by using the following commands:

```
Use R ::= Q[xyzt];
Set Indentation;
Elim(t, Ideal(t^31+t^6-x, t^8-y, t^10-z ) );
```

The system automatically changes the ordering to an elimination term-ordering for $t$, performs the computation, and, finally, restores the original ordering and gives the result:

```
Ideal( y^5 - z^4,
       z^8 + 2xy^3 - x^2yz - z^3,
       xy^4z^4 + 1/2yz^7 + 3/2x^2y^2 - x^3z - 1/2yz^2,
       y^4z^5 - y^4 + 2xy^2z - x^2z^2,
       y^2z^6 + 1/2xz^7 - 1/2x^2 3y - y^2z + 3/2xz^2,
       x^2y^4z^3 + 3y^3z^5 + 2xyz^6 - x^4 - 3y^3 + 4xyz )
-------------------------------
```

## Minimal Free Resolutions

*Example 9.* In this example we compute the minimal free resolution of the ideal $I$ generated by the 2 by 2 minors of a catalecticant matrix $A$, using the interactive environment.

We define the ideal $I$ and start the computation of its minimal free resolution using the Hilbert-driven algorithm described in [Capani et al. 1997].

```
Use R ::= Z/(32003)[z[0..3,0..3,0..3]];

A := Mat[
  [z[3,0,0], z[2,1,0], z[2,0,1]],
  [z[2,1,0], z[1,2,0], z[1,1,1]],
  [z[2,0,1], z[1,1,1], z[1,0,2]],
  [z[1,2,0], z[0,3,0], z[0,2,1]],
  [z[1,1,1], z[0,2,1], z[0,1,2]],
  [z[1,0,2], z[0,1,2], z[0,0,3]]
];

I := Ideal(Minors(2,A));
Res(I);
0 --> R(-9) --> R^27(-7) --> R^105(-6) --> R^189(-5) --> R^189(-4)
--> R^105(-3) --> R^27(-2)
-------------------------------
```

**Online Help** CoCoA's online help is roughly divided into two parts: a manual and a list of commands. The manual includes a tutorial which can be started by entering

```
H.Tutorial();
```

If you are a new user of CoCoA, the tutorial is a good place to start.

Each section of the manual and each command is uniquely identified by a set of keywords. The set of keywords always includes the title of the section or the title of the command. The online help command `Man` takes a string from the user and searches for a (case insensitive) match among the keywords. For instance, `Man('gbasis')` will display information about the function 'GBasis'.

### 4.2.12 CREP

CREP stands for Combinatorial REPresentation theory of finite-dimensional algebras. The name reflects the fact, that the study of finite-dimensional algebras has led to a rich variety of combinatorial methods and algorithms.

The development of the CREP system is coordinated at the working group "representation theory of algebras" at the University of Bielefeld/Germany. However CREP consists of a collection of functions from different places (Bayreuth, Berlin, Chemnitz, Düsseldorf, Essen, München, Paderborn (Germany), Kiew (Ukraine), Toruń (Poland), Zürich (Switzerland) ). Everyone working in the field is invited to contribute to the system.

CREP started as a collection of several mutually rather independent programs written in Pascal. However, soon we felt the need for communication between these different programs as well as for a common user interface for interactive work. Also one wished for an interface to a general algebra system.

Thus we started to use the Maple system to call the various programs collected inside CREP and to handle communication between these as well as the dialogue with the user. This way, CREP turned into a system, which in many respects behaves like a Maple library.

Currently CREP consists of the following parts:

The basic CREP package containing functions for completely separating algebras, unit forms and partially ordered sets. These include

- Computation of preprojective components and tubes.
- Computation of quadratic forms associated to an algebra (Tits form, Euler form).
- Testing a quadratic form for weak positivity and weak non-negativity.
- Determining the representation type of a finite partially ordered set.

- Testing a completely separating algebra for finite type, wild type, and tameness of polynomial growth by means of posets of thin startmodules.
- Test, whether a given completely separating algebra is is a tubular algebra.
- A database each of tame concealed algebras and of minimal wild concealed algebras.

The `heralg`-package containing functions for hereditary algebras, including

- Computation of fundamental invariants of hereditary algebras (Cartan matrix, Coxeter matrix, growth number) and one point extensions.
- Computation of right perpendicular categories.

The `tubular`-package containing functions implementing:

K-theoretic methods for canonical algebras and weighted projective lines including:
- Operations on the Grothendieck group.
- Performing Auslander-Reiten translations.
- Evaluating the Euler characteristic.

Special emphasis lies on tubular canonical algebras and the evaluation of 'telescoping functors'.

Further there are three packages each built around a database of a certain class of algebras. Functions inside each package give access to this database, allow to test, whether a given algebra is a member of the class in question. These are

The `esrd`-package for exceptional sincere representation directed algebras.

The `repetit`-package for repetitive algebras of type $E_6$, $E_7$, and $E_8$ and iteratedly tilted algebras of type $E_6$, $E_7$, and $E_8$ (which are closely related).

The `tpe`-package for completely separating tame poset algebras containing algebras of type $\tilde{E}_6$, $\tilde{E}_7$ or $\tilde{E}_8$.

While the heralg- and the tubular-package are written completely in Maple, the other packages also make use of external programs in Pascal, some of programs in C. So that, besides the Maple system, the corresponding compilers are needed for a full installation.

Plans for the future development of CREP include, besides the completion of the material already existing, the implementation of algorithms on the following topics:

- Computing the Auslander-Reiten translate of an indecomposable module of an algebra over a finite field.
- Calculation of Hall polynomials.

- Universal coverings of algebras.
- Degenerations of algebras and modules.
- Computing the quiver and relations of a given algebra.

CREP can be obtained via ftp from the server ftp.uni-bielefeld.de where it can be found in the directory pub/math/f-d-alg/crep

For questions or remarks on CREP, one may use the e-mail address fdowner@mathematik.uni-bielefeld.de or contact Peter Dräxler, Fakultät für Mathematik, Universität Bielefeld, Postfach 100 131, D-33501 Bielefeld, e-mail: draexler@mathematik.uni-bielefeld.de.

Additional references: [Dräxler and Nörenberg 1996, 1997; Dräxler et al. 1999; Dräxler and Nörenberg 1999; Dräxler 1993].

Peter Dräxler (Bielefeld) and Rainer Nörenberg (Essen)

### 4.2.13 The Desir Project and Its Continuation

**Formal Solutions of Linear Differential Equations**
The DESIR* project has started in the 80's. At the beginning, it was centered around the local analysis in the complex plane of scalar $n$th order linear differential equations $L(y) = 0$, where

$$L(y) = a_n(x)y^{(n)} + a_{n-1}(x)y^{(n-1)} + \cdots + a_1(x)y' + a_0(x)y,$$

the coefficients $a_i$'s belonging to $\mathbb{C}[x]$. For this type of equations, a Newton Polygon has been defined, and based on it, a Newton Algorithm has been build for computing a basis of formal solutions of $L(y) = 0$. In the general case, when the origin is an irregular singularity, the solutions can be written under the following form:

$$y(t) = e^{Q(1/t)} t^\lambda \Phi(t), \qquad (38)$$

where $x = t^r$, $r \in \mathbb{N}^*$, $\lambda \in \mathbb{C}$, $Q$ is a polynomial with coefficients in $\mathbb{C}$ and $\Phi \in \mathbb{C}[[t]][\log t]$ [Della Dora et al. 1982].
This general form covers several particular cases: if the origin is a regular singularity $r = 1, Q = 0$, and if the origin is an ordinary point $r = 1, Q = 0, \lambda = 0, \Phi \in \mathbb{C}\{t\}$.
This study has led to an effective implementation of the symbolic resolution of linear differential equations in the DESIR solver [Tournier 1987], written in Reduce.
The software DESIR–II [Pflügel 1997b] is an improved version of this first one, and has been developed in MAPLE.

---

* The softwares DESIR, DESIR-II and ISOLDE are accessible on the Web at the address http://www-lmc.imag.fr/CF/logiciel.html

### From Formal Solutions to Numerical and Graphical Solutions

For a long time, we are interested in illustrating how the formal solutions obtained by the former algorithms can give very precise informations on the *actual* solutions, even if the series appearing in 38 are generally divergent.

For this purpose, following previous work on the graphical representation of the solutions [Richard-Jung 1988], the COMPAS package is now developed in our team.

Given

- a *problem*: a linear differential operator $L$ and initial conditions determining a particular solution of $L(y) = 0$,

- a *path* in the complex plane,

the aim is to calculate the particular solution defined by the problem on the path, using different methods (formal and numerical) and to give its graphical representation.

This package is implemented in ALDOR. It is able to communicate data with MAPLE: it passes the differential operator to DESIR-II and recovers the basis of formal solutions. It contains various numerical methods for summing divergent series: from the elementary *summation at the leading term* to more sophisticated algorithms, well adapted to $k$-summable or multi-summable series [Jung et al. 1996].

### Linear Differential Systems

Recent research in the domain of symbolic integration of linear differential systems has led to the development of the package ISOLDE, written in MAPLE. The ISOLDE package handles systems of first–order linear differential equations

$$\frac{dY}{dx} = A(x)Y + b(x)$$

where $A \in \mathcal{M}_n(K)$ and $b \in K^n$. Here $K$ is a differential field of characteristic zero with constant field $C \supset \mathbb{Q}$. The problems solved by ISOLDE can be divided into two parts: *local problems* and *global problems*. In the former case we have $K = C[[x]][x^{-1}]$, the field of meromorphic formal power series, or an extension of this field, and in the latter $K = C(x)$, the field of rational functions in $x$ over $C$ or one of its extension. The local problems are:

1. find the exponential parts of the system,
2. find a basis of the regular formal solution space,
3. find a basis of formal solutions,
4. compute local factorisations of the system.

The algorithms used for problem 1 are described in [Barkatou 1997] and an improved method in [Pflügel 1998]. In [Barkatou 1998] and more in detail in [Barkatou and Pflügel 1997] the computation of regular formal solutions is considered, and an efficient implementation is available. The algorithms in [Pflügel

1998] for problem 1 together with problem 2 are based on local factorisation, and they solve the problem of computing a basis of formal solutions.
A typical application of the local methods is the computation of global solutions. We treat the following problems:

1. find the rational solutions,
2. find the exponential solutions,
3. decide whether two differential systems are equivalent,
4. compute a factorisation of completely reducible systems.

The first problem is solved in [Barkatou 1989], the second in [Pflügel 1997a]. The remaining problems are considered in [Barkatou and Pflügel Rostock, Germany, 1998].

Claire Di Crescenzo, Françoise Jung and Eckhard Pflügel (Grenoble)

### 4.2.14 DISCRETA: A Tool for Constructing $t$-Designs

The computer algebra package DISCRETA arouse of a project for the construction of discrete combinatorial structures especially designs at the University of Bayreuth. Its primary aim is the construction of $t$-designs with prescribed automorphism group. A $t$-design or better a $t$-$(v, k, \lambda)$ design $\mathcal{D} = (V, \mathcal{B})$ is an incidence system on a set $V = \{1, 2, \ldots, v\}$ of *points* and a distinguished set $\mathcal{B}$ of $k$-subsets of $V$ which are called *blocks*. The system is called $t$-design if for each $t$-subset $T \subseteq V$ there are exactly $\lambda$ blocks in the design containing $T$. A design with no repeated blocks is called simple. The construction of $t$-designs with large $t$ is a big challenge for combinatorics starting with the famous 5-$(24, 8, 1)$ design described by Witt with $M_{24}$ as group of automorphisms. (An automorphism of a design is a bijective mapping $\pi : V \to V$ with $\mathcal{B}^\pi = \{b^\pi \mid b \in \mathcal{B}\}$ inducing an action on the blocks which is supposed to map $\mathcal{B}$ onto $\mathcal{B}$.) $t$-designs with $\lambda = 1$ are called *Steiner systems*. They are of particular interest, a 2-$(v, 3, 1)$ design is a Steiner Triple System, or $STS(v)$.

In order to describe small examples of designs we consider the 2-dimensional projective space over a finite field of order $q$. One starts by labelling projective points and hyperplanes in a specific manner: So, $\langle (x_1, x_2, x_3) \rangle$ with $x_1, x_2, x_3 \in GF(q)$, not all three zero, denotes a one-dimensional subspace as well as the hyperplane which is orthogonal to it. A point $(x_1, x_2, x_3)$ is contained in the hyperplane orthogonal to $(y_1, y_2, y_3)$ if and only if $\sum_{i=0}^{3} x_i y_i = 0$. Such a design is called a *projective plane*. The projective plane for $q = 2$ is also a $STS(7)$, the well known Fano plane:

### 4.2.14 DISCRETA: A Tool for Constructing $t$-Designs

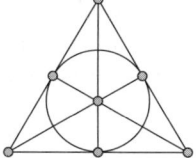

All these designs from projective planes have the remarkable property that there are equally many points and blocks. Such a design is called *symmetric* (in fact, Dembowski in his influential book [Dembowski 1997] calls these designs projective). An interesting but difficult question is the construction of designs with the parameters of a projective plane but which are different from a projective plane design.

There is a unique $STS(9)$, which is the affine plane of order 3. The next smallest Steiner Triple Systems are on 13 points. There are two different designs, cf. Figure 19.

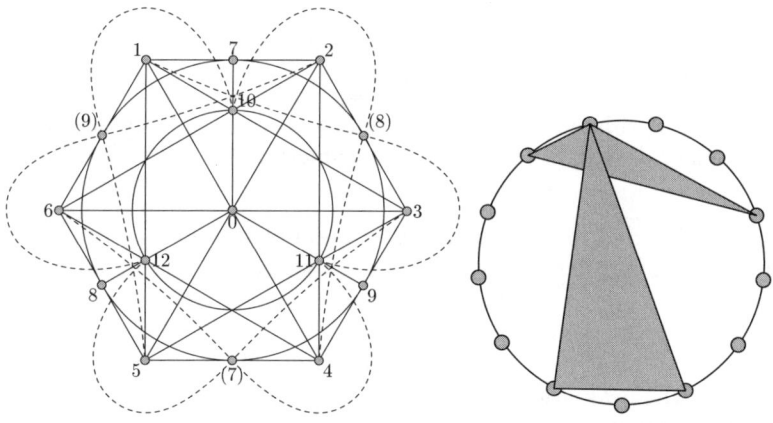

**Fig. 19.** The two $STS(13)$

In the first drawing, the blocks are indicated by lines. The points 7, 8 and 9 are doubled. The second system has a cyclic automorphism group with two orbits on blocks. We have chosen one block of each orbit and drawn them as triangles. The complete set of blocks is obtained by letting the blocks rotate 13 times along the circle. The number of Steiner triple systems grows rapidly with the number of points. There are exactly 80 $STS(15)$ and the number of $STS(19)$ is known to be in the millions.

All designs which we have seen so far were 2-designs. In order to solve the difficult problem of finding $t$-designs with large $t$ one may apply the following method: *prescribe a certain group which the design should possess as a group of automorphisms.* Then, one proceeds as follows: This group – call it $A$ – induces orbits on the set of $t$-subsets and $k$-subsets of $V$ and in order to construct a $t$-design which is invariant under $A$ collect orbits of $A$ on $k$-sets. The condition to form a $t$-design is equivalent to solutions of a linear system of equations with non-negative integral coefficients. This Diophantine System is called the Kramer-Mesner System according to Kramer and Mesner who described that method in [Kramer and Mesner 1976]. The 0/1-solutions of that system allow to describe the design as a collection of $A$-orbits.

DISCRETA is designed to easily apply this method to sets of moderate parameters and in particular to a variety of different permutation groups. One starts with choosing an appropriate group from a large list of groups including infinite series of permutation groups like projective groups. Then one fixes the parameters of the design (the degree of the group coincides with the number of points $v$). DISCRETA allows to compute the Kramer-Mesner matrix and to solve the corresponding Diophantine system yielding designs. Internally, the program uses double cosets in groups to evaluate transversals for the $A$-orbits on subsets. The Diophantine equations are solved by a refined LLL-algorithm which even allows to enumerate all solutions for systems of reasonable size (cf. [Wassermann 1998]). In typical applications, the size of the matrices which can be handled are up to some hundred rows and columns.

A variation of this construction allows to search for large sets of designs. These are sets of designs with the same parameters that partition the set of all $k$-subsets of $V$.

A far-reaching and important question is the determination of isomorphism types of designs constructed in the manner described above. In small cases, the system provides a direct approach to isomorphism checking by computing a canonical labelling of the incidence matrix for each of the designs. However, for large designs one must switch to other methods. One approach supported by DISCRETA is the evaluation of invariants of the designs like intersection numbers. Often one can prove that all designs have distinct invariants and thus must be non-isomorphic. Another method of checking isomorphism involves group theoretic methods using Sylow $p$-subgroups of the full group of automorphisms. These methods allow to search for isomorphisms in subgroups which usually are small compared to the full group of permutations of the complete point set $V$ (cf. [Betten et al. 1998c] for both methods of isomorphism checking).

DISCRETA provides a graphical user interface and is therefore easy to use. No programming skills are required. It runs on Unix systems including Linux PCs and is written in C++. The program is freely available from the web pages of the authors, cf. [Betten et al. 1998d]. On that page, one may also find screenshots of typical DISCRETA sessions. There is also a database of design parameter sets with large $t$. This page collects results obtained with DISCRETA and results from other research teams. Outstanding results obtained with DISCRETA are the first 7- and 8-designs with small parameters and a 5-$(36, 6, 1)$ Steiner system.

In addition to the graphical user interface there is also an interface to the computer-algebra system GAP [GAP 1997]. The whole functionality of DISC-RETA is available via GAP commands from the GAP command prompt. This allows writing programs making use of DISCRETA for example if one wants to loop about a large set of parameters of groups avoiding tedious manual processing. This interface to GAP provides all important data from the construction process to the GAP-side. Thus, using this interface, one may incorporate group theoretic algorithms and the DISCRETA functionality to a very powerful tool for the construction and classification of designs.

Additional references: [Assmus, Jr. and Key 1992; Beth et al. 1986; Colbourn and Dinitz 1996; Magliveras and Leavitt 1984; Kreher 1996; Betten et al. 1995]

Anton Betten, Reinhard Laue, Alfred Wassermann (Bayreuth, Germany)

## 4.2.15 FELIX

**4.2.15.1 Introduction.** The core part of FELIX is a library of tools for developing and testing programs in symbolic computing.

Main focus of the authors was the implementation of algorithms from the theory of commutative and non-commutative ideals. Therefore, the system features a rather comprehensive library of modules written in the FELIX-language as well as compiled object files for that particular area. However, as a LISP-like system, FELIX is far more universal.

In the first paragraph, we present the basic structure of the system, and it's hardware requirements; followed by a brief description of the programming language and environment in the second paragraph. In the third paragraph, we take a closer look at the built-in algebraic algorithms of FELIX. Finally, we conclude with a short sample session.

**4.2.15.2 System Overview.** FELIX consists of three layers. The kernel of the system provides the basic data structures

- nodes (as basic objects of lists),
- short and long integers,
- names,
- integer and bit vectors, and
- (one-dimensional) arrays

as well as functions operating on these basic objects. Moreover, the kernel contains simple control structures, system interfaces, and last but not least memory management routines (especially *garbage-collection*).

Since the kernel was intentionally kept as small as possible the user is provided with only a primitive, LISP-like language which is not intended to be a user interface, although it can be called from any application. In the spirit of the C-language paradigm, the actual programming language, as well as various development tools, were written in FELIX itself. This concludes the description of the second system level.

The third layer consists of a set of algorithms for symbolic computing. For the time being, it is limited to the area of commutative and non-commutative ideals.

Currently, FELIX runs on personal computers under operating systems which provide the DOS Protected Mode Interface (DPMI), e.g. WINDOWS or OS/2, and on workstations under various versions of the UNIX operation system. Under UNIX there are included special built-in functions for interprocess communication and networking. These functions cover establishing connections and transferring data. The implemented interface is mainly based on stream sockets and includes a special protocol for the data exchange. So, it supports the distributed computation on heterogeneous, loosely coupled systems.

**4.2.15.3 Programming Environment.** From the data structures point of view FELIX is considered to be LISP-like. The "nodes" - mentioned in the first paragraph - constitute the basic building blocks of lists. In contrast to LISP, the set of atomic data types was enlarged by integer vectors, bit strings, and arrays. These data types, including basic functions operating on them, allow a representation of polynomials which matches the requirements of Gröbner basis computations.

Although FELIX inherited its functional programming characteristics - together with its data structures - from LISP, the syntax of the programming language is also PASCAL-like, adding procedural elements to it.

FELIX can be used interactively, like a pocket calculator. Additionally, algorithms can be defined as operators. The system does not distinguish between subroutines and functions. One typical functional characteristic is the convention that each operator call returns a value which can be substituted into complicated expressions. However, the return value could be less important than the side effects of the function call in which case the operator would take on procedural attributes.

In principal, FELIX's operators are global, i.e., there is no scoping of function names, and nested declarations are not permitted. The language supports explicit and implicit recursive function calls. Besides its formal parameters, an operator can have two kinds of local variables, which differ with respect to their scope rules. Where the first class - the LOCAL variables - have static binding. Those variables are only visible within the operator body. The so-called FLUID variables are subject to dynamic binding, i.e., they appear to operators called within that operator like global variables. Their binding is depending on the sequence of function calls.

Within an operator definition, the following control directives are provided:

- **Loops :**
  LOOP, FOR-DO, REPEAT-UNTIL, WHILE-DO
- **Branches :**
  WHEN, IF-THEN-ELSE, CASE
- **Block :**
  BEGIN-END
- **Exit :**
  RETURN

Labels and unconditional jumps were intentionally left out.

For any experienced programmer, the semantics of the commands is quite self explanatory, we are not going into the details here. Only LOOP, the conditional jump WHEN, and the block and return instructions are part of the system kernel. The WHEN instruction executes a sequence of statements pending on the value of a boolean expression, and at the same time effects a jump out of the current block. Such a block structure either starts with a LOOP or a BEGIN statement, or is given by an operator definition. BEGIN-END blocks are needed to contain jumps within the function body. The LOOP instruction is implemented as infinite loop; it encloses a sequence of statements which are repeated cyclically. It can be exited via the WHEN instruction; the current operator structure is exited via the RETURN instruction, where the jump can break out from multiply nested blocks. The higher level loop and branch constructs are translated into the smaller set of kernel instructions by a preprocessor at input time.

If there is need, one can define additional language constructs quite easily. The only requirement is that they start with a keyword; the definition of the subsequent syntax is at the discretion of the user.

Reading in an operator definition causes an interpreter object being created, and bound to the operator's name. Upon completion of this transaction, the operator becomes available immediately. Especially during the development and testing phase this form of an operator has its advantages.

Later in the process such an interpreted operator might become a burden. Execution is relatively slow, and loading from source code adds to the running time. Therefore, FELIX has a built-in compiler which allows translation of a set of operators into machine code, and to combine them in a library of modules. Each module in turn, can contain local operators which are only callable from operators within the same module. This concept of data hiding has the advantage that only operators which are visible from the outside ("exported operators") have to be defined in the module interface.

Compilation and binding of modules is an integral part of the interactive mode; compiler and interpreter are equivalent apart from peculiarities imposed by the modularization. Users do not have to concern about the process of binding modules. However, means are provided to control it manually. Whenever an undefined operator is called, automatically all module files are searched for an exported operator having that name. Upon success, the module is added and processed; on failure, it will generate an unresolved operator reference.

The system provides tools for debugging. Users can specify watchdogs for particular functions, either to just count the number of calls, or to print the function's name together with its actual parameters upon entry, the name and its return value on exit. The most rigourous method of testing is the setting of a break point. In this case, execution stops immediately before entering the function, and the user is put into - compared to the basic system slightly degraded - interactive mode. At that point, utilities are available to trace back the chain of function calls in terms of the names of the called operators, their actual parameters, and their LOCAL variables. Moreover, the values of global variables can be displayed, and even the values of parameters on the stack may be changed interactively.

For compiled functions, the options for monitoring execution is more limited in the sense, that calls to operators within the same module, and calls to kernel functions cannot be traced.

Besides explicit setting of break points, the user can interrupt the computation at (almost) any time and enter the debugging mode. This is especially of advantage during time consuming computations, in case one should have doubts regarding the correctness of the results early on.

Users can define a text editor of their choice to be used during an interactive session for editing of sample or program files. There exist functions for input and output of source code and sample files, where file input may be nested. In case syntax errors should occur during reading into the system, one can immediately jump to the respective file location in the text editor (if positioning is supported by the editor). Furthermore, a shell can be initiated in order to execute operating system commands during computations.

For an overview about data types and programming facilities we refer to [Apel and Klaus 1992].

**4.2.15.4 Algebraic Algorithms.** FELIX is specially designed for computations in rings and modules. Besides operations over the basic algebraic structures, it can also perform operations in sub-structures, especially in ideals and modules, as well as operations related to morphisms between the structures.

So far, the following algebraic structures are part of the system:

– Rings of polynomials,

- Algebras of solvable type [Kandry-Rody and Weispfennning 1990],
- free non-commutative algebras, and
- finitely generated modules over rings mentioned above.

For coefficients, one can choose from

- integers,
- rationals,
- complex numbers,
- residue classes of integers, and
- rational functions.

Besides basic arithmetic in above structures, additional operations are available for some algebraic objects. Naturally, they are limited to particular classes of rings, e.g., GCD and factorization of polynomials.

Operations on ideals play a central role in the system. They are based to a large extent on Buchberger's algorithm for Gröbner Basis computation [Buchberger 1965]. That algorithm and its various generalizations to non-polynomial rings (e.g., [Möller and Mora 1986; Kandry-Rody and Weispfennning 1990; Mora 1986; Apel and Laßner 1988; Apel 1992]) is used to compute elimination ideals (which relates to solving systems of non-linear algebraic equations), modules and chains of syzygies, as well as intersections and quotients of ideals. A treatise of these application can be found in e.g., [Buchberger 1985b] and [Gianni et al. 1988].

Special attention was paid to flexibility of Buchberger's algorithm. Several options for term ordering, and the selection strategy for critical pairs and reductions are available [Apel and Klaus 1991a]. Also the influence of various parallelization approaches were investigated [Klaus 1996]. Statistics is maintained in order to provide insights into basis growth during run-time, and questions regarding the choice of critical pairs and their processing. The level of statistics can be controlled by flags.

Quite some time was spent on experiments to find an optimal internal representation of polynomials and elements of free non-commutative algebras [Apel and Klaus 1991b]. In principal, a distributed representation of polynomials in Gröbner Basis computations is preferable over a recursive representation. The main advantage of recursive representations is the property that during any session, new variables can be added continuously without affecting previously computed objects. The design of FELIX does not support this feature, the complete definition of the working structure is imperative. Therefore, the system uses only the distributed representation for polynomials anymore. However, this leads to the problem of representing monomials and words (non-commutative products of variables).

For this purpose we introduced the basic data types *integer vector* and *bit string*. Kernel functions implement the semi-group operations, subword matching, LCM computation, and linear semi-group ordering on these data structures. Integer vectors and bit strings are uniquely stored as balanced binary trees. Their favourable properties allow fairly compact storage of monomials, assuming sparsely populated integer vectors.

#### 4.2.15.5 Sample Session.

```
> select rat[a,x,y]/{y*x==x*y+a}_
> k := y^3+x^2*y+x*y$
@ := X^2*Y+Y^3+X*Y
> h := x^2+x$
@ := X^2+X
> standard(leftideal(k,h))$
******************************************************************
LEFTIDEAL of 4 elements
   X^2+X
   A^2
   Y^3-2*A*X-A
   A*Y^2
******************************************************************
> standard(rightideal(k,h))$
******************************************************************
RIGHTIDEAL of 5 elements
   X^2+X
   Y^3
   A*Y^2
   A^2*Y
   A^3
******************************************************************
>standard(ideal(k,h))$
******************************************************************
IDEAL of 3 elements
   A
   X^2+X
   Y^3
******************************************************************
```

Joachim Apel and Uwe Klaus (Leipzig)

### 4.2.16 *Fermat*

**4.2.16.1 Introduction** *Fermat* is a special-purpose system oriented toward polynomial and matrix algebra over the integers $\mathbb{Z}$, the rationals $\mathbb{Q}$ and finite fields. It has been developed by Robert H. Lewis of Fordham University, New York. It is among the class of Computer Algebra Systems which are specialized in doing computations on polynomials and matrices, such as for instance Macaulay [Bayer and Stillman 1992b], Macaulay 2 [Grayson and Stillman 1996], CoCoA [Giovini and Niesi 1990] and Singular [Greuel et al. 1995].

The architecture of *Fermat* is designed to offer flexibility through *efficiency* in both time and space, *portability*, *programmability*, and *choices of coefficient arithmetics*.

Currently *Fermat* runs on Macintosh and on Windows 95/NT. Versions for Linux are expected sometime in 1999. The system can be obtained from the URLs:

http://www.bway.net/~lewis/

and

http://www.fordham.edu/lewis/

Besides the main, or *rational version* that will be described here, there is also a *floats version* of *Fermat* for graphics on the Macintosh, in which the basic data type is real (or complex) number with 18 significant digits of accuracy. Though one can define and manipulate polynomials and matrices much as in the rational version, the floating point data type does not permit the sophisticated polynomial and matrix algorithms of the main version. However, it allows the user to create arbitrarily complex geometric *objects* consisting of lines, points, and polygons. These objects can be viewed from arbitrary angles, rotated, translated, cut up, or strung together to form *animation sequences*, or movies. For example, numerical solutions of differential equations, such as planetary motions, can be displayed as movies. Surfaces can be graphed as *hidden-line objects* or *shaded objects*. Every point, line, or polygon can be colored.

In the documentation for *Fermat* and on the web pages, the main (rational) version is referred to as *QFermat* and the float version as *FFermat*.

**4.2.16.2 History and Use** *Fermat* began as a project in a compiler design course in 1986. It was therefore a programming language from the very beginning, and has always gone by the name "Fermat." It grew into a computer algebra system gradually. It flourished under three years of R. E. U. (Research Experiences for Undergraduates) projects in which teams of undergraduates used it to search for nilpotent topological spaces [Lewis and Moore 1997]. It has also been used to study certain infinite almost-free groups [Lewis and Liriano 1994], to study elliptic curves [Brumer 1992], and in image analysis [Lewis 1997], [Lewis and Nakos 1998].

**4.2.16.3 Features** *Fermat* performs both simple and sophisticated operations on multivariate polynomial rings, rational functions, and matrices containing polynomials and rational functions. Arbitrarily large integers are available. In

*Fermat* one may choose to work modulo a specified integer $n$, thereby changing the *ground ring* from $\mathbb{Z}$ to $\mathbb{Z}/p\mathbb{Z}$. On top of this ground ring $F$ may be attached any number of symbolic variables $t_1, t_2, \ldots, t_n$, thereby creating the polynomial ring $F[t_1, t_2, \ldots, t_n]$ and its quotient field, the field of rational functions. Further, monic polynomials $p, q, \ldots$ can be chosen in a certain way to mod out, creating the quotient ring $F(t_1, t_2, \ldots) \,/<p, q, \ldots>$. *Fermat* is especially effective when this quotient ring is a field. Finally, it is possible to allow *Laurent polynomials*, those with negative as well as positive exponents. Once the computational ring is established in this way, all computations are of elements of this ring.

The system is capable of performing basic operations such as: sums, products, quotients, powers, derivatives, gcd, content, and lcm of polynomials; sums, products, quotients, powers, and derivatives of rational functions; sums, products, powers, determinants, adjoints, inverses, of matrices. Many determinant algorithms are implemented, such as expansion by minors, Gaussian elimination, Gauss-Bareiss, and Lagrangian interpolation. The user may choose among them or let *Fermat* choose.

More advanced operations are square-free factoring, complete factoring of univariate polynomials over finite fields; various degree, term, and coefficient functions; characteristic polynomial and minimal polynomial of matrices; various normalizations and diagonalizations of matrices. *Fermat* is especially good at Smith Normal Form.

In *Fermat* one may create *sparse* matrices. This is an alternate storage structure in which only the nonzero entries only are stored, in list structures. The matrix functions may be applied to sparse matrices as well as "ordinary" matrices.

One of the main features of *Fermat* is a Pascal-like programming language. The syntax is flexible and intuitive, allowing one to write expressions like

[x]*[y]^3 - 2[z], where [x], [y], and [z] are matrices. The language contains the usual features: assignment, conditionals, for-loops, while-loops, procedures (functions), recursion. A compiler is available for execution speed. Certain other features of the language speed up procedures in which all variables are of simple data types.

Another important feature of *Fermat* is the ability to read and save data and programs to ordinary ASCII files, which may be created and edited with any word processor. This should permit easy communication with other systems.

*Fermat* is written in Pascal, about 65,000 lines in the main (rational) version, and about 35,000 more in the float version.

**4.2.16.4 Examples** *Fermat* has proven to be extremely efficient in both time and space. One reason for this may be that the multivariate polynomial gcd algorithm is very elaborate, involving for example modular techniques and Hensel's lemma. Another reason may be efficient data structures and garbage collection techniques.

To give complete examples here of this speed and efficiency is difficult, because problems on which *Fermat* excels are inevitably large. The *Fermat* web pages contain several test problems or "challenges," summarized here. They involve resultants, Smith Normal Form, and the evaluation of complex expressions of rational functions.

*Example 10.* Consider three homogeneous polynomials $f$, $g$, and $h$ of degree two in three variables,

```
f := a6*y^2 + a5*y*z + a4*y*w + a3*z^2 + a2*z*w + a1*w^2;
g := b6*y^2 + b5*y*z + b4*y*w + b3*z^2 + b2*z*w + b1*w^2;
h := c6*y^2 + c5*y*z + c4*y*w + c3*z^2 + c2*z*w + c1*w^2;
```

Each polynomial has six coefficients that are independent parameters, for a total of 18 parameters. Van der Waerden shows in [van der Waerden 1950] how to create three $15 \times 15$ matrices $d1$, $d2$, and $d3$ such that the resultant of $f, g,$ and $h$ is the $GCD(Det[d1], Det[d2], Det[d3])$. *Fermat* easily computes the answer, a homogeneous polynomial in all 18 parameters with 21894 terms.

*Example 11.* Four rational functions $s_1, s_2, s_3, s_4$ and a polynomial $res$ are given in 16 variables, $q_1, q_2, q_3, q_4$, and 12 others. $res$ has 239 terms, and each $s_i$ has around 40 terms in both numerator and denominator. The test is to verify that $res$ evaluated at $q_1 = s_1, q_2 = s_2, q_3 = s_3$, and $q_4 = s_4$ is 0. This is an actual problem that came up in image analysis.

*Example 12.* Compute the Smith Normal Form of a $51975 \times 13860$ sparse matrix. It has 4 nonzero entries in each row, and has rank 12440. This is an actual problem that came up in the computation of the homology of simplicial complexes.

Robert H. Lewis (Fordham University, New York)

### 4.2.17 FoxBox and Other Blackbox Systems

In the second half of the 1980s, Erich Kaltofen invented algorithms for computing with symbolic mathematical objects in straight-line program and black box representation [Kaltofen 1988, 1989; Kaltofen and Trager 1990; Kaltofen and Saunders 1991]. Thus is the genesis of black box computer algebra. The main ingredient is the idea that a program, which can be a straight-line single assignment program or a procedure expressed in a programming language, that computes the value of a multivariate polynomial when substituting elements from a given field for the variables, implicitly defines the polynomial. Interpolation or sparse interpolation [Zippel 1990; Lakshman Y. N. and Saunders 1995; Grigoriev and Lakshman 2000; Kaltofen et al. 2000] can be employed to reconstruct the polynomial in conventional representations. However, such conversion can cause exponential space growth, or in Knuth's words [Knuth 1998], "would fill the universe." The black box algorithms adopt the implicit representation for the results of the operations: for instance, an irreducible factor of a black box polynomial is again a procedure for evaluating that factor. Hence black box computer algebra can efficiently compute a sparse irreducible factor of polynomial that is itself dense in standard representation and whose other factors are also dense.

Efficiency can be measured both theoretically and practically. For polynomials and rational functions, our algorithms for GCD, factoring, and retrieval of numerators and denominators are of polynomial-time complexity in the number of variables. For linear algebra, where the black box model of a matrix is a procedure that computes the product of the matrix with an input vector, our algorithms are of quadratic running time in the dimensions of the matrix, at least when the coefficient field is finite (see the subsection on Linear Algebra on page 2.3). Many of our algorithms at this stage of knowledge require randomization.

The black box model has been put to practice by several computer implementations. We have built the DAGWOOD [Freeman et al. 1988], FOXBOX [Díaz and Kaltofen 1998], WILISYS [Kaltofen and Lobo 1999b], and PROTOBOX [Kaltofen et al. 2000] systems* and are building the LINBOX library. DAGWOOD is a system written in Lisp that implements algorithms for manipulating polynomials and rational functions in straight-line representation. Experiments with DAGWOOD showed that the straight-line outputs can have excessive length. That drawback is completely ameliorated by our FOXBOX system written in C++ that implements algorithms for manipulating polynomials and rational functions in black box representation. We know of efforts elsewhere to utilize the straight-line program representation: John Canny's and Ashu Rege's APU (available from Canny jfc@cs.berkeley.edu) and the TERA project for equation solution http://tera.medicis.polytechnique.fr/. WILISYS is a system written in C that implements the block Wiedemann algorithm for solving sparse linear systems over a finite field. David Saunders and we have joined efforts with the Givaro designers [Gautier et al. 1999] to build a new library for black box linear algebra, LINBOX. Initial results are reported in [Dumas et al. 2000]. See http://www.cis.udel.edu/~caviness/linbox/ for progress on the project.

Both FOXBOX and WILISYS can be executed on parallel systems. Moreover, generic programming methodology is employed both in FOXBOX and LINBOX. We use a clearly defined object interface for the underlying coefficient domain

---

* The programs are available in source code from links in Erich Kaltofen's Internet homepage, see http://www.kaltofen.net.

and polynomial or matrix arithmetic. We plug in several libraries, such as NTL, Saclib, GnuMP, Givaro for performing basic arithmetic operations.

PROTOBOX is a system written in Maple that implements algorithms for manipulating polynomials in black box representation. It is used primarily for prototyping our black box algorithms for later implementation in C++. A similar prototype Maple program was developed in [Gut 1999].

Erich Kaltofen (Raleigh, USA)

## 4.2.18 GAP

### 4.2.18.1 Introduction
GAP (Groups, Algorithms and Programming) is a system for computational discrete algebra with particular emphasis on, but not restricted to computational group theory. GAP was developed at Lehrstuhl D für Mathematik (LDFM), RWTH Aachen, Germany from 1986 to 1997. After the retirement of J. Neubüser from the chair of LDFM, the development and maintenance of GAP is coordinated by the School of Mathematical and Computational Sciences at the University of St Andrews, Scotland. Several users have contributed to the system via share packages which can be loaded into the system and then act as extensions to the main library. Some share packages also provide interfaces to separate standalones, such as the ANU quotient algorithms (see section 4.3.1).

The current (March 2000) version of GAP is GAP 4.2, released February 2000.

GAP and its sources, including share packages, data library and the manual, are distributed freely, subject to "copyleft" conditions (which are detailed on the web pages mentioned below). Please cite the GAP manual if you use GAP in research (again see the web pages). GAP runs on any Unix system (in particular Linux), under Windows 9x/NT, and on Macintosh systems. It requires a minimum of 20 MB disk space, the full distribution (including all share packages and extensive data library) takes about 180 MB. To run GAP one needs a minimum of 20 MB of main memory, 64 MB permits decent usage, for most purposes 128 MB are sufficient.

GAP 4 supersedes the older version GAP 3.4.4. While this older version is still available (a few share packages have not yet been translated and only run under the older version) we strongly encourage everybody to upgrade to GAP 4 which provides much new functionality and is the version that will be developed further.

More details about the system, the software itself, and instructions for obtaining and installing it can be found at http://www-gap.dcs.st-and.ac.uk/~gap. If you have no web access you can get this information in another form. Please contact "The GAP Group, Mathematical Institute, St Andrews, Fife, KY16 9SS, Scotland ", Tel. +44/1334/463251, Fax 463278, e-mail gap@dcs.st-and.ac.uk.

#### 4.2.18.2 A Brief Overview of the Structure of GAP
GAP consists of a kernel (a C program), the library of functions (GAP programs), the data library and the documentation. Distributed with it are a number of user contributed share packages.

**4.2.18.2.1 Kernel Features:** The kernel implements a programming language, also called GAP. Among the features of this language are:

- Pascal-like control structures.
- Automatic memory management. This includes garbage collection, it automatically throws away objects which are no longer accessible.
- Built-in data types for key algebraic objects.
- Flexible list and record data types.

The language is interpreted and can be compiled.

The kernel also implements the environment that allows one to interact with GAP in the so–called read–eval–print loop, supports debugging of GAP programs and provides online access to the reference manual. This manual (which consists of over 1000 pages in total) is also provided as an HTML document for browsing, and in printable form.

**4.2.18.2.2 Library Features:** The library of functions contains implementations of various algebraic algorithms written in the GAP language. Besides the possibility of using them, you can also look at them to find out how they work, and use them as examples for your own programs. Altogether the library contains over 300 000 lines of GAP code, providing several thousand functions to the user.

**4.2.18.2.3 Share Packages:** There is a growing collection of user contributed share packages. Some of these are written in the GAP language, others include external C programs. All of these can be called directly from GAP as if they were library functions.

**4.2.18.2.4 Data Library:** The data library currently takes up about 120 MB, its content will be summarized below.

#### 4.2.18.3 Capabilities of GAP

In this description we do not always distinguish between the functionality provided by the body of GAP and by its share libraries.

**4.2.18.3.1 Basic Functionality:** Long integer and rational arithmetic, cyclotomic fields, finite fields, residue class rings, $p$-adic numbers, permutations, polynomials (multivariate polynomials and rational functions), vectors and matrices, various combinatorial functions, elementary number theory, a wide variety of list operations. Input/Output streams for interfacing to other software.

**4.2.18.3.2 Groups and Group Elements:** In short, almost all the functionality described in Section 2.7 is available in GAP.

To give an idea of capabilities, GAP has been used to classify all groups of order up to 2000 (with the exception of 1024), to determine the transitive permutation groups of degree up to 31, and to find the automorphism groups of compact Riemann surfaces up to genus 48.

Some examples of the use of GAP and a bibliography of work involving GAP can be found at http://www-gap.dcs.st-and.ac.uk/~gap/Info/examples.html, we mention in particular the surveys [Neubüser 1995; Hulpke and Linton 1999].

Groups can be given in various forms: for example as permutation or matrix groups (by generating elements), as finitely presented groups (see Section 2.7.9) or as polycyclicly presented groups (see Section 2.7.8). GAP knows how to construct a number of well-known groups such as symmetric and classical groups. There is a wide variety of functions for the investigation of groups that will compute e.g. order, conjugacy classes of elements, derived series, composition series (including identification of the composition factors), Sylow subgroups, certain characteristic subgroups, maximal subgroups, normal subgroups, subgroup lattice, automorphism group, cohomology groups, character table and table of marks. Of course the range of applicability of the particular functions depends very much on the order and structure of the group.

Some of these functions just build on the concept of a group while others (usually the more efficient ones, for instance newly developed nearly linear methods for permutation groups used in GAP 4) utilize the way in which a particular group is given. GAP tries automatically to select a good method but the user can take over full control of this selection of methods.

To give a small sample of special methods: The nearly linear time methods for permutation groups include functions to compute a stabilizer chain, $p$-core, radical, centre and composition series. There are also tasks for which no polynomial time methods are known and for which GAP relies on partition backtrack methods, for example centralizer, normalizer or intersection of subgroups. For finitely presented groups there is the family of coset table methods such as Todd-Coxeter, Reidemeister-Schreier, low index, as well as Tietze transformations. There also are share packages for a double coset enumerator, a Knuth-Bendix and automatic groups methods. Further there are the so-called quotient methods, building in one form or other on "collection" techniques, in particular $p$-quotient, nilpotent quotient, several solvable quotient and a polycyclic quotient algorithm. Polycyclic generating systems can also be used for groups given as permutation groups. There also is again a wide variety of functions utilizing polycyclic presentations, being able to compute efficiently conjugacy classes, centralizers, normalizers, intersections, maximal subgroups, formation theoretic subgroups, and complements of solvable subgroups.

#### 4.2.18.3.3 Representations and Characters of Groups: Again, much of the functionality described in Section 2.8 is implemented:

Ordinary representations (over fields of characteristic 0) are mainly investigated via their characters. GAP provides methods to compute character tables automatically from concretely given groups (e.g. permutation groups) as well as a large set of tools for calculating with (partial) characters for the interactive construction of character tables. It also admits deduction of group theoretic properties from character tables, which is possible for example using the data base of character tables. Modular representations (over fields whose characteristic divides the group order) can be studied via Brauer characters or by calculations with matrices representing the generators of a group by the whole complex of methods known as MEAT-AXE (see 4.2.10), vector enumeration and condensation techniques. A link to the Atlas of Group Representations (see http://www.mat.bham.ac.uk/atlas) provides access to many permutation and matrix representations of almost simple groups and related information.

The share package **CHEVIE** (see 4.2.9) allows one to work with generic characters of groups of Lie type and Hecke algebras.

GAP also contains functions for the creation and manipulation of tables of marks. (Such a table represents information about the subgroup lattice of a group. See [Pfeiffer 1997] for further information.)

#### 4.2.18.3.4 Vector Spaces, Modules and Algebras: Compared to GAP 3, GAP 4 substantially extends the capabilities in this area. Vector spaces over all available fields and modules over all available rings can be defined. E.g., there are algorithms for the efficient calculation of Hermite and Smith normal forms.

Algebras (associative ones as well as Lie algebras) can be given by structure constants, by generating matrices or by a finite presentation (The latter case is handled by share packages.) There are routines for computing the structure of finite dimensional Lie algebras, e.g., for computing Cartan subalgebras, the direct sum decomposition, a Levi decomposition, the solvable radical and nil radicals (cf. 2.6.6). For associative algebras, GAP provides basic functionality, however, so far not much of the functionality described in Section 2.6 has been implemented.

The share package **SPECHT** permits computations concerning special modules (e.g. Specht modules) arising in representation theory.

#### 4.2.18.3.5 Graphs and Codes: The package **GRAPE** (see 4.3.10) allows one to classify certain graphs and designs, to determine the invariants of graphs and to calculate their automorphism groups. A special feature of this package is

the use of (parts of) the automorphism group (and hence of permutation group methods) of a graph to reduce the amount of data needed for the description of the graph. The package GUAVA can construct many classes of codes, derive invariants such as minimal distance and weight distribution and find their automorphism groups.

**4.2.18.3.6  Further Functionalities:** These include share packages for crystallographic groups, semigroups, near rings, crossed modules and cat-1 groups. There is a graphical user interface XGAP (in function similar to QUOTPIC, see 4.2.32) that permits for example the display of subgroup lattices.

**4.2.18.4  Data Library**
The GAP data library includes a number of databases with GAP language interfaces allowing them to be searched and studied efficiently. It currently includes:

- Primitive permutation groups, distributed in various data bases, covering all primitive groups of degree up to 255 and non-affine groups up to degree 999. [Dixon and Mortimer 1988; Short 1992; Theißen 1997]
- Transitive groups of degree up to 23 (soon to be extended to degree 31). [Hulpke 1996b]
- All groups of order up to 2 000, excluding 1 024. [Besche and Eick 1998; Eick and O'Brien 1999]
- Perfect groups up to order $10^6$ (with some specified omissions). [Holt and Plesken 1989]
- Solvable irreducible matrix groups on $GF(p)$ vector spaces up to order 255. [Short 1992]
- Space groups up to dimension 4 via the CRYSTCAT share package [Brown et al. 1978], and up to dimension 6 via a link to CARAT (see 4.2.7).
- Maximal finite irreducible subgroups of $GL_n(Z)$ (classified partly up to $Z$- and partly up to $Q$-conjugacy) up to dimension 24 (soon to be extended to dimension 31). [Nebe 1996; Nebe and Plesken 1995]
- Ordinary character tables, including all of those in the ATLAS of Finite Groups. [Conway et al. 1985]
- Modular character tables, including all of those in the ATLAS of Brauer Characters. [Jansen et al. 1995]
- Tables of marks for various groups. [Pfeiffer 1991; Merkwitz 1998]

**4.2.18.5  GAP 4 Versus GAP 3**
For interactive use or simple programming GAP 4 looks a lot like the older version GAP 3, especially the GAP language has not changed. A few commands have changed names and some tasks will be performed more efficiently. Nevertheless the GAP 4 kernel has been rebuilt from the ground up (Martin Schönert, Frank Celler). It now has more efficient memory management, faster function calling, save/load workspace facilities, fast vector arithmetic for finite fields, and streams. It is easier to extend and 64 bit clean. There is now a GAP compiler that produces human-readable C code. This C code can be compiled and loaded dynamically

(UNIX only) or compiled into a (new) kernel. Compiled code is automatically loaded when it exists.

In GAP 4, a user can design new objects (similar to the records of GAP 3). Similar to kernel objects, such as permutations and elements in polycyclic groups, these new elements can "hide" their innards to be protected against accidental changes. These new features are intended to make it easier than in GAP 3 to extend the system consistently by new data types. Examples for new data structures using this facility in GAP 4 are enumerators (special kinds of lists) and iterators (which admit to loop over virtual lists). Also the representation of algebraic structures via records in GAP 3 has been replaced in GAP 4 by one that uses these new objects. At the same time, the operations records of GAP 3 have been replaced by a more flexible system [Breuer and Linton 1998]. Every GAP 4 object has a type, which is used in the choice of methods for an upcoming computation. Part of this type is known information about the object. For example, when GAP is asked to compute the conjugacy classes of a group, different methods are available. One of these is a method for solvable groups. This method can be chosen if the group is known to be solvable, which would be part of its type. In particular this mechanism is used to utilize mathematical implications; for example, a group that is known to be nilpotent automatically knows that it is solvable.

#### 4.2.18.6 Support Structure and Acknowledgements
A mailing list "GAP-Forum" has been established where interested users can discuss topics by e-mail. In particular information about new versions and bugfixes is distributed via this forum. See the web page
http://www-gap.dcs.st-and.ac.uk/~gap/Info/miles.html for information
on how to subscribe.

If you have technical or installation problems, we suggest that you write to the support address `gap-trouble@dcs.st-and.ac.uk` instead (not to the GAP Forum), as such discussions are usually not very interesting for a larger audience. Your e-mail will be read by several people who shall try to provide support.

GAP is intended to be an international enterprise in which people interested in computational group theory and related topics can cooperate. Hence its further development depends also on outside contributions. A "GAP-Council" of experts in various areas of computational group theory has been established which not only advises the further development of GAP, but in particular, acting as an editorial board, will formally acknowledge acceptance of contributions to GAP in a similar way as editorial boards of journals accept publications. Details on this scheme are again found on the GAP web pages.

We acknowledge with gratitude the support of the development of GAP in Aachen by grants of the Deutsche Forschungsgemeinschaft in the frame of the Schwerpunkt "Algorithmen in Zahlentheorie und Algebra", by the EU Human Capital and Mobility Network, grant number ERBCHRXCT930418 "Computational Group Theory", by the Engineering and Physical Sciences Research Council Grant GR/L21013, by the Leverhulme trust grant F/268/Q, and by the British Council and the Deutscher Akademischer Austauschdienst under the ARC programme.

Thomas Breuer (Aachen), Alexander Hulpke (Columbus)

### 4.2.19 GiNaC

GiNaC is a relatively new library that was designed to overcome certain design problems encountered and accumulated during the development of xloops [Lars Brücher et al. 1998], a package for automatic calculation of Feynman diagrams in quantum field theories. GiNaC (which stands for *GiNaC is Not a CAS*) is entirely written in C++ and provides classes for efficient handling of symbolic expressions. The user interacts with it in compiled C++-programs but other interfaces are easily conceivable, like an interactive shell (a sample implementation is actually included with the sources) or even an interface to the C++ interpreter Cint by Masaharu Goto [Masaharu Goto 2000].

GiNaC provides methods for the efficient handling of multivariate polynomials, quotients thereof (heuristic and subresultant PRS GCD algorithms), truncated power series, some symbolic linear algebra and special functions as well as algebras needed in high energy physics.

The philosophy behind GiNaC is driven by the observation that sometimes large-scale scientific applications call for a combined approach: computer algebraic methods drive some numerical simulation the results of which get displayed graphically. It is not uncommon to see the whole process itself being controlled via a graphical user interface. The design of such combined symbolical/numerical/graphical systems is largely complicated by technical odds and ends. While many of the more advanced computer algebra systems have powerful built-in graphical capabilities, numerically they cannot compete with compiled code thus establishing a need to use a second system. The interfaces are frequently constructed using fragile pipes, shell scripts, etc. The situation doesn't get any better when one considers the linguistic capabilities of computer algebra systems. To give an idea of what we mean we refer to the section called "Mathematics versus Computer Science" in Wester's chapter "A Critique of the Mathematical Abilities of CA Systems" in [Wester 1999]: it has an interesting section where some violations of scope are illustrated.

Here is a complete example how symbolic computation works in C++ using the GiNaC framework:

```
#include <ginac/ginac.h>
using namespace GiNaC;
```

```
ex HermitePoly(const symbol & x, int n)
{
    const ex HKer = exp(-pow(x,2));
    // uses the identity H_n(x) == (-1)^n exp(x^2) (d/dx)^n exp(-x^2)
    return normal(pow(-1,n) * diff(HKer, x, n) / HKer);
}

int main(int argc, char **argv)
{
    if (argc != 2) {
        cout << "usage:\n" << argv[0] << " <degree>" << endl;
        return -1;
    }
    int degree = atoi(argv[1]);
    symbol z("z");
    cout << "H_" << degree << "(z)=="
         << HermitePoly(z,degree) << endl;
    return 0;
}
```

The compiled program takes an integer argument $n$ from the command line and computes the $n$th Hermite polynomial in $z$ using straightforward differentiation. The `normal()` function therein cancels the generators `HKer` in numerator and denominator. The result is returned and then printed like this: "H_5(z)==120*z-160*z^3+32*z^5". Note that operators +, -, * and / have been overloaded to allow construction of composite expressions while `pow(b,e)` has to be used for exponentiation (overloading operator ^ in C++ would lead to surprises, since it has lower precedence than *). Generally, in order to be easy to understand, GiNaC tries to follow the spirit of the standardized language C++ as far as possible.

As a package for long integers and arbitrary precision arithmetic GiNaC uses Bruno Haible's extraordinary CLN library (cf. 4.6.11). It is compatible in the design since it is also written in C++ (and it is compatible in the license).

GiNaC is free software, covered by the GNU General Public License. It is available from http://www.ginac.de/. It should run on any machine where CLN has successfully been ported to (we have positive results from Linux, Solaris and Tru64 using GCC as C++ compiler).

Christian Bauer, Alexander Frink, Richard Kreckel (Mainz)

### 4.2.20 Kan/sm1

**4.2.20.1 Introduction** Kan/sm1 is a special-purpose system for doing *Computations in algebraic analysis* especially $D$-modules based on Gröbner basis.

## 4.2.20 Kan/sm1

The ring of differential operators (or the sheaf of them) is denoted by $D$ and modules over $D$ is called $D$-modules. "Kan" is the name of the project and also the name of the total architecture. "sm1" is a Gröbner engine for $D$ in the project kan. The system sm1 is designed to be used as a backend engine, but it has a small interpreter like Postscript and is possible to use it in a standalone system. It was also used in a course to teach what is a stack machine.

Currently sm1 runs on the following platforms as far as I checked: PC (under Linux and Free BSD), Sun workstations.

The system relies on two other packages:

1. gc 4.10 : Copyright 1988, 1989 Hans-J. Boehm, Alan J. Demers, Copyright (c) 1991-1995 by Xerox Corporation. All rights reserved.
2. gmp 2.0.2 (GNU MP).

The system can be obtained by anonymous ftp from the URL: ftp://ftp.math.kobe-u.ac.jp/pub/kan or http://www.math.kobe-u.ac.jp/KAN

The latest version of kan/sm1 is 2.981031. The first version of kan/sm1 was released on 1991. The second version was released on 1994. The design was completely changed in the second version; all computation is done in the homogenized Weyl algebra and weight vector specifying orders was introduced. Some algorithms in the system and applications are explained in the book [Saito et al. 2000].

The system is capable of performing basic operations such as sums, products, powers of differential operators, Gröbner basis in the Weyl algebra (ring of differential operators with polynomial coefficients) and in the homogenized Weyl algebra.

Besides these, more advanced operations for $D$-modules are available. For example, holonomic rank; indicial equations; $b$-functions and annihilating ideal of $f^s$ [Oaku 1997a]; fundamental operations for $D$-modules such as restrictions, integrals, tensor products, localizations, algebraic local cohomology groups [Oaku 1997b]; de Rham cohomology groups [Oaku and Takayama 1999]. Packages for these operations are written in sm1 Postscript language with a collaboration with Toshinori Oaku.

The main objects in sm1 are machine integers, numbers (bignum), strings, arrays, executable strings (encoded program), stacks, errors, polynomials, differential operators, and rings.

The related projects with kan are

1. D-Macaulay : "D-Macaulay" is a system for doing algebraic analysis by computer based on computations of Groebner bases. It is a system built on "Macaulay (classic)" by D.Bayer and M.Stillman [Bayer and Stillman

1992b], so users of Macaulay classic can easily use it. It computes Gröbner bases and syzygies in the ring of homogennized differential operators $D^{(h)}$. It shares codes for arithmetic in $D$ with kan/sm1. The latest version is 1998,10/31 with a new document with Will Traves, but the program itself is stable since 1994, July.

2. Open xxx (a protocol between mathematical programs): The program open sm1 (ox_sm1) is a version of sm1 that works as a server with the open xxx protocol. The same protocol is also implemented in the general computer algebra system Risa/Asir [Noro and et al. 1993, 1995] and the server version of asir with the open xxx protocol is called open asir (ox_asir). This project is under a work in progress.

The web cite of the projects above is as same as that of kan.

### 4.2.20.2 Some Examples

*Example 13.* The Gröbner basis of the GKZ hypergeometric system

$$f_1 = \partial_2 \partial_3 - \partial_1 \partial_4, f_2 = \theta_1 - \theta_4 + 1 - c, f_3 = \theta_2 + \theta_4 + a, f_4 = \theta_3 + \theta_4 + b$$

with $a = b = c = 0$ and the weight vector $(-1, 0, 0, 0; 1, 0, 0, 0)$ can be computed as follows. Here $\theta_i = x_i \partial_i$. We can find the indicial polynomial (or, $b$-function) along $x_1 = 0$ of the system.

```
(1)     [(x1,x2,x3,x4) ring_of_differential_operators
(2)       [[(x1) -1 (Dx1) 1]] weight_vector
(3)       0] define_ring
(4)     [(Dx2 Dx3 - Dx1 Dx4).
(5)       (x1 Dx1 - x4 Dx4 + h^2).
(6)       (x2 Dx2 + x4 Dx4 ).
(7)       (x3 Dx3 + x4 Dx4 ).
(8)     ] /ff set
(9)     [ff] groebner  ::
```

In the lines (1), (2), (3), the homogenized Weyl algebra of $2 \times 4$ variables is defined. The variable x1 stands for the variable $x_1$, Dx1 stands for the variable $\partial_1$ and so on. In the line (2), a weight vector $(u, v) = (-1, 0, 0, 0, 1, 0, 0, 0)$ is given and all terms are sorted by the weight vector. Generators are given in the lines (4) – (7) and the reduced Gröbner basis in the homogenized Weyl algebra is obtained in the line (9) and is printed. From the output, we can see that the indicial equation along $x_1 = 0$ is $s(s+1)$. See the book [Saito et al. 2000] for details on GKZ hypergeometric systems.

*Example 14.* Let us consider the ring of polynomials in two variables. Suppose that we are given two random polynomials of two variables $f$ and $g$. Then, $f$ and $g$ generically generate a zero dimensional ideal. However, if we consider the left ideal $I$ in the Weyl algebra generated by two random elements, then the chance of getting the trivial ideal is almost 100 percent. Let us compute the Gröbner basis of random generators

$$x\partial_x + \partial_y^2 + y, \quad y\partial_y^2 + x$$

with kan/sm1. The output is a constant, which means that they generate the trivial ideal.

```
(1)     [(x,y) ring_of_differential_operators 0] define_ring /R set
(2)     [(Homogenize) 0] system_variable
(3)     /ff [ (x Dx + Dy^2 + y). (y Dy^2+x).] def
(4)     [ff] groebner_sugar ::
```

The lines (1) and (2) define the Weyl algebra $\mathbf{Q}\langle x, y, \partial_x, \partial_y\rangle$. The ring structure is put in the variable R. $\partial_x$ is denoted by Dx and $\partial_y$ is denoted by Dy. The elements will be ordered by a default order (the degree reverse lexicographic order) in this example. The line (2) means that we will not use the homogenized Weyl algebra and $h$ should be regarded 1 in the sequel. In the lines (3) and (4), the reduced Gröbner basis is computed and is printed.

*Example 15.* Computing $b$-functions of polynomials: $b$-function is a polynomial in one variable associated to a polynomial in several variables. It is known that the roots of the $b$-function are rational numbers. Let us compute the $b$-function of $x^3 - y^2$.

```
(bfunc) run
[(x^3-y^2) [(x) (y)]] bfunction ::

216*s^3+648*s^2+642*s+210
```

The output is factored to

$$6(6s+5)(6s+7)(s+1).$$

See [Oaku 1997a] for details.

*Example 16.* Computation of the cohomology groups of $X = \mathbf{C}^2 \setminus V(x^3 - y^2)$ by $D$-module restriction (inverse image).

```
(restriction) run
[[(- 2 x Dx - 3 y Dy +1) (3 y Dx^2 - 2 x Dy)]
    [(x) (y)] [[(x) (y)] [ ]]] restriction  ::


[  [  0 , [  ] ]  , [   1 , [  ] ]  , [   1 , [  ] ] ]
```

The input system

$$-2x\partial_x - 3y\partial_y + 1, \; 3y\partial_x^2 - 2x\partial_y$$

is the Fourier transformation of the annihilating differential operators for $1/(x^3 - y^2)$. The output means that

$$H^2(X, \mathbf{C}) = 0, \; H^1(X, \mathbf{C}) = \mathbf{C}, \; H^0(X, \mathbf{C}) = \mathbf{C}.$$

This computation is done by using a free resolution of input system. See [Oaku and Takayama 1999] for details.

Nobuki Takayama (Kobe)

### 4.2.21 KANT V4

**4.2.21.1 Survey** KANT V4 is a program package for computations in algebraic number fields. The emphasis is on the interaction of elements of several such fields. Algebraic integers are considered as elements of a specified order of an appropriate number field $\mathcal{F}$. Arbitrary algebraic numbers are presented by an algebraic integer and a denominator, usually chosen as a natural number. The available algorithms make it possible to compute all invariants of $\mathcal{F}$ and also to solve tasks like calculating the solutions of Diophantine equations related to $\mathcal{F}$. Subfields of $\mathcal{F}$ can be generated and $\mathcal{F}$ can be embedded into overfields. The potential of moving elements between different fields (orders) is a significant feature of our system.

KANT V4 was developed at the University of Düsseldorf from 1987 until 1993 and at the Technical University Berlin afterwards. During these years the performance of existing algorithms and their implementations grew dramatically. While calculations in number fields of degree 4 and up were nearly impossible before 1970 and number fields of degree more than 10 were beyond reach until 1990, it is now possible to compute in number fields of degree well over 20, and – in special cases – even beyond 1000. Members of the KANT group have contributed considerably to this progress. Our philosophy is to overcome bottlenecks of existing implementations by improvements of the theory rather than by efforts to write better code.

KANT V4 consists of a C–library of more than 1000 functions for doing arithmetic in number fields. Of course, the necessary auxiliaries from linear algebra over rings, especially lattices, are also included. The set of these functions is based on the computer algebra system MAGMA [Bosma et al. 1997] from which we adopt our storage management, arithmetic for (long) integers and arbitrary precision floating point numbers, arithmetic for finite fields, polynomial arithmetic and a variety of other tools. Essentially, all of the public domain part of

MAGMA is contained in KANT V4 . In return, almost all KANT V4 routines are included in MAGMA. At present, the only other computer algebra system allowing similiar extensive calculations with algebraic numbers and number fields is PARI (Bordeaux) [Batut et al. 2000].

A user of KANT V4 routines needs to write his own header programs which requires some knowledge of the storage handling in MAGMA. To make KANT V4 easier to use we developed a shell called KASH . This shell is based on that of the group theory package GAP [GAP 1997] and the handling is similar to that of MAPLE. We put great effort into the handling of number theoretical objects in a similiar way as a mathematician does with pencil and paper. For example, there is just one command Factor for the factorization of elements from a factorial monoid like rational integers, polynomials over a field, or ideals from a Dedekind ring.

In the subsequent sections we discuss in some detail:

- the realization of number theoretical objects in KANT V4 and the corresponding data types;
- the most important algorithms contained in KANT V4;
- the KANT shell KASH ;
- distributed computations via PVM;
- the integrated SQL–database for number fields.

These sections are followed by an illustrative example. The development of KANT V4 as well as KASH is continued in view of providing the user with the most advanced tools for computations in algebraic number fields.

**4.2.21.2 Basic Concepts and Data Structures** The design of KANT V4 is based on the mathematical structures of global fields. Nearly all mathematical objects in KANT V4 can have several representations at the same time. The following objects are the most important ones: orders, algebraic numbers, ideals in orders, lattices.

In KANT V4 an order $\mathcal{O}$ is a free module with another order $\mathcal{O}'$ as its coefficient ring:

$$\mathcal{O} = \mathcal{O}'\omega_1 + \cdots + \mathcal{O}'\omega_n. \tag{1}$$

There are essentially two ways to create orders:

- $\mathcal{O}$ can be an equation order generated by the successive powers of a zero $\alpha$ of a monic irreducible polynomial $f(t) \in \mathcal{O}'[t]$.
- In the case that $\mathcal{O}$ is given as in (1) an overorder $\mathcal{M}$ of $\mathcal{O}$ can be defined by a transformation $(d, T) \in \mathbb{N} \times \mathcal{O}'^{n \times n}$ such that $(\eta_1, \ldots, \eta_n) = \frac{1}{d}(\omega_1, \ldots, \omega_n) T$ is an $\mathcal{O}'$-basis of $\mathcal{M}$.

We note that the rational integers $\mathbb{Z}$ are an order in **KANT V4** .

**KANT V4** supports two different presentations for an algebraic number $\alpha$. The first one is the basis presentation of $\alpha$ with respect to the basis of a given order $\mathcal{O}$ in the form: $(d, (a_1, \ldots, a_n)) \in \mathbb{N} \times \mathcal{O}'^n$ such that $\alpha = \frac{1}{d}(\alpha_1 \omega_1 + \cdots + \alpha_n \omega_n)$ and $\mathcal{O}'$ is the coefficient ring for $\mathcal{O}$. The second method is only available if $\mathcal{O}$ is a $\mathbb{Z}$–order, e.g. if the coefficient ring $\mathcal{O}'$ equals $\mathbb{Z}$. In this case **KANT V4** can represent $\alpha$ by a vector of the conjugates of $\alpha$.

Fractional ideals can be represented in several ways. The most important ones are as a $\mathbb{Z}$-module or as a greatest common divisor of two principal ideals (2-element representation). There is a complete ideal calculus as described in [Pohst and Zassenhaus 1989].

The final important number theoretical structure we shortly discuss are lattices. In **KANT V4** a lattice $\Lambda$ can generally be defined either by giving its Gram matrix or by specifying linearly independent vectors $\mathbf{b}_1, \ldots, \mathbf{b}_k \in \mathbb{R}^n$ with $\Lambda = \mathbb{Z}\mathbf{b}_1 + \cdots + \mathbb{Z}\mathbf{b}_k$. Additionally, a lattice can be obtained from the canonical scalar product of a $\mathbb{Z}$-order.

**4.2.21.3 Library Functions and Algorithms** We give a short overview of the **KANT V4** library and the most important algorithms. Many of the library functions can be accessed by using the shell. This will be discussed later.

For a given number field $\mathcal{F}$, represented by an order $\mathcal{O}$, **KANT V4** offers the complete arithmetic of elements in $\mathcal{F}$. For ideals of arbitrary $\mathbb{Z}$ orders addition and multiplication are possible. Ideals of maximal orders can be inverted and factorized into prime ideals; this includes the decomposition law of rational primes in $\mathcal{F}$. For arbitrary orders this is possible only, if the ideal under consideration is coprime to the index of that order. All elementary operations for polynomials and matrices are adopted from MAGMA.

According to H. Zassenhaus, one of the founders of computational algebraic number theory, the main tasks in this area are the computation of an integral basis, the unit group, the class group and the Galois group.

- *Integral Bases* An integral basis for the maximal order $o_\mathcal{F}$ of a number field $\mathcal{F}$ is computed by a combination of the Round–2 and Round–4 algorithm [Cohen 1996b; Pohst 1993; Pohst and Zassenhaus 1989].
- *Unit Groups* There are several methods for the computation of the unit group of an order. A maximal system of independent units can either be obtained by a "generalized" continued fraction algorithm [Pohst 1993] or by finding relations in the context of class group computations (see below). Then the full unit group is calculated by a stepwise enlargement of the already determined subgroup.
- *Class Groups* For class group computations **KANT V4** offers a variety of options. We can only describe the general ideas. Let us assume that $\mathbf{p}_1, \ldots, \mathbf{p}_s$ are prime ideals representing ideal classes $\mathbf{p}_i H_\mathcal{F}$ which generate the class

group $Cl_\mathcal{F}$. They form a so-called *factor basis*. By calculating sufficiently many *relations* $\alpha o_\mathcal{F} = \prod_{i=1}^{s} \mathbf{p}_i^{a_i}$ we already get a subgroup $U$ of $Cl_\mathcal{F}$. Computing an Euler product with sufficient accuracy we can check $U = Cl_\mathcal{F}$ if we assume the Generalized Riemann Hypothesis. That check can be done also without GRH but is then much more tedious.

- *Galois Groups* Let $f(t) \in \mathbb{Z}[t]$ denote an irreducible polynomial of degree $n$. Its Galois group $\Gamma$ is a transitive subgroup of $S_n$, the symmetric group of $n$ letters. For an approximation of $\Gamma$ from below, the modulo $p$ factorizations of $f$ for a few primes $p$ give a rather precise idea of the cycle types to be expected in $\Gamma$, excluding all transitive subgroups of $S_n$ which are *too small*. Using factorization methods over number fields, the inductive construction of $\mathbb{Q}(\rho_1, \ldots, \rho_s)$ for some roots $\rho_1, \ldots, \rho_s$ of $f$ will usually produce an approximation of $\Gamma$ from above. In the case that both approximations still admit several candidates for $\Gamma$ we need to decide whether $\Gamma$ is contained in one of certain conjugate subgroups of $S_n$. This decision is possible through the use of indicator functions. Presently, **KANT V4** enables the computation of Galois groups for polynomials up to degree 12 [Pohst and Zassenhaus 1989; Eichenlaub and Olivier 1995].

Finally, we list more sophisticated and specialized features which are currently realized. Implementations were partly done using the programming language of the shell.

- *Subfields and Automorphisms* Let $\mathcal{F} = \mathbb{Q}(\alpha)$ be an algebraic number field which is given by a zero $\alpha$ of the corresponding minimal polynomial $f \in \mathbb{Z}[t]$. Each subfield $\mathcal{E} = \mathbb{Q}(\beta)$ of $\mathcal{F}$ can be described by a pair $(h, g)$ where $g$ is the minimal polynomial of $\beta$ and $h \in \mathbb{Q}[t]$ in an *embedding polynomial* with $h(\alpha) = \beta$. In **KANT V4** subfields are described in this way and can be computed by a generalized and improved version [Klüners and Pohst 1997] of Dixon's method [Dixon 1990]. Additionally **KANT V4** provides functions for the computation of automorphisms in arbitrary normal extensions over $\mathbb{Q}$ and relative abelian extensions.
- *Relative Norm Equations* Let $\mathbb{Q} \subseteq \mathcal{F} \subseteq \mathcal{E}$ be algebraic number fields and let $M \subseteq \mathcal{E}$ be a free $o_\mathcal{F}$ module. For a non–zero $\beta \in o_\mathcal{F}$ **KANT V4** provides routines for the computation of all non-associate $\gamma \in M$ with $N_{\mathcal{E}/\mathcal{F}}(\gamma) = \beta$. These routines are based on a generalization [Fieker et al. 1997] of the algorithm by Fincke and Pohst [Fincke and Pohst 1983].
- *Kummer Extensions* Let $\mathcal{F}$ be an algebraic number field containing the $n$-th roots of unity. We consider an extension $\mathcal{E}/\mathcal{F}$ such that $\mathcal{E} = \mathcal{F}(\sqrt[n]{\mu})$ where $\mu \in o_\mathcal{F}$. **KANT V4** is able to compute the relative discriminant $\mathbf{d}_{\mathcal{E}/\mathbb{F}}$ of the extension $\mathcal{E}/\mathcal{F}$ and a small set $\{\xi_1, \ldots, \xi_m\} \subseteq o_\mathcal{E}$ such that $o_\mathcal{E} = \xi_1 o_\mathcal{F} + \cdots + \xi_m o_\mathcal{F}$. Additionally **KANT V4** can calculate an integral basis of $o_\mathcal{E}$ from $\{\xi_1, \ldots, \xi_m\}$ [Daberkow 1995; Daberkow and Pohst 1995, 1996].
- *Hilbert Class Fields and Ray Class Fields* Let $\mathcal{F}$ be an algebraic number field and let $\mathbf{m}$ be a congruence module of $\mathcal{F}$. Making extensive use of compu-

tations in Kummer extensions and of the Artin map, **KANT V4** can arithmetically compute the ray class field $\varGamma^\mathbf{m}$ of $\mathcal{F}$ which corresponds to the ray class group $Cl^\mathbf{m}$ [Daberkow and Pohst 1998]. If $\mathcal{F}$ is an imaginary quadratic field, **KANT V4** enables the computation of the ray class field $\varGamma^\mathbf{m}$ and the integral basis of $\varGamma^\mathbf{m}$ by analytical methods [Schertz 1990].
- *Thue Equations* One of the classical objects of number theory is the Diophantine equation of Thue $f(X,Y) = a$, where $f(X,Y) \in \mathbb{Z}[X,Y]$ is an irreducible form of degree $\geq 3$ and $a$ is an integer. In **KANT V4** such equations are solved by the methods in [Bilu and Hanrot 1996].
- *Unit Equations* Let $\mathcal{F}$ be an algebraic number field and let $\alpha, \beta \in \mathcal{F}^\times$. By using methods from the geometry of numbers **KANT V4** can compute all units $\varepsilon, \eta \in o_\mathcal{F}$ which satisfy the unit equation $\alpha\varepsilon + \beta\eta = 1$ [Wildanger 1997].
- *Index Form Equations* Let $\mathcal{F}$ be an algebraic number field which is normal or has degree $\leq 4$. **KANT V4** can compute all non-equivalent $\alpha \in o_\mathcal{F}$ whose index $(o_\mathcal{F} : \mathbb{Z}[\alpha])$ equals a given positive integer ([Gaál et al. 1993; Wildanger 1997]).
- *Integral Points of Mordell's equation* By making use of cubic index form equations **KANT V4** enables the computation of all integral points of Mordell's equation $y^2 = x^3 + k$ ([Wildanger 1997]).
- *Global Function Fields* A global function field is a field extension $F/k$ of transcendence degree 1, where $k$ denotes a finite field. Hence we can assume the existence of an element $y \in F$ and of an irreducible bivariate polynomial $f$ over $k$, which is separable and monic in the second variable, such that $f(x,y) = 0$. There is the ability for computing the integral closure of $k[x]$ and of the degree valuation ring of $k(x)$ in $F$, called finite resp. infinite maximal order. Arithmetic operations of algebraic functions and of ideals of the maximal orders are provided, valuations of these at arbitrary places can be computed. There are functions for reducing the basis and for computing fundamental units of a finite maximal order [Pohst and Schörnig 1996]. Furthermore, the genus of the function field and Riemann-Roch spaces for divisors consisting only of places which extend the degree valuation can be determined.

**4.2.21.4 The Shell** A recent and extremely important part of our software is KASH – the KAnt SHell. For a proper use of **KANT V4**, the user needs to have some experience with programming in C and an understanding of the memory management in MAGMA. Because of this disadvantage, we started to build a shell around the C–library **KANT V4**, which combines the functionality of **KANT V4** with a comfortable user interface based on GAP, a software package for group theory [GAP 1997].

Within the shell, the user can do arithmetical operations with integers, rationals, real and complex numbers (with arbitrary precision), matrices or — after the definition of an order — with algebraic numbers, ideals, etc.. Of course, all results can be assigned to variables for later use.

There are two different types of functions available, the "internal functions" and the "user functions". The first are built-in functions of the internal function library, i.e. they are written in C, linked to **KASH** and cannot be changed. With these at hand, most of the algorithms mentioned above can be performed.

In contrast to these, the user can create his own (user) functions: With the PASCAL–like programming language, he can create loops, conditional branches, functions etc. and use all internal and user functions. In this environment, he can even write sophisticated programs. All user functions and programs can be stored as (external) text files which build a user function library (in contrast to the internal function library).

Presently, there are more than 350 internal functions installed, 200 additional predefined user functions and comprehensive references are available. Because **KASH** grows weekly, updates will be made more often than for the **KANT V4**–library.

Additionally, **KASH** possesses an interface to the public domain PVM–software (see next section), which allows distributed computing and enables users to write programs in **KASH** that utilize PVM to perform certain tasks concurrently.

**4.2.21.5 Distributed Computing** It is possible, both in **KANT V4** and **KASH**, to benefit from distributed computing. The requirement for this (in addition to **KASH**) is a network of workstations running PVM3 (at least version 3.3, see [Geist 1994]). Based on the PVM-protocol we provide a high-level interface for process communication and exchange of **KANT V4** data. We support two different modes of communication: one is based on **KANT V4**, providing C-functions, and the other is based on **KASH**, consisting of several **KASH** commands.

Our (virtual) parallel computer is hierarchical, consisting of one master, the **KASH** session running in the foreground, and an arbitrary number of slaves, running on different machines in the background. Additional slaves can be added at run time if the current task is able to use them. It is possible to give arbitrary time spots for each machine, e.g. allowing the slaves only to run at night or on the weekend, in order to be able to do really large jobs at times which are convenient.

As an example we discuss the processing of the computation of the maximal order by the Round-2 algorithm. After the factorization of the discriminant, the $p$-maximal overorders for all primes $p$ whose square divides the discriminant must be computed. Since those calculations are independent, we use different slaves for different primes. The combination of the results is done on the master afterwards. Especially when "large" computations (involving many primes) are carried out, a lot of time can be saved in this way.

**4.2.21.6 The Database** Accessible from **KASH** is an SQL–database for number fields [Daberkow and Weber 1996]. The database is designed to give easy

and fast access to several hundreds of thousands of number fields. Currently the
following invariants are stored (if known) and can be used as keys in a selection:

- a generating polynomial together with its signature
- an integral basis, the field discriminant
- the unit group and regulator
- the class group with structural information
- the Galois group

Isomorphy can be tested with **KASH**. In a first step one can check some invariants and if all tests are successful there is the possibility to choose between several algorithms for proving the isomorphy.

The underlying SQL–database (currently Postgres95) is public domain and available for every system supported by **KASH**.

**4.2.21.7  Example** The following is an example of **KASH** for the computation of the Hilbert class field for $\mathcal{F} := \mathbb{Q}(\rho)$ where $\rho^3 + \rho^2 - 42\rho - 107 = 0$. We start by reading the equation order of $\mathcal{F}$.

```
kash> f := Poly(Zx,[1,1,-42,-107]);;   # f(x)=x^3+x^2-42x-107
kash> F := Order(f);;                  # Create the equation order
                                       # of the polynomial f
kash> F := OrderMaximal(F);;
kash> OrderUnitsFund(F);;
kash> OrderClassGroup(F,"euler");;     # option "euler" is necessary
                                       # for the function
                                       # OrderHilbertClassField
kash> F;
Generating polynomial:   x^3 + x^2 - 42*x - 107
Discriminant: 70313
Regulator: 21.20506
Units:
[3, 1, 0]        [9, 12, 2]
class number 2
class group structure C2
cyclic factors of the class group:
<5, [3, 0, 1]>
```

The discriminant is always the discriminant of the order. The cyclic factors are given in a 2 element normal representation.

We apply the user function `OrderHilbertClassField` to it.

```
kash> Y := OrderHilbertClassField (F);
```

```
Starting Class Field Computation
   Degree      : 3
   Signature   : [ 3, 0 ]
   Class Group : [ 2, [ 2 ]]
=======================================
Checking cyclic group C2
----------------------------------------------------------
----------------------------------------------------------
Computing Class Field for cyclic subgroup C2
```

We obtain the following 4 elements $\alpha_1, \ldots, \alpha_4$, a power product of which yields a generating element.

```
List of Generators :
[ [1299, 255, -62], -1, [3, 1, 0], [9, 12, 2] ]
```

We compute a generating element $\mu = \alpha_1^{e_1} \cdot \ldots \cdot \alpha_4^{e_4}$ for $(e_1, \ldots, e_4) \in (\mathbb{Z}/2\mathbb{Z})^4$. Only unramified extensions $\mathcal{F}(\sqrt{\mu})$ of $\mathcal{F}$ are processed further.

```
Exponent Vector [ 1, 1, 0, 1 ] -->[79, 5, -2]
```

It corresponds to the element $\mu = 79 + 5\rho - 2\rho^2$. Since we obtain just one unramified extension, it has to be the class field. Of course, there is also a built in checking routine.

**4.2.21.8 Availability** KASH is freely available via

   ftp.math.tu-berlin.de   at   pub/algebra/Kant/Kash.

It has been ported to the following architectures:

- HP 700 / HP–UX 9.01,
- IBM RS 6000 / AIX 3.2.5,
- Intel 486 / IBM OS/2 Warp 3.0, Linux, MS DOS 5.0, MS Windows95,
- Silicon Graphics / IRIX 6.2,
- Sun SPARC / SunOS 4.1.3,
- Sun SPARC / SunOS 5.5.

## 4.2.22 LiDIA

LiDIA* is a powerful C++ library for computational number theory which provides a collection of highly optimized implementations of various multiprecision data types and time-intensive algorithms. LiDIA is developed by the LiDIA group at the Technische Universität Darmstadt. Although not in the public domain, LiDIA can be used freely for non commercial purposes.

---

* LiDIA(tm) is a registered trademark of Dr. Thomas Papanikolaou.

### 4.2.22.1 Introduction

In early 1994, the research group for computational number theory at J. Buchmann's department decided to reorganize their software packages, that had been developed separately until then and for which many basic routines had been written over and over again, into a single library which was called LiDIA[Biehl et al. 1995]. Most of the design, and the implementation of the basic components like the *kernel*, the *interface* and the majority of the components for doing arithmetic was done by Thomas Papanikolaou [Papanikolaou 1997].

Since *object oriented programming* allows modelling of mathematical objects and implementation of algorithms in a very natural way, and since C and C++ belong to the most accepted programming languages in scientific computing, C++ was chosen as the *implementation language* for LiDIA. As a side effect, the C++ class concept with its data hiding and interface definitions supports modularity of the resulting code. One of the major design goals for the library was high *portability* while preserving *efficiency*. This lead to a layered structure with a modularized small kernel.

### 4.2.22.2 Structure

The **kernel** of LiDIA, the lowest layer, is the only part of the system which may contain machine dependent code; it is consequently kept as small as possible. It consists of code for the multiprecision integer package 2.1.1 and the memory manager, which are used by all LiDIA classes. The official release LiDIA-2.0.1* supports the multiprecision integer packages `lip`[Lenstra 1995] by Arjen Lenstra, `libI`[Dentzer 1991] by Ralf Dentzer, `gmp`[Granlund 1996] by Torbjörn Granlund, and `cln`[Haible 1998] by Bruno Haible, as well as the memory manager `boehm`[Boehm and Weiser 1988] by Hans-J. Boehm and Alan J. Demers, and the `free list manager`[Dentzer 1991] by Ralf Dentzer.

In order to guarantee that the kernel can easily be adapted to specific hardware, the functions of the kernel can only be called through the **interface**. The interface of the long integer arithmetic consists of a C++ class, called `bigint`, that can be used in the same way as the built-in type `int`. The `bigint` class provides additional functions, e.g. for computing gcd's, square roots and logarithms to base 2.

The **simple classes** are basically the multiprecision arithmetics for the rational, real, and complex numbers as well as a modular arithmetic.

The **main part** of the library is a steadily growing number of classes which are being built on top of the interface and the simple classes.

The LiDIA library uses **parametrized classes** (template classes) for generic objects like vectors, matrices, polynomials and power series.

---

* Released in April 2000

**4.2.22.3 User Interface** The top level of LIDIA consists of the **user interface LC**, which is an interpreter, that makes the functionality and documentation of LIDIA interactively available. LC implements a subset of the C++ language and supports function overloading and automatic coercions in addition to standard programming facilities. Functions and statements are treated as ordinary objects and may be manipulated at run-time. Because of the interpreted language, LC functions can be easily transformed to C++ programs which can then be compiled.

**4.2.22.4 Contents**
The current release of LIDIA includes

- **Arithmetic Interfaces** to various multiprecision integer packages
- **Basic arithmetic** over $\mathbb{Z}$, $\mathbb{Q}$, $\mathbb{R}$, $\mathbb{C}$, interval arithmetic, $GF(2^n)$, $GF(p^n)$
- **Factorization**: Integer Factorization (see 2.4.2): Trial Division, the Elliptic Curve Method, Self-Initializing Multipolynomial Quadratic Sieve with Lanczos algorithm; Factorization of Polynomials over finite fields (see 2.2.2): V. Shoup's algorithms; Factoring ideals of algebraic number fields
- **Lattice Basis Reduction**: various versions of LLL/MLLL (Schnorr-Euchner, Benne de Weger)
- **Linear Algebra over** $\mathbb{Z}$: basic operations, normal forms of matrices (G. Havas algorithms)
- **Number Fields**: Quadratic Number Fields including a new implementation of Buchmann's subexponential algorithm for computing classgroups using MPQS techniques, Higher-Degree Number Fields arithmetic and maximal order (see 2.4.3)
- **Polynomials**: template classes for univariate polynomials with special algorithms for different domains
- **Other Generic Data Types**: vectors, matrices, power series, and hash tables implemented as templates.

There is an implementation of the Atkin/Elkies algorithm for counting the number of points on an elliptic curve over a finite prime field with large characteristic [Lehmann et al. 1994] (see 2.4.5. This package uses the LIDIA library, but is not part of the free distribution. For details, please contact the LIDIA group directly.

**4.2.22.5 Applications**
LIDIA has been widely used by graduate students in Saarbrücken and Darmstadt and has proved a very productive tool for research. Various projects at the Universität des Saarlandes and the Technische Universität Darmstadt have benefited from the ease of use and the efficient implementation of algorithms for computational number theory. Besides, LIDIA was an integral part of several seminars at different universities as it covers the mathematics for cryptographic algorithms. On the other hand, the LIDIA library was used for finding practical

solutions to cryptographic problems such as McCurley's 129-digit discrete log challenge [McCurley 1990; Denny and Weber 1998].

A LiDIA users conference, which took place in Darmstadt in November 1996, helped bringing together users' and programmers' experiences and resulted in intensified cooperation. Contributions vary from special floating point arithmetic to an extensive elliptic curve package. Further applications developed in Darmstadt were numerous lattice reduction algorithms [Wetzel 1998], polynomial factorization routines over finite fields [Shoup 1994; Pfahler 1997], and a very efficient linear algebra package [Theobald 1999]. Recently, the research focus has been on algebraic number fields, especially on ideal arithmetic in quadratic number fields [Jacobson 1998] and class group computations over arbitrary number fields [Buchmann et al. 1999]. All these programs have or will become part of the LiDIA library.

Among the applications that have been reported by LiDIA users from other research intitutions since the first release in 1995 are problem solving in algebraic geometry and diophantine approximation [Keyser et al. 1998], and elliptic curve computations. LiDIA has also proved a succesful base system for the mwrank package by John Cremona as well as for the Groebner base system Opal. There even exists a communication interface between SAC-2 and LiDIA [Simon 1996], and there are modules for replacing functions in Mathematica (via MathLink), e.g. to make use of the very efficient integer factorization routines in LiDIA.

### 4.2.22.6  How to Use LiDIA

The philosophy, the usage and the entire functionality of LiDIA is described in great detail in the documentation that is delivered with the package [LiDIA-Group 2000]. The 600-page manual is available in DVI, PS or HTML or UNIX man pages format. After having installed LiDIA (which should be straightforward due to the installation routines), it can be used like any other C++-library. A short sample program for computing with fractional ideals of quadratic orders might give an impression of how LiDIA classes and functions are used in C++:

```
#include <LiDIA/quadratic_order.h>

main()
{
  quadratic_order QO;
  quadratic_ideal A,B,C;
  bigint D,p,x;

  do {
    cout << "Please enter a quadratic discriminant: "; cin >> D;
  } while (!QO.assign(D));
  cout << "\n";
```

```
/* compute 2 prime ideals */
p = 3;
while (!generate_prime_ideal(A,p,QO))
   p = next_prime(p);
p = next_prime(p);
while (!generate_prime_ideal(B,p,QO))
   p = next_prime(p);

cout << "A = " << A << "\n"
     << "B = " << B << "\n";

power(C,B,3);
C *= -A;
square(C,C);
cout << "C = (A^-1*B^3)^2 = " << C << "\n";

cout << "|<C>| = " << C.order_in_CL() << "\n";

C.reduce();
cout << "C reduced = " << C << "\n";

cout << "ring of multipliers of C:\n"
     << C.ring_of_multipliers() << "\n";

if (A.DL(B,x))
   cout << "log_B A = " << x << "\n" << flush;
else
   cout << "no discrete log:   |<B>| = " << x << "\n" << flush;
}
```

#### 4.2.22.7 Availability

LiDIA is free for non commercial use and comes with the full source code. Version 2.0.1 has been successfully compiled on Sun SparcStations (under SunOS and Solaris), HP Workstations (under HPUX-9.xx and HPUX-10.xx), DEC Alphas (running Linux), and Intel PC running Linux or OS/2 or Windows NT using GNU's g++-2.8.1, egcs-1.1, g++-2.95.2, and MS VisualC++ 6.0, respectively. It should compile on any system providing long filenames and a modern C++ compiler.

#### 4.2.22.8 Homepage & Download

The LiDIA hompage is located at http://www.informatik.tu-darmstadt.de/TI/LiDIA, where you can find the most current release of LiDIA as well as the online documentation. Alternatively, LiDIA can be retrieved via anonymous FTP from ftp.informatik.tu-darmstadt.de in the directory pub/TI/systems/LiDIA. Mirror FTP sites are ftp.math.uni-hamburg.de/pub/soft/math/LiDIA and unix.hensa.ac.uk/ftp/mirrors/LiDIA.

For commercial licences and further information on LiDIA please contact

**The LiDIA Group**
Technische Universität Darmstadt
Arbeitsgruppe J. Buchmann
Fachbereich 20 - Informatik
D - 64283 Darmstadt
email: lidia@cdc.informatik.tu-darmstadt.de


Johannes Buchmann and Thomas Pfahler (Darmstadt)

### 4.2.23 Lie

LiE, a software package for Lie group computations

LiE is a specialised computer algebra package for computations concerning Lie groups and algebras, and their finite dimensional representations. LiE was developed at CWI in Amsterdam, but development currently continues at TUE (Eindhoven) and at the Université de Poitiers. The source code of the package is freely available on the internet. The package offers a large number of algorithms performing computations related to Lie groups and algebras. Areas in which built-in algorithms are provided include root systems, the Weyl group and its action on the root and weight lattices, symmetric group characters, semisimple elements and their centralizers, highest weight modules and general finite dimensional representations, their characters and decompositions of tensor products, restrictions to subgroups (branching), symmetric and alternating tensor powers, and more general plethysms. Furthermore LiE provides an interactive programming environment, allowing not only easy and flexible handling of input and output data, but also enabling customization of the package by the user to particular applications, by means of variables, control structures and functions. A library of examples is supplied, showing how LiE can be applied to various kinds of problems. LiE comes with a manual ([van Leeuwen et al. 1992] of over 100 pages (as a file in DVI-format) which provides an introduction to the use of LiE, background information about the mathematical field covered by the package, full documentation on all functions of the package and the language provided by the programming environment, and a discussion of the library of examples. In addition to this LiE provides on-line help on all its features, as well as theoretical information about Lie groups and related topics.

### Background

The theory of reductive Lie groups and algebras and their finite dimensional representations is very well suited for a computational approach. This is mainly due to their clear classification and the uniform and explicit description of their representations, and to many results such as character formulae and rules for computing tensor product decompositions. While these calculations lie in the realm of exact (as opposed to numerical) computation, they are very time consuming when performed by manipulations of general symbolic expressions, as found for instance in a computer algebra system like Maple; this is due to the combinatorial complexity of the calculations. On the other hand, the calculations can be expressed in terms of very simple data structures, such as vectors and matrices with integer entries, and this fact can be used to gain efficiency. Based on these ideas the software package LiE was conceived, in which the central algorithms, such as computation of characters, traversal of Weyl group orbits, and tensor product decompositions, are implemented by algorithms that are compiled into the package, and which can therefore exploit direct access to the internal representation of the simple data structures. The resulting efficiency gain is illustrated by the comparison of character computations in LiE with those using a Maple package written for the same purpose: a speedup factor of approximately 100 was found.

This extensive library of specialised functions is turned into a practical tool for everyday use by embedding it in a user environment consisting of an interpreted, typed programming language providing operator and function overloading. Thus various ways of using LiE are facilitated, from entering simple commands to writing sophisticated new algorithms at a high level. Furthermore input, output and editing facilities are included, and the combination of all these aspects makes LiE into a powerful and versatile stand-alone program.

**Data Types**

LiE provides the user with a few simple data types: integers, integer vectors and matrices, (multivariate) Laurent polynomials with integer coefficients, values that symbolically describe the types of reductive complex Lie groups (or algebras), and texts. With objects of these types however, a large number of concepts pertaining to the study of Lie groups or algebras can be described. For instance, LiE can perform calculations in root systems (computing Cartan matrices, sets of positive roots, inner products, closures of subsets of roots, etc.), in which case roots are represented by integer vectors, and sets of roots by matrices. LiE may also compute in the Weyl group (finding for instance reduced expressions and elements of minimal length in (double) cosets for Weyl subgroups), in which case vectors represent expressions for Weyl group elements as product of simple reflections. Alternatively Weyl group elements may be used to act upon the root lattice or the weight lattice, in which case they can be represented by square matrices (conversions between these different representations are provided in LiE). In the study of the symmetric groups (which is needed for instance to define plethysms) vectors may stand for permutations, partitions or Young tableaux, and polynomials may be used to represent symmetric group characters. Finally, in the operations pertaining to the representation theory of reductive Lie groups—the computationally hardest type of algorithms provided by LiE—the predominant types are vector (representing weights) and polynomial (for sets of weights with multiplicities, representing either reducible modules as decomposed into highest weight modules, or characters of modules).

With so many algorithms related to Lie groups and their representations available in one package, and with the flexible programming interface provided by the simple but useful data types, LiE provides a powerful tool for studying Lie groups, both for students and for academic researchers in the field.

## History and Current Status

Early versions of LiE have been available and used in many places since 1990, and a new major version 2.0 was released in January 1992. In July 1996, version 2.1 of LiE was made publically available for free, and the current version number is 2.2.

Roughly speaking, version 1.3 already contained many of the key algorithms, but in a somewhat rudimentary form. With version 2.0, everything was integrated in a way that is much more practical for users, with the following important improvements:

- introduction of the polynomial data type, which is especially suited to represent characters of modules or decompositions of reducible modules,
- systematisation of the interface to functions, so that each handles the most general case in which the corresponding computation makes sense,
- improvement of key internal routines, leading to a considerable increase in overall performance,
- inclusion of new functions, for instance dealing with Young tableaux, the Bruhat ordering and Kazhdan-Lusztig polynomials,
- greatly improved documentation: an extensive manual, many instructive examples, and improved on-line help facility.

After version 2.0, the external characteristics of LiE have been relatively stable. Nevertheless, there have been, and will continue to be, considerable improvements beneath the surface, in terms of efficiency, portability, and in particular of documentation of the source code. Making LiE freely available was intended not only to provide public access to the computations of LiE, but also to the sources for the algorithms being used. For this to work in practice, the source code should be optimally documented, and for this the technique of literate programming is being applied; most of the core mathematical algorithms have already been so documeted, and it is intended that eventually all of the sources will be.

## Availability

The package is written in portable ANSI C, and can therefore be compiled any computer system with a ANSI C compiler. The package is distributed via the internet a well documented collection of source files; the basic URL is http://wwwmathlabo.univ-poitiers.fr/~maavl/LiE/. The form in which LiE is distributed has been tailored to UNIX platforms (for instance by supporting the use of make), but the program itself makes no essential use of any UNIX-specific features. The entire program can be (and has been) compiled on other platforms; this may require some adaptation to the particular properties of the compilation environment. For those who only wish to perform some computations with LiE, or want to try out the program before deciding whether to download the sources, there is an interactive interface to LiE via the World Wide Web. It is reachable via the above URL, and allows a collection of the most useful functions in LiE to be invoked by means of filling in a form; the answer delivered as a tailor-made Web page.

References are [van Leeuwen et al. 1992], [Cohen and de Man 1996], and [Littelmann 1996].

Marc A. A. van Leeuwen (Poitiers)

## 4.2.24  LIE

LIE is a self-contained stand-alone program for Lie symmetry analysis of differential equations that runs under DOS on any PC and does not require any additional computer algebra system. It covers Lie-Backlund, contact, generalised and approximate symmetries besides conventional point symmetries, see also 2.11 and [Head 1993, 1996].

The zipped distribution of LIE contains the operating program, instruction and information files and a number of test data files that cover its various capabilities. To save space the source code is no longer included but is available on request. BIGLIE is a new program that can handle very large problems that are beyond LIE. Both programs are freely available from the many Simtelnet archives ( in /msdos/math ) or on the web at http://www.cmst.csiro.au/LIE/LIE.htm where also the README files can be browsed.

Alan Head (Melbourne)

## 4.2.25  A Brief Introduction to Macaulay 2

*Macaulay 2* is a software system devoted to supporting research in algebraic geometry and commutative algebra, implemented by Michael Stillman and Daniel

Grayson, starting in 1992. The current version of *Macaulay 2* is always available to the public at the web site, http://www.math.uiuc.edu/Macaulay2. There you may view the hypertext documentation for *Macaulay 2*, download the complete source code, or download executable versions that will run under Linux, various other Unix systems, Windows-95/98/NT, and Macintosh PowerPC.

*Macaulay 2* supports high level mathematical concepts as first-class objects. This includes polynomial rings, quotient rings, ideals, modules, homomorphisms of rings, homomorphisms of modules, graded modules, maps of graded modules, chain complexes, maps of chain complexes, chain homotopies, sets, multi-sets, vectors, complex numbers, Galois fields, product rings, monoids, free resolutions, algebraic varieties (affine or projective), and coherent sheaves. These high-level types of objects are all defined by code written in our interpreted language, and thus the user has many examples available as a guide for the introduction of new data types.

The core algorithms handle (homogeneous or inhomogeneous) ideals and modules over graded (and multi-graded) polynomial rings, skew-commutative algebras, and Weyl algebras. All this is available over various ground rings, including finite fields, the rational numbers, and the integers. Users may compute Gröbner bases (see 2.2.5), *SAGBI* bases, syzygies, projective resolutions, kernels, cokernels, homology, Hilbert functions, saturations, symmetric powers, exterior powers, Pfaffians, *Ext*, *Tor*, radicals of ideals, integral closures, irreducible decomposition, and coherent sheaf cohomology. Further development of the code is underway for primary decomposition, Gröbner bases of noncommutative algebras, and Gröbner bases of inhomogeneous ideals over $\mathbb{Z}$.

We have settled upon the following efficient and versatile paradigm for representing modules in *Macaulay 2*. Every module is presented as a submodule of a quotient module of a free module, presented internally as a pair of matrices with the same number of rows. The columns of one matrix are the relations, and the columns of the other matrix are the generators.

The implementation of resolutions of modules has been highly optimized. The algorithm we use is based on Schreyer's method, with improvements due to Roberto La Scala, and has appeared as a joint paper with La Scala [La Scala and Stillman 1998]. The implementation is significantly faster than *Macaulay*, the forerunner of *Macaulay 2* which was written by David Bayer and Michael Stillman. Computation of resolutions is fully interruptable and may be continued later, and various termination conditions on the computation may be given. The algorithm and implementation we use has been designed to work also in the case when the ring is an exterior algebra or a Weyl algebra modulo an ideal.

We have incorporated some code from other authors. The package FACTORY [Greuel and Stobbe 1993-1997] by Gert-Martin Greuel and Ruediger Stobbe provides for factorization of polynomials, and LIBFAC [Messollen 1996] by Michael Messollen provides the ability to compute characteristic sets and decomposition of ideals. We are grateful to them for their contribution.

Operations on integer matrices are available, including the Lenstra-Lenstra-Lovasz lattice basis reduction algorithm. This algorithm is immensely useful when computing the kernel of a map between free abelian groups, because it often is able to locate a basis whose vectors have small coefficients.

*Macaulay 2* includes the basic facilities required by any symbolic algebra system. This includes functions for handling strings, arbitrary precision integers and rational numbers, lists, hash tables, files, and two-dimensional formatting of data. The ability to save the entire state of the system and to restore it later is provided, and allows the program to start up quickly, even though much of its code is interpreted.

The programming language provided by *Macaulay 2* to the users is a versatile object-oriented language with inheritance, so that introduction of new types of objects is easily accomplished in such a way that the standard operators of the language apply to the new types (see 2.17). For example, the system code that implements ordered monoid rings contains the statement `OrderedMonoidRing = new Type of Ring`. Methods that apply to all rings will then also apply to all ordered monoid rings. Here's the way this works. The new type `OrderedMonoidRing` will be a hash table with a pointer to its *class*, `Type`, and to its *parent*, `Ring`. Later on, when we create a ordered monoid ring $R$, its class will be `OrderedMonoidRing`. Unary and binary methods concerned with ordered monoid rings will be stored in `OrderedMonoidRing`; the search for methods applicable to a ring such as $R$ starts in `OrderedMonoidRing`, and continues in the parent, i.e., in `Ring`, where methods that are more widely applicable may reside. The parent relationship is intended to mimic the mathematical notion of being a subset of something, and the class relationship mimics the mathematical notion of being an element of something.

The language adheres as much as possible to customary mathematical notation and conventions about typing mathematics in ASCII. For example, to compute $Ext^i(M, N)$ the user types `Ext^i(M,N)`, and to construct the polynomial ring $\mathbf{Z}/101[x, y, z]$ one types `ZZ/101[x,y,z]`. The latter construction involves building the ring $\mathbf{Z}/101$ and the array $[x, y, z]$ and then locating the binary method to be used whenever a ring occurs adjacent to an array. In the system's code, that binary method is installed with a statement of the following form.

```
Ring Array := (R,args) -> ( ... )
```

The system is probably unique among computer algebra systems in providing the user with functions that are *closures*, as in the *Scheme* language. For example, the user can define a function that composes two other functions by typing the following assignment.

```
compose = (f,g) -> x -> f g x
```

Here are a few examples of the use of *Macaulay 2*.

We start with a Gröbner basis computation. We'd like to find a polynomial relation between $f = t^3 + t^2 + 1$ and $g = t^4 - t$, so we create a ring suitable for elimination of the variable $t$.

```
i1 : R = QQ[t,f,g,MonomialOrder => Eliminate 1]

o1 = R

o1 : PolynomialRing
```

The first line above is the user input, the second line is the output value, and the third signals the output type.

```
i2 : compactMatrixForm = false;

i3 : transpose gens gb ideal(f - (t^3 + t^2 + 1), g - (t^4 - t))

       | 4       3       2       2      3      2                  2                |
o3 =   | f   - 7f  - 2f g - 4f*g   - g  + 18f  + 3f*g + 6g   - 21f - g + 9         |
       |                                                                            |
       |              2                    3      2             2                   |
       |         t*g   - t*f + 6t*g + 5t - f  + 3f  + 3f*g + 3g  + g - 2            |
       |                                                                            |
       |                                              2         2                   |
       |         t*f*g + t*f - 4t*g - 3t + f  - f*g - g  - 4f - g + 3               |
       |                                                                            |
       |                2                     2                                     |
       |         t*f  - 4t*f + t*g + 5t - f  - f*g + 3f + 3g - 2                    |
       |                                                                            |
       |                        2                                                   |
       |                   t  + t*f - 2t - f - g + 1                                |

                  5       1
o3 : Matrix R  <--- R
```

The topmost entry in the matrix above is the desired relation.

Let's investigate the trace of a general nilpotent 3 by 3 matrix. We need variables for each of its entries, so let's make a polynomial ring with nine variables in it.

```
i4 : R = QQ[a..i];
```

Here is the general matrix.

```
i5 : m = genericMatrix(R,a,3,3)

o5 = | a  d  g |
     |         |
     | b  e  h |
     |         |
     | c  f  i |

             3      3
o5 : Matrix R  <--- R
```

To make the matrix nilpotent, we may work modulo the ideal generated by the entries of $m^3$, so we make a quotient ring. Results of algebraic operations in this quotient ring will be reduced to normal form using a Gröbner basis of the ideal, automatically computed for us.

```
i6 : S = R/(ideal m^3);
```

Let's reduce the elements of the matrix modulo the ideal, and call the new matrix $m'$.

```
i7 : m' = substitute(m,S);

             3      3
o7 : Matrix S  <--- S
```

Its cube is zero, as we verify.

```
i8 : m'^3

o8 = 0

             3      3
o8 : Matrix S  <--- S
```

Its trace is not zero, of course, since it's linear.

```
i9 : t = trace m'

o9 = a + e + i

o9 : S
```

The trace is a nilpotent element of $S$ because over a field, the trace of a nilpotent matrix is zero. After a bit of experimentation at the keyboard, we find which power of the trace is zero.

```
i10 : t^6
```

$$o10 = 90a^2e^2i^2 + 90b*d*e^2i^2 + 90a*e^3i^2 + 90e^4i^2 + 90c*e^2g*i^2 + 90a^2f*h^2 \ldots$$

```
o10 : S

i11 : t^7

o11 = 0

o11 : S
```

Now let's compute a big projective resolution. As an example, we'll take the Grasmmannian variety parameterizing projective planes inside $\mathbb{P}^5$. We start by setting up a ring with enough variables.

```
i12 : clearAll

i13 : R = ZZ/101[a..t];
```

We construct the graded coordinate ring of our variety as an $R$-module using a built-in routine.

```
i14 : M = R^1/Grassmannian(2,5,R)

o14 = cokernel | c*e - b*f + a*h   d*e - b*g + a*i   d*f - c*g + a*j   c*k ...

                             1
o14 : R - module, quotient of R
```

Now we compute the projective resolution, and display the graded Betti numbers.

```
i15 : time C = res M
     -- used 19.57 seconds

            1        35       140       301       735      1080       735       301 ...
o15 = R  <-- R  <-- R  <-- R  <-- R  <-- R  <-- R  <-- R    ...
                                                                                  ...
            0        1         2         3         4         5         6         7  ...

o15 : ChainComplex
```

```
i16 : betti C

o16 = total: 1 35 140 301  735 1080  735 301 140 35 1
         0:  1  .   .   .    .    .    .   .   .  .  .
         1:  .  35 140 189   .    .    .   .   .  .  .
         2:  .   .   . 112  735 1080  735 112   .  .  .
         3:  .   .   .   .    .    .    . 189 140 35  .
         4:  .   .   .   .    .    .    .   .   .  .  1

o16 : Net
```

Now let's illustrate how one computes cohomology of a coherent sheaf on a projective variety. We take the K3 surface defined by the equation $a^4 + b^4 + c^4 + d^4 = 0$. We begin with the homogeneous coordinate ring.

```
i17 : R = QQ[a,b,c,d]/(a^4+b^4+c^4+d^4)

o17 = R

o17 : QuotientRing
```

We use `Proj` to make a projective variety from it.

```
i18 : X = Proj R

o18 = Proj R

o18 : ProjectiveVariety
```

Now let's compute the cotangent sheaf $\Omega^1_X$ of $X$ and call it Omega.

```
i19 : Omega = cotangentSheaf(1,X)

                   |                  3    3             |
o19 = sheaf(cokernel | a  0  0  b   c  - d   0    0  |)
                   |                                    |
                   |                       3    3       |
                   | 0  0  a  d   0  b   - c    0  |
                   |                                    |
                   |                 3        3         |
                   | 0  a  0  - c  b  0   - d    0  |
                   |                                    |
                   |                            3    3  |
```

```
               |  0    d    c    0    0    0    a    b |
               |                                        |
               |                      3              3  |
               |  d    0   -b    0    0    a    0    c  |
               |                                        |
               |                      3              3  |
               | -c   -b    0    0    a    0    0    d  |
```

o19 : CoherentSheaf

The answer appears as the coherent sheaf associated to an $R$-module presented as the cokernel of a certain homogeneous matrix. We may compute the direct sum $\bigoplus_i H^1(X, \Omega^1(i))$ as an $R$-module with HH.

```
i20 : M = HH^1 Omega

o20 = cokernel |  0    0    0    0    a    b    c    d |
               |                                        |
               |  3    3    3    3                      |
               |  a    b    c    d    0    0    0    0  |

                              2
o20 : R - module, quotient of R
```

Notice that the module is evidently a direct sum of two simpler modules, since its presentation matrix is in block diagonal form, up to a permutation matrix. We can compute the Betti number $h^{1,1}$ by asking for the value of the Hilbert function of $M$ at 0, which is equivalent to asking for the dimension of the degree 0 part of $M$.

```
i21 : hilbertFunction(0,M)

o21 = 20
```

The answer, 20, is what we expected for a K3 surface, and the direct sum decomposition observed above writes 20 as $19 + 1$.

Let's take a look at a package written by Gregory Smith [Smith 1998], a graduate student working with David Eisenbud. It computes the global Ext-group for coherent sheaves on a projective variety. Let's use Grothendieck duality to see whether it's working properly.

```
i22 : k = ZZ/101; R = k[v,w,x,y,z]/(w*x, y*z); X = Proj R

o24 = Proj R
```

o24 : ProjectiveVariety

We know the dualizing sheaf $\omega$ will turn out to be $O_X(-1)$.

```
i25 : omega = OO_X(-1)
```

```
               1
o25 = sheaf(R )
```

```
o25 : CoherentSheaf
```

Here's the sheaf we will use.

```
i26 : G = sheaf coker matrix {{v,w},{w,x}}
```

```
o26 = sheaf(cokernel | v  w |)
                     |      |
                     | w  x |
```

```
o26 : CoherentSheaf
```

Let's load Smith's package.

```
i27 : load "globalExt.m2"
```

And now let's use it to compute $Ext^2(G, \omega)$.

```
i28 : globalExt(2,G,omega)
```

```
        2
o28 = k
```

```
o28 : k - module, free
```

By duality it ought to have the same dimension as $H^0(X, G)$, namely 2.

```
i29 : hilbertFunction(0,HH^0 G)
```

```
o29 = 2
```

And it does.

Now let's take a look at the computation of $\bigoplus_i Ext^i_R(M, N)$ where $R$ is a complete intersection, following an algorithm developed by Shamash [Shamash 1969] and Eisenbud [Eisenbud 1980].

```
i30 : compactMatrixForm = true;
```

```
i31 : R = ZZ/103[x,y]/(x^3,y^2);
```

Notice that this ring is not regular, so our modules may have infinitely many nonvanishing Ext-modules. Let's create a module $N$ as the cokernel of a random $1 \times 2$ matrix.

```
i32 : N = cokernel random (R^1, R^{-2,-2})

o32 = cokernel {0} | -9x2-44xy -13x2-33xy |

                           1
o32 : R - module, quotient of R
```

Now let's compute the total Ext-module.

```
i33 : E = Ext(N,N)

o33 = cokernel {2, 2} | 0  0  0  0 0 0 0 0 0 0 0    0      0     0    ...
               {1, 1} | 0  0  0  0 0 0 x 0 0 0 y 0  0      0     $X_1 ...
               {1, 1} | 0  0  0  0 0 0 x 0 0 0 y 0  0      $X_1  0    ...
               {1, 1} | 0  0  0  0 0 x 0 0 y 0 0 0  0      0     0    ...
               {1, 1} | 0  0  0  x 0 0 0 0 y 0 0 0  0      0     0    ...
               {0, 0} | x2 xy y2 0 0 0 0 0 0 0 $X_1x $X_1y 0     0    ...

               ZZ                                                     ...
o33 : --- [$X , $X , x, y, Degrees => {{1, 2}, {2, 2}, {1, 0}, {1, 0}}] ...
          103   0    1                                                ...
```

The answer is expressed as a finitely presented module over a polynomial ring over $R$ in two new variables, $\$X_0$ and $\$X_1$. The pairs of numbers labelling each row illustrate a useful feature of *Macaulay 2*, the ability to handle rings and modules which are bi-graded, i.e., graded by $\mathbf{Z} \times \mathbf{Z}$. (The second member of the pair is $i$.)

We have a system which has started to attract serious users. In the development of *Macaulay 2* we have paid close attention to minimizing the use of time and space in our implementation. We have paid close attention to making the language adhere as closely as possible to customary mathematical notation and concepts, and we anticipate that users will find the environment a natural and easy one.

Daniel R. Grayson (Urbana) and Michael E. Stillman (Ithaca)

## 4.2.26 MAS

MAS, the Modula-2 Algebra System

MAS, the Modula-2 Algebra System, is an experimental computer algebra system. MAS combines imperative programming facilities with algebraic specification capabilities for design and study of algebraic algorithms. It contains a large library of implemented up-to-date Gröbner basis algorithms for nearly all algebraic structures where such methods exist. MAS further includes algorithms for real quantifier elimination, parametric real root counting, and for computing in (commutative and noncommutative) polynomial rings. Most algorithms of the ALDES / SAC-2 library [Collins and Loos 1982b; Loos 1976] and of the DIP library [Böge et al. 1986] are contained.

This article describes MAS including the (at the time of this writing) most recent version 1.01, which has been released in March 1998.

We survey the development of MAS, the system design and the mathematical libraries.

**4.2.26.1 History and Development of MAS** The development of MAS began in 1985. At this time, the Fortran-based ALDES / SAC-2 system (see section 4.2.35) had one of the most comprehensive libraries of algebraic algorithms available. There were about 650 algorithms in ALDES / SAC-2 and in addition H. Kredel had about 450 algorithms developed on top of ALDES / SAC-2.

MAS was intended to be a computer algebra system with an up-to-date language and a design which allows to reuse the existing ALDES / SAC-2 algorithm libraries, to avoid the immense work that a re-implementation of these libraries would cause.

The wish for reusing existing software in an interactive environment with specification capabilities contributes most to the evolution of MAS.

Modula-2 had been chosen as the programming language since it was one of the most modern programming languages in 1985 and incorporated the concept of modular programming which allows to easily build reusable program libraries. It is easily possible to integrate function written in the C programming language into MAS. In fact, a small part of the recent version of MAS is written in C.

In 1987 the MAS kernel had been completed and most of the ALDES / SAC-2 libraries had been automatically translated to Modula-2. Since then a lot of new algorithms have been implemented in MAS, often by students as "Diplomarbeit".

Four versions of MAS have since then been released to the public, always with greatly increased functionality. The size of the recent version of MAS is a triple of the size of the first release of MAS, see figure 20.

| Version | Release Date | Lines | Bytes | Procedures |
|---|---|---|---|---|
| 0.3 | 07.09.1989 | 30 896 | 1 095 647 | 1 083 |
| 0.6 | 04.04.1991 | 36 374 | 1 209 994 | 1 270 |
| 0.7 | 13.05.1993 | 67 103 | 2 262 993 | 2 154 |
| 1.0 | 27.10.1996 | 101 027 | 3 362 577 | 2 849 |

**Fig. 20.** MAS Source Code Sizes

The first version was version 0.3 [Kredel 1990c,b], released on September 7th, 1989. Binaries were available for Commodore Amiga, Atari ST, and IBM PC compatible machines.

Version 0.6 [Kredel 1991a,b] was released on April 4th, 1991. Binaries were available for Commodore Amiga, Atari ST, and IBM PC compatible machines. Also, there was a translation to the C programming language available for compilation on UNIX machines.

On May 13th, 1993 version 0.7 [Kredel 1993a,b] was released. Binaries were available for IBM PC compatible machines running MS-DOS or OS/2 and IBM RS/6000 running AIX. Also, there was a translation to the C programming language available for compilation on UNIX machines. Support for Commodore Amiga and Atari ST was dropped.

The most recent version, version 1.0 [Kredel and Pesch 1996a,b] was released on October 27th, 1996. There exist binaries for HP 9000, running HP-UX, IBM RS/6000, running AIX, Intel PC, running Linux, Intel PC, running OS/2, all architectures running Nextstep and Sun Sparc, running SunOS.

**4.2.26.2 Design Overview** MAS is an experimental computer algebra system combining imperative programming facilities with algebraic specification capabilities for design and study of algebraic algorithms. MAS brings together specifications of abstract data types, imperative programs, controlled polymorphism, compiled libraries and term rewriting.

The *goal* of MAS is to provide:

1. an *interactive* computer algebra system
2. comprehensive algorithm *libraries*, including the ALDES / SAC-2 [Collins and Loos 1982b], and the DIP system
3. a familiar program *development* system with an efficient compiler,
4. an algebraic *specification* component for data structure and algorithm design
5. algorithm *documentation* open to the users.

Key *attributes* of MAS are:

1. *Portability*: MAS is easily portable to a new computer architecture. Machine dependencies are usually handled automatically by the use of GNU autoconf.
2. *Extensibility*: It is possible to add and interface to external algorithm libraries.
3. *Transparent low level facilities*:
   - Storage management: Garbage collection is provided without the need for user cooperation.
   - Stable error handling: No system break down on misspelled expressions and runtime exceptions.
   - Input / output with streams: No changes are required to existing libraries to redirect I/O.
4. *Efficiency*: Critical parts can be compiled and still be accessed interactively.
5. *Expressiveness*: MAS offers the possibility to specify abstract algebraic concepts like rings or fields.

The goals and attributes have been achieved by the following main *design concepts*:

MAS replaces the ALDES language [Loos 1976] and the FORTRAN implementation system of SAC-2 by the Modula-2 language [Wirth 1985]. Modula-2 is well suited for the development of large program libraries. The language is powerful enough to implement all parts of a computer algebra system. Modula-2 compilers have easy to use program development environments. Moreover there are Modula to C translators freely available. So MAS can be compiled on every system providing a C compiler. C components can be easily integrated in MAS.

To provide an interactive calculation system, a LISP interpreter is implemented in Modula-2 with full access to the library modules. For better usability a Modula-2 like imperative (interaction) language was defined, including a type system and function overloading capabilities. To increase expressiveness, high-level specification language constructs have been included together with conditional term rewriting capabilities. They resemble facilities known from algebraic specification languages like ASL [Wirsing 1986] and are discussed in detail in [Kredel 1991c].

**4.2.26.3 System Components** The MAS components are identified in figure 21. Active components (programs) are enclosed in square boxes and passive components (data) are enclosed in oval boxes. Arrows indicate flow of data and lines between boxes show that the components are related in some way.

As already mentioned MAS itself is a Modula-2 program. Thus the MAS program can be recompiled and linked together with other symbolic and numerical libraries by a suitable Modula-2 compiler. This is shown as an arrow from the compiler box on the right to the enclosing MAS box on the left.

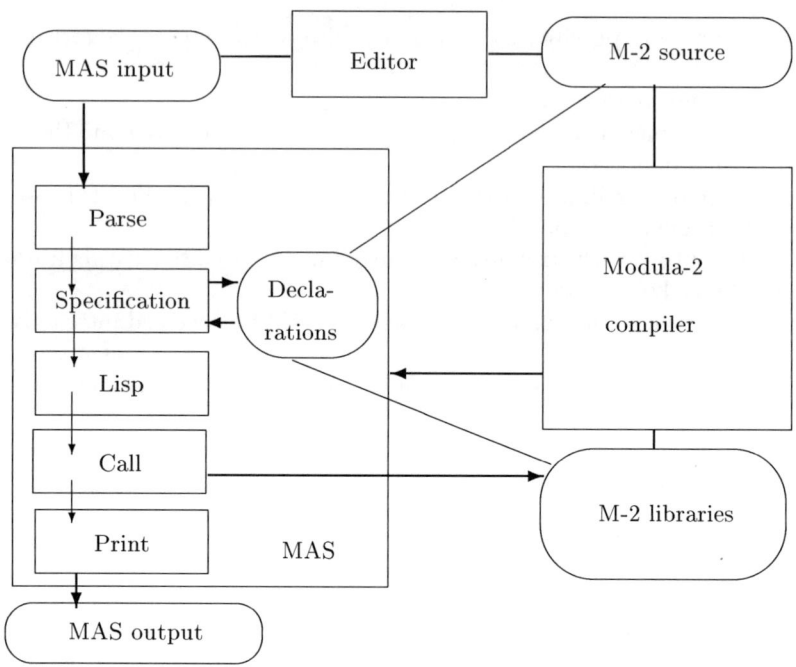

**Fig. 21.** MAS Components

On the top line the editor box both acts on the Modula-2 source code (on the right) and the MAS input data (on the left). The input is processed by the following internal components:

1. The *parser* for the MAS language (Parse box): character strings in concrete syntax are transformed into abstract syntax trees. Static syntax check together with variable scope analysis is performed.
2. The *specification processor* (Specification box) with an attached *data base of declarations* (Declarations box): declarations are extracted from the parse tree and stored in the declaration base, information is retrieved during interpretation. The declarations reflect the Modula-2 source code and the library structure.
3. The *LISP interpreter* (LISP box): according to the type or the function name of an S-expression inner most (that is eager) evaluation is performed.
4. The *interface to the libraries* (Call box): if external functions are encountered then corresponding compiled procedures from the libraries is called with the appropriate parameters.

5. The *pretty printing part* (Print box): the results of a computation are displayed by the pretty printing part.

**4.2.26.4 Mathematical Libraries** Besides kernel libraries, which handle memory management, error handling, input / output, etc., the interpreter, consisting of the parser, LISP, specification component, etc. and the list processing libraries, one of the most important parts of MAS is the mathematical libraries.

The mathematical libraries comprise most of the ALDES / SAC-2 libraries by Collins and Loos [Collins and Loos 1982b], and the DIP system by Gebauer and Kredel (see e.g. [Böge et al. 1986]).

Moreover the mathematical libraries include many new algorithms originally implemented in MAS.

One of the most prominent features of MAS is that it provides up-to-date Gröbner basis algorithms for almost all algebraic structures, where such methods exists.

In the following, we describe the mathematical libraries in more detail. "SAC" indicates parts taken from the ALDES / SAC-2 system, "DIP" indicates parts taken from the DIP system. Everything else has been originally developed for MAS.

**4.2.26.5 Arithmetic** On the mathematical side, arithmetic is the basic part of every computer algebra system. MAS contains algorithms for integer, modular integer, rational, rational complex, rational quaternion and rational octonion arithmetic. Algorithms for arbitrary precision floating point arithmetic and the handling of sets are available.

Apart from addition, multiplication, subtraction, division andd exponentiation these algorithms include integer greatest common divisor and least common multiple computation, integer random number generation, prime number generation, integer factorization and floating point root computation.

Further procedures for handling sets and procedures for combinatorics are included.

**4.2.26.6 Linear Algebra** MAS contains many algorithms from linear algebra including Gaussian elimination, LU-decomposition and determinant computation. Vectors and matrices with integral or rational entries can be handled.

**4.2.26.7 Polynomial Arithmetic** The basic requirement for the implementation of algorithms from the field of commutative algebra and algebraic geometry is a polynomial arithmetic. MAS can handle polynomials from polynomial rings over a large number of different coefficient rings. These include integers, rational numbers and modular numbers and polynomials with integral or rational coefficients.

**4.2.26.8 Arbitrary Domain Polynomials** MAS contains a mechanism that allows to implement algorithms on polynomials that can handle polynomials over arbitrary coefficient domains. Implemented domains are

- integers,
- modular integers,
- rational numbers,
- rational complex numbers,
- rational quaternion numbers,
- rational octonion numbers,
- finite fields,
- algebraic number fields,
- integral polynomials,
- rational polynomials,
- rational functions and
- arbitrary precision floating point numbers.

**4.2.26.9 Non-commutative Polynomials & Solvable Polynomial Rings**
MAS contains algorithms for computations with non-commutative polynomials, see [Kredel 1992], also see [Rody and Weispfennning 1990; Becker et al. 1993; Pesch 1997]. These include computation of the center, one-sided and two-sided Gröbner bases.

**4.2.26.10 Gröbner Bases** MAS contains implementations of Gröbner basis algorithms (see [Becker et al. 1993] and section 2.2.5) for almost all structures where such algorithms exists. Different versions with different selection strategies (including an improved exploitations of Buchberger's second criterion and the "sugar"-strategy [Rose 1995]) are implemented for commutative polynomial rings over various coefficient fields.

There are algorithms for D-Gröbner bases in commutative polynomial rings over integral domains [Mark 1992].

Arbitrary term orders can be specified by linear forms [Becker et al. 1993]. Moreover, the termorders which are usually needed, are implemented directly.

The algorithms for noncommutative polynomial rings are described above.

**4.2.26.11 Primary Ideal Decomposition** MAS contains algorithms for the primary ideal decomposition of zero-dimensional ideals [Kredel 1989], see also section 2.5.1.

**4.2.26.12 Factorized Gröbner Bases** A package to compute factorized Gröbner bases has been developed [Pfeil 1994].

**4.2.26.13  Comprehensive Gröbner Bases** Comprehensive Gröbner are finite sets of polynomials with parameterized coefficient that are Gröbner bases for all possible values of these coefficients, see [Weispfenning 1992a; Becker et al. 1993; Pesch 1997]. MAS contains algorithms to compute Gröbner systems or parts of such and comprehensive Gröbner bases [Pesch 1994]. Many applications and related algorithms are available.

**4.2.26.14  Universal Gröbner Bases** Universal Gröbner Bases are sets of polynomial which are Gröbner bases for every admissible term order. An algorithm to compute these is implemented in MAS (see [Belkahia 1992; Weispfenning 1989a; Becker et al. 1993]).

**4.2.26.15  Modules of Syzygies** One application of Gröbner bases is the computation of modules of syzygies, i.e. the computation of a module-basis of the set of solutions of a polynomial linear equation, see [Becker et al. 1993]. This has been implemented (see [Philipp 1991]) for commutative and solvable polynomial rings.

**4.2.26.16  Involutive Bases** Involutive bases are a special kind of (non-reduced) Gröbner bases, which can be computed by an algorithm different from Buchberger's. MAS contains a package for computing involutive bases and factorized involutive bases [Grosse-Gehling 1995].

**4.2.26.17  Invariant Polynomials** MAS contains a package for reduction of permutation invariant polynomials [Göbel 1992, 1996], see also 2.2.8.

**4.2.26.18  Ring Theory, Algebraic Geometry** Apart from Gröbner bases, MAS contains other algorithms from ring theory and algebraic geometry like GCD computation and factorization of univariate and multivariate polynomials.

**4.2.26.19  Module Arithmetic** MAS contains algorithms for solving linear Diophantine equation systems.

**4.2.26.20  Logic Formulae** A package for representation of logic formulae was included. It makes methods for formula simplification available on a new data structure for logic formulae.

**4.2.26.21  Real Root Counting** MAS contains an algorithm for counting real roots of multivariate polynomial algorithms based on Hermite's method [Lippold 1993]. Special algorithms for the univariate case are included.

**4.2.26.22  Real Quantifier Elimination** Elimination of quantifiers in formulas over the real numbers based on the methods of comprehensive Gröbner bases and parametric real root counting [Dolzmann 1994; Weispfenning 1998] (see also 2.15.2) is implemented in MAS.

**4.2.26.23 Availability** MAS (binaries and sources), more information and a bibliography are freely available (for non-commercial purposes) on the WWW from URL:

$$\text{http://alice.fmi.uni-passau.de/mas.html}$$

MAS is also available by anonymous-ftp from the URL:

$$\text{ftp://alice.fmi.uni-passau.de/pub/ComputerAlgebraSystems/mas.}$$

Binaries are available for

- HP 9000, running HP-UX,
- IBM RS/6000, running AIX,
- Intel PC, running Linux,
- Intel PC, running OS/2,
- Sun Sparc, running SunOS and
- multiple architectures, running Nextstep.

Please direct questions, remarks, bug-reports, etc. to:

$$\text{mas@alice.fmi.uni-passau.de}$$

Additional references: [Kredel 1990a; Buchberger et al. 1982; Buchberger 1985b].

Heinz Kredel (Mannheim), Michael Pesch (Passau)

## 4.2.27 MASYCA

The system MASYCA (**ma**thematics **sy**mbolic **ca**lculations) was developed by W.L.F. Degen and implemented on a Cyber 174 machine at the computer centre of the University of Stuttgart (Germany). The system's cernel consists of a manipulator for symbolic polynomial expressions with coefficients in $\mathbf{Z}$ or in $\mathbf{Z}/(m)$. Variables may stand for themselves or denote expressions assigned to them. All variables may have the meaning of an implicit function of one or two variables; thus derivatives are very simple to handle (e. g. $f'$, $sin' = cos$, $a' = 0$ etc.).

Two more packages are included. The first one serves to manipulate index expressions, containing among others the faculty function, the binomials and integers modulo $m$ (however only within the machine's range). The second one allows calculations with vectors, matrices and tensors (with three indices).

A special feature that can not be found in other computer algebra systems (as to the auther's knowledge even in nowaday's systems) is the possibility to assign a value to a *product* of two or to a *power* (higher than one) of one variable. Thus, not only scalar products become very simple but all kinds of calculations within arbitrary residue classes are accessible; e. g. to work with complex numbers one has only to introduce the rule $i^2 = -1$.

Frequently used command sequences may be composed to subroutines (with global variables only), stored on files and recalled at any place of a command. A control structure and an editor, both only with poor capabilities, are included for that purpose.

The system must be considered in relation to the time of its creation: REDUCE was the only interactive system available in Europe at that time; furthermore there were strong restrictions of memory for a normal user which forced the designer to keep code and data within the limit of about 300kB. Nevertheless the system has the advantages of being easy to learn and to apply on not too large calculations as for example substitutions, differentiations, asymptotic expansions of symbolic polynomial expressions as well as vectors and matrices with elements of that kinds. By the way it was written in FORTRAN (with data encoded bit-wise) and run very fast (compared with REDUCE up to ten times faster).

**Contact:**
Prof. Dr. W. Degen, Universität Stuttgart, Mathematisches Institut B, Pfaffenwaldring 57, 70569 Stuttgart
Tel.: 0711-685-5317, Fax: -5304, E-mail: degen@mathematik.uni-stuttgart.de

Reference: [Degen 1980]

Wendelin Degen (Stuttgart)

## 4.2.28 MOC

The MOC system (whose name is derived from Modular Characters) is a package to support the computation of modular character tables and decomposition matrices of finite groups. It consists of a collection of FORTRAN stand alone programs, each solving a simple task (e.g. matrix multiplication). More comprehensive and complex tasks are achieved by UNIX shell scripts linking these programs. This design makes it easy for users to extend MOC by their own commands.

MOC has its own long integer arithmetic as well as an arithmetic for abelian number fields, based upon integral bases and their multiplication tables. By $p$-adic approximation MOC solves a given system of integral linear equations, provided it has a unique solution over the integers. An extension of this method can be used if the system has a unique solution over the rational numbers. Finally, MOC makes use of methods of Integer Linear Programming. In contrast to the

Meat-Axe (see 4.2.10), MOC only operates with characters rather than matrix representations. Thus the restrictions of the Meat-Axe through the degrees of the representations disappear. A detailed description of MOC and its algorithms is given in [Hiss et al. 1993].

From the many results which were obtained with substantial help of MOC let us just mention [Hiss and Lux 1989] and [Jansen et al. 1995]. The latter reference contains in particular all modular character tables of the sporadic groups of order less than $10^9$. Since then, many more modular tables of sporadic groups have been computed with the assistance of MOC, Meat-Axe, and GAP (for GAP see [GAP 1997] and 4.2.18). For example, the three systems were used to determine all modular tables for the second Conway group. The known modular character tables are available through GAP. The web-site http://www.math.rwth-aachen.de/~MOC/ contains the latest results on modular character tables and the corresponding decomposition matrices.

MOC was developed by Gerhard Hiss (Aachen), Christoph Jansen (Aachen), Klaus Lux (Tucson), and Richard Parker (Cambridge) from 1984 to 1991. We plan to incorporate MOC into GAP in the form of GAP library functions.

Gerhard Hiss (Aachen), Klaus Lux (Tucson)

### 4.2.29 NTL: A Library for Doing Number Theory

NTL is a C++ library for basic polynomial and matrix operations that are implemented with intricate fast algorithms, like FFT-based polynomial multiplication. It was written largely by Victor Shoup with contributions from several researchers. We cite from NTL's the Internet web site http://shoup.net/ntl/, where the code is also available in source under the usual GNU free software license agreements.

"NTL is a high-performance, portable C++ library providing data structures and algorithms for arbitrary length integers; for vectors, matrices, and polynomials over the integers and over finite fields; and for arbitrary precision floating point arithmetic.

NTL provides high quality implementations of state-of-the-art algorithms for:

- arbitrary length integer arithmetic and arbitrary precision floating point arithmetic;
- polynomial arithmetic over the integers and finite fields including basic arithmetic, polynomial factorization, irreducibility testing, computation of minimal polynomials, traces, norms, and more;
- lattice basis reduction, including very robust and fast implementations of Schnorr-Euchner, block Korkin-Zolotarev reduction, and the new Schnorr-Horner pruning heuristic for block Korkin-Zolotarev;

– basic linear algebra over the integers, finite fields, and arbitrary precision floating point numbers.

NTL's polynomial arithmetic is one of the fastest available anywhere, and has been used to set "world records" for polynomial factorization and determining orders of elliptic curves.

NTL's lattice reduction code is also one of the best available anywhere, in terms of both speed and robustness, and one of the few implementations of block Korkin-Zolotarev reduction with the Schnorr-Horner pruning heuristic. It has been used to "crack" several cryptosystems.

NTL can be easily installed in a matter of minutes on just about any platform, including virtually any 32- or 64-bit machine running any flavor of Unix, as well as PCs running Windows 95, 98, or NT, and Macintoshes. NTL achieves this portability by avoiding esoteric C++ features, and by avoiding assembly code; it should therefore remain usable for years to come with little or no maintenance, even as processors and operating systems continue to change and evolve. However, NTL can be used in conjunction with GMP (the GNU Multi-Precision library) for enhanced performance.

NTL provides a clean and consistent interface to a large variety of classes representing mathematical objects. It provides a good environment for easily and quickly implementing new number-theoretic algorithms, without sacrificing performance."

The editors

### 4.2.30 PARI

The PARI-GP package

The PARI system is a package which is capable of doing formal computations on recursive types at high speed; it is primarily aimed at number theorists, but can be used by anybody whose primary needs are speed and ease of use.

**Comparison with other systems**: although quite an amount of symbolic manipulation is possible in PARI, it does badly compared to more sophisticated systems like Axiom, Macsyma, Maple, Mathematica or Reduce on such manipulations (e.g. multivariate polynomials, formal integration, etc...). On the other hand, the three main advantages of the system are its speed, the possibility of using directly data types which are familiar to number theorists (e.g. element of a number field, binary quadratic form...), and its extensive algebraic number theory module (with particular emphasis on explicit class field theory over number fields) which has no equivalent in the above-mentioned systems.

**Interpreter vs. Library mode**: it is possible to use PARI in two different ways:

1. as a library, which can be called from a high-level language application, for instance written in C or C++;
2. as a sophisticated programmable calculator, named GP, which uses a scripting language containing most of the control instructions of a standard language like C (there are interfaces with other popular interpreted languages: CLisp, perl and python; also custom routines from external shared libraries are easy to add at runtime to the GP core).

**Basic mathematical reference**: Henri Cohen's book [Cohen 1996b] describes the PARI core number theoretic algorithms (the multiprecision kernel and transcendental functions were omitted). Since then, a number of functions centered around constructive class field theory and relative extensions of number fields have been added.

**History**: PARI was born in the mid-eighties in the mathematics department of Bordeaux I University, written by Christian Batut, Dominique Bernardi, Henri Cohen and Michel Olivier. It is now moving towards a more open, internet-based development model. It has been used by number theorists worldwide but also in various other fields ranging from physics to topology or statistics. More advanced Computer Algebra Systems like Magma and, to a lesser extent, MuPAD are partly based on the PARI source code.

Of course, improvements and additions are done continuously and at the time of this writing (April 2000), the current version of PARI/GP is 2.0.19.beta. But a stable version will shortly be released, and the system is as fully operational as any other.

**Supported architectures**: although kernel multiprecision routines are partly programmed in assembly language, a portable kernel is provided so that all architectures are supported, some of them (Intel/x86, Motorola, Sun/SPARC and, to a lesser degree, DEC/alpha) being especially optimized. Unix is the preferred operating system to run GP, but DOS (needing an i386 processor or better), OS/2, Windows and PowerMacintosh versions, stripped down of a few (non-mathematical!) features are also available (and so is Linux which provides an excellent Unix system on the corresponding architectures).

**How to get it**: PARI/GP is copyrighted but free. The complete sources, as well as some binaries, are available by anonymous ftp from the URL:

    ftp://megrez.math.u-bordeaux.fr/pub/pari

General information, including mailing lists archive, can be obtained from the PARI Web site at the address

    http://www.parigp-home.de/

**Example**: let $a$ be a root of the polynomial $x^7 - x - 1$, and $b = (4a - 1)/a^3$. Let's compute the minimal polynomial of $b$. A brute-force approach is simply to compute a floating point approximation of (a random conjugate of) $b$, then look for linear relations of degree up to 7 among powers of $b$, using a variant of the LLL algorithm:

```
gp > \p 100
   realprecision = 105 significant digits (100 digits displayed)
gp > l = polroots(x^7-x-1); a = l[1]; b = (4*a-1)/a^3
%1 = 2.50458056791561956479270161537513606477154443358689870719\
   79354866878184631688907785789587138408324154 + 0.E-105*I
gp > algdep(b, 7)
%2 = x^7 - 5*x^6 + 147*x^4 + 640*x^3 - 31*x^2 + 2176*x - 20479
```

Of course, since we already know an *algebraic* expression for $b$, it is much more efficient to compute directly its characteristic polynomial:

```
gp > a = Mod(x, x^7-x-1); b = (4*a-1)/a^3
%3 = Mod(-5*x^6 + 5*x^5 - x^4 + 5, x^7 - x - 1)
gp > charpoly(b)
%4 = x^7 - 5*x^6 + 147*x^4 + 640*x^3 - 31*x^2 + 2176*x - 20479
```

(here $a$ is represented as $\bar{x} \in \mathbb{Z}[x]/(x^7 - x - 1)$, and $b$ is computed from this algebraic expression). Let's factor the characteristic polynomial of $b$ in the original number field. This can be done in a number of ways in PARI, one of them being:

```
gp > lift( factornf(%4, y^7-y-1) )
%5 =
[x + (5*y^6 - 5*y^5 + y^4 - 5) 1]
[x^6 + (-5*y^6 + 5*y^5 - y^4)*x^5 + \
   (-20*y^4 + 25*y^3 - 9*y^2 + y)*x^4 + \
   (14*y^6 - y^5 - 80*y^2 + 120*y + 72)*x^3 + \
   (320*y^6 + 240*y^5 - 16*y^4 - 18*y^3 + y^2)*x^2 + \
   (-y^6 + 1280*y^4 + 640*y^3 - 304*y^2 - 56*y - 8)*x + \
   (64*y^6 + 16*y^5 + 4*y^4 + y^3 + 5120*y^2 + 1280*y + 256) 1]
```

The `lift` function was used to simplify the way the coefficients are written down, otherwise the polynomials in $y$ would be given as elements of $\mathbb{Z}[y]/(y^7 - y - 1)$. The first line gives us back the expression for $b$ we computed in %3 above and we see that $b$ is the only root of its characteristic polynomial lying in the (abstract) number field $\mathbb{Q}[x]/(x^7 - x - 1)$. It must indeed be so since

```
gp > polgalois(x^7-x-1)
%6 = [5040, -1, 1]
```

tells us that the Galois group is the full symmetric group $S_7$ (of order 5040). Note that we can compute many more non-trivial invariants of the corresponding number field. We first store in a *big number field* structure some technical data needed for subsequent computations in our number field:

```
gp > bnf = bnfinit(x^7-x-1);      (; = no printed output)
gp > bnf.zk                        (maximal order)
%7 = [1, x, x^2, x^3, x^4, x^5, x^6]
gp > bnf.disc                      (discriminant)
%8 = -776887
gp > bnf.clgp                      (class group is trivial)
%9 = [1, [], []]
gp > bnf.fu                        (fundamental units)
%10 = [x, x^4 - x, x^6 - x^5]
```

Namely the maximal order admits a power basis (discriminant $-776887$), has trivial class group, and admits a system of fundamental units including $a$. Note that $a$ is an exceptional unit ($1 - a$ is also a unit):

```
gp > bnfisunit(bnf, 1-a)
%11 = [-5, 0, 1, Mod(1, 2)]~
```

The output gives the coordinates of $1 - a$ on the generating system given by the fundamental units above and the torsion unit ($-1$, of order 2). Now

```
gp > idealfactor(bnf, b)
%12 = [[20479, [-5120, 1, 0, 0, 0, 0, 0]~, 1, 1, \
       [4, 20, 80, 320, 1280, 5120, 1]~] 1]
```

tells us that $b$ generates a prime ideal $\wp$ above 20479 (the one containing $a-5120$). Indeed, we can check directly whether $\wp$ is principal

```
gp > P = %12[1, 1]; bnfisprincipal(bnf, P).gen
%13 = [-1, 4, 0, 0, 0, 0, 0]~
```

We obtain a generator of $\wp$ (namely $4 * a - 1$, given on the integral basis %7), which differs from $b$ by a unit (namely $a^3$):

```
gp > bnfisunit(bnf, nfeltdiv(bnf, %13, b))
%14 = [3, 0, 0, Mod(0, 2)]~
```

The whole series of computation in this example was carried out in 5 seconds user time on a slow workstation (2s on a 333MHz Pentium II).

Reference: [Batut et al. 2000].

Karim Belabas (Orsay)

### 4.2.31 Parsac

PARSAC [Küchlin 1991c, 1995] is a parallel Computer Algebra library written in C and based on the concept of parallel *threads of control*. PARSAC originated, in 1990, in the attempt to parallelize the first C version of SAC-2, which became

SACLIB a few years later. From the start, PARSAC was designed for networks of shared-memory parallel workstations running a version of UNIX (only minicomputer servers existed at the time).

PARSAC relies on the existence of lightweight processes *(threads)* in the operating system. In 1990, only MACH had threads, but they have since become a POSIX standard. The S-threads programming environment [Küchlin 1992] of PARSAC extends the operating system threads in two essential ways to make them fit for symbolic computation: (a) per-thread synchronized access to a heap with list-cells is provided together with parallel garbage collection [Küchlin and Nevin 1991], and (b) V-threads [Küchlin and Ward 1992], an ultra light-weight user-level threads layer, is added which allows the programmer to fork tens of thousands of threads without regard for the limitations of hardware and operating system.

The parallelization concept encouraged by virtual S-threads is parallel divide-and-conquer. Heavy-weight VS-threads created first high up in the call-tree are grabbed by empty processors and executed as real operating system threads. Excess light-weight VS-threads created further down in the call-tree are dynamically converted back to procedure calls when the parent tries to join them.

All but a few low-level SACLIB routines execute unmodified on VS-threads, providing the core of algebraic functionality in PARSAC. Parallel algebraic algorithms can now be written both by inserting thread forks into existing SACLIB code and by writing new parallel code with calls to sequential SACLIB routines. A parallel PARSAC program may thus consist of calling two SACLIB functions in parallel, such as for multiplying numerators and denominators of two rational numbers. Many sequential SACLIB algorithms already follow a divide-and-conquer scheme, such as Karatsuba long integer multiplication, polynomial real root isolation or the multiple homomorphic image algorithms for multivariate polynomial g.c.d. and resultant computations. These were the first to be parallelized, in 1990–91 [Küchlin 1991c; Küchlin et al. 1991; Collins et al. 1990; Küchlin 1991a,b]. Parallel integer multiplication showed that S-threads work efficiently down to a very low grain-size. The experience with high-level algorithms showed that the concept supports programming techniques which are manageable on a daily basis—no tricky low-level code needs to be written. The work with multivariate polynomial g.c.d. computation also provided the motivation for the virtual threads concept, because pure S-threads could not handle the amount of parallelism generated by larger examples.

In recent years, research has proceeded in 3 directions. First, on the system side, VS-threads were ported to Solaris 2.x and POSIX threads, and the concept was extended to the network. Our DTS programming environment supports the threads abstraction over a network of workstations running the same set of algorithms. Under suitable restrictions, an algorithm parallelized under S-threads can be distributed simply by replacing some S-thread forks by analogous DTS forks. The DTS system has been applied to the distributed computation of modular resultants [Bubeck et al. 1995]. DTS was originally implemented as an extension of PVM and has recently been re-implemented as a class library on top of the ACE environment.

Second, the parallel equational theorem prover PaReDuX was developed after combining VS-threads and the ReDuX term-rewriting system [Bündgen et al. 1996b]. PaReDuX contains parallel Knuth-Bendix completion, Peterson-Stickel AC-completion, and unfailing completion. A distributed unfailing completion procedure after the Teamwork concept has been constructed using DTS and combined with shared memory parallel completion using S-threads; the system runs on a network of parallel workstations [Bündgen et al. 1996a].

Third, a new parallelization concept for Buchberger's Gröbner Basis algorithm has been developed and implemented in a new system, again using the S-threads concept. In each completion cycle, the algorithm converts $w$ many critical pairs to S-polynomials which it reduces in parallel. When $k$ normal forms have been obtained, the best is selected and introduced into the basis. This favors short reductions over long ones and often leads to large super-linear speedups [Amrhein et al. 1996a]. Our implementation, together with a sequential implementation of the Gröbner Walk basis conversion method, is accessible via a Java GUI over the internet under URL http://www-sr.informatik.uni-tuebingen.de/projects/pareqs. Current work extends the algorithm over the network, using DTS.

For further information see http://www-sr.informatik/uni-tuebingen.de/

Wolfgang Kuechlin (Tübingen)

### 4.2.32 Quotpic

QUOTPIC is an interactive graphics display program, which, given a finite presentation for a group G, constructs quotients of the group and displays the output graphically as a lattice on a screen. Although QUOTPIC is by no means the first program to construct quotients of finitely presented groups, and in specific cases will certainly be outperformed by more specialist programs, it is the first such program to use interactive graphics; the use of the graphics enables the user to obtain very easily an overall picture of the quotient structure of the group. The user merely has to input a presentation of the group, and then select options, using a mouse and menus. The program runs on UNIX machines with X-Windows display.

QUOTPIC builds quotients via descending series of normal subgroups. In general the factors of each series are finitely generated abelian, but the top factor (and in fact other factors too, although the size of calculations usually limits this option far from the top of a series) might be any finite group which possesses a permutation representation of relatively small degree (usually less than 100) or alternatively any user-specified finite group of relatively small order. For these purposes the program has direct access to data on all groups with trivial Fitting subgroups and order less than 10000, and to all simple groups with order less than 1000000, and can receive data on further groups from the user. Thus the quotients which are built are usually polycyclic by finite, where the size of the insoluble part is normally restricted to having moderately small order by practical, rather than theoretical, considerations. The quotients are usually finite, but some information about infinite abelian by finite and nilpotent by finite quotients can be obtained.

A number of standard algorithms, programmed in C, and called via system calls from QUOTPIC, are used to build the quotients downwards. This loose connection between the graphics program QUOTPIC and the programs which do the actual computations of the quotients mean that it has been possible to import fast implementations from elsewhere, currently from Aachen, the University of Queensland, and the Australian National University.

QUOTPIC and the programs it calls are freely available by *anonymous ftp* across the internet from either ftp.ncl.ac.uk, where it currently sits in the subdirectory pub/local/nser in compressed tar files isom$i$.tar.Z, for

$i = 1, 2, 3, \ldots$, or similarly from ftp.maths.warwick.ac.uk in the subdirectory people/dfh/isom_quotpic. Alternatively it can be accessed from http://www.maths.warwick.ac.uk/~dfh/. Documentation is included with the source code. Further information can be obtained by contacting the authors of the program.

Contact addresses:
Dr. Derek Holt, Mathematics Institute, University of Warwick, Coventry CV4 7AL, UK.
Tel: 01203 523480 e-mail: dfh@maths.warwick.ac.uk.
Dr. Sarah Rees, Department of Mathematics and Statistics, University of Newcastle, Newcastle NE1 7RU, UK.
Tel: 0191 222 7236, e-mail: sarah.rees@ncl.ac.uk.

Reference: [Holt and Rees 1993]

Sarah Rees (Newcastle)

## 4.2.33 ReDuX

The ReDuX system provides a computing environment both for the developer of term rewriting software and for the user who wants to apply term rewriting systems to a variety of theorem proving fields like theorem proving in finitely presented algebraic structures, inductive theorem proving for algebraic specifications and hardware verification.

ReDuX supports term rewriting in many sorted term algebras. It allows for a very flexible definition of term notations (including pre-, in-, postfix operators, LISP and function notation and parentheses operators). ReDuX features efficient Knuth-Bendix completion procedures, Peterson-Stickel completion for commutative and/or associative-commutative theories, inductive completion procedures based on positional ground reducibility tests, tools to analyze the set of irreducible variable free terms, an unfailing completion procedure based on ordered rewriting, rewriting with built-in operations and a random term generator.

Descriptions of ReDuX can be found in [Küchlin 1982; Bündgen 1993; Bündgen et al. 1996c] and under the URL

http://www-sr.informatik.uni-tuebingen.de/~buendgen/redux.html

For academic purposed ReDuX is available free of charge via FTP under

ftp://ftp.informatik.uni-tuebingen.de/pub/SR/ReDuX

ReDuX is distributed together with all sources, ALDES/SAC-2 (see 4.2.35) program development tools, documentation and many pre-assembled term rewriting and theorem proving tools. It runs under most UNIX operating systems.

Reinhard Bündgen (Böblingen)

### 4.2.34  RepTiles   A Program for Interactively Generating Periodic Tilings

REPTILES is a program for systematically enumerating and interactively designing periodic tilings of the Euclidean plane [Delgado Friedrichs et al. 1995; Huson 1997]. It is based on the concept of *Delaney symbols*, which serve as a fundamental data-structure for periodic tilings [Dress 1984, 1987] and provide a symbolic language for formulating operations on such tilings.

In its simplest application, the program reads Delaney symbols from a file, *interprets* them and displays the encoded tilings. The user can interactively modify the appearance of the tiling, e.g. set edge widths or face colors.

More advanced features of the program allow the user to apply geometric operations to existing tilings and thus produce *new tilings from old*. Such operations include vertex-truncation, edge-contraction, dualization, and splitting or gluing of tiles. These transformations are implemented on a symbolic level in terms of the corresponding Delaney symbols. Further, the program provides tools for changing the symmetry group of a given tiling, such as symmetry-breaking and its inverse, "symmetry-making".

Additionally, the program includes algorithms for systematically enumerating *all possible* periodic tilings of a certain type up to a given degree of complexity [Dress and Scharlau 1984; Huson 1993]. REPTILES also contains a search engine for finding tilings of a given type in large files of Delaney symbols.

The theory of Delaney symbols generalizes to higher dimensions and has been used to classify three-dimensional periodic tilings [Dress et al. 1993a; Delgado Friedrichs and Huson 1999; Delgado Friedrichs et al. 1999].

At present, REPTILES is only available for the Macintosh. However, within the near future, the third named author plans to release a new program 2DTILER, that will generate and visualize periodic tilings of all three two-dimensional geometries, i.e. the plane, the sphere and the hyperbolic plane. This program will be compiled for a number of different platforms, including UNIX, Mac OS and Windows.

The first named author is currently developing a similar program 3DTILER for the analysis and visualization of three-dimensional tilings.

REPTILES is available from:

ftp://ftp.uni-bielefeld.de/pub/math/tiling/reptiles

Olaf Delgado Friedrichs, Andreas W.M. Dress (Bielefeld)
and Daniel H. Huson (Rockville, USA)

## 4.2.35  SAC-1, Aldes/SAC-2, Saclib

Saclib, the symbolic algebraic computation library, is a collection of functions written in the programming language C that performs polynomial operations with exact coefficients. Saclib (see also [Buchberger et al. 1993]) is the current state of a project begun in the early 1960s by George Collins for the implementation of cylindrical algebraic decomposition of a semi-algebraic set, thus performing logic decisions in real geometry (see section 2.5.3). The original library, SAC-1, was written in FORTRAN and pioneered garbage collection by reference counting. R. Loos designed the computing language Aldes (algorithm design language) for programming the library in a Knuth-like higher manner. The compiled Aldes programs constitute the SAC-2 library, which is available in ANSI Fortran, LISP, Modula, and C. Many of Collin's and Loos's collaborators and students have made contributions to the libraries.

The editors

### 4.2.36 SciNapse: Software that Writes PDE Software

SciNapse(TM) is a software system that automatically transforms a concise specification of a mathematical modeling problem specification into an executable numerical program. It can be thought of as a very smart optimizing compiler with optional directives. Although the framework is quite general, SciNapse focuses on using finite difference methods to solve systems of partial differential equations. A specialized version of the system, SciFinance(TM) (a commercially available product), generates options and derivatives pricing codes for the financial services industry. In addition, SciNapse has been used to generate numerical codes for applications in areas such as modeling seismic- and sonic-wave propagation. Specifications typically range from 10 to 50 lines, from which SciNapse will generate about 50 times the number of lines at a rate of about 3000 lines per minute.

The input to SciNapse is a specification in a concise but formal language that describes regions, equations, outputs, and inputs. Macros and equation generators can be defined for commonly used patterns. Specifications of solution techniques (solvers, evolution algorithms, interpolations, and so on) are optional. The system will apply its constraints and heuristics to make any solver choices or other decisions not indicated by the specification. SciNapse has an extensive knowledge-base (including information about boundary value problems, discretization, and optimizing transformations) that is implemented with objects, transformation rules, a reasoning system, and computer algebra.

SciNapse's close integration with a computer algebra system, Mathematica(TM), confers more powerful code-generation abilities than libraries or conventional expert systems. SciNapse uses computer algebra for a wide variety of purposes ranging from code generation to error analysis to optimization. For example, some transformation rules apply substantial algebraic manipulations, enabling arbitrary-order discretization rules and n-dimensional coordinate transformations. The ability to write such code generation rules means that the equations and templates can be abstracted away from the coordinate system or discretization method. Computer algebra is also used in producing codes with general coordinates and dimensionless parameters. SciNapse automatically calculates truncation errors, inserting this information as comments in the generated code, and (optionally) using the error to determine grid sizes. Computer algebra allows derived quantities to be computed as well as data structures and data-based operations to be optimized for a particular problem (e.g. dependent on the stencil structure resulting from an arbitrary discretization).

SciNapse is implemented on top of Mathematica and therefore runs under Windows, Unix, or any other system supported by Mathematica.

A minimum RAM of 96Mb and processor speed of about 200 is recommended.

Installation simply requires loading a set of files on a CD or in tar file; a root load file or notebook handles the details.

References: [Randall et al. 1998; Akers et al. 1997, 1998]

Web site: www.scicomp.com

Email: info@scicomp.com

> Elaine Kant, Stan Steinberg, Curt Randall, Larry Akers, and Bob Young
> (SciComp Inc.)

### 4.2.37 SENAC

**SENAC** stands for Software Environment for Numeric and Algebraic Computation. SENAC is a modular software system for scientists, applied mathematicians and engineers. It integrates numeric, symbolic and graphical features, it interfaces to existing libraries, and is portable. The SENAC modules are:

- Sencore - An interactive computer algebra host language for SENAC modules,
- Senpack - An interactive library including Numerical Recipes,
- Sengraph - A graphics library with PostScript output,
- Numlink - A fully automated symbolic numeric interface to the NAG Library,
- Graflink - A symbolic graphic interface to the NAG Graphics Library.

**SENCORE** is a special purpose symbolic manipulation language which provides an interactive, easy to use software host within which the symbolic form of mathematical modelling problems can be easily represented and numeric and graphical routines invoked.

#### 4.2.37.1 SENCORE features

- an easy to learn and use stand-alone natural user interface with extensive error checking,
- interactive input of symbolic and numeric expressions,
- symbolic expression output formatting,
- full featured floating point number formatting,
- probabilistic and deterministic algorithms for integer primality testing and factoring,
- list, matrix and sparse matrix data types,
- a wide range of matrix transformation functions,
- monomial and recursive polynomial representations,
- Zippel's probabilistic GCD and multinomial-factorisation,
- canonical rational simplification of SENAC expressions,
- user defined functions and programs for symbolic and numeric transformation,

- differentiation of expressions and undefined functions,
- symbolic integration by pattern matching or the Risch algorithm,
- translation of user defined functions into FORTRAN for SENAC calculation or external use,
- access to common blocks in FORTRAN code from SENAC,
- callbacks to SENAC from FORTRAN code for function evaluation,
- command timing and interruption,
- internal form of SENAC expressions,
- compiler for SENAC functions [Jan 1993],
- array data types [Jan 1993],
- a commentary feature to describe the progress of algorithms,
- optional interfaces to graphical and numeric subroutine libraries,
- online help for the host language and each interface included as a standard feature.

**4.2.37.2 Interfaces and Modules** In addition to the host module Sencore, there are available with SENAC two optional modules and two optional interfaces: an interactive numeric library Senpack (including 170 Numerical Recipes functions) and graphics library Sengraph, and the interfaces Numlink to the NAG Library of numeric subroutines and Graflink to the NAG Graphics Library.

**4.2.37.3 Availability** For further information in the UK/Europe: Dr. Minaz Punjani, SENAC Coordinator, University of London Computer Centre (ULCC), 20 Guilford Street, London WC1N 1DZ. Fax +44(71)242-1845 Tel +44(71)405-8400 e-mail: `senac@ulcc.ac.uk`.

All other regions: Associate Professor Kevin Broughan, Mathematical Software Project Director, Department of Mathematics and Statistics, University of Waikato, Private Bag 3105, Hamilton 2001, New Zealand. Telephone: -64(7)856-2889x8330 or -64(7)856-6358 Fax: -64(7)838-4155 e-mail: `senac@waikato.ac.nz`.

Minaz Punjani (London)<br>(Original 1993 contribution [Eds.])

### 4.2.38 Simath - Algorithms in Number Theory

SIMATH, that is *SInix–MATHematik*, is a computer algebra system focusing mainly on *algebraic number theory*. It is being developed at the Universität des Saarlandes in Saarbrücken (Germany), partially supported by the Siemens AG/Munich.

The main characteristics of SIMATH are

- the main area of application: *algebraic number theory*.

## 4.2.38 SIMATH - Algorithms in Number Theory

- the concept of the system: SIMATH is *a transparent system* which means that all sources are part of the system so the user can adapt existing general algorithms to specific problems and integrate her/his own algorithms at any point within th e system.
- the programming language: SIMATH is written in C; likewise, the user works in C. SIMATH functions are integrated into a C or C++ program simply by function calls.

SIMATH may also be accessed via the interactive calculator *simcalc* which features

- most of the existing SIMATH algorithms,
- comprehensive error checking,
- detailed "help facilities".

This makes *simcalc* particularly suitable for a quick calculation on the side and users with little programming experience.

**Simath**

SIMATH consists of

- an *interface* between the operating system and SIMATH;
- the C-library giving acces to more than 2000 functions;
- the *basic* system which consists of modified input/output functions, and a *list* system with an automatic *garbage collector* and dynamic memory administration;
- a *multiple precision arithmetic* package for computations over $\mathbb{Z}$, $\mathbb{Q}$, $\mathbb{R}$, $\mathbb{C}$, $\mathbb{Z}/m\mathbb{Z}$, $\mathbb{Q}_p$, finite fields, and global fields, i.e. algebraic number fields and function fields;
- a *polynomial* package for computations with polynomials in any number of unknowns over any of the structures contained in the arithmetic package;
- a *matrix–vector* package for matrix/vector computations over the structures contained in the arithmetic package and over polynomial rings;
- an *elliptic curves* package with elliptic-curve-specific functions over the rational numbers, prime fields, finite fields of characteristic 2 and algebraic number fields (see also 2.4.5);
- the interactive SIMATH calculator *simcalc*.

The *number theory* package contains higher algorithms for algebraic number theory (see Section 2.4) such as

- integral bases, extension of valuations, and the decomposition law for number fields and congruence function fields;

- for quadratic congruence function fields: regulators, unit groups, divisor and ideal class number, and generators and type of isomorphism of the ideal class group and the zero class group;
- conductor, minimal model, and an algorithm for finding the rank and a basis of an elliptic curve over rational numbers, as well as Tate's algorithm over the rational numbers and quadratic number fields;
- combined Schoof-Shanks algorithm for counting points on elliptic curves over prime fields and finite fields of characteristic 2;
- an algorithm for constructing elliptic curves with a given number of points over a given prime field;
- LLL-algorithm.

The system libraries each consist of a package of functions which - - except for some internal initialization and managing procedures – can be integrated into any C or C++ program by a simple function call.

### simcalc

The calculator *simcalc* is a user interface for solving problems to specific need. Giving full access to the SIMATH C-library, it enables the user to perform calculations in an extensive range and allows her/him the use of standard mathematical notation in a fully interactive environment.

*simcalc* handles calculations in

- $\mathbb{Z}, \mathbb{Q}, \mathbb{R}$ (with arbitrary precision), $\mathbb{C}, \mathbb{Z}/m\mathbb{Z}$, algebraic number fields $\mathbb{Q}(\alpha)$ and Galois-fields $\mathbb{F}_{p^n}$,
- polynomial rings over all these structures,
- matrices and vectors over all these structures,
- elliptic curves (and their points) over $\mathbb{Q}, \mathbb{F}_p, \mathbb{Q}(\alpha)$ and $\mathbb{F}_{2^n}$.

*simcalc* is easy to use because of its built-in system facilities, e.g.

- user-defined functions
- loop constructions and if-statements
- substitution of variables in polynomial structures
- extensive and comprehensive on-line help
- complete set of on-line documentation
- input errors are intercepted by self-explanatory error messages
- possibility to edit the input line and use the history with the usual keys of emacs
- arrays as variable names
- variable store with the possibility to list it entirely or partly and to delete in it
- overwrite protection that can be switched on and off

- user-defined configuration of simcalc
- data input from files
- data output on files
- statistical functions
- interrupt an output or a computation
- enter shell-commands and branch into a subshell.

### Availability

Currently, we are running SIMATH on

- HP 9000 series 700 under HP-UX 9.0x and HP-UX 10.x
- SGI machines under IRIX 5.3
- Sun SPARCstation under SunOS 4.1.1
- Intel based PCs under Linux 1.x and 2.x

It should not be difficult to compile and run SIMATH on most 32 bit UNIX platforms. For example, SIMATH is known to run on SPARCstations under Solaris 2.x and Intel based PCs under FreeBSD 2.x

The latest version of SIMATH may be obtained by anonymous ftp from ftp.math.uni-sb.de (134.96.32.23) in the directory /pub/simath. If you have any questions or problems, please contact the following address:

>SIMATH-Gruppe
>Lehrstuhl Prof. Dr. H.G. Zimmer
>Fachbereich 9 Mathematik
>Universität des Saarlandes
>Postfach 151150
>D-66041 Saarbrücken
>
>e-mail: simath@math.uni-sb.de
>www: **http://emmy.math.uni-sb.de/~simath/**

Marc Conrad and Susanne Schmitt (Saarbrücken)

## 4.2.39 SINGULAR – A Computer Algebra System for Polynomial Computations

### Overview of SINGULAR

**Main functionality**

SINGULAR is a computer algebra system for polynomial computations with emphasize on the special needs of commutative algebra, algebraic geometry, and singularity theory.

SINGULAR's main computational objects are polynomials, ideals and modules over a large variety of rings. SINGULAR features one of the fastest and most general implementations of various algorithms for computing standard bases. Furthermore, it provides multivariate polynomial factorizations, resultant, characteristic set and gcd computations, syzygy and free-resolution computations, and many more related functionalities.

Based on an easy-to-use interactive shell and C-like programming language, SINGULAR's internal functionality is augmented and user-extendible by libraries written in the SINGULAR programming language. A general and efficient implementation of links as endpoints of communications allows SINGULAR to make its functionality available to and be easily incorporated into other programmes.

**Background**

SINGULAR's development started in 1984 with an implementation of Mora's Tangent Cone algorithms in Modula-2 on an Atari computer (K.P. Neuendorf, G. Pfister, H. Schönemann; Humboldt-Universität zu Berlin). Using this first implementation, the existence of complete intersection singularities which are not quasi-homogeneous, but Poincare-complex exact could be shown [Pfister and Schönemann 1989]. In the early 1990s SINGULAR's "home-town" moved to Kaiserslautern and it was ported Unix, MS-DOS, and MacOS. Continuous extensions and refinements led in 1997 to the release of SINGULAR version 1.0 [Greuel et al. 1997] and in June 1998 to the release of SINGULAR version 1.2 (the current version).

SINGULAR was and is still being developed by G.-M. Greuel, G. Pfister, and H. Schönemann at the University of Kaiserslautern (Germany). SINGULAR's current developer group consists, further, of O. Bachmann, K. Krüger, T. Siebert, T. Wichmann, and W. Pohl. Many more people have made contributions to SINGULAR: W. Decker, C. Gorzel, H. Grassmann, A. Heydtmann, C. Jung, U. Klein, K. Krüger, B. Martin, M. Messollen, W. Neumann, T. Nüssler, and R. Stobbe.

**Availability**

SINGULAR is publicly available as a binary program for all common Unix platforms, for Windows 95/NT and for MacOS. The current version number is 1.2. It can be downloaded by anonymous ftp from

ftp://www.mathematik.uni-kl.de/pub/Math/Singular

See the README file in the frp directory for further instructions on how to download and install SINGULAR.

### 4.2.39 SINGULAR – A Computer Algebra System for Polynomial Computations

Besides the executable SINGULAR program, the distribution contains the source code of all SINGULAR libraries and the user manual (resp. tutorial) in various formats (PostScript, info, and, HTML).

For more and always up-to-date information, SINGULAR's home page im WWW can be reached at

$$\text{http://www.mathematik.uni-kl.de/}{\sim}\text{zca/Singular}$$

**Mathematical Features**

SINGULAR primary computational objects are ideals resp. modules which are generated by polynomials resp. polynomial-vectors over polynomial rings or, more generally, over the localization of a polynomial ring with respect to any ordering on the set of monomials which is compatible with the semigroup structure.

Supported baserings include:

- polynomial rings with a large variety of polynomial orderings (common simple, block, elimination, weighted, and general matrix orderings),
- localization of a polynomial ring at a prime ideal generated by a subset of the variables
- factor rings by an ideal of one of the above,
- rings of tensor products of one of the above.

Moreover, with a specially compiled version, SINGULAR supports

- exterior algebras,
- tensor products of one of the above rings with an exterior algebra, and
- Weyl algebras, and, D-modules.

Supported ground fields for these rings include:

- rational numbers $\mathbb{Q}$,
- finite fields $\mathbb{Z}/p$ (where $p$ is a prime $\leq 32003$),
- Galois fields (finite fields with $q = p^n \leq 2^{15}$ elements),
- transcendental extensions ($K(A, B, C, \ldots)$, $K = \mathbb{Q}$ or $\mathbb{Z}/p$),
- algebraic extensions ($K[t]$/minimal-polynomial, $K = \mathbb{Q}$ or $\mathbb{Z}/p$), and
- floating point real numbers with single precision.

The main algorithms implemented in SINGULAR are:

- General standard basis algorithm for *any* monomial ordering which is compatible with the natural semi-group structure of the exponents. This includes well-orderings (Buchberger algorithm) and tangent cone orderings (Mora algorithm) as special cases.

- Hilbert–driven Gröbner basis algorithms, weighted–ecart–method and high–corner–method, FGLM algorithm for change of ordering.

- Factorizing Buchberger algorithm.

- Intersection, quotient, elimination and saturation of ideals.

- Schreyer's and La Scala's algorithm for computations of syzygies and free resolutions of modules.

- Combinatorial algorithms for computations of dimensions of factor rings, Hilbert series and multiplicities of modules.

- Multivariate polynomial gcd, resultant, and factorization algorithms.

- Wang's algorithm to compute characteristic sets.

- Primary decomposition of ideals and modules based on Gianni/Traeger/Zacharias's and on Shimoyama/Yokoyama's algorithms.

- Puiseux development (over arbitrary characteristic) of plane curve singularities.

- Arnold's classifier of singularities.

- Normalization (integral closure) of affine rings.

- Ring invariants under certain group actions (finite groups, special unipotent groups).

- Invariants of singularities of affine or projective varieties such as Milnor, Tjurina and discriminate numbers, singular locus, T1 and T2 modules, semiuniversal deformations, etc.

- Algorithms for homological algebras as, for example, for $Ext$, $Tor$.

See also [Grassmann et al. 1996; Greuel and Pfister 1996; Greuel et al. 1997; Greuel and Pfister 1998] for more details on the implemented algorithms.

**Computational Features**

SINGULAR has a convenient and intuitive interactive user interface (shell) which has key-bindings similar to those of Unix' `tcsh` shell. SINGULAR's user interface provides both, access to SINGULAR's mathematical functionality and a convenient, powerful, and C-like programming language (strongly typed, and lexicographically scoped) which includes all the usual programming constructs (like loops, procedures, local/global variables, etc). Based on this programming language, users may extend SINGULAR's functionality by writing their own libraries.

At the moment, the SINGULAR distribution includes the following libraries:

### 4.2.39 SINGULAR – A Computer Algebra System for Polynomial Computations

classify.lib   Arnold's classifier of singularities
deform.lib     $T^1, T^2$, deformations of isolated singularities
elim.lib       elimination, saturation and blowing up
finvar.lib     invariant rings of finite groups
hnoether.lib   Hamburger-Noether (Puiseux) developments
homolog.lib    $Ext^i$, Massey-products, versal deformation of modules
invar.lib      invariants of $G_a$ acting on $K[x_1, \ldots, x_n]$
normal.lib     normalization of a ring
primdec.lib    primary decomposition of ideals
sing.lib       invariants of singularities: dimensions, Milnor, Tjurina numbers, critical locus, etc.
latex.lib      LaTeX typesetting of SINGULAR output.

SINGULAR furthermore features links as general endpoints of communications, i.e. as something, SINGULAR can read from or write to. To this point, the following link types are implemented:

- Ascii text:
  Output can conveniently be viewed and manipulated. Read/write is not the fastest.
- DBM (Standard Unix database):
  Provides access to data stored in a data base.
- MP file:
  Stores data in the binary Multi Protocoll (MP) format [Bachmann et al. 1996]. Read/write is very fast.
- MP TCP (Socket based, Client/Server links):
  Exchanges data in binary MP format between processes (on the same or different computers); data exchange is very efficient

The functionality of theses links is provided to the user by a general, consistent and convenient link interface.

Based on MP TCP links, SINGULAR can very efficiently communicate with itself which opens the door for implementations of parallel/distributed algorithms. Furthermore, the same links can be used to communicate with other Computer Algebra programs which have an MP interface. At the moments, there are MP interfaces for MuPAD, and Mathematica, enabling the use of one system from the others (e.g. one can use SINGULAR's functionality from within Mathematica or MuPAD).

**Implementation**

SINGULAR's kernel is implemented in C/C++. SINGULAR's main implementation design goal is speed. Therefore, all time-consuming operations like standard bases computations or factorization are implemented in its kernel. As another consequence, SINGULAR has the concept of a global ring which needs to be defined prior to any polynomial operations. Arbitrary precision integer arithmetic is accomplished by linking SINGULAR with the GNU multiple precision library gmp and modulo arithmetic is accomplished using look-up tables. Polynomials are internally represented as linked lists of monomials, where a monomial consists of a coefficient and an exponent vector.

To illustrate SINGULAR's speed of Gröbner basis computations, the table below shows the timings of various systems for solving problem 6 of the ISSAC'97 system challenge (computation of a lexicographical Gröbner basis). All timings were taken on a Pentium P90 with 32 MB of RAM running Linux:

| MapleV R4, MuPAD 1.3, Reduce, CoCoA (3.0.2), Macaulay 2 (0.8.14) | Mathematica 3.0.1 | GB v3[3] | | SINGULAR[4] | |
|---|---|---|---|---|---|
| | | groebner | fglm | stdfglm | groebner |
| _[1] | 362 sec[2] | 207 sec | 34 sec | 6 sec | 2 sec |

1): interrupted after appr. 15 CPU hours
2): on Pentium Pro 200
3): GB is the program develloped by Faugère
4): stdfglm, is the SINGULAR commands for the FGLM-based standard basis algorithm, grobner is a command which heuristically chooses the "best" way to compute a standard bases.

**Future Work**

SINGULAR's development is an actively ongoing project. Currently, the following features are under development:

- user interfaces (Emacs mode, Tcl/TK based user interface)
- faster computations of standard bases based on better implementation techniques and advanced algorithms
- implementation of parallel/distributed algorithms for computing standard bases
- dynamic modules [Sorgatz 1996] and library compilation
- Newton polyeder
- Standardbases for subalgebras of polynomial rings

Gert-Martin Greuel, Gerahrd Pfister, Hans Schönemann (Kaiserslautern)

## 4.2.40 SymbMath

SymbMath (an abbreviation for Symbolic Mathematics) is a symbolic calculator that can solve symbolic math problems.

SymbMath is a computer algebra system that can perform exact numeric, symbolic and graphic computations. It manipulates complicated formulas and returns answers in terms of symbols, formulas, exact numbers, tables and graphs.

SymbMath is an expert system that is able to learn from a user's input. If the user only inputs one formula without writing any code, it will automatically learn many problems related to this formula (e.g. it learns many integrals involving an unknown function $f(x)$ from one derivative $f'(x)$).

SymbMath is a symbolic, numeric and graphics computing environment where you can set up, run and document your calculation, draw your graph, and use external functions in the same way as standard functions since the external functions are auto-loaded.

SymbMath is a programming language in which you can define conditional, case, piecewise, recursion, multi-value functions and procedures, derivatives, intergrals and rules.

SymbMath is database where you can search your data.

SymbMath is a multi-windowed editor in which you can copy-and- paste anywhere in a file and between files, even from the Help file.

It runs on IBM PCs (80x86) with 400 KB free memory under MS-DOS. It can provide analytical and numeric answers for:

- Differentiation: regular or higher order, partial or total, mixed and implicit differentiation, one-sided derivatives.
- Integration: indefinite or definite integration, multiple integration, infinity as a bound, parametric or iterated integration, line or surface integrals, discontinuous or implicit integration.
- Solution of equations: roots of a polynomial, systems of algebraic or differential equations.
- Manipulation of expressions: simplification, factoring or expansion, substitution, evaluation.
- Calculation: exact and floating-point numeric computation of integer, rational, real and complex numbers in the range from minus to plus infinity, even with different units.
- Limits: real, complex or one-sided limits.
- Complexnumbers: calculation, functions, derivatives, integration.
- Sum and product: partial, finite or infinite.
- Others: series, lists, arrays, vectors, matrices, tables.
- Also included are:
  - Plot: functions, polar, parametric, data, and list.

- Draw: lines, arcs, ellipse, cirlces, ovals.
- Auto-loaded library in the source code.
- Pull-down and pop-up menus, resizeable and colorable windows.
- Procedural, conditional, iterational, recursive, functional, rule-based, logic, pattern-matching and graphic programming.
- Searching database.

Its two versions (Shareware and Advanced) are available from the author. The shareware version is available from anonymous FTP sites (e.g. ftp://ftp.unsw.edu.au/pub/UNSW/symbmath/sm332a.zip).

If you get SymbMath in ZIP format (e.g. `sm332a.zip`), you should `unzip` it with parameter -d by

```
pkunzip -d sm332a c:\symbmath
```

Email: huang@emphasys.com.au, w.huang@unsw.edu.au

URL: http://acsusun.acsu.unsw.edu.au/~s9300078/

Weiguang Huang (Sydney, Australia)

### 4.2.41 Symmetrica

SYMMETRICA a computer algebra package for the symmetric group

SYMMETRICA is a computer algebra package for representation theory, invariant theory and the combinatorial theory of the finite symmetric group and certain related groups (like the alternating group and wreath products of symmetric groups and the general linear group). This also includes the theory of symmetric function.

SYMMETRICA is a collection of routines written in C, so it is possible to use the system on nearly every computer platform. The system is independent of the hardware and the operating system.

The development started with the implementation of the Littlewood-Richardson-rule, where we used Schubert polynomials for the computation. The next step was the implementation of the character theory of the symmetric group (Murnaghan-Nakayama-rule) and methods for the computation of matrix representations.

The source code was written in an object oriented style. This allows easy extension of the source code. The user receives the source code, so he can implement his own extensions and also can use system tools like profilers for improvement of his own routines.

SYMMETRICA provides at the moment the following routines:

- ordinary irreducible characters for symmetric group, alternating group, and wreathproduct of symmetric groups.
- decomposition numbers and Brauer characters of the symmetric group
- decomposition of characters of the symmetric group (Littlewood Richardson, plethysm)
- ordinary and modular irreducible matrix representations of the symmetric group
- symmetry adapted bases, this includes irreducible polynomial matrix representations of the general linear group $GL_m(\mathbb{C})$,
- symmetric functions and related polynomials (zonal polynomials, Schubert polynomials)
- cycle indices
- computations in the group algebra of the symmetric group

To do these computations there is a long intger arithmetik and the cyclotomic arithmetic (which is needed in the character theory of the alternating group).

SYMMETRICA was developed together with A. Lascoux and his coworkers (Paris VII, Marne la Valle), with A.O. Morris and T. McDonough (Aberystwyth), with H. Fripertinger (Graz) and W. Müller (Bayreuth) and his coworkers.

SYMMETRICA is public domain, and can be fetched via the hompage at "www.mathe2.uni-bayreuth.de".

Adalbert Kerber and Axel Kohnert (Bayreuth)

## 4.2.42 Theorema: Computation and Deduction in Natural Style

The *Theorema* system provides an easy-to-use intelligent environment for the working scientist, by combining the computing capabilities of a computer algebra system with the deduction capabilities of automatic theorem provers. This system can be used interactively for bulding and checking mathematical models (which also means designing and verifying new algorithms) as well as for writing (interactive) textbooks in mathematics, computer science, and engineering, and additionally for the training of students in mathematical logic and computer science.

The system is based on early ideas of Bruno Buchberger [Buchberger 1996] and is developed by a team of researchers at RISC-Linz under his leadership [Buchberger et al. 1997b, 1998]. The distinctive features of *Theorema* are:

**Natural language** presentation using *Mathematica* notebooks. Both for input and for output, the description of mathematical models and of the algorithms uses the language of high-order predicate logic – the formulae being rendered in the usual two-dimensional mathematical notation. The proofs produced by our system are presented in natural language, similar to the one which is used in scientific textbooks.

**Natural style** deduction and integration of provers. The system comprises several provers: for high-order predicate-logic, for equational simplification, for induction on lists and on natural numbers, for sets, for mathematical analysis, and for functor-built domains. (New provers are constantly added.) The inference mechanisms used by the various provers are modeled upon the proving style of working mathematicians. Moreover, combining these provers is also done in the same style: depending on the domain to which the formulae refer, a prover incorporating specific knowledge about this domain will be used. *Theorema* uses a general mechanism for integrating these domain–specific provers according to a specified strategy.

**Combining proving and computing.** The provers can be used for defining and proving properties of mathematical models and algorithms, while a specially provided "computing engine" can execute the logical description of these algorithms directly.

A demo of the current version of the system is available at www.theorema.org.

Bruno Buchberger and Tudor Jebelean (RISC–Linz)

### 4.2.43  THEORIST—a User Interface for Symbolic Algebra

*The following is the contribution to original report in 1993. Theorist was bought by Waterloo Maple Inc—the editors.*

**4.2.43.1  Introduction**  Theorist is a symbolic algebra program that emphasizes a new graphical user interface. The user interface is graphical and interactive and has been well received. In particular, many people find they can use it after merely watching someone else use it, as opposed to the usual learning curves typically associated with such systems.

**4.2.43.2  Equation Entry**  Equations can be entered into Theorist either through the keyboard or via a palette, which has pictures of each variable, constant, function and operator available.

Theorist requires that all names of variables and functions have appropriate declarations. As new names are encountered in a notebook, Theorist will query the user as to the nature of the name and will allow the user to choose whether the name is declared to be a constant, a variable, a linear operator, or a function. The option is also given to not declare the name, in case it was misspelled. This system avoids the problems inherent in systems that don't use declarations, while avoiding the inconvenience of maintaining declarations by hand.

### 4.2.43.3 Computer Algebra

The fundamental principle behind interaction in Theorist is a system called the "Propositional Schema".

Actually, "Propositional Schema" is a new name for an old idea, used for centuries, where mathematical knowledge is described by equations or inequalities (propositions) and new propositions are derived based upon assumed propositions. It is obvious and intuitive to those who have been educated in the twentieth century. Theorist is usually easier to understand than the typical symbolic algebra program where the user enters a question, in the form of a function enclosing argument expressions that are to be manipulated, and the system displays an answer. In practice, some users who are very familiar with certain command-oriented systems have some trouble adapting to Theorist; this problem has not been encountered with new users.

In Theorist, there are no questions and answers. Instead, there are assumptions and conclusions. Assumptions are equations that the user types in; they have a square icon on the left. Conclusions are equations that Theorist generates when the user does a manipulation; they have a triangular icon instead. Both are somewhat equivalent in that they can be used in subsequent manipulations. Also, graphs can be drawn using either of them.

For example, the user can type in this equation:

$$y = \sqrt{\tan(x)}$$

The user can then select the $x$ and, holding down the Command (Clover) modifier key, drag the $x$ to the far left of the equation. When the appropriate highlighting displays, the user releases the mouse button, and a conclusion is generated:

$$x = \arctan\left(y^2\right)$$

Another kind of manipulation that can be done graphically is substitution. For example, the user can select the resulting equation from the last example, and can hold down the Command modifier key again and drag it on top of an expression such as this:

$$\int \sqrt{\tan(x)}\, dx$$

to yield this result:

$$2 \int \frac{y^2}{y^4 + 1} dy$$

(Note in this example that the differential operator $d$ actually differentiated $\arctan(y^2)$ in the process of simplification.) Any equation can be used in this way and any expression that matches the left hand side will be transformed into the right hand side.

Theorist also supports wildcard variables with which the user can make expression patterns that can match a wide variety of expression targets.

**4.2.43.4  Graphics**  In Theorist, the user can select any equation and request that a graph be created for it. Once created, the graph is an interactive object. The user can scroll or rotate the graph for a different viewpoint. The user can also zoom in or out, or increase or decrease the resolution (number of panels or line segments) in the objects in the graph, by way of buttons around the periphery of the graph.

The simplest transformation of a graph is by clicking and dragging on the face of the graph. The mouse pointer turns into a spread hand, ready to scroll the image. In the case of a two dimensional graph, the graph bounds are scrolled sideways as though it were a large sheet of paper under the user's hand. The new graph image is drawn such that the axes remain fixed relative to the edge of the graph, but the domain and range have been changed.

In the case of a three dimensional graph, when the user clicks down, a cube is displayed showing the bounds of the graphic space of primary attention. Dragging motions from the user result in rotations of this cube. When the mouse button is released, the three dimensional objects are redrawn as seen in the new orientation.

There are many different kinds of graphs available by menu command. In addition, the graphs themselves are programmable in that the user can delete or add more parametrically defined graphical objects to those that make up the graph as created by menu command.

For instance, if the user enters the equation $z = \sin(x)\sin(y)$, and then chooses "Color 3D" from the Graph menu, a three dimensional graph will be created containing a surface object, plus some axis and grid objects. The surface object draws a parametrically defined surface in the three dimensional space. Each axis object draws one axis line; there are three by default in this kind of graph. Each grid object draws a set of parallel grid lines (they are parallel in the space of their parameters but needn't be parallel graphically). The axis and grid objects are merely annotation but work similarly to the way the surface object works. The user can delete any of these, or can add more of any kind, and can change their expressions and other properties by simply clicking and typing.

**4.2.43.5  Calculations**  With Theorist, the user can do calculations by using the Calculate manipulation. As with any other manipulation, the user can calculate an expression and the result is shown below in a conclusion equation.

In addition, calculations are performed as part of drawing a graph. Drawing a graph merely involves repeatedly calculating expressions that are inside the graph. Those calculations reference variables that are defined by equations outside the graph, such as the original equation the graph was requested to use.

Any equation of a specific form can become a Working Statement, which defines the value of a variable, the definition of a function, or other properties. A Working Statement has a dot in its icon to indicate its special status. The user can choose to turn on or off Working Statement status. In addition, the program itself will turn it on or off in an effort to resolve undefined or multiply defined variables. For instance, when the user creates a graph of $y = \sin(x)$, the equation becomes a Working Statement that defines $y$. Theorist then evaluates the vector $(x, y)$ repeatedly in order to plot the points of the line. When it needs to evaluate $y$, it uses the Working Statement.

Users can create notebooks with any number of Working Statements that are defined in terms of each other, effectively forming numerical pipelines that are evaluated during graph drawing. Users who change any Working Statement will trigger the redrawing of any graphs that depend upon it. Finite recursion is allowed; a conditional function is supplied to prevent such recursion from becoming infinite.

**4.2.43.6 Theories** A theory is a set of propositions that are segregated from their surrounding propositions both graphically (with an enclosing box) and mathematically.

A theory represents a limited set of possibilities. Propositions inside of a theory are only assumed to be true within that theory, although propositions outside the theory are valid within the theory. The integrity of theories is enforced throughout Theorist. Propositions in different theories can be mixed only if one theory is a direct descendent of the other. For instance, a substitution mixing propositions cannot be done between propositions in sibling theories, although they can be done between propositions in a theory and propositions in the theory's grandparent theory.

Theories can be made manually by the user or they can be made automatically as the result of some operation. For instance, Theorist has a numerical root finder that works in conjunction with two dimensional graphs. A user might enter the equation $y = x^3 - 5x + 1$ and then draw a graph of it and find one or more roots. Each root is placed as an equation in its own theory, for instance, $x = 2.1284$. This equation can subsequently be substituted into other equations to learn more about the mathematical system under investigation in the vicinity of that point. All resulting propositions, however, will be placed in that theory because they are not valid unless $x = 2.1284$ is also valid. Sibling theories may have similar investigations underway for other roots such as $x = -2.33006$ or $x = 0.20163$. It is impossible, however, for the user to mix these mutually exclusive situations without manually copying and pasting expressions.

Theories may be nested to any required depth. At the top level, the main theory encloses everything in the notebook.

**4.2.43.7 Notebooks and Outlining** Theorist data is stored in "notebooks". Each notebook holds equations and graphs. In addition, notebooks can contain textual and picture comments provided by the user that are ignored by all mathematical processes. All such data is displayed in a window on the screen.

Theorist notebooks are organized with a topical outliner system. The user can open a topic to see its subtopics by double clicking on its icon (on the left).

Many facilities in Theorist work with the outliner system to keep a notebook orgainized. For instance, all name declarations are hidden under the first proposition of a notebook, which is usually a comment. Another example is that when conclusions are generated from other equations, the conclusions are indented underneath the equation they were derived from, so that the dependency tree can be easily visualized. Whole trains of thought can be hidden with a single double-click or eliminated with a single deletion keystroke.

**4.2.43.8 Limitations** Theorist runs on any Macintosh with at least one megabyte of memory. Implementations for other graphical user interfaces are planned for the future but there are no such versions at present.

For the most part, Theorist's symbolic capabilities are greater than Power-Math and Milo, roughly comparable to Derive, but inferior to Maple, Mathematica, Macsyma and Scratchpad II. For instance, it can invert matrices symbolically but it cannot find eigenvalues either symbolically or numerically (without the user going through the steps for finding roots of the characteristic polynomial). As is always the case, it is hard to compare systems and there are features in all of these programs that are unavailable on "more powerful" systems. In particular, many interactive features of Theorist increase the power of the user in unexpected ways. This article is merely a brief overview.

Theorist has no procedural programmability. Theorist can be programmed, though, in certain nonprocedural ways:

- algebraic rules can be added to its manipulation facilities
- graphical objects can be added to its graphs
- variables and functions can be defined for numerical calculations

Allan Bonadio (San Francisco)

## 4.3 Packages

### 4.3.1 ANU Polycyclic Quotient Programs

**4.3.1.1 Background** The study of groups defined by finite presentations is one of the classical areas in group theory. Despite the fact that it has been shown that many general questions about finitely presented groups are algorithmically unsolvable, there exist algorithms and their computer implementations for studying these groups (see also 2.7.3 and 2.7.9). For example, one family of algorithms investigates groups given by finite presentations by computing presentations for polycyclic factor groups. These include the ANU quotient algorithms designed to compute presentations for quotients of a finitely presented group that have prime-power order, are nilpotent or are finite soluble.

A *polycyclic group* has a descending series of subgroups, such that each is normal in the previous one and the quotient of two successive subgroups is cyclic (see Segal [Segal 1983], Sims [Sims 1994] or 2.7.8). It can be described by a *polycyclic presentation* which exhibits a polycyclic series of the group. The class of polycyclic groups includes the class of finite soluble and finitely generated nilpotent groups and hence also the class of $p$-groups. The central importance of polycyclic presentations is that they facilitate practical computation in polycyclic groups. In practice this is achieved by collection algorithms (see Leedham-Green & Soicher [Leedham-Green and Soicher 1990] and Sims [Sims 1994]).

We describe programs that compute polycyclic presentations for groups of prime-power order, nilpotent groups or finite soluble groups described as factor groups of finitely presented groups. In each case the input to the algorithm is a group $G$ given by a finite presentation. The output is a polycyclic presentation for $G/N$ together with an epimorphism $G \longrightarrow G/N$ where $N$ is a subgroup in a normal series $N_0 = G \geq N_1 \geq \ldots \geq N_i \geq \cdots$ associated to each algorithm. The basic step in each algorithm is to compute a polycyclic presentation for $G/N_{i+1}$ and an epimorphism $G \longrightarrow G/N_{i+1}$ from a polycyclic presentation for $G/N_i$ and an epimorphism $G \twoheadrightarrow G/N_i$. During this step the algorithms profit from an algorithm described in Newman, Nickel & Niemeyer [Newman et al. 1998] which was generalised to the nilpotent case in [Nickel 1996].

Note that there are several alternative algorithms to the ones described here and references to these can be found in the papers listed below.

**4.3.1.2 The ANU $p$-Quotient Program** Details of the $p$-quotient algorithm can be found in Newman [Newman 1976] and Havas & Newman [Havas and Newman 1980]. In addition to the $p$-quotient algorithm, the ANU $p$-Quotient Program offers access to implementations of the following:

- the $p$-group generation algorithm described in Newman [Newman 1977] and O'Brien [O'Brien 1990];

- an algorithm to decide isomorphism of $p$-groups described in O'Brien [O'Brien 1994];
- an algorithm to compute the automorphism group of a $p$-group described in O'Brien [O'Brien 1995].

The program has been used to obtain results such as:

- there are 56092 groups of order 256 (O'Brien [O'Brien 1991]);
- the largest finite group generated by 3 elements in which every element has order dividing 5 has order $5^{2822}$ (Newman & O'Brien [Newman and O'Brien 1996]);
- and (using variants for Lie and associative algebras which are not yet in the public domain) that the largest finite group generated by 2 elements in which the order of every element divides 7 has order dividing $7^{20418}$ [Newman and Vaughan-Lee 1998].

The program is implemented in C and is available as stand-alone or as part of the systems **GAP**, see 4.2.18, MAGMA, see 4.1.5, and QUOTPIC, see 4.2.32.

For a recent description of the program and some of its applications, see Newman & O'Brien [Newman and O'Brien 1996].

### 4.3.1.3 The ANU Nilpotent Quotient Program
Details of the algorithm implemented in the ANU Nilpotent Quotient Program can be found in Nickel [Nickel 1996]. In addition to the computation of nilpotent factor groups, the program provides facilities for the computation of nilpotent groups that satisfy an Engel identity or are have Engel elements as generators.

The program has been used to obtain insight into certain classes of infinite nilpotent groups (see Newman & Nickel [Newman and Nickel 1994] and Nickel [Nickel 1999]) and to compute polycyclic presentations for infinite nilpotent Engel groups such as the free nilpotent 2-generator $n$-Engel group for $n = 4, 5, 6$.

The implementation is written in C and is available as stand-alone via anonymous ftp and as part of the systems **GAP**, QUOTPIC and MAGMA.

### 4.3.1.4 The ANU Soluble Quotient Program
Details of the algorithm can be found in Niemeyer [Niemeyer 1994b, 1995].

The program has been used to compute a polycyclic presentation for the Burnside group $B(2,6)$, the freest group on two generators with exponent 6. It was also used to find small sets of sixth power relations that define certain groups of exponent six, see Havas et al. [Havas et al. 1999].

The program is implemented in C and is available as stand-alone, or as part of the system **GAP**. An alternative soluble quotient algorithm described in [Plesken 1987] and [Brückner 1998] is available in MAGMA.

**4.3.1.5  Availability** Latest versions of the software described here can be found at http://wwwmaths.anu.edu.au/research.groups/algebra

M.F. Newman (Canberra), Werner Nickel (Darmstadt),<br>
Alice C. Niemeyer (Perth, Australia), E.A. O'Brien (Auckland)

## 4.3.2  AREP

The special purpose library **AREP** (Abstract REPresentations) provides data structures and algorithms for calculating symbolically with structured matrix representations of finite groups, [Egner and Püschel 1998]. For example, it is possible to work with inductions or tensor products of representations without constructing large matrices explicitly. The main distinction of **AREP** from other libraries for representation theory is that representations are manipulated up to *equality* and not just up to equivalence (for which character theory is the adequate device). See Section 2.8 for an overview on representation theory in computer algebra. **AREP** is a reviewed **GAP** share package and distributed together with the **GAP3** system (see Section 4.2.18).

**AREP** consists of four major building blocks: Structured matrices (datatype `AMat`), structured matrix representations of finite groups (datatype `ARep`), functions for decomposing monomial representations into irreducibles, and functions for combinatorial search of certain types of symmetries in matrices. A brief overview of these building blocks follows.

**Structured Matrices:** The recursive data type `AMat` implements a term algebra for structured matrices. It was created to represent matrix expressions like $2 \cdot (A \oplus B)^C \otimes D \cdot E^2$ in symbolic form. The basic building blocks for `AMat`s are permutation matrices, monomial matrices (one non-zero entry in each row and column) and, of course, flat matrices (lists of lists of field elements). Functions are provided to simplify `AMat`s, convert them into flat matrices, or evaluate trace, determinant, transposition and other well-known functions efficiently by recursion over the structure of the expression. An `AMat` can also be exported to LaTeX yielding an expression as the one shown below.

**Structured Representations:** The recursive data type `ARep` implements a term algebra for structured matrix representations of finite groups. For example, the expression $(\phi \uparrow_T G)^A$ represents a conjugated induction of a representation $\phi$ to a supergroup $G$ with transversal $T$. Basic building blocks of `ARep`s are permutation, monomial and flat matrix representations. These can be constructed, e.g., by giving the images on a fixed set of generators (possibly adding degree and base field). Constructors of `AReps` are induction, restriction, inner and outer tensor product, direct sum, conjugation, extension and galois conjugation of the base field. Functions are provided that convert `AReps` into flat matrix representations or monomial representations (when possible), determine the character, compute intertwining spaces and other well-known functions by recursion over the structure.

**Decomposing Monomial Representations:** On top of the basic functions for AReps, the sophisticated function DecompositionMonRep takes a monomial representation of a solvable group under Maschke's condition and decomposes it stepwise into a direct sum of irreducibles. The function returns an ARep that represents the same representation as its argument but the expression makes the decomposition explicit. Moreover, the conjugating matrix is an AMat that kept track of all operations applied to decompose the monomial representation, [Püschel 1999b]. This allows to read off the irreducibles, the decomposition of the complex group ring into a direct sum of full matrix algebras, the isomorphism of this decomposition and a fast algorithm for the evaluation of this isomorphism. In the special case of a regular representation, a fast Fourier transform for the group is obtained.

**Finding Symmetries of Matrices:** AREP provides functions to find certain "symmetries" of matrices. For example, the function MonMonSymmetry performs a combinatorial search to find all pairs $(L, R)$ of monomial matrices satisfying $LM = MR$ for a given matrix $M$, [Egner 1997]. The result is returned as a pair of monomial AReps that represent a common group. Other functions in AREP search for a pair of permutation representations or for pairs $(\mu, \phi)$ where $\mu$ is monomial and $\phi$ is a permuted direct sum of irreducibles, [Püschel 1999a; Egner 1997]. The symmetry of a matrix (if one exists) can be used to factorize the matrix as explained below.

**Fast Discrete Signal Transforms:** The original motivation to develop AREP was to construct fast algorithms for given discrete linear signal transforms, like the fast Fourier transform, automatically, [Püschel 1999b,a, 1998; Egner 1997; Minkwitz 1995, 1993]. It involves all four building blocks of AREP. In a first step, a given matrix $M$, representing the discrete linear signal transform $x \mapsto Mx$, is studied for symmetries in the sense mentioned above. The resulting AReps are then decomposed into irreducibles using the decomposition function. The conjugating matrices obtained in this way are extracted, combined and a correction matrix is inserted. This way, the matrix $M$ is factored into a product of sparse and highly structured matrices. The result, expressed as an AMat, represents a fast algorithm for applying the transform $M$. The ARep can then, for example, be exported into the TPL-system, [Moura et al. 1998], in which rule-based optimizations can be applied and Fortran code can be produced. This makes algorithms generated by AREP interesting for applications.

The following example shows a factorization of an $8 \times 8$ discrete cosine transform (DCT, type II) generated entirely automatic (even the LaTeX-expression to print it). The corresponding algorithm needs 13 additions and 29 multiplications and is among the best ones known.

$$\begin{aligned}\text{DCT}_8 = \\ [(2,5)(4,7)(6,8),8] \cdot \tfrac{1}{2} \cdot (\tfrac{1}{\sqrt{2}} \cdot \mathbf{1}_2 \oplus \text{R}_{\frac{3}{8}\pi} \oplus \text{R}_{\frac{15}{16}\pi} \oplus \text{R}_{\frac{21}{16}\pi}) \cdot \\ [(2,4,7,3,8),8] \cdot ((\text{DFT}_2 \otimes \mathbf{1}_3) \oplus \mathbf{1}_2) \cdot [(5,6),8] \cdot \\ (\mathbf{1}_4 \oplus \tfrac{1}{\sqrt{2}} \cdot \text{DFT}_2 \oplus \mathbf{1}_2) \cdot [(2,3,4,5,8,6,7),8] \cdot \\ (\mathbf{1}_2 \otimes ((\text{DFT}_2 \oplus \mathbf{1}_2) \cdot [(2,3),4] \cdot (\mathbf{1}_2 \otimes \text{DFT}_2))) \cdot \\ [(1,8,6,2)(3,4,5,7),8],\end{aligned}$$

$[\sigma, n]$ denotes the $n \times n$ permutation matrix of permutation $\sigma$, $\mathbf{1}_k$ is the $k \times k$ identity matrix, $\text{DFT}_2 = \begin{pmatrix} 1 & 1 \\ 1 & -1 \end{pmatrix}$, $\oplus, \otimes$ denotes the direct sum resp. Kronecker product of matrices, and $\text{R}_\alpha$ is the $2 \times 2$ rotation matrix for angle $\alpha$.

Sebastian Egner (Eindhoven), Markus Püschel (Pittsburgh), and Thomas Beth (Karlsruhe)

### 4.3.3 CALI

CALI is a REDUCE package that contains algorithms for computations in commutative algebra closely related to the Gröbner algorithm for ideals and modules. Its heart is a new implementation of the Gröbner algorithm, see 2.2.5, that is more flexible than the original Gröbner implementation of REDUCE. It allows also for the computation of syzygies and is applicable to submodules of free modules with generators represented as rows of a matrix.

Based on this central knowledge the package offers implementations of a wide variety of algorithms for commutative algebra problems as described, e.g., in 2.5.1 or [Becker et al. 1993]. As main topics CALI contains facilities for

- defining rings, ideals and modules,
- computing Gröbner bases and local standard bases, see 2.2.6,
- computing syzygies, resolutions and (graded) Betti numbers,
- computing (also weighted) Hilbert series, multiplicities, independent sets, and dimensions, see [Gräbe 1993],
- computing normal forms and representations,
- computing sums, products, intersections, quotients, stable quotients, elimination ideals etc.,
- primality tests, computation of radicals, unmixed radicals, equidimensional parts, primary decompositions etc. of ideals and modules, see [Gräbe 1997],

- advanced applications of Gröbner bases (blowup, associated graded ring, analytic spread, symmetric algebra, monomial curves etc.),
- applications of linear algebra techniques to zero dimensional ideals, as, e.g., the FGLM change of term orders, border bases and affine and projective ideals of sets of points,

– splitting polynomial systems of equations mixing the Gröbner algorithm with factorization, triangular systems, and different versions of the extended Gröbner factorizer, see [Gräbe 1995c,b].

CALI contains also facilities for local computations, using a modern implementation of Mora's standard basis algorithm, see [Mora et al. 1992] and [Gräbe 1994], that works for arbitrary term orders. A unified approach to both the local and the global cases is based on reduction with bounded ecart for non Noetherian term orders as described in [Gräbe 1995a]. This allows for a common driver for the Gröbner algorithm in both cases.

For a more detailed description of CALI the reader may consult the manual and also an extended test file with many different examples. Both files are contained in the package that is part of the Contributed Software Section in the REDUCE distribution. They may also be obtained from the URL http://www.informatik.uni-leipzig.de/~compalg

Hans-Gert Gräbe (Leipzig)

### 4.3.4 CLN

CLN is a C++ library for doing multiple-precision computations with high efficiency. CLN is easy to use because it supports algebraic syntax and provides for automatic memory management. It is free software, covered by the GNU General Public License.

CLN's data types are integers, rational numbers, floating-point numbers, complex numbers, modular integers, and univariate polynomials, all of them with unlimited procision. CLN implements elementary, logical and transcendental functions.

Due to C++ as implementation language, programs written to use CLN are type-safe and efficient at the same time, and easy to understand because of CLN's algebraic syntax.

CLN's extraordinary efficiency comes 1. from the Karatsuba and Schönhage-Strassen multiplication algorithms it implements, 2. from consequent use of the binary splitting technique for transcendental function evaluation [Haible and Papanikolaou 1997], and 3. from the GMP kernel which it integrates [Granlund 1996].

CLN is currently used by researchers in the domains of number theory, cryptography, complex analysis, and physics.

CLN is available from http://clisp.cons.org/~haible/packages-cln.html. It runs on Unix and Windows, using GCC as C/C++ compiler.

Bruno Haible

### 4.3.5 CRACK, LIEPDE, APPLYSYM and CONLAW

**General description**
To investigate non-linear differential equations (DEs) or partial differential equations (PDEs), for which no general solution techniques are known, one either relies on numerics or, if one is interested in exact results, one investigates special properties, like symmetries or conservation laws. Any such results may give a better understanding of the problem, provide the basis for using more adequate numerical algorithms or even solve the problem analytically.

The circumstance that special properties are investigated (symmetries,...), which may but need not be present, has the consequence that the conditions for these properties lead to an overdetermined system of equations, usually PDEs. To solve overdetermined PDE systems, the package CRACK contains about a dozen modules for the

– integration of different types of PDEs,
– direct and indirect separation of PDEs,
– computation of a Pseudo Differential Gröbner Basis,
– substitution of equations into each other,
– length reduction of equations,
– solution of underdetermined linear ODEs/PDEs,
– reduction of redundant arbitrary constants and functions in solutions of DEs, and
– performing point transformations.

Most of them can work together automatically, all of then can be used interactively driven by a menu. The programs LIEPDE, APPLYSYM and CONLAW use the package CRACK to solve the PDE-systems they generate.

CONLAW is a package of four programs implementing four different approaches for the computation of first integrals of single or systems of ordinary DEs (ODEs) or conservation laws for single or systems of PDEs [Wolf et al. 1999], [Wolf 1998b].

LIEPDE is a program for the determination of infinitesimal point-, contact- and generalized higher order symmetries of single DEs and DE-systems. For a description see [Wolf 1993] and for an application [Sokolov and Wolf 1999].

With the program APPLYSYM point symmetries, computed with LIEPDE, can be integrated to yield symmetry- and similarity variables, i.e. a symmetry reduction [Wolf 1995]. After a transformation of the DEs into these variables, symmetry variables themselves disappear which allows the lowering of the order of ODEs or the reduction of the number of variables in the search of special solutions to PDEs.

**System requirements**
The required system is REDUCE, version 3.6 or 3.7. The *.rlg files (see below) are produced from test files *.tst in a 8MB session running REDUCE, version 3.7 under LINUX.

**Availability**
For each of the four packages CRACK, LIEPDE, APPLYSYM and CONLAW there are test files *.tst, log files *.rlg, manual descriptions *.tex and source files *.red freely available from ftp.maths.qmw.ac.uk. Currently (Sep. 1998) two web demos allow the use of CRACK and CONLAW for problems of restricted size, accessible from http://cathode.maths.qmw.ac.uk/demos.html. A publication list related to these programs and algorithms can be found at http://www.maths.qmw.ac.uk/~tw.

**Target topics**
CRACK and related application programs are applicable whenever smooth analytic properties of differential equations or geometric objects including manifolds are investigated. With "smooth" we mean properties that can be described by differential equations which are valid in an open neighbourhood in contrast to problems from Discrete Mathematics like "Which integers satisfy ...". In this sense CONLAW, LIEPDE and APPLYSYM are just examples for applications of CRACK.

**Comparisons with other systems**
The strength of the package CRACK in comparison with other packages also aiming at the solution of overdetermined PDE systems is based on

– a wide range of algorithms and modules mentioned above,
– the possibility to run batch jobs as well as to work interactively with full access to inspect, add and change data and flags/parameters of the algorithms,
– a high flexibility in the way the above mentioned modules interact.

What distinguishes the programs in CONLAW from other programs is that they allow the computation of

– conservation laws (CLs) for ODEs as well as PDEs of a non-restricted number of independent variables,
– CLs which are non-polynomial, even non-rational in the integrating factors and conserved quantities,
– CLs which have an explicit dependence on the independent variables,
– CLs for which an ansatz for the integrating factor and/or the first integral/conserved current can be made.

Examples of computations that become possible with CONLAW including new conservation laws are given in [Wolf et al. 1999], [Wolf 1998b]. A characterization of other programs for the computation of conservation laws is given in [Göktaş and Hereman 1997].

A comprehensive collection of references related to the computation of symmetries is given in [Hereman 1996]. The high efficiency of LIEPDE in comparison with other programs comes from

- using CRACK for solving the symmetry conditions,
- the ability to formulate the overdetermined symmetry conditions in portions and to solve them successively before the next conditions are formulated which are reduced in size and easier to solve as a consequence.

**User base**
CRACK and related programs are available for about a decade in the Reduce network library and have been distributed as part of Reduce in recent years.

**Applications**
The programs CONLAW, LIEPDE, APPLYSYM are applicable to investigate any kind of differential equation or system of differential equations (apart from complexity issues of course, when PDEs with many functions of many independent variables are to be investigated with respect to symmetries or conservation laws of high order). Examples are given in the *.tst files and in [Wolf et al. 1999], [Wolf 1998b].

These programs are themselves applications of CRACK and many other applications of CRACK can be thought of, for example, 1. the inverse variational problem, i.e. the question whether for a given system of DEs there is an equivalent variational principle, 2. the consistency of additional constraints added to a set of equations.

CRACK has been used in General Relativity either to investigate a special ansatz to find exact solutions [Stephani and Wolf 1996], [Mars and Wolf 1997] of the field equations or to characterise space times by determining their symmetries [Wolf and Gebot 1994], [Wolf 1998c].

LIEPDE has been used for the classification of PDE-systems in [Sokolov and Wolf 1999].

**Acknowledgement**
Andreas Brand is the author of a number of core modules of CRACK. The currently used data structure and program structure of the kernel of CRACK are due to him.

Francis Wright contributed a module that provides simplifications of expressions involving symbolic derivatives and integrals. Also, CRACK makes extensive use of the REDUCE program ODESOLVE written by Malcolm MacCallum and Francis Wright.

Arrigo Triulzi contributed in supporting the use of different total orderings of derivatives in doing pseudo differential Gröbner basis computations.

Work on this package has been supported by the Konrad Zuse Institute / Berlin through a fellowship of T.W.. Winfried Neun and Herbert Melenk are thanked for many discussions and constant support.

Anthony Hearn provided free copies of REDUCE to us as a REDUCE developers group which also is thankfully acknowledged.

Thomas Wolf (QMW/London)

### 4.3.6 DIMSYM

Dimsym is a REDUCE ([Hearn 1996], also see section(4.1.9)) package primarily for the determination of symmetries of differential equations , see 2.11.3. It also can be used to compute symmetries of distributions of vector fields or differential forms on finite dimensional manifolds, symmetries of geometric objects (eg, isometries) , and also to solve linear partial differential equations . Dimsym was initially an attempt to port ALan Head's LIE 4.2.24 program to REDUCE to overcome LIE's memory limitations but it quickly became an independent system.

To use its primary function the user specifies a system of ordinary and/or partial differential equations and the type of symmetry to be found (Lie point, Lie–Bäcklund or some user-provided ansatz) . Dimsym then produces the corresponding determining equations (a system of linear partial differential equations for the generator of the generic symmetry). It proceeds to solve these equations, reporting any special conditions required to produce a solution. Finally, Dimsym gives the generators of the symmetry group (which may of course be infinite dimensional).

The program allows the user to compute Lie brackets , directional derivatives and so on and it has an interface with the REDUCE package EXCALC [Schrüfer 1987] so that all the machinery of calculus on manifolds can be utilised from within the program. Its use can be interactive or batch and there are extensive tracing options.

Dimsym is a freely distributed package which includes the program (rlisp) source code, a user manual and an extensive set of example files. The latest version can be downloaded from the dimsym web site at
http://www.latrobe.edu.au/www/mathstats/Maths/Dimsym/

A more detailed description of the program and its performance can be found in [Sherring et al. 1997], while [O'Connor and Prince 1998; Edwards and Broadbridge 1995; Aldridge and Prince 1998; Jerie and Prince 1998, 1999; Kovalev 1996] give the reader some idea of the program's use as a research tool.

G. E. Prince and M. Jerie (Bundoora)

### 4.3.7  EinS

EinS is a package for MATHEMATICA intended for calculations with indexed objects (which may be in particular tensors). The package automatically handles dummy indices and Einstein's summation notation, enables one to define new indexed objects and to assign symmetries to that objects. EinS has an efficient simplification algorithm based on pattern matching technique which takes full account for the symmetries of the objects and the possibility to rename dummy indices. Other important features of EinS are the ability to perform automatically "3+1" split of implicit summations, printing expressions in a natural 2-dimensional form and exporting them into plain TeX or LaTeX with user-controllable alignment commands. Unlike some other packages of this kind EinS does not employ tensor properties of the objects. This allows the package to work effectively with any kind of objects possessing some kind of indices. A detailed description of the package can be found in [Klioner 1998]. EinS runs under any version of MATHEMATICA 4.1.7 starting from 2.1. The package together with User Guide and examples can be obtained from the URL:

http://rcswww.urz.tu-dresden.de/ klioner/eins.html

A typical (but not the only) application field of EinS is various calculations in the post-Newtonian approximation scheme of metric gravity theories (see, also 3.1.2.1). A number of important problems have been tackled with the use of EinS in this area. EinS has been used to compute the Landau-Lifshits energy-momentum pseudotensor of the gravitational field in a rotating reference system in the first post-Newtonian approximation of general relativity. This result has been used to define a post-Newtonian angular velocity of rotation of a massive extended non-isolated body [Klioner 1996]. The problem of constructing of a local reference system for a massive extended body in the framework of the Parametrized Post-Newtonian formalism [Klioner and Soffel 1998] has been also treated with the help of EinS.

Sergei Klioner (St.Petersburg, Russia – Dresden, Germany)

### 4.3.8  *FeynArts* and *FormCalc*

#### 4.3.8.1  Generating and Visualizing Feynman Diagrams and Amplitudes
Feynman diagrams are a powerful and intuitive technique of field theory

for evaluating perturbative expansions of Green functions and observables, usually the $S$ matrix (see Sect. 3.1.1.1). The accuracy of a calculation is linked to the number of loops in the Feynman diagrams, and already a one-loop calculation can easily involve several hundreds of diagrams, particularly so in models with many particles (see also Sect. 3.1.1.3).

*FeynArts* is a system for the generation and visualization of Feynman diagrams and amplitudes based on *Mathematica*. It works in three steps. First, the topologically inequivalent diagrams are created for a given number of loops and external legs. Then, fields are inserted into these topologies on the basis of a user-definable model. Lastly, the amplitudes are generated by applying the Feynman rules to the diagrams. The results are completely symbolic and not specific to a certain class of model. In addition, the diagrams can be drawn with output directly in PostScript, or in a LaTeX picture environment.

The insertion of fields and subsequent generation of amplitudes is possible at three levels, generic fields, classes of fields, or specific particles. *FeynArts* takes the information about the model from two special files: the representation of the kinematical quantities like spinors or vector fields is set up in a generic model file, and the particle content is given in the classes model file. Since the user can create own model files, the applicability of *FeynArts* is virtually unlimited within perturbative quantum field theory. The following model files are supplied with *FeynArts*: As generic model the Lorentz formalism (`Lorentz.gen`) and as classes model the Standard Model in several variations: the electroweak part (`SM.mod`), the same including QCD (`SMQCD.mod`), and in the background-field formulation (`SMbgf.mod`). A model file for the MSSM is in preparation.

The current version 2.2 is a much-expanded version of *FeynArts* 1 [Küblbeck et al. 1990] which can handle in particular fermion-number violating interactions (important for supersymmetric theories), vertices of arbitrary degree (important for effective theories), and mixing propagators. *FeynArts* is an open-source package and can be obtained from `http://www.feynarts.de`.

**4.3.8.2 Simplifying One-Loop Feynman Diagrams** As already mentioned, the accuracy of a Feynman-diagrammatic calculation increases with the number of loops in each diagram. Over the last decade, experimental data have reached such a high accuracy that in order to draw significant conclusions from the comparison with theoretical predictions, a one-loop calculation has become the lowest acceptable approximation in many cases.

Incidentally, for the computation of higher-loop diagrams no generic algorithms are known, hence the fully automatic calculation of an arbitrary diagram is possible only up to one loop.

*FormCalc* is a *Mathematica*-based program that simplifies one-loop Feynman diagrams. It reads amplitudes generated with *FeynArts* and returns the results in a way well suited for further numerical (or analytical) evaluation. The latter is straightforward with the utility programs provided by *FormCalc* (e.g. for

### 4.3.9 FeynCalc – Tools and Tables for Elementary Particle Physics

translation into Fortran code) and the implementations of the one-loop integrals in the associated package *LoopTools*.

Internally, *FormCalc* delegates the hard work (e.g. working out fermionic traces) to FORM by Jos Vermaseren [Vermaseren 1991]. Thus *FormCalc* is merely a driver that threads the *FeynArts* amplitudes through FORM in an appropriate way. The idea is rather simple: *FormCalc* prepares the symbolic expressions of the diagrams in an input file for FORM, runs FORM, and retrieves the results. This interaction is transparent to the user.

*FormCalc* combines the speed of FORM with the powerful instruction set of *Mathematica* and the latter greatly facilitates further processing of the results. Owing to FORM's speed, *FormCalc* can process, for example, the 1000-odd one-loop diagrams of W–W scattering in the Standard Model in about 10 minutes on an ordinary Pentium PC.

One important aspect of *FormCalc* is that it automatically gathers spinor chains, scalar products of vectors, and antisymmetric tensors contracted with vectors, and introduces abbreviations for them. In calculations with non-scalar external particles where such objects are ubiquitous, code produced from the *FormCalc* output (say, in Fortran) can be significantly shorter and faster than without the abbreviations.

*FormCalc* can work either in dimensional regularization ("$D$ dimensions") or in constrained differential renormalization ("4 dimensions"). At the one-loop level, the latter technique is equivalent to dimensional reduction. This means that *FormCalc* can process also supersymmetric diagrams. Details on both regularization methods and their implementation can be found in [Hahn and Pérez-Victoria 1999].

The associated package *LoopTools* provides the numeric implementations of the one-loop functions that appear in the *FormCalc* output. The one-loop functions in the *LoopTools* library are accessible in Fortran, C++, and *Mathematica*.

Both *FormCalc* and *LoopTools* are open-source packages and can be obtained from http://www.feynarts.de/{formcalc,looptools}.

Thomas Hahn (Karlsruhe), Hagen Eck and Sepp Küblbeck (formerly Würzburg)

### 4.3.9 FeynCalc – Tools and Tables for Elementary Particle Physics

FeynCalc 3.0 is a *Mathematica* 4.1.7 package for algebraic calculations in elementary particle physics. FeynCalc 3.0 is partially based on an earlier FeynCalc version [Mertig et al. 1991c] and on algorithms described in [Denner 1993]. Calculational and database tools are provided for frequently occuring tasks. Some of those are:

- Lorentz index contraction (`Contract`), Dirac algebra manipulation (`Calc`)
- color factor calculation and simplification for SU(N) (`SUNSimplify`)
- automatic Feynman rule derivation (`FeynRule`)
- automatic 1-loop diagram simplification up to 4 external legs (`OneLoop`, `OneLoopSum`, `OneLoopSimplify`)
- general noncommutative algebra and special noncommutative operator algebra (`DotSimplify`, `ExpandPartialD`)
- tables for Feynman parameter integrals and Mellin transforms (`Integrate2`), convolutions (`Convolute`)* and Feynman rules (`Twist2GluonOperator`)
- special translation facilities to change FeynCalc syntax to and from FORM syntax (`FeynCalc2FORM`, `FORM2FeynCalc`)
- optimized Fortran generation with `Isolate` and `Write2`.

Many more functions like, e.g., `SimplifyPolyLog` for simplification of polylogs and logs of real arguments, `Series2` for optimized Taylor expansion of Gamma functions, are implemented in FeynCalc in a modular and extendable way.

FeynCalc 3.0 is useful for small and medium sized research calculations. The new version is well integrated into the *Mathematica* 3.0 notebook environment with automatic TeX- like typesetting.

FeynCalc is completely programmed in the *Mathematica* language. This enables the user to take advantage of the many possibilities of *Mathematica* like graphics and numerics. However, as with all general purpose systems, a speed penalty is unavoidable. For example the 4-loop QCD $\beta$-function can and should be only calculated with FORM, but the author and others have calculated without any problem 2-loop QCD and Standard Model self-energies with FeynCalc 3.0 and TARCER [Mertig and Scharf 1998], a FeynCalc compatible program to calculate massive 2-loop propagator-type integrals. Also it is rather easy to automatically write and call existent FORM programs from within FeynCalc. This hybrid approach of using the best of both worlds has been used for the calculation of the 2-loop spin splitting functions [Mertig and van Neerven 1996], which is after [Cho and Leibovich 1996a,b] the most important work (measured in number of citations) where FeynCalc has been used. Further applications of FeynCalc were the calculation of 1-loop radiative corrections to $e^+e^- \to HZ$, high energy approximation of $e^+e^- \to W^+W^-$, $gg \to t\bar{t}$, and more.

FeynCalc 3.0 is still compatible with FeynArts 2.2**, a program to generate Feynman diagrams, if the FeynArts function `ToFA1Convention` is used on the generated amplitudes.

The latest commercial version FeynCalc 3.0, available from the author, is an add-on to *Mathematica* 3.0. The manual has a lot of examples and is freely available, e.g., from the author's web-page, http://www.mertig.com, or from the CD of this book. FeynCalc 3.0 works on any system running *Mathematica* 3.0.

---

* `Convolute[f, g, x]` does: $f(x) \otimes g(x) = \int_0^1 dx_1 \int_0^1 dx_2 \, \delta(x - x_1 x_2) \, f(x_1) \, g(x_2)$
** for updated versions see http://www-itp.physik.uni-karlsruhe.de/~hahn

It is advised to use *Mathematica* on a computer running Linux or Unix with as much main memory as possible ($> 32$ MB) for more complicated calculations.

The user base of FeynCalc consists of theoretical physicists around the world working in QCD, the Standard Model and minimal extensions, chiral perturbation theory and similar theories. FeynCalc can also be used in Quantum Field Theory courses, since it enables the students to perform more difficult calculations in seconds or minutes, freeing their time to understand concepts instead of getting lost in technicalities of calculations.

### 4.3.10 GRAPE

**GRAPE** is a **GAP** (4.2.18) share package for computing with finite graphs endowed with groups of automorphisms. The package is designed primarily for constructing and analysing graphs related to groups, designs and finite geometries. Special emphasis is placed on the determination of regularity properties and subgraph structure. Research applications of **GRAPE** have included the discovery and analysis of certain distance-regular graphs, the analysis of vertex-transitive graphs for low-rank representations of sporadic groups, and the discovery, analysis and classification of designs and finite geometries of various types.

The **GRAPE** philosophy is that a graph $G$ always comes together with a known subgroup $A$ of the automorphism group of $G$ (the group $A$ usually comes from the construction of $G$). Then $A$ is used to store $G$ efficiently and to speed up computations with $G$.

**GRAPE** includes functions to construct graphs (usually using group actions), to determine connected components, diameter and girth, to compute induced subgraphs and geodesics in graphs, to determine regularity parameters of graphs, to determine complete subgraphs of given weight-sum in a vertex-weighted graph, to calculate automorphism groups and test for graph isomorphism, to classify distance-regular graphs with a given vertex-transitive group of automorphisms, and to classify partial linear spaces with given point graph and parameters.

The vast majority of **GRAPE** functions are written entirely in the **GAP** language, except for the automorphism group and isomorphism testing functions, which use B. D. McKay's *nauty* package [McKay 1990]. The **GAP**-only parts of **GRAPE** run on any system that runs **GAP**, but the functions making use (invisible to the user) of *nauty* run under Unix only.

The web-page for **GRAPE** is

http://www-gap.dcs.st-and.ac.uk/~gap/Share/grape.html

from which the latest version of **GRAPE** can be downloaded (currently version 4.0 for use with **GAP** 4.2 [GAP Group 1997]) and the HTML-manual for **GRAPE** can be browsed.

Except for the *nauty* package, **GRAPE** was designed and written by L.H. Soicher, who you should email if you install **GRAPE**. If you use **GRAPE** to solve a problem then also tell him about it, and reference [Soicher 1993]. If you use the automorphism group and graph isomorphism testing functions of **GRAPE** then you should also reference *nauty* [McKay 1990].

Leonard H. Soicher (London)

### 4.3.11  Recognising Matrix Groups over Finite Fields

**4.3.11.1  Background**  The general linear group $GL(d, q)$ is the group of all non-singular $d \times d$ matrices with entries in the field with $q$ elements. A major research effort over the past decade has been directed at developing algorithms to answer questions about the subgroup $G$ generated by a subset of $GL(d, q)$ supplied as input. The analogous problem for subgroups of the symmetric groups has long been a very successful area of computational group theory. The main goal is to produce a composition series for $G$. We are still some way from this goal. From the standpoint of complexity analysis we would like to have algorithms that require a number of bit operations polynomial in the size of the input, and hence polynomial in $d \log q$; but some key algorithms, such as the explicit recognition algorithms for classical groups, appear to be ineluctably polynomial in $dq$. This causes a problem, as large extension fields of the original field may arise, for example when dealing with non-absolutely irreducible modules.

The main theoretical underpinning of the project comes from Aschbacher's eight families of subgroups of $GL(d, q)$; see [Aschbacher 1984]. Much of the effort that has been put into this project has been devoted to design algorithms to determine whether or not $G$ belongs to some Aschbacher family. The second phase of the project is to determine, in case $G$ is simple (modulo scalars), a pair of efficiently computable inverse isomorphisms between $G$ and a standard copy of this simple group.

A feature of the project is its strong dependence on statistical methods. Although results returned are proved to be correct, many steps on the way use one-sided Monte Carlo algorithms; that is to say, algorithms that return a Yes – No answer, when only one of these answers is reliable. Often the algorithms involve selecting independent uniformly distributed random elements from the group. For example, at some points we will proceed by making conjectures based on sampling group elements, and then try to confirm these conjectures.

As there is no efficient algorithm known to select independent uniformly distributed random elements from a group $G$ in time equivalent to a matrix multiplication, the complexity analysis of the algorithms contains a variable counting the number of such selections. Thus to achieve polynomial runtime in practice, a major component of the project has been to produce a pseudo random element generator which, while imperfect, performs very well in practice. As an indication of the difficulty of the matrix case as opposed to the permutation group case, we mention the fact that even efficiently computing the order of a single element of a matrix group is quite a subtle business.

Another technique that should be mentioned is the search for a faithful permutation representation of $G$ of manageably small degree arising from the action of $G$ on the set of subspaces of the natural module. Although asymptotically hopeless, this is a surprisingly effective tool in practice.

The aim of the matrix package is to provide integrated and comprehensive access to implementations of this collection of algorithms.

**4.3.11.2 Contents of the Matrix Package** The matrix group package contains algorithms to

- test whether a $G$-module is irreducible or absolutely irreducible and to test whether two irreducible $G$-modules are isomorphic; see [Parker 1984], [Holt and Rees 1994] and [Neumann and Praeger 2001].
- decide whether a matrix group has certain decompositions with respect to a normal subgroup; see [Holt et al. 1996a].
- decide whether a subgroup of $GL(d,q)$ acts primitively on the underlying vector space $V$; see [Holt et al. 1996b].
- decide whether a subgroup of $GL(d,q)$ contains $SL(d,q)$ or a classical group in its natural representation. Here we provide access to the algorithms of
  - Celler and Leedham-Green [Celler and Leedham-Green 1997b]; and
  - Niemeyer and Praeger [Niemeyer and Praeger 1997, 1998, 1999], which are modelled on the SL-recognition algorithm of Neumann and Praeger [Neumann and Praeger 1992].
- constructively recognise the special linear group; see [Celler and Leedham-Green 1998].
- decide whether a subgroup of $GL(d,q)$ preserves a symplectic, unitary or quadratic form on $V$.
- select random elements from a group; see [Celler et al. 1995].
- compute the order of an invertible matrix; see [Celler and Leedham-Green 1997a].
- select base points for the Random Schreier-Sims algorithm for matrix groups; see [Murray and O'Brien 1995].
- decide whether a subgroup of $GL(d,q)$ preserves a tensor decomposition; see [Leedham-Green and O'Brien 1997b,a].

The algorithms described here are available both as a share package library of GAP, see 4.2.18 and in MAGMA [Bosma et al. 1997], see 4.1.5.

C.R. Leedham-Green (London, UK), Alice C. Niemeyer (Perth, Australia),
E.A. O'Brien (Auckland), Cheryl E. Praeger (Perth, Australia)

### 4.3.12 MOLGEN

MOLGEN a computer algebra package for the generation of molecular graphs for chemical structure elucidation

MOLGEN is a computer algebra package for the generation of structural formulae of chemical molecules, i.e. of molecular graphs or connectivity isomers. It generates, efficiently and redundancy free, i.e. without doublettes, all the mathematically existing molecular graphs that correspond to given data from spectroscopy. Typical cases are given chemical formula, an interval for the possible ring sizes, and prescribed as well as forbidden substructures of the molecule in question. MOLGEN is applied (in research as well as in industry) to molecular structure elucidation as well as in combinatorial chemistry, where a library of molecules has to be generated from (optional) a central part and further building blocks and reactions according to which the building blocks react with the central molecule.

Its mathematical tools are taken both from combinatorics and algebra, in particular orderly generation, group actions, double coset methods (for the evaluation of a transversal of the orbits) and the homomorphism principle are used.

MOLGEN is a collection of routines written in $C^{++}$. Several versions are available which differ in their user shell, the features and the platform.

There are several extensions of MOLGEN, devoted to special purposes:

- MOLGEN 4.0 allows in particular to use information that stems from NMR-spectroscopy.
- MOLGEN–MS is devoted to structure elucidation using mass spectroscopy.
- MOLGEN–COMB is intended for combinatorial chemistry.

The basic generators contained in the various versions of MOLGEN were developed over many years and in several research products, in particular by D. Moser, R. Grund, Th. Grüner. Emphasis has changed from building the fastest to the most flexible generator choosing its strategy according to the constraints of the actual problem. The methods used overlap with those applied in DISCRETA (4.2.14, see also 2.9).

There is also a multimedia product MOLiS under development that contains MOLGEN and which is intended for schools in order to teach the notion of isomerism. Isomerism is the name for the fact that there are different molecules with the same chemical formula, i.e. with the same atomic constituents but with different properties (e.g. different boiling points, solvability, stability, and so on). To the chemical formula $C_6H_6$, for example, there exist 217 different structural formulae, at least 70 of them correspond to chemical substances that have been found or produced, and there are many more if we take into account that molecules are not only topological entities but geometric ones.

MOLGEN has a graphical user interface and presents its results with automatically generated 2D- or 3D-drawings according to the rules of chemists. A standard file format is provided for data exchange with other packages. MOLGEN is also included in SpecInfo (Chemical Concepts, Weinheim) which combines it with a large database of spectra and which is commercially used in world wide industry.

Information about MOLGEN can be obtained from its hompage at "www.mathe2.uni-bayreuth.de ".

Adalbert Kerber and Reinhard Laue (Bayreuth)

## 4.3.13 ORME

*The following is the contribution to original report in 1993—the editors.*

For the purpose of making clear and nice proofs of their completeness, completion procedures are described by transition rules. The philosophy of ORME is to use for implementation the same paradigm that was shown so useful for proofs. This way one gets readable programs and a good view on high level optimizations. ORME also provides a tool box for easily build prototypes in equational reasoning. The paper entitled "Implementation of Completion by Transition Rules + Control: ORME", invited lecture at second international conference on Algebraic and Logic Programming [Lescanne 1990], describes an implementation of an associative and commutative completion.

ORME contains also tools for proving termination based on polynomial and elementary (functions described with plus, times and exponential) described in [Lescanne 1992].

Pierre Lescanne (Vandœuvre-les-Nancy)

## 4.3.14 Ratappr

For presentation of complicated function and quick calculation of its values in the concrete points, this function is often replaced by simpler expression. It is often required that the absolute or relative error on the interval $[a, b]$ be not larger than a certain value $\mu$. If the number $m + 1$ of approximation parameters is to be minimized, we come to a problem of the best Chebyshev or minimax approximation.

Some computer systems have procedures for the minimax approximation of the function by rational expression [Geddes 1993]

$$R_{k,l}(x) = \frac{\sum_{i=0}^{k} a_i x^i}{\sum_{i=0}^{l} b_i x^i}, \ k+l = m.$$

However, more complicated rational expression $V_{k,l}(x) = x^s R_{k,l}(x^p)$, where $s$ and $p$ are integers, may be widely used to approximate functions.

If there is a need to reduce the approximation error $\mu$, it is convenient to use the approximation on subintervals of the interval $[a, b]$, i.e. an approximation by splines [Popov 1989]. If the maximum approximation error is the same on each subinterval (spline link), then the approximation is called a "balanced" spline approximation. If each spline link is a best Chebyshev approximation, then such spline approximation is called the Chebyshev spline. Balanced Chebyshev spline approximation is an optimal one [Meinardus 1967]. This means that if the boundaries of the spline links (knots of spline) are selected from the condition of the balanced approximation, we get the least possible error with the fixed number of links or the least possible number of links with the fixed maximum error.

The balanced spline approximation usage may greatly reduce the evaluation time and it makes possible to approximate functions on large intervals. The approximation efficiency may be greatly increased if different approximating expressions are used on different subintervals. The process of finding such optimal spline approximations is more time consuming than other methods of approximation, but the resulting function evaluation will be quicker.

The *ratappr* is the special package for numerical minimax approximations of functions by rational expressions ($R_{k,l}(x)$ and $V_{k,l}(x)$) and balanced rational splines. Package *ratappr* has been made for MAPLE V RELEASE 5, see 4.1.6 [Monagan et al. 1998].

> with(ratappr);

[*balance*, *balanceP*, *choice*, *minimaxC*, *minimaxN*, *minimaxP*, *optappr*, *optapprP*, *rat*, *ratio*]

Methods for building the balanced approximation with given error $\mu$ by Chebyshev rational splines are done by the procedures *balance* and *balanceP*. The output of the procedures is a Chebyshev spline. Moreover, it is possible to print the global sequence *er*, that contains: the interval of approximation, the number of subintervals $r$, the coefficient $\kappa$ and maximum errors $\mu_i$ of the best Chebyshev approximation on the subintervals. The coefficient $\kappa$ is gained while dividing maximum error by average value of errors $\mu_i, i = \overline{1, r}$.

*Example 1.* Balanced approximation of the function $f(x) = e^x$ with weight $w(x) = 1$ on the interval $[0, 2]$ by Chebyshev spline with links — rational expressions $R_{1,1}(x)$ with given error $\mu = 0.001$ is achieved by procedure:

```
> Digits:=8:exr:=balance(exp,0..2,.001,[1,1],1):'exr'=exr(x);er;
```

$$exr = \begin{cases} 0 & x < 0 \\ \dfrac{-2.3567607 + x}{-2.3544041 - 1.4100893\,x} & x \le .65197472 \\ \dfrac{-2.9402326 + x}{-2.7349278 - 2.5379810\,x} & x \le 1.1869567 \\ \dfrac{-3.4288848 + x}{-2.4219463 - 4.1474258\,x} & x \le 1.6408641 \\ \dfrac{-3.8491909 + x}{-1.0149995 - 6.3240642\,x} & x \le 2.0349192 \\ 0 & 2.0349192 < x \end{cases}$$

$[0..2.0349192]$, $r = 4$, $\kappa = 1.0008$, $\mu = (.0009999, .001000, .001001, .001001)$

The balanced approximation with different rational expressions on different subintervals is done by procedures *optappr* and *optapprP*. The approximation by $R_{k_i,l_i}(x)$ or $V_{k_i,l_i}(x)$, with $k_i + l_i = m$, is used on each subinterval.

*Example 2.* Optimal approximation for $f(x) = \Gamma(x)$, $w(x) = 1$, $x \in [1, 3]$, $m = 2$ with given error $\mu = .00119$ is

```
>g:=optappr(GAMMA,1..3,.00119,2,1):'g'=g(x);er;
```

$$g = \begin{cases} 0 & x < 1. \\ -1.7089967\,\dfrac{1}{.22842843 + (-2.9394596 + x)\,x} & x \le 1.7039553 \\ 1.9046179 + (-1.3352704 + .44126821\,x)\,x & x \le 2.4220359 \\ 15.404227\,\dfrac{1}{38.508260 + (-13.270062 + x)\,x} & x \le 3.0031028 \\ 0 & 3.0031028 < x \end{cases}$$

$[1...3.0031028]$, $r = 3$, $\kappa = 1.0003$, $\mu = (.001189, .001189, .001189)$

The basic concept of the algorithms in these procedures is to make use of representations of approximation error by special analytical expressions, which cannot be possible without using computer algebra systems.

Bogdan Popov and Oksana Laushnyk (Lviv)

### 4.3.15 TTC: Tools of Tensor Calculus

TTC (Tools of Tensor Calculus) [Castellví et al. 1994, 1995; Balfagón and Jaén 1998] is a MATHEMATICA[Wolfram 1991] package for doing tensor and exterior calculus on differentiable manifolds 3.1.4.1.

One of the main purposes of TTC is to have an easy way to write inputs, i.e. to have a close to textbook notation for inputs. As an example of this feature consider the possible input:

```
In[]:= O[i,j]+ (T[s,.<<j,.;-n] Q[i>>.,-s])[.;n]    //Index[gn]
```

where gn is the name of the metric tensor in use and T and Q are explicit tensors. The symbol .; stands for covariant derivative and .<< >>. enclose the indexes to be antisymmetrize. Once the input is entered TTC perfoms the indicated operations over the explicit tensors giving as output a single tensor.

Due to the generic character of TTC their applications are the ones differential geommetry has. Some of the typical fields of application are: General relativity, electromagnetism, continuum mechnanics...

The easy way to link inputs and outputs, and the abovementioned easy way to write inputs, can be interesting also for teaching purposes.

In this short and *academic* session we will show the flavour of TTC. The session is part of a MATHEMATICAnotebook. The inputs are indicated by In[]:=. The outputs have no prefix. We start using symbolic index utilities, then we make some explicit calculus.

```
In[]:=<<ttc.m
-----------------------------------------
|  Tools  of  Tensor  Calculus 4.1.0    |
|  A.Balfagon,P.Castellvi and X.Jaen    |
|     http://baldufa.upc.es/ttc         |
|     e-mail:Xavier.Jaen@upc.es         |
|   version: november, 23,1999          |
-----------------------------------------
|     Session started on                |
|     April,      12, 2000              |
|     at 9   h 57 min 6 s               |
-----------------------------------------

In[]:=InputCoordinates[sw,{r,th,ph,t}];
InputSMetric[gn,sw,"g",g];InputSRiemann[gn,sw,"R",Rie,Ric,Rs];
InputTensor[M,sw,{1,1,1}];
InputSymmetries[M[a,b,c],Cyclic[a,b,c][2],{{a,b}}[1]];
```

## 4.3.15 TTC: Tools of Tensor Calculus

```
InputTensor[V,sw,{1}];

In[]:=InputIndex[{a,b,c,d,e,f,h,i,j}];

In[]:=(M[a,b,c]V[-a]V[-d])[.;-b,.;-c]-
(M[a,b,c]V[-a]V[-d] )[.;-c,.;-b]//Index[gn]
```

$$0_a + (M^{bcd}{}_a V_b V_{.;c .;d}) - (M^{bcd}{}_a V_b V_{.;d .;c})$$

```
In[]:=%//Index[gn,SuperIndexExpand]
```

$$0_a + M^{bcd}{}_{.;c .;d} V_a V_b - M^{bcd}{}_{.;d .;c} V_a V_b + M^{bcd}{}_b V_a V_{.;c .;d} -$$

$$M^{bcd}{}_b V_a V_{.;d .;c} + M^{bcd}{}_a V_b V_{.;c .;d} -$$

$$M^{bcd}{}_a V_b V_{.;d .;c}$$

```
In[]:=MVVR=%//Index[gn,SimplifyAllIndex[2]]
```

$$0_a - \frac{M_{bcd} V^d V^{bce} R_e{}^a}{2}$$

```
In[]:=Compact[On];
```

(*This is the way TTC compact explicit calculus and outputs.
  TTC[nnn] is the symbol for compacted scalars. If Off TTC
  does not compact outputs and the times might increase *)

```
In[]:=InputMetric[gn,sw,
-(1-2 m/r) sw[t]^2 + 1/(1-2 m/r) sw[r]^2 +
r^2 (sw[th]^2 + Sin[th]^2 sw[ph]^2 )];

In[]:=(V=>r^2 sw[-r]+ r^2 Sin[th] sw[-th];

    M=>r Sin[th] 1/2(sw[1,3,2]-sw[3,1,2])+
    2 r Sin[th] 1/2 (sw[2,1,3]-sw[1,2,3])+
    3 r   Sin[th] 1/2(sw[2,3,1]-sw[3,2,1])
    );

(* => is the way to input compact tensors (or scalars).
Only to be used in explicit calculus.*)

In[]:=M[i,j,k]+M[k,i,j]+M[j,k,i]//Index[gn]

0

In[]:=Rie[-i,-j,-k,-l]M [l,-o,j] V[i,.;k] //Index[gn]

-2 TTC[347] dph

In[]:=MVVR=MVVR//Index[gn];    Compact[Off];    MVVR

        2     2
-5 m r  Sin[th]
---------------- dph
        2
```

*Tools of Tensor Calculus (TTC)*

web address: http://baldufa.upc.es/ttc

e-mail Xavier.Jaen@upc.es

# Acknowledgments

X.J. would like to thank the Comisión Asesora de Investigación Científica y Técnica for partial financial support, under Contract No. PB96-0384. The authors would also like to thank the Laboratori de Física Matemàtica at the Societat Catalana de Física for partial financial support.

A. Balfagón (Universitat Ramón Llull, Spain), X. Jaén (Universitat Politecnica de Catalunya, Spain)

# 5 Meetings and Publications

## 5.1 Conferences and Proceedings

In the following we list a selection of international conferences and workshops in computer algebra with printed or electronic proceedings. For a supplementary list of further conferences, meetings and workshops in computer algebra compare [Wester 1999].

- 1979
    - Algorithm in Modern Mathematics and Computer Science. September 16-22, Urgench, Uzbek SSR. [Ershov and Knuth 1979]
- 1981
    - SYMSAC'81. 1981 ACM Symposium on Symbolic and Algebraic Computation, August 5-7, Snowbird, Utah, USA. [Wang 1981]
- 1982
    - EUROCAM'82. European Computer Algebra Conference, April 5-7, Marseille, France. [Calmet 1982]
- 1983
    - EUROCAL'83, European Computer Algebra Conference, March 28-30, London, England. [van Hulzen 1983]
    - AAECC-1. Applied Algebra, Algebraic Algorithms and Error-Correcting Codes. 1st International Conference, June 1983, Toulouse, France. [Poli 1985]

- 1984
    - EUROSAM'84. International Symposium on Symbolic and Algebraic Computation, July 9-11, Cambridge, England. [Fitch 1984]
    - AAECC-2. Applied Algebra, Algebraic Algorithms and Error-Correcting Codes. 2nd International Conference, October 1-5, Toulouse, France. [Poli 1986]

- 1985
    - EUROCAL'85. European Conference on Computer Algebra, April 1-3, Linz, Austria. [Buchberger 1985a]
    - RTA'85. 1st Conference on Rewriting Techniques and Applications, May 20-22, 1985, Dijon, France. [Jouannaud 1985]
    - AAECC-3. Applied Algebra, Algebraic Algorithms and Error-Correcting Codes. 3rd International Conference, July 15-19, Grenoble, France. [Calmet 1986]
- 1986
    - SYMSAC '86. Symposium on Symbolic and Algebraic Manipulation, ACM, July 21-23, 1986, Waterloo, Ontario, Canada. [Char 1986]
    - International Conference on Computers and Mathematics. July 30 - August 1, 1986, Stanford, U.K. [Chudnovsky 1990]

- AAECC-4. Applied Algebra, Algebraic Algorithms and Error-Correcting Codes. 4th International Conference, September 23-26, Karlsruhe, Germany. [Beth and Clausen 1988]

− 1987
- NASA-Ames Workshop on the Use of Symbolic Methods to Solve Algebraic and Geometric Problems Arising in Engineering. January 15-16, 1987, Moffett Field, Califorina, USA. [Grossman 1989]
- International Symposium on Trends in Computer Algebra, May 19-21, Bad Neuenahr, Germany. [Janšen 1988]
- RTA'87. 2nd Conference on Rewriting Techniques and Applications, May 25-27, Bordeaux, France. [Lescanne 1987]
- EUROCAL'87. European Conference on Computer Algebra, June 2-5, Leipzig, Germany. [Davenport 1989]
- AAECC-5. Applied Algebra, Algebraic Algorithms and Error-Correcting Codes. 5th International Conference, June 15-19, Menorca, Spain. [Huguet and Poli 1989]

− 1988
- CADE'88. First International Workshop on Computer Algebra and Differential Equations, May 24-27, Grenoble, France. [Tournier 1988]
- Conference on Algebraic Logic and Universal Algebra in Computer Science. June 1-4, Ames, Iowa, USA. [Bergman et al. 1990]
- ISSAC'88. International Symposium on Symbolic and Algebraic Computation, July 4-8, Rome, Italy. [Gianni 1989]
- AAECC-6. Applied Algebra, Algebraic Algorithms and Error-Correcting Codes. July 4–8, Rom, Italy. [Mora 1989a]

− 1989
- RTA'89. 3rd International Conference on Rewriting Techniques and Applications, April 3-5, Chapel Hill, North Carolina, USA. [Dershowitz 1989]
- Conference on Computers and Mathematics. June 13-17, 1989, Mass, USA. [Kaltofen and Watt 1989]
- ISSAC'89. International Symposium on Symbolic and Algebraic Computation, July 17-19, Portland, Oregon, USA. [Gonnet 1989]

− 1990
- DISCO'90. International Symposium on Design and Implementation of Symbolic Computation Systems. April 10-12, Capri, Italy. [Miola 1990b]
- MEGA'90. Effective Methods in algebraic Geometry, April 17-21, Livorno, Italy. [Mora and Traverso 1991]
- CADE'90. International Workshop on Computer Algebra and Differential Equations, May 6-9, Cornell University, Ithaca, NY, USA. [Singer 1990b]
- 2nd International Workshop on Computer Algebra and Parallelism. May 9-11, Ithaca, USA. [Zippel 1992]
- International Conference on Computer Algebra in Physical Research. May 22-26, 1990, Dubna, UdSSR. [Shirkov et al. 1991]

- ISSAC'90. International Symposium on Symbolic and Algebraic Computation, ACM, August 20-24, Tokyo, Japan. [Watanabe and Nagata 1990]
- AAECC-8. Applied Algebra, Algebraic Algorithms and Error-Correcting Codes, August 20–24, 1990, Tokyo, Japan. [Sakata 1991]

– 1991
  - RTA'91. 4th Conference on Rewriting Techniques and Applications, April 10-12, Como, Italy. [Book 1991]
  - Cortona Conference on Computational Algebraic Geometry and Commutative Algebra. June 17-21, 1991, Cortona, Italy. [Eisenbud and Robbiano 1991]
  - ISSAC'91. International Symposium on Symbolic and Algebraic Computation, ACM, July 15-17, Bonn, Germany. [Watt 1991]
  - AAECC-9. Applied Algebra, Algebraic Algorithms and Error-Correcting Codes. 9th International Conference, October 7-11, New Orleans, LA, USA. [Mattson et al. 1991]

– 1992
  - ERCIM'92. Advanced course on Partial Differential Equations and Group Theory, April 6-10, Bonn, Germany. [INRIA, GMD 1992]
  - DISCO'92. International Symposium on Design and Implementation of Symbolic Computation Systems, April 13-15, Bath, U.K.. [Fitch 1992]
  - MEGA'92. Computational Algebraic Geometry, April 21-25, Nice, France. [Eysette and Galligo 1993]
  - International Conference on Formal Power Series and Algebraic Combinatorics. June 15-19, 1992, Québec, Montréal, Canada. [Leroux and Reutenauer 1992]
  - ISSAC'92. International Symposium on Symbolic and Algebraic Computation, ACM, July 27-29, Berkeley, CA, USA. [Wang 1992]

– 1993
  - AAECC-10. Applied Algebra, Algebraic Algorithms and Error-Correcting Codes. May 10-14, San Juan de Puerto Rico, Puerto Rico. [Cohen et al. 1993]
  - RTA'93. Conference on Rewriting Techniques and Applications, June 16-18, Montreal, Canada. [Kirchner 1993]
  - ISSAC'93. International Symposium on Symbolic and Algebraic Computation, ACM, July 6-8, Kiev, Ukraine. [Bronstein 1993]
  - DISCO'93. International Symposium on Design and Implementation of Symbolic Computation Systems, September 15-17, Gmunden, Austria. [Miola 1993]
  - Quantifier Elimination and Cylindrical Algebraic Decomposition, October 6-8, RISC Linz, Austria. [Caviness and Johnson 1998]

– 1994
  - MEGA'94. Effektive Methods in algebraic Geometry. April 5-9, Santander, Spain. [Gonzàles-Vega and Recio 1996]
  - ISSAC'94. International Symposium on Symbolic and Algebraic Computation, ACM, July 20-22, Oxford, UK. [Giesbrecht 1994b]

- PASCO'94. Parallel Symbolic Computation, September 26-28, Castle of Hagenberg, Austria. [Hong 1994]
- 1995
  - RTA'95. 6th International Conference on Rewriting Techniques and Applications, April 5-7, Kaiserslautern, Germany. [Xiang 1995]
  - Symbolic Rewriting Techniques, April 30 - May 4, 1995, Monte Verita, Ascona, Switzerland. [Bronstein et al. 1998]
  - IMACS-ACA'95. Conference on Applications of Computer Algebra, May 16-20, Albuquerque, New Mexico, USA. For electronic proceedings see: http://math.unm.edu/ACA/1995.html
  - ISSAC'95. International Symposium on Symbolic and Algebraic Computation, ACM, July 10-12, Montreal, Canada. [Levelt 1995]
  - AAECC-11. 11th International Symposium on Applied Algebra Algebraic Algorithms and Error-Correcting Codes, July 17-22, Paris, France. [Cohen et al. 1995b]
  - SCAN-95. International Symposium on Scientific Computing, Computer Arithmetic and Validated Numerics, September 26-29, Wuppertal, Germany. [Alefeld 1995]
- 1996
  - ISSAC'96. International Symposium on Symbolic and Algebraic Computation, ACM, July 24-26, Z"urich, Switzerland. [Lakshman 1995]
  - RTA'96. 7th International Conference, July 27-30, New Brunswick, NJ, USA. [Ganzinger 1996]
  - ASCM'96. Asian Symposium on Computer Mathematics, August 20-22, Kope, Japan. [Koboyashi 1996]
  - ADG'96. International Workshop on Automated Deduction in Geometry, Toulouse, France, September 1996. [Wang 1997]
- 1997
  - RTA'97. 8th International Conference on Rewriting Techniques and Applications, June 2-5, Sitges, Spain. [Comon 1997]
  - First International Theorema Workshop. June 9-10, 1997, Castle of Hagenberg, Austria. [Buchberger et al. 1997a]
  - AAECC-12. The 12th AAECC Conference, June 23-27, IUT-A Rangueil, Univ. P. Sabatier-Toulouse, Toulouse, France. [Mora 1997]
  - PASCO'97. Second International Symposium on Parallel Symbolic Computation, ACM, July 20-22, 1997, Maui, HI, USA. [Hitz and Kaltofen 1997]
  - ISSAC'97. International Symposium on Symbolic and Algebraic Computation, ACM, July 21-23, Maui, HI, USA. [Küchlin 1997]
  - IMACS-ACA'97. The 1997 IMACS Conference on Applications of Computer Algebra, July 24-26, Maui, Hawaii, USA. For electronic proceedings see: http://math.unm.edu/ACA/1997.html

- 1998
  - International Conference on Gröbner Bases: 33 Years of Gröbner Bases. February 2-4, RISC-Linz, Austria. [Buchberger and Winkler 1998]

- RTA'98. 9th Conference on Rewriting Techniques and Applications, March 30 - April 1, Tsukuba, Japan. [Nipkow 1998]
- MEGA'98. The Fifth International Symposium on Effective Methods in Algebraic Geometry, June 22-27, St. Malo, France. [Lombardi and Roy 1999]
- ADG'98. Second International Workshop on Automated Deduction in Geometry. August 1-3, 1998, Beijing, China. [Gao et al. 1999]
- IMACS-ACA'98. The Fourth International IMACS Conference on Applications of Computer Algebra, August 9-11, Prague, Czech Republic. For electronic proceedings see: http://math.unm.edu/ACA/1998/index.html
- ISSAC'98. International Symposium on Symbolic and Algebraic Computation, Aug. 13-15, Rostock, Germany. [Gloor 1998]

– 1999
- CASC-99. The Second Workshop on Computer Algebra in Scientific Computing, May 31 - June 4, Munich, Germany. [Ganzha et al. 1999]
- IMACS-ACA'99. Applications of Computer Algebra Conference Euroforum, June 24-27, El Escorial, Madrid, Spain. For electronic proceedings see:
http://math.unm.edu/ACA/1999.html
- RTA'99. 10th International Conference on Rewriting Techniques and Applications, July 2-4, Trento, Italy. [Narendran and Rusinowitch 1999]
- ISSAC'99. International Symposium on Symbolic and Algebraic Computation, July 29-31, 1999, Simon Fraser University, British Columbia, Vancouver, Canada. [Dooley 1999]
- AAECC-13. Symposium on Applied Algebra Algebraic Algorithms Error-Correcting Codes, November 14-19, Hawaii, USA. [Fossorier et al. 1999]

– 2000
- MEGA'2000, Effective Methods in Algebraic Geometry, June 20-24, University of Bath, England. [Singer 2001].
- IMACS-ACA'2000. 6th IMACS Conference on Applications of Computer Algebra, June, 25-28, 2000, St. Petersburg, Russia. For electronic proceedings see: http://math.unm.edu/ACA/2000/index.html
- RTA'2000. International Conference on Rewriting Techniques and Applications, July 10-12, Norwich, U.K.. [Bachmair 2000].
- ISSAC'2000. International Symposium on Symbolic and Algebraic Computation, August 7-9, 2000, University of St. Andrews, Scotland. [Traverso 2000].
- EACA'2000, Encuentro de Algebra Computacional y Applicaciones, Sept. 6-8, Barcelona, Spain. [Montes 2000].
- ADG 2000. 3rd International Workshop on Automated Deduction in Geometry, September 25-27, Zürich, Switzerland. [Richter-Gebert and Wang 2000].
- CASC'2000. The Third Workshop on Computer Algebra in Scientific Computing, October 5-9, 2000, Sarmarkand, Uzbekistan. [Ganzha et al. 2000].

- ASCM'2000. The 4th Asian Symposium on Computer Mathematics, December 17-21, 2000, Chiang Mai, Thailand. [Gao and Wang 2000a].
- 2001
  - IMACS-ACA'2001. 7th IMACS Conference on Applications of Computer Algebra, May 31-June 3, Technical Vocational Institute, Albuquerque, New Mexico, USA. For electronic proceedings see: http://math.unm.edu/ACA/2001/Proceedings/MainPage.html
  - ISSAC 2001, International Symposium on Symbolic and Algebraic Computation, July 22-25, 2001, University of Western Ontario, London, Ontario, Canada. [Mourrain 2001].
  - CASC'2001. The Fourth Workshop on Computer Algebra in Scientific Computing, September 22 - 26, 2001, Konstanz, Germany. [Ganzha et al. 2001].

## 5.2 Books on Computer Algebra

The following list of books is intended as a guide to computer algebra for the non-expert. It is not intended to be exhaustive. In particular we have excluded the conference proceedings cited above and books in mathematics or computer science related to computer algebra but without a significant portion in computer algebra proper. For reference we have also included some books providing historical sources of computer algebra methods.

1. General textbooks on computer algebra.
[Fachgruppe Computeralgebra der GI and DMV and GAMM 1993; Caviness and Johnson 1998; Davenport et al. 1989; von zur Gathen and Gerhard 1999; Geddes et al. 1992; Mignotte 1989, 1992]
2. Monographs and textbooks on specific topics related to computer algebra.
[Abhyankar 1988; Adams and Loustaunau 1994b; Aho et al. 1974; Adleman and Huang 1994; Aït-Kaci and Nivat 1989a,b; Adleman and Huang 1992; Akritas 1989; Atkinson 1984a; Avenhaus 1995; Seress 1999; Assmus, Jr. and Key 1992; Baader and Nipkow 1998; Bachmair 1991; Buell 1989; Hiss and Lux 1989; Brown et al. 1978; Beth and Clausen 1987; Buchberger et al. 1986; Barbeau 1989; Bachmair 1991; Bach and Shallit 1996; Beck and Kolman 1977; Becker et al. 1993, 1998; Benson 1993; Berlekamp 1984; Beth 1991; Beth et al. 1986; Betten et al. 1998a; Blum et al. 1998; Bini and Pan 1994; Blahut 1987; Bluman and Kumei 1989; Bochnak et al. 1987; Borodin and Munroe 1975; Bressoud 1989; Butler 1991; Bürgisser et al. 1996; Bündgen 1998b; Bürgisser 2000; Canny 1987; Clausen and Baum 1993; Chou 1988; Cohen 1996b; Cohn 1966; Colbourn and Dinitz 1996; Conway et al. 1985; Conway and Sloane 1988; Cox et al. 1992, 1998a; Craig 1986; Cremona 1992; Devroye 1986; Davenport 1981; Graham and Phillip 1985; Epstein et al. 1992; Ehrig and Mahr 1985; Bykov et al. 1998; Eisenbud 1995; Fakler 1997a; Ferrante and Rackoff 1979; Fripertinger and Kerber 1995; Froeberg 1997;

Ganzha and Vorozhtsov 1996; Gallager 1963; Gantmacher 1966; Gibbons and Rytter 1988; Gielen and Sansen 1991; Goppa 1988; Gröbner 1968, 1970; Hardy and Wright 1979; Hayes-Roth et al. 1983; Hensel 1908; Holt and Plesken 1989; Hoffmann 1989b; Jacobson 1985; Janet 1929; Jantzen 1988; Johnson 1990; Jones and Lins 1996; Klimov and Rudenko 1989a; Kahrs and Brandenburg 1998; Kamke 1961; Kaplansky 1957; Kerber 1991; Klimov and Rudenko 1989b; Kluge 1996; Knuth 1998; Koblitz 1987, 1998; Kolchin 1973; Kraft 1984; Kranakis 1986; Kreuzer and Robbiano 2000; Krull 1935; Lenstra Jr. and Tijdeman 1982; Lausch and Noebauer 1973; LeChenedac 1986; Lehmann 1992; Lenstra and Lenstra Jr. 1993; Lie 1881; Lipson 1981; Lidl and Niederreiter 1986; Lyndon and Schupp 1977; Lovász 1986; Lüneburg 1987; Marden 1966; Morgan 1987; Meyer and Schmidt 1990; MacCallum et al. 1994; Magid 1994; Macaulay 1916; Marden 1949, 1966; Matzat 1987; Malle and Matzat 1999; MacWilliams and Sloane 1988; Menezes 1993; Mines et al. 1988; Mishra 1993; Moore 1979; Jansen et al. in Vorbereitung; Mumford et al. 1994; Magnus et al. 1966; Netto 1896/1900; Newman 1972; Olver 1993; Pomerance 1984; Pavelle 1985; Flannery et al. 1992; Pohst and Zassenhaus 1989; Perron 1954; Petkovšek et al. 1996; Pommaret 1978, 1994; Pohst 1993; Pohst and Zassenhaus 1997; Praeger and Soicher 1997; Rand 1994; Riesel 1985; Rice 1988; Robbiano 1989; Ritt 1950; Rand and Armbruster 1987; Renschuch 1976; Richter 1989; Riquier 1910; Robinson 1956; Rosenstein 1982; Saito et al. 2000; Sedgewick 1983; Segal 1983; Sambandham 1986; Stauffer 1989; Stauffer et al. 1988; Vetter et al. 1994; Schinzel 1982; Schlesinger 1895; Segal 1983; Shparlinski 1992; Silverman 1986; Silverman and Tate 1992; Sims 1984, 1994; Singer and van der Put 1997; Springer 1977; Smith 1995; Stephani 1989; Stauffer et al. 1993; Sturmfels 1993; Tarski 1951; Thomas 1937; Tsfasman and Vladut 1991; Uspensky 1948; Winograd 1980; Vasconcelos 1998; van Tilborg 1988; Wang 2001; Weber 1898/1899/1908; Wegener 1987; Wicker and Bhargava 1994; Winkler 1996; Winograd 1980; Zimmer 1972; Benninghofen et al. 1987; Jansen et al. in Vorbereitung; van der Waerden 1931; Zariski and Samuel 1958,1960; Zippel 1993]

3. Monographs and textbooks on computer algebra systems and their applications.

The number of books on a specific popular computer algebra systems and their applications amounts to several hundred. So they can not all be listed here. Please consult the home pages of the systems for additional information.

[Veigneau 1997; Bauldry and Fiedler 1995; Blachmann 1991; Baumann 1998; Ben-Israel and Gilbert 1999; Boggess et al. 1995; Braden et al. 1992; Brackx and Constales 1991; Braun and Hauser 1995; Braun and Meise 1995; Char et al. 1991a,b; Caviness et al. 1989; Crandall 1991; Davis et al. 1994; Devitt 1993; Ellis et al. 1992; Ellis and Lodi 1990; Fell 1998; Finch and Lehmann 1992; Fuchssteiner et al. 1993; GAP 1997; Char et al. 1992; Lidl et al. 1987; Gray and Glynn 1991; Hearn 1996; Parker and Christensen 1994; Soleng 1993; Grozin 1991; Harper et al. 1991; Hehl et al. 1992; Heller 1991; Heugl et al. 1996; Gloor et al. 1995; Jenks and Sutor 1992; Koepf et al. 1993;

Koepf 1994, 1996, 1998a; Kutzler et al. 1990; Maeder 1990; Kutzler 1997a,c,b; MacCallum and Wright 1991; Fuchssteiner and the MuPAD Group 1996; Oevel et al. 2000; Batut et al. 2000; Macsyma 1977; Melenk 1995; Metzner et al. 1999; Monagan et al. 1998; Rayna 1987; Rand 1984; Redfern et al. 1998; Hearn 1993; Skiena 1990; Stauffer et al. 1988; Steeb and Lewien 1992; Stroeker and Kaashoek 1999; Ueberberg 1992; Vardi 1991; Vermaseren 1991; Wagon 1990; Watt et al. 1994a; Wester 1999; Wolfram 1988, 1992, 1991, 1996, 1998; Hehl et al. 1992]

4. Collections of articles in computer algebra.

[Arnoni and Buchberger 1988; Buchberger et al. 1982; Bose 1985; Della Dora and Fitch 1989; Caviness and Johnson 1998; Chudnovsky and Jenks 1990; Cohen 1993; Cohen et al. 1995a; Cox and Sturmfels 1998; Ershov and Knuth 1981; Fleischer et al. 1994; Gao and Wang 2000b; Griewank and Corliss 1991; Herzberger 1994; Cannon 1990, 1991; Kaltofen and Buchberger 1990; Kraft et al. 1987; Deuflhard and Engquist 1987; Matzat et al. 1999; Miola 1990a; Micali 1989; Michalski et al. 1983; Pavelle 1985; Pohst 1987b; Pethö et al. 1990; Caviness and Johnson 1998; Singer 1991a; Specker and Strassen 1976; Tangora 1988, 1989; Gonzalez-Vega et al. 1998b; Tournier 1990; Yeh 1978]

# Cited References

Abbott, J. Univariate factorization over the integers. Preprint, 1998. Cited on page(s) 364.

Abbott, J., Bigatti, A., Kreuzer, M., and Robbiano, L. Computing ideals of points. *Journal of Symbolic Computation*, 30(4):341–356, 2000a. Cited on page(s) 364.

Abbott, J., Kreuzer, M., and Robbiano, L. Computing zero-dimensional schemes. Preprint, 2000b. Cited on page(s) 364.

Abbott, J., van Leeuwen, A., and Strotmann, A. Objectives of OpenMath. Technical Report 12, RIACA, 1996. Cited on page(s) 151, 153.

Abdallah, Chaouki T., Dorato, Peter, Yang, Wei, Liska, Richard, and Steinberg, Stanly. Applications of quantifier elimination theory to control system design. In *Proceedings of the 4th IEEE Mediterranean Symposium on Control and Automation*, pages 340–345. IEEE, 1996. Cited on page(s) 139.

Abhyankar, S. S. *Enumerative Combinatorics of Young Tableaux*. Marcel Dekker, New York, 1988. Cited on page(s) 490.

Ablowitz, M. J. and Clarkson, P. A. *Solitons, Nonlinear Evolution Equations and Inverse Scattering*, volume 149 of *London Math. Soc. Lec. Note Ser.* Cambridge University Press, London, 1991. Cited on page(s) 99, 101.

Ablowitz, M. J., Ramani, A., and Segur, H. A connection between nonlinear evolution equations and ordinary differential equations of P-type. I & II. *J. Math. Phys.*, 21:715–721; 1006–1015, 1980. Cited on page(s) 99.

Abramov, S. A. On the summation of rational functions. *Zh. vychisl. Mat. math. Fiz. 4*, 11:1071–1075, 1971. English translation in *USSR Comput. Math. Phys.* Cited on page(s) 91.

Abramov, S. A. The rational component of the solution of a first-order linear recurrence relation with a rational side. *Zh. vychisl. Mat. math. Fiz. 4*, 15: 1035–1039, 1975. English translation in *USSR Comput. Math. Phys.* Cited on page(s) 91.

Abramowitz, M. and Stegun, I. A. *Handbook of Mathematical Functions*. Dover Publ., New York, 1964. Cited on page(s) 214.

Adams, W. W. and Loustaunau, P. *An introduction to Gröbner bases*, volume 3 of *Graduate Studies in Math*. AMS, Oxford University Press, Providence, R.I., 1994a. Cited on page(s) 365.

Adams, William W. and Loustaunau, Philippe. *An Introduction to Groebner Basis*, volume 3 of *Graduate Studies in Mathematics*. AMS, Providence, Rhode Island, 1994b. Cited on page(s) 28, 490.

Adem, A. and Milgram, R. J. *Cohomology of Finite Groups*, volume 309 of *Grundlehren der Mathematischen Wissenschaften*. Springer-Verlag, Berlin-Heidelberg-New York, 1994. Cited on page(s) 34, 87, 197.

Adleman, L. M. and Huang, M.-D., editors. *Algorithmic Number Theory Symposium, Proceedings ANTS I*, volume 677 of *Lecture Notes in Computer Science*, Berlin-Heidelberg-New York, 1994. Springer-Verlag. Cited on page(s) 45, 490.

Adleman, L. M. and Huang, M.-D. A. *Primality Testing and Abelian Varieties over Finite Fields*, volume 1512 of *Lecture Notes in Mathematics*. Springer-Verlag, Berlin-Heidelberg-New York, 1992. Cited on page(s) 50, 490.

Ahmed, M. O. and Corless, R. M. The method of modified equations in Maple. Technical report, Dept. of Applied Mathematics, University of Western Ontario, London (Canada), 1997. Cited on page(s) 108.

Aho, A. V., Hopcroft, J. E., and Ullman, J. D. *The Design and Analysis of Computer Algorithms.* Addison-Wesley, 1974. Cited on page(s) 490.

Ahuja, S., Carriero, N., and Gelernter, D. Linda and friends. *IEEE Computer*, pages 26–34, August 1986. Cited on page(s) 147.

Aiba, A., Sakai, K., Sato, Y., Hawley, D. J., and Hasegawa, R. Constraint logic programming language CAL. In Institute for New Generation Computer Technology (ICOT), Tokyo, Japan, editor, *FGCS'88, Fifth Generation Computer Systems 1988, Proceedings of the International Conference on Fifth Generation Computer Systems 1988, Tokyo, Japan, November 28 – December 2, 1988*, volume 1, pages 263–276, Berlin-Heidelberg-New York, 1988. Springer-Verlag. Cited on page(s) 141.

Aït-Kaci, H. and Nivat, M., editors. *Algebraic Techniques*, volume 1 of *Resolution of Equations in Algebraic Structures.* Academic Press, New York, 1989a. ISBN 0-12-046370-9. Cited on page(s) 490.

Aït-Kaci, H. and Nivat, M., editors. *Rewriting Techniques*, volume 2 of *Resolution of Equations in Algebraic Structures.* Academic Press, New York, 1989b. ISBN 0-12-046371-7. Cited on page(s) 490, 568, 606.

Akers, R., Baffes, P., Kant, E., Randall, C., Steinberg, S., and Young, R. Automatic synthesis of numerical codes for solving partial differential equations. In *Special Issue on Non-Standard Applications of Computer Algebra*, volume 45 of *Mathematics and Computers in Simulation*, pages 3–22. Elsevier Science Publishers, North-Holland, January 1998. Nos. 1-2. Cited on page(s) 441.

Akers, R., Kant, E., Randall, C., Steinberg, S., and Young, R. Scinapse: A problem-solving environment for partial differential equations. *IEEE Computational Science and Engineering*, 4(3):32–42, 1997. July-Sept. Cited on page(s) 441.

Akritas, A. G. *Elements of Computer Algebra with Applications.* John Wiley&Sons, New York, 1989. Cited on page(s) 490.

Akritas, A. G. and Bavel, Z. Teaching great ideas of Mathematics with Mathematica. *Mathematica in Education and Research*, 7(4):39–45, 1998. Cited on page(s) 255.

Alagar, V. S. and Thanh, M. Fast polynomial decomposition algorithms. In *Proc. EUROCAL85*, pages 150–153. Springer-Verlag Lect. Notes in Comput. Sci. 204, 1985. Cited on page(s) 27.

Albouy, A. Integral manifolds of the N-body problem. *Inventiones mathematicae*, 114:463–488, 1993. Cited on page(s) 177.

Albouy, A. Symétrie des Configurations Centrales de Quatre Corps. *C. R. Acad. Sci. Paris*, 320:217–220, 1995. Cited on page(s) 177.

Albouy, A. The symmetric central configurations of four equal masses. *Contemporary Mathematics*, 198:131–135, 1996. Cited on page(s) 177, 179.

Albouy, A. and Chenciner, A. Le problème des $N$ corps et les distances mutuelles. *Inventiones mathematicae*, 131:151–184, 1998. Cited on page(s) 177.

Aldaz, M., Heintz, J., Matera, G., Montaña, J. L., and Pardo, L. M. Combinatorial hardness proofs for polynomial evaluation (extended abstract). In Brim, L., Gruska, J., and Zlatuska, J., editors, *MFCS'98*, Lecture Notes in Computer Science 1450, pages 167–175, Berlin, 1998a. Springer. Cited on page(s) 15.

Aldaz, M., Heintz, J., Matera, G., Montaña, J. L., and Pardo, L. M. Time–space tradeoffs for polynomial evaluation. *C. R. Acad. Sci., Paris*, 327:907–912, 1998b. Cited on page(s) 15.

Aldaz, M., Heintz, J., Matera, G., Montaña, J. L., and Pardo, L. M. Time–space tradeoffs in algebraic complexity theory. *J. of Complexity*, 16:2–49, 2000. Cited on page(s) 15.

Aldaz, M., Matera, G., Montaña, J. L., and Pardo, L. M. A new method to obtain lower bounds for polynomial evaluation. *Theor. Comput. Sci.*, 259 (1-2):577–596, 2001. Cited on page(s) 15.

Aldridge, J. E. and Prince, G. E. Computer algebra solution of the inverse problem in the calculus of variations. *Computer Physics Communications*, 115:489–509, 1998. Cited on page(s) 469.

Alefeld, Götz, editor. *Proceedings of the International Symposium on Scientific Computing, Computer Arithmetic and Validated Numerics*, number 90 in Mathematical Research, 1995. Akad.-Verlag. Cited on page(s) 488.

Alefeld, G. and Herzberger, J. *Introduction to Interval Computations*. Academic Press, New York, 1983. Cited on page(s) 111.

Alonso, C., Gutiérrez, J., and Recio, T. An implicitization algorithm with fewer variables. *Computer Aided Geometric Design*, 12:251–258, 1995a. Cited on page(s) 235.

Alonso, C., Gutierrez, J., and Recio, T. A rational function decomposition algorithm by near-separated polynomials. *J. Symbolic Computation*, 19:527–544, 1995b. Cited on page(s) 28.

Alvarez-Sobreviela, Luis. REDUCE-MathML Interface. Technical report, Konrad-Zuse-Zentrum Berlin (ZIB), 1998. Distributed with Reduce 3.7. Cited on page(s) 150.

Amrhein, Beatrice and Gloor, Oliver. The fractal walk. In Buchberger, Bruno and Winkler, Franz, editors, *Gröbner Bases and Applications*, volume 251 of *LNS*, pages 305–322, Cambridge, UK, February 1998. LMS, Cambridge University Press. ISBN 0-521-63298-6. Cited on page(s) 30, 52, 353.

Amrhein, Beatrice, Gloor, Oliver, and Küchlin, Wolfgang. A case study of multi-threaded Gröbner basis completion. In Lakshman [1995], pages 95–102. Cited on page(s) 352, 353, 436.

Amrhein, Beatrice, Gloor, Oliver, and Küchlin, Wolfgang. Walking faster. In Calmet and Limoncelli [1996], pages 150–161. Cited on page(s) 353.

Amrhein, Beatrice, Gloor, Oliver, and Küchlin, Wolfgang. On the Walk. *Theoretical Computer Science*, 187(187):179–202, 1997a. Cited on page(s) 30, 52, 353.

Amrhein, B., Gloor, O., and Maeder, R. E. Visualizations for Mathematics courses based on a computer algebra system. *JSC*, 23(5/6):447–452, 1997b. Cited on page(s) 258.

Anderson, J. R. *Cognitive Psychology and its Implications*. Freeman, New York, 4th edition, 1995. Cited on page(s) 257.

Andrews, G. E. On a conjecture of Peter Borwein. *Journal of Symbolic Computation*, 20(5/6):487–502, November/December 1995. Cited on page(s) 264.

Anick, David J. On the homology of associative algebras. *Trans. Amer. Math. Soc.*, 296(2):641–659, 1986. Cited on page(s) 210.

Apel, J. *Gröbnerbasen in nichtkommutativen Algebren und ihre Anwendung*. PhD thesis, Universität Leipzig, 1988. Cited on page(s) 59, 61, 62.

Apel, J. A relationship between Gröbner bases of ideals and modules of G-algebras. *Contemporary Mathematics*, 131:195–203, 1992. Cited on page(s) 62, 379.

Apel, J. A Gröbner approach to involutive bases. *J. Symb. Comp.*, 19:441–457, 1995. Cited on page(s) 31, 103.

Apel, J. *Zu Berechenbarkeitsfragen der Idealtheorie*. Habilitation, Universität Leipzig, 1997. Cited on page(s) 62.

Apel, J. Computational ideal theory in finitely generated extension rings. Informatik-Report 12, Universität Leipzig, Leipzig, 1998a. Cited on page(s) 59.

Apel, J. The theory of involutive divisions and an application to Hilbert function computations. *J. Symb. Comp.*, 25:683–704, 1998b. Cited on page(s) 31.

Apel, J. and Klaus, U. FELIX—an assistant for algebraists. In Watt [1991], pages 382–389. ISBN 0-89791-437-6. Cited on page(s) 379.

Apel, J. and Klaus, U. Implementation aspects for non-commutative domains. In Shirkov et al. [1991], pages 127–132. ISBN 981-02-0687-9. Cited on page(s) 379.

Apel, J. and Klaus, U. Data representation and in-built compilation in the computer algebra program FELIX. In Fitch, J., editor, *Design and Implementation of Symbolic Computation Systems, International Symposium DISCO '92, Bath, U.K., April 13–15, 1992, Proceedings*, volume 721 of *Lecture Notes in Computer Science*, pages 173–192, Berlin-Heidelberg-New York, 1992. Springer-Verlag. ISBN 3-540-57272-4, 0-387-57272-4. Cited on page(s) 378.

Apel, J. and Lassner, W. An algorithm for calculations in enveloping fields of Lie algebras. In *Proc. Int. Conf. on Computer Algebra and its Application in Theoretical Physics, Dubna 1985*, D11-85-791, pages 231–241. JINR, Dubna, 1985. Cited on page(s) 61.

Apel, J. and Laßner, W. An extension of Buchberger's algorithm and calculations in enveloping fields of Lie algebras. *Journal of Symbolic Computation*, 6(3): 361–370, 1988. Cited on page(s) 61, 379.

Apel, J. and Schmüdgen, K. Classification of three-dimensional covariant differential calculi on Podleś quantum spheres. *Lett. Math. Phys.*, 32:25–36, 1994. Cited on page(s) 213, 214.

Apiola, Heikki, Laone, Marko, and Valkeila, Esko, editors. *Proceedings of the Workshop on Symbolic and Numeric Computing, Helsinki University of Technology 1993*, Helsinki, 1994. Rolf Nevanlinna Institute. ISBN 952-9528-27-2. Cited on page(s) 525, 561.

Arais, E. A., Shapeev, V. P., and Yanenko, N. N. Realization of Cartan's method of exterior differential forms on an electronic computer. *Sov. Math. Dokl.*, 15: 203–205, 1974. Cited on page(s) 102.

Arhrib, A. and Moultaka, G. Radiative corrections to $e^+e^- \to H^+H^-$: THDM versus MSSM. *Nucl. Phys.*, B558:3, 1999. Cited on page(s) 170.

Arndt, J. The Hfloat package. http://www.spectracom.de/~arndt/joerg.html, 1990. Cited on page(s) 12.

Arnon, D. and McCallum, S. A polynomial time algorithm for the topological type of a real algebraic curve. *Journal of Symbolic Computation*, 5:213–236, 1988. Cited on page(s) 238.

Arnoni, D. S. and Buchberger, B., editors. *Algorithms in Real Algebraic Geometry*. Academic Press, New York, 1988. Cited on page(s) 492.

Arora, S. Reductions, codes, PCPs and inapproximability. In *Proceedings of the 36th Conference on Foundations of Computer Science*, pages 404–413, 1995. Cited on page(s) 129.

Arora, S. and Sudan, M. Improved low-degree testing and its applications. In *Proceedings of the the 29th Symposium on Theory of Computing*, pages 485–495, 1997. Cited on page(s) 130.

Arvind and Culler, David E. Resource requirements of dataflow programs. In *Proceedings of the 15th Annual International Symposium on Computer Architectures*, pages 141–150, Honolulu, Hawai, May 1988. Cited on page(s) 148.

Arvind, Nikhil, Rishiyur S., and Pingali, Keshav K. I-structures: Data structures for parallel programming. *ACM Transactions on Programming Languages and Systems*, 11(4):598–632, 1989. Cited on page(s) 149.

Aschbacher, M. On the maximal subgroups of the finite classical groups. *Invent. Math.*, 76:469–514, 1984. Cited on page(s) 474.

Askey, R. and Gasper, G. Jacobi polynomial expansions of Jacobi polynomials with non-negative coefficients. *Proc. Camb. Phil. Soc.*, 70:243–255, 1971. Cited on page(s) 214.

Askey, R. and Wilson, J. A. *Some Basic Hypergeometric Polynomials that Generalize Jacobi Polynomials*, volume 319 of *Memoirs of the Amer. Math. Soc.* American Mathematical Society, 1985. Cited on page(s) 214.

Askey, R. A., Koepf, W., and Koornwinder, T. H., editors. *Orthogonal Polynomials and Computer Algebra*, volume 28 (6) of *Journal of Symbolic Computation*, 1999. Cited on page(s) 215.

Assmus, E. F. and Key, J. D. *Designs and their Codes*, volume 103 of *Cambridge Tracts in Mathematics*. Cambridge University Press, 1993. ISBN 0-521-45839-0. Cited on page(s) 129.

Assmus, Jr., E. F. and Key, J. D. *Designs and their codes*, volume 103 of *Cambridge Tracts in Mathematics*. Cambridge University Press, Cambridge, 1992. ISBN 0-521-41361-3. Cited on page(s) 375, 490.

Astesiano, Egidio, Kreowski, Hans-Jörg, and Krieg-Brückner, Bernd. *Algebraic Foundations of System Specification*. Springer, 1999. Cited on page(s) 218.

Atkin, A. O. L. and Morain, F. Elliptic curves and primality proving. *Mathematics of Computation*, 61(203):29–68, July 1993. Cited on page(s) 48.

Atkins, D., Graff, M., Lenstra, A. K., and Leyland, P. C. The magic words are squeamish ossifrage. In *Advances in Cryptology, ASIACRYPT'94*, number 917 in Lecture Notes in Computer Science, pages 263–277, 1995. Cited on page(s) 45.

Atkinson, Michael D., editor. *Computational Group Theory*, London, New York, 1984a. Academic Press. Cited on page(s) 70, 490.

Atkinson, M. D., editor. *Computational Group Theory, Proceedings of a 1982 LMS Symposium, Durham*, New York, 1984b. Academic Press. Cited on page(s) 88, 197, 585, 589.

Aubry, P., Lazard, D., and Maza, M. Moreno. On the Theories of Triangular Sets. *Journal of Symbolic Computation*, 28:105–124, 1999. Cited on page(s) 178, 179.

Aubry, P. and Maza, M. Moreno. Triangular Sets for Solving Polynomial Systems: a Comparative Implementation of Four Methods. *Journal of Symbolic Computation*, 28:125–154, 1999. Cited on page(s) 178.

Ausbrooks, R., Buswell, S., Dalmas, S., Devitt, S., Diaz, A., Hunter, R., Smith, B., Soiffer, N., Sutor, R., and Watt, S. *Mathematical Markup Language (MathML) Version 2.0*. W3C Recommendation 21-February-2001. World Wide Web Consortium, 2001. D. Carlisle, P. Ion, N. Poppelier and R. Miner, editors, http://www.w3.org/TR/2001/REC-MathML2-20010221. Cited on page(s) 154.

Auslander, M., Reiten, I., and Smalø, S. O. *Representation Theory of Artin Algebras*. CUP, ACUP, 1995. Cited on page(s) 64.

Austrian Experiment. Austrian experiment. *The International DERIVE Journal*, 3(1), 1996. Cited on page(s) 250.

Auzinger, W. and Stetter, H. J. An elimination algorithm for the computation of all zeros of a system of multivariate polynomial equations. In Agarwal, Ravi P., Chow, Y. M., and Wilson, S. J., editors, *Numerical Mathematics*, volume 86 of *ISNM*, pages 11–30. Birkhäuser, 1988. Cited on page(s) 52, 113, 122.

Avdeev, L. V. *Comp. Phys. Comm.*, 98:15, 1996. Cited on page(s) 166.

Avenhaus, Jürgen. *Reduktionssysteme*. Springer-Verlag, 1995. Cited on page(s) 136, 490.

Avenhaus, J. and Denzinger, J. Distributing equational theorem proving. In Kirchner, Claude, editor, *Rewriting Techniques and Applications (LNCS 690)*, pages 62–76. Springer-Verlag, 1993. (Proc. RTA'93, Montreal, Canada, June 1993). Cited on page(s) 136.

Avitzur, R., Bachmann, O., and Kajler, N. From Honest to Intelligent Plotting. In Levelt [1995], pages 32–41. Cited on page(s) 150.

Ax, James and Kochen, Simon. Diophantine problems over local fields. *American Journal of Mathematics*, 87:605–648, 1965. Parts I and II. Cited on page(s) 138.

Ax, James and Kochen, Simon. Diophantine problems over local fields. *Annals of Mathematics*, 83:437–456, 1966. Part III. Cited on page(s) 138.

Baader, Franz and Nipkow, Tobias. *Term Rewriting and All That*. Cambridge University Press, 1998. Cited on page(s) 136, 490.

Babai, Laszlo. Computational complexity in finite groups. In *Proceedings of the International Congress of Mathematicians, Kyoto, Japan, 1990*, page what pages. The Mathematical Society of Japan, 1991. Cited on page(s) 66.

Bach, Eric and Shallit, Jeffrey. *Algorithmic Number Theory, Volume 1*. MIT Press, 1996. Cited on page(s) 42, 490.

Bachmair, L. *Canonical Equational Proofs*. Birkhäuser Verlag, Basel, 1991. Cited on page(s) 490.

Bachmair, Leo, editor. *Rewriting Techniques and Applications/ 11th International Conference*, number 1833 in Lecture Notes in Computer Science, July 2000. Springer. Cited on page(s) 489.

Bachmann, O., Gray, S., and Schönemann, H. MPP:A Framework for Distributed Polynomial Systems Based on MP. In *Proc. of the International Symposium on Symbolic and Algebraic Computation (ISSAC'96)*, Zurich, Switzerland, July 1996. ACM Press. Cited on page(s) 449.

Bachmann, O., Schönemann, H., and Gray, S. MPP: A Framework for Distributed Polynomial Computations. In Lakshman [1995], pages 103–111. Cited on page(s) 153.

Bachmann, O., Schönemann, H., and Gray, S. A Proposal for Syntactic Data Integration for Math Protocols. In Hitz and Kaltofen [1997], pages 165–175. ISBN 0-89791-951-3. Cited on page(s) 153.

Bachmann, O., Schönemann, H., and Sorgatz, A. Connecting MuPAD and Singular with MP. *Maple Tech*, 5(2/3):117–121, July 1999. Cited on page(s) 330.

Backelin, Jörgen. BERGMAN. ftp://ftp.matematik.su.se/pub/src/bergman, 1998. For more information: joeb@matematik.su.se. Cited on page(s) 210.

Backelin, J. and Fröberg, R. How we prove that there are exactly 924 cyclic 7–roots. In Watt [1991], pages 103–111. ISBN 0-89791-437-6. Cited on page(s) 351.

Backelin, J. and Ufnarovski, V., 1998. Personal communication. Cited on page(s) 210.

Backstrom, Gunnar. *Fields of Physics on the PC by Finite Element Analysis*. Studentlitteratur, Lund, Sweden, second edition edition, 1996. ISBN 91-44-00293-9. Cited on page(s) 283.

Baddoura, J. *Integration in Finite Terms with Elementary Functions and Dilogarithms*. PhD thesis, MIT, Mathematics, 1994. Cited on page(s) 95.

Bailey, D. H. MPFUN: A portable high performance multiprecsion package. http://www.nas.nasa.gov/NAS/TechReports/RNRreports/dbailey/RNR-90-022/RNR-90-022.html, 1990. Cited on page(s) 12.

Balfagón, A., Castellví, P., and Jaén, X. TTC: *Symbolic Tensor Calculus with Index*, 2001. http://baldufa.upc.es/ttc/ (documentation and source code). Cited on page(s) 175.

Balfagón, A. and Jaén, X. TTC: Symbolic tensor calculus with indexes. *Comput. Phys.*, 12,3, 1998. Cited on page(s) 480.

Bank, B., Giusti, M., Heintz, J., and Mbakop, G. Polar varieties, real equation solving and data-structures: the hypersurface case. *J. of Complexity*, 13(1): 5–27, 1997. Cited on page(s) 54.

Barbeau, E. J. *Polynomials*. Springer-Verlag, Berlin-Heidelberg-New York, 1989. Cited on page(s) 490.

Barbier, C., Bettess, P., and Bettess, J. A. Automatic generation of mapping functions for infinite elements using REDUCE. *Journal of Symbolic Computation*, 14(5):523–534, November 1992. Cited on page(s) 151.

Barbier, C., Clark, P., Bettess, P., and Bettess, J. Automatic Generation of Shape Functions for Finite Element Analysis Using REDUCE. *Engineering Computations*, 7(4):349–358, 1990. Cited on page(s) 151.

Bardis, L. and Patrikalakis, M. Approximate conversion of rational B–spline patches. *Computer Aided Geometric Design*, 6:189–204, 1989. Cited on page(s) 241.

Bareiss, E. H. Sylvester's identity and multistep integers preserving Gaussian elimination. *Mathematics of Computation*, 22:565–578, 1968. Cited on page(s) 36.

Barkatou, M.A. and Pflügel, E. An Algorithm Computing the Regular Formal Solutions of a System of Linear Differential Equations, 1997. RR 988, LMC–IMAG, *Also in JSC 28(4-5), pp. 569-587 (1999)*. Cited on page(s) 371.

Barkatou, M.A. and Pflügel, E. On the Equivalence Problem of Linear Differential Systems and its Application for Factoring Completely Reducible Systems. *ACM Press, In Oliver Gloor, editor, proceedings of ISSAC '98*, pages 268–275, Rostock, Germany, 1998. Cited on page(s) 372.

Barkatou, M. A. Contribution à l'étude des équations différentielles et de différences dans le champ complexe. PhD thesis, INPG - Grenoble, 1989. Cited on page(s) 372.

Barkatou, M. A. An algorithm to compute the exponential part of a formal fundamental matrix solution of a linear differential system near an irregular singularity. *App. Alg. Eng. Comm. Comp.*, 8:1–23, 1997. Cited on page(s) 97, 371.

Barkatou, M. A. A fast algorithm to compute the rational solutions of systems of linear differential equations, 1998. RR 973, IMAG Grenoble. Cited on page(s) 371.

Barkatou, M. A. On rational solutions of linear differential systems. *Journal of Symbolic Computation*, 28:547–567, 1999. Cited on page(s) 97.

Barnes, Donald W. and Lambe, Larry A. A fixed point approach to homological perturbation theory. *Proc. Amer. Math. Soc.*, 112(3):881–892, 1991. ISSN 0002-9939. Cited on page(s) 210, 211.

Barton, D. R. and Zippel, R. E. Polynomial decomposition algorithms. *J. Symb. Comp.*, 1:159–168, 1985. Cited on page(s) 27.

Basu, Saugata, Pollack, Richard, and Roy, Marie-Françoise. On the combinatorial and algebraic complexity of quantifier elimination. In Goldwasser, Shafi, editor, *Proceedings of the 35th Annual Symposium on Foundations of Computer Science*, pages 632–641, Santa Fe, NM, November 1994. IEEE Computer Society Press, Los Alamitos, CA. Cited on page(s) 53, 57, 138, 204.

Basu, Saugata, Pollack, Richard, and Roy, Marie-Francoise. Computing a set of points meeting every cell defined by a family of polynomials on a variety. In et al. Goldberg, Ken, editor, *Algorithmic foundations of robotics.*, pages 537–555, Wellesley, MA, 1995. A. K. Peters. ISBN 1-56881-045-8/hbk. Proceedings of the workshop on the algorithmic foundations of robotics, WAFR '94, held in San Francisco, CA, USA, 17-19 February, 1994. Cited on page(s) 57.

Basu, Saugata, Pollack, Richard, and Roy, Marie-Francoise. Computing roadmaps of semi-algebraic sets (extended abstract). In *Twenty-Eighth ACM Symp. on Theory of Computing*, pages 168–173, 1996a. Cited on page(s) 57, 234.

Basu, Saugata, Pollack, Richard, and Roy, Marie-Francoise. On the combinatorial and algebraic complexity of quantifier elimination. *J. ACM*, 43(6):1002–1045, 1996b. Cited on page(s) 57.

Basu, Saugata, Pollack, Richard, and Roy, Marie-Francoise. On the number of cells defined by a family of polynomials on a variety. *Mathematika*, 43(1): 120–126, 1996c. Cited on page(s) 57.

Basu, Saugata, Pollack, Richard, and Roy, Marie-Francoise. Computing roadmaps of semi-algebraic sets on a variety. (extended abstract). In et al. Cucker, Felipe, editor, *Foundations of computational mathematics.*, pages 1–15, Berlin, 1997a. Springer. ISBN 3-540-61647-0. Selected papers of a conference, held at IMPA in Rio de Janeiro, Brazil, January 1997. Cited on page(s) 57.

Basu, Saugata, Pollack, Richard, and Roy, Marie-Francoise. On computing a set of points meeting every cell defined by a family of polynomials on a variety. *J. Complexity*, 13(1):28–37, 1997b. Cited on page(s) 57.

Basu, Saugata, Pollack, Richard, and Roy, Marie-Francoise. Complexity of computing semi-algebraic descriptions of the connected components of a semi-algebraic set. In Gloor, O., editor, *Proc. 98 Int. Symp. on Symb. and Alg. Comp.*, pages 25–29. ACM, 1998a. Cited on page(s) 57.

Basu, Saugata, Pollack, Richard, and Roy, Marie-Francoise. A new algorithm to find a point in every cell defined by a family of polynomials. In Caviness, B. and Johnson, J., editors, *Quantifier Elimination and Cylindrical Algebraic Decomposition*, Texts and Monographs in Symbolic Computation, pages 341–349. Springer, Wien, New York, 1998b. Cited on page(s) 57.

Batut, C., Belabas, K., Bernardi, D., Cohen, H., and Olivier, M. *User's Guide to PARI-GP.* by anonymous ftp from ftp://megrez.math.u-bordeaux.fr/pub/pari, 2000. see also http://www.parigp-home.de/. Cited on page(s) 12, 45, 397, 434, 492.

Bauberger, S. and Weiglein, G. Calculation of two-loop top-quark and Higgs-boson corrections in the electroweak Standard Model. *Nucl. Instrum. Meth.*, A389:318, 1997. Cited on page(s) 171.

Bauberger, S. and Weiglein, G. Higgs-mass dependence of two-loop corrections to $\Delta r$. *Phys. Lett.*, B419:333, 1998. Cited on page(s) 171.

Bauer, A. and Petkovšek, M. Mixed multibasic and hypergeometric Gosper-type algorithms. *Journal of Symbolic Computation*, 28:711–736, 1999. Cited on page(s) 92.

Bauldry, W. C. and Fiedler, J. R. *Calculus Laboratories with MAPLE, A Tool not an Oracle.* Symbolic Computation Series. Brooks/Cole Publishing Company, Pacific Grove, California 93950, 1995. Cited on page(s) 491.

Baum, U. and Clausen, M. Computing irreducible representations of supersolvable groups. *Math. Comp.*, 63:351–359, 1994. Cited on page(s) 85.

Baumann, G. *Symmetry Analysis of Differential Equations with* Mathematica. TELOS/Springer, 1998. Cited on page(s) 190, 192, 194, 195, 491.

Baumslag, Gilbert, Cannonito, Frank B., Robinson, Derek J. S., and Segal, Dan. The algorithmic theory of polycyclic-by-finite groups. *Journal of Algebra*, 142: 118–149, 1991. Cited on page(s) 78.

Baumslag, G. and et al., editors. *Geometric and Computational Perspectives on Infinite Groups*, volume 25 of *DIMACS Series in Discrete Mathematics and Theoretical Computer Science*, Providence, 1996. AMS. Cited on page(s) 71, 586.

Baur, W. Simplified lower bounds for polynomials with algebraic coefficients. *J. of Complexity*, 13(1):38–41, 1997. Cited on page(s) 15.

Baur, W. and Strassen, V. The complexity of partial derivatives. *Theoret. Comput. Sci.*, 22:317–330, 1983. Cited on page(s) 14.

Bayer, D., Galligo, A., and Stillman, M. Gröbner bases and extension of scalars. In *Proc. Comput. Algebr. Geom. and Commut. Algebra*, 1991. Cortona, Italy. Cited on page(s) 31.

Bayer, Dave and Stillman, Mike. Computation of Hilbert functions. *J. Symbolic Comput.*, 14(1):31–50, 1992a. ISSN 0747-7171. Cited on page(s) 209.

Bayer, D. and Stillman, M. *Macaulay: A system for computation in algebraic geometry and commutative algebra.* Available via anonymous ftp from math.harvard.edu, 1992b. Available via anonymous ftp from math.harvard.edu. Cited on page(s) 209, 210, 381, 393.

Beals, Robert. Computing blocks of imprimitivity for small-base groups in nearly linear time. In Finkelstein and Kantor [1993], pages 17–26. Cited on page(s) 83.

Beals, Robert. Towards polynomial time algorithms for matrix groups. In Finkelstein and Kantor [1997], pages 31–54. Cited on page(s) 75.

Beck, R. E. and Kolman, B., editors. *Computers in Nonassociative Rings and Algebras.* Academic Press, 1977. Cited on page(s) 347, 490.

Becken, O. Algorithmen zum Lösen einfacher Differentialgleichungen. Rostocker Informatik-Berichte 17, Universität Rostock, 1995. http://www.informatik.uni-rostock.de/~obecken/. Cited on page(s) 110.

Becker, E., Marinari, M. G., Mora, T., and Traverso, C. The shape of the shape lemma. In von zur Gathen, J. and Giesbrecht, M., editors, *ISSAC'94*, pages 129–133. ACM Press, 1994. Cited on page(s) 52.

Becker, E. and Neuhaus, R. Computation of real radicals of polynomial ideals. In Eysette, Frederic and Galligo, Andre, editors, *Computational Algebraic Geometry*, volume 109 of *Progress in mathematics*, pages 1–20, Boston, Basel, Berlin, 1993. Birkhäuser. ISBN 0-8176-3678-1. Cited on page(s) 56.

Becker, Eberhard and Wörmann, Thorsten. On the trace formula for quadratic forms. In Jacob, William B., Lam, Tsit-Yuen, and Robson, Robert O., editors, *Recent Advances in Real Algebraic Geometry and Quadratic Forms*, volume 155 of *Contemporary Mathematics*, pages 271–291. American Mathematical Society, American Mathematical Society, Providence, Rhode Island, 1994. ISBN 0-8218-5154-3. Proceedings of the RAGSQUAD Year, Berkeley, 1990–1991. Cited on page(s) 57, 139.

Becker, Th. Standard bases and some computations in rings of power series. *Journal of Symbolic Computation*, 10(2):165–178, 1990. Cited on page(s) 32.

Becker, Thomas. Gröbner bases versus D-Gröbner bases, and Gröbner bases under specialization. *AAECC*, 5:1–8, 1994. Cited on page(s) 31.

Becker, Th. and Weispfenning, V. The Chinese remainder problem, multivariate interpolation and Gröbner bases. In Watt, S. M., editor, *ISSAC'91*, pages 64–69, 1991. Cited on page(s) 30.

Becker, Thomas, Weispfenning, Volker, and Kredel, Heinz. *Gröbner Bases, a Computational Approach to Commutative Algebra*, volume 141 of *Graduate Texts in Mathematics*. Springer, New York, corrected second printing edition, 1998. Cited on page(s) 28, 30, 32, 33, 52, 53, 205, 490.

Becker, Th., Weispfenning, V., and with cooperation from H. Kredel. *Gröbner Bases, A Computational Approach to Commutative Algebra*, volume 141 of *Graduate Texts in Mathematics*. Springer-Verlag, New York, Berlin, Heidelberg, 1993. Cited on page(s) 4, 426, 427, 463, 490.

Beckermann, Bernhard and Labahn, George. When are two polynomials relatively prime? *Journal of Symbolic Computation*, 26:677–689, 1998. Cited on page(s) 121.

Beckermann, B., Labahn, G., and Villard, G. Shifted Normal Forms of Polynomial Matrices. In *International Symposium on Symbolic and Algebraic Computation, Vancouver, Canada*, pages 189–196. ACM Press, July 1999. Cited on page(s) 39.

Beenakker, W., Denner, A., and Kraft, A. $e^+e^-$ annihilation into heavy fermion pairs in the two Higgs-doublet model. *Nucl. Phys.*, B410:219, 1993. Cited on page(s) 170.

Beke, E. Die Irreducibilität der homogenen linearen Differentialgleichungen. *Math. Ann.*, 45:278–294, 1894a. Cited on page(s) 215.

Beke, E. Die symmetrischen Functionen bei linearen homogenen Differentialgleichungen. *Math. Ann.*, 45:295–300, 1894b. Cited on page(s) 215.

Belkhahia, Tijani. Implementierung eines Algorithmus zur Konstruktion universeller Gröbner basen in SAC-2/ALDES. Diplomarbeit, Universität Passau, Passau, July 1992. Cited on page(s) 427.

Beltzer, A. I. *Variational and Finite Element Methods*. Springer–Verlag, Berlin Heidelberg, 1990. ISBN 3-540-51598-4 and 0-387-51598-4. Cited on page(s) 283.

Ben-Israel, Adi and Gilbert, Robert P. *Calculus with Macsyma*. Springer–Verlag, Vienna, 1999. in preparation. Cited on page(s) 283, 491.

Ben-Or, M., Kozen, D., and Reif, J. The complexity of elementary algebra and geometry. *Journal Comput. System Sci.*, 32:251–264, 1986. Cited on page(s) 139.

Bendezu, A. Barrientos and Kniehl, B. $W^{\pm}H^{\pm}$ associated production at the Large Hadron Collider. *Phys. Rev.*, D59:015009, 1999. Cited on page(s) 170.

Benninga, S. and Wiener, Z. Financial engineering. *Mathematica in Education and Research*, 6(3):27–34, 6(4):11–14, 7(1):12–16, 7(2):13–21, 7(3):11–19, 7(4):39–45, 1997–1998. Cited on page(s) 255.

Benninghofen, B., Kemmerich, S., and Richter, M. M. *Systems of Reductions*, volume 277 of *Lecture Notes in Computer Science*. Springer-Verlag, Berlin-Heidelberg-New York, 1987. Cited on page(s) 491.

Benson, David J. *Polynomial Invariants of Finite Groups*. Number 190 in Lond. Math. Soc. Lecture Note Ser. Cambridge Univ. Press, Cambridge, 1993. Cited on page(s) 33, 490.

Bergman, C. H., Maddux, R. D., and Pigozzi, D. L., editors. *Algebraic Logic and Universal Algebra in Computer Science, Conference, Ames, Iowa, USA, June 1–4, 1988, Proceedings*, volume 425 of *Lecture Notes in Computer Science*, Berlin-Heidelberg-New York, 1990. Springer-Verlag. ISBN 3-540-97288-9, 0-387-97288-9. Cited on page(s) 486.

Berlekamp, E. R. Factoring polynomials over finite fields. *Bell System Techn. Journal*, 46:1853–1859, 1967. Cited on page(s) 24.

Berlekamp, E. R. Factoring polynomials over large finite fields. *Mathematics of Computation*, 24:713–735, 1970. Cited on page(s) 24.

Berlekamp, E. R. *Algebraic Coding Theory, revised 1984 edition*. Aegean Park Press, Laguna Hills, Ca., 1984. Cited on page(s) 490.

Berry, J., Graham, E., and Watkins, A. J. P. Integrating the DERIVE program into the teaching of mathematics. *International DERIVE Journal*, 1(1):83–96, 1994. Cited on page(s) 251.

Bertrand, D. Lectures on differential Galois theory by A.R. Magid. *Bull. Amer. Math. Soc.*, 33:289–294, 1996. (Review). Cited on page(s) 97.

Bertrand, Laurent. Computing a hyperelliptic integral using arithmetic in the jacobian of the curve. *Applicable Algebra in Engineering, Communication and Computing*, 6:275–298, 1995. Cited on page(s) 95.

Besche, H. U. and Eick, B. The groups of order at most 1000 except 512 and 768. *J. Symb. Comput.*, 27:405–413, 1998. Cited on page(s) 197, 389.

Beth, Th. *Public-Key Cryptography: State of the Art and Future Directions*, volume 578 of *Lecture Notes in Computer Science*. Springer-Verlag, Berlin-Heidelberg-New York, 1991. Cited on page(s) 490.

Beth, T. and Clausen, M. Computer-Algebra und Komplexität. Vorlesungsmanuskript Universität Karlsruhe, 1987. Cited on page(s) 490.

Beth, T. and Clausen, M., editors. *Applicable Algebra, Error-Correcting Codes, Combinatorics and Computer Algebra, 4th International Conference, AAECC-4, Karlsruhe, FRG, September 23–26, 1986, Proceedings*, volume 307 of *Lecture Notes in Computer Science*, Berlin-Heidelberg-New York, 1988. Springer-Verlag. ISBN 3-540-19200-X, 0-387-19200-X. Cited on page(s) 486.

Beth, Thomas and Grassl, Markus. Improved Decoding of Quantum Error Correcting Codes from Classical Codes. In Toffoli, Tommaso, Biafore, Michael, and Leao, Joao, editors, *PhysComp96*, pages 28–31, Boston, November 1996. Cited on page(s) 130.

Beth, Thomas, Jungnickel, Dieter, and Lenz, Hanfried. *Design theory*. Cambridge University Press, Cambridge, 1986. ISBN 0-521-33334-2. Cited on page(s) 375, 490.

Betten, A., Fripertinger, H., Kerber, A., Wassermann, A., and Zimmermann, K.-H. *Codierungstheorie, Konstruktion und Anwendung Linearer Codes*. Springer-Verlag, Berlin-Heidelberg-New York, 1998a. Cited on page(s) 90, 490.

Betten, Anton, Kerber, Adalbert, Kohnert, Axel, Laue, Reinhard, and Wassermann, Alfred. The discovery of simple 7-designs with automorphism group $P\Gamma L(2,32)$. In *Applied algebra, algebraic algorithms and error-correcting codes (Paris, 1995)*, volume 948 of *Lecture Notes in Comput. Sci.*, pages 131–145. Springer, Berlin, 1995. Cited on page(s) 375.

Betten, A., Kerber, A., Laue, R., and Wassermann, A. Simple 8-designs with small parameters. *Designs, Codes, Cryptography*, 15:5–27, 1998b. Cited on page(s) 90.

Betten, Anton, Kerber, Adalbert, Laue, Reinhard, and Wassermann, Alfred. Simple 8-designs with small parameters. *Designs, Codes, Cryptography*, 15: 5–27, 1998c. Cited on page(s) 374.

Betten, A., Klin, M. C., Laue, R., and Wassermann, A. Graphical $t$-Designs via Polynomial Kramer-Mesner Matrices. *Discrete Mathematics*, 197/198:83–109, 1999. Cited on page(s) 90.

Betten, A., Laue, R., and Wassermann, A. Some simple 7-designs. In Hirschfeld, J. W. P., Magliveras, S. S., and de Resmini, M. J., editors, *Geometry, Combinatorial Designs and Related Structures, Proceedings of the First Pythagorean Conference*, number 245 in LMS Lecture Notes, pages 15–25. Cambridge University Press, 1997. Cited on page(s) 90.

Betten, A., Laue, R., and Wassermann, A. *DISCRETA – A tool for constructing t-designs*. Lehrstuhl II für Mathematik, Universität Bayreuth, http://www.mathe2.uni-bayreuth.de/betten/DISCRETA/Index.html, 1998d. Cited on page(s) 374.

Beukers, F. The maximal differential ideal is generated by its invariants. *Indag. Mathem. N.S.*, 11:13–18, 2000. Cited on page(s) 97.

Beyer, W. A., Fawcett, L. R., Mauldin, R. D., and Swartz, B. K. The volume common to two congruent circular cones whose axes intersect symmetrically. *Journal of Symbolic Computation*, 4:381–390, 1987. Cited on page(s) 222.

Bialkowski, Stephen. *Photothermal Spectroscopy Method for Chemical Analysis*, volume 134 of *Chemical Analysis Series*. John Wiley & Sons, 1995. Cited on page(s) 283.

Biehl, I., Buchmann, J., and Papanikolaou, Th. LiDIA: a library for computational number theory. SFB 124 report, Universität des Saarlandes, 1995. Cited on page(s) 404.

Bigatti, A. M. Computations of Hilbert-Poincaré Series. *Journal for Pure and Applied Algebra*, 119:237–253, 1997. Cited on page(s) 364.

Bigatti, A. M., LaScala, R., and Robbiano, L. Computing toric ideals. *Journal of Symbolic Computation*, 27:351–365, 1999. Cited on page(s) 364.

Bilu, Y. and Hanrot, G. Solving Thue equations of high degree. *Journal of Number Theory*, 60:373–392, 1996. Cited on page(s) 400.

Binder, F. Fast computations in the lattice of polynomial rational fucntion fields. In *Proc. ISSAC-96*. ACM Press, 1995. Cited on page(s) 27.

Bini, D. and Pan, V. *Polynomial and matrix computations*. Birkhäuser, 1994. Cited on page(s) 146, 490.

Bini, D. A. and Fiorentino, G. Design, analysis, and implementation of a multiprecision polynomial rootfinder. *Numerical Algorithms*, 23:127–173, 2000. Cited on page(s) 26.

Birch, B. J. and Swinnerton-Dyer, H. P. F. Notes on elliptic curves I. *J. Reine Angew. Math.*, 212:7–25, 1963. Cited on page(s) 49.

Birkhoff, Garret. On the structure of abstract algebras. *Proceedings of the Cambridge Philosophical Society*, 31:433–454, 1935. Cited on page(s) 133.

Blachmann, N. *Mathematica: A Practical Approach*. Prentice Hall, Englewood Cliffs, N.J. 07632, 1991. Cited on page(s) 491.

Blahut, Richard E. *Fast Algorithms for Digital Signal Processing*. Addison-Wesley, 1987. Cited on page(s) 228, 490.

Blelloch, Guy E., Gibbons, Phillip B., Matias, Yossi, and Narlikar, Girija J. Space efficient scheduling of Parallelism with Synchronization Variables. In *Proceedings of the 9th Symposium on Parallel Algorithms and Architectures*. ACM Press, June 1997. Cited on page(s) 148.

Blömer, J. Computing sums of radicals in polynomial time. In *Proc. 32nd IEEE Symposium on Foundations of Computer Science*, pages 670–677. IEEE, 1991. Cited on page(s) 18.

Blömer, J. How to denest ramanujan's nested radicals. In *Proc. 33rd IEEE Symposium on Foundations of Computer Science*, pages 447–456. IEEE, 1992. Cited on page(s) 18.

Blömer, J. A probabilistic zero test for expressions involving roots of rational numbers. In *Sixth European Symposium on Algorithms*, pages 151–162, 1998. Cited on page(s) 18.

Blum, L., Cucker, F., Shub, M., and Smale, S. *Complexity and Real Computation*. Springer, 1998. Cited on page(s) 126, 128, 490.

Bluman, G. W. and Kumei, S. *Symmetries and Differential Equations*. Applied Mathematical Sciences 81. Springer-Verlag, New York, 1989. Cited on page(s) 98, 99, 490.

Blumofe, Robert D. and Leiserson, Charles E. Space-efficient scheduling of multithreaded computations. *SIAM Journal on Computing*, 27(1):202–229, 1998. Cited on page(s) 147, 148.

Bocharov, A. V. Will DELiA grow into an expert system? In Miola [1990b], pages 266–267. ISBN 3-540-52531-9, 0-387-52531-9. Cited on page(s) 141.

Bochnak, J., Coste, M., and Roy, M.-F. *Géométrie Algébrique Réelle*, volume 12 of *Ergebnisse der Mathematik und ihrer Grenzgebiete, 3. Folge*. Springer-Verlag, Berlin-Heidelberg-New York, 1987. ISBN 3-540-16951-2, 0-387-16951-2. Cited on page(s) 56, 490.

Boehm, H. and Weiser, M. Garbage collection in an uncooperative environment. *Software Practice & Experience*, pages 807–820, September 1988. Cited on page(s) 404.

Böge, W., Gebauer, R., and Kredel, H. Some examples for solving systems of algebraic equations by calculating Gröbner bases. *Journal of Symbolic Computation*, 2(1):83–98, 1986. Cited on page(s) 421, 425.

Boggess, A. et al. *CalcLabs with Maple V*. Brooks/Cole, Pacific Grove, CA, 1995. Cited on page(s) 257, 491.

Böing, H. and Koepf, W. Reduce package for the indefinite and definite summation of $q$-hypergeometric terms. Technical Report 97-04, Konrad-Zuse-Zentrum Berlin (ZIB), 1997. Cited on page(s) 93.

Böing, H. and Koepf, W. Algorithms for $q$-hypergeometric summation in computer algebra. *Journal of Symbolic Computation*, 28:777–199, 1999. Cited on page(s) 93.

Bonini, C., Nischke, K.-P., and Traverso, C. Computing gröbner bases numerically: some experiments. Manuscript, 1998. Cited on page(s) 242.

Bonorden, Olaf, von zur Gathen, Joachim, Gerhard, Jürgen, Müller, Olaf, and Nöcker, Michael. Factoring a binary polynomial of degree over one million. *ACM SIGSAM Bulletin*, 35(1), March 2001. Cited on page(s) 25.

Book, R. V., editor. *Rewriting Techniques and Applications, Como, Proceedings*, volume 488 of *Lecture Notes in Computer Science*, Berlin-Heidelberg-New York, 1991. Springer-Verlag. Cited on page(s) 487.

Boos, E. E. and et al. SNUTP 94-116 (1994); (hep-ph/9503280), 1994. Cited on page(s) 166.

Borodin, A. and Munroe, I. *The Computational Complexity of Algebraic and Numeric Problems*. Elsevier Science Publishers, New York, 1975. Cited on page(s) 490.

Borst, W. N., Goldman, V. V., and van Hulzen J. A. Gentran90: A Reduce Package for the Generation of Fortran-90 Code. In Giesbrecht [1994b], pages 45–51. Cited on page(s) 151.

Bose, N. K., editor. *Multidimensional Systems Theory*. K. Reidel Publishing Company, Dordrecht, 1985. Cited on page(s) 492.

Bosma, W., Cannon, J., and Playoust, C. The Magma algebra system I: The user language. *J. Symb. Comp.*, 24(3/4):235–265, 1997. Cited on page(s) 46, 84, 88, 296, 396, 475.

Bosma, W. and Cannon, J. J., editors. *Handbook of Magma Functions*. Sydney, 1995–2000. 2000 pages. Cited on page(s) 296.

Bosma, W. and Cannon, J. J., editors. *Solving Problems with Magma*. Sydney, 2000. 190 pages. Cited on page(s) 296.

Bosma, W., Cannon, J. J., and Matthews, G. Programming with algebraic structures: design of the Magma language. In *Proceedings of the 1994 International Symposium on Symbolic and Algebraic Computation*, pages 52–57. Association for Computing Machinery, 1994. Cited on page(s) 296.

Boston, N. A use of computers to teach group theory and introduce students to research. *J. Symbolic Computation*, 23:453–458, 1997. Cited on page(s) 253.

Boston, N., Dabrowski, W., Foguel, T., Gies, P., Jackson, D., Leavitt, J., and Ose, D. The proportion of fixed-point-free elements of a transitive permutation group. *Comm. in Alg.*, 21(9):3259–3275, 1993. Cited on page(s) 253.

Boulanger, J.-L. Object oriented method for Axiom. *ACM SIGPLAN Notices*, 30(2):33–41, February 1995. Cited on page(s) 264.

Boulier, F., Lazard, D., Ollivier, F., and Petitot, M. Representation for the radical of a finitely generated differential ideal. In Levelt, A. H. M., editor, *Proc. ISSAC '95*, pages 158–166. ACM Press, New York, 1995. Cited on page(s) 103, 105.

Boulier, F., Lazard, D., Ollivier, F., and Petitot, M. Computing representations for radicals of finitely generated differential ideals. Submitted to Journal of Symbolic Computation, 1997. Cited on page(s) 105.

Bourbaki, N. *Algèbre I: Chapitres 1 à 3 (Éléments de Mathématique)*. Paris: Hermann, 1970. Nouvelle Édition. Cited on page(s) 296.

Boute, R. The euclidean definitions of the functions div and mod. *ACM Transactions on Programming Languages and Systems*, 14:127–144, 1992. Cited on page(s) 13.

Bouziane, D., Kandi-Rody, A., and Maârouf, H. Computing representations for radicals of finitely generated differential ideals. Submitted to Journal of Symbolic Computation, 1998. Cited on page(s) 105.

Brackx, F. and Constales, D. *Computer Algebra with Lisp and REDUCE*, volume 72 of *Mathematics and Its Applications*. Kluwer Academic Publishers, Dordrecht, Boston, London, 1991. Cited on page(s) 491.

Braden, B., Krug, D. K., McCartney, Ph. W., and Wilkinson, St. *Discovering Calculus with Mathematica*. John Wiley and Sons, New York, 1992. Cited on page(s) 257, 491.

Brandt, S. R. and Seidel, E. The evolution of distorted rotating black holes I: Methods and tests. *Phys.Rev. D*, 52:856–69, 1995. Also available at: http://xxx.lanl.gov/abs/gr-qc/9412072. Cited on page(s) 176.

Brans, C. H. Computer algebra and general relativity. In Fleischer, J., Grabmeier, J., Hehl, F. W., and Küchlin, W., editors, *Computer Algebra in Science and Engineering*, pages 183–195. World Scientific Publishing, 1995. Cited on page(s) 172.

Braun, R. and Meise, R. *Analysis mit MAPLE*. Vieweg, Wiesbaden, 1995. Cited on page(s) 257, 491.

Braun, Stefan and Hauser, Harald. *Macsyma Version 2 - Systematische und praxisnahe Einfuhrung mit Anwendungsbeispielen (Macsyma 2.0 - A Systematic and Practical Introduction with Applications)*. Addison-Wesley (Deutschland), 1995. ISBN 3-89319-751-6. Cited on page(s) 283, 491.

Bray, Tim, Paoli, Jean, and Sperberg-McQueen, C. M. Extensible Markup Language 1.0. Technical report, Worldwide Web Consortium, 1998. Available at http://www.w3.org/TR/1998/REC-xml-19980210. Cited on page(s) 150.

Bremner, M. On the Z-module structure of a free semialternative ring. *Communications in Algebra*, 27:1951–1965, 1999. Cited on page(s) 348.

Brent, R. ECM champs. available via ftp://ftp.comlab.ox.ac.uk/pub/Documents/techpapers/Richard.Brent/champs.txt, October 2001. Cited on page(s) 44.

Brent, R. P. Brent's multiple precision, ACM algorithm 524. netlib@ornl.gov, 1981. Cited on page(s) 13.

Bressoud, D. M. *Factorization and Primality Testing.* Undergraduate Texts in Mathematics. Springer-Verlag, Berlin-Heidelberg-New York, 1989. Cited on page(s) 490.

Breuer, Thomas and Linton, Steve. The GAP 4 type system – Organising algebraic algorithms. In Gloor [1998], pages 38–45. ISBN 1-58113-002-3. Cited on page(s) 390.

Broadbery, P. A., Gómez-Díaz, T., and Watt, S. M. On the implementation of dynamic evaluation. In Levelt [1995], pages 77–84. Cited on page(s) 264.

Broadhurst, D. J. *Z. Phys.*, C54:599–606, 1992. Cited on page(s) 168.

Bronstein, Manuel. Integration of elementary functions. *Journal of Symbolic Computation*, 9(2):117–173, 1990. Cited on page(s) 95.

Bronstein, Manuel. The Risch differential equation on an algebraic curve. In Watt [1991], pages 241–246. ISBN 0-89791-437-6. Cited on page(s) 264.

Bronstein, Manuel. Integration and differential equations in computer algebra. *Programmirovanie*, 18(5), 1992a. Cited on page(s) 264.

Bronstein, Manuel. Linear ordinary differential equations: Breaking through the order 2 barrier. In Wang [1992], pages 42–48. Cited on page(s) 264.

Bronstein, Manuel, editor. *Proceedings of the 1993 International Symposium on Symbolic and Algebraic Computation*, volume 8, Kiev, Ukraine, July 1993. ACM Press. Cited on page(s) 487.

Bronstein, Manuel. An improved algorithm for factoring linear ordinary differential operators. In Giesbrecht [1994b], pages 336–347. Cited on page(s) 264.

Bronstein, M. $\Sigma^{IT}$: A strongly-typed embeddable computer algebra library. In Calmet, J. and Limongelli, C., editors, *Proc. DISCO '96*, Lecture Notes in Computer Science 1128, pages 22–33, Karlsruhe, 1996. Springer. Cited on page(s) 98, 331.

Bronstein, Manuel. *Symbolic Integration I – Transcendental Functions*, volume 1 of *Algorithms and Computation in Mathematics*. Springer, Heidelberg, Berlin-Heidelberg-New York, 1997. Cited on page(s) 4, 95, 96.

Bronstein, M., Davenport, J. H., and Trager, B. M. Symbolic integration is algorithmic! Tutorial Notes, Conference on Computers and Mathematics, MIT, June 1989. Cited on page(s) 264.

Bronstein, M., Grabmeier, J., and Weispfenning, V., editors. *Symbolic Rewriting Techniques*, volume 15 of *Progress in Computer Science and Applied Logic*, Basel, 1998. Birkhäuser Verlag. Cited on page(s) 137, 488, 561, 578, 590, 618.

Bronstein, M. and Petkovsek, M. An introduction to pseudo-linear algebra. *Theor. Comp. Sci.*, 157:3–33, 1996. Cited on page(s) 97.

Broughan, K. A., Dewar, M. C., Keady, G., Robb, T., and Richardon, M. G. Some Symbolic Computing Links to the NAG Numeric Library. *ACM SIGSAM Bulletin*, 1991. Cited on page(s) 152.

Brown, H., Bülow, R., Neubüser, J., Wondratschek, H., and Zassenhaus, H. *Cristallographic Groups of Fourdimensional Space*. John Wiley&Sons, New York, 1978. Cited on page(s) 243, 389, 490.

Brown, R. The twisted Eilenberg-Zilber theorem. In *Simposio di Topologia (Messina, 1964)*, pages 33–37. Edizioni Oderisi, Gubbio, 1965. Cited on page(s) 210.

Brown, R. and Tonks, A. Calculations with simplical and cubical groups in AXIOM. *Journal of Symbolic Computation*, 17(2):159–180, February 1994. Cited on page(s) 264.

Brücher, L. and et al. *Nucl.Instrum. Meth*, A389:323, 1997. Cited on page(s) 166.

Brücher, L., Franzkowski, J., and Kreimer, D. Xloops: Automated Feynman diagram calculation. *Comput. Phys. Commun.*, 115:140, 1998. Cited on page(s) 170.

Brückner, H. *Algorithmen für endliche auflösbare Gruppen und Anwendungen*, volume 22 of *Aachener Beiträge zur Mathematik*. Verlag der Augustinus Buchhandlung, Aachen, 1998. Dissertation. Cited on page(s) 85, 460.

Brumer, A. The rank of $J_0(N)$. *Asterisque*, 228:41–68, 1992. Cited on page(s) 381.

Bryant, Robert L., Chern, S. S., Gardner, Robert B., schmidt, Hubert L. Gold, and Griffiths, P. A. *Exterior differential systems*. Number 18 in Publications, Mathematical Sciences Research Institute. Springer-Verlag, New York, 1991. ISBN 0-387-97411-3. Cited on page(s) 207.

Bubeck, T., Hiller, M., Küchlin, W., and Rosenstiel, W. Distributed symbolic computation with DTS. In Ferreira, Afonso and Rolim, José, editors, *Parallel Algorithms for Irregularly Structured Problems, 2nd Intl. Workshop, IRREGULAR'95*, volume 980 of *LNCS*, pages 231–248, Lyon, France, September 1995. Cited on page(s) 436.

Buchberger, Collins, Encarnación, Hong, Johnson, Krandick, Loos, Mandache, Neubacher, and Vielhaber. SACLIB User's Guide, 1993. On-line software documentation. Cited on page(s) 352, 439.

Buchberger, Bruno. *Ein Algorithmus zum Auffinden der Basiselemente des Restklassenringes nach einem nulldimensionalen Polynomideal*. Doctoral dissertation, Mathematical Institute, University of Innsbruck, Innsbruck, Austria, 1965. Cited on page(s) 28, 135, 205, 379.

Buchberger, B. Ein algorithmisches Kriterium für die Lösbarkeit eines algebraischen Gleichungssystems. *Aequ. Math.*, 4:374–383, 1970. Cited on page(s) 28.

Buchberger, B., editor. *EUROCAL '85, European Conference on Computer Algebra, Linz, Austria, April 1–3, 1985, Proceedings Vol. 1: Invited Lectures*, volume 203 of *Lecture Notes in Computer Science*, Berlin-Heidelberg-New York, 1985a. Springer-Verlag. ISBN 3-540-15983-5, 0-387-15983-5. Cited on page(s) 485.

Buchberger, B. Gröbner bases: An algorithmic method in polynomial ideal theory. In Bose, N. K., editor, *Multidimensional Systems Theory*, pages 184–232. D. Reidel, Dordrecht, 1985b. Cited on page(s) 28, 229, 379, 428.

Buchberger, B. Should students learn integration rules? *RISC-Linz Series*, 89-07.0, 1989. Cited on page(s) 251.

Buchberger, B. Using Mathematica for doing simple mathematical proofs. In *Proceedings of the 4th Mathematica Users' Conference, Tokyo*, pages 80–96. Wolfram Media Publishing, 1996. Invited paper. Cited on page(s) 453.

Buchberger, B., Collins, G. E., and Loos, R., editors. *Computer Algebra, Symbolic and Algebraic Computation*. Springer-Verlag, Wien - New York, first edition, 1982. Cited on page(s) 261, 428, 492.

Buchberger, B., Collins, G. E., and Loos, R., editors. *Computer Algebra, Symbolic and Algebraic Computation*. Springer-Verlag, Wien - New York, second edition, 1983. Cited on page(s) 558, 569, 576.

Buchberger, Bruno, Ida, Tetsuo, and Vasaru, Daniela, editors. *First International Theorema Workshop*, 1997a. RISC-Linz. Cited on page(s) 488.

Buchberger, B., Jebelean, T., Kriftner, F., Marin, M., Tomuta, E., and Vasaru, D. A survey of the *Theorema* project. In Kuechlin, W., editor, *Proceedings of ISSAC'97*, pages 384–391. ACM Press, 1997b. Cited on page(s) 453.

Buchberger, B., Jebelean, T., and Vasaru, D. Theorema: A system for formal scientific training in natural language presentation. In *ED-MEDIA '98*. AACE Press, 1998. Cited on page(s) 453.

Buchberger, B., Kutzler, B., and Feilmeier, M. *Rechnerorientierte Verfahren*. B.G. Teubner, Stuttgart, 1986. Cited on page(s) 490.

Buchberger, Bruno and Winkler, Franz, editors. *Gröbner Bases and Applications*, volume 251 of *Lecture Note Series*, 1998. London Mathematical Society, Cambridge University Press. Cited on page(s) 488, 578, 612.

Buchmann, J., Jacobson, M. J., Neis, S., Theobald, P., and Weber, D. *Sieving Methods for Class Group Computation*, pages 3–10. Springer Verlag, 1999. in B.H. Matzat, G.-M. Greuel, G. Hiss (Eds.): Algorithmic alegera and number theory. Cited on page(s) 406.

Buchmann, J., Loho, J., and Zayer, J. An implementation of the general number field sieve. In *Proceedings of Crypto 93*, number 773 in LNCS, pages 159–165, 1994. Cited on page(s) 44.

Buchmann, J. and Neis, S. Algorithms for linear algebra problems over principal ideal rings. Technical report, Technische Hochschule Darmstadt, 1996. Cited on page(s) 39.

Bücken, Hans. Reduktionssysteme und Wortproblem. Technical Report 3, RWTH Aachen, 1979. Cited on page(s) 134.

Buckheit, J., Chen, S., Donoho, D. L., and Johnstone, I. About WaveLab, 1996. ftp://playfair.stanford.edu/~wavelab. Cited on page(s) 229.

Buell, D. A. *Binary Quadratic Forms Classical Theory and Modern Computations*. Springer-Verlag, Berlin-Heidelberg-New York, 1989. Cited on page(s) 490.

Buhler, Joe, Amin Shokrollahi, M., and Stemann, Volker. Fast and Precise Computations of Discrete Fourier Transforms using Cyclotomic Integers. *STOC'97, El Paso, Texas, USA*, pages 40–47, 1997. Cited on page(s) 228.

Buium, A. *Differential Algebra and Diophantine Geometry*. Actualités mathématiques. Hermann, Paris, 1994. Cited on page(s) 104.

Bündgen, Reinhard. Reduce the redex → ReDuX. In Kirchner, Claude, editor, *Rewriting Techniques and Applications (LNCS 690)*, pages 446–450. Springer-Verlag, 1993. (Proc. RTA'93, Montreal, Canada, June 1993). Cited on page(s) 438.

Bündgen, Reinhard. Buchberger's algorithm: the term rewriter's point of view. *Theoretical Computer Science*, 159(2):143–190, 1996. Cited on page(s) 135.

Bündgen, Reinhard. Symmetrization based completion. In Bronstein, M., Grabmeier, J., and Weispfenning, V., editors, *Symbolic Rewriting Techniques*, pages 47–70. Birkhäuser Verlag, 1998a. Cited on page(s) 134.

Bündgen, Reinhard. *Termersetzungssysteme, Theorie, Implementierung, Anwendung*. Vieweg-Verlag, 1998b. Cited on page(s) 136, 490.

Bündgen, R., Göbel, M., and Küchlin, W. A master-slave approach to parallel term rewriting on a hierarchical multiprocessor. In Calmet and Limoncelli [1996], pages 183–194. Cited on page(s) 436.

Bündgen, Reinhard, Göbel, Manfred, and Küchlin, Wolfgang. Strategy compliant multi-threaded term completion. *Journal of Symbolic Computation*, 21(4–6):475–505, 1996b. Cited on page(s) 136, 436.

Bündgen, Reinhard, Sinz, Carsten, and Walter, Jochen. ReDuX 1.5: New facets of rewriting. In Ganzinger, Harald, editor, *Rewriting Techniques and Applications (LNCS 1103)*, pages 412–415. Springer-Verlag, 1996c. (Proc. RTA'96, New Brunswick, NJ, USA, July 1996). Cited on page(s) 438.

Burge, W. and Watt, S. Infinite structures in SCRATCHPAD II. Technical report, IBM Thomas J. Watson Research Center, Bos 218, Yorktown Heights, NY 10598, USA, 1987. Cited on page(s) 16.

Burge, William H. Scratchpad and the Rogers-Ramanujan identities. In Watt [1991], pages 189–190. ISBN 0-89791-437-6. Cited on page(s) 264.

Bürgi, U. Charged pion pair production and polarizabilities to two loops. *Nucl. Phys.*, B479:392, 1996. Cited on page(s) 170.

Bürgisser, P. *Completeness and Reduction in Algebraic Complexity Theory*. Number 7 in Algorithms and Computation in Mathematics. Springer-Verlag, 2000. Cited on page(s) 126, 128, 490.

Bürgisser, P., Clausen, M., and Shokrollahi, M. A. *Algebraic Complexity Theory*, volume 315 of *Grundlehren der mathematischen Wissenschaften*. Springer Verlag, 1996. Cited on page(s) 126, 128, 129, 490.

Burhenne, Klaus-Dieter. Implementierung eines Algorithmus zur Quantorenelimination für lineare reelle Probleme. Diploma thesis, Universität Passau, D-94030 Passau, Germany, December 1990. Cited on page(s) 139.

Burrus, C. S. Notes on the FFT. *Rice University, Houston, Texas*, 1995. Cited on page(s) 228.

Butler, Gregory. *Fundamental Algorithms for Permutation Groups*, volume 559 of *Lecture Notes in Computer Science*. Springer-Verlag, Berlin-Heidelberg-New York, 1991. Cited on page(s) 4, 71, 490.

Butler, G. and Cannon, J. J. Cayley version 4: The user language. In Gianni, P., editor, *Proceedings of the 1988 International Symposium on Symbolic and Algebraic Computation*, number 358 in Lecture Notes in Computer Science, pages 456–466, New York, 1989. Springer-Verlag. Cited on page(s) 295.

Butler, G. and Cannon, J. J. The design of Cayley, a language for modern algebra. In Miola, A., editor, *Design and Implementation of Symbolic Computation System*, number 429 in Lecture Notes in Computer Science, pages 10–19. Springer-Verlag, 1990. Cited on page(s) 295.

Bykov, V., Kytmanov, A., Lazman, M., and Passare, M. *Elimination methods in Polynomial Computer Algebra.* Kluwer, Dordrecht, 1998. Cited on page(s) 490.

Caboara, M., Conti, P., and Traverso, C. Yet another algorithm for ideal decomposition. In *Proceedings of AAECC-12*, number 1255 in Lecture Notes in Computer Science, pages 39–54. Springer Verlag, 1997. Cited on page(s) 364.

Caboara, M. and Traverso, C. Efficient algorithms for ideal operations. In *ISSAC 98: Proceedings of the 1998 International Symposium on Symbolic and Algebraic Computation*, pages 147–152. ACM Press, 1998. Extended Abstract. Cited on page(s) 364.

Cabral, H. On the integral manifolds of the N-body problem. *Inventiones mathematicae*, 20:59–72, 1973. Cited on page(s) 177.

Cade, J. J. A public key cipher which allows signatures. In *Proc. 2nd Conf. on Appl. Linear Algebra*. SIAM, 1985. Cited on page(s) 27.

CalculusLive. *Calculus Live.* John Wiley and Sons, New York, 1999. Cited on page(s) 258.

Calderbank, A. R., Rains, E. M., Shor, P. W., and Sloane, N. J. A. Quantum Error Correction and Orthogonal Geometry. *Physical Review Letters*, 78(3):405–208, 20. January 1997a. LANL preprint quant–ph/9605005. Cited on page(s) 130.

Calderbank, A. R., Rains, E. M., Shor, P. W., and Sloane, N. J. A. Quantum Error Correction via Codes over $GF(4)$. In *Proceedings ISIT 97*, page 292, 1997b. Cited on page(s) 130.

Calderbank, A. R. and Shor, Peter W. Good quantum error–correcting codes exist. *Physical Review A*, 54(2):1098–1105, August 1996. Cited on page(s) 130.

Callaham, T. K. and Knobloch, E. Symmetry-breaking bifurcations on cubic lattices. *Nonlinearity*, 10:1179–1216, 1997. Cited on page(s) 107, 215.

Calmet, J., editor. *Computer Algebra, EUROCAM'82, European Computer Algebra Conference, Marseille, France, 5–7 April 1982*, volume 144 of *Lecture Notes in Computer Science*, Berlin-Heidelberg-New York, 1982. Springer-Verlag. ISBN 3-540-11607-9, 0-387-11607-9. Cited on page(s) 485, 524, 582.

Calmet, J., editor. *Algebraic Algorithms and Error-Correcting Codes, 3rd International Conference, AAECC-3, Grenoble, France, July 15–19, 1985, Proceedings*, volume 229 of *Lecture Notes in Computer Science*, Berlin-Heidelberg-New York, 1986. Springer-Verlag. ISBN 3-540-16776-5, 0-387-16776-5. Cited on page(s) 485, 582.

Calmet, J. and Campbell, J. A., editors. *Proceedings of the 1st Conference on Artificial Intelligence and Symbolic Mathematical Computing, Karlsruhe, August 1992*, volume 737 of *Lecture Notes in Computer Science*, Berlin-Heidelberg-New York, 1993. Springer-Verlag. Cited on page(s) 142.

Calmet, J. and Campbell, J. A., editors. *Proceedings of the 2nd Conference on Artificial Intelligence and Symbolic Mathematical Computing, Cambridge, August 1994*, volume 958 of *Lecture Notes in Computer Science*, Berlin-Heidelberg-New York, 1994. Springer-Verlag. Cited on page(s) 142.

Calmet, J. and Campbell, J. A., editors. *Special issue of the Annals of Mathematics and Artificial Intelligence*, volume 19, Amsterdam, 1997. Balzer Pub. Cited on page(s) 142.

Calmet, J., Campbell, J. A., and J., Pfalzgraf, editors. *Proceedings of the 3rd Conference on Artificial Intelligence and Symbolic Mathematical Computing, Steyr, September 1996*, volume 1138 of *Lecture Notes in Computer Science*, Berlin-Heidelberg-New York, 1996. Springer-Verlag. Cited on page(s) 142.

Calmet, J. and Limoncelli, J., editors. *Proceedings of DISCO '96*, Heidelburg, 1996. Springer Verlag. Cited on page(s) 495, 512.

Caniglia, L., Galligo, A., and Heintz, J. Some new effectivity bounds in computational geometry. In Mora [1989a], pages 131–151. ISBN 3-540-51083-4, 0-387-51083-4. Cited on page(s) 54.

Cannon, John, editor. *Computational Group Theory I*, volume 9/5&6 of *Journal of Symbolic Computation*. Academic Press, London, New York, 1990. Cited on page(s) 66, 71, 492.

Cannon, John, editor. *Computational Group Theory II*, volume 12/5&5 of *Journal of Symbolic Computation*. Academic Press, London, New York, 1991. Cited on page(s) 66, 71, 492.

Cannon, J. J. The basis of a computer system for modern algebra. In *Proceedings of the 1981 ACM Symposium on Symbolic and Algebraic Computation, (SYMSAC '81)*, pages 1–5, New York, 1981. Cited on page(s) 295.

Cannon, J. J. A Language for Group Theory, 1982. Sydney, 340 pages. Cited on page(s) 295.

Cannon, J. J. An introduction to the group theory language Cayley. In *Computational Group Theory*, pages 145–183. Academic Press, London, 1984. Cited on page(s) 295.

Cannon, J. J. Overview of Magma V2.7 features.
http://www.maths.usyd.edu.au:8000/u/magma/, 2000. Sydney, 50 pages. Cited on page(s) 298.

Cannon, J. J. and Playoust, C. An introduction to algebraic programming in Magma. Sydney: School of Mathematics and Statistics, University of Sydney, 1996. Also see: http://www.maths.usyd.edu.au:8000/u/magma/. Cited on page(s) 211.

Cannon, J. J. and Playoust, C. An introduction to algebraic programming with Magma: The categories, 2001a. (Springer, to appear), 500 pages. Cited on page(s) 296.

Cannon, J. J. and Playoust, C. An introduction to algebraic programming with Magma: The language, 2001b. (Springer, to appear), 350 pages. Cited on page(s) 296.

Canny, J. *The Complexity of Motion Planning*. MIT Press, Cambridge, Mass., 1987. Cited on page(s) 234, 490.

Canny, J. and Manocha, D. The implicit representation of rational parametric surfaces. *Journal of Symbolic Computation*, 13:485–510, 1992. Cited on page(s) 235, 236.

Cantor, D. G. and Zassenhaus, H. A new algorithm for factoring polynomials over finite fields. *Mathematics of Computation*, 36:587–592, 1981. Cited on page(s) 25.

Capani, A., DeDominicis, G., Niesi, G., and Robbiano, L. Computing minimal finite free resolutions. *Journal for Pure and Applied Algebra*, 117–118:105–117, 1997. Special volume: Algorithms for Algebra. Cited on page(s) 364, 367.

Capani, A., Niesi, G., and Robbiano, L. *CoCoA, a system for doing Computations in Commutative Algebra*. Available via anonymous ftp from cocoa.dima.unige.it, 3.7 edition, 1999. Cited on page(s) 210.

Caprasse, H., Demaret, J., and Schrüfer, E. Can EXCALC be used to investigate high-dimensional cosmological models with nonlinear lagrangians? In Gianni [1989], pages 116–124. ISBN 3-540-51084-2, 0-387-51084-2. Cited on page(s) 180.

Caprotti, Olga, Carlisle, David P., and Cohen, Arjeh M., editors. *The OpenMath Standard Version 1.0*. The OpenMath Esprit Consortium, February 2000. http://www.openmath.org/standard/omstd.pdf. Cited on page(s) 158.

Cariello, F. and Tabor, M. Painlevé expansions for nonintegrable evolution equations. *Physica D*, 39:77–94, 1989. Cited on page(s) 101.

Carlson, J. F. Calculating group cohomology: Tests for completion, 1996. Preprint. Cited on page(s) 197.

Carlson, J. F., Green, E. L., and Schneider, G. J. A. Computing Ext algebras for finite groups. *J. Symb. Comput.*, 24:317–325, 1997. Cited on page(s) 64, 87.

Carrá-Ferro, Giuseppa and Gallo, Giovanni. A procedure to prove geometrical statements. Technical report, Dip. Mathematica Univ. Catania, Italy, 1987. Cited on page(s) 203.

Cartan, E. *Les Systèmes Différentielles Extérieurs et leurs Applications Géométriques*. Hermann, Paris, 1945. Cited on page(s) 102.

Cartan, Henri and Eilenberg, Samuel. *Homological algebra*. Princeton University Press, Princeton, N. J., 1956. Cited on page(s) 207, 208, 209.

Caruso, F. A Macsyma implementation of Zeilberger's fast algorithm. Technical Report 99-16, SFB F103, University of Linz, Austria, 1999. Cited on page(s) 93.

Casperson, D., Ford, D., and MacKay, J. An ideal decomposition algorithm. *J. Symbolic Computation*, 21(2):133–137, 1996. Cited on page(s) 27.

Casperson, D. and McKay, J. Symmetric functions, m-sets and Galois groups. *Math.Comput.*, 63:749–757, 1994. Cited on page(s) 47.

Cassels, J. W. S. and Flynn, E. V. *Prolegomena to a Middlebrow Arithmetic of Curves of Genus 2*, volume 230 of *Lect. Notes of the London Math. Soc.* Cambridge Univ. Press, Cambridge, 1996. Cited on page(s) 50.

Castellví P., Jaén, X., and LLanta, E. TTC: Symbolic tensor and exterior calculus. *Comput. Phys.*, 8,3, 1994. Cited on page(s) 480.

Castellví P., Jaén, X., and LLanta, E. Symbolic tensor calculus using index notation. *Comput. Phys.*, 9,3, 1995. Cited on page(s) 480.

Castro, D., Giusti, M., Heintz, J., Matera, G., and Pardo, L. M. Data structures and smooth interpolation procedures in elimination theory. Manuscript École Polytechnique, 1999. Cited on page(s) 54.

Castro, D., Hägele, K., Morais, J. E., and Pardo, L. M. Kronecker's and Newton's approaches to solving : A first comparison. *J. of Complexity*, 17(1):212–203, 2001. Cited on page(s) 15.

Cattaneo, G. *Atti. Sem. Mat. Fis. Univ. Modena*, 3:83, 1948. Cited on page(s) 192.

Caviness, B. and Johnson, J., editors. *Quantifier Elimination and Cylindrical Algebraic Decomposition*. Texts and Monographs in Symbolic Computation. Springer, Berlin, Heidelberg, New York, 1998. Cited on page(s) 57, 487, 490, 492, 618.

Caviness, B. F., Gilbert, Robert P., and Shtokhamer, Roman. An introduction to applied symbolic computation using Macsyma. University of Delaware, 1989. Cited on page(s) 491.

Celler, Frank and Leedham-Green, C. R. Calculating the order of an invertible matrix. In *Groups and Computation II*, volume 28 of *Amer. Math. Soc. DIMACS Series*. (DIMACS, 1995), 1997a. Cited on page(s) 475.

Celler, Frank and Leedham-Green, C. R. A non-constructive recognition algorithm for the special linear and other classical groups. In Finkelstein, L. and Kantor, W. M., editors, *Groups and Computation II*, volume 28 of *Amer. Math. Soc. DIMACS Series*, pages 61–67. (DIMACS, 1995), 1997b. Cited on page(s) 475.

Celler, F. and Leedham-Green, C. R. A constructive recognition algorithm for the special linear group. In *The atlas of finite groups: ten years on (Birmingham, 1995)*, pages 11–26. Cambridge Univ. Press, Cambridge, 1998. Cited on page(s) 475.

Celler, Frank, Leedham-Green, Charles R. Murray, Scott H. Niemeyer, Alice C. and O'Brien, E. A. Generating random elements of a finite group. *Comm. Algebra*, 23:4931–4948, 1995. Cited on page(s) 475.

Cellini, P., Gianni, P., and Traverso, C. Algorithms for the shape of semialgebraic sets: a new approach. In Mattson et al. [1991], pages 1–18. ISBN 3-540-54522-0, 0-387-54522-0. Cited on page(s) 238, 239.

Çetin, A. Enis, Gerek, Ömer N., and Yardimci, Yasemin. Equiripple FIR Filter Design by the FFT Algorithm. *IEEE Signal Processing Magazine*, pages 60–64, 1997. Cited on page(s) 228.

Char, B., Johnson, J., Saunders, D., and Wack, A. P. Some experiments with parallel Bignum arithmetic. In *First International Symposium on Parallel Symbolic Computation (PASCO'94)*, Lecture Notes Series in Computing – Vol. 5, pages 94–103. World Scientific Publishing, 1994. Cited on page(s) 147, 148.

Char, B. W., editor. *1986 ACM Symposium on Symbolic and Algebraic Computation, University of Waterloo, Ontario*, New York, 1986. Academic Press. Cited on page(s) 485.

Char, B. W., Geddes, K. O., Gonnet, G. H., Leong, B., Monagan, M. B., and Watt, S. M. *Maple V Language Reference Manual*. Springer-Verlag, New York, first edition, 1991a. Cited on page(s) 3, 491.

Char, B. W., Geddes, K. O., Gonnet, G. H., Leong, B., Monagan, M. B., and Watt, S. M. *Maple V Library Reference Manual*. Springer-Verlag, New York, 1991b. Cited on page(s) 491.

Char, B. W., Geddes, K. O., Gonnet, G. H., Leong, B., Monagan, M. B., and Watt, S. M. *First Leaves: Tutorial Introduction to Maple*. Springer-Verlag, New York, 1992. Cited on page(s) 359, 491.

Chen, L., Eberly, W., Kaltofen, E., Saunders, B. D., Turner, W. J., and Villard, G. Efficient matrix preconditioners for black box linear algebra. *Linear Algebra and Appl.*, page to appear, 2001. Special issue on *Infinite Systems of Linear Equations Finitely Specified.* Cited on page(s) 38.

Chenadec, Philippe Le. *Canonical Forms in Finitely Presented Algebras.* Pitman, London, 1986. Cited on page(s) 134.

Cherlin, Gregory and Francoise, Point. On extensions of Presburger arithmetic. In *Proceedings Fourth Easter Conference on Model Theory,Berlin*, pages 17–34, 1986. Cited on page(s) 138.

Cherry, G. W. Integration in finite terms with special functions: the error function. *Journal of Symbolic Computation*, 1(3):283–302, 1985. Cited on page(s) 95.

Cherry, G. W. Integration in finite terms with special functions: the logarithmic integral. *SIAM Jour. Comp.*, 15:1–21, 1986. Cited on page(s) 95.

Chetyrkin, K. G. *Teor. Math. Phys.*, 75:26, 1988. Cited on page(s) 166.

Chetyrkin, K. G. and Tkachov, F. V. *Nucl. Phys.*, B192:159–204, 1981. Cited on page(s) 168.

Chin, Paulina, Corless, Robert M., and Corliss, George F. Optimization strategies for the approximate GCD problem. In Gloor [1998], pages 228–235. ISBN 1-58113-002-3. Cited on page(s) 118.

Chionh, E. W. *Base Points, Resultants, and the Implicit Representation of Rational Surfaces.* Dissertation, University of Waterloo, Waterloo, Canada, 1990. Cited on page(s) 235.

Cho, Peter and Leibovich, Adam K. Color octet quarkonia production. *Phys. Rev.*, D53:150–162, 1996a. Cited on page(s) 472.

Cho, Peter and Leibovich, Adam K. Color octet quarkonia production. 2. *Phys. Rev.*, D53:6203–6217, 1996b. Cited on page(s) 472.

Chou, Shang-Ching. *Mechanical Geometry Theorem Proving.* K. Reidel Publishing Company, Dordrecht, 1988. Cited on page(s) 32, 33, 104, 201, 203, 205, 490.

Chou, Shang-Ching. Automatic reasoning in geometries, using the characteristic set method and Gröbner basis method. In S., Watanabe and M., Nagata, editors, *ISSAC'90*, pages 255–260. ACM Press, 1990. Cited on page(s) 32, 33.

Chou, Shang-Ching and Yiao-Shan, Gao. Methods for mechanical geometry formula deriving. In S., Watanabe and M., Nagata, editors, *ISSAC'90*, pages 265–270. ACM Press, 1990a. Cited on page(s) 32, 33.

Chou, Shang-Ching and Yiao-Shan, Gao. Ritt-Wu's decomposition algorithm and geometry theorem proving. In *Proceedings of the 10th International Conference on Automated Deduction*, volume 449 of *LNCS*, pages 207–220. Springer, 1990b. Cited on page(s) 33.

Chudnovsky, David, editor. *Computers in Mathematics*, volume 7 of *Lecture notes in pure and applied mathematics*, New York u.a., 1990. Dekker. Cited on page(s) 485.

Chudnovsky, D. V. and Jenks, R. D., editors. *Computers in Mathematics*, volume 7 of *Lecture notes in pure and applied mathematics.* Marcel Dekker, New York, 1990. Cited on page(s) 492, 558.

Chyzak, F. *Fonctions holonomes en calcul formel.* PhD thesis, Ecole Polytechnique, Palaiseau, France, 1998. Cited on page(s) 92, 221.

Chyzak, F. and Salvy, B. Non-commutative elimination in Ore algebras proves multivariate identities. *Journal of Symbolic Computation*, 26:187–227, 1998. Cited on page(s) 92.

Ciucu, M. and Krattenthaler, C. The number of centered lozenge tilings of a symmetric hexagon. *Journal of Combinatorial Theory (A)*, 86:103–126, 1999. Cited on page(s) 93.

Clarkson, Kenneth L. Safe and efficient determinant evaluation. In *Proc. 33rd Ann. Symp. Foundations Comput. Sci.*, pages 387–395, Los Alamitos, California, 1992. IEEE Computer Society Press. Cited on page(s) 114.

Clarkson, P. A., editor. *Applications of Analytic and Geometrical Methods to Nonlinear Differential Equations*, volume 413 of *NATO Adv. Sci. Inst. Ser. C.* Kluwer, Dortrecht, 1993. Cited on page(s) 101.

Clausen, Michael and Baum, Ulrich. *Fast Fourier Transforms.* Bibliographisches Institut, 1993. Cited on page(s) 228, 490.

CoFI. The Common Framework Initiative for algebraic specification and development, electronic archives. Notes and Documents accessible from http://www.brics.dk/Projects/CoFI, 2001. Contains a summary of CASL. Cited on page(s) 218, 597.

Cohen, A. M., editor. *Computer algebra in industry - Problem Solving in Practice - Proc. 1991 SCAFI Seminar*, 1993. CWI, Amsterdam, John Wiley & Sons. Cited on page(s) 226, 492.

Cohen, A. M., Cuypers, H., and Sterk, H., editors. *Some tapas of computer algebra*, volume 4 of *Advance in Computational Mathematics*. Springer-Verlag, Berlin-Heidelberg-New York, 1998. Cited on page(s) 60, 63.

Cohen, A. M. and de Man, R. Computational evidence for deligne's conjecture regarding exceptional lie groups. *Comptes Rendus Acad. Sci. Paris, t. 322, Série I*, pages 427–432, 1996. Cited on page(s) 411.

Cohen, A. M. and Heck, A. Applied computer-algebra: Experience from cam design. In Fleischer, J., Grabmeier, J., Hehl, F. W., and Küchlin, W., editors, *Computer-Algebra in Science and Engineering*, pages 303–316. World Scientific Publishers, 1995. Cited on page(s) 234.

Cohen, A. M., Ivanyos, G., and Wales, D. B. Finding the radical of an algebra of linear transformations. *Journal for Pure and Applied Algebra*, 117:177–193, 1997. Cited on page(s) 63.

Cohen, Arejh M., van Gastel, Leendert, and Lunel, Sjoerd V., editors. *Computer Algebra in Industry 2 - Problem Solving in Practice.* John Wiley and Sons, 1995a. ISBN ISBN 0-471-95529-9. Cited on page(s) 492.

Cohen, G., Giusti, M., and Mora, T., editors. *Proceedings of the 11th International Symposium on Applied Algebra, Algebraic Algorithms and Error-Correcting Codes*, volume 948 of *Lecture Notes in Computer Science*, Berlin, 1995b. Springer. Cited on page(s) 488.

Cohen, G., Mora, T., and Moreno, O., editors. *Proceedings of the 10th International Symposium on Applied Algebra, Algebraic Algorithms and Error-Correcting Codes*, volume 673 of *Lecture Notes in Computer Science*, May 1993. Springer. Cited on page(s) 487.

Cohen, H., editor. *Algorithmic Number Theory Symposium, Proceedings ANTS II, Bordeaux*, volume 1122 of *Lecture Notes in Computer Science*, Berlin-Heidelberg-New York, 1996a. Springer-Verlag. Cited on page(s) 45, 47.

Cohen, H. *A Course in Computational Algebraic Number Theory*. Number 138 in Graduate Texts in Mathematics. Springer-Verlag, Berlin-Heidelberg-New York, 1996b. Cited on page(s) 4, 16, 17, 18, 42, 43, 47, 49, 398, 432, 490.

Cohen, H. Hermite and smith normal form algorithms over Dedekind domains. *Mathematics of Computation*, 65(216):1681–1699, 1996c. Cited on page(s) 39.

Cohen, H. and Lenstra, H. W. Primality testing and jacobi sums. *Math. Comp.*, 42:297–330, 1984. Cited on page(s) 42.

Cohen, J. K. Arclength—an example of the New Calculus. *Mathematica in Education and Research*, 3(2):23–28, 1994. Cited on page(s) 255.

Cohen, P. J. Decision procedures for real and p-adic fields. *Comm. Pure Appl. Math.*, 22:131–151, 1969. Cited on page(s) 138.

Cohn, R. M. *Difference Algebra*. Interscience, New York, 1966. Cited on page(s) 104, 490.

Colbourn, Charles J. and Dinitz, Jeffrey H., editors. *The CRC handbook of combinatorial designs*. CRC Press Series on Discrete Mathematics and its Applications. CRC Press, Boca Raton, FL, 1996. ISBN 0-8493-8948-8. Cited on page(s) 375, 490.

Colin, A. Solving a System of Algebraic Equations with Symmetries. *Journal of Pure and Applied Algebra*, 117–118:195–215, 1997a. Cited on page(s) 178.

Colin, A. *Théorie des invariants effective. Applications à la théorie de Galois et à la résolution de systémes algébriques. Implantanation en AXIOM*. PhD thesis, École polytechnique, 1997b. Cited on page(s) 48.

Collart, S., M., Kalkbrener, and Mall, D. Converting bases with the Gröbner walk. *J. Symb. Computation*, 24:465–469, 1997. Cited on page(s) 30, 52.

Collins, George and Loos, Rüdiger. Real zeros of polynomials. In Buchberger, Bruno, Collins, George, and Loos, Rüdiger, editors, *Computer algebra, symbolic and algebraic computation*, pages 83–94. Springer, Wien, 1982a. Cited on page(s) 19.

Collins, George E. Quantifier elimination for the elementary theory of real closed fields by cylindrical algebraic decomposition. In Brakhage, H., editor, *Automata Theory and Formal Languages. 2nd GI Conference*, volume 33 of *Lecture Notes in Computer Science*, pages 134–183. Gesellschaft für Informatik, Springer-Verlag, Berlin, Heidelberg, New York, 1975. Cited on page(s) 138, 206.

Collins, George E. Quantifier elimination by cylindrical algebraic decomposition - twenty years of progress. In Caviness, B. F. and Johnson, J. R., editors, *Quantifier Elimination and Cylindrical Algebraic Decomposition*, Texts and Monographs in Symbolic Computation, pages 8–23. Springer, Wien, New York, 1998a. Cited on page(s) 138.

Collins, Georg E. Quantifier elimination for real closed fields by cylindrical algebraic decomposition. In Caviness, B. F. and Johnson, J. R., editors, *Quantifier Elimination and Cylindrical Algebraic Decomposition*, Texts and Monographs in Symbolic Computation, pages 85–121. Springer, Wien, New York, 1998b. Cited on page(s) 138.

Collins, George E. and Hong, Hoon. Partial cylindrical algebraic decomposition for quantifier elimination. *Journal of Symbolic Computation*, 12(3):299–328, September 1991. Cited on page(s) 138, 204, 206.

Collins, G. E., Johnson, J., and Küchlin, W. W. Parallel real root isolation using the coefficient sign variation method. In Zippel [1992], pages 71–88. ISBN 3-540-55328-2, 0-387-55328-2. Cited on page(s) 435.

Collins, G. E. and Loos, R. G. ALDES and SAC-2 now available. *ACM SIGSAM Bulletin*, 12(2):19, 1982b. Cited on page(s) 421, 422, 425.

Colmerauer, A. An introduction to PROLOG-III. *Communications of the ACM*, 33(7):69–90, 1990. Cited on page(s) 141.

Comet, S. Improved methods to calculate the characters of the symmetric group. *Math. Comput.*, 14:104–117, 1960. Cited on page(s) 84.

Comon, Hubert, editor. *Proceedings of the 8th International Conference on Rewriting Techniques and Applications*, volume 1232, Berlin, 1997. Springer. Cited on page(s) 488.

Compoint, E. and Singer, M. F. Computing Galois groups of completely reducible differential equations. *Journal of Symbolic Computation*, 28:473–499, 1999. Cited on page(s) 97.

Conlon, S. B. Calculating characters of $p$-groups. *J. Symb. Comput.*, 9:535–550, 1990. Cited on page(s) 84.

Conte, R. Singularities of differential equations and integrability. In *Introduction to Methods of Complex Analysis and Geometry for Classical Mechanics and Non-Linear Waves*, pages 49–143, D. Benest and C. Frœschlé, eds. Éditions Frontières, Gif-sur-Yvette, France, 1993. Cited on page(s) 101.

Conte, R., editor. *The Painlevé Property, One Century Later*. CRM Series in Mathematical Physics. Springer, Berlin, 1998. Cited on page(s) 101.

Conte, R., Fordy, A. P., and Pickering, A. A perturbative Painlevé approach to nonlinear differential equations. *Physica D*, 69:33–58, 1993. Cited on page(s) 101.

Conway, J. H., Delgado Friedrichs, O., Huson, D. H., and Thurston, W. P. On three-dimensional space groups. submitted, 2000. Cited on page(s) 243.

Conway, J. H. et al. *Atlas of Finite Groups*. Clarendon Press, 1985. Cited on page(s) 84, 196, 389, 490.

Conway, J. H. and Sloane, N. J. A. *Sphere Packings, Lattices, and Groups*. Springer-Verlag, 1988. Cited on page(s) 129, 490.

Coppersmith, D. Solving linear systems over GF(2): block Lanczos algorithm. *Linear Algebra and Appl.*, 192:33–60, 1993. Cited on page(s) 38.

Coppersmith, D. Solving homogeneous linear equations over GF(2) via block Wiedemann algorithm. *Mathematics of Computation*, 62(205):333–350, 1994. Cited on page(s) 38.

Corless, R., Gianni, P., and Trager, B. A reordered Schur factorization method for zero-dimensional polynomial systems with multiple roots. In Küchlin [1997], pages 133–140. Cited on page(s) 241.

Corless, Robert M., Gianni, Patrizia M., and Trager, Barry M. A reordered Schur factorization method for zero-dimensional polynomial systems with multiple roots. In Küchlin [1997], pages 133–140. Cited on page(s) 122.

Corless, Robert M., Gianni, Patrizia M., and Trager, Barry M. Approximate GCD, or, schönhage's algorithm revisited. Technical report, Ontario Research Centre for Computer Algebra, 2001a. Cited on page(s) 119.

Corless, Robert M., Gianni, Patrizia M., Trager, Barry M., and Watt, Stephen M. The Singular Value Decomposition for polynomial systems. In Levelt [1995], pages 195–207. Cited on page(s) 113, 117, 119, 120, 241, 264.

Corless, Robert M., Giesbrecht, Mark W., Jeffrey, David J., and Watt, Stephen M. Approximate polynomial decomposition. In Dooley [1999], pages 213–219. Cited on page(s) 115, 123.

Corless, Robert M., Giesbrecht, Mark W., Kotsireas, Ilias S., and Watt, Stephen M. Towards factoring bivariate approximate polynomials. In Mourrain [2001], pages 85–92. Cited on page(s) 118, 124.

Corless, Robert M. and Jeffrey, David J. Well ... it isn't quite that simple. *ACM SIGSAM Bulletin*, 26(3):2–6, August 1992. Cited on page(s) 139.

Coste, M. and Roy, M.-F. Thom's lemma, the coding of real algebraic numbers and the topology of semi-algebraic sets. *Journal of Symbolic Computation*, 5 (1):121–129, 1988. Cited on page(s) 19, 56.

Cox, D., Little, J., and O'Shea, D. *Ideals, Varieties and Algorithms: An Introduction to Computational Algebraic Geometry and Commutative Algebra*. Undergraduate Texts in Mathematics. Springer-Verlag, New York, 1992. Cited on page(s) 4, 28, 33, 53, 235, 365, 490.

Cox, David, Little, John, and O'Shea, Donald. *Using algebraic geometry*. Graduate Texts in Mathematics. Springer-Verlag, New York, Berlin, Heidelberg, 1998a. ISBN 0-387-98487-9, 3-540-98492-5. Cited on page(s) 28, 490.

Cox, D. A., Sedeberg, T. W., and Chen, F. The moving line ideal basis of planar rational curves. *Computer Aided Geometric Design*, 15:803–827, 1998b. Cited on page(s) 237.

Cox, David A. and Sturmfels, Bernd, editors. *Applications of Computational Algebraic Geometry*. American Mathematical Society, January 1997/ San Diego (California) 1998. ISBN ISBN 0-8218-0750-1. Proceedings of Symposia in Applied Mathematics/ Volume 53. Cited on page(s) 492.

Craig, J. J. *Introduction to Robotics, Mechanics and Control*. Addison-Wesley, Reading, Mass., 1986. Cited on page(s) 232, 490.

Crandall, R. *Mathematica for the Sciences*. Addison-Wesley, Reading, Massachusetts, 1991. Cited on page(s) 491.

Cremona, J. E. *Algorithms for Modular Elliptic Curves*. Cambridge University Press, Cambridge, 1992. Cited on page(s) 49, 490.

Creutzburg, R., Hatz, V., Nückel, A., and D., Zerfowski. A new design environment for scalable Fermat arithmetic hardware. In *Proceedings of the Int. Symp. on DSP for Comm. Sys., Univ. of Warwick, 1992*, 1992. Cited on page(s) 161.

Crippen, G. and Havel, T. *Distance Geometry and Molecular Conformation*. Research Studies Press Ltd., 1988. Cited on page(s) 243.

Csendes, T. and Pintér, J. The Impact of Accelerating Tools on the Interval Subdivision Algorithm for Global Optimization. *European Journal of Operational Research*, 65:314–320, 1993. Cited on page(s) 111.

Cucker, F., Gonzalez-Vega, L., and Roselló, F. On algorithms for real algebraic plane curves. In Mora and Traverso [1991], pages 63–88. Cited on page(s) 238.

Curtis, R. and Wilson, R., editors. *The Atlas of Finite Groups: Ten Years On*, volume 249 of *London Mathematical Society Lecture Note Series*, Cambridge, 1998. Cambridge University Press. Cited on page(s) 88, 197, 589, 591, 619.

Cushman, R. and Sanders, J. A. A survey of invariant theory applied to normal forms of vector fields with nilpotent linear part. In Stanton, D., editor, *Invariant Theory and Tableaux*, IMA Volumes in Mathematics and its Applications 19, pages 82–106. Springer-Verlag, New York, 1987. Cited on page(s) 106.

Daberkow, M. Über die Berechnung der ganzen Elemente in Radikalerweiterungen algebraischer Zahlkörper. Dissertation, Technische Universität Berlin, 1995. Cited on page(s) 399.

Daberkow, M., Fieker, C., Klüners, J., Pohst, M. E., Roegner, K., Schörnig, M., and Wildanger, K. KANT V4. *J. Symb. Comp.*, 24:267–283, 1997. Cited on page(s) 45, 300, 306.

Daberkow, M. and Pohst, M. Computations with relative extensions of number fields with an application to the construction of Hilbert class fields. In *Proceedings of ISAAC'95*, pages 68–76. ACM Press, 1995. Cited on page(s) 399.

Daberkow, M. and Pohst, M. On integral bases in relative quadratic extensions. *Mathematics of Computation*, 65:319–329, 1996. Cited on page(s) 399.

Daberkow, M. and Pohst, M. On the computation of Hilbert class fields. *J. Number Theory*, 69:213–230, 1998. Cited on page(s) 400.

Daberkow, M. and Weber, A. A database for number fields. In Calmet, J. and Limongelli, C., editors, *Design and Implementation of Symbolic Computation Systems*, volume 1128 of *Lecture Notes in Computer Science*, pages 320–330. Springer-Verlag, 1996. Cited on page(s) 401.

Dalmas, S., Gaëtano, M., and Watt, S. An OpenMath 1.0 Implementation. In Küchlin [1997], pages 241–248. Cited on page(s) 151, 153, 158.

Daubechies, I. Orthonormal bases of compactly supported wavelets. *Comm. Pure Applied Math.*, XLI(41):909–996, 1988. Cited on page(s) 228.

Davenport, J. On the parallel Risch algorithm (III): Use of tangents. *ACM SIGSAM Bulletin*, 16(3):3–6, 1982. Cited on page(s) 95.

Davenport, J. H. *On the Integration of Algebraic Functions*, volume 102 of *Lecture Notes in Computer Science*. Springer-Verlag, Berlin-Heidelberg-New York, 1981. Cited on page(s) 95, 490.

Davenport, J. H., editor. *EUROCAL '87, European Conference on Computer Algebra, Leipzig, GDR, June 2-5, 1987, Proceedings*, volume 378 of *Lecture Notes in Computer Science*, Berlin-Heidelberg-New York, 1989. Springer-Verlag. ISBN 3-540-51517-8, 0-387-51517-8. Cited on page(s) 486, 566, 617.

Davenport, J. H. The AXIOM system. AXIOM Technical Report ATR/3, NAG Ltd., Oxford, 1992a. Cited on page(s) 264.

Davenport, J. H. How does one program in the AXIOM system? AXIOM Technical Report ATR/4, NAG Ltd., Oxford, 1992b. Cited on page(s) 264.

Davenport, J. H. Primality testing revisited. In Wang [1992], pages 123–129. also in: AXIOM Technical Report, ATR/6, NAG Ltd., Oxford, 1993. Cited on page(s) 264.

Davenport, J. H. and Faure, C. R. The "unknown" in computer algebra. *Programmirovanie*, pages 4–10, January 1994. Cited on page(s) 264.

Davenport, J. H., Gianni, P., and Trager, B. M. Scratchpad's view of algebra II: A categorical view of factorization. In Watt [1991], pages 32–38. ISBN 0-89791-437-6. also in: AXIOM Technical Report, ATR/2, NAG Ltd., Oxford, 1992. Cited on page(s) 264.

Davenport, James H. and Heintz, Joos. Real quantifier elimination is doubly exponential. *Journal of Symbolic Computation*, 5(1–2):29–35, February–April 1988. Cited on page(s) 204.

Davenport, J. H., Siret, Y., and Tournier, E. *Computer Algebra: Systems and Algorithms for Algebraic Computation*. Academic Press, London, 1989. Cited on page(s) 4, 490.

Davenport, J. H. and Smith, G. Fast recognition of symmetric and alternating Galois groups. *Journal of pure and applied algebra*, 153(1):17–, October 2000. Cited on page(s) 47.

Davenport, J. H. and Trager, B. M. Scratchpad's view of algebra I: Basic commutative algebra. In Miola [1990b], pages 40–54. ISBN 3-540-52531-9, 0-387-52531-9. also in: AXIOM Technical Report, ATR/1, NAG Ltd., Oxford, 1992. Cited on page(s) 264.

Davies, G. Second-Order Black Hole Perturbations: A Computer Algebra Approach, I – The Schwarzschild Spacetime. http://xxx.lanl.gov/abs/gr-qc/9810056, 1998. Cited on page(s) 176.

Davis, B., Porta, H., and Uhl, J. J. *Calculus & Mathematica*. Addison-Wesley, Redwood City, CA, 1994. Cited on page(s) 257, 491.

de Jagher, P. C. *Physica*, 101A:629, 1980. Cited on page(s) 192.

de Weger, B. M. M. Algorithms for diophantine equations. CWI Tract 65, Stichting Mathematisch Centrum, Amsterdam, 1989. Cited on page(s) 16.

Degen, W. MASYCA – eine interaktive Kommandosprache zur symbolischen Manipulation mathematischer Formeln. *Angewandte Informatik*, 1:18–26, 1980. Cited on page(s) 429.

Dehn, M. Ueber unendliche diskontinuierliche Gruppen. *Math. Ann.*, 71:116–144, 1911. Cited on page(s) 69.

DeJong, M. L. Some thoughts on calculus laboratories. *Mathematica in Education and Research*, 3 (3):37–41, 1994. Cited on page(s) 255.

Delgado Friedrichs, O., Dress, A. W. M., and Huson, D. H. Tilings and symbols - A report on the uses of symbolic calculation in tiling theory. In *Computer Algebra in Science and Engineering*, pages 273–286. World Scientific, 1995. Cited on page(s) 438.

Delgado Friedrichs, O., Dress, A. W. M., Huson, D. H., Klinowski, J., and Mackay, A. L. Systematic enumeration of crystalline networks. *Nature*, 400: 644–647, 1999. Letters to Nature. Cited on page(s) 243, 439.

Delgado Friedrichs, O. and Huson, D. H. Tiling space by platonic solids I. *Discrete and Computational Geometry*, 21:299–315, 1999. Cited on page(s) 439.

Deligne, P. and Husemöller, D. Survey of Drinfeld modules. *Contemporary Mathematics*, 67:25–91, 1987. Cited on page(s) 50.

Della Dora, J., Dicrescenzo, C., and Tournier, E. An algorithm to obtain formal solutions of a linear homogeneous differential equation at an irregular singular point. In Calmet [1982]. ISBN 3-540-11607-9, 0-387-11607-9. Cited on page(s) 370.

Della Dora, J. and Fitch, J., editors. *Computer Algebra and Parallism*. Academic Press, 1989. Cited on page(s) 146, 492.

Della Dora, J. and Stolovitch, L. Normal forms of differential systems. In Tournier, E., editor, *Computer Algebra and Differential Equations*, London Mathematical Society Lecture Note Series 193, pages 143–184. Cambridge University Press, 1994. Cited on page(s) 105.

Demana, F. and Waits, B. K. A zero-based technology enhanced mathematics curriculum for secondary mathematics. In Ralston, A. and Burkhardt, H., editors, *Proceedings of WG 11, ICME 8*. England, 1997. Cited on page(s) 247.

Dembowski, Peter. *Finite geometries*. Classics in Mathematics. Springer-Verlag, Berlin, 1997. ISBN 3-540-61786-8. Reprint of the 1968 original. Cited on page(s) 373.

Demmel, J. On condition numbers and the distance to the nearest ill-posed problem. *Numerische Mathematik*, 51:251–289, 1987. Cited on page(s) 116.

Denef, J. and van den Dries, L. p-adic and real subanalyitic sets. *Annals of Mathematics*, 128:79–138, 1988. Cited on page(s) 138.

Denis, M. *Image et cognition*. Presse Universitaire de France, Paris, 1989. Cited on page(s) 257.

Denner, A. Techniques for calculation of electroweak radiative corrections at the one loop level and results for W physics at LEP200. *Fortschr. Phys.*, 41: 307–420, 1993. Cited on page(s) 471.

Denner, A., Dittmaier, S., and Schuster, R. Radiative corrections to $\gamma\gamma \to W^+W^-$ in the electroweak Standard Model. *Nucl. Phys.*, B452:80, 1995a. Cited on page(s) 170.

Denner, A., Dittmaier, S., and Weiglein, G. Application of the background-field method to the electroweak Standard Model. *Nucl. Phys.*, B440:95, 1995b. Cited on page(s) 170.

Denner, A. and Hahn, T. Radiative corrections to $W^+W^- \to W^+W^-$ in the electroweak Standard Model. *Nucl. Phys.*, B525:27, 1998. Cited on page(s) 170.

Denner, A., Küblbeck, J., Mertig, R., and Böhm, M. Electroweak radiative corrections to e+ e- $\to$ H Z. *Z. Phys.*, C56:261–272, 1992. Cited on page(s) 170.

Denny, Th. F. and Weber, D. The solution of McCurley's discrete log challenge. In *CRYPTO'98*, volume 1462 of *Lecture Notes in Computer Science*. Springer-Verlag, 1998. Cited on page(s) 406.

Dentzer, R. libI: eine lange ganzzahlige Arithmetik, 1991. Cited on page(s) 404.

Dentzer, R. libI: Eine lange ganzzahlige Arithmetik. University of Heidelberg, ftp.iwr.uni-heidelberg.de/pub/IntArith/, 1993. Cited on page(s) 12, 13.

Dentzer, R. On geometric embedding problems and semiabelian groups. *Manuscripta Math.*, 86:199–216, 1995. Cited on page(s) 48.

Derksen, Harm. Computation of invariants for reductive groups. *Adv. Math.*, 141:366–384, 1999. Cited on page(s) 35.

Dershowitz, N. Termination of rewriting. *Journal of Symbolic Computation*, 3 (1):69–116, 1987. Cited on page(s) 136.

Dershowitz, N., editor. *Rewriting Techniques and Applications, 3rd International Conference, RTA-89, Chapel Hill, North Carolina, USA, April 3–5, 1989, Proceedings*, volume 355 of *Lecture Notes in Computer Science*, Berlin-Heidelberg-New York, 1989. Springer-Verlag. ISBN 3-540-51081-8, 0-387-51081-8. Cited on page(s) 486.

Dershowitz, Nachum and Jouannaud, Jean-Pierre. Rewrite systems. In van Leeuven, Jan, editor, *Formal Models and Semantics*, volume B of *Handbook of Theoretical Computer Science*, chapter 6, pages 243–320. Elsevier, 1990. Cited on page(s) 136.

Dettweiler, M. and Reiter, S. An algorithm of Katz and its application to the inverse Galois problem. *Journal of Symbolic Computation*, 30(6):761–798, 2000. Cited on page(s) 48.

Deuflhard, P. and Engquist, B., editors. *Large Scale Scientific Computing*, volume 7 of *Progress in Scientific Computing*. Birkhäuser Verlag, Boston, 1987. Cited on page(s) 492.

Devitt, J. S. *Calculus with MAPLE V*. Brooks/Cole, Pacific Grove, CA, 1993. Cited on page(s) 257, 491.

Devroye, L. *Lecture Notes on Bucket Algorithms*. Birkhäuser Verlag, Basel, 1986. Cited on page(s) 490.

Dewar, M. C. *Interfacing Algebraic and Numeric Computation*. PhD thesis, The University of Bath, Bath, 1991. Cited on page(s) 152.

Dewar, M. C. Using Computer Algebra To Select Numerical Algorithms. In Wang [1992], pages 1–8. Cited on page(s) 152.

Dewar, Michael C. Manipulating Fortran Code in AXIOM and the AXIOM-NAG Link. In Apiola et al. [1994], pages 1–12. ISBN 952-9528-27-2. Cited on page(s) 151, 263.

Díaz, A., Hitz, M., Kaltofen, E., Lobo, A., and Valente, T. Process scheduling in DSC and the large sparse linear systems challenge. *Journal of Symbolic Computation*, 19(1–3):269–282, 1995. Cited on page(s) 147, 148.

Díaz, A. L. and Kaltofen, E. FoxBox a system for manipulating symbolic objects in black box representation. In Gloor [1998], pages 30–37. ISBN 1-58113-002-3. Cited on page(s) 384.

Dickerson, M. Polynomial decomposition algorithms for multivariate polynomials. Technical Report TR87-826, Comput. Sci., Cornell Univ., April 1987. Cited on page(s) 27.

Dicrescenzo, C. and Duval, D. Algebraic extensions and algebraic closure in SCRATCHPAD II. In Gianni [1989], pages 440–446. ISBN 3-540-51084-2, 0-387-51084-2. Cited on page(s) 53.

D'Inverno, R. *Introducing Einstein's Relativity*. Calderon Press, 1992. Cited on page(s) 194.

Dixon, A. L. The eliminant of three quantics in two independent variables. *Proceedings of London Mathematical Society*, 6:49–69, 209–236, 1908. Cited on page(s) 239, 240.

Dixon, J. Exact solution of linear equations using $p$-adic expansions. *Numer. Math.*, 40(1):137–141, 1982. Cited on page(s) 36.

Dixon, J. Computing subfields in algebraic number fields. *Journal Aust. Math. Soc., Ser. A*, 49:434–448, 1990. Cited on page(s) 399.

Dixon, John D. and Mortimer, Brian. The primitive permutation groups of degree less than 1000. *Math. Proc. Cambridge Philos. Soc.*, 103:213–238, 1988. Cited on page(s) 389.

Dixon, J. D. and Wilf, H. S. The random selection of unlabeled graphs. *Journal of Algorithms*, 4:205–213, 1983. Cited on page(s) 89.

Djouadi, A., Gambino, P., Heinemeyer, S., Hollik, W., Jünger, C., and Weiglein, G. Supersymmetric contributions to electroweak precision observables: QCD corrections. *Phys. Rev. Lett.*, 78:3626, 1997. Cited on page(s) 170, 171.

Djouadi, A., Gambino, P., Heinemeyer, S., Hollik, W., Jünger, C., and Weiglein, G. Leading QCD corrections to scalar quark contributions to electroweak precision observables. *Phys. Rev.*, D57:4179, 1998. Cited on page(s) 171.

Dold, Albrecht and Puppe, Dieter. Non-additive functors, their derived functions, and the suspension homomorphism. *Proc. nat. Acad. Sci. USA*, 44: 1065–1068, 1958. Cited on page(s) 208.

Dolzmann, Andreas. Reelle Quantorenelimination durch parametrisches Zählen von Nullstellen. Diploma thesis, Universität Passau, D-94030 Passau, Germany, November 1994. Cited on page(s) 139, 427.

Dolzmann, Andreas. Solving scheduling problems with REDLOG. Extended abstract at the Rhine Workshop on Computer Algebra, April 1998. Cited on page(s) 139.

Dolzmann, Andreas. Solving geometric problems with real quantifier elimination. In Gao, Xiao-Shan, Wang, Dongming, and Yang, Lu, editors, *Automated Deduction in Geometry*, volume 1669 of *Lecture Notes in Artificial Intelligence (Subseries of LNCS)*, pages 14–29. Springer-Verlag, Berlin Heidelberg, 1999a. ISBN 3-540-66627-9. Cited on page(s) 54.

Dolzmann, Andreas. Solving geometric problems with real quantifier elimination. Technical Report MIP-9903, FMI, Universität Passau, D-94030 Passau, Germany, January 1999b. Cited on page(s) 139, 206.

Dolzmann, Andreas, Gloor, Oliver, and Sturm, Thomas. Approaches to parallel quantifier elimination. In Gloor [1998], pages 88–95. ISBN 1-58113-002-3. Cited on page(s) 353.

Dolzmann, Andreas and Sturm, Thomas. *Redlog User Manual*. FMI, Universität Passau, D-94030 Passau, Germany, October 1996. Edition 1.0 for Version 1.0. Cited on page(s) 139.

Dolzmann, Andreas and Sturm, Thomas. Guarded expressions in practice. In Küchlin, Wolfgang W., editor, *Proceedings of the 1997 International Symposium on Symbolic and Algebraic Computation (ISSAC 97)*, pages 376–383, New York, NY, July 1997a. ACM, ACM Press. Cited on page(s) 139, 206.

Dolzmann, Andreas and Sturm, Thomas. Redlog: Computer algebra meets computer logic. *ACM SIGSAM Bulletin*, 31(2):2–9, June 1997b. ISSN 0163-5824. Cited on page(s) 54, 139, 204.

Dolzmann, Andreas and Sturm, Thomas. Simplification of quantifier-free formulae over ordered fields. *Journal of Symbolic Computation*, 24(2):209–231, August 1997c. Cited on page(s) 206.

Dolzmann, Andreas and Sturm, Thomas. Redlog user manual. Technical Report MIP-9905, FMI, Universität Passau, D-94030 Passau, Germany, April 1999. Edition 2.0 for Version 2.0. Cited on page(s) 205.

Dolzmann, Andreas, Sturm, Thomas, and Weispfenning, Volker. Real quantifier elimination in practice. Technical Report MIP-9720, FMI, Universität Passau, D-94030 Passau, Germany, December 1997. Cited on page(s) 57.

Dolzmann, Andreas, Sturm, Thomas, and Weispfenning, Volker. A new approach for automatic theorem proving in real geometry. *Journal of Automated Reasoning*, 21(3):357–380, 1998b. ISSN 0168-7433. Cited on page(s) 57, 139, 201, 205, 207.

Dolzmann, Andreas, Sturm, Thomas, and Weispfenning, Volker. Real quantifier elimination in practice. In Matzat, B. H., Greuel, G.-M., and Hiss, G., editors, *Algorithmic Algebra and Number Theory*, pages 221–247. Springer, Berlin, 1998c. ISBN 3-540-64670-1. Cited on page(s) 57, 139.

Domich, P. D., Kannan, R., and Trotter, Jr., L. E. Hermite normal form computation using modulo determinant arithmetic. *Math. of Operations Research*, 12(1):50–59, 1987. Cited on page(s) 39.

Dooley, S., editor. *ISSAC 99 Proc. 1999 Internat. Symp. Symbolic Algebraic Comput.*, New York, N. Y., 1999. ACM Press. Cited on page(s) 489, 521, 542, 552, 556, 583, 618.

Dorato, Peter, Yang, Wei, and Abdallah, Chaouki. Robust multi-objective feedback design by quantifier elimination. *Journal of Symbolic Computation*, 24 (2):153–159, August 1997. Special issue on applications of quantifier elimination. Cited on page(s) 139.

Dorey, F. and Whaples, G. Prime and composite polynomials. *J. Algebra*, 28: 88–101, 1974. Cited on page(s) 26, 27.

Draexler, P. and Noerenberg, R. CREP Manual - Version 1.0 using Maple as surface. Technical Report E 96-002, SFB 343, Bielefeld, 1996. Cited on page(s) 65.

Draexler, P. and Noerenberg, R. Classification problems in the combinatorial representation theory of finite-dimensional algebras. Technical Report E 97-008, SFB 343, Bielefeld, 1997. Computational methods for representations of groups and algebras, Proceedings of the Euroconference at the University of Essen, April 1-5, 1997 (edited by P. Dräxler, C.M. Ringel, G.O. Michler). Cited on page(s) 65.

Dräxler, Peter. Computational aspects in the representation theory of finite-dimensional algebras. In *Computational Algebra*, 1993. Cited on page(s) 370.

Dräxler, P., Michler, G. O., and Ringel, C. M., editors. *Computational Methods for Representations of Groups and Algebras*. Birkhäuser Verlag, Basel, 1999. Cited on page(s) 370.

Dräxler, Peter and Nörenberg, Rainer. *CREP Manual, Version 1.0*, 1996. Ergänzungsreihe 96 – 002, SFB 343 'Diskrete Strukturen in der Mathematik', Universität Bielefeld. Cited on page(s) 370.

Dräxler, Peter and Nörenberg, Rainer. *CREP Manual, Part2*, 1997. Ergänzungsreihe 97 – 009, SFB 343 'Diskrete Strukturen in der Mathematik', Universität Bielefeld. Cited on page(s) 370.

Dräxler, Peter and Nörenberg, Rainer. Classification problems in the representation theory of finite-dimensional algebras. In *Computational Methods for Representations of Groups and Algebras*, 1999. Cited on page(s) 370.

Drescher, K. Axioms, categories and domains. Automath Technical Report, University of Paderborn, August 1997. http://www.sciface.com/support/papers/PS/dom-pack.ps.gz. Cited on page(s) 332.

Dress, Andreas and Wenzel, Walter. Grassmann-Plücker relations and matroids with coefficients. *Adv. in Math.*, 86:68–110, 1991. Cited on page(s) 34.

Dress, A. W. M. Regular polytopes and equivariant tessellations from a combinatorial point of view. In *Algebraic Topology*, pages 56–72. SLN 1172, Göttingen, 1984. Cited on page(s) 438.

Dress, A. W. M. Presentations of discrete groups, acting on simply connected manifolds, in terms of parametrized systems of coxeter matrices—a systematic approach. *Advances in Mathematics*, 63:196–212, 1987. Cited on page(s) 243, 438.

Dress, A. W. M., Dreiding, A., and Haegi, H. Classification of mobile molecules by category theory. In Maruani, J. and Serre, J., editors, *Studies in Physical and Theoretical Chemistry 23, Symmetries and Properties of Non-Rigid Molecules: A Comprehensive Survey, International Symposium, Paris, July, 1-7. 1982*, pages 39–58, 1982. Cited on page(s) 242.

Dress, A. W. M., Huson, D., and Molnár, E. The classification of face-transitive 3D-tilings. *Acta Crystallographica*, A49:806–817, 1993a. Cited on page(s) 243, 439.

Dress, A. W. M., Müller, A., and Pope, M. Polyoxometalates: A class of compounds with remarkable topology. In Müller, A. and Pope, M., editors, *Polyoxometalates: From Platonic Solids to Antiretroviral Activity*. Kluwer Academic Publishers, Dordrecht, Boston, London, 1993b. Cited on page(s) 243.

Dress, A. W. M. and Scharlau, R. Zur Klassifikation äquivarianter Pflasterungen. *Mitteilungen aus dem Math. Seminar Giessen*, 164:83–136, 1984. Cited on page(s) 439.

Drijvers, P. Graphics calculators and computer algebra systems: Differences & similarities. *The International Derive Journal*, 1(1):71–82, 1994. Cited on page(s) 248.

Drijvers, P. What issues do we need to know more about: Questions for future educational research concerning CAS. In et al., J. Berry, editor, *The state of computer algebra in mathematics education*. Chartwell-Bratt, Bromley, 1997a. Cited on page(s) 249.

Drijvers, P. You never forget your first love ... : The TI-92 in teacher education. *The International Journal on Computer Algebra in Mathematics Education*, 4(1):69–76, 1997b. Cited on page(s) 249.

Drijvers, P. Assessment and new technologies: Different policies in different countries. *The International Journal on Computer Algebra in Mathematics Education*, 5(2):81–93, 1998. Cited on page(s) 248.

Drijvers, P. and Doorman, M. The graphics calculator in mathematics education. *The Journal of Mathematical Behaviour*, 14(4):425–440, 1997. Cited on page(s) 248.

Drijvers, P., Verweij, A., and Winsen, E. Van. Mathematics lessons with Derive developed by the CAVO working group. *Zentralblatt für Didaktik der Mathematik*, 97(4):118–123, 1997. Cited on page(s) 248.

Dumas, J.-G., Saunders, B. D., and Villard, G. Integer Smith form via the valence: experience with large sparse matrices from homology. In Traverso [2000], pages 95–105. Cited on page(s) 384.

Dupée, B. J. and Davenport, J. H. An intelligent interface to numerical routines. In Calmet, J. and Limongelli, C., editors, *Proc. DISCO '96*, Lecture Notes in Computer Science 1128, pages 252–262. Springer, Heidelberg, 1996. Cited on page(s) 108, 152, 263.

Durham II. *Proc. Symposium Finite Simple Groups II, Durham 1978*, New York, 1980. Academic Press. Cited on page(s) 197, 604.

Duval, A. and Loday-Richaud, M. Kovaçic's algorithm and its application to some families of special functions. *Appl. Alg. Eng. Comm. Comp.*, 3:211–246, 1992. Cited on page(s) 97.

Duval, D. Algebraic numbers : an example of dynamic evaluation. *Journal of Symbolic Computation*, 18(5):429–445, November 1994. Cited on page(s) 40, 264.

Duval, D. and Jung, F. Examples of problem solving using computer algebra. *IFIP Transactions. A. Computer Science and Technology*, A-2:133–141, 143, 1992. Cited on page(s) 264.

Dziobek, O. Über einen merkwürdigen Fall des Vielkörperproblems. *Astron. Nach.*, 152:32–46, 1900. Cited on page(s) 177.

Eberly, W. Black box Frobenius decompositions over small fields. In Traverso [2000], pages 106–113. Cited on page(s) 40.

Eberly, W., Giesbrecht, M., and Villard, G. On computing the determinant and Smith form of an integer matrix. In *Proc. 41st IEEE Symposium on Foundations of Computer Science (FOCS)*, 2000. Cited on page(s) 40.

Eberly, W. and Kaltofen, E. On randomized Lanczos algorithms. In Küchlin [1997], pages 176–183. Cited on page(s) 37.

Ebner-Altunay, P. Standardbasen in Potenzreihenringen. Diplomarbeit, Universität Passau, Passau, 1991. Cited on page(s) 32.

Eckart, C. and Young, G. The approximation of one matrix by another of lower rank. *Psychometrika*, 1(3):211–218, September 1936. Cited on page(s) 119.

Edelman, Alan, Elmroth, Erik, and Kågström, Bo. A geometric approach to perturbation theory of matrices and matrix pencils. part I: Versal deformations. *SIAM J. Matrix Anal. Appl.*, 18(3):653–692, 1997. Cited on page(s) 125.

Edneral, V. F. Computer evaluation of cyclicity in planar cubic system. In Küchlin, W. W., editor, *Proc. ISSAC '97*, pages 305–309. ACM Press, New York, 1997. Cited on page(s) 108.

Edwards, M. P. and Broadbridge, P. Exceptional symmetry reductions of Burgers' equation in two and three spatial dimensions. *Journal of Applied Mathematics and Physics*, 46(4):595–622, 1995. Cited on page(s) 468.

Egner, S. *Zur Algorithmischen Zerlegungstheorie Linearer Transformationen mit Symmetrie*. PhD thesis, Universität Karlsruhe, Informatik, 1997. Cited on page(s) 218, 462.

Egner, S. and Püschel, M. *AREP – Constructive Representation Theory and Fast Signal Transforms*. GAP share package, 1998. http://www-gap.dcs.st-and.ac.uk/~gap/Share/arep.html and http://avalon.ira.uka.de/home/pueschel/arep/arep.html. Cited on page(s) 85, 461.

Ehrig, H. and Mahr, B. *Fundamentals of Algebraic Specifications I*. Springer-Verlag, Berlin-Heidelberg-New York, 1985. Cited on page(s) 142, 490.

Eichenlaub, Y. *Problèmes effectifs de théorie de Galois en degrés 8 à 11*. Thèse, Université Bordeaux 1, 1996. Cited on page(s) 48.

Eichenlaub, Y. and Olivier, M. Computation of Galois groups for polynomials with degree up to eleven. Preprint, Université Bordeaux I, 1995. Cited on page(s) 48, 399.

Eick, B. and O'Brien, E. A. The groups of order 512. In Matzat et al. [1999], pages 379–380. ISBN ISBN 3-540-64670-1. Selected Papers from a Conference Held at the University of Heidelberg in October 1997. Cited on page(s) 197, 389.

Eisenbud, D. Homological algebra on a complete intersection, with an application to group representations. *Trans. AMS*, 260:35–64, 1980. Cited on page(s) 419.

Eisenbud, David. *Commutative Algebra with a view Toward Algebraic Geometry*. Springer, 1995. ISBN ISBN 0-387-94269-6. Graduate Texts in Mathematics 150. Cited on page(s) 490.

Eisenbud, D., Huneke, C., and Vasconcelos, W. Direct methods for primary decomposition. *Inventiones Mathematicae*, 2:207–236, 1992. Cited on page(s) 53.

Eisenbud, David and Robbiano, Lorenzo, editors. *Computational Algebraic Geometry and Commutative Algebra*, volume 34 of *Symposia Mathematica*, 1991. Cambridge University Press. Cited on page(s) 487.

Ekedahl, T., Grabmeier, J., and Lambe, L. Data structures and algorithms for algebraic computations with applications to the cohomology of finite $p$-groups. Preprint, University Stockholm, Stockholm, 1995. Cited on page(s) 59.

Ellis, W., Johnson, E. W., Lodi, E., and Schwalbe, D. *Maple V Flight Manual: Tutorials for Calculus, Linear Algebra, and Differential Equations*. Brooks/Cole Publishing Company, Pacific Grove, California 93950, 1992. Cited on page(s) 491.

Ellis, W. and Lodi, E., editors. *A Tutorial Introduction to Mathematica*. Brooks/Cole Publishing Company, Pacific Grove, California 93950, 1990. Cited on page(s) 491.

Emiris, I. On the complexity of sparse elimination. *J. Complexity*, 12:134–166, 1996. Cited on page(s) 52, 54.

Emiris, Ioannis, Galligo, André, and Lombardi, Henri. Numerical univariate polynomial GCD. In Renegar, J., Shub, M., and Smale, S., editors, *Proc.*

*1995 AMS-SIAM Summer Seminar on the Mathematics of Numerical Analysis*, volume 32 of *Lectures in Applied Math*, pages 323–343, Park City, Utah, 1996. Cited on page(s) 121.

Emiris, Ioannis, Galligo, André, and Lombardi, Henri. Certified approximate univariate GCD. *Journal of Pure and Applied Algebra*, 117 & 118:229–251, 1997. Cited on page(s) 121.

Emiris, Ioannis Z. *Symbolic-numeric algebra for polynomials*, volume 39, pages 261–281. Marcel Dekker, New York, 1999. Cited on page(s) 113, 122, 123.

Epstein, D. B. A., Cannon, J. W., Holt, D. F., Levy, S. V. F., Paterson, M. S., and Thurston, W. P. *Word Processing in Groups*. Academic Press, New York, 1992. Cited on page(s) 83, 490.

Erdélyi, A., Magnus, W., Oberhettinger, F., and Tricomi, F. G. *Higher Transcendental Functions*, volume 1–3. McGraw-Hill, New York, 1953–1955. Cited on page(s) 214.

Erlingsson, Ú, Kaltofen, E., and Musser, D. Generic Gram-Schmidt orthogonalization by exact division. In *ISSAC 96 Proc. 1996 Internat. Symp. Symbolic Algebraic Comput.*, pages 275–282, 1996. Cited on page(s) 13.

Ernst, F. J., Garcia, A. D., and Hauser, I. Colliding gravitational waves with noncollinear polarzation. I. *Journal Math. Phys.*, 28:2155–2161, 1987. Cited on page(s) 174.

Ershov, A. P. and Knuth, D. E., editors. *Proceedings of the Conference on Algorithms in Modern Mathematics and Computer Science*, volume 122 of *Lecture Notes in Computer Science*, Urgench, Uzbek SSR, September 1979. Springer. Cited on page(s) 485.

Ershov, A. P. and Knuth, D. E., editors. *Algorithms in Modern Mathematics and Computer Science*, volume 122 of *Lecture Notes in Computer Science*. Springer-Verlag, Berlin-Heidelberg-New York, 1981. Cited on page(s) 492, 621.

Ershov, Juri L. On elementary theories of local fields. *Algebra i Logika Sem.*, 4 (2):5–30, 1965. Cited on page(s) 138.

Espinola, J., Gonzalez-Vega, L., and Necula, I. Generic implicitation of low degree rational surfaces. Available at http://frisco.matesco.unican.es, 1999. Cited on page(s) 236.

Eysette, F. and Galligo, A., editors. *Proceedings MEGA 1992*, number 109 in Progress in Mathematics, Basel, 1993. Birkhäuser Verlag. Cited on page(s) 487.

Fachgruppe Computeralgebra der GI and DMV and GAMM, editor. *Computeralgebra in Deutschland*. Fachgruppe Computeralgebra der GI, DMV, GAMM, Passau und Heidelberg, 1993. Cited on page(s) 197, 490.

Fakler, W. *Algebraische Algorithmen zur Lösung von linearen Differentialgleichungen*. MuPAD Reports. Teubner, Stuttgart, 1997a. Cited on page(s) 4, 490.

Fakler, W. On second order homogeneous linear differential equations with Liouvillian solutions. *Theor. Comp. Sci.*, 187:27–48, 1997b. Cited on page(s) 97.

Fakler, W. *Algebraische Algorithmen zur Lösung von linearen Differentialgleichungen.* MuPAD Reports. B.G. Teubner, Stuttgart, 1999. Cited on page(s) 331.

Faradžev, I. A. and Klin, M. H. Computer package for computations with coherent configurations. In Watt [1991], pages 219–223. ISBN 0-89791-437-6. Cited on page(s) 91.

Faradzev, I. A. Constructive enumeration of combinatorial objects. *CMRS*, 1977. Cited on page(s) 89.

Farkas, D., Feustel, D., and Green, E. L. Synergy in the theories of Gröbner bases and path algebras. *Canadian Journal of Mathematics*, 45(4):727–739, 1993. Cited on page(s) 61, 64.

Fateman, R. Honest Plotting, Global Extrema, and Interval Arithmetic. In Wang [1992], pages 216–223. Cited on page(s) 150.

Faugere, J., Gianni, P., Lazard, D., and Mora, T. Efficient computation of zero-dimensional gröbner bases by change of ordering. *J. Symb. Computation*, 16: 329–344, 1993. Cited on page(s) 30.

Faugère, J.-C. http://posso.lip6.fr/~jcf/ Gb and FGb hypertext documentation on the WWW, 1994a. Cited on page(s) 178.

Faugère, J. C. *Résolution des Systèmes d'équations Algébriques.* PhD thesis, The Université Paris VI, Paris, 1994b. Cited on page(s) 153.

Faugère, J.-C. A new efficient algorithm for computing Gröbner bases (F4). *Journal of Pure and Applied Algebra*, 139:61–88, 1999. Cited on page(s) 178.

Faugère, J. C., Gianni, P., Lazard, D., and Mora, T. Efficient computation of zero-dimensional Gröbner bases by change of ordering. *Journal of Symbolic Computation*, 16(4):329–344, October 1993. Cited on page(s) 52.

Faugère, J.-C. and Kotsireas, I. Symmetry theorems for the Newtonian 4- and 5-body problems with equal masses. In et al., Ganzha, editor, *CASC'99 Proceedings, TUM Munich*, pages 81–92. Springer Verlag, 1999. Cited on page(s) 177.

Fell, Richard N. *Macsyma Tutorial for Calculus.* Jones and Bartlett Publishers, Sudbury, Mass., 1998. ISBN 0-7637-0622-1. Cited on page(s) 283, 491.

Feng, H. *Decomposition and computation of the topology of plane real algebraic curves.* Dissertation, The Royal Institute of Technology, Stockholm, Sweden, 1992. Cited on page(s) 238.

Feo, J. T., Cann, D. C., and Oldehoeft, R. R. A report on the SISAL language project. *Journal of Parallel and Distributed Computing*, 10(4):349–366, 1990. Cited on page(s) 148, 149.

Fermigier, S. Construction of High-Rank Elliptic Curves over **Q** and **Q**($t$). In *Lect. Notes in Comp. Sci.*, volume 1122, pages 115–120, Berlin-Heidelberg-New York, 1996. Springer-Verlag. Cited on page(s) 49.

Ferrante, Jeanne and Rackoff, Charles W. *The Computational Complexity of Logical Theories.* Number 718 in Lecture Notes in Mathematics. Springer-Verlag, Berlin, 1979. Cited on page(s) 138, 206, 490.

Ferro, G. Carrá. Gröbner bases and differential ideals. In *AAECC'5*, number 356 in Lect. Notes in Comp. Science, pages 129–140, Menorca, 1987. Springer Verlag. Cited on page(s) 104.

Feuillebois, F. and Lasek, A. Computer aided application of the principle of least degeneracy. *J. Appl. Math. Phys. (ZAMP)*, 28(6):1142–1146, 1977. Cited on page(s) 222.

Fieker, C., Jurk, A., and Pohst, M. On solving relative norm equations in algebraic number fields. *Mathematics of Computation*, 66:399–410, 1997. Cited on page(s) 399.

Finch, J. K. and Lehmann, M. *Exploring Calculus with Mathematica*. Addison-Wesley, Redwood City, CA, 1992. Cited on page(s) 257, 491.

Fincke, U. and Pohst, M. A procedure for determining algebraic integers of given norm. volume 162 of *Lecture Notes in Computer Science*, pages 194–202. Springer-Verlag, 1983. Cited on page(s) 399.

Finkelstein, Larry and Kantor, William M., editors. *Groups and Computation*, volume 11 of *DIMACS Series in Discrete Mathematics and Theoretical Computer Science*, Providence, 1993. AMS. Cited on page(s) 70, 502.

Finkelstein, Larry and Kantor, William M., editors. *Groups and Computation II*, volume 28 of *DIMACS Series in Discrete Mathematics and Theoretical Computer Science*, Providence, 1997. AMS. Cited on page(s) 70, 502, 573, 586, 588.

Fitch, J., editor. *EUROSAM 84, International Symposium on Symbolic and Algebraic Computation, Cambridge, England, July 9–1, 1984*, volume 174 of *Lecture Notes in Computer Science*, Berlin-Heidelberg-New York, 1984. Springer-Verlag. ISBN 3-540-13350-X, 0-387-13350-X. Cited on page(s) 485.

Fitch, John, editor. *Design and Implementation of Symbolic Computation Systems*, volume 721 of *Lecture Notes in Computer Science*, Berlin, 1992. Springer. Cited on page(s) 487.

Fitchas, N. and Galligo, A. Nullstellensatz effectif et conjecture de Serre (theoréme de Quillen-Suslin). *Mathematische Nachrichten*, 149:231–253, 1990. Cited on page(s) 51.

Fitchas, N., Galligo, A., and Morgenstern, J. Precise sequential and parallel complexity bounds for quantifier elimination over algebraically closed fields. *Journal for Pure and Applied Algebra*, 67:1–14, 1990. Cited on page(s) 138.

Fitchas (J. Heintz and J. Sabia L. M. Pardo and P. Solernò), N., Giusti, M., and Smietansky, F. Sur la complexitè du thèoréme des Zèros. In Florenzano, M., Guddat, J., Jimenez, M., Jongen, H. T., Lagomasino, G. L., and Marcellán, F., editors, *Approximationa and Optimization in the Caribbean II*, Approximation and Optimization 8, pages 274–329, La Habana, 1995. Peter Lange. Cited on page(s) 54.

Flajolet, P. and Salvy, B. Computer algebra libraries for combinatorial structures. *Journal of Symbolic Computation*, 20:653–671, 1995. Cited on page(s) 220.

Flajolet, P., Salvy, B., and Zimmermann, P. Lambda-Upsilon-Omega: The 1989 Cookbook. Technical Report RR 1073, INRIA, 1989. Cited on page(s) 220.

Flajolet, P., Salvy, B., and Zimmermann, P. Automatic average-case analysis of algorithms. *Theoretical Computer Science, Series A*, 79:37–109, 1991. Cited on page(s) 220.

Flajolet, P. and Sedgewick, R. *An Introduction to the Analysis of Algorithms.* Addison-Wesley, Reading, Massachusetts, 1996. Cited on page(s) 220.

Flannery, B., Press, W., Teukolsky, S., and Vetterling, W. *Numerical Recipes in C, The Art of Scientific Computing.* Cambridge University Press, New York, Melbourne, 2nd edition, 1992. Cited on page(s) 12, 491.

Flaschka, H., Newell, A. C., and Tabor, M. Integrability. In *What is Integrability*, pages 73–114, V. E. Zakharov, ed. Springer, New York, 1991. Cited on page(s) 99, 101.

Fleischer, J., Grabmeier, J., Hehl, F. W., and Küchlin, W., editors. *Computer Algebra in Scinece and Engineering.* World Scientific, Singapore, 1994. Cited on page(s) 492.

Fleischer, J., Grabmeier, J., Hehl, F. W., and Küchlin, W., editors. *Computer Algebra in Science and Engineering*, 1995. World Scientific Publishers, Singapore. Cited on page(s) 226, 227, 545.

Fleischer, J. and Kalmykov, M. Yu. ON-SHELL2: FORM based package for the calculation of two-loop self-energy single scale Feynman diagrams occurring in the Standard Model. *Comp. Phys. Comm.*, 128:531–549, 2000. Cited on page(s) 165.

Fleischer, J. and Tarasov, O. V. SHELL2. *Comp. Phys. Comm.*, 71:193, 1992. Cited on page(s) 165.

Fleischer, J. and Tarasov, O. V. *Z. Phys. C*, 64:413, 1994. Cited on page(s) 166.

Ford, B. and Chatelin, F., editors. *Problem Solving Environments for Scientific Computing*, Amsterdam, 1987. North-Holland. Cited on page(s) 153.

Fortenbacher, A. Efficient type inference and coercion in computer algebra. In Miola [1990b], pages 56–60. ISBN 3-540-52531-9, 0-387-52531-9. Cited on page(s) 143, 263.

Fossorier, M., Hideki, I., Lin, Shu, and Poli, A., editors. *Proceedings of the 13th International Symposium on Applied Algebra, Algebraic Algorithms and Error-Correcting Codes*, volume 1719 of *Lecture Notes in Computer Science*, Berlin, 1999. Springer. Cited on page(s) 489.

Fowler, D. Farewell to Greek-bearing GIFs Part I: MathML. *Math&Stats.*, 9(3): 38–39, 1998. Cited on page(s) 256.

Fowler, D. *The Fractal Dimension of the Blues.* Wolfram Media, Champaign, Illinois, under development. Cited on page(s) 256.

Freeman, T. S., Imirzian, G., Kaltofen, E., and Lakshman Yagati. DAGWOOD: A system for manipulating polynomials given by straight-line programs. *ACM Trans. Math. Software*, 14(3):218–240, 1988. Cited on page(s) 384.

Freire, E., Gamero, E., and Ponce, E. Normal forms for planar systems with nilpotent linear part. In Seydel, R., Schneider, F. W., Küpper, T., and Troger, H., editors, *Bifurcation and Chaos: Analysis, Algorithms, Applications*, ISNM 97, pages 123–128. Birkhäuser, Basel, 1991. Cited on page(s) 106.

Freire, E., Gamero, E., Ponce, E., and Franquelo, L. G. An algorithm for symbolic computation of center manifolds. In Gianni, P., editor, *Proc. ISSAC '88*, pages 218–230. Springer-Verlag, Berlin, 1988. Cited on page(s) 106.

Fried, M. D. and MacRae, R. E. On curves with separated variables. *Math. Ann.*, 180:220–226, 1969. Cited on page(s) 26.

Frink, A., Körner, J. G., and Tausk, J. B. *Comp.Phys.Comm.*, 115:140, 1998. hep-ph/9709490. Cited on page(s) 166.

Fripertinger, H. and Kerber, A. *Isometry Classes of Indecomposable Linear Codes*, volume 948 of *LNCS*. Springer-Verlag, Berlin-Heidelberg-New York, 1995. Cited on page(s) 90, 490.

Froeberg, Ralf. *An Introduction to Gröbner Bases*. Pure and Applied Mathematics. John Wiley and Sons, New York, NY10158-0012, USA, 1997. Cited on page(s) 28, 490.

Fuchssteiner, B. et al. *MuPAD Benutzerhandbuch*. Birkhäuser Verlag, Basel, 1993. Cited on page(s) 491.

Fuchssteiner, B. and the MuPAD Group. *MuPAD User's Manual - MuPAD Version 1.2.2*. John Wiley&Sons, New York, 1996. Revised edition in preparation. Cited on page(s) 332, 492.

Gaál, I., Pethö, A., and Pohst, M. On the resolution of index form equations in quartic number fields. *Journal of Symbolic Computation*, (16):563–584, 1993. Cited on page(s) 400.

Gabriel, P. Unzerlegbare Darstellungen I. *Manuscr. Math.*, 6:71–103, 1972. Cited on page(s) 64.

Gabriel, P. and Roiter, A. V. *Algebra VIII: Representations of Finite-Dimensional Algebras*, volume 73 of *Encyclopedia of the Mathematical Sciences*. Springer-Verlag, New York, 1992. Editors: Kostrikin, A.I. and Shafarevich, I.V. Cited on page(s) 64.

Gaffney, P. W. and Houstis, E. N., editors. *Programming Environments for High-Level Scientific Problem Solving*, Amsterdam, 1992. North-Holland. Cited on page(s) 153.

Galile, Franois, Roch, Jean-Louis, Cavalheiro, Gerson G. H., and Doreille, Mahtias. A general modular specification for distributed schedulers. In *Pact'98*, Paris, France., October 1998. Cited on page(s) 147, 148.

Gallager, R. G. *Low Density Parity-Check Codes*. MIT Press, Cambridge, MA, 1963. Cited on page(s) 129, 491.

Galligo, A. Some algorithmic questions on ideals of differential operators. In *Proc. EUROCAL'85*, volume 204 of *LNCS*, pages 413–421. Springer, 1985. Cited on page(s) 61.

Galligo, André, Gonzalez-Vega, Laureano, and Lombardi, Henri. Continuity properties for flat families of polynomials (I). *Journal of Pure and Applied Algebra*, page to appear, 2001. Cited on page(s) 121.

Galligo, André and Rupprecht, David. Semi-numerical determination of irreducible branches of a reduced space curve. In Mourrain [2001], pages 137–142. Cited on page(s) 121.

Galligo, A. and Vorobjov, N. Complexity of finding irreducible components of a semialgebraic set. *Journal pure appl. algebra*, 11:174–193, 1995. Cited on page(s) 56, 57.

Galligo, André and Watt, Stephen M. A numerical absolute primality test for bivariate polynomials. In Küchlin [1997], pages 217–224. Cited on page(s) 124.

Gallo, G., Mishra, B., and Ollivier, F. Some constructions in rings of differential polynomials. In *AAECC-9*, number 539 in Lecture Notes in Computer Science, pages 171–182, New Orléans, 1991. Springer Verlag. Cited on page(s) 104.

Gantmacher, F. R. *The Theory of Matrices*, volume 2. Chelsea Publ. Co., New York, N. Y., 1960. Cited on page(s) 122.

Gantmacher, F. R. *Théorie des matrices*. Dunod, Paris, France, 1966. Cited on page(s) 39, 491.

Ganzha, V. G., Mayr, E. W., and Vorozhtsov, E. V., editors. *Computer algebra in scientific computing*, 1999. Springer. Cited on page(s) 489.

Ganzha, V. G., Mayr, E. W., and Vorozhtsov, E. V., editors. *Computer Algebra in Scientific Computing*, October 2000. Springer. Cited on page(s) 489.

Ganzha, V. G., Mayr, E. W., and Vorozhtsov, E. V., editors. *Computer Algebra in Scientific Computing*, September 2001. Springer-Verlag. In Konstanz, Germany. Cited on page(s) 490.

Ganzha, V. G. and Vorozhtsov, E. V. *Computer-Aided Analysis of Difference Schemes for Partial Differential Equations.* John Wiley, New York, 1996. Cited on page(s) 108, 490.

Ganzinger, Harald, editor. *Proceedings of the 7th International Conference on Rewriting Techniques and Applications*, volume 1103, 1996. Springer. Cited on page(s) 488.

Gao, Shuhong. Factoring multivariate polynomials via PDE, 2001. Available from sgao@math.clemson.edu. Cited on page(s) 26.

Gao, S. and Lenstra Jr., H. W. Optimal normal basis. *Design, Codes and Cryptography*, 2:315–323, 1992. Cited on page(s) 161.

Gao, X. S. and Chou, S. C. Implicitization of rational parametric equations. *Journal of Symbolic Computation*, 14:459–470, 1992. Cited on page(s) 235.

Gao, Xiao-Shan and Wang, Dongming, editors. *ASCM'2000, The 4th Asian Symposium on Computer Mathematics*, Singapore, 2000a. World Scientific Publishing Co. Cited on page(s) 490.

Gao, Xiao-Shan and Wang, Dongming, editors. *Mathematics Mechanization and Applications*. Academic Press, London, 2000b. Cited on page(s) 492.

Gao, Xiao-Shan, Wang, Dongming, and Yang, Lu, editors. *Automated Deduction in Geometry*, volume 1669 of *Lecture Notes in Artificial Intelligence*, 1999. Springer. Cited on page(s) 489.

GAP. *GAP – Groups, Algorithms, and Programming, Version 4*. The GAP Group, Lehrstuhl D für Mathematik, RWTH Aachen, Germany and School of Mathematical and Computational Sciences, U. St. Andrews, Scotland, 1997. Also see http://www-gap.dcs.st-and.ac.uk/~gap/. Cited on page(s) 84, 88, 196, 211, 359, 375, 397, 400, 430, 491.

*GAP – Groups, Algorithms, and Programming, Version 4*. The GAP Group, Lehrstuhl D für Mathematik, RWTH Aachen, Germany and School of Mathematical and Computational Sciences, U. St. Andrews, Scotland, 1997. Also see http://www-gap.dcs.st-and.ac.uk/~gap/. Cited on page(s) 473.

Gatermann, K. Mixed symbolic-numeric solution of symmetrical nonlinear systems. In Watt [1991], pages 431–432. ISBN 0-89791-437-6. Cited on page(s) 217.

Gatermann, K. SYMCON. Available by anonymous ftp from elib at ZIB, 1991b. Cited on page(s) 217.

Gatermann, K. SYMMETRY. REDUCE network library, available via electronic mail to netlib@rand.org, 1991c. Cited on page(s) 217.

Gatermann, K. Computation of Bifurcation Graphs. In Allgower, E., Georg, K., and Miranda, R., editors, *Exploiting Symmetry in Applied and Numerical Analysis*, AMS Lectures in Applied Mathematics Vol. 29, pages 187–201. AMS, 1993. Cited on page(s) 217.

Gatermann, K. The moregroebner package version 2.1 - an improvement of the grobner package, 1996a. A Maple program. Available by www by address http://www.zib.de/gatermann/moregroebner.html. Cited on page(s) 216.

Gatermann, K. Semi-invariants, equivariants and algorithms. *Applicable Algebra in Engineering, Communications and Computing*, 7:105–124, 1996b. Cited on page(s) 34, 107, 216.

Gatermann, K. SYMCON, symbolic part updated for REDUCE 3.6. Available by http://www.zib.de/gatermann/rsymcon.html, 1997. Cited on page(s) 217.

Gatermann, K. *Computer Algebra Methods for Equivariant Dynamical Systems*, volume 1728 of *Lecture Notes in Mathematics*. Springer, Berlin, 2000. Cited on page(s) 179, 216.

Gatermann, K. and Guyard, F. The symmetry package in Maple, 1996. Available by www by address http://www.zib.de/gatermann/symmetry.html. Cited on page(s) 35, 216.

Gatermann, K. and Guyard, F. Gröbner bases, invariant theory and equivariant dynamics. *J. Symb. Comp.*, 28:275–302, 1999. Cited on page(s) 216.

Gatermann, K. and Hohmann, A. Hexagonal lattice dome - illustration of a nontrivial bifurcation problem. Preprint SC 91-8, Konrad-Zuse-Zentrum, Berlin, 1991a. Cited on page(s) 216.

Gatermann, K. and Hohmann, A. Symbolic exploitation of symmetry in numerical pathfollowing. *Impact of Computing in Science and Engineering*, 3: 300–365, 1991b. Cited on page(s) 217.

Gatermann, K. and Lauterbach, R. Automatic classification of normal forms. *Nonlin. Anal.*, 34:157–190, 1998a. Cited on page(s) 106.

Gatermann, K. and Lauterbach, R. Automatic classification of normal forms. *Nonlinear Analysis, Theory, Methods, and Applications*, 34:157–190, 1998b. Cited on page(s) 216.

Gatermann, K. and Werner, B. Group theoretical mode interactions with different symmetries. *International Journal on Bifurcation and Chaos*, 4:177–191, 1994. Cited on page(s) 217.

Gatermann, K. and Werner, B. Secondary Hopf bifurcation caused by steady-state steady-state mode interaction. In Chaddam, J., Golubitsky, M., Langford, W., and Wetton, B., editors, *Pattern Formations: Symmetry Methods and Applications*, volume 5 of *Fields Institute Communications*, pages 209–224. AMS, Providence, R.I., 1996. Cited on page(s) 216.

Gates, B. L. A Numerical Code Generation Facility for REDUCE. In Jenks, R. D., editor, *SYMSAC'76, 1976 ACM Symposium on Symbolic and Algebraic Computation*, pages 94–99, New York, 1976. Academic Press. Cited on page(s) 151.

Gates, B. L. GENTRAN: An Automatic Code Generation Facility for REDUCE. *ACM SIGSAM Bulletin*, 19:24–42, 1985. Cited on page(s) 151.

Gates, B. L. and Wang, P. S. A LISP-based RATFOR Code Generator. In *Third MACSYMA Users' Conference, Schenectady, N.Y*, July 1984. Cited on page(s) 151.

von zur Gathen, J. Parallel algorithms for algebraic problems. *SIAM Journal on Computing*, 13:802–824, 1984. Cited on page(s) 146.

von zur Gathen, J. Irreducibility of multivariate polynomials. *Journal Comput. System Sci.*, 31:225–264, 1985. Cited on page(s) 14.

von zur Gathen, J. Algebraic complexity theory. *Ann. Review of Comp. Sci.*, 3: 317–347, 1988. Cited on page(s) 126, 128.

von zur Gathen, J. Functional decomposition of polynomials: the tame case. *Journal of Symbolic Computation*, 9(3):281–299, 1990a. Cited on page(s) 27, 123, 124.

von zur Gathen, J. Functional decomposition of polynomials: the wild case. *J. Symb. Comput.*, 10:437–452, 1990b. Cited on page(s) 27.

von zur Gathen, Joachim and Gerhard, Jürgen. *Modern Computer Algebra*. Cambridge University Press, 1999. Cited on page(s) 24, 25, 490.

von zur Gathen, J. and Kaltofen, E. Factorization of multivariate polynomials over finite fields. *Math. Comp.*, 45:251–261, 1985. Cited on page(s) 130.

von zur Gathen, Joachim and Panario, Daniel. Factoring polynomials over finite fields: a survey. *Journal of Symbolic Computation*, 31(1/2):3–17, 2001. Cited on page(s) 24.

von zur Gathen, J. and Shoup, V. Computing Frobenius maps and factoring polynomials. *Computational Complexity*, 2:187–224, 1992. Cited on page(s) 25.

von zur Gathen, J. and Weiss, J. Homogeneous bivariate decompositions. *J. Symbolic Computation*, 19:409–434, 1995. Cited on page(s) 27, 123, 233.

Gauthier, B. *Calcul symbolique sur les séries hypergéométriques*. PhD thesis, Université Marne-la-Vallée, France, 1999. Cited on page(s) 93.

Gautier, T. and Roch, J. L. PAC++ system and parallel algebraic numbers computations. In Hong, Hoon, editor, *PASCO'94*, volume 5 of *Lecture Notes Series on Computing*, pages 145–153. World Scientific Publishing, 1994. Cited on page(s) 148.

Gautier, T., Roch, J.-L., and Villard, G. Givaro. http://www-apache.imag.fr/software/givaro/, January 1999. Cited on page(s) 384.

Gebauer, R., Kalkbrener, M., Wall, B., and Winkler, F. CASA: A computer algebra package for constructive algebraic geometry. In Watt, S. M., editor, *Proceedings of ISSAC '91*, pages 403–410, Bonn, Germany, July 1991. ACM Press. Cited on page(s) 357.

Gebel, J. *Bestimmung aller ganzen und S-ganzen Punkte auf elliptischen Kurven über den rationalen Zahlen mit Anwendung auf die Mordellschen Kurven*. PhD thesis, Universität des Saarlandes, Saaarbrücken, 1996. Cited on page(s) 49.

Gebel, J., Pethö, A., and Zimmer, H. G. Computing integral points on elliptic curves. *Acta Arith.*, 68:171–192, 1994. Cited on page(s) 49.

Gebel, J., Pethö, A., and Zimmer, H. G. Computing $S$-integral points on elliptic curves. In *Lect. Notes in Comp. Sci.*, number 1122, pages 157–171, Berlin-Heidelberg-New York, 1996. Springer-Verlag. Cited on page(s) 49.

Gebel, J. and Zimmer, H. G. Computing the Mordell-Weil group of an elliptic curve. In Kisilevsky, H. and Murty, M. Ram, editors, *Elliptic Curves and Related Topics*, volume 4 of *Proc. and Lect. Notes*, pages 61–83. Centre Rech. Math., Montréal, Amer. Math. Soc., 1993. Cited on page(s) 49.

Geck, M., Hiss, G., Lübeck, F., Malle, G., and Pfeiffer, G. CHEVIE—A system for computing and processing generic character tables. *AAECC*, 7:175–210, 1996. Cited on page(s) 88, 196, 359.

Geddes, K. O. A package for numerical approximation. *Maple Technical Newsletter*, 10:28–36, 1993. Cited on page(s) 478.

Geddes, K. O., Czapor, S. R., and Labahn, G. *Algorithms for Computer Algebra*. Kluwer Academic Publishers, Boston, 1992. Cited on page(s) 4, 25, 490.

Geiselmann, W. A Note on the Hash Function of Tillich and Zémor. In Boyd, C., editor, *Cryptography and Coding*, Lecture Notes in Computer Science 1025, pages 257–263. Springer, 1995. Cited on page(s) 131.

Geiselmann, W. and Gollmann, D. Correlation attacks on cascades of clock controlled shift registers. In Kim, K. and Matsumoto, T., editors, *Advances in Crytology - ASIACRYPT 1996*, Lecture Notes in Computer Science 1163, pages 346–359. Springer, 1989. Cited on page(s) 131.

Geiselmann, W. and Gollmann, D. VLSI-design for exponentiation in $GF(2^n)$. In Seberry, J. and Pieprzyk, J., editors, *Proc. AUSCRYPT 1990*, number 453 in Lecture Notes in Computer Science, pages 398–405, Berlin-Heidelberg-New York, 1990. Springer-Verlag. Cited on page(s) 161.

Geiselmann, W. and Ulmer, F. Constructing a third order differential equation. *Theor. Comp. Sci.*, 187:3–6, 1997. Cited on page(s) 98.

Geissler, K. and Klüners, J. Galois group computation for rational polynomials. *Journal of Symbolic Computation*, 30(6):653–674, 2000. Cited on page(s) 48.

Geist, A. *PVM 3 user's guide and reference manual*. Oak Ridge National Laboratory, Oak Ridge, Tennessee 37831, 1994. Cited on page(s) 401.

Gekeler, E.-U. Automorphe Formen über $\mathbf{F}_q(T)$ mit kleinem Führer. *Abh. Math. Sem. Univ. Hamburg*, 55:111–146, 1985. Cited on page(s) 50.

Gentleman, W. M. and Johnson, S. C. The evaluation of determinants by expansion by minors and the general problem of substitution. *Mathematics of Computation*, 28:543–548, 1974. Cited on page(s) 37.

Gentleman, W. M. and Johnson, S. C. Analysis of algorithms, a case study: Determinants of matrices with polynomial entries. *ACM Trans. Math Software*, 2:232–241, 1976. Cited on page(s) 37.

Gerasoulis, A. and Yang, T. PYRROS : Static Task Scheduling and Code Generation for Message-Passing Multi-Processor. In *Proceedings of the Sixth ACM International Conference on Supercomputing*, 1992. Cited on page(s) 148.

Gerdt, Vladimir. Gröbner bases and involutive methods for algebraic and differential equations. In *Proceeding of the workshop:"Computer Algebra in Science and Engineering"*, Bielefeld, Aug'94, pages 117–138, Singapore, 1994. World Scientific. Cited on page(s) 31.

Gerdt, Vladimir. Involutive bases of polynomial ideals. *Mathematics and Computers in Simulation*, 45:519–542, 1998a. Cited on page(s) 31.

Gerdt, Vladimir. Involutive division technique: some generalizations and optimizations. Technical report, Laboratory of Computing techniques and Automation, Dubna, 1998b. Cited on page(s) 31.

Gerdt, Vladimir. Minimal involutive bases. *Mathematics and Computers in Simulation*, 45:543–560, 1998c. Cited on page(s) 31.

Gerdt, Vladimir. Completion of linear differential systems to involution. In Ganzha, V. G., Mayr, E. W., and Vorozhtov, E. V., editors, *Computer Algebra in Scientific Computation, CASC'99*, pages 115–137, Berlin, 1999. Springer. Cited on page(s) 31.

Gerdt, V. P. Gröbner bases and involutive methods for algebraic and differential equations. In Fleischer, J., Grabmeier, J., Hehl, F. W., and Küchlin, W., editors, *Computer Algebra in Science and Engineering*, pages 117–137. World Scientific, Singapore, 1995. Cited on page(s) 103.

Gerdt, V. P. and Blinkov, Yu.A. Involutive bases of polynomial ideals. *Math. Comp. Simul.*, 45:519–542, 1998a. Cited on page(s) 103.

Gerdt, V. P. and Blinkov, Yu.A. Minimal involutive bases. *Math. Comp. Simul.*, 45:543–560, 1998b. Cited on page(s) 103.

Gerdt, V. P., Kornyak, V. V., Berth, M., and Czichowski, G. Construction of involutive monomial sets for different involutive divisions. In Ganzha, V. G., Mayr, E. W., and Vorozhtov, E. V., editors, *Computer Algebra in Scientific Computation, CASC'99*, pages 147–157, Berlin, 1999. Springer. Cited on page(s) 31, 103.

Gerdt, V. P. and Lassner, W. Isomorphism verification for complex and real Lie algebras by Gröbner basis technique. In et al., N. H. Ibragimov, editor, *Modern Group Analysis: Advanced Analytical and Computational Methods in Mathematical Physics*, pages 245–254, Amsterdam, 1993. Kluwer Academic Publishers. Cited on page(s) 63.

Gessel, I. Finding identities with the WZ-method. *Journal of Symbolic Computation*, 20:537–566, 1995. Cited on page(s) 93.

Gianni, P. Properties of Gröbner bases under specialization. In Davenport, J. H., editor, *EUROCAL'87*, number 378 in LNCS, pages 293–297. Springer, 1987. Cited on page(s) 31.

Gianni, P., editor. *Symbolic and Algebraic Computation, International Symposium ISSAC '88, Rome, Italy, July 4–8, 1988, Proceedings*, volume 358 of *Lecture Notes in Computer Science*, Berlin-Heidelberg-New York, 1989. Springer-Verlag. ISBN 3-540-51084-2, 0-387-51084-2. Cited on page(s) 486, 515, 525, 583.

Gianni, P., Seppälä, M., Silhol, R., and Trager, B. Riemann surfaces, plane algebraic curves and their period matrices. *Journal of Symbolic Computation*, 26(6):789–803, 1998. Special issue on Symbolic Numeric Algebra for Polynomials S. M. Watt and H. J. Stetter, editors. Cited on page(s) 119.

Gianni, P., Trager, B., and Zacharias, G. Gröbner bases and primary decomposition of polynomial ideals. *Journal of Symbolic Computation*, 6(2):149–168, 1988. Cited on page(s) 379.

Gianni, P. and Traverso, C. Shape determination of real curves and surfaces. *Ann. Univ. Ferrara Sez VII Sec. Math.*, XXIX:87–109, 1983. Cited on page(s) 238, 240.

Gibbon, J. D., Radmore, P., Tabor, M., and Wood, D. The Painlevé property and Hirota's method. *Stud. Appl. Math.*, 72:39–63, 1985. Cited on page(s) 101.

Gibbons, A. M. and Rytter, W. *Efficient parallel algorithms.* Cambridge University Press, 1988. Cited on page(s) 146, 491.

Gielen, G. and Sansen, W. *Symbolic Analysis for Automated Design of Analog Integrated Circuits.* Kluwer Academic Publishers, Boston, 1991. Cited on page(s) 491.

Giesbrecht, M. Fast algorithms for rational forms of integer matrices. In *International Symposium on Symbolic and Algebraic Computation, Oxford, England*, pages 305–311. ACM Press, 1994a. Cited on page(s) 40.

Giesbrecht, M., editor. *ISSAC '94, International Symposium on Symbolic and Algebraic Computation, Oxford, United Kingdom*, New York, 1994b. ACM. Cited on page(s) 487, 507, 509, 545, 616.

Giesbrecht, M. Fast computation of the Smith normal form of an integer matrix. In *International Symposium on Symbolic and Algebraic Computation, Montreal, Canada*, pages 110–118, 1995a. Cited on page(s) 39.

Giesbrecht, M. Nearly optimal algorithms for canonical matrix forms. *SIAM Journal on Computing*, 24(24):948–969, 1995b. Cited on page(s) 40, 41.

Giesbrecht, M. Fast computation of the Smith normal form of a sparse integer matrix. In *Algorithms in Number Theory Symposium*, LNCS 1122, pages 173–186, 1996. Cited on page(s) 40.

Giesbrecht, M. Efficient parallel solutions of sparse systems of linear diophantine equations. In Hitz and Kaltofen [1997], pages 1–10. ISBN 0-89791-951-3. Cited on page(s) 37.

Giesbrecht, M. Fast computation of the smith form of a sparse integer matrix. *Computational Complexity*, 2001. Accepted for publication. Cited on page(s) 40.

Giesbrecht, M., Lobo, A., and Saunders, B. D. Certifying inconsistency of sparse linear systems. In Gloor [1998], pages 113–119. ISBN 1-58113-002-3. Cited on page(s) 38.

Giesbrecht, M. and Storjohann, A. Computing rational forms of integer matrices. Submitted to Journal of Symbolic Computation., 2000. Cited on page(s) 40.

Gil, Isabelle. Computation of the Jordan canonical form of a square matrix (using the AXIOM programming language). In Wang [1992], pages 138–145. Cited on page(s) 264.

Gilfoyle, G. P. A new teaching approach to quantum mechanical tunneling. *Mathematica in Education and Research*, 4(1):19–23, 1995. Cited on page(s) 255.

Giovine, A., Mora, T., Niesi, G., Robbiano, L., and Traverso, C. "One sugar cube, please" or selection strategies in the Buchberger algorithm. In M., Watt Stephen, editor, *ISSAC'91*, pages 49–54. ACM Press, 1991. Cited on page(s) 29.

Giovini, A. and Niesi, G. CocoA: A user-friendly system for commutative algebra, in A. miola, editor, design and implementation of symbolic computation systems, international symposium DISCO '90, capri, italy. In *Lecture Notes in Computer Science, vol. 429*, pages 20–29, Berlin-Heidelberg-New York, 1990. Springer-Verlag. Cited on page(s) 381.

Giusti, M., Hägele, K., Heintz, J., Morais, J. E., Montaña, J. L., and Pardo, L. M. Lower bounds for diophantine approximation. *J. of Pure and Appl. Agebra*, 117 & 118:277–317, 1997a. Cited on page(s) 54.

Giusti, M. and Heintz, J. Algorithmes - disons rapides - pour la décomposition d' une variété algébrique en composantes irréductibles et équidimensionelles. In Mora, T. and Traverso, C., editors, *MEGA'90*, Progress in Mathematics 94, pages 169–194, Castiglioncello, 1991. Birkhäuser. Cited on page(s) 54.

Giusti, M. and Heintz, J. La détermination des points isolés et de la dimension d'une variété algébrique peut se faire en temps polynomial. In Eisenbud, D. and Robbiano, L., editors, *Computational Algebraic Geometry and Commutative Algebra*, Symposia Matematica XXXIV, pages 216–256, Cortona, 1993. Cambridge University Press. Cited on page(s) 14, 54.

Giusti, M., Heintz, J., Morais, J. E., Morgenstern, J., and Pardo, L. M. Straight–line programs in geometric elimination theory. *J. of Pure and App. Algebra*, 124:101–146, 1998. Cited on page(s) 54.

Giusti, M., Heintz, J., Morais, J. E., and Pardo, L. M. When polynomial equation systems can be "solved" fast? In Cohen, G., Giusti, M., and Mora, T., editors, *AAECC-11*, Lecture Notes in Computer Science 948, pages 205–231, Paris, 1995. Springer. Cited on page(s) 54.

Giusti, M., Heintz, J., Morais, J. E., and Pardo, L. M. Le Rôle des structures des donnèes dans les problèmes d'èlimination. *C. R. Acad. Sci. Paris*, 325: 1223–1228, 1997b. Cited on page(s) 54.

Giusti, M., Lecerf, G., and Salvy, B. A Gröbner free alternative for polynomial system solving. *Journal of Complexity*, 17(1):154–211, March 2001. Cited on page(s) 54.

Giusti, M. and Schost, E. Solving some over–determined systems. In Dooley [1999], pages 1–8. Cited on page(s) 54.

Gloor, O., editor. *ISSAC 98 Proc. 1998 Internat. Symp. Symbolic Algebraic Comput.*, New York, N. Y., 1998. ACM Press. ISBN 1-58113-002-3. Cited on page(s) 489, 509, 517, 525, 526, 541, 552, 601, 608, 610.

Gloor, O., Amrhein, B., and Maeder, R. E. *Illustrierte Mathematik: Visualisierung von mathematischen Gegenständen*. Birkhäuser, 1994. ISBN 3-7643-5100-4. CD-ROM mit Begleitheft. Cited on page(s) 258.

Gloor, O., Amrhein, B., and Maeder, R. E. *Illustrated Mathematics*. TELOS/Springer, 1995. ISBN 0-387-14222-3. CD-ROM with Booklet. Cited on page(s) 258, 491.

Gloor, Oliver and Müller, Stefan. PARCAN—a parallel computer algebra nucleus. In preparation, 1998. Cited on page(s) 351.

Göbel, Manfred. Reduktion G-symmetrischer Polynome für beliebige Permutationsgruppen G. Diplomarbeit, Universität Passau, Passau, 1992. Cited on page(s) 427.

Göbel, Manfred. Computing bases for rings of permutation-invariant polynomials. *J. Symbolic Computation*, 19:285–291, 1995. Cited on page(s) 35.

Göbel, Manfred. *Computing Bases for Permutation-Invariant Polynomials*. Dissertation, Universität Tübingen, 1996. Cited on page(s) 427.

Göbel, Manfred. The Invariant Package of MAS. In Comon, Hubert, editor, *Rewriting Techniques and Applications, 7th Intl. Conf., RTA-96*, number 1232 in LNCS, pages 327–330, Berlin, Heidelberg, 1997. Springer-Verlag. Cited on page(s) 35.

Goguen, J. A, Thatcher, J. W, and Wagner, E. G. An initial algebra approach to the specification, correctness and implementation of abstract data types. In Yeh [1978], pages 80–144. Cited on page(s) 142.

Göktaş, Ü. and Hereman, W. Symbolic computation of conserved densities for systems of nonlinear evolution equations. *Journal of Symbolic Computation*, 24:591–621, 1997. Cited on page(s) 189, 467.

Gollan, H. and Grabmeier, J. Algorithms in representation theory and their realization in the computer algebra system Scratchpad. *Bayreuther Mathematische Schriften*, 33:1–23, 1990. Cited on page(s) 264.

Gomez, C. MACROFORT: A FORTRAN Code Generator for MAPLE. Technical report, Institut National de Recherche en Informatique et en Automatique, 1990. Cited on page(s) 151.

Gonnet, G., editor. *Proceedings of the ACM-SIGSAM 1989 International Symposium on Symbolic and Algebraic Computation, ISSAC '89, July 17-19, 1989, Portland, Oregon*, New York, 1989. ACM. ISBN 0-89791-325-6. Cited on page(s) 486, 543, 613.

Gonzàles-Vega, L. and Recio, T., editors. *Algorithms in Algebraic Geometry and Applications*, volume 143 of *Progress in Mathematics*, 1996. Birkhäuser. Cited on page(s) 487.

Gonzàlez, L., Lombardi, H., Recio, T., and Roy, M.-F. Sturm-Habicht sequence. In Gonnet [1989], pages 136–146. ISBN 0-89791-325-6. Cited on page(s) 57.

Gonzàlez, L., Lombardi, H., Recio, T., and Roy, M.-F. Sous-résultants et spécialisation de la suite de Sturm I. *Informatique théorique et applications*, 24 6:561–588, 1990. Cited on page(s) 57.

Gonzalez-Lopez, M. J. and Recio, T. On the symbolic insimplification of the general 6R-manipulator kinematic equations. In *ISSAC 94*, pages 354–358. ACM, 1994. Cited on page(s) 233.

González-Vega. A combinatorial algorithm solving some quantifier elimination problems. In Caviness, B. F. and Johnson, J. R., editors, *Quantifier Elimination and Cylindrical Algebraic Decomposition*, Texts and Monographs in Symbolic Computation, pages 365–375. Springer, Wien, New York, 1998. Cited on page(s) 139.

González-Vega, Laureano. Applying quantifier elimination to the Birkhoff interpolation problem. *Journal of Symbolic Computation*, 22(1):83–103, July 1996. Cited on page(s) 139.

Gonzalez-Vega, L. Implicitization of parametric curves and surfaces by using multidimensional newton formulae. *Journal of Symbolic Computation*, 23:137–151, 1997. Cited on page(s) 235.

Gonzalez-Vega, L. and Kahoui, M. El. An improved upper complexity bound for the topology computation of a real algebraic plane curve. *Journal of Complexity*, 12:527–544, 1996. Cited on page(s) 20, 238.

Gonzalez-Vega, L. and Necula, I. Tracing, algebraically driven, of planar curves implicitely defined. Available at http://frisco.matesco.unican.es, 1999. Cited on page(s) 238.

Gonzalez-Vega, L., Rouillier, F., and Roy, M.-F. Symbolic recipes for polynomial system solving. In Cohen, A., Cuypers, H., and Sterk, H., editors, *Some tapas of computer algebra*. Springer-Verlag, Berlin-Heidelberg-New York, 1998a. Cited on page(s) 57.

Gonzalez-Vega, L., Roy, M.-F., Rouillier, F., and Trujillo, G. Symbolic recipes for real solutions. In Cohen, A., Cuypers, H., and Sterk, H., editors, *Some tapas of computer algebra*. Springer, Berlin, 1998b. Cited on page(s) 20, 57, 492.

Goppa, V. D. *Geometry and Codes*. Kluwer Academic Publishers, Boston, 1988. Cited on page(s) 491.

Gosper jr., R. W. Decision procedures for indefinite hypergeometric summation. *Proc. Natl. Acad. Sci. USA*, 75:40–42, 1978. Cited on page(s) 91, 215.

Goss, D. *Basic Structures of Function Field Arithmetic*, volume 35 of *Erg. Math.* Springer-Verlag, Berlin-Heidelberg-New York, 1996. Cited on page(s) 50.

Goss, D., Hayes, D. R., and Rosen, M. I., editors. *The Arithmetic of Function Fields*. W. De Gruyter, Berlin, 1992. Cited on page(s) 50.

Gottesman, Daniel. *Stabilizer Codes and Quantum Error Correction*. PhD thesis, California Institute of Technology, Pasadena, California, 1997. Cited on page(s) 130.

Göttsch, Dörte. Entscheidbare Theorien der Form Th(N,+f) für einstellige Funktionen f. Master's thesis, Universität Kiel, 1988. Advisor: K. Potthoff. Cited on page(s) 138.

Graaf, W. De. *Algorithms for the Structure Determination of Lie Algebras*. Dissertation, Eindhoven University of Technology, Eindhoven, 1997. Cited on page(s) 63.

Gräbe, Hans-Gert. On lucky primes. *J. Symb. Computation*, 15:199–209, 1993. Cited on page(s) 30.

Gräbe, H.-G. Two remark on independent sets. *J. Alg. Comb.*, 2:137–145, 1993. Cited on page(s) 463.

Gräbe, H.-G. The tangent cone algorithm and homogenization. *J. Pure Applied Algebra*, 97:303–312, 1994. Cited on page(s) 32, 464.

Gräbe, H.-G. Algorithms in local algebra. *J. Symb. Comp.*, 19:545–557, 1995a. Cited on page(s) 464.

Gräbe, H.-G. On factorized Groebner bases. In Fleischer, J., Grabmeier, J., Hehl, F. W., and Küchlin, W., editors, *Computer Algebra in Science and Engineering*, pages 77–89, Singapore, 1995b. World Scientific. Cited on page(s) 464.

Gräbe, H.-G. Triangular systems and factorized Groebner bases. In G., Cohen, M., Giusti, and T., Mora, editors, *AAECC-11*, number 948 in LNCS, pages 248–261. Springer, 1995c. Cited on page(s) 464.

Gräbe, H.-G. Minimal primary decomposition and factorized Groebner bases. *J. AAECC*, 8:265–278, 1997. Cited on page(s) 52, 463.

Gräbe, H.-G. and W., Lassner. A parallel gröbner factorizer. In Hong, Hoon, editor, *PASCO'94*, pages 174–180. World Scientific, 1994. Cited on page(s) 52.

Grabmeier, J. An application of a symbolic version of the Raleigh-Ritz method to the design of aircraft turbines. In Fleischer et al. [1995], pages 317–330. Cited on page(s) 227, 264.

Grabmeier, J., Huber, K., and Krieger, U. Das Computeralgebra-System AXIOM bei kryptologischen und verkehrstheoretischen Untersuchungen des Forschungsinstituts der Deutschen Bundespost TELEKOM. Technischer Report TR 75.91.20, IBM Wissenschaftliches Zentrum, Heidelberg, 1991. Cited on page(s) 264.

Grabmeier, Johannes and Lambe, Larry. Computing resolutions over finite $p$-groups. In Betten, A., Kohnert, A., Laue, R., and Wassermann, A., editors, *Algebraic Combinatorics and Applications*, pages 157–195, Berlin-Heidelberg-New York, 2001. Springer-Verlag. ISBN 3-540-41110-0. Cited on page(s) 211.

Grabmeier, J. and Scheerhorn, A. Finite fields in AXIOM. AXIOM Technical Report ATR/5, NAG Ltd., Oxford, 1992. Cited on page(s) 22, 264.

Grabmeier, J. and Wisbauer, R. Computations in algebras of finite rank. In Fischer, K. G., Loustaunau, P., Shapiro, J., Green, E. L., and Farkas, D., editors, *Computational Algebra*, volume 151 of *Lectures Notes in Pure and Applied Mathematics*, pages 131–165, New York, 1993. Marcel Dekker. Cited on page(s) 63, 64, 264.

Graham, E. E. and Phillip, G. *Syzygies*. Cambridge University Press, Cambridge, 1985. Cited on page(s) 490.

Graham, R. L. Bounds on multiprocessing timing anomalies. *SIAM J. Appl. Math.*, 17(2):416–426, 1969. Cited on page(s) 147.

Graham, R. L., Knuth, D. E., and Patashnik, O. *Concrete Mathematics. A foundation for Computer Science*. Addison-Wesley, Reading, Massachussets, second edition, 1994. Cited on page(s) 4.

Granlund, T. *The GNU Multiple Precision Arithmetic Library, gmp 2.0.2*. Free Software Foundation, 1996. 45 pages. Cited on page(s) 12, 13, 351, 352, 404, 464.

Grassl, Markus, Beth, Thomas, and Pellizzari, Thomas. Codes for the Quantum Erasure Channel. *Physical Review A*, 56(1):33–38, July 1997. LANL preprint quant–ph/9610042. Cited on page(s) 130.

Grassmann, H., Greuel, G. M., Martin, B., Neumann, W., Pfister, G., Pohl, W., Schönemann, H., and Siebert, T. On an implementation of standard bases and syzygies in SINGULAR. *Computational methods in Lie Theory, AAECC*, 7:235–249, 1996. Cited on page(s) 448.

Gray, S., Kajler, N., and Wang, P. MP: A Protocol for Efficient Exchange of Mathematical Expressions. In Giesbrecht [1994b], pages 330–335. Cited on page(s) 153.

Gray, T. and Glynn, J. *Exploring Mathematics with Mathematica*. Addison-Wesley, Reading, Massachusetts, 1991. Cited on page(s) 257, 491.

Grayson, D. and Stillman, M. *Macaulay 2*. Available via anonymous ftp from math.uiuc.edu, 1996. Available via anonymous ftp from math.uiuc.edu. Cited on page(s) 210, 381.

Green, D. J. *Constructing projective resolutions for p-groups*, volume 24 of *Vorlesungen aus dem Fachbereich Mathematik der Universität Essen*. Universität Essen, Fachbereich Mathematik, Essen, 1997. Cited on page(s) 197.

Greuel, G.-M. Constant Milnor number implies constant multiplicity for quasihomogeneous singularities. *Manuscripta Math*, 56:159–166, 1986. Cited on page(s) 200.

Greuel, G.-M., Lossen, C., and Shustin, E. Plane curves of minimal degree with prescribed singularities. *Inventiones mathematicae*, 133:539–580, 1998. Cited on page(s) 201.

Greuel, G.-M., Martin, B., and Pfister, G. Numerische Charakterisierung quasihomogener Gorenstein-Kurvensingularitäten. *Math. Nachr.*, 124:123–131, 1985. Cited on page(s) 198.

Greuel, G.-M. and Pfister, G. On moduli spaces of semiquasihomogeneous singularities. Preprint, Universität Kaiserslautern, Kaiserslautern, 1993. Cited on page(s) 199.

Greuel, G. M. and Pfister, G. Advances and improvements in the theory of standard bases and szyzygies. *Arch. d. Math.*, 66:163–176, 1996. Cited on page(s) 32, 55, 199, 200, 448.

Greuel, G. M. and Pfister, G. Groebner bases and algebraic geometry. In Buchberger, B. and Winkler, F., editors, *Groebner bases and applications*, volume 251 of *London Mathematical Society Lecture Notes Series*, pages 109–143. Cambridge University Press, 1998. Cited on page(s) 32, 55, 448.

Greuel, G. M., Pfister, G., and Schönemann, H. *Singular:, A System for Computation in Algebraic Geometry and Singular Theory*. Available via anonymous ftp from helios.mathematik.uni-kl.de, 1995. Available via anonymous ftp from helios.mathematik.uni-kl.de. Cited on page(s) 210, 381.

Greuel, G.-M., Pfister, G., and Schönemann, Hans. Singular Reference Manual. In *Reports On Computer Algebra*, number 12. Centre for Computer Algebra, University of Kaiserslautern, version 1.0, May 1997. http://www.mathematik.uni-kl.de/~zca/Singular. Cited on page(s) 446, 448.

Greuel, G.-M. and Stobbe, R. FACTORY, a subroutine library for factorization, 1993-1997. Cited on page(s) 412.

Griewank, A. and Corliss, G. F., editors. *Automatic Differentiation of Algorithms: Theory, Implementation,and Application*, volume 53 of *Proceedings in Applied Mathematics*. SIAM, 1991. Cited on page(s) 492.

Grigor'ev, D. Complexity of deciding Tarski algebra. *Journal of Symbolic Computation*, 5(1):65–108, 1988. Cited on page(s) 57.

Grigor'ev, D. and Vorobjov, N. Counting connected components of semialgebraic set in subexponential time. *Computational Complexity*, 2:133–186, 1992. Cited on page(s) 57.

Grigor'ev, D. Yu. Complexity of quantifier elimination in the theory of ordinary differential equations. In Davenport, J. H., editor, *Proc. EUROCAL '87*, Lecture Notes in Computer Science 378, pages 11–25, Leipzig, 1987. Springer. Cited on page(s) 105.

Grigoriev, D. Yu. and Lakshman, Y. N. Algorithms for computing sparse shifts for multivariate polynomials. *Journal of Appl. Alg. in Eng. Comm. and Comp.*, 11(1):43–67, 2000. Cited on page(s) 384.

Gröbner, W. *Algebraische Geometrie*, volume I. Bibliographisches Institut, Mannheim, 1968. Cited on page(s) 491.

Gröbner, W. *Algebraische Geometrie*, volume II. Bibliographisches Institut, Mannheim, 1970. Cited on page(s) 491.

Grosse-Gehling, Rainer. Konstruktion involutiver Basen im Computeralgebra-System MAS. Diplomarbeit, Universität Passau, Passau, May 1995. Cited on page(s) 427.

Grossman, Robert, editor. *Symbolic Computation - Applications to Scientific Computing*, volume 5 of *Frontiers in Applied Algebra*, Philadelphia, 1989. SIAM Press. Cited on page(s) 486.

Grozin, A. G. *Using REDUCE in High Energy Physics*. Cambridge University Press, 1991. Cited on page(s) 491.

GRTensor. GRTensorII: Online documentation and information, 2001. http://grtensor.phy.queensu.ca. Cited on page(s) 175.

Gruntz, D. Symbolic computation of explicit Runge-Kutta formulas. In Gander, W. and rebíček, J. H, editors, *Solving Problems in Scientific Computing Using Maple and MATLAB*, chapter 19, pages 267–283. Springer-Verlag, Berlin, 1995. Cited on page(s) 108.

Gugenheim, V. K. A. M. On the chain-complex of a fibration. *Illinois J. Math.*, 16:398–414, 1972. Cited on page(s) 210.

Gugenheim, V. K. A. M. and Milgram, J. On successive approximations in homological algebra. *Trans. Amer. Math. Soc.*, 150:157–181, 1970. Cited on page(s) 212.

Gurevich, Yuri. Elementary properties of ordered abelian groups. *Translations AMS*, 46:165–192, 1965. Cited on page(s) 137.

Gurevich, Yuri. Expanded theory of ordered abelian groups. *Annal of Math. Logic*, 12:193–228, 1977. Cited on page(s) 137, 138.

Gut, C. *Faktorisierung von multivariaten Polynomen über endliche n Körpern unter Verwendung der Blackbox-Darstellung.* Term project, ETH, Zurich, Switzerland, January 1999. Cited on page(s) 385.

Gutierrez, J. A polynomial decomposition algorithm over factorial domains. *Comptes Rendus Mathematiques de l'Academie des Sciences*, 13(2–3):81–86, April-June 1991. Cited on page(s) 27.

Gutierrez, J. and Recio, T. Advances on the simplification of sine-cosine equations. *J. Symbolic Computation*, 26:31–70, 1998. Cited on page(s) 28.

Gutierrez, J., Recio, T., and de Velasco, C. Ruiz. A polynomial decomposition algorithm of almost quadratic complexity. In *Proc. AAECC-6/88*, volume 357 of *Lect. Notes in Comput. Sci.*, pages 471–476. Springer-Verlag, 1989. Cited on page(s) 27.

Hafner, J. L. and McCurley, K. S. Asymptotically fast triangularization of matrices over rings. *SIAM Journal on Computing*, 20(6):1068–1083, December 1991. Cited on page(s) 39.

Hägele, K., Morais, J. E., Pardo, L. M., and Sombra, M. The intrinsic complexity of the arithmetic nullstellensatz. *J. of Pure and App. Algebra*, 146:103–183, 2000. Cited on page(s) 54.

Hahn, T. and Pérez-Victoria, M. Automatized one-loop calculations in 4 and $D$ dimensions. *Comput. Phys. Commun.*, 118:153–165, 1999. Cited on page(s) 170, 471.

Haible, B. CLN, a class library for numbers. http://clisp.cons.org/~haible/packages-cln.html, 1998. 66 pages. Cited on page(s) 12, 13, 404.

Haible, Bruno and Papanikolaou, Thomas. Fast multiprecision evaluation of series of rational numbers. *Technical Report, Fachbereich Informatik, Technische Hochschule Darmstadt*, 1997. Available from http://www.informatik.tu-darmstadt.de/TI/Veroeffentlichung/TR/Welcome.html, http://clisp.cons.org/~haible/papers/ca/binsplit-slides1997/. Cited on page(s) 464.

Halstead, Robert H. Parallel symbolic computing. *IEEE Computer*, 19(8):35–43, 1986. Cited on page(s) 147.

Hansen, E.R. *Global Optimization using Interval Analysis*. Marcel Dekker, New York, Basel, Hong Kong, 1992. Cited on page(s) 111.

Hantzschmann, K. Implementierbare Fehlerabschätzungen für Näherungslösungen von Randwertaufgaben bei Systemen gewöhnlicher Differentialgleichungen. Studientexte Weiterbildungszentrum für Mathematische Kybernetik und Rechentechnik / Informationsverarbeitung 73, Technische Universität Dresden, 1984. Cited on page(s) 109.

Hantzschmann, K. Probleme und Algorithmen der Computeranalytik. In *Mitteilungen der MGDDR/Mathematikerkongress, Bd. 2*, pages 61–67. MGDDR, 1990. Cited on page(s) 109.

Hantzschmann, K. and Jung, A. Adaptive Approximation zur näherungsweisen Lösung von Differentialgleichungen. Rostocker Informatik-Berichte 18, Universität Rostock, 1995. http://www.informatik.uni-rostock.de/~ajung/. Cited on page(s) 110.

Hantzschmann, K. and Thinh, N. X. Analytical approximate solution of singular ordinary differential equations. In Shirkov et al. [1991], pages 169–174. ISBN 981-02-0687-9. Cited on page(s) 110.

Hardy, G. H., Littlewood, J. E., and Polya, G. A. *Inequalities*. Cambridge University Press, 2nd edition, 1951. Cited on page(s) 117.

Hardy, G. H and Wright, E. M. *An Introduction to the Theory of Numbers*. Clarendon Press, Oxford, fifth edition, 1979. Cited on page(s) 491.

Harlander, R. *Ph. D. thesis, University of Karlsruhe, to be published*, 1998. Cited on page(s) 168.

Harlander, R., Seidensticker, T., and Steinhauser, M. *Phys. Lett.*, B426:125–132, 1998. Cited on page(s) 168.

Harlander, R. and Steinhauser, M. Automatic computation of Feynman diagrams. *Progress in Particle and Nuclear Physics*, 43(1):167–228, 1999. Cited on page(s) 165.

Harper, David, Wooff, Chris, and Hodgkinson, David. *A Guide to Computer Algebra Systems*. John Wiley&Sons, New York, 1991. Cited on page(s) 491.

Harper, John F. and Dyer, Charles C. *Tensor Algebra with REDTEN*, 1994. http://www.scar.utoronto.ca/~harper/redten/root.html. Cited on page(s) 175.

Hartley, D. EDS: A Reduce package for exterior differential systems. *Computer Physics Communications*, 100:177–194, 1996a. Cited on page(s) 182.

Hartley, D. Overview of computer algebra in relativity. In Hehl et al. [1996], pages 173–191. Cited on page(s) 172, 174.

Hartley, D. Involution analysis for nonlinear exterior differential systems. *Math. Comput. Modelling*, 25(8/9):51–62, 1997. Cited on page(s) 182.

Hartley, D. H. and Tucker, R. W. A constructive implementation of the Cartan-Kähler theory of exterior differential systems. *Journal of Symbolic Computation*, 12:655–667, 1991. Cited on page(s) 102.

Havas, G. and Majewski, B. S. Integer matrix diagonalization. *Journal of Symbolic Computation*, 24:399–408, 1997. Cited on page(s) 39, 51, 76.

Havas, G., Majewski, B. S., and Matthews, K. R. Extended Gcd and Hermite normal form algorithms via lattice basis reduction. *Experimental Mathematics*, 7(2):125–135, 1998. Cited on page(s) 39.

Havas, George and Newman, M. F. Application of computers to questions like those of Burnside. In *Burnside Groups*, volume 806 of *Lecture Notes in Math.*, pages 211–230, Berlin, Heidelberg, New York, 1980. (Bielefeld, 1977), Springer-Verlag. Cited on page(s) 459.

Havas, George, Newman, M. F. Niemeyer, Alice C. and Sims, Charles. Groups with Exponent Six. *Communications in Algebra*, 28:3619–3638, 1999. Cited on page(s) 460.

Hayes-Roth, F., Waterman, D. A., and Lenat, D. B., editors. *Building Expert Systems*. Addison-Wesley, Reading, Massachusetts, 1983. ISBN 0-201-10686-8. Cited on page(s) 141, 142, 491.

Head, Alan K. LIE: a PC program for Lie analysis of differential equations. *Computer Physics Communications*, 77,2:241–248, 1993. Cited on page(s) 411.

Head, Alan K. LIE: a PC program for Lie analysis of differential equations. *Computer Physics Communications*, 96,2:311–313, 1996. Cited on page(s) 411.

Hearn, A. *Reduce User's Manual*. The Rand Corp., Santa Monica, CA, 1985. Cited on page(s) 169.

Hearn, A. C. *REDUCE user's manual Version 3.5*. RAND publication, Santa Monica, 1993. Cited on page(s) 165, 492.

Hearn, A. C. *REDUCE User's Manual, Version 3.6*. The Rand Corporation, Santa Monica (CA), 1996. Cited on page(s) 186, 468, 491.

Heckenberger, I. and Schüler, A. Exterior algebras related to the quantum group $\mathcal{O}(O_q(3))$. *Czechoslovak Journal of Physics*, 48:1355–1362, 1998. Cited on page(s) 214.

Heckler, C., Metzner, T., and Zimmermann, P. Progress report on parallelism in MUPAD, 1997. MUPAD Report, Dpt. Math. Comp. Sc., Univ. Paderborn, Germany. Cited on page(s) 147.

Heegard, C., Little, J., and Saints, K. Systematic encoding via Gröbner bases of geometric Goppa codes. *IEEE Transactions on Information Theory*, 41:1752–1761, 1995. Cited on page(s) 129.

Hehl, F. W., Puntigam, R. A., and Ruder, H., editors. *Relativity and scientific computing*, Berlin-Heidelberg-New York, 1996. Springer-Verlag. Cited on page(s) 549, 611, 612, 619.

Hehl, F. W., Winkelmann, V., and Meyer, H. *Computer-Algebra. Ein Kompaktkurs über die Anwendung von REDUCE*. Springer-Verlag, Berlin-Heidelberg-New York, 1992. Cited on page(s) 491, 492.

Heid, K., Blume, G., Flanagan, K., Iseri, L., and Kerr, K. The impact of CAS on nonroutine problem solving by college mathematics students. *The International Journal of Computer Algebra in Mathematics Education*, 5(4):215–250, 1998. Cited on page(s) 250.

Heinemeyer, S., Hollik, W., and Weiglein, G. Precise prediction for the mass of the lightest Higgs boson in the MSSM. *Phys. Lett.*, B440:296, 1998a. Cited on page(s) 171.

Heinemeyer, S., Hollik, W., and Weiglein, G. QCD corrections to the masses of the neutral $CP$-even Higgs bosons in the MSSM. *Phys. Rev.*, D58:091701, 1998b. Cited on page(s) 170, 171.

Heinemeyer, S., Hollik, W., and Weiglein, G. *FeynHiggs*: a program for the calculation of the masses of the neutral $CP$-even Higgs bosons in the MSSM. *Comput. Phys. Commun.*, 124:76, 2000. Cited on page(s) 171.

Heintz, J., Matera, G., Pardo, L. M., and Wachenchauzer, R. On the intrinsic complexity of parametric elimination methods. *Electronic J. of SADIO*, 1: 38–51, 1998. Cited on page(s) 54.

Heintz, J., Recio, T., and Roy, M.-F. Algorithms in real algebraic geometry and applications to computational geometry. *Dimacs Series in Discrete Mathematics and Theoretical Computer Science*, 6:137–163, 1991a. Cited on page(s) 126, 128.

Heintz, J., Roy, M.-F., and Solernó, P. On the complexity of semi-algebraic sets. In Ritter, G. X., editor, *Proceedings IFIP 89*, pages 293–298, Amsterdam, 1989. North-Holland. Cited on page(s) 57.

Heintz, J., Roy, M.-F., and Solernó, P. Sur la complexité du principe de Tarski-Seidenberg. *Bull. Soc. Math. France*, 118:101–126, 1990. Cited on page(s) 57.

Heintz, J., Roy, M.-F., and Solernó, P. Description de composantes connexes d'un ensamble semi-algébrique en temps simplement exponentiel. *C.R. Acad. Sci.*, 311:167–70, 1991b. Cited on page(s) 57.

Heintz, J., Roy, M.-F., and Solernó, P. Single exponential path finding in semi-algebraic sets I: The case of smooth compact hypersurface. In Sakata [1991], pages 180–196. ISBN 3-540-54195-0, 0-387-54195-0. Cited on page(s) 57.

Heintz, J., Roy, M.-F., and Solernó, P. Single exponential path finding in semi-algebraic sets II: The general case. In Bajaj, C., editor, *Algebraic Geometry and Its Applications: A Collection of Papers from Shreeram S. Abhyankar's 60th Birthday Conference*, Berlin-Heidelberg-New York, 1995. Springer-Verlag. Cited on page(s) 57.

Heintz, J. and Schnorr, C. P. Testing polynomials which are easy to compute. In *Proceedings of the 12th Ann. ACM Symp. on Ther. of Comput.*, pages 262–272, New York, 1980. ACM Press. Cited on page(s) 14.

Heller, Barbara. *Macsyma for Statisticians.* John Wiley & Sons, 1991. ISBN 0-471-62590-6. Cited on page(s) 283, 491.

Hendriks, P. and van der Put, M. Galois action on solutions of linear differential equations. *Journal of Symbolic Computation*, 19:559–576, 1995. Cited on page(s) 97, 98.

Hensel, K. *Theorie der Algebraischen Zahlen.* B.G. Teubner, Leipzig, 1908. Cited on page(s) 491.

Hentzel, I. R. and Jacobs, D. P. A dynamic programming method for building free algebras. *Computers & Mathematics with Applications*, 22:61–66, 1991. Cited on page(s) 64, 347.

Hentzel, I. R., Jacobs, D. P., and Kleinfeld, E. Rings with $(a, b, c) = (a, c, b)$ and $(a, [b, c], d) = 0$: A case study using albert. *Int. J. of Computational Mathematics*, 49:19–27, 1993a. Cited on page(s) 348.

Hentzel, I. R., Jacobs, D. P., and Muddana, S. V. Experimenting with the identity $(xy)z = y(zx)$. *Journal of Symbolic Computation*, 16:289–293, 1993b. Cited on page(s) 348.

Hentzel, I. R., Jacobs, D. P., Peresi, L. A., and Sverchkov, S. R. Solvability of the ideal of all weight zero elements in bernstein algebras. *Communications in Algebra*, 22:3265–3275, 1994. Cited on page(s) 348.

Hereman, W. Review of symbolic software for the computation of Lie symmetries of differential equations. *Euromath Bull.*, 2:45–82, 1994. Cited on page(s) 97, 99, 103.

Hereman, W. Symbolic software for lie symmetry analysis. In Ibragimov, N. H., editor, *CRC Handbook of Lie Group Analysis of Differential Equations, Vol. 3: New trends in Theoretical Developments and Computational Methods*, chapter 13, pages 367–413. CRC Press, Boca Raton, Florida, 1996. Cited on page(s) 97, 99, 103, 467.

Hereman, W., Goktaş, U., Colagrosso, M., and Miller, A. Algorithmic integrability tests for nonlinear differential and lattice equations. *Comp. Phys. Comm.*, 115:428–446, 1998. Cited on page(s) 101.

Hereman, W. and Zhuang, W. Symbolic software for soliton theory. *Acta Appl. Math.*, 39:361–378, 1995. Cited on page(s) 101.

Hermite, E. Sur l'intégration des fractions rationelles. *Nouvelles Annales de Mathématiques ($2^{\text{eme}}$ série)*, 11:145–148, 1872. Cited on page(s) 94.

Herzberger, J., editor. *Topics in Validated Computations — Studies in Computational Mathematics.* Elsevier, AElsevier, 1994. Cited on page(s) 111, 492.

Heugl, H., Klinger, W., and Lechner, J. *Mathematikunterricht mit Computeralgebra-Systemen.* Addison-Wesley, Bonn, 1996. Cited on page(s) 246, 491.

Heydtmann, Agnes Eileen. `finvar.lib`: A Singular-Library to Compute Invariant Rings and More, 1997. http://www.mathematik.uni-kl.de/ftp/pub/Math/Singular/bin/Singular-1.1-share.tar.gz. Cited on page(s) 35.

Higham, Nicholas J. *Accuracy and Stability of Numerical Algorithms.* SIAM, 1996. Cited on page(s) 116.

Hilbert, David. Über die Theorie der algebraischen Formen. *Math. Ann.*, 36: 473–534, 1890. Cited on page(s) 34.

Hillebrand, R. Flexible mathematical GUI controls for mathematical electronical documents. Talk at the IAMC workshop, ISSAC'99, Vancouver, Canada, July 1999. http://www.mupad.de/CONF/issac99.ps.gz. Cited on page(s) 328.

Hiss, G., Jansen, C., Lux, K., and Parker, R. *Computational Modular Character Theory*. http://www.math.rwth-aachen.de/~MOC/CoMoChaT/, 1993. Cited on page(s) 430.

Hiss, G. and Lux, K. *Brauer Trees of Sporadic Groups*. Clarendon Press, Oxford, 1989. Cited on page(s) 196, 430, 490.

Hitz, M. and Kaltofen, E., editors. *Second Internat. Symp. Parallel Symbolic Comput. PASCO '97*, New York, 1997. ACM. ISBN 0-89791-951-3. Cited on page(s) 146, 488, 499, 541.

Hitz, M. A. and Kaltofen, E. Efficient algorithms for computing the nearest polynomial with constrained roots. In Gloor [1998], pages 236–243. ISBN 1-58113-002-3. Cited on page(s) 117, 121.

Hitz, M. A., Kaltofen, E., and Lakshman Y. N. Efficient algorithms for computing the nearest polynomial with A real root and related problems. In Dooley [1999], pages 205–212. Cited on page(s) 117, 121, 124, 125.

Hoffmann, C. M. Algebraic and numerical techniques for offsets and blends. In *Computation of Curves and Surfaces*, volume 307 of *NATO ASI Series*, pages 499–528. Kluwer Academic Publishers, 1989a. Cited on page(s) 240.

Hoffmann, C. M. *Geometric and Solid Modelling: An Introduction*. Morgan Kaufmann Publishers, 1989b. Cited on page(s) 235, 240, 491.

Hoffmann, C. M. and Vermeer, P. J. Eliminating extraneous solutions in curve and surface operations. *International Journal of Computational Geometry and Applications*, 1(1):47–66, 1991. Cited on page(s) 240.

Høholdt, T. and Pellikaan, R. On the decoding of algebraic-geometric codes. *IEEE Transactions on Information Theory*, 41:1589–1614, 1995. Cited on page(s) 129.

Holdener, J. Calculus&Mathematica: Instructors' perspectives on continuing controversies. *Mathematica in Education and Research*, 6(4):6–10, 1997. Cited on page(s) 255.

Holt, D. F. The mechanical computation of first and second cohomology groups. *J. Symb. Comput.*, 1:351–361, 1985. Cited on page(s) 87, 197.

Holt, D. F. and Hurt, D. F. Computing automatic coset systems and subgroup presentations. *J. Symb. Comput.*, 27:1–19, 1998. Cited on page(s) 196.

Holt, Derek F. Leedham-Green, C. R. O'Brien, E. A. and Rees, Sarah. Computing matrix group decompositions with respect to a normal subgroup. *J. Algebra*, 184:818–838, 1996a. Cited on page(s) 475.

Holt, Derek F. Leedham-Green, C. R. O'Brien, E. A. and Rees, Sarah. Testing matrix groups for primitivity. *J. Algebra*, 184:795–817, 1996b. Cited on page(s) 475.

Holt, D. F. and Plesken, W. *Perfect Groups*. Clarendon Press, Oxford, 1989. Cited on page(s) 87, 197, 389, 491.

Holt, Derek F. and Rees, Sarah. A graphics system for displaying finite quotients of finitely presented groups. In Finkelstein, Larry and Kantor, William M., editors, *Groups and Computation*, volume 11 of *DIMACS Series in Discrete*

*Mathematics and Theoretical Computer Science*, pages 113–126. AMS-ACM, 1993. Cited on page(s) 437.

Holt, D. F. and Rees, S. Testing modules for irreducibility. *J. Austral. Math. Soc. Ser. A*, 57:1–16, 1994. Cited on page(s) 86, 363, 475.

Holzinger, A. Computer-aided Mathematics instruction with Mathematica 3.0. *Mathematica in Education and Research*, 6(4):37–40, 1997. Cited on page(s) 256.

Hommenl, G. and Kovács, P. Simplification of symbolic inverse kinematic transformations through functional decomposition. In *Proc. of the Conference Adv. in Robotics*, pages 88–95, 1992. Cited on page(s) 27.

Hong, H. Non-linear constraint solving over real numbers in constraint logic programming. Technical Report 92-08, RISC-Linz, 1992a. Cited on page(s) 139.

Hong, H. Non-linear real constraints in constraint logic programming. In Kirchner and Levi [1992], pages 201–212. Cited on page(s) 141.

Hong, Hoon, editor. *First International Symposium on Parallel Symbolic Computation (PASCO'94)*, Lecture Notes Series in Computing, Linz, Austria, September 1994. World Scientific Publishing. Cited on page(s) 146, 488.

Hong, H. An efficient method for analyzing the topology of plane real algebraic curves. *Mathematics and Computers in Simulation*, 42(4–6):571–582, 1996. Cited on page(s) 238.

Hong, Hoon, Liska, Richard, and Steinberg, Stanly. Testing stability by quantifier elimination. *Journal of Symbolic Computation*, 24(2):161–187, August 1997. Special issue on applications of quantifier elimination. Cited on page(s) 139.

Hong, H., Neubacher, A., and Schreiner, W. The design of the SACLIB / PACLIB kernels. In Miola, A., editor, *DISCO'93, Gmunden, Austria*, LNCS 722. Springer-Verlag, 1993. Cited on page(s) 147.

Hong, H., Neubacher, A., Schreiner, W., and Stahl, V. The C++ interface to the STURM multi-processor kernel. Technical Report TR 94-31, RISC Institute, Linz, Austria, 1994. Cited on page(s) 147.

Hong, Hoon and Weispfenning, Volker. Algorithmic theory of admissible term orders. Technical report, North Carolina State University, Department of Mathematics and University of Passau, Department of Mathematics, 1999. Submitted for publication. Cited on page(s) 29.

Hopcroft, J., Borodina, A., Fagin, R., and Tompa, M. Decreasing the nesting depth of expressions involving square roots. *J. Symb. Comput.*, 1:169–188, 1985. Cited on page(s) 18.

Horng, G. and Huang, M.-D. Solving polynomials by radicals with roots of unity in minimum depth. *Mathematics of Computation*, 68:881–885, 1999. Cited on page(s) 18.

Hosten, Serkan and Thomas, Rekha. Gröbner bases and integer programming. In Buchberger, Bruno and Winkler, Franz, editors, *Gröbner Bases and Applications*, volume 251 of *LNS*, pages 144–158, Cambridge, UK, 1998. LMS, Cambridge University Press. ISBN 0-521-63298-6. Cited on page(s) 30.

Howell, J. A. Spans in the module $(\mathbb{Z}_m)^s$. *Linear and Multilinear Algebra*, 19: 67–77, 1986. Cited on page(s) 39.

Hsiang, Jieh. Refutational theorem proving using term-rewriting systems. *Artificial Intelligence*, 25:255–300, 1985. Cited on page(s) 135.

Huang, M. and Ierardi, D. Efficient algorithms for the Riemann-Roch problem and for addition in the Jacobian of a curve. In *Proceedings of the 32nd Conference on Foundations of Computer Science*, pages 678–687, 1991. Cited on page(s) 129.

Huang, Xiaohan and Pan, Victor Y. Fast rectangular matrix multiplication and applications. *Journal of Complexity*, 14:257–299, 1998. Cited on page(s) 25.

Huang, Yuzhen, Wu, Wenda, Stetter, Hans J., and Zhi, Lihong. Pseudofactors of multivariate polynomials. In Traverso [2000], pages 161–168. Cited on page(s) 118, 124.

Hubert, E. The general solution of an ordinary differential equation. In Lakshman, Y. N., editor, *ISSAC'96*, pages 196–203, Zürich, Switzerland, 1996. Cited on page(s) 105.

Hubert, E. Essential components of an algebraic differential equation. *Journal of Symbolic Computation*, 28(4-5):657–680, 1999. Cited on page(s) 105.

Huet, Gérard and Hullot, Jean-Marie. Proofs by induction in equational theories with constructors. In *Proc. 21st FoCS*, pages 96–107, Los Angeles, CA, 1980. Cited on page(s) 136.

Huguet, L. and Poli, A., editors. *Applied Algebra, Algebraic Algorithms and Error-Correcting Codes, 5th International Conference, AAECC-5, Menorca, Spain, June 15–19, 1987, Proceedings*, volume 356 of *Lecture Notes in Computer Science*, Berlin-Heidelberg-New York, 1989. Springer-Verlag. ISBN 3-540-51082-6, 0-387-51082-6. Cited on page(s) 486.

Hulpke, A. *Konstruktion transitiver Permutationsgruppen*. PhD thesis, Rheinisch-Westfälische Technische Hochschule, Aachen, Germany, 1996a. Cited on page(s) 47.

Hulpke, A. *Konstruktion transitiver Permutationsgruppen*, volume 18 of *Aachener Beiträge zur Mathematik*. Verlag der Augustinus Buchhandlung, Aachen, 1996b. Cited on page(s) 197, 389.

Hulpke, Alexander and Linton, Steve. Construction of Co3. an example of the use of a CGT system. In Campbell, C. M., Robertson, E. F., Ruskuc, N., and Smith, G. C., editors, *Groups St Andrews 1997 in Bath*, volume 260/261 of *LMS Lecture Notes Series*, pages 394–409. Cambridge University Press, 1999. Cited on page(s) 387.

Hulsbergen, W. W. J. *Conjectures in Arithmetic Algebraic Geometry. A Survey*. Aspects of Math. Vieweg, Braunschweig/Wiesbaden, 1992. Cited on page(s) 49.

Huson, D. H. The generation and classification of tile-$k$-transitive tilings of the euclidean plane, the sphere and the hyperbolic plane. *Geometriae Dedicata*, 47:269–296, 1993. Cited on page(s) 439.

Huson, D. H. Visualization of periodic tilings. In Hege, Hans-Christian and Polthier, Konrad, editors, *Visualization and Mathematics*, pages 135–139. Springer Verlag, Heidelberg, 1997. Cited on page(s) 438.

Husty, M. L. An algorithm for solving the direct kinematics of general stewart-platforms. *J. Mech. and Mach. Th.*, 31:365–380, 1996. Cited on page(s) 233.

IBM. ACRITH High-Accuracy Arithmetic Subroutine Library, Program Description and User's Guide. Technical Report SC 33-6164-3, IBM Publications, 1986. Cited on page(s) 111.

Ifrah, G. *Universalgeschichte der Zahlen.* Campus Verlag, Frankfurt, New York, 1991. ISBN 3-593-34192-1. Cited on page(s) 1.

Ikl'e, M. O. *Exact Solutions to a Discrete Model For Coagulation-Fragmentation.* PhD thesis, University of Wisconsin, Madison, 1993. Cited on page(s) 223.

Iliopoulos, C. S. Worst-case complexity bounds on algorithms for computing the canonical structure of finite abelian groups and the Hermite and Smith normal forms of an integer matrix. *SIAM Jour. Comp.*, 18(4):658–669, 1989. Cited on page(s) 39.

Ince, E. L. *Ordinary Differential Equations.* Dover, New York, 1956. Cited on page(s) 99.

*Advanced Course on Partial Differential Equations and Group Theory*, Bonn, 1992. INRIA, GMD. Cited on page(s) 487.

Institute, American National Standards. American National Standard Programming Language Fortran. ANSI X3.9, American National Standards Institute, 1978. Cited on page(s) 151.

International Standard. Information technology - programming language C. ISO/IEC 9899, WG14/N843, 1990. Cited on page(s) 13.

International Standard. Information technology - language independent arithmetic - part 1 integer and floating point arithmetic. ISO/IEC 10967-1:1994(E), 1994. 92 pages. Cited on page(s) 13.

International Standard. Information technology - language independent arithmetic - part 2 elementary numerical functions. ISO/IEC 10967-2:1995(E), 1995. 79 pages. Cited on page(s) 13.

Ion, Patrick and Miner, Robert. Mathematical Markup Language. Technical report, Worldwide Web Consortium, 1998. Available at http://www.w3.org/TR/1998/REC-MathML. Cited on page(s) 150.

Ischtwan, Joseph and Collins, Michael A. Symmetry-invariant reaction path potentials. *J. Chem. Phys.*, 94, 1991. Cited on page(s) 34.

Ishikawa, T. Minami-Taeya group 'GRACE manual", KEK-92-19, 1993. Cited on page(s) 166.

Ivanov, A. A., Linton, S. A., Lux, K., Saxl, J., and Soicher, L. H. Distance-transitive representations of the sporadic groups. *Comm. Algebra*, 23:3379–3427, 1995. Cited on page(s) 90.

Ivanov, A. V. Constructive enumerations of incidence systems. *Ann. Discrete Math.*, 26:227–246, 1985. Cited on page(s) 89.

Jacobs, D. P. The Albert nonassociative algebra system: A progress report. In *Proceedings of International Symposium on Symbolic and Algebraic Computation*, pages 41–44, Oxford, July 1994. ACM. Cited on page(s) 64, 348.

Jacobs, David P. A course in computational nonassociative algebra. *Journal of Symbolic Computation*, 23(5/6):497–502, May/June 1997. Cited on page(s) 264.

Jacobson, M. J. Experimental results on class groups of real quadratic fields (extended abstract). In *Algorithmic Number Theory - ANTS III*, volume 1423 of *Lecture Notes in Computer Science*, pages 463–474, 1998. Cited on page(s) 406.

Jacobson, N. *Basic Algebra*, volume 1 and 2. Freeman, W.H., New York, 1985. Cited on page(s) 491.

Jacquemard, A. and Teixeira, M. A. Reversible normal forms of degenerate cusps for planar diffeomorphisms. *Bull. Sci. Math.*, 122:409–425, 1998. Cited on page(s) 106.

Jaffar, J. and Maher, M. J. Constraint logic programming: A survey. *Journal Logic Programming*, 19/20:503–581, 1994. Cited on page(s) 139.

Jaffar, J., Maher, M. J., Stuckey, P. J., and Yap, R. H. C. Beyond finite domains. In *Principles and Practice of Constraint Programming PPCP'94*, 1994. Cited on page(s) 139.

Jaffar, J., Michaylov, S., Stuckey, P. J., and Yap, R. H. C. The CLP(R) language and system. *ACM Transactions on Programming Languages and Systems*, 14 (3):339–395, 1992. Cited on page(s) 139, 141.

Janet, M. Sur les Systèmes d'Équations aux Dérivées Partielles. *J. Math. Pure Appl.*, 3:65–151, 1920. Cited on page(s) 102, 192.

Janet, M. *Leçons sur le Systèmes d'Équations aux Derivées Partielles*. Gauthier-Villars, Paris, 1929. Cited on page(s) 104, 491.

Jansen, C., Lux, K., Parker, R., and Wilson, R. *An Atlas of Brauer Characters*. Clarendon Press, Oxford, 1995. Cited on page(s) 86, 196, 389, 430.

Jansen, C., Lux, K., Parker, R., and Wilson, R. A. *An Atlas of Modular Character Tables*. Cambridge University Press, Cambridge, in Vorbereitung. Cited on page(s) 491.

Janßen, R., editor. *Trends in Computer Algebra, International Symposium Bad Neuenahr, May 19-21, 1987, Proceedings*, volume 296 of *Lecture Notes in Computer Science*, Berlin-Heidelberg-New York, 1988. Springer-Verlag. ISBN 3-540-18928-9, 0-387-18928-9. Cited on page(s) 486.

Jansson, C. On Self-Validating Methods for Optimization Problems. In Herzberger, J., editor, *Topics in Validated Computations — Studies in Computational Mathematics 5*, pages 381–438, Amsterdam, 1994. North-Holland. Cited on page(s) 111.

Jantzen, M. *Confluent String Rewriting*. Number 14 in EATCS Monographs on Theoretical Computer Science. Springer-Verlag, Berlin, 1988. Cited on page(s) 135, 491.

Jarić, M. V., Michel, L., and Sharp, R. T. Zeros of covariant vector fields for the point groups: Invariant formulation. *J. Physique*, 45:1–27, 1984. Cited on page(s) 34.

Jayant, N. S. and Noll, Peter. *Digital Coding of Waveforms*. Englewood Cliffs, N.J. 07632, 1984. Cited on page(s) 227.

Jeanerrod, C.-P. and Pflügel, E. A reduction algorithm for matrices depending on a parameter. In Dooley [1999], pages 121–128. Cited on page(s) 125.

Jebelean, T. An algorithm for exact division. *jsc*, 15:169–180, 1993. Cited on page(s) 12.

Jeffrey, D. J. Rectifying transformations for the integration of rational trigonometric functions. *Journal of Symbolic Computation*, 24(5):563–574, November 1997. Cited on page(s) 96.

Jeffrey, D. J., Labahn, G., von Mohrenschildt, M., and Rich, A. D. Integration of the signum, piecewise and related functions. In *Proceedings of ISSAC'97*, pages 324–330. ACM Press, 1997. Cited on page(s) 96.
Jeffrey, D. J. and Rich, A. D. Recursive integration of piecewise-continuous functions. In *Proceedings of ISSAC'98*, pages 290–294. ACM Press, 1998. Cited on page(s) 96.
Jenks, R. D. and Sutor, R. S. *AXIOM, The Scientific Computation System*. Springer-Verlag, Berlin-Heidelberg-New York, 1992. ISBN 0-387-97855-0, 3-540-97855-0. Cited on page(s) 3, 211, 242, 263, 491.
Jenks, R. D., Sutor, R. S., and Watt, S. M. Scratchpad II: An abstract datatype system for mathematical computation. In Janssen, R., editor, *Trends in Computer Algebra, International Symposium Bad Neuenahr, May 19-21, 1987, Proceedings*, volume 296 of *Lecture Notes in Computer Science*, pages 12–37, Berlin-Heidelberg-New York, 1988a. Springer-Verlag. ISBN 3-540-18928-9, 0-387-18928-9. Cited on page(s) 264.
Jenks, R. D., Sutor, R. S., and Watt, S. M. Scratchpad II: An abstract datatype system for mathematical computation. In Rice [1988]. Cited on page(s) 266.
Jenks, R. D. and Trager, B. M. A language for computational algebra. In Wang [1981], pages 6–13. Cited on page(s) 266.
Jerie, J. E. R. O'Connor M. and Prince, G. E. Computer algebra determination of symmetries in general relativity. *Computer Physics Communications*, 115: 363–380, 1998. Cited on page(s) 469.
Jerie, J. E. R. O'Connor M. and Prince, G. E. Spacetime symmetries for the Kerr metric. *Classical and Quantum Gravity*, 16:2885–2887, 1999. Cited on page(s) 469.
Jirstrand, Mats. Nonlinear control system design by quantifier elimination. *Journal of Symbolic Computation*, 24(2):137–152, August 1997. Special issue on applications of quantifier elimination. Cited on page(s) 139.
Joerg, C. F. *The Cilk system for parallel multithreaded computing*. PhD thesis, Massachussets Institute of Technology, January 1996. Cited on page(s) 148.
Johnson, D. L. *Presentations of Groups*, volume 15 of *London Mathematical Society Student Texts*. Cambridge University Press, Cambridge, 1990. Cited on page(s) 4, 491.
Jones, Richard and Lins, Rafael. *Garbage Collection*. Wiley, 1996. ISBN 0-471-94148-4. Cited on page(s) 352, 491.
Jouannaud, J. P., editor. *Rewriting Techniques and Applications, RTA-85, Dijon, France, May 20–22, 1985*, volume 202 of *Lecture Notes in Computer Science*, Berlin-Heidelberg-New York, 1985. Springer-Verlag. ISBN 3-540-15976-2, 0-387-15976-2. Cited on page(s) 485.
Jouannaud, Jean-Pierre and Kirchner, Hélène. Completion of a set of rules modulo a set of equations. *SIAM J. on Computing*, 14(4):1155–1194, 1986. Cited on page(s) 133.
Jouannaud, Jean-Pierre and Kounalis, Emmanuel. Proofs by induction in equational theories without constructors. *Information and Computation*, 82:1–33, 1989. Cited on page(s) 136.

Jung, F., Naegele, F., and Thomann, J. An algorithm of multisummation of formal power series solutions of linear ODE equations. *Proceedings of IMACS'93, Lille, Mathematics and Computers in Simulation*, 42:409–425, 1996. Cited on page(s) 371.

Kadlecsik, J. *RICCIR: Ricci calculus package in REDUCE*, 1996. http://www.kfki.hu/(html2,hu)/cnc/szhkpub/riccir/riccir.html. Cited on page(s) 175.

Kahrs, M. and Brandenburg, K., editors. *Applications of Digital Signal Processing to Audio and Acoustics*. Kluwer Academic Publishers, 1998. Cited on page(s) 227, 491.

Kajler, N. Building a Computer Algebra Environment by Composition of Collaborative Tools. In Miola [1990b]. ISBN 3-540-52531-9, 0-387-52531-9. Cited on page(s) 152.

Kajler, N. CAS/PI: a Portable and Extensible Interface for Computer Algebra Systems. In Wang [1992], pages 376–386. Cited on page(s) 152.

Kalkbrener, M. Implicitization of rational parametric curves and surfaces. In Sakata [1991], pages 249–259. ISBN 3-540-54195-0, 0-387-54195-0. Cited on page(s) 235.

Kalkbrener, M. On the stability of Gröbner bases under specializations. *J. Symb. Computation*, pages 51–58, 1997. Cited on page(s) 31.

Kalouti, H., Lazic, D. E., and Beth, Th. On the Relation Between Distance Distributions of Linear Block Codes and the Binomial Distribution. *Annales des Télécommunications*, 50(4):762–778, 1995. Cited on page(s) 129.

Kaltofen, E. Factorization of polynomials. In Buchberger et al. [1983], pages 95–113. Cited on page(s) 24.

Kaltofen, E. Fast parallel absolute irreducibility testing. *Journal of Symbolic Computation*, 1(1):57–67, 1985a. Misprint corrections: *J. Symbolic Comput.* vol. 9, p. 320 (1989). Cited on page(s) 113, 124.

Kaltofen, E. Polynomial-time reductions from multivariate to bi- and univariate integral polynomial factorization. *SIAM Journal of Computing*, 14(2):469–489, 1985b. Cited on page(s) 130.

Kaltofen, E. Greatest common divisors of polynomials given by straight-line programs. *Journal of the ACM*, 35(1):231–264, 1988. Cited on page(s) 14, 383.

Kaltofen, E. Factorization of polynomials given by straight-line programs. In Micali [1989], pages 375–412. Cited on page(s) 15, 25, 383.

Kaltofen, E. Polynomial factorization 1982-1986. In Chudnovsky and Jenks [1990], pages 285–300. Cited on page(s) 24.

Kaltofen, E. On computing determinants of matrices without divisions. In Wang [1992], pages 342–349. Cited on page(s) 14, 37, 40.

Kaltofen, E. Polynomial factorization 1987-1991. In Simon, I., editor, *Proc. Latin '92*, volume 583 of *Lecture Notes in Computer Science*, pages 294–313, Berlin-Heidelberg-New York, 1992b. Springer-Verlag. Cited on page(s) 15, 24.

Kaltofen, E. Polynomial factorization 1987-1991. In Simon, I., editor, *Proc. LATIN '92*, volume 583 of *Lect. Notes Comput. Sci.*, pages 294–313, Heidelberg/New York, 1992c. Springer Verlag. Cited on page(s) 124.

Kaltofen, E. Analysis of Coppersmith's block Wiedemann algorithm for the parallel solution of sparse linear systems. *Mathematics of Computation*, 64 (210):777–806, 1995a. Cited on page(s) 38.

Kaltofen, E. Effective Noether irreducibility forms and applications. *Journal Comput. System Sci.*, 50(2):274–295, 1995b. Cited on page(s) 26.

Kaltofen, E., June 2000. Personal communication. Cited on page(s) 40.

Kaltofen, E. and Buchberger, B., editors. *Computational Algebraic Complexity*. Academic Press, New York, 1990. Cited on page(s) 492.

Kaltofen, E., Krishnamoorthy, M. S., and Saunders, B. D. Fast parallel computation of Hermite and Smith forms of polynomials matrices. *SIAM J. Alg. Disc. Meth.*, 8:683–690, 1987. Cited on page(s) 41.

Kaltofen, E., Krishnamoorthy, M. S., and Saunders, B. D. Parallel algorithms for matrix normal forms. *Linear Algebra and its Applications*, 136:189–208, 1990. Cited on page(s) 40, 41.

Kaltofen, E., Lee, W.-s., and Lobo, A. A. Early termination in Ben-Or/Tiwari sparse interpolation and a hybrid of Zippel's algorithm. In Traverso [2000], pages 192–201. Cited on page(s) 384.

Kaltofen, E. and Lobo, A. Distributed matrix-free solution of large sparse linear systems over finite fields. *Algorithmica*, 24(3–4):331–348, July–August 1999a. Special Issue on "Coarse Grained Parallel Algorithms". Cited on page(s) 38.

Kaltofen, E. and Lobo, A. Distributed matrix-free solution of large sparse linear systems over finite fields. *Algorithmica*, 24(3–4):331–348, July–August 1999b. Cited on page(s) 384.

Kaltofen, E. and Pan, V. Processor efficient parallel solution of linear systems over an abstract field. In *Proc. SPAA '91 3rd Ann. ACM Symp. Parallel Algor. Architecture*, pages 180–191, New York, N.Y., 1991. ACM Press. Cited on page(s) 38.

Kaltofen, E. and Pan, V. Processor-efficient parallel solution of linear systems II: the positive characteristic and singular cases. In *Proc. 33rd Annual Symp. Foundations of Comp. Sci.*, pages 714–723, Los Alamitos, California, 1992. IEEE Computer Society Press. Cited on page(s) 38.

Kaltofen, E. and Saunders, B. D. On Wiedemann's method of solving sparse linear systems. In *Proc. AAECC-9*, LNCS 539, Springer Verlag, pages 29–38, 1991. Cited on page(s) 38, 40, 383.

Kaltofen, Erich and Shoup, Victor. Fast polynomial factorization over high algebraic extensions of finite fields 1997. In *ISSAC 97 Proc. 1997 Internat. Symp. Symbolic Algebraic Comput.*, pages 184–188. ACM, 1997. Cited on page(s) 25.

Kaltofen, Erich and Shoup, Victor. Subquadratic-time factoring of polynomials over finite fields. *Mathematics of Computation*, 67(223):1179–1197, 1998. Cited on page(s) 25.

Kaltofen, Erich and Trager, Barry M. Computing with polynomials given by black boxes for their evaluations: Greatest common divisors, factorization, separation of numerators and denominators. *Journal of Symbolic Computation*, 9:301–320, 1990. Cited on page(s) 25, 37, 383.

Kaltofen, E. and Watt, S. M., editors. *Computer and Mathematics'89*, 1989. Springer-Verlag. Cited on page(s) 486.

Kamke, E. *Differentialgleichungen. Lösungsmethoden und Lösungen I.* Akademische Verlagsanstalt, Leipzig, 1961. Cited on page(s) 98, 224, 283, 491.

Kandry-Rody, A. and Weispfennning, V. Non-commutative Gröbner bases in algebras of solvable type. *Journal of Symbolic Computation*, 9(1):1–26, 1990. Cited on page(s) 59, 61, 379.

Kannan, R. Polynomial-time algorithms for solving systems of linear equations over polynomials. *Theor. Computer Science*, 39:69–88, 1985. Cited on page(s) 39.

Kannan, R., Lenstra, A. K., and Lovasz, L. Polynomial factorization and non-randomness of bits of algebraic and some transcendental numbers. In *Proc. of the 16th Ann. ACM Symp. on Theor. of Comput.*, pages 191–200, New York, 1984. ACM Press. Cited on page(s) 15.

Kant, E., Daube, F., MacGregor, B., and Wald, J. MathCode: A Code Generation Package for Mathematica. Technical report, Schlumberger Technologies Corporation, September 1990. Cited on page(s) 151.

Kantor, William M. Simple groups in computational group theory. In *Proceedings of the International Congress of Mathematicians, Berlin, 1998*, pages 77–86. Deutsche Mathematiker-Vereinigung, 1998. Cited on page(s) 66, 71, 75.

Kaplansky, I. *Introduction to Differential Algebra*. Hermann, Paris, 1957. Cited on page(s) 104, 491.

Kapur, Deepak. A refutational approach to geometry theorem proving. Presented at Workshop Geometric Reasoning, Oxford, England, June 30–July 3, 1986a. Cited on page(s) 33.

Kapur, Deepak. Using Gröbner bases to reason about geometry problems. *Journal of Symbolic Computation*, 2(4):399–408, December 1986b. ISSN 0747-7171. Cited on page(s) 201, 203.

Kapur, Deepak. Comparison of various multivariate resultant formulations. In Levelt, A. H. M., editor, *ISSAC'95*, pages 187–194. ACM-Press, 1995. Cited on page(s) 52.

Kapur, Deepak. Automated geometric reasoning: Dixon resultants, Gröbner bases, and characteristic sets. In Wang, Dongming, editor, *Automated Deduction in Geometry*, volume 1360 of *LNAI*. Springer, 1998. Cited on page(s) 32, 33.

Kapur, D. and Lakshman, Y. N. Elimination methods: An introduction. In Donald, B., Kapur, D., and Mundy, J., editors, *Symbolic and Numerical Computation for Artificial Intelligence*. Academic Press, 1992. Cited on page(s) 33, 52.

Kapur, D. and Madlener, Klaus. A completion procedure for comuting a canonical basis for a $k$-subalgebra. In Kaltofen, E. and M., Watt S., editors, *Computers and Mathematics*, pages 1–11. Springer, 1989. Cited on page(s) 31.

Kapur, Deepak and Mundy, Joseph L. Wu's method and its application to perspective viewing. *Artificial Intelligence*, 37(1-3):15–36, December 1988. Special volume on geometric reasoning. Cited on page(s) 32, 33.

Kapur, D. and Narendran, P. An equational approach to theorem proving in first-order predicate calculus. In Joshi, A. K., editor, *Proc. IJCAI 1985*, pages 1146–1153, Los Angeles, 1985. Morgan Kaufmann. Cited on page(s) 30.

Karmarkar, N. K. and Lakshman Y. N. Approximate polynomial greatest common divisors and nearest singular polynomials. In Lakshman [1995], pages 35–42. Cited on page(s) 117, 118, 119, 120, 121, 124.

Karp, R. M. and Ramachandran, V. Parallel algorithms for shared-memory machines. In van Leuwen, J., editor, *Handbook of Theoretical Computer Science Vol. A*, pages 869–941. North-Holland, 1990. Cited on page(s) 41.

Karr, M. Summation in finite terms. *Journal of the ACM*, 28:305–350, 1981. Cited on page(s) 91.

Karr, M. Theory of summation in finite terms. *Journal of Symbolic Computation*, 1:303–315, 1985. Cited on page(s) 91.

Karypis, George and Kumar, Vipin. Multilevel k-way partitionning scheme for irregular graphs. *Journal of Parallel and Distributed Computing*, 48(1): 96–129, 1998. Cited on page(s) 148.

Keady, G. GENTRANs from REDUCE and from MACSYMA. Technical report, University of Waikato Mathematics Department, Waikato, September 1991. Cited on page(s) 151.

Keady, G. and Nolan, G. Production of Argument SubPrograms in the AXIOM-NAG Link: examples involving non-linear systems. In Apiola et al. [1994], pages 13–31. ISBN 952-9528-27-2. Cited on page(s) 263.

Kearfott, B. and Du, K. The Cluster Problem in Global Optimization. *Computing Suppl.*, 9:117–127, 1993. Cited on page(s) 111.

Kearfott, R.B., Dawande, M., Du, K., and Hu, C. INTLIB: A portable Fortran-77 interval standard function library. *ACM Trans. Math. Software*, 20:447–459, 1994. Cited on page(s) 111.

Keller, B. J. Alternatives in implementing noncommutative Gröbner basis systems. In Bronstein et al. [1998], pages 105–126. Cited on page(s) 61.

Keller-Gehrig, W. Fast algorithms for the characteristic polynomial. *Theor. Computer Science*, 36:309–317, 1985. Cited on page(s) 40.

Kemper, Gregor. The *invar* package for calculating rings of invariants. IWR Preprint **93-34**, Heidelberg, 1993. Cited on page(s) 35, 216.

Kemper, G. Calculating invariant rings of finite groups over arbitrary field. *J. Symb. Comput.*, 21:351–366, 1996. Cited on page(s) 34, 197.

Kemper, G. and Malle, G. The finite irreducible linear groups with polynomial ring of invariants. *Transf. Groups*, 2:57–89, 1997. Cited on page(s) 197.

Kemper, G. and Malle, G. Invariant rings and fields of finite groups. In Matzat et al. [1999], pages 265–281. ISBN ISBN 3-540-64670-1. Selected Papers from a Conference Held at the University of Heidelberg in October 1997. Cited on page(s) 87.

Kemper, Gregor and Steel, Allan. Some algorithms in invariant theory of finite groups. In Dräxler, P., Michler, G. O., and Ringel, C. M., editors, *Proceedings of the Euroconference on Computational Methods for Representations of Groups and Algebras*, number 173 in Progress in Mathematics, Basel, 1999. Birkhäuser. Cited on page(s) 34, 35.

Kempfert, H. On sign determination in real algebraic number fields. *Numerische Mathematik*, 11:170–174, 1968. Cited on page(s) 56.

Kerber, A. *Algebraic Combinatorics via Finite Group Actions*. Bibliographisches Institut, Mannheim, 1991. Cited on page(s) 90, 491.

Kerber, A., Kohnert, A., and Lascoux, A. SYMMETRICA, an object oriented computer-algebra system for the symmetric group. *Journal of Symbolic Computation*, 14:195–203, 1992. Cited on page(s) 88, 346.

Kerber, A. and Laue, R. Group actions, double cosets, and homomorphisms: unifying concepts for the constructive theory of discrete structures. *Acta Applicandae Mathematicae*, 1998. Cited on page(s) 89.

Keyser, J., Krishnan, S., and Manocha, D. Efficient and accurate B-rep generation of low degree sculptured solids using exact arithmetic. In *Proceedings of ACM Solid Modeling '97*, pages 42–55, Baltimore, 1997. ACM Press. Cited on page(s) 240.

Keyser, J., Krishnan, S., Manocha, D., and Culver, T. Efficient and reliable computation with algebraic numbers for geometric algorithms. Technical Report Technical Report TR98-012, Department of Computer Science, University of N. Carolina, Chapel Hill, 1998. available via http://www.cs.unc.edu/\~geom/intersect.html. Cited on page(s) 406.

Kharitononv, V. L. Asymptotic stability of an equilibrium of a family of systems of linear differential equations. *Differential Equations*, 14:1483–1485, 1979. Cited on page(s) 113, 122.

Kilgenstein, P. *Ausbildung der dreidimensionalen Grenzschicht im abgebremsten Ellipsoid*. Dissertation, Darmstadt University of Technology, Technische Universität Darmstadt, 1984. Cited on page(s) 223.

Kimura, M. *Tensor N.S.*, 30:27–43, 1976. Cited on page(s) 188.

Kinoshita, T., editor. *Quantum Electrodynamics*, Advanced Series on Directions in High Energy Physics, Vol. 7, Singapore, 1990. World Scientific. Cited on page(s) 167.

Kipnis, A. and Shamir, A. Cryptoanalysis of the Oil and Vinegar Scheme. In Krawczyk, H., editor, *Advances in Cryptology Crypto'98*, number 1462 in Lecture Notes in Computer Science, pages 257–266, Boston, August 1998. Cited on page(s) 131.

Kirchner, Claude, editor. *5th International Conference on Rewriting Techniques and Applications*, volume 690, Montereal, Canada, June 1993. Springer. Cited on page(s) 487.

Kirchner, Claude, Kirchner, Hélène, and Michaël, Rusinowitch. Deduction with symbolic constraints. *Revue d'Intelligence Artificielle*, 4(3):9–52, 1990. Cited on page(s) 136.

Kirchner, H. and Levi, G., editors. *Algebraic and Logic Programming, Third International Conference, Volterra, Italy, September 1992, Proceedings*, volume 632 of *Lecture Notes in Computer Science*, Berlin-Heidelberg-New York, 1992. Springer-Verlag. Cited on page(s) 553, 574.

Kirrinnis, P. Partial fraction decomposition in C(z) and simultaneous newton iteration for factorization in C[z]. *J. Complexity*, 14:378–444, 1998. Cited on page(s) 26.

Kirwan, W. E. et al. *Moving Beyond Myths: Revitalizing Undergraduate Mathematics*. National Academy Press, Washington, D. C., 1991. Cited on page(s) 254.

Klappenecker, A., Nückel, A., and May, F. U. Lossless image compression using wavelets over finite rings and related architectures. In Unser, M. A., Aldroubi, A., and Laine, A. F., editors, *Wavelet Applications in Signal and Image Processing V*. SPIE, 1997. Cited on page(s) 218.

Klatte, R., Kulisch, U., Neaga, M., Ratz, D., and Ullrich, Ch. *PASCAL-XSC — Sprachbeschreibung mit Beispielen*. Springer, 1991. Cited on page(s) 111.

Klaus, U. Parallel standard basis computations. In Carrière, A. and Oudin, L. R., editors, *Proc. 5 th Rhine workshop*, pages 10.1–10.14, Saint Louis, 1996. ISL PR 801/96. Cited on page(s) 379.

Kleczka, M., Kleczka, W., and Kreuzer, E. Bifurcation analysis: A combined numerical and analytical approach. In Roose, D., De Dier, B., and Spence, A., editors, *Continuation and Bifurcations: Numerical Techniques and Applications*, pages 123–137. Kluwer, Dordrecht, 1990. Cited on page(s) 107.

Klimov, D. M. and Rudenko, V. M. *Computer Algebra Methods for Mechanics Problems*. Moskva, 1989a. Cited on page(s) 491.

Klimov, D. M. and Rudenko, V. M. *Metodyi kompyuternoi algebryi v zadachakh mekhaniki (Methods of computer algebra in Problems of Mechanics)*. Nauka, 1989b. Cited on page(s) 226, 491.

Klimyk, A. and Schmüdgen, K. *Quantum Groups and Their Representations*. Springer-Verlag, Heidelberg, 1997. Cited on page(s) 213.

Klioner, S. A. Angular velocity of extended bodies in general relativity. In Ferraz-Mello, S., Morando, B., and Arlot, J. E., editors, *Dynamics, ephemerides and astrometry in the solar system*, pages 309–320. Kluwer, Dordrecht, 1996. Cited on page(s) 469.

Klioner, S. A. New System for indicial computation and its applications in gravittaional physics. In G., McLenaghan R. and Cheb-Terrab, E. S., editors, *Thematic Issue of Computer Physics Communication on "Computer Algebra in Physics Research"*, volume 115, pages 231–244. 1998. Cited on page(s) 469.

Klioner, Sergei A. *EinS: A Mathematica package for calculations with indexed objects. Information*, 1999. http://rcswww.urz.tu-dresden.de/~klioner/eins.html. Cited on page(s) 175.

Klioner, S. A. and Soffel, M. Nordtvedt effect in rotational motion. *Physical Review D*, 58:084023, 1998. Cited on page(s) 469.

Kluge, O. *Entwicklung einer Programmierumgebung für die Parallelverarbeitung in der Computer-Algebra*. MuPAD Reports. B.G. Teubner, Stuttgart, 1996. Cited on page(s) 331, 491.

Klüners, J. On computing subfields - A detailed description of the algorithm. *Journal de Théorie des Nombres de Bordeaux*, 10:243–271, 1998. Cited on page(s) 48.

Klüners, J. On polynomial decompositions. *J. Symb. Comput.*, 27:261–269, 1999. Cited on page(s) 27.

Klüners, J. and Malle, G. Explicit Galois realization of transitive groups of degree up to 15. *Journal of Symbolic Computation*, 30(6):675–716, 2000. Cited on page(s) 48.

Klüners, J. and Pohst, M. On computing subfields. *J. Symb. Comput.*, 24(3): 385–397, 1997. Cited on page(s) 27, 48, 399.

Knowles, P. H. Integration of a class of transcendental Liouvillian functions with error-functions, Part I. *Journal of Symbolic Computation*, 13(5):525–543, 1992. Cited on page(s) 95.

Knowles, P. H. Integration of a class of transcendental Liouvillian functions with error-functions, Part II. *Journal of Symbolic Computation*, 16(3):227–242, 1993. Cited on page(s) 95.

Knüppel, O. PROFIL / BIAS — A Fast Interval Library. *Computing 53*, pages 277–287, 1994. Cited on page(s) 111.

Knuth, Donald E. *The Art of Computer Programming*, volume 2: Seminumerical Algorithms. Addison-Wesley, 2nd edition, 1981a. Cited on page(s) 112.

Knuth, D. E. *Seminumerical Algorithms*, volume 2 of *The Art of Computer Programming*. Addison-Wesley, Reading, Massachusetts, second edition, 1981b. Cited on page(s) 17, 21, 23.

Knuth, Donald E. *The Art of Computer Programming, vol. 1, Fundamental Algorithms*. Addison-Wesley, Reading MA, 3rd edition, 1997. Cited on page(s) 25.

Knuth, D. E. *Seminumerical Algorithms*, volume 2 of *The Art of Computer Programming*. Addison-Wesley, Reading, Massachusetts, third edition, 1998. Cited on page(s) 4, 11, 384, 491.

Knuth, Donald E. and Bendix, Peter B. Simple word problems in universal algebra. In Leech, John, editor, *Computational Problems in Abstract Algebra*, pages 169–183. Pergamon Press, 1970. (Proc. of a conference held in Oxford, England, 1967). Cited on page(s) 80, 133.

Koblitz, N. *p-adic Numbers, p-adic Analysis, and Zeta-Functions*. Springer-Verlag, Berlin-Heidelberg-New York, second edition, 1984. Cited on page(s) 21.

Koblitz, N. *A Course in Number Theory and Cryptography*, volume 114 of *Graduate Texts in Mathematics*. Springer-Verlag, Berlin-Heidelberg-New York, 1987. Cited on page(s) 4, 491.

Koblitz, N. *Algebraic Aspects of Cryptography*, volume 3 of *Algorithms and Computation in Mathematics*. Springer, 1998. Cited on page(s) 131, 491.

Koboyashi, Hidetsune, editor. *Asian Symposium on Computer Mathematics*, Tokyo, 1996. Scientists Incorporated. Cited on page(s) 488.

Koekoek, R. and Swarttouw, R. F. The Askey-scheme of hypergeometric orthogonal polynomials and its $q$-analogue. Technical Report 98–17, Delft University of Technology, Faculty of Information Technology and Systems, Department of Technical Mathematics and Informatics, http://aw.twi.tudelft.nl/~koekoek/research.html, 1998. Cited on page(s) 214.

Koepf, W. *Höhere Analysis mit Derive*. Vieweg, Braunschweig/Wiesbaden, 1994. Cited on page(s) 244, 491.

Koepf, W. Algorithms for $m$-fold hypergeometric summation. *Journal of Symbolic Computation*, 20:399–417, 1995a. Cited on page(s) 93, 215.

Koepf, W. REDUCE package for indefinite and definite summation. *SIGSAM Bulletin*, 29:14–30, 1995b. Cited on page(s) 93.

Koepf, W. *Derive für den Mathematikunterricht.* Vieweg, Braunschweig/Wiesbaden, 1996. Cited on page(s) 244, 492.

Koepf, W. The algebra of holonomic equations. *Mathematische Semesterberichte*, 44:173–194, 1997. Cited on page(s) 215.

Koepf, W. *Hypergeometric Summation. An Algorithmic Approach to Summation and Special Function Identities.* Vieweg, Braunschweig/Wiesbaden, 1998a. Cited on page(s) 4, 92, 93, 215, 492.

Koepf, W. Numeric versus symbolic computation. *International Journal of Computer Algebra in Mathematics Education*, 5 (1):29–54, 1998b. Cited on page(s) 244.

Koepf, W., Ben-Israel, A., and Gilbert, R. P. *Mathematik mit Derive.* Vieweg, Braunschweig/Wiesbaden, 1993. Cited on page(s) 244, 257, 491.

Koepf, W. and Schmersau, D. Representations of orthogonal polynomials. *J. Comput. Appl. Math.*, 90:57–94, 1998. Cited on page(s) 215.

Kohnert, A. and Veigneau, S. Using Schubert basis to compute with multivariate polynomials. *Advances in Applied Mathematics*, 19:45–60, 1997. Cited on page(s) 345.

Kolchin, E. R. *Differential Algebra and Algebraic Groups.* Academic Press, New York, 1973. Cited on page(s) 104, 491.

Koornwinder, T. H. On Zeilberger's algorithm and its $q$-analogue: a rigorous description. *Journal of Computational and Applied Mathematics*, 48:91–111, 1993. Cited on page(s) 93, 215.

Köppl, Ch. Eine REDUCE-Implementierung eines Quantoreneliminationsverfahrens für die Presburger Arithmetik. Master's thesis, Universität Passau, Universität Passau, FMI, 1991. Cited on page(s) 139.

Kotsireas, I. *Algorithms for solving polynomial systems: application to central configurations in the N-body problem of Celestial Mechanics.* PhD thesis, Université Paris 6, France, December 1998a. Cited on page(s) 177, 178.

Kotsireas, I. Using Gröbner Bases and Invariant Theory to check a conjecture in Celestial Mechanics. IMACS-ACA'98, Prague, Submitted to Journal of Symbolic Computation, Special Volume on Applications of Gröbner Bases, 1998b. Cited on page(s) 179.

Kotsireas, I. and Lazard, D. Central configurations of the 5-body problem with equal masses in three-dimensional space. In *Zap. Nauchn. Sem.*, volume 258, pages 292–317, St.-Petersburg, 1999. Otdel. Mat. Inst. Steklov (POMI). CASC'98. Cited on page(s) 177, 178.

Kotsireas, I. and Schicho, J. A computer algebra solution to a planar newtonian 4-body problem with unequal masses. MEGA'2000 Bath, 2000. Cited on page(s) 179.

Kovács, P. and Hommel, G. Fast functional decomposition of SC-polynomials. In *Proc. Conf. Robotics and Automation*, pages 980–987. IEEE, 1993. Cited on page(s) 233.

Kovács, P., Kockelkorn, M., and Hommel, G. A nonuniversal symbolic solution technique for the inverse kinematics problem applied to the cycloheptane molecule. *J. Mech. and Mach. Th.*, 2000. submitted. Cited on page(s) 232.

Kovalev, V. F. Group and renormgroup symmetry of quasi-Chaplygin media. *J. Nonlinear Math. Phys.*, 3:351–356, 1996. Cited on page(s) 469.

Kozen, D. and Landau, S. Polynomial decomposition algorithms. *J. Symb. Comput.*, 7:445–456, 1989a. Cited on page(s) 27.

Kozen, D. and Landau, S. Polynomial decomposition algorithms. *Journal of Symbolic Computation*, 7:445–456, 1989b. Cited on page(s) 123.

Kozen, D., Landau, S., and Zippel, R. Decomposition of algebraic functions. *Journal of Symbolic Computation*, 22(3):235–246, September 1996. Cited on page(s) 28.

Kozen, Dexter and Stefansson, Kjartan. Computing the newtonian graph. *Journal of Symbolic Computation*, 24(2):125–136, 1997. Cited on page(s) 139.

Kraft, Hanspeter. *Geometrische Methoden in der Invariantentheorie*. Vieweg, Braunschweig, 1984. Cited on page(s) 33, 491.

Kraft, Hanspeter, Slodowy, Peter, and Springer, Tonny A., editors. *Algebraische Transformationsgruppen und Invariantentheorie*. Number 13 in DMV Seminar. Birkhäuser, Basel, 1987. Cited on page(s) 33, 492.

Kramer, E. S. and Mesner, D. M. $t$-designs on hypergraphs. *Discrete Math.*, 15: 263–296, 1976. Cited on page(s) 374.

Kranakis, E. *Primality and Cryptography*. Wiley-Teubner, Stuttgart, 1986. Cited on page(s) 491.

Krattenthaler, C. HYP and HYPQ. *Journal of Symbolic Computation*, 20:737–744, 1995. Cited on page(s) 93.

Krattenthaler, C. and Zeilberger, D. Proof of a determinant evaluation conjectured by Bombieri, Hunt and van der Poorten. *New York Journal of Mathematics*, 3:54–102, 1997. Cited on page(s) 93.

Krauss, F. and Soff, G. Next-to-leading order QCD corrections to $B\bar{B}$ mixing and $\epsilon(K)$ within the MSSM. *hep-ph/9807238*, 1998. Cited on page(s) 170.

Krawczyk, R. Newton-Algorithmen zur Bestimmung von Nullstellen mit Fehlerschranken. *Computing 4*, pages 187–201, 1969. Cited on page(s) 111.

Kredel, H. Primary ideal decomposition. In Davenport [1989], pages 270–281. ISBN 3-540-51517-8, 0-387-51517-8. Cited on page(s) 426.

Kredel, H. MAS Modula-2 algebra system. In Miola [1990b], pages 270–271. ISBN 3-540-52531-9, 0-387-52531-9. Cited on page(s) 428.

Kredel, Heinz. *MAS, Modula-2 Algebra System, Interactive Usage, Version 0.3*. Universität Passau, D-94030 Passau, May 1990b. Available by anonymous-ftp from alice.fmi.uni-passau.de. Cited on page(s) 422.

Kredel, Heinz. *MAS, Modula-2 Algebra System, Library Description, Version 0.30*. Universität Passau, D-94030 Passau, January 1990c. Available by anonymous-ftp from alice.fmi.uni-passau.de. Cited on page(s) 422.

Kredel, Heinz. *MAS, Modula-2 Algebra System, Interactive Usage, Version 0.6*. Universität Passau, D-94030 Passau, March 1991a. Available by anonymous-ftp from alice.fmi.uni-passau.de. Cited on page(s) 422.

Kredel, Heinz. *MAS, Modula-2 Algebra System, Specifications, Definition Modules, Indexes, Version 0.6*. Universität Passau, D-94030 Passau, March 1991b. Available by anonymous-ftp from alice.fmi.uni-passau.de. Cited on page(s) 422.

Kredel, H. The MAS specification component. In Maluszynski, J. and Wirsing, M., editors, *Programming Language Implementation and Logic Programming, PLILP'91, Passau, Germany, August 26–28, 1991*, volume 528 of *Lecture Notes in Computer Science*, pages 39–50, Berlin-Heidelberg-New York, 1991c. Springer-Verlag. Cited on page(s) 423.

Kredel, H. *Solvable Polynomial Rings*. Dissertation, Universität Passau, Passau, 1992. also: Verlag Shaker, Aachen 1993. Cited on page(s) 61, 426.

Kredel, Heinz. *MAS, Modula-2 Algebra System, Interactive Usage, Version 0.7*. Universität Passau, D-94030 Passau, February 1993a. Available by anonymous-ftp from alice.fmi.uni-passau.de. Cited on page(s) 422.

Kredel, Heinz. *MAS, Modula-2 Algebra System, Specifications, Definition Modules, Indexes, Version 0.7*. Universität Passau, D-94030 Passau, February 1993b. Available by anonymous-ftp from alice.fmi.uni-passau.de. Cited on page(s) 422.

Kredel, Heinz and Pesch, Michael. *MAS, Modula-2 Algebra System, Interactive Usage, Version 1.0*. Universität Passau, D-94030 Passau, October 1996a. Available by anonymous-ftp from alice.fmi.uni-passau.de. Cited on page(s) 422.

Kredel, Heinz and Pesch, Michael. *MAS, Modula-2 Algebra System, Specifications, Definition Modules, Indexes, Version 1.0*. Universität Passau, D-94030 Passau, October 1996b. Available by anonymous-ftp from alice.fmi.uni-passau.de. Cited on page(s) 422.

Kreher, D. L. $t$-designs, $t \geq 3$. In *The CRC handbook of combinatorial designs*, pages 47–66. CRC Press, Boca Raton, FL, 1996. ISBN 0-8493-8948-8. Cited on page(s) 375.

Kreuzer, Martin and Robbiano, Lorenzo. *Computational Commutative Algebra 1*. Springer, 2000. ISBN ISBN 3-540-67733-X. Cited on page(s) 365, 491.

Krick, T. and Pardo, L. M. Une approche informatique pour l'approximation diophantienne. *C. R. Acad. Sci., Paris*, 318:407–412, 1994. Cited on page(s) 54.

Krick, T. and Pardo, L. M. A computational method for diophantine approximation. In Recio, T. and Vega, L. G., editors, *MEGA-94*, Progress in Mathematics 143, pages 193–254, Basel, 1996. Birkhäuser. Cited on page(s) 14, 54.

Krull, W. *Idealtheorie*. Ergebnisse der Mathematik und ihrer Grenzgebiete. Springer-Verlag, Berlin, 1935. Reprint Chelsea Publ, Co., New York, 1950. Cited on page(s) 491.

Kruskal, M. D. Flexibility in applying the Painlevé test. In Levi, D. and Winternitz, P., editors, *Painlevé Transcendents*, pages 187–195. Plenum, New York, 1992. Cited on page(s) 101.

Kruskal, M. D. and Clarkson, P. A. The Painlevé-Kowalevski and poly-Painlevé tests for integrability. *Stud. Appl. Math.*, 86:87–165, 1992. Cited on page(s) 101.

Kruskal, M. D., Joshi, N., and Halburd, R. Analytic and asymptotic methods for nonlinear singularity analysis: a review and extensions of tests for the Painlevé property. In Grammaticos, B. and Tamizhmani, K. M., editors, *Integrability of Nonlinear Systems. Proc. CIMPA Winter School on Nonlinear Systems, Pondicherry, India*, pages 171–205. Springer-Verlag, Berlin, 1997. Cited on page(s) 101.

Kruskal, M. D., Ramani, A., and Grammaticos, B. Singularity analysis and its relation to complete, partial and non-integrability. In *Partially Integrable Evolution Equations in Physics*, pages 321–372, R. Conte and N. Boccara, 1990. Kluwer, Dortrecht. Cited on page(s) 100, 101.

Küblbeck, J., Böhm, M., and Denner, A. *FeynArts*: Computer algebraic generation of Feynman graphs and amplitudes. *Comp. Phys. Comm.*, 60:165–180, 1990. Cited on page(s) 166, 170, 470.

Küchlin, Wolfgang. An implementation and investigation of the Knuth-Bendix completion algorithm. Master's thesis, Informatik I, Universität Karlsruhe, D-7500 Karlsruhe, W-Germany, 1982. (Reprinted as Report 17/82.). Cited on page(s) 438.

Küchlin, Wolfgang. Inductive completion by ground proof transformation. In Aït-Kaci and Nivat [1989b], chapter 7. ISBN 0-12-046371-7. Cited on page(s) 136.

Küchlin, W. PARSAC-2: a parallel SAC-2 based on threads. In *Eights Int. Symposium on Applied Algebra, Algebraic Algorithms and Error-Correcting Codes, Tokyo, Japan.*, LNCS 508. Springer-Verlag, 1990. Cited on page(s) 147.

Küchlin, W. W. On the multi-threaded computation of integral polynomial greatest common divisors. In Watt [1991], pages 333–342. ISBN 0-89791-437-6. Cited on page(s) 435.

Küchlin, W. W. On the multi-threaded computation of modular polynomial greatest common divisors. In Zima, H., editor, *Parallel Computation, 1st International ACPC Conference, Proceedings*, volume 591 of *Lecture Notes in Computer Science*, pages 369–384, Berlin-Heidelberg-New York, 1991b. Springer-Verlag. Cited on page(s) 435.

Küchlin, W. W. PARSAC-2: A parallel SAC-2 based on threads. In Sakata [1991], pages 341–353. ISBN 3-540-54195-0, 0-387-54195-0. Cited on page(s) 434, 435.

Küchlin, Wolfgang W. The S-threads environment for parallel symbolic computation. In Zippel, Richard, editor, *Computer Algebra and Parallelism*, volume 584 of *LNCS*, pages 1–18, Ithaca, NY, March 1992. Springer-Verlag. (Proc. CAP'90, Ithaca, NY, May 1990). Cited on page(s) 352, 435.

Küchlin, W. W. PARSAC-2: Parallel computer algebra on the desk-top. In Fleischer, J., Grabmeier, J., Hehl, F., and Küchlin, W., editors, *Computer Algebra in Science and Engineering*, pages 24–43, Singapore, 1995. World Scientific. ISBN 981-02-2319-6. Proc. of a ZiF workshop, Bielefeld, Germany, Aug. 28–31, 1994. Cited on page(s) 434.

Küchlin, W. W., editor. *ISSAC '97, International Symposium on Symbolic and Algebraic Computation, Maui, Hawaii*, New York, 1997. ACM. Cited on page(s) 488, 520, 522, 529, 535, 591, 607, 614, 622.

Küchlin, W. W., Lutz, D., and Nevin, N. J. Integer multiplication in PARSAC-2 on stock microprocessors. In Mattson et al. [1991], pages 206–217. ISBN 3-540-54522-0, 0-387-54522-0. Cited on page(s) 435.

Küchlin, W. W. and Nevin, N. J. On multi-threaded list-processing and garbage collection. In *Third IEEE Symposium on Parallel and Distributed Processing, Proceedings*, pages 894–897. IEEE Press, 1991. Cited on page(s) 435.

Küchlin, W. W. and Ward, J. A. Experiments with virtual C threads. In *Fourth IEEE Symposium on Parallel and Distributed Processing*, pages 50–55. IEEE Press, 1992. Cited on page(s) 435.

Kühnle, Klaus and Mayr, Ernst W. Exponential space computation of Gröbner bases. In Lakshman, Y. N., editor, *ISSAC'96*, pages 60–71. ACM Press, 1996. Cited on page(s) 29.

Kurth, Frank and Clausen, Michael. Adaptive FBT and $M$-band wavelet packet algorithms in audio signal processing. *IEEE Transactions on Signal Processing*, 47(2):549–554, 1999. Cited on page(s) 229.

Kutzler, B. *Improving Mathematics Teaching with* DERIVE. Chartwell-Bratt, 1997a. ISBN 0-86238-442-2. Cited on page(s) 277, 492.

Kutzler, B. *Introduction to* DERIVE *for Windows*. Soft Warehouse, 1997b. Cited on page(s) 277, 492.

Kutzler, B. *Introduction to the TI-92*. bk teachware Lehrmittel GmbH&CoKG, 1997c. ISBN 3-901769-02-1. Cited on page(s) 282, 492.

Kutzler, B., Lichtenberger, F., and Winkler, F. *Softwaresysteme zur Formelmanipulation*. Expert Verlag, 1990. Cited on page(s) 492.

Kutzler, Bernhard A. *Algebraic Approaches to Automated Theorem Proving*. PhD thesis, Johannes Kepler Universität Linz, 1988. RISC-Linz series no. 88-74.0. Cited on page(s) 30, 33, 201, 203, 204, 205, 206.

Kutzler, Bernhard A. and Stifter, Sabine. On the application of Buchberger's algorithm to automated geometry theorem proving. *Journal of Symbolic Computation*, 2(4):389–397, December 1986. ISSN 0747-7171. Cited on page(s) 30, 201, 203, 205, 206.

La Scala, R. and Stillman, M. Strategies for computing minimal free resolutions. *Journal of Symbolic Computation*, 26(4):409–431, 1998. Cited on page(s) 412.

Labhalla, S. E., Lombardi, H., and Marlin, R. Algorithmes de calcul de la réduction d'Hermite d'une matrice à coefficients polynomiaux. *Theoretical Computer Science*, 161(1–2):69–92, 1995. Cited on page(s) 39.

Labonte, G. An algorithm for the construction of matrix representations for finitely presented non-commutative algebras. *Journal of Symbolic Computation*, 9(1):27–38, 1990. Cited on page(s) 59, 85.

Lafon, J. C. Summation in finite terms. In Buchberger et al. [1983], pages 71–77. Cited on page(s) 91.

Lagarias, J. C. and Odlyzko, A. M. Effective versions of the Chebotarev density theorem. In Fröhlich, A., editor, *Algebraic Number Fields (L-functions and Galois properties)*, pages 409–464. Academic Press, 1977. Cited on page(s) 47.

Lake, K. GR 15 Proceedings A5(ii) Computer Methods in GR: Algebraic computing. http://xxx.lanl.gov/abs/gr-qc/9803072, 1998. Cited on page(s) 172.

Lake, K. and et al. GRTensorII Demonstration Page General Relativity & Geometry, 2001. http://www.grtensor.org. Cited on page(s) 173.

Lakshman, Y. N., editor. *ISSAC '96, International Symposium on Symbolic and Algebraic Computation, Zurich, Switzerland*, New York, 1995. ACM. Cited on page(s) 488, 495, 499, 561, 598.

Lakshman Y. N. and Saunders, B. D. Sparse polynomial interpolation in nonstandard bases. *SIAM J. Comput.*, 24(2):387–397, 1995. Cited on page(s) 384.

Lakshmanan, M. and Sahadevan, R. Painlevé analysis, Lie symmetries, and integrability of coupled nonlinear oscillators of polynomial type. *Physics Reports*, 224:1–93, 1993. Cited on page(s) 101.

LaMacchia, B. A. and Odlyzko, A. M. Solving large sparse linear systems over finite fields. In Menezes, A. J. and Vanstone, S., editors, *Advances in Cryptology: CRYPTO '90*, volume 537 of *Lecture Notes in Computer Science*, pages 109–133, Berlin-Heidelberg-New York, 1991. Springer-Verlag. Cited on page(s) 37.

Lambe, Larry. Next generation computer algebra systems AXIOM and the scratchpad concept: applications to research in algebra. In *Analysis, algebra, and computers in mathematical research (Luleå, 1992)*, volume 156 of *Lecture Notes in Pure and Appl. Math.*, pages 201–222. Dekker, New York, 1994. Cited on page(s) 210, 211.

Lambe, Larry and Löfwall, Clas. The cohomology ring of the free loop space of a wedge of spheres and cyclic homology. *C. R. Acad. Sci. Paris Sér. I Math.*, 320(10):1237–1242, 1995. ISSN 0764-4442. Cited on page(s) 210.

Lambe, Larry and Stasheff, Jim. Applications of perturbation theory to iterated fibrations. *Manuscripta Math.*, 58(3):363–376, 1987. ISSN 0025-2611. Cited on page(s) 210.

Lambe, L. A. Resolutions via homological perturbation. *Journal of Symbolic Computation*, 12(1):71–87, July 1991. Cited on page(s) 210, 211, 264.

Lambe, Larry A. Homological perturbation theory, Hochschild homology, and formal groups. In *Deformation theory and quantum groups with applications to mathematical physics (Amherst, MA, 1990)*, volume 134 of *Contemp. Math.*, pages 183–218. Amer. Math. Soc., Providence, RI, 1992. Cited on page(s) 210, 211.

Lambe, Larry A. Resolutions which split off of the bar construction. *J. Pure Appl. Algebra*, 84(3):311–329, 1993. ISSN 0022-4049. Cited on page(s) 210, 211.

Lambert, R. *Computational Aspects of Discrete Logarithms*. PhD thesis, University of Waterloo, 1996. Cited on page(s) 37.

Landau, S. Polynomial time algorithms for Galois groups. In *Proceedings of the International Symposium on Symbolic and Algebraic Computation*, volume 174 of *Lecture Notes in Computer Science*, pages 225–236. Springer, 1984. Cited on page(s) 47.

Landau, S. Factoring polynomials over algebraic number fields. *SIAM J. Comput.*, 14:184–195, 1985. Cited on page(s) 18, 47.

Landau, S. Simplification of nested radicals. *SIAM J. Comput.*, 21:85–110, 1992. Cited on page(s) 18.

Landau, S. and Miller, G. L. Solvability by radicals is in polynomial time. *J. of Computer and System Sciences*, 30:179–208, 1985. Cited on page(s) 47.

Lane, Saunders Mac. *Homology*. Classics in Mathematics. Springer-Verlag, Berlin, 1995. ISBN 3-540-58662-8. Reprint of the 1975 edition. Cited on page(s) 207, 208.

Laporta, S. and Remiddi, E. *Phys. Lett. B*, 379:283, 1996. Cited on page(s) 166.

Lari-Lavassani, A., Langford, W., and Huseyin, K. Symmetry-breaking bifurcations on multidimensional fixed point subspaces. *Dynamics and Stability of Systems*, 9:345–373, 1994. Cited on page(s) 107.

Lari-Lavassani, A., Langford, W. F., Huseyin, K., and Gatermann, K. Steady-state mode interactions for $D_3$ and $D_4$-symmetric systems. *Dynamics of Continuous, Discrete and Impulsive Systems*, 6:169–209, 1999. Cited on page(s) 216.

Larin, S. A., Tkachov, F. V., and Vermaseren, J. A. M. *Preprint NIKHEF-H/91-18*, 1991. Cited on page(s) 166, 169.

Lars Brücher, Johannes Franzkowski, and Dirk Kreimer. XLoops: Automated Feynman diagram calculation. *Comput. Phys. Commun.*, 115:140–160, 1998. Cited on page(s) 391.

Lassner, W. Symbol representations of noncommutative algebras. In *EUROCAL'85*, LNCS 204, pages 99–115. Springer, 1985. Cited on page(s) 61.

Lassner, W. Operator ordering and symbol representations of enveloping algebras. In Shirkov et al. [1991], pages 296–305. ISBN 981-02-0687-9. Cited on page(s) 59.

Laudal, O. A. and Pfister, G. *Local Moduli and Singularities*, volume 1310 of *Lecture Notes in Mathematics*. Springer-Verlag, Berlin-Heidelberg-New York, 1988. Cited on page(s) 199.

Laue, R. Construction of combinatorial objects – A tutorial. *Bayreuther Math. Schr.*, 43:53–96, 1993. Cited on page(s) 90.

Laue, R. Halvings on small point sets. Technical Report CDMTCS-078, Center of Discrete Mathematics and Theoretical Computer Science, University of Auckland, New-Zealand, 1997. Cited on page(s) 90.

Laure, P. and Demay, Y. Symbolic computation and equation on the center manifold: Application to the Couette-Taylor problem. *Comp. Fluids*, 16:229–238, 1988. Cited on page(s) 106.

Lausch, H. and Noebauer, W. *Algebra of Polynomials*. North-Holland, Amsterdam, 1973. Cited on page(s) 491.

Lawo, C. C-XSC, a programming environment for verified scientific computing and numerical data processing. *In: Adams, E., Kulisch, U. (eds.) Scientific computing with automatic result verification: Academic Press*, pages 71–86, 1992. Cited on page(s) 111.

Lay, G. and Zimmer, H. G. Constructing elliptic curves with given group order over large finite fields. In *Lect. Notes in Comp. Sci.*, volume 877, pages 250–263, Berlin-Heidelberg-New York, 1994. Springer-Verlag. Cited on page(s) 49.

Lazard, Daniel. Gröbner bases, Gaussian elimination, and resolution of systems of algebraic equations. In van Hulzen, J. A., editor, *EUROCAL'83*, volume 162 of *LNCS*, pages 146–156. Springer, 1983. Cited on page(s) 32.

Lazard, D. A new method for solving algebraic systems of positive dimension. *Discrete Applied Mathematics*, 33:147–160, 1991. Cited on page(s) 53, 178, 179.

Lazard, D. Solving zero-dimensional algebraic systems. *Journal of Symbolic Computation*, 13(2):117–131, 1992. Cited on page(s) 52, 53.

Lazard, D. On the representation of rigid-body motions and its application to generalized platform manipulators. In Angeles, Hommel, and Kovács, editors, *Computational Kinematics*, pages 175–182. Kluwer, Dordrecht, 1993. Cited on page(s) 232.

LeChenedac, Ph. *Canonical Forms in Finitely Presented Algebras*. Pitman, London, 1986. Cited on page(s) 491.

Lee, H. R. and Saunders, B. D. Fraction free Gaussian elimination for sparse matrices. *Journal of Symbolic Computation*, 19:393–402, 1995. Cited on page(s) 36.

Lee, H. Y. and Liang, C. G. Displacement analysis of the general spatial 7-link 7R mechanism. *J. Mech. and Mach. Th.*, 3:219–226, 1988. Cited on page(s) 232.

Lee, John M. *Ricci: A Mathematica package for doing tensor calculations in differential geometry. Documentation and source code*, 2000. http://www.math.washington.edu/~lee/Ricci/. Cited on page(s) 175.

Leech, J., editor. *Computational Problems in Abstract Algebra, Proceedings of a Conference Held at Oxford Under the Auspices of the Science Research Council, Atlas Computer Laboratory, 29. Aug. to 2. Sept. 1967*, Oxford, 1970. Pergamon Press. ISBN 0-08-012975-7. Cited on page(s) 70.

Leedham-Green, C. R. and O'Brien, E. A. Recognising tensor products of matrix groups. *Internat. J. Algebra Comput.*, 7:541–559, 1997a. Cited on page(s) 475.

Leedham-Green, C. R. and O'Brien, E. A. Tensor products are projective geometries. *Journal of Algebra*, 189:514–528, 1997b. Cited on page(s) 475.

Leedham-Green, C. R. and Soicher, L. H. Collection from the left and other strategies. *J. Symbolic Comput.*, 9:665–675, 1990. Cited on page(s) 459.

Leeuwen, M. A. A. Van and Roelofs, G. H. M. Termination for a class of algorithms for constructing algebras by generators and relations. *Journal of Pure and Applied Algebra*, 117 & 118:431–445, 1997. Cited on page(s) 60.

Lehmann, F., editor. *Semantic Networks in Artificial Intelligence*. Pergamon Press, Oxford, 1992. Cited on page(s) 491, 619.

Lehmann, F., Maurer, M., Müller, V., and Shoup, V. Counting the number of points on elliptic curves over finite fields of characteristic greater than three. In *Lect. Notes in Comp. Sci.*, volume 877, pages 60–70, Berlin-Heidelberg-New York, 1994. Springer-Verlag. Cited on page(s) 49, 405.

Lehmann, N. J. Fehlerschranken für Näherungslösungen bei Differentialgleichungen. *Numerische Mathematik*, 10:261–288, 1967. Cited on page(s) 110.

Lehmann, N. J. Computer algebra and practical analysis. In Buchberger, B., editor, *EUROCAL '85, European Conference on Computer Algebra, Linz, Aus-*

*tria, April 1–3, 1985, Proceedings Vol. 1: Invited Lectures*, volume 203 of *Lecture Notes in Computer Science*, pages 102–113, Berlin-Heidelberg-New York, 1985a. Springer-Verlag. ISBN 3-540-15983-5, 0-387-15983-5. Cited on page(s) 109.

Lehmann, N. J. Die Analytische Maschine - Grundlagen einer Computer-Analytik. In *Sitzungsberichte der Sächsischen Akademie der Wissenschaften zu Leipzig, Mathematisch- naturwissenschaftliche Klasse*, volume 118, 4. Akademie Verlag, Berlin, 1985b. Cited on page(s) 109.

Lehmann, N. J. Adaptive Approximation und Anwendungen. Studientexte Weiterbildungszentrum Computermathematik 105, Technische Universität Dresden, 1989. Cited on page(s) 109.

Lemaire, Jean-Michel. *Algèbres connexes et homologie des espaces de lacets.* Springer-Verlag, Berlin, 1974. Lecture Notes in Mathematics, Vol. 422. Cited on page(s) 209.

Lenstra, A. K. Polynomial factorization by root approximation. In Fitch, John, editor, *Proc. of the 3rd Int. Symp. on Symb. and Alg. Comput. EUROSAM 84*, Lecture Notes in Computer Science 174, pages 272–276, Berlin-Heidelberg-New York-London-Paris-Tokyo-Hong Kong, 1984. Springer. Cited on page(s) 15.

Lenstra, A. K. *LIP, long integer package.* Bellcore, lenstra@bellcore.com, 1995. 45 pages. Cited on page(s) 12, 13, 404.

Lenstra, A. K. and Lenstra Jr., H. W. Algorithms in number theory. In van Leeuwen, J., editor, *Handbook of theoretical computer science, Volume A, Algorithms and Complexity*, chapter 12. Elsevier, 1990. Cited on page(s) 44.

Lenstra, A. K., Lenstra Jr., H. W., and Lovász, L. Factoring polynomials with rational coefficients. *Mathematische Annalen*, 261:515–534, 1982. Cited on page(s) 15, 25, 47, 51.

Lenstra, A. K., Lenstra, Jr., H. W., Manasse, M. S., and Pollard, J. M. The factorization of the ninth Fermat number. *Math. Comp.*, 61:319–349, 1993. Cited on page(s) 45.

Lenstra, A. K. and Lenstra Jr., H. W. (eds). *The development of the number field sieve.* Lecture Notes in Math. Springer-Verlag Berlin, 1993. Cited on page(s) 44, 491.

Lenstra Jr., H. W. Factoring with elliptic curves. *Annals of Mathematics*, 126: 649–673, 1987. Cited on page(s) 13, 48.

Lenstra Jr., H. W. and Tijdeman, R. *Computational Methods in Number Theory I,II*. Mathematisch Centrum, Amsterdam, 1982. Cited on page(s) 491.

Leon, Jeffrey S. Partitions, refinements, and permutation group computations. In Finkelstein and Kantor [1997], pages 123–158. Cited on page(s) 83.

Leroux, Pierre and Reutenauer, Christoph, editors. *4th Conference on Formal Power Series and Algebraic Combinatorics*, Québec à Montréal, June 1992. Université du Québec à Montréal. Cited on page(s) 487.

Lescanne, P., editor. *Rewriting Techniques and Applications, RTA-87, Bordeaux, Proceedings*, volume 256 of *Lecture Notes in Computer Science*, Berlin-Heidelberg-New York, 1987. Springer-Verlag. Cited on page(s) 486.

Lescanne, P. Implementation of completion by transition rules + control: ORME. In Kirchner, H. and Wechler, W., editors, *Algebraic and Logic Programming, Second International Conference, Nancy, France, October 1990, Proceedings*, volume 463 of *Lecture Notes in Computer Science*, pages 262–269, Berlin-Heidelberg-New York, 1990. Springer-Verlag. Cited on page(s) 477.

Lescanne, P. Termination of rewrite systems by elementary interpretations. In Kirchner and Levi [1992], pages 21–36. Cited on page(s) 477.

Levelt, A. H. M., editor. *ISSAC '95, International Symposium on Symbolic and Algebraic Computation, Montreal, Canada*, New York, 1995. ACM. Cited on page(s) 488, 498, 509, 521, 606.

Levelt, T. Invited lecture ISSAC 97. unpublished, 1997. Cited on page(s) 232.

Levi, D. and Winternitz, P., editors. *Painlevé Transcendents: Their Asymptotics and Physical Applications*, volume 278 of *NATO Adv. Sci. Inst. Ser. B (Phys.)*. Plenum, New York, 1992. Cited on page(s) 99.

Lewis, R. H. The six line problem and resultants. Presented to the "Grand Challenges" session at IMACS-ACA, Hawaii, July 1997, 1997. Cited on page(s) 381.

Lewis, R. H. and Liriano, S. Isomorphism classes and derived series of almost-free groups. *Experimental Mathematics*, 3:255–258, 1994. Cited on page(s) 381.

Lewis, R. H. and Moore, G. D. Computer search for nilpotent complexes. *Experimental Mathematics*, 6:239–246, 1997. Cited on page(s) 381.

Lewis, R. H. and Nakos, G. Solving the six line problem with resultants. Presented to the "Grand Challenges" session at IMACS-ACA, Prague, August 1998, 1998. Cited on page(s) 381.

Lewis, R. H. and Wester, M. Comparison of polynomial-oriented computer algebra systems. http://www.fordham.edu/lewis/cacomp.html, 1999. Poster at ISSAC '99. Cited on page(s) 306.

Leyland, P. freelip: adapted from the RSA-129 package. Oxford university, pcl@oucs.ox.ac.uk, 1997. Cited on page(s) 12, 13.

Li, T. Y., Sauer, T., and Yorke, J. A. The cheater's homotopy. *SIAM J. Num. Anal.*, 26(5):1241–1251, 1989. Cited on page(s) 123.

*LiDIA, a library for computational number theory*. LiDIA-Group, Technical University Darmstadt, 1997. 419 pages. Cited on page(s) 13.

LiDIA-Group. LiDIA 1.3.1 – *a library for computational number theory*. Technische Universität Darmstadt, 1998. Available via anonymous FTP from ftp.informatik.tu-darmstadt.de:/pub/TI/systems/LiDIA or via WWW from http://www.informatik.tu-darmstadt.de/TI/LiDIA. Cited on page(s) 45.

LiDIA-Group. LiDIA 2.0.1 – *a library for computational number theory*. Technische Universität Darmstadt, 2000. Available via anonymous FTP from ftp.informatik.tu-darmstadt.de:/pub/TI/systems/LiDIA or via WWW from http://www.informatik.tu-darmstadt.de/TI/LiDIA. Cited on page(s) 406.

Lidl, R., Matthews, R., and Wells, R. *Galois software and manual*. University of Tasmania, Hobart, Australia, 1987. Cited on page(s) 491.

Lidl, R. and Niederreiter, H. *Introduction to Finite Fields.* Cambridge University Press, Cambridge, 1986. Cited on page(s) 17, 23, 491.

Lie, S. *Verh. Gesell. d. Wissenschaften zu Christiania,* 1874. Cited on page(s) 190.

Lie, S. *Vorlesungen über Differentialgleichungen mit bekannten infinitesimalen Transformationen.* Teubner, Leipzig, 1881. (reprint Chelsea Publishing Company, New York, 1967). Cited on page(s) 98, 491.

Lie, S. In Engel, F. and Hergard, editors, *Gesammelte Werke, Vol. 1-7,* Leipzig, 1899. Teubner. Cited on page(s) 190.

Lie, S. and Scheffers, G. *Vorlesungen über Differentialgleichungen mit bekannten infinitesimalen Transformationen.* Teubner, 1891. Cited on page(s) 192, 194.

Linton, S. On vector enumeration. *Linear Algebra and Appl.,* 192:235–248, 1993. Cited on page(s) 59.

Linton, S. A. Constructing matrix representations of finitely presented groups. *J. Symb. Comput.,* 12:427–438, 1991. Cited on page(s) 85.

Linton, S. A., Parker, R. A., Walsh, P. G., and Wilson, R. A. Computer construction of the Monster. *J. Group Theory,* 1:307–337, 1998. Cited on page(s) 87, 196.

Liouville, J. Premier (et second) mémoire sur la détermination des intégrales dont la valeur est algébrique. *Jour. École Polytechnique,* 14 c.22:124–148, 149–193, 1833. Cited on page(s) 94.

Liouville, J. Mémoire sur l'intégration d'une classe de fonctions transcendentes. *Journal für reine und angewandte Mathematik,* 13, 1835. Cited on page(s) 94.

Lippold, Frank. Implementierung eines Verfahrens zum Zählen reeller Nullstellen multivariater Polynome. Diplomarbeit, Universität Passau, Passau, September 1993. Cited on page(s) 427.

Lipson, J. D. *Elements of Algebra and Algebraic Computing.* Addison-Wesley, Reading, Massachusetts, 1981. Cited on page(s) 491.

Liska, R. and Steinberg, S. Applying quantifier elimination to stability analysis of difference schemes. *The Computer Journal,* 36:497–509, 1993. Cited on page(s) 139.

Liska, Richard and Wendroff, Burton. Where numerics can benefit from computer algebra in finite difference modelling of fluid flows. In Ganzha, V. G., Mayr, E. W., and e.v. Vorzhtsov, editors, *Computer algebra in scientific computing, CASC'99,* Berlin, Heidelberg, 1999. Springer. Cited on page(s) 139.

Littelmann, P. An effective method to classify nilpotent orbits. In *Algorithms in algebraic geometry and applications (Santander 1994),* Progress in Math. 143, pages 255–269. Birkhäuser Verlag, 1996. Cited on page(s) 411.

Lloyd, N. G. and Pearson, J. M. REDUCE and the bifurcation of limit cycles. *Journal of Symbolic Computation,* 9:215–224, 1990. Cited on page(s) 107.

Lo, Eddie H. Finding intersections and normalizers in finitely generated nilpotent groups. *Journal of Symbolic Computation,* 25:45–59, 1998a. Cited on page(s) 78.

Lo, Eddie H. A polycyclic quotient algorithm. *Journal of Symbolic Computation,* 25:45–59, 1998b. Cited on page(s) 82.

Lohner, R. *Einschließung der Lösung gewöhnlicher Anfangs- und Randwertaufgaben und Anordnungen*. PhD thesis, University of Karlsruhe, 1988. Cited on page(s) 111.

Lombardi, H. Effective real Nullstellensatz and variants. In Mora and Traverso [1991], pages 263–288. Cited on page(s) 56.

Lombardi, H. and Roy, M.-F., editors. *Effective Methods in Algebraic Geometry*, volume 139 of *Journal of pure and applied algebra*, 1999. North-Holland. vol. 139, Nos. 1-3, special issue. Cited on page(s) 489.

Loos, R. The algorithm description language ALDES (Report). *ACM SIGSAM Bulletin*, 14(1):15–39, 1976. Cited on page(s) 421, 423.

Loos, R. Generalized polynomial remainder sequences. In Buchberger et al. [1983], pages 115–137. Cited on page(s) 23.

Loos, Rüdiger and Weispfenning, Volker. Applying linear quantifier elimination. *The Computer Journal*, 36(5):450–462, 1993. Special issue on computational quantifier elimination. Cited on page(s) 54, 139, 204, 206.

Lossen, C. New asymptotics for the existence of plane curves with prescribed singularities. *Comm. in Alg.*, 27:3263–3282, 1999. Cited on page(s) 201.

Lovász, L. *An Algebraic Theory of Numbers, Graphs, and Convexity*. Society for Industrial and Applied Mathematics, Philadelphia, 1986. Cited on page(s) 491.

Lübeck, F. *Charaktertafeln für die Gruppen $CSp_6(q)$ mit ungeradem q und $Sp_6(q)$ mit geradem q*. Universität Heidelberg, Heidelberg, 1993. Dissertation. Cited on page(s) 88.

Luby, M., Mitzenmacher, M., Shokrollahi, M. A., Spielman, D., and Stemann, V. Practical loss-resilient codes. In *Proceedings of the the 29th Symposium on Theory of Computing*, pages 150–159, 1997. Cited on page(s) 129.

Luks, Eugene M. Isomorphism of graphs of bounded valence and be tested in polynomial time. *Journal of Computer and System Sciences*, 25:42–65, 1982. Cited on page(s) 70.

Lüneburg, H. *On the Rational Normal Form of Endomorphisms, A Primer to Constructive Algebra*. Bibliographisches Institut, Mannheim, 1987. Cited on page(s) 491.

Lunter, G. *Bifurcations in Hamiltonian systems, computing singularities by Gröbner bases*. PhD thesis, Rijksuniversiteit Groningen, 1999. Cited on page(s) 106.

Lyndon, Roger C. and Schupp, Paul E. *Combinatorial Group Theory*. Springer Verlag, 1977. Cited on page(s) 78, 491.

Maârouf, H. *Étude de quelques problèmes effectifs en algèbre différentielle*. Thèse de troisième cycle, Université de Marrakech, Faculté des sciences Semlalia, 1996. Cited on page(s) 105.

Maârouf, H., Kandri-Rody, A., and Ssafini, M. Triviality and dimension of a system of algebraic differential equations. *J. Autom. Reason.*, 20:365–385, 1998. Cited on page(s) 105.

Macaulay, F. S. *Algebraic Theory of Modular Systems*, volume 19 of *Cambridge Tracts in Mathematics*. Cambridge University Press, Cambridge, 1916. Cited on page(s) 491.

MacCallum, M. A. H. *Sheep: Information and source code*, 1995a. http://www.maths.qmw.ac.uk/hyperspace/#ftp. Cited on page(s) 175.

MacCallum, M. A. H. Using computer algebra to solve ordinary differential equations. In Cohen, A. M., van Gastel, L., and Lunel, S. M. Verduyn, editors, *Computer Algebra in Industry 2*, pages 19–41. John Wiley, New York, 1995b. Cited on page(s) 97.

MacCallum, M. A. H. Computer algebra and applications in relativity and gravity. In Macias, A., Matos, T., Obregon, O., and Quevedo, H., editors, *Recent Developments in Gravitation and Mathematical Physics: Proceedings of the First Mexican School on Gravitation and Mathematical Physics*. World Scientific Publishing, 1996. Cited on page(s) 172.

MacCallum, M. A. H., Skea, J. E. F., McCrea, J. D., and McLenaghan, R. G. *Algebraic Computing in General Relativity*. Clarendon Press, 1994. Cited on page(s) 172, 175, 491.

MacCallum, M. A. H. and Wright, F. J. *Algebraic Computing with REDUCE*, volume 1 of *Lecture Notes from the First Brasilian School on Computer Algebra*. Clarendon Press, Oxford, 1991. Cited on page(s) 492.

Macintyre, A. Model completeness. In Barwise, Jon, editor, *Handbook of Mathematical Logic*, volume 90 of *Studies in Logic and the Foundations of Mathematics*, pages 139–180. North Holland Publishing Company, 1977. Cited on page(s) 138.

Macintyre, A. and Wilkie, A. J. On the decidability of the real exponential field. In *Kreiseliana: About and around Georg Kreisel*, pages 441–467. A.K. Peters, 1996. Cited on page(s) 138.

MacLane, Saunders. *Categories for the working mathematician. 4th corrected printing*, volume 5 of *Graduate Texts in Mathematics*. Springer-Verlag, Berlin, Heidelberg, New York, 1988. Cited on page(s) 207.

Macsyma. MACSYMA reference manual version 9, 1977. Cited on page(s) 165, 492.

Macsyma. Macsyma product information, 1998. http://www.macsyma.com/ (inoperational on Oct 16, 2001). Cited on page(s) 175.

Macsyma Inc. *Macsyma Mathematics and System Reference Manual*. Arlington, Mass., sixteenth edition edition, March 1996a. Cited on page(s) 283, 286, 287.

Macsyma Inc. *Macsyma User's Guide*. Arlington, Mass., second edition edition, 1996b. Cited on page(s) 283, 287.

Macsyma Inc. *Introduction to Macsyma*. Arlington, Mass., March 1997. Cited on page(s) 283, 287.

Macsyma Inc. *Introduction to PDEase2D*. Arlington, Mass., November 1998a. on-line version. Cited on page(s) 283.

Macsyma Inc. *Macsyma Scientific Graphics Reference Manual*. Arlington, Mass., sixteenth edition edition, 1998b. Cited on page(s) 283, 287.

Macsyma Inc. *PDEase2D Reference Manual*. Arlington, Mass., third edition edition, May 1998c. Cited on page(s) 283.

Macsyma Inc. *Scientific Notebook Interface Reference Manual*. Arlington, Mass., 1998d. Cited on page(s) 283, 287.

Macsyma Inc. and Three's Company. *Macsyma Mathematics and System Reference Manual (in Japanese)*, sixteenth edition edition, 1996. Cited on page(s) 283.

MacWilliams, F. J. and Sloane, N. J. A. *The Theory of Error-Correcting Codes*. North-Holland, 1988. Cited on page(s) 129, 491.

Madlener, K. and Reinert, B. Computing Gröbner bases in monoid and group rings. In Bronstein, M., editor, *Symbolic and Algebraic Computation, International Symposium ISSAC '93, Kiew, 1993, Proceedings*, pages 254–263, New York, 1993. ACM. Cited on page(s) 62.

Madlener, K. and Reinert, B. Gröbner bases in non-commutative reduction rings. In Buchberger and Winkler [1998], pages 408–420. Cited on page(s) 63.

Madlener, K. and Reinert, B. Relating rewriting techniques on monoids and rings: Congruences on monoids and ideals in monoid rings. *Theoretical Computer Science*, 208(1-2):3–31, 1998b. Cited on page(s) 62.

Madlener, K. and Reinert, B. String rewriting and Gröbner bases – A general approach to monoid and group rings. In Bronstein et al. [1998], pages 127–180. Cited on page(s) 62.

Maeder, R. E. *Programming in Mathematica*. Addison-Wesley, Reading, Massachusetts, 1990. Cited on page(s) 492.

Magid, A. *Lectures on Differential Galois Theory*. University Lectures Series 7. Amer. Math. Soc., Providence, RI, 1994. Cited on page(s) 97, 491.

Magliveras, Spyros S. and Leavitt, David W. Simple 6-(33, 8, 36) designs from $P\Gamma L_2(32)$. In *Computational group theory (Durham, 1982)*, pages 337–352. Academic Press, London, 1984. Cited on page(s) 375.

Magnus, W., Oberhettinger, F., and Soni, R. P. *Formulas and Theorems for the Special Functions of Mathematical Physics*. Springer, Berlin–Heidelberg–New York, 1966. Cited on page(s) 214, 491.

Mall, Daniel. Covers and fans of polynomial ideals. *Theor. Comp. Science*, 187: 167–178, 1997. Cited on page(s) 30.

Malle, G. Darstellungstheoretische Methoden bei der Realisierung einfacher Gruppen vom Lie Typ als Galoisgruppen. In Michler and Ringel [1991], pages 443–459. Cited on page(s) 196.

Malle, G. and Matzat, B. H. *Inverse Galois Theory*. Springer Verlag, Heidelberg, 1999. Cited on page(s) 48, 491.

Mandache, A. Applications of polynomial systems solving in geometric modelling and robotics: An annotated bibliography. Technical report, RISC-Linz, 1993. Cited on page(s) 235.

Manin, Yu.I. Cyclotomic fields and modular curves. *Russian Mathematical Surveys*, 26:7–78, 1971. Cited on page(s) 49.

Manocha, D. *Algebraic and numeric techniques for modelling and robotics*. Dissertation, University of California, Berkeley, USA, 1992. Cited on page(s) 236, 239, 240.

Manocha, D. and Canny, J. Real time inverse kinematics for general 6R manipulators. In *Proc. IEEE Robotics and Automation*, pages 383–389, 1992. Cited on page(s) 232.

Manocha, D. and Demmel, J. Algorithms for intersecting parametric and algebraic curves II: multiple intersections. *Computer Vision, Graphics and Image Processing: Graphical Models and Image Processing*, pages 81–100, 1995. Cited on page(s) 117, 122.

Maple. Maple product information, 2001. http://www.maplesoft.com/. Cited on page(s) 175.

Marché, Claude. Normalized rewriting: an alternative to rewriting modulo a set of equations. *Journal of Symbolic Computation*, 11(1), 1996. Cited on page(s) 134.

Marden, M. *The Geometry of the Zeros of a Polynomial in a Complex Variable*, volume 3 of *AMS Math. Surv.* 1949. Cited on page(s) 491.

Marden, M. *Geometry of Polynomials*. AMS, Providence, RI, second edition, 1966. Cited on page(s) 491.

Mark, Wolfgang. Gröbnerbasen über Hauptidealringen und euklidischen Ringen. Diplomarbeit, Universität Passau, Passau, 1992. Cited on page(s) 426.

Mars, M. and Wolf, T. Diagonal and non-diagonal $G_2$ perfect-fluid cosmologies with a proper conformal killing vector. *Class. Quantum Grav.*, 14:1–28, 1997. Cited on page(s) 467.

Marti, J., Hearn, A., Griss, M., and Griss, C. Standard lisp report. *SIGPLAN Notes, ACM, N.Y.*, 14:48, 1979. Cited on page(s) 186.

Masaharu Goto. The CINT C/C++ Interpreter, 2000. http://root.cern.ch/root/Cint.html. Cited on page(s) 391.

Mathematica. Mathematica product information, 2001. http://www.wri.com/. Cited on page(s) 175.

Mathews, S. and McCallister, L. Mathematica in the middle school and high school: An aid to solving word problems in Algebra I. *Mathematica in Education and Research*, 4(3):14–20, 1995. Cited on page(s) 256.

Mathtensor. Mathtensor online information, 2001. http://smc.vnet.net/MathSolutions.html. Cited on page(s) 175.

Mattman, T. and McKay, J. Computation of Galois groups over function fields. *Math.Comput.*, 66:823–831, 1997. Cited on page(s) 47.

Mattson, H. F., Mora, T., and Rao, T. R. N., editors. *Applied Algebra, Algebraic Algorithms and Error-Correcting Codes, 9th International Symposium, AAECC-9, New Orleans, LA, USA, October 7–11, 1991, Proceedings*, volume 539 of *Lecture Notes in Computer Science*, Berlin-Heidelberg-New York, 1991. Springer-Verlag. ISBN 3-540-54522-0, 0-387-54522-0. Cited on page(s) 487, 516, 569.

Matz, O., Miller, A., Potthoff, A., Thomas, W., and Valkema, E. Report on the program AMoRE. Technical Report 9507, Institut für Informatik und Praktische Mathematik, Universität Kiel, 1995. Cited on page(s) 349.

Matzat, B. H. *Konstruktive Galoissche Theorie*, volume 1284 of *Lecture Notes in Mathematics*. Springer-Verlag, Berlin-Heidelberg-New York, 1987. Cited on page(s) 491.

Matzat, B. H., Greuel, G.-M. W., and Hiss, G., editors. *Algorithmic Algebra and Number Theory*, Berlin-Heidelberg-New York, 1999. Springer-Verlag. ISBN ISBN 3-540-64670-1. Selected Papers from a Conference Held at the University of Heidelberg in October 1997. Cited on page(s) 88, 197, 492, 530, 561, 591.

Mayes, R. ACT in Algebra: Student attitude and belief. *The International Journal of Computer Algebra in Mathematics Education*, 5(1):3–14, 1998. Cited on page(s) 250.

Mayr, Ernst W. Membership in polynomial ideals over Q is exponential space complete. In Monien, B. and Cori, R., editors, *STACS'89*, volume 349 of *LNCS*, pages 400–406. Springer, 1989. Cited on page(s) 29.

Mayr, Ernst W. and Meyer, Albert R. The complexity of the word problem for commutative semigroups and polynomial ideals. *Advances in Mathematics*, 46:305–329, 1982. Cited on page(s) 29.

McClellan, M. T. The exact solution of systems of linear equations with polynomial coefficients. *J. ACM*, 20:563–588, 1973. Cited on page(s) 36.

McCurley, K. S. The discrete logarithm problem. In *Cryptology and Computational Number Theory*, number 42 in Proc. Symp. in Applied Mathematics, pages 49–74. American Mathematical Society, 1990. Cited on page(s) 406.

McKay, B. D. *nauty users guide*. Computer Science Department, Australian National University, 1990. Technical report TR-CS-90-02. Also see http://cs.anu.edu.au/people/bdm/nauty/. Cited on page(s) 473, 474.

Mei, Chiang C. *Mathematical Analysis in Engineering*. Cambridge University Press, 1994. ISBN 0-521-46053-0. Cited on page(s) 283.

Meinardus, G. *Approximation of Functions. Theories and Numerical Methods*. Springer, Berlin, 1967. Cited on page(s) 478.

Melenk, H. *REDUCE Symbolic Mode Primer, Version 1*. Konrad-Zuse-Zentrum Berlin, 1995. Cited on page(s) 492.

Melenk, H., Möller, H. M., and Neun, W. Symbolic solution of large stationary chemical kinetic problems. *Impact of Computing in Science and Engineering*, 1(2):138–167, 1989. Cited on page(s) 244.

Menezes, A. J. *Elliptic curve public key cryptosystems*. Kluwer Academic Publishers, 1993. Cited on page(s) 43, 491.

Merkwitz, Thomas. Markentafeln endlicher Gruppen. Diplomarbeit, Lehrstuhl D für Mathematik, Rheinisch Westfälische Technische Hochschule, 1998. Cited on page(s) 389.

Merriman, J. R., Siksek, S., and Smart, N. P. Explicit 4-descents on an elliptic curve. *Acta Arith.*, 77:385–404, 1996. Cited on page(s) 49.

Mertig, R., Böhm, M., and Denner, A. *Comp. Phys. Comm.*, 64:345, 1991a. Cited on page(s) 166.

Mertig, R., Böhm, M., and Denner, A. FeynCalc: Computer-algebraic calculation of Feynman amplitudes. *Comput. Phys. Commun.*, 64:345, 1991b. Cited on page(s) 170.

Mertig, R., Böhm, M., and Denner, A. FeynCalc: Computer algebraic calculation of Feynman amplitudes. *Comput. Phys. Commun.*, 64:345, 1991c. Cited on page(s) 471.

Mertig, R. and Scharf, R. TARCER: A Mathematica program for the reduction of two loop propagator integrals. *Comput. Phys. Commun.*, 111:265, 1998. Cited on page(s) 472.

Mertig, R. and van Neerven, W. L. The Calculation of the Two-Loop Spin Splitting Functions $P_{ij}^{(1)}(x)$. *Z. Phys.*, C70:637–654, 1996. Cited on page(s) 472.

Messollen, M. LIBFAC, a subroutine library for characteristic sets and irreducible decomposition, 1996. Cited on page(s) 412.

Mestre, J.-F. Un exemple de courbe elliptique sur Q de rang $\geq 15$. *Comptes Rendue Academie Sciences*, 314:453–455, 1992. Cited on page(s) 49.

Metzner, T., Radimersky, M., Sorgatz, A., and Wehmeier, S. *User's Guide to Macro Parallelism in MuPAD 1.4.1*. B.G. Teubner, Stuttgart, January 1999. Cited on page(s) 330, 492.

Meyer, K. R. Lie transform tutorial II. In Meyer, K. R. and Schmidt, D. S., editors, *Computer Aided Proofs in Analysis*, IMA Volumes in Mathematics and its Applications 28, pages 190–210. Springer-Verlag, New York, 1991. Cited on page(s) 106.

Meyer, K. R. and Schmidt, D. S. *Computer Aided Proofs in Analysis*. Springer-Verlag, Berlin-Heidelberg-New York, 1990. Cited on page(s) 491.

Micali, S., editor. *Randomness and Computation*. JAI Press, Greenwhich CT, 1989. Cited on page(s) 492, 558.

Michalski, R. S., Carbonell, J. G., and Mitchell, T. M., editors. *Machine Learning*. Tioga Publ. Co, Palo Alto, CA, 1983. Cited on page(s) 492, 582.

Michler, G. O. and Ringel, C. M., editors. *Representation Theory of Finite Groups and Finite-Dimensional Algebras*, volume 95 of *Progress in Mathematics*, Basel, 1991. Birkhäuser Verlag. Cited on page(s) 197, 578.

Mignotte, M. *Mathématiques pour le calcul formel*. PUF, 1989. Cited on page(s) 490.

Mignotte, M. *Mathematics for Computer Algebra*. Springer-Verlag, Berlin-Heidelberg-New York, 1992. ISBN 0-387-97675-2. Cited on page(s) 4, 17, 490.

Mihalescu, P. Cyclotomy primality proving – recent developments. In Buhler, J. P., editor, *ANTS III*, number 1423 in LNCS, pages 95–110, 1998. Cited on page(s) 42.

Milne, P. S. On the solutions of a set of equations. In Donald, B., editor, *Symbolic and Numerical Computation for Artificial Intelligence*, pages 89–102. Academic Press, 1992. Cited on page(s) 240.

Mines, R., Richman, F., and Ruitenburg, W. *A Course in Constructive Algebra*. Springer-Verlag, New York, 1988. Cited on page(s) 491.

Minkwitz, T. *Algorithmensynthese für lineare Systeme mit Symmetrie*. PhD thesis, Universität Karlsruhe, Informatik, 1993. Cited on page(s) 462.

Minkwitz, T. Algorithms Explained by Symmetry. *Lecture Notes on Computer Science*, 900:157–167, 1995. Cited on page(s) 462.

Minnichelli, R. J., Anagnost, J. J., and Desoer, C. A. An elementary proof of Kharitonov's stability theorem with extensions. *IEEE Trans. Automatic Control*, 34(9):995–998, 1989. Cited on page(s) 122.

Miola, A., editor. *Computing Tools for Scientific Problem Solving*. Academic Press, New York, 1990a. Cited on page(s) 492.

Miola, A., editor. *Design and Implementation of Symbolic Computation Systems, International Symposium DISCO '90, Capri, Italy, April 10–12, 1990, Proceedings*, volume 429 of *Lecture Notes in Computer Science*, Berlin-Heidelberg-New York, 1990b. Springer-Verlag. ISBN 3-540-52531-9, 0-387-52531-9. Cited on page(s) 486, 506, 523, 534, 558, 566, 596.

Miola, A., editor. *International Symposium on Design and Implementation of Symbolic Computation Systems*, volume 722 of *Lecture Notes in Computer Science*, September 1993. Springer. Cited on page(s) 487.

Mishra, Bhubaneswar. *Algorithmic Algebra*. Texts and Monographs in Computer Science. Springer Verlag, New York, 1993. ISBN 0-540-94090-1. Cited on page(s) 28, 32, 33, 205, 491.

Mitchell, T. M., Utgoff, P. E., and Banerji, R. B. Learning by experimentation: Acquiring and refining problem-solving heuristics. In Michalski et al. [1983]. Cited on page(s) 142.

Mitschi, C. and Singer, M. F. Connected linear groups as differential Galois groups. *J. Algebra*, 184:333–361, 1996. Cited on page(s) 98.

Mnafeg, O. El. Quantorenelimination für die additive Arithmetik mit Exponentialfunktion. Master's thesis, Universität Passau, Mai 1992. Cited on page(s) 139.

Moenck, R. On computing closed forms for summations. In *Proceedings of the 1977 MACSYMA Users' Conference, held at Berkeley, Calif., July 27–29, 1977*, number 2012 in NASA CP, pages 225–236, Washington, D.C., 1977. NASA Scientific and Technical Information Service. Cited on page(s) 91.

Moenck, R. T. and Carter, J. H. Approximate algorithms to derive exact solutions to systems of linear equations. In *Proc. EUROSAM '79*, volume 72 of *Lecture Notes in Computer Science*, pages 65–73, Berlin-Heidelberg-New York, 1979. Springer-Verlag. Cited on page(s) 36.

Möller, H. M. and Mora, T. Upper and lower bounds for the degree of gröbner bases. In Fitch, J., editor, *EUROSAM'84*, volume 174 of *LNCS*, pages 172–183. Springer, 1984. Cited on page(s) 29.

Möller, H. M. and Mora, T. New constructive methods in classical ideal theory. *J. of Algebra*, 100:138–178, 1986. Cited on page(s) 30, 210, 379.

Monagan, M. B., Geddes, K. O., Heal, K. M., Labahn, G., and Vorkoetter, S. M. *Maple V. Programming Guide*. Springer-Verlag, New York, 1998. Cited on page(s) 478, 492.

Montes, Antonio, editor. *Actas del 6º Encuentro de Algebra Computacional y Aplicaciones*, September 2000. UPC, Universitat Politecnica de Catalunya. Cited on page(s) 489.

Montgomery, P. L. A block Lanczos algorithm for finding dependencies over GF(2). In *Proc. Eurocrypt 1995*, volume 921 of *Lecture Notes in Computer Science*, pages 106–120, Berlin-Heidelberg-New York, 1995. Springer-Verlag. Cited on page(s) 38.

Moore, R.E. *Interval Analysis*. Prentice-Hall, Englewood Cliffs, N.J., 1966. Cited on page(s) 111.

Moore, R.E. *Methods and Applications of Interval Analysis*. SIAM, Philadelphia, 1979. Cited on page(s) 111, 491.

Mora, F. Gröbner bases for non-commutative polynomial rings. In Calmet [1986], pages 353–362. ISBN 3-540-16776-5, 0-387-16776-5. Cited on page(s) 61, 379.

Mora, T. An algorithm to compute the equations of tangent cones. In Calmet [1982], pages 158–165. ISBN 3-540-11607-9, 0-387-11607-9. Cited on page(s) 32.

Mora, T. Seven variations on standard bases. Preprint 45, Universität Genua, Genua, 1988. Cited on page(s) 62.

Mora, T., editor. *Applied Algebra, Algebraic Algorithms and Error-Correcting Codes, 6th International Conference, AAECC-6, Rome, Italy, July 4–8, 1988, Proceedings*, volume 357 of *Lecture Notes in Computer Science*, Berlin-Heidelberg-New York, 1989a. Springer-Verlag. ISBN 3-540-51083-4, 0-387-51083-4. Cited on page(s) 486, 514.

Mora, T. Gröbner bases for non-commutative algebras. In Gianni [1989], pages 150–161. ISBN 3-540-51084-2, 0-387-51084-2. Cited on page(s) 61.

Mora, T. An introduction to commutative and non-commutative Gröbner bases. *Theoretical Computer Science*, 134:131–173, 1994. Cited on page(s) 61.

Mora, Teo, editor. *Proceedings of the 12th International Symposium on Applied Algebra, Algebraic Algorithms and Error-Correcting Codes*, Lecture Notes in Computer Science, Berlin, 1997. Springer. Cited on page(s) 488.

Mora, T., Pfister, G., and Traverso, C. An introduction to the tangent cone algorithm. In Hoffman, C. M., editor, *Issues in non-linear geometry and robotics*. JAI Press, 1992. Cited on page(s) 32, 464.

Mora, T. and Robbiano, L. Gröbner fan of an ideal. *Journal of Symbolic Computation*, 6:183–208, 1988. Cited on page(s) 30.

Mora, T. and Traverso, C., editors. *MEGA'90, Effective Methods in Algebraic Geometry, Castiglioncello, Livorno, Italy*, volume 94 of *Progress in Mathematics*, Basel, 1991. Birkhäuser Verlag. Cited on page(s) 486, 522, 576.

Morain, F. Primality proving using elliptic curves – an update. In Buhler, J. P., editor, *ANTS III*, number 1423 in LNCS, pages 111–127, 1998. Cited on page(s) 42, 43.

Morgan, A. *Solving Polynomial Systems Using Continuation for Engineering and Scientific Problems*. Prentice Hall, Englewood Cliffs, N.J. 07632, 1987. Cited on page(s) 491.

Morrison, M. A. and Brillhart, J. A method of factoring and the factorization of $F_7$. *Mathematics of Computation*, 29:183–205, 1975. Cited on page(s) 16.

Moses, J. *Symbolic Integration*. PhD thesis, MIT, Cambridge, USA, 1967. Cited on page(s) 141.

Moura, J. M. F., Johnson, J., Johnson, R., Padua, D., Prasanna, V., and Veloso, M. M. SPIRAL: Portable Library of Optimized Signal Processing Algorithms, 1998. http://www.ece.cmu.edu/~spiral/. Cited on page(s) 462.

Mourrain, B., editor. *ISSAC 2001 Proc. 2001 Internat. Symp. Symbolic Algebraic Comput.*, New York, N. Y., 2001. ACM Press. Cited on page(s) 490, 521, 535, 600.

Mourrain, B. and Pan, V. Asymptotic acceleration of solving multivariate polynomial systems of equations. In *30th Ann. ACM Symposium on Theory of Computing (STOC)*, pages 302–308, Baltimore, 1998. ACM Press. Cited on page(s) 242.

Mulders, T. and Storjohann, A. Fast algorithms for linear algebra modulo $N$. In *Algorithms — ESA '98*, LNCS 1461, 1998. Cited on page(s) 39.

Mulders, T. and Storjohann, A. Diophantine linear system solving. In Dooley [1999]. Cited on page(s) 36, 37.

Mulders, T. and Storjohann, A. On lattice reduction for polynomial matrices. Technical Report 356, Departement Informatik, ETH Zürich, December 2000. Cited on page(s) 39.

Mulmuley, K. A fast parallel algorithm to compute the rank of a matrix over an arbitrary field. *Combinatorica*, 7:101–104, 1987. Cited on page(s) 38.

Mumford, David, Fogarty, John, and Kirwan, Frances. *Geometric Invariant Theory*. Number 34 in Ergebnisse der Math. und ihrer Grenzgebiete. Springer-Verlag, Berlin, Heidelberg, New York, 3 edition, 1994. Cited on page(s) 33, 491.

Mund, E. H. Computer algebra and finite element methods in engineering. In Cohen, A. M., van Gastel, L., and Lionel, S. Verduyn, editors, *Computer Algebra in Industry 2*, pages 81–100. John Wiley, New York, 1995. Cited on page(s) 108.

Murray, Scott H. and O'Brien, E. A. Selecting base points for the Schreier-Sims algorithm for matrix groups. *Journal of Symbolic Computation*, 19:577–584, 1995. Cited on page(s) 475.

Musser, David R. Proving inductive properties of abstract data types. In *Proc. 7th PoPL*, pages 154–162, Las Vegas, Nevada, January 1980. ACM. Cited on page(s) 136.

Mňuk, M. and Winkler, F. CASA—a system for computer aided constructive algebraic geometry. In *Proceedings of the International Symposium DISCO'96*, number 1128 in LNCS, pages 297–307. Springer, 1996. Cited on page(s) 357.

Myers, N. A new and useful template technique: "traits". In Lippman, St. B., editor, *C++ Gems*, SIGS Books, page 24. Prentice-Hall, 1996. http://www.cantrip.org/traits.html. Cited on page(s) 13.

NAG Ltd. *The NAG Fortran Library – Mark 19*, 1999. Cited on page(s) 152.

Nagao, K. and Kouya, T. An example of an elliptic curve over **Q** with rank $\geq 21$. *Proc. Japan Acad.*, 70(Ser. A, No. 4):104–105, 1994. Cited on page(s) 49.

Nagata, M. *Lectures on the Fourteenth Problem of Hilbert*. Tata Institute of Fundamental Research, Bombay, 1965. Cited on page(s) 34.

Nakao, M.R. A Numerical Verification Method for the Existence of Weak Solutions for Nonlinear Boundary Value Problems. *Journal of Mathematical Analysis and Applications*, 164:489–507, 1992. Cited on page(s) 111.

Narendran, P. and Rusinowitch, M., editors. *Proceedings of the 10th International Conference on Rewriting Techniques and Applications*, volume 1631, Berlin, 1999. Springer. Cited on page(s) 489.

Narlikar, Girija J. and Blelloch, Guy E. Space-efficient implementation of nested parallelism. In *Proceedings of the Sixth ACM SIGPLAN Symposium on Principles and Practice of Parallel Programming*, pages 25–36, June 1997. Cited on page(s) 148.

Nauheim, R. Systems of algebraic equations with bad reduction. *J.Symb.Comput*, 25:619–641, 1998. Cited on page(s) 48.

Naundorf, H. *Ein denotationales Modell für parallele objektbasierte Systeme*. MuPAD Reports. B.G. Teubner, Stuttgart, 1996. Cited on page(s) 331.

Naundorf, H. *MAMMUT - Eine verteilte Speicherverwaltung für symbolische Manipulation.* MuPAD Reports. B.G. Teubner, Stuttgart, 1997. Cited on page(s) 331.

Naylor, W. N. and Watt, S. M. On the relationship between OpenMath and MathML. In *Workshop on Internet Accessible Mathematical Communication.* ICM, 2001. http://icm.mcs.kent.edu/research/iamc01/proceedings.html. Cited on page(s) 158.

Nebe, G. Finite subgroups of $GL_n(\mathbf{q})$ for $25 \leq n \leq 31$. *Exp. Math.*, 5:163–195, 1996. Cited on page(s) 389.

Nebe, G. and Plesken, W. Finite rational matrix groups. *Mem. Am. Math. Soc.*, 556, 1995. Cited on page(s) 85, 197, 389.

Nemes, I., Petkovšek, M., Wilf, H. S., and Zeilberger, D. How to do Monthly problems with your computer. *Amer. Math. Monthly*, 104:505–1997, 199. Cited on page(s) 93.

Netto, E. *Vorlesungen über Algebra*, volume 1 und 2. B.G. Teubner, Leipzig, 1896/1900. Cited on page(s) 491.

Neubüser, J. Untersuchungen des Untergruppenverbandes endlicher Gruppen auf einer programmgesteurten elektronischen Dualmaschine. *Numer. Math.*, 2:280–292, 1960. Cited on page(s) 69.

Neubüser, J. An invitation to computational group theory. In *Groups '93, Galway/St. Andrews*, volume 212 of *London Math. Soc. Lecture Note Series*, pages 457–475. Cambridge University Press, 1995. Cited on page(s) 66, 387.

Neubüser, J. An elementary introduction to coset table methods in computational group theory. In Campbell, C. M. and Robertson, E. F., editors, *Groups – St. Andrews 1981*, number 71 in LMS Lecture Notes, pages 1–45. Cambridge University Press, 1982. Cited on page(s) 90.

Neubüser, J., Pahlings, H., and Plesken, W. CAS; Design and use of a system for the handling of characters of finite groups. In Atkinson [1984b], pages 195–247. Cited on page(s) 84.

Neuhaus, R. Computation of real radicals of polynomial ideals - II. *Journal pure appl. algebra*, 124:261–280, 1998. Cited on page(s) 56.

Neumaier, A. *Interval Methods for Systems of Equations, Encyclopedia of Mathematics and its Applications.* Cambridge University Press, 1990. Cited on page(s) 111.

Neumann, Peter M. and Praeger, Cheryl E. A recognition algorithm for special linear groups. *Proc. London Math. Soc. (3)*, 65:555–603, 1992. Cited on page(s) 475.

Neumann, Peter M. and Praeger, Cheryl E. Cyclic matrices and the meataxe. In Kantor, W. M. and Ákos Seress, editors, *Groups and Computation* III, *Conference Proceedings at the Ohio State University, June* 15-19, 1999, OSU Mathematical Research Institute Publications, pages 291–300, Berlin, New York, 2001. de Gruyter. Cited on page(s) 475.

Newell, A. C., Tabor, M., and Zeng, Y. B. A unified approach to Painlevé expansions. *Physica D*, 29:1–68, 1987. Cited on page(s) 101.

Newman, M. *Integral Matrices.* Academic Press, 1972. Cited on page(s) 39, 491.

Newman, M. F. Calculating presentations for certain kinds of quotient groups. In SYMSAC '76, *Proc. ACM Sympos. symbolic and algebraic computation*, pages 2–8, New York, 1976. (New York, 1976), Association for Computing Machinery. Cited on page(s) 459.

Newman, M. F. Determination of groups of prime-power order. In *Group Theory*, volume 573 of *Lecture Notes in Math.*, pages 73–84, Berlin, Heidelberg, New York, 1977. (Canberra, 1975), Springer-Verlag. Cited on page(s) 459.

Newman, M. F. and Nickel, Werner. Engel elements in groups. *Journal for Pure and Applied Algebra*, 96:39–45, 1994. Cited on page(s) 460.

Newman, M. F. Nickel, Werner, and Niemeyer, Alice C. Descriptions of groups of prime-power order. *Journal of Symbolic Computation*, 25:665–682, 1998. Cited on page(s) 459.

Newman, M. F. and O'Brien, E. A. Application of computers to questions like those of Burnside, II. *Internat. J. Algebra Comput.*, 6:593–605, 1996. Cited on page(s) 460.

Newman, M. F. and Vaughan-Lee, Michael. Some Lie rings associated with Burnside groups. *Electron. Res. Announc. Amer. Math. Soc.*, 4:1–3 (electronic), 1998. ISSN 1079-6762. Cited on page(s) 460.

Nguyen, Truong Q. Digital filter bank design quadratic-constrained formulation. *IEEE Transactions on Signal Processing*, 43:2103–2108, 1995. Cited on page(s) 229.

Nickel, Werner. Computing nilpotent quotients of finitely presented groups. In Baumslag and et al. [1996], pages 175–191. Cited on page(s) 82, 459, 460.

Nickel, Werner. Some groups with right Engel elements. In *Groups St Andrews 1997 in Bath*, volume 261 of *London Math. Soc. Lecture Note Ser.*, pages 571–578. Cambridge University Press, 1999. Cited on page(s) 460.

Nickel, W., Niemeyer, A. C., O'Keefe, C. M., and Praeger, C. E. The block-transitive point-imprimitive 2-(729,8,1) designs. *Communication and Computing*, 3:47–61, 1992. Cited on page(s) 90.

Niemeyer, Alice C. A finite soluble quotient algorithm. *Journal of Symbolic Computation*, 18:541–561, 1994a. Cited on page(s) 82.

Niemeyer, Alice C. A soluble quotient algorithm. *Journal of Symbolic Computation*, 18:541–561, 1994b. Cited on page(s) 460.

Niemeyer, Alice C. Computing finite soluble quotients. In Bosma, Wieb and van der Poorten, Alf, editors, *Proc. of CANT '92*, pages 75–82. Kluwer Academic Publishers, 1995. Cited on page(s) 460.

Niemeyer, Alice C. and Praeger, Cheryl E. Implementing a recognition algorithm for classical groups. In Finkelstein and Kantor [1997], pages 273–296. Cited on page(s) 75, 475.

Niemeyer, Alice C. and Praeger, Cheryl E. A recognition algorithm for classical groups over finite fields. *Proc. London Math. Soc.* (3), 77:117–169, 1998. Cited on page(s) 475.

Niemeyer, Alice C. and Praeger, Cheryl E. A recognition algorithm for non-generic classical groups over finite fields. *J. Austral. Math. Soc. (A)*, 67:223–253, 1999. Cited on page(s) 475.

Nikiforov, A. F., Suslov, S. K., and Uvarov, V. B. *Orthogonal Polynomials of a Discrete Variable*. Springer, Berlin–Heidelberg–New York, 1991. Cited on page(s) 214.

Nipkow, Tobias, editor. *Proceedings of the 9th International Conference on Rewriting Techniques and Applications*, volume 1379, Berlin, 1998. Springer. Cited on page(s) 489.

Nipkow, Tobias and Prehofer, Christian. Higher-order rewriting and equational reasoning. In Bibel, W. and Schmitt, P. H., editors, *Automated Deduction – A Basis for Applications*, volume 1. Kluwer Academic Publishers, 1998. Cited on page(s) 136.

Noll, Peter. MPEG Digital Audio Coding. *IEEE Signal Processing Magazine*, pages 59–81, 1997. Cited on page(s) 228.

Nonnenmacher, T. F. *J. Non-Equilib. Thermodyn.*, 5:361, 1980. Cited on page(s) 192.

Noro, T. and et al. *Risa/Asir*. Available via anonymous ftp from endeavor.fujitsu.co.jp/pub/isis/asir, 1993, 1995. Cited on page(s) 394.

Novikov, P. S. On the algorithmic unsolvability of the word problem in group theory. *Trudy Math. Inst. im. Steklov*, 44:1–143, 1955. In Russian. Cited on page(s) 69.

Oaku, T. An algorithm of computing $b$-functions. *Duke Mathematical Journal*, 87:115–132, 1997a. Cited on page(s) 393, 395.

Oaku, T. Algorithms for $b$-functions, restrictions, and algebraic local cohomology groups of $D$-modules. *Advances in Applied Mathematics*, 19:61–105, 1997b. Cited on page(s) 393.

Oaku, T. and Takayama, N. An algorithm for de Rham cohomology groups of the complement of an affine variety via D-module computation. *Journal of Pure and Applied Algebra*, 139:201–233, 1999. Cited on page(s) 393, 396.

O'Brien, E. A. The $p$-group generation algorithm. *J. Symbolic Comput.*, 9: 677–698, 1990. Cited on page(s) 459.

O'Brien, E. A. The groups of order 256. *J. Algebra*, 143:219–235, 1991. Cited on page(s) 67, 197, 460.

O'Brien, E. A. Isomorphism testing for $p$-groups. *Journal of Symbolic Computation*, 17:133–147, 1994. Cited on page(s) 460.

O'Brien, E. A. Computing automorphism groups of $p$-groups. In Bosma, Wieb and van der Poorten, Alf, editors, *Computational Algebra Number and Number Theory*, pages 83–90. (Sydney, 1992), Kluwer Academic Publishers, Dordrecht, 1995. Cited on page(s) 460.

O'Connor, J. E. R. and Prince, G. E. Finding collineations of Kimura metrics. *General Relativity and Gravitation*, 1:69–82, 1998. Cited on page(s) 468.

Odlyzko, A. M. Discrete logarithms in finite fields and their cryptographic significance. In Beth, T., Cot, N., and Ingmarsson, I., editors, *Advances in Cryptology, Proceedings Eurocrypt 1984*, volume 209 of *Lecture Notes in Computer Science*, pages 224–314, Berlin-Heidelberg-New York, 1984. Springer-Verlag. Cited on page(s) 22.

Oettli, W. and Prager, W. Compatibility of approximate solution of linear equations with given error bounds for coefficients and right-hand sides. *Numerische Mathematik*, 6:405–409, 1964. ISSN 0029-599X. Cited on page(s) 115.

Oevel, W. Numerical computations in MuPAD 1.4. mathPAD *journal*, 8(1): 57–66, March 1998. See http://www.mupad.de/mathpad.shtml. Cited on page(s) 332.

Oevel, W. MuPAD 1.4: An overview, 1999. http://www.sciface.com/support/papers/PS/qref_en.ps.gz. Cited on page(s) 331.

Oevel, W., Postel, F., Wehmeier, S., and Gerhard, J. *The MuPAD Tutorial*. Springer-Verlag, Berlin-Heidelberg-New York, 2000. German version: W. Oevel, F. Postel, G. Rüscher, and S. Wehmeier. *Das MuPAD Tutorium*. Springer, 1999. Cited on page(s) 331, 492.

Ollivier, F. *Le problème de l'identifiabilité structurelle globale: étude théorique, méthodes effectives et bornes de complexité*. PhD thesis, École polytechnique, Palaiseau, France, 1990a. Cited on page(s) 104.

Ollivier, F. Standard bases of differential ideals. In Sakata, S., editor, *Proc. AAECC-8*, Lecture Notes in Computer Science 508, pages 304–321, Tokyo, 1990b. Springer. Cited on page(s) 104.

Olshevsky, V. Pivoting on structured matrices with applications. *Linear Algebra and Appl.*, to appear, 2001. See also http://www.cs.gsu.edu/~matvro/papers.html. Cited on page(s) 38.

Olver, P. J. *Applications of Lie Groups to Differential Equations*, volume 107 of *Graduate Texts in Mathematics*. Springer-Verlag, New York-Berlin-Heidelberg-Tokyo, 1986. Cited on page(s) 98, 99.

Olver, P. J. *Applications of Lie groups to differential equations*. Springer, 2nd edition, 1993. Cited on page(s) 191, 491.

Opgenorth, J., Plesken, W., and Schulz, T. Crystallographic algorithms and tables. *Acta Cryst.*, 54(5):517–531, 1998. Sect. A. Cited on page(s) 85, 243, 355.

O'Shea, D. B. Topologically trivial deformations of isolated hypersurface singularities are equimultiple. *Proc. AMS*, 101:260–262, 1987. Cited on page(s) 200.

Ostheimer, Gretchen. Algorithms for polycyclic-by-finite matrix groups. In Finkelstein and Kantor [1997], pages 297–308. Cited on page(s) 75.

Ostrogradski, M. W. De l'intégration des fractions rationelles. *Bull. Acad. Imp. Sci. St. Petersbourg*, 4:146–167, 286–300, 1845. Cited on page(s) 94.

Ostrowski, A. M. On two problems in abstract algebra connected with horner's rule. In *Studies in Math. and Mech. presented to Richard von Mises*, Studies in Math. and Mech., pages 40–48. Academic Press, 1954. Cited on page(s) 14.

Ozello, P. *Calcul exact des formes de Jordan et de Frobenius d'une matrice*. PhD thesis, Université Scientifique Technologique et Médicale de Grenoble, France, 1987. Cited on page(s) 40.

Pan, V. Y. Solving a polynomial equation: Some history and recent progress. *SIAM Review*, 39(2):187–220, 1997. Cited on page(s) 26.

Pan, Victor Ya. Approximate polynomial GCDs, Pade approximation, polynomial zeros, and bipartite graphs. In *Proc. 9th ACM-SIAM Symposium on Discrete Algorithms*, pages 68–77, Philadelphia, 1998. SIAM Publications. Cited on page(s) 120.

Pan, Victor Y. *Structured Matrices and Polynomials: Unified Superfast Algorithms*. Birkhäuser Verlag, Basel, 2001. Cited on page(s) 38.

Papanikolaou, Th. *Entwurf und Entwicklung einer objektorientierten Bibliothek für algorithmische Zahlentheorie*. PhD thesis, Universität des Saarlandes, 1997. Cited on page(s) 404.

Pardo, L. M. How lower and upper complexity bounds meet in elimination theory. In Cohen, G., Giusti, M., and Mora, T., editors, *AAECC-11*, Lecture Notes in Computer Science 948, pages 33–69, Berlin-Heidelberg-New York-London-Paris-Tokyo-Hong Kong, 1995. Springer. Cited on page(s) 54.

Park, H., Kalker, T., and Vetterli, M. Groebner Bases and Multidimensional Perfect Reconstruction FIR Systems. *Journal of Multidimensional Systems and Signal Processing*, 8:11–30, 1997. Cited on page(s) 228.

Parker, L. and Christensen, S. M. *MathTensor: A system for doing tensor analysis by computer*. Addison-Wesley, Redwood, 1994. Cited on page(s) 173, 175, 176, 183, 491.

Parker, R. A. The computer calculation for modular characters (the Meat-Axe). In Atkinson [1984b], pages 267–274. Cited on page(s) 85, 196, 363, 475.

Parker, R. A. An integral Meat-axe. In Curtis and Wilson [1998], pages 215–228. Cited on page(s) 85.

Pasqualina, Conti and Carlo, Traverso. Buchberger algorithm and integer programming. In *AAECC-9*, volume 539 of *LNCS*, pages 130–139. Springer, 1991. Cited on page(s) 30.

Pasqualina, Conti and Carlo, Traverso. A case of automatic theorem proving in Euclidean geometry: the MacLane $8_3$ theorem. In G., Cohen, M., Giusti, and T., Mora, editors, *AAECC-11*, number 948 in LNCS, pages 183–193. Springer, 1995. Cited on page(s) 33.

Pauer, F. On lucky ideals for Gröbner basis computation. *J. Symb. Computation*, 14:471–482, 1992. Cited on page(s) 30.

Paul, Etienne. Equational methods in first order predicate calculus. *Journal of Symbolic Computation*, 1(1):7–29, 1985. Cited on page(s) 135.

Paule, P. Short and easy computer proofs of the Rogers-Ramanujan identities and of identities of similar type. *Electronic Journal of Combinatorics*, 1 (R10): 9p., 1994. Cited on page(s) 93.

Paule, P. Greatest factorial factorization and symbolic summation. *Journal of Symbolic Computation*, 20:235–268, 1995. Cited on page(s) 92.

Paule, P. and Riese, A. A Mathematica $q$-analogue of Zeilberger's algorithm based on an algebraically motivated approach to $q$-hypergeometric telescoping. In et al., M. E. H. Ismail, editor, *Fields Proceedings of the Workshop "Special Functions, q-Series and Related Topics"*, volume 14 of *Fields Institute Communications*, pages 179–210. American Mathematical Society, 1997. Cited on page(s) 92, 93.

Paule, P. and Schorn, M. A Mathematica version of Zeilberger's algorithm for proving binomial coefficient identities. *Journal of Symbolic Computation*, vol. 20:673–698, 1995. Cited on page(s) 93.

Paule, P. and Strehl, V. Symbolic summation — some recent developments. In et al., J. Fleischer, editor, *Computer Algebra in Science and Engineering*, pages 138–162. World Scientific, 1995. Cited on page(s) 92.

Pavelle, Richard, editor. *Application to Computer Algebra*. Kluwer Academic Publishers, Boston, 1985. Cited on page(s) 283, 491, 492.

Pearson, J. M., Lloyd, N. G., and Christopher, C. J. Algorithmic derivation of centre conditions. *SIAM Rev.*, 38:619–636, 1996. Cited on page(s) 107.

Peckman, K. One mile wide and one inch deep: Giving the secondary Mathematics curriculum more depth with Mathematica. *Mathematica in Education and Research*, 7(4):29–38, 1998. Cited on page(s) 256.

Pecquet, L. *A first course in Magma - the computer algebra system*. Springer-Verlag, 2000. Cited on page(s) 307.

Pedersen, H. *Counting Real Zeros*. PhD thesis, Courant Institute, New York, 1991. Cited on page(s) 57.

Pedersen, P., Roy, M.-F., and Szpirglas, A. Counting real zeroes in the multivariate case. In Eysette, F. and Galigo, A., editors, *Computational Algebraic Geometry*, volume 109 of *Progress in Mathematics*, pages 203–224. Birkhäuser, Boston, Basel; Berlin, 1993. ISBN 3-7643-3678-1. Proceedings of the MEGA 92. Cited on page(s) 57, 139.

Péladan-Germa, A. Testing identities of series defined by algebraic partial differential equations. In Cohen, G., Giusti, M., and Mora, T., editors, *AAECC-11*, Lecture Notes in Computer Science 948, pages 393–407, Paris, 1995. Springer. Cited on page(s) 105.

Pellizzari, Thomas, Beth, Thomas, Grassl, Markus, and Müller-Quade, Jörn. Stabilization of Quantum States in Quantum Optical Systems. *Physical Review A*, 54(4):2698–2702, October 1996. Cited on page(s) 130.

Pemberton, Steve, Altheim, Murray, and et al. *XHTML[tm] 1.0: The Extensible HyperText Markup Language*. W3C Recommendation 26 January 2000. World Wide Web Consortium, 2000. http://www.w3.org/TR/2000/REC-xhtml1-20000126. Cited on page(s) 160.

Penttila and Williams, B. Regular packings of PG(3,q). preprint, December 1997. Cited on page(s) 90.

Perron, O. *Die Lehre von den Kettenbrüchen*. B.G. Teubner, Stuttgart, 1954. Cited on page(s) 16, 491.

Pesch, Michael. Die MAS-Implementation des Algorithmus zur Berechnung umfassender Gröbner-Basen. Internal Report, April 1994. Cited on page(s) 427.

Pesch, Michael. *Gröbner Bases in Skew Polynomial Rings*. Dissertation, Universität Passau, Passau, 1997. Cited on page(s) 426, 427.

Pesch, M. Two-sided Gröbner bases in iterated Ore extensions. In Bronstein et al. [1998], pages 227–243. Cited on page(s) 61, 62.

Peterson, G. and Stickel, M. Complete sets of reductions for some equational theories. *Journal of the ACM*, 28:223–264, 1981. Cited on page(s) 133.

Pethö, A., Zimmer, H. G., Pohst, M. E., and Williams, H. C., editors. *Computational Number Theory, Conference Proceedings Univ. Debrecen, (Hungary), 04-08.09.1989*. W. De Gruyter, Berlin, 1990. Cited on page(s) 492.

Petitot, M. Experience with Axiom. In Jacob, G., Oussous, N. E., and Steinberg, S., editors, *Proceedings SC 93. International IMACS Symposium on Symbolic Computation. New Trends and Developments*, page 240, Lille, France, 1993. LIFL Univ. Lille. Cited on page(s) 264.

Petkovšek, M. Hypergeometric solutions of linear recurrences with polynomial coefficients. *Journal of Symbolic Computation*, 14:243–264, 1992. Cited on page(s) 92.

Petkovšek, M. and Wilf, H. S. A high-tech proof of the Mills-Robbins-Rumsey determinant formula. *Electronic Journal of Combinatorics*, 3 (R19):4p., 1996. Cited on page(s) 93.

Petkovšek, M., Wilf, H. S., and Zeilberger, D. $A = B$. A. K. Peters, Wellesley, MA, 1996. Cited on page(s) 4, 92, 93, 215, 491.

Pfahler, Th. Factoring univariate polynomials over finite fields with lidia. In *3rd International IMACS Conference on Applications of Computer Algebra*, 1997. Cited on page(s) 406.

Pfeiffer, G. Von Permutationscharakteren und Markentafeln. Diplomarbeit, Lehrstuhl D für Mathematik, Rheinisch Westfälische Technische Hochschule, 1991. Cited on page(s) 389.

Pfeiffer, G. The Subgroups of $M_{24}$, or How to Compute the Table of Marks of a Finite Group. *Exp. Math.*, 6(3):247–270, 1997. Cited on page(s) 388.

Pfeil, Jörn. Implementierung von Varianten des Buchberger-Algorithmus mit Polynomfaktorisierung. Diplomarbeit, Universität Passau, Passau, August 1994. Cited on page(s) 426.

Pfister, G. and Schönemann, Hannes. Singularities with exact Poincare complex but not quasihomogeneous. *Revista Mathematica de la Universidad Complutense de Madrid*, 2(2&3):161–171, 1989. Cited on page(s) 198, 446.

Pflügel, E. An Algorithm for Computing Exponential Solutions of First Order Linear Differential Systems. In Küchlin [1997], pages 164–171. Cited on page(s) 97, 98, 372.

Pflügel, E. On the Latest Version of DESIR. *Theoretical Computer Science*, 187: 81–86, 1997b. Cited on page(s) 370.

Pflügel, E. *Résolution symbolique des systèmes différentiels linéaires*. PhD thesis, Universit J. Fourier - Grenoble, 1998. Cited on page(s) 371.

Philipp, Jörg. Syzygien Berechnung im Computeralgebra System MAS. Diplomarabeit, Universität Passau, Passau, June 1991. Cited on page(s) 427.

Pin, J. E. *Varieties of Formal Languages*. Oxford Univ. Press, Oxford, 1986. Cited on page(s) 349.

Pirastu, R. and Strehl, V. Rational summation and the Gosper-Petkovšek representation. *Journal of Symbolic Computation*, 20:627–635, 1995. Cited on page(s) 92.

Plaisted, D. A. Some polynomial and integer divisibility problems are NP-hard. *SIAM J. Comp.*, 7:458–464, 1978. Cited on page(s) 14.

Plesken, W. Towards a soluble quotient algorithm. *Journal of Symbolic Computation*, 4:111–122, 1987. Cited on page(s) 82, 85, 460.

Plesken, W. Finite rational matrix groups: a survey. In Curtis and Wilson [1998], pages 229–248. Cited on page(s) 85.

Plesken, W. Presentations and representations of groups. In Matzat et al. [1999], pages 423–434. ISBN ISBN 3-540-64670-1. Selected Papers from a Conference Held at the University of Heidelberg in October 1997. Cited on page(s) 85, 196.

Plesken, W., Opgenorth, J., and Schulz, T. CARAT - a package for mathematical crystallography. *J. Appl. Cryst.*, 31:827–828, 1998. Cited on page(s) 197.

Plum, M. Computer Assisted Existence Proofs for two Point Boundary Value Problems. *Computing*, 46:19–34, 1991. Cited on page(s) 111.

Plum, M. Numerical Existence Proofs and Explicit Bounds for Solutions of Nonlinear Elliptic Boundary Value Problems. *Computing*, 49:25–44, 1992. Cited on page(s) 111.

Podleś, P. Quantum spheres. *Lett. Math. Phys.*, 14:193–202, 1987. Cited on page(s) 212.

Podleś, P. Differential calculus on quantum spheres. *Lett. Math. Phys.*, 18:107–119, 1989. Cited on page(s) 212.

Podleś, P. The classification of differential structures on quantum 2-spheres. *Comm. Math. Phys.*, 150:167–179, 1992. Cited on page(s) 212, 213.

Pohst, M. A modification of the LLL reduction algorithm. *Journal of Symbolic Computation*, 4:123–127, 1987a. Cited on page(s) 51.

Pohst, M. *Computational algebraic number theory*. DMV Lecture Notes. Birkhäuser Verlag, 1993. Cited on page(s) 398, 491.

Pohst, M. and Schörnig, M. On integral basis reduction in global function fields. In Cohen, Henri, editor, *ANTS II*, volume 1122 of *Lecture Notes in Computer Science*, pages 273–282. Springer-Verlag, 1996. Cited on page(s) 400.

Pohst, M. E., editor. *Algorithmic Methods in Algebra and Number Theory*. Academic Press, New York, 1987b. Cited on page(s) 492.

Pohst, M. E. and Zassenhaus, H. *Algorithmic Algebraic Number Theory*. Cambridge University Press, Cambridge, 1989. Cited on page(s) 4, 17, 21, 47, 398, 399, 491.

Pohst, M. E. and Zassenhaus, H. *Algorithmic Algebraic Number Theory, Paperback Ed.* Cambridge University Press, Cambridge, 1997. Cited on page(s) 491.

Poli, A. Editorial. *Discrete Mathematics*, 56(2-3):93–303, 1985. volume 56, (2-3) constitutes the proceedings of AAECC-1. Cited on page(s) 485.

Poli, A., editor. *Applied Algebra, Algorithmics, and Error-Correcting Codes, 1 2nd International Conference, AAECC-2, Toulouse, France, October 1–5, 1984, Proceedings*, volume 228 of *Lecture Notes in Computer Science*, Berlin-Heidelberg-New York, 1986. Springer-Verlag. ISBN 3-540-16767-6, 0-387-16767-6. Cited on page(s) 485.

Poljak, S. and Rohn, J. Checking robust nonsingularity is NP-hard. *Math. Control Signals Systems*, 6:1–9, 1993. Cited on page(s) 125.

Pollack, Richard and Roy, Marie-Francoise. On the number of cells defined by a set of polynomials. *C. R. Acad. Sci., Paris*, 316(6):573–577, 1993. Cited on page(s) 57.

Polyanin, A. D. and Zhurov, A. I. Algebraicheskii metod integrirovaniya differentsialnyikh uravnenii nelineinoi mekhaniki (algebraic method for the integration of differential equations of nonlinear mechanics). *Dokl. Akad. Nauk, T.*, 337(2):196–199, 1994. Cited on page(s) 223.

Pomerance, C. *Lecture Notes on Primality Testing and Factoring*. The Mathematical Association of America, 1984. Cited on page(s) 491.

Pommaret, J.-F. *Systems of Partial Differential Equations and Lie Pseudogroups.* Gordon&Breach, New York, London, Paris, 1978. Cited on page(s) 103, 491.

Pommaret, J. F. *Partial Differential Equations and Group Theory.* Kluwer, Dordrecht, 1994. Cited on page(s) 102, 103, 491.

Popov, B. A. *Balanced Approximation by Splines.* Naukova Dumka, Kyiv, 1989. Cited on page(s) 478.

Popov, V. L. and Vinberg, E. B. Invariant theory. In Parshin, N. N. and Shafarevich, I. R., editors, *Algebraic Geometry IV*, number 55 in Encyclpaedia of Mathematical Sciences, Berlin, Heidelberg, 1994. Springer-Verlag. Cited on page(s) 33.

Postel, F. MuPAD as a tool, tutee and tutor. Talk at the ACDCA Summer Academy "Recent Research on Derive/TI-92-Supported Mathematics Education", August 1999. http://www.mupad.de/CONF/ACDCA99/paper_acdca99.zip. Cited on page(s) 332.

Postel, F. and Zimmermann, P. Solving ordinary differential equations. In Wester [1999], chapter 11, pages 191–210. Cited on page(s) 332.

Praeger, C. E. and Soicher, L. H. L. H. *Low rank representations and graphs for sporadic groups.* Number 8 in Aust. Math. Soc. Lecture Series. Cambridge University Press, 1997. Cited on page(s) 90, 491.

Presburger, M. Über die Vollständigkeit eines gewissen Systems der Arithmetik. In *Comptes rendues du 1er Congres des Mathematique des Pays Slaves*, volume 395, pages 92–101, Warsaw, 1929. Cited on page(s) 138.

Prevost, S. Exploring the $\epsilon$-$\delta$ definition of limit with Mathematica. *Mathematica in Education and Research*, 3(4):17–21, 1994. Cited on page(s) 255.

Prosper, V. and Lascoux, A. $\mu$-EC, a MuPAD Environment for Combinatorics, 1998. http://phalanstere.univ-mlv.fr/~muec. Cited on page(s) 346.

Prudnikov, A. P., Brychkov, Yu. A., and Marichev, O. I. *Integrals and Series*, volume 3: *More Special Functions*. Gordon and Breach Publ., New York, 1990. Cited on page(s) 215.

Püschel, M. *Konstruktive Darstellungstheorie und Algorithmengenerierung.* PhD thesis, Universität Karlsruhe, Informatik, 1998. Translated in [Püschel 1999a]. Cited on page(s) 218, 462, 593.

Püschel, M. Constructive Representation Theory and Fast Signal Transforms. Technical Report Drexel-MCS-1999-1, Drexel Univ., Philadelphia, 1999a. Translation of [Püschel 1998]. Cited on page(s) 462, 593.

Püschel, M. Decomposing Monomial Representations of Solvable Groups. Technical Report Drexel-MCS-1999-2, Drexel Univ., Philadelphia, 1999b. Submitted for publication. Cited on page(s) 462.

Rabin, M. O. Recursive unsolvability of group theoretic problems. *Ann. of Math.*, 67:172–194, 1958. Cited on page(s) 69.

Rabin, Michael O. Decidable theories. In Barwise, Jon, editor, *Handbook of Mathematical Logic*, volume 90 of *Studies in Logic and the Foundations of Mathematics*, pages 595–629. North Holland Publishing Company, 1977. Cited on page(s) 138.

Raghavan, M. and Roth, B. A general solution for the inverse kinematics of all series chains. In *Proc. 8th Symp. on Rob. and Manip. (RoManSy-90)*, pages 24–31. CISM-IFToMM, 1992. Cited on page(s) 232.

Ragot, J. F., van Hoij, M., Ulmer, F., and Weil, J. A. Liouvillian solutions of linear differential equations of order three and higher. *Journal of Symbolic Computation*, 28:589–609, 1999. Cited on page(s) 97.

Rains, Eric M. Polynomial Invariants of Quantum Codes. LANL preprint quant–ph/9704042, April 1997. Cited on page(s) 131.

Ramani, A., Grammaticos, B., and Bountis, T. The Painlevé property and singularity analysis of integrable and non-integrable systems. *Physics Reports*, 180:159–245, 1989. Cited on page(s) 100, 101.

Ramis, J. P. and Martinet, J. Théorie de Galois différentielle et resommation. In Tournier, E., editor, *Computer Algebra and Differential Equations*. Academic Press, New York, 1990. Cited on page(s) 97.

Rand, Richard H. *Computer Algebra in Applied Mathematics: An Introduction to Macsyma*. Pitman Publishing, Marshfield, Mass., 1984. ISBN 0-273-08632-4. Cited on page(s) 283, 492.

Rand, Richard H. *Topics in Nonlinear Dynamics with Computer Algebra*, volume 1 of *Computation in Education: Mathematics, Science and Engineering*. Gordon and Breach Science Publishers S.A., 1994. ISBN 2-88449-113-9 and 2-88449-114-7. Cited on page(s) 283, 491.

Rand, Richard H. and Armbruster, Dieter. *Perturbation Methods, Bifurcation Theory and Computer Algebra*, volume 65 of *Applied Mathematical Sciences*. Springer–Verlag, New York, 1987. ISBN 0-387-96589-0 and 3-540-96589-0. Cited on page(s) 223, 283, 491.

Randall, C., Kant, E., and Chhabra, A. Using program synthesis to price derivatives. *Journal of Computational Finance*, 1(2):97–128, 1998. Cited on page(s) 441.

Rayna, G. *Reduce, Software for Algebraic Computation*. Springer-Verlag, Berlin-Heidelberg-New York, 1987. Cited on page(s) 492.

Rayward-Smith, V. J., Burton, F. W., Fujimoto, R. M., and Janacek, G. J. Load balancing strategies for the implementation of parallel programs. Technical report, Simon Fraser University Centre for System Science, Vancouver, Canada, 1990. Cited on page(s) 148.

Read, R. C. Every one a winner. *Ann. Discrete Math.*, 2:107–120, 1978. Cited on page(s) 89.

Redfern, Darren, Chandler, Edgar, and Fell, Richard N. *The Macsyma ODE Lab Book*. Jones and Bartlett Publishers, Sudbury, Mass., 1998. ISBN 0-7637-0532-2. Cited on page(s) 283, 492.

Reduce. Reduce online documentation and information, 1999. http://www.uni-koeln.de/REDUCE/index.html. Cited on page(s) 175.

Reid, G. J. Algorithms for reducing a system of PDEs to standard form, determining the dimension of its solution space and calculating its Taylor series solution. *Eur. J. Appl. Math.*, 2:293–318, 1991a. Cited on page(s) 103.

Reid, G. J. Finding abstract Lie symmetry algebras of differential equations without integrating determining equations. *Eur. J. Appl. Math.*, 2:319–340, 1991b. Cited on page(s) 103.

Reid, G. J., Wittkopf, A. D., and Boulton, A. Reduction of systems of nonlinear partial differential equations to simplified involutive form. *Euro. J. Appl. Math.*, 7:635–666, 1996. Cited on page(s) 103.

Reid, Miles. Chapters on algebraic surfaces. In Kollár, J., editor, *Complex algebraic varieties, IAS/Park City Mathematics Series 3*, pages 1–154. AMS, Providence R.I., 1997. Cited on page(s) 302, 303.

Reinert, B., Mora, T., and Madlener, K. A note on Nielsen reduction and coset enumeration. In *ISSAC1998*, pages 171–178, 1998. Cited on page(s) 63.

Renegar, J. On the computational complexity and geometry of the first-order theory of the reals. Part I: Introduction. Preliminaries. The geometry of semialgebraic sets. The decision problem for the existential theory of the reals. *Journal of Symbolic Computation*, 13(3):255–300, 1992a. Cited on page(s) 57, 138, 204.

Renegar, J. On the computational complexity and geometry of the first-order theory of the reals. Part II: The general decision problem. Preliminaries for quantifier elimination. *Journal of Symbolic Computation*, 13(3):301–328, 1992b. Cited on page(s) 57, 138, 204.

Renegar, J. On the computational complexity and geometry of the first-order theory of the reals. Part III: Quantifier elimination. *Journal of Symbolic Computation*, 13(3):329–330, 1992c. Cited on page(s) 57, 138, 204.

Renschuch, B. *Elementare und praktische Idealtheorie*. VEB Deutscher Verlag der Wissenschaften, Berlin, 1976. Cited on page(s) 491.

Reutenauer, C. *Free Lie Algebras*. Clarendon Press, Oxford, 1993. Cited on page(s) 58.

Rice, J. R., editor. *Mathematical Aspects of Scientific Software*, volume 14 of *IMA Volumes in Mathematics and Its Applications*. Springer-Verlag, Berlin-Heidelberg-New York, 1988. Cited on page(s) 491, 557.

Richard-Jung, F. *Représentations graphiques de solutions d'équations différentielles dans le champ complexe*. PhD thesis, Université de Strasbourg, 1988. Cited on page(s) 371.

Richardson, D. Towards computing non algebraic cylindrical decompositions. In Watt, Stephen M., editor, *Proceedings of the 1991 International Symposium on Symbolic and Algebraic Computation*, pages 247–255, Bonn, Germany, July 1991a. Cited on page(s) 138.

Richardson, D. Wu's method and the Khovanskii finiteness theorem. *Journal of Symbolic Computation*, 12:127–141, 1991b. Cited on page(s) 138.

Richardson, D. The elementary constant problem. In Wang, Paul S., editor, *Proceedings of the 1992 International Symposium on Symbolic and Algebraic Computation*, pages 108–116, Bercely, California, July 1992. Cited on page(s) 138.

Richardson, D. A zero structure theorem for exponential polynomials. In Bronstein, Manuel, editor, *Proceedings of the 1993 International Symposium on Symbolic and Algebraic Computation*, pages 144–151, Kiev, Ukraine, July 1993. Cited on page(s) 138.

Richardson, D. A simplified method of recognizing zero among elementary constants. In Levelt, A. H., editor, *Proceedings of the 1995 International Sympo-*

*sium on Symbolic and Algebraic Computation ; July 10 - 12*, New York, NY, 1995. ACM Press. Cited on page(s) 138.

Richardson, D. Local theories and cylindrical decompositon. In Caviness, B. F. and Johnson, J. R., editors, *Quantifier Elimination and Cylindrical Algebraic Decomposition*, Texts and Monographs in Symbolic Computation, pages 351–364. Springer, Wien, New York, 1998. Cited on page(s) 138.

Richter, M. M. *Prinzipien der Künstlichen Intelligenz*. B.G. Teubner, Stuttgart, 1989. Cited on page(s) 142, 491.

Richter-Gebert, J. and Wang, Dongming, editors. *Automated Deduction in Geometry*, volume 2061 of *Lecture Notes in Artificial Intelligence*, 2000. Springer-Verlag. Cited on page(s) 489.

Riemann. Maple package Riemann: Documentation and source code, 1998. http://www.astro.queensu.ca/~portugal/Riemann.html. Cited on page(s) 175.

Riese, A. A generalization of Gosper's algorithm to bibasic hypergeometric summation. *Electronic Journal of Combinatorics*, 3 (R19):16p., 1996. Cited on page(s) 92.

Riesel, H. *Prime Numbers and Computer Methods for Factorization*. Birkhäuser Verlag, Boston, 1985. Cited on page(s) 491.

Rimey, K. Template-based Formula Editing in Kaava. In Miola [1990b]. ISBN 3-540-52531-9, 0-387-52531-9. Cited on page(s) 152.

Rinard, Martin. The design, implementation and evaluation of jade. *ACM Transactions on Programming Languages and Systems*, 20(3):483–545, 1998. Cited on page(s) 148.

Ringel, C. M. *Tame Algebras and Integral Quadratic Forms*. lnm, New York, 1984. Cited on page(s) 64.

Rioboo, Renaud. *Quelques aspects du calcul exact avec des nombres réels*. Thèse de doctorat, Université de Paris 6, Informatique, 1991. Cited on page(s) 96.

Rioboo, Renaud. Real algebraic closure of an ordered field, implementation in Axiom. In Wang [1992], pages 206–215. Cited on page(s) 264.

Riordan, M. List of arbitrary precision C packages. http://www.csc.fi/math_topics/Mail/FAQ/msg00015.html, 1994. 6 pages. Cited on page(s) 13.

Riquier, F. *Les systèmes d'équations aux dérivées partielles*. Gauthier-Villars, Paris, 1910. Cited on page(s) 102, 192, 491.

Risch, Robert. On the integration of elementary functions which are built up using algebraic operations. Research Report SP-2801/002/00, System Development Corporation, Santa Monica, CA, USA, 1968. Cited on page(s) 94.

Risch, Robert. Further results on elementary functions. Research Report RC-2402, IBM Research, Yorktown Heights, NY, USA, 1969a. Cited on page(s) 94.

Risch, Robert. The problem of integration in finite terms. *Transactions of the American Mathematical Society*, 139:167–189, 1969b. Cited on page(s) 94.

Risch, Robert. The solution of the problem of integration in finite terms. *Bulletin of the American Mathematical Society*, 76:605–608, 1970. Cited on page(s) 94.

Ritt, J. F. Prime and composite polynomials. *Trans. Amer. Math. Soc.*, 23: 51–66, 1922. Cited on page(s) 26, 27.

Ritt, J. F. *Differential Equations from the Algebraic Standpoint*, volume 14 of *Amer. Math. Soc. Colloq. Publ.* AMS, New-York, 1932. Cited on page(s) 104.

Ritt, J. F. *Differential Algebra*, volume 33 of *Amer. Math. Soc. Colloq. Publ.* AMS, Providence, 1950. Cited on page(s) 32, 104, 491.

Roach, K. Maple and orthogonal polynomials. Talk at the *First ISAAC Conference, Newark, Delaware, June 3-7, 1997, session on Orthogonal Polynomials*, 1997. Cited on page(s) 215.

Robbiano, Lorenzo. Term orderings on the polynomial ring. In Caviness, B. F., editor, *EUROCAL'85, European conference on computer algebra, vol. II*, volume 357 of *Springer LNCS*, pages 31–44. Springer, 1985. Cited on page(s) 29.

Robbiano, L. *Computational Aspects of Commutative Algebra*. Academic Press, New York, 1989. Cited on page(s) 491.

Robbiano, L. Gröbner Bases and Statistic. In Buchberger, B. and Winkler, F., editors, *Gröbner Bases and Applications (Proc. of the Conf. 33 Years of Gröbner Bases)*, volume 251 of *London Mathematical Society Lecture Notes Series*. Cambridge University Press, 1998. Cited on page(s) 365.

Robbiano, Lorenzo and Sweedler, M. Subalgebra bases. In Burns, W. and Simis, A., editors, *Proc Commutative Algebra Salvador*, volume 1430 of *LNM*, pages 61–87. Springer, 1988. Cited on page(s) 31.

Robidoux, Nicolas. Computer algebra and interpolation: a lesson plan. *Journal of Symbolic Computation*, 23(5/6):551–576, May/June 1997. Cited on page(s) 264.

Robinson, A. *Complete theories*. North-Holland, Amsterdam, 1956. Cited on page(s) 138, 491.

Robinson, A. and Zakon, E. Elementary properties of ordered abelian groups. *Trans. AMS*, 96:222–236, 1960. Cited on page(s) 138.

Robinson, J. A. A machine-oriented logic based on the resolution principle. *Journal of the ACM*, 12(1):23–41, January 1965. Cited on page(s) 135.

Roch, J. L. and Villard, G. Fast parallel computation of the Jordan normal form of matrices. *Parallel Processing Letters*, 6(2):203–212, 1996. Cited on page(s) 41.

Roch, Jean-Louis and Villard, Gilles. Parallel computer algebra. Technical report, Lecture notes for a tutorial in ISSAC'97, Hawaii, July 1997. URL http://www-apache.imag.fr/~jlroch/ps/97-issac.ps.gz. Cited on page(s) 146.

Rody, A. Kandri and Weispfennning, V. Non-commutative Gröbner bases in algebras of solvable type. *Journal of Symbolic Computation*, 9(1):1–26, 1990. Cited on page(s) 426.

Roesner, K. G. Verified solutions for parameters of an exact solution for non-Newtonian liquids using computer algebra. *Zeitschrift fur Angewandte Mathematik und Physik*, 75(suppl. 2):S435–S438, 1995. Cited on page(s) 264.

Roesner, K. G. and Zikanov, O. Taylor-couette instability of polymer solutions. *Comp. Fluid Dyn. Journ.*, 6(1):1–22, 1997. Cited on page(s) 222.

Roggenbach, Markus, Mossakowski, Till, and Schröder, Lutz. Basic datatypes in CASL. Note L-12 (version 0.4), in [CoFI 2001], March 2000a. Cited on page(s) 219.

Roggenbach, Markus, Schröder, Lutz, and Mossakowski, Till. Specifying real numbers in CASL. In Choppy, Christine and Bert, Didier, editors, *Recent Developments in Algebraic Development Techniques, 14th International Workshop, WADT'99*, volume 1827 of *Lecture Notes in Computer Science*, pages 146–161. Springer-Verlag, 2000b. Cited on page(s) 219.

Ronveaux, A., Zarzo, A., and Godoy, E. Recurrence relations for connection coefficients between two families of orthogonal polynomials. *J. Comput. Appl. Math.*, 62:67–73, 1995. Cited on page(s) 214, 215.

Roos, Jan-Erik. On computer-assisted research in homological algebra. *Math. Comput. Simulation*, 42(4-6):475–490, 1996. Cited on page(s) 210.

Rose, Claus. The MAS module DIPAGB: An implementation of the normal with sugar selection strategy in the Buchberger algorithm. Praktikum, Universität Passau, Passau, January 1995. Cited on page(s) 426.

Rosenstein, J. G. *Linear Orderings*, volume 98 of *Pure and Applied Mathematics*. Academic Press, London, 1982. Cited on page(s) 491.

Rothstein, M. A new algorithm for the integration of exponential and logarithmic functions. In *Proceedings of the 1977 MACSYMA Users Conference*, pages 263–274. NASA Pub. CP-2012, 1977. Cited on page(s) 95.

Rouiller, Fabrice. *Algorithmes efficaces pour l'étude des zéros réels des systèmes polynomiaux*. PhD thesis, Université Rennes, 1996. Cited on page(s) 139.

Rouiller, Fabrice. Solving zero-dimensional systems through the rational univariate representation. *AAECC*, 9:433–461, 1999. Cited on page(s) 52.

Rouillier, F. *Algorithmes efficaces pour l'étude des zéros réels des systèmes polynomiaux*. Dissertation, Université de Rennes I, Rennes, France, 1996. Cited on page(s) 178, 241.

Roy, M.-F. Computation of the topology of a real algebraic curve. *Asterisque*, 192:17–34, 1991. Cited on page(s) 238.

Roy, M.-F. Basic algorithms in real algebraic geometry: from Sturm theorem to the existential theory of real numbers. In *Lectures on Real Geometry in memoriam of Mario Raimondo,*, volume 23 of *Expositions in Mathematics*, pages 1–67. de Gruyter, 1996. Cited on page(s) 20, 57, 139.

Roy, M.-F. and Vorobjov, N. Computing the complexfication of semialgebraic sets. In Lakshman [1995], pages 26–34. Cited on page(s) 56, 57.

Ruch, E. and Klein, D. J. Double cosets in chemistry and physics. *Theoretica Chimica Acta*, 63:447–472, 1983. Cited on page(s) 90.

Rump, S.M. Validated Solution of Large Linear Systems. In Albrecht, R., Alefeld, G., and Stetter, H.J., editors, *Computing Supplementum 9, Validation Numerics*, pages 191–212. Springer, 1993. Cited on page(s) 111.

Rump, S.M. Verification Methods for Dense and Sparse Systems of Equations. In Herzberger, J., editor, *Topics in Validated Computations — Studies in Computational Mathematics*, pages 63–136, Elsevier, Amsterdam, 1994. Cited on page(s) 111.

Rump, S.M. INTLAB - INTerval LABoratory. In Csendes, Tibor, editor, *Developements in Reliable Computing*, pages 77–104. Kluwer Academic Publishers, 1999. Cited on page(s) 111.

Ruppert, W. M. Reduzibilität ebener Kurven. *J. reine angew. Math.*, 369: 167–191, 1986. Cited on page(s) 26.

Ruppert, W. M. Reducibility of polynomials $f(x,y)$ modulo $p$. *J. Number Theory*, 77:62–70, 1999. Cited on page(s) 26.

Rupprecht, David. An algorithm for computing certified approximate GCD of $n$ univariate polynomials. *Journal of Pure and Applied Algebra*, 139, 1999. Cited on page(s) 121.

Rupprecht, David. Semi-numerical absolute factorization of polynomials with integer coefficients. *Journal of Symbolic Computation*, page to appear, 2001. Cited on page(s) 124.

Rust, C. J. and Reid, G. J. Rankings of partial derivatives. In Küchlin, W., editor, *Proc. ISSAC '97*, pages 9–16. ACM Press, 1997. Cited on page(s) 102.

Rust, C. J., Reid, G. J., and Wittkopf, A. D. Existence and uniqueness theorems for formal power series solutions of analytic differential systems. In Dooley, S., editor, *Proc. ISSAC '99*, pages 105–112. ACM Press, 1999. Cited on page(s) 103.

Saab, E. First in computer-based classes. *http://www.math.missouri.edu/~news/issue2/front2.html*, 1997. Cited on page(s) 254.

Sadik, B. The complexity of formal resolution of linear partial differential equations. In Cohen, M. Giusti G. and Mora, T., editors, *AAECC-11*, Lecture Notes in Computer Science 948, pages 408–414, Paris, France, 1995. Springer Verlag. Cited on page(s) 105.

Sadik, B. A bound for the order of characteristic set elements of an ordinary prime differential ideal and some applications. *Appl. Alg. Eng. Comm. Comp.*, 10:251–268, 2000a. Cited on page(s) 105.

Sadik, Brahim. Une note sur les algorithmes de décomposition en algèbre différentielle. (A note on the decomposition algorithms in differential algebra). *C. R. Acad. Sci., Paris, Sr. I, Math.*, 330(8):641–646, 2000b. Cited on page(s) 105.

Saints, K. and Heegard, C. Algebraic-geometric codes and multidimensional cyclic codes: A unified theory and algorithms for decoding using Gröbner bases. *IEEE Transactions on Information Theory*, 42, 1995. Cited on page(s) 129.

Saito, K. Quasihomogene isolierte Singularitäten von Hyperflächen. *Inv. Math.*, 14:123–142, 1971. Cited on page(s) 197.

Saito, Mutsumi, Sturmfels, Bernd, and Takayama, Nobuki. *Gröbner Deformations of Hypergeometric Differential Equations*. Springer, 2000. ISBN ISBN 3-540-66065-8. Algorithms and Computation in Mathematics, Volume 6. Cited on page(s) 393, 394, 491.

Sakata, S. Finding a minimal set of linear recurring relations capable of generating a given two-dimensional array. *Journal of Symbolic Computation*, 5: 321–337, 1988. Cited on page(s) 129.

Sakata, S. Extension of the Berlekamp-Massey algorithm to $N$ dimensions. *Information and Computation*, 84:207–239, 1990. Cited on page(s) 129.

Sakata, S., editor. *Applied Algebra, Algebraic Algorithms and Error-Correcting Codes, 8th International Conference, AAECC-8, Tokyo, Japan, August 20–24, 1990, Proceedings*, volume 508 of *Lecture Notes in Computer Science*,

Berlin-Heidelberg-New York, 1991. Springer-Verlag. ISBN 3-540-54195-0, 0-387-54195-0. Cited on page(s) 487, 550, 558, 568.

Sakata, S., Jensen, H. Elbrønd, and Høholdt, T. Generalized Berlekamp-Massey decoding of algebraic geometric codes up to half the Feng-Rao bound. *IEEE Transactions on Information Theory*, 41:1762–1768, 1995a. Cited on page(s) 129.

Sakata, S., Justesen, J., Madelung, Y., Jensen, H. E., and Høholdt, T. Fast decoding of algebraic-geometric codes up to the designed minimum distance. *IEEE Transactions on Information Theory*, 41:1672–1677, 1995b. Cited on page(s) 129.

Salvy, B. and Zimmermann, P. Gfun: a Maple package for the manipulation of general holonomic functions of one variable. *ACM Transactions on Mathematical Software*, 20:163–177, 1994. Cited on page(s) 220.

Sambandham, M. Bharucha. *Random Polynomials*. Academic Press, New York, 1986. Cited on page(s) 491.

Sasaki, Tateaki. Approximate multivariate polynomial factorization based on zero-sum relations. In Mourrain [2001], pages 284–291. Cited on page(s) 124.

Sasaki, T. and Murao, H. Efficient Gaussian elimination method for symbolic determinants and linear systems. *ACM Trans. Math. Software*, 8(3):277–289, 1982. Cited on page(s) 36.

Sasaki, Tateaki, Saito, Tomokatsu, and Hilano, Teruhiko. Analysis of approximate factorization algorithm I. *Japan J. Industrial and Applied Math*, 9(3):351–368, October 1992. Cited on page(s) 124.

Sasaki, Tateaki and Sasaki, Mutsuko. A unified method for multivariate polynomial factorization. *Japan J. Industrial and Applied Math*, 10(1):21–39, February 1993. Cited on page(s) 26, 124.

Sasaki, Tateaki, Suzuki, Masayuki, r, Miroslav Kolá, and Sasaki, Mutsuko. Approximate factorization of multivariate polynomials and absolute irreducibility testing. *Japan J. Industrial and Applied Math*, 8(3):357–375, October 1991. Cited on page(s) 26, 124.

Satir, A. Duff-Inami-Pope-Sezgin-Stelle bosonic membrane equations as an involutary system. *Prog. of Theoret. Phys.*, 100(6):1273–1280, 1998. Cited on page(s) 181.

Sattinger, D. H. *Group Theoretic Methods in Bifurcation Theory*. Number 762 in Lecture Notes in Math. Springer-Verlag, Berlin, Heidelberg, New York, 1979. Cited on page(s) 34.

Scheen, C. Implementation of the Painlevé test for ordinary differential equations. *Theor. Comp. Sci.*, 187:87–104, 1997. Cited on page(s) 101.

Scheerhorn, A. Trace- and norm-compatible extensions of finite fields. *Journal of Appl. Alg. in Eng. Comm. and Comp.*, 3:199–209, 1992. Cited on page(s) 22.

Scheerhorn, A. *Darstellungen des algebraischen Abschlusses endlicher Körper*. Dissertation, Universität Erlangen, 1993. Cited on page(s) 22.

Schemmel, K.-P. An extension of Buchberger's algorithm to compute all reduced gröbner bases of a polynomial ideal. In Davenport, J. H., editor, *EUROCAL'87, European conference on computer algebra*, volume 378 of *Springer LNCS*, pages 300–310. Springer, 1987. Cited on page(s) 30.

Schertz, R. Zur expliziten Berechnung von Ganzheitsbasen in Strahlklassenkörpern über imaginär-quadratischen Zahlkörpern. *Journal of Number Theory*, 34:41–53, 1990. Cited on page(s) 400.

Schijo, J. Rational parameterization of real algebraic surfaces. In Gloor [1998], pages 302–308. ISBN 1-58113-002-3. Cited on page(s) 237.

Schinzel, A. *Selected Topics on Polynomials*. University of Michigan Press, Ann Arbor, MI., 1982. Cited on page(s) 27, 491.

Schirokauer, O., Weber, D., and Denny, Th. *Discrete Logarithms: The effectiveness of the Index Calculus Method*, pages 337–361. Springer-Verlag, 1996. Cited on page(s) 22.

Schirra, S., Burnikel, C., Fleischer, R., and Mehlhorn, K. A strong and easily computable separation bound for arithmetic expressions involving square roots. In *Proc. 8th ACM-SIAM Symposium on Discrete Algorithms*, New York, 1997. Association for computing machinery. Cited on page(s) 19.

Schlesinger, L. *Handbuch der Theorie der linearen Differentialgleichungen*. B.G. Teubner, Leipzig, 1895. Cited on page(s) 491.

Schlomiuk, D., Guckenheimer, J., and Rand, R. Integrability of plane quadratic vector fields. *Expo. Math.*, 8:3–25, 1990. Cited on page(s) 107.

Schneider, C. An implementation of Karr's summation algorithm in Mathematica. Technical Report 99-15, SFB F103, University of Linz, Austria, 1999. Cited on page(s) 93.

Schnupp, P. and Leibrandt, U. *Expertensysteme – Nicht nur für Informatiker*. Springer-Verlag, Berlin-Heidelberg-New York, 1986. Cited on page(s) 142.

Schöbel, C. and Schöbel, F. The symbolic classification of real four-dimensional Lie algebras. In *Proceedings of the Conference Physics Computing '92, Prag*, 1992. Cited on page(s) 65.

Schönhage, A. Quasi-gcd computations. *Journal of Complexity*, 1:118–137, 1985. Cited on page(s) 113, 119, 125.

Schönhage, A. Equation solving in terms of computational complexity. In Gleason, A. M., editor, *Proceedings of the International Congress of Mathematicians 1986, Vol. I*, pages 131–153, Providence, 1987. AMS. Cited on page(s) 25, 126, 128.

Schönhage, A. and Vetter, E. A new approach to resultant computations and other algorithms with exact division. In *Lecture Notes in Computer Science*, volume 855, pages 448–459, 1994. Cited on page(s) 12.

Schoof, R. Elliptic curves over finite fields and the computation of square roots mod $p$. *Mathematics of Computation*, 44:483–494, 1985. Cited on page(s) 49.

Schrüfer, E. EXCALC user's manual. Technical report, RAND Corp., Santa Monica, California, 1987. Cited on page(s) 180, 468.

Schrüfer, E. Excalc: A system for doing calculations in the calculus of modern differential geometry. GMD-SCAI: St. Augustin, 1994. Cited on page(s) 183.

Schrüfer, E., Hehl, F. W., and McCrea, J. D. Exterior calculus on a computer: The REDUCE-package EXCALC applied to general relativity and to the Poincaré Gauge theory. *Gen. Rel. Grav.*, 19:197–218, 1987. Cited on page(s) 180, 183.

Schü, J., Seiler, W. M., and Calmet, J. Algorithmic methods for Lie pseudogroups. In Ibragimov, N., Torrisi, M., and Valenti, A., editors, *Proc. Modern Group Analysis: Advanced Analytical and Computational Methods in Mathematical Physics*, pages 337–344. Kluwer, Dordrecht, 1993. Cited on page(s) 103.

Schwartz, J. T. Fast probabilistic algorithms for verification of polynomial identities. *Journal of the ACM*, 27:701–717, 1980. Cited on page(s) 14.

Schwartz, J. T., Dewar, R. B. K., Dubinsky, E., and Schonberg, E. *Programming with Sets: An Introduction to SETL*. Springer-Verlag, New York, 1986. Cited on page(s) 297.

Schwartz, N. Stability of Gröbner bases. *J. pure and appl. Algebra*, 53:171–186, 1988. Cited on page(s) 30.

Schwarz, F. The Riquier-Janet theory and its application to nonlinear evolution equations. *Physica*, 11D:243–251, 1984. Cited on page(s) 102.

Schwarz, F. An algorithm for determining the size of symmetry groups. *Comp.*, 49:95–115, 1992. Cited on page(s) 102.

Schwarz, F. Janet bases for $2^{nd}$ order ordinary differential equations. In Lakshman, editor, *Proc. ISSAC'96*. ACM Press, 1996. Cited on page(s) 99.

Schwarz, F. *Einführung in die Elementare Zahlentheorie*. MuPAD Lectures. B.G. Teubner, Stuttgart, 1998. Cited on page(s) 323, 327.

Schwarz, F. Differentialgleichungen Lösen mit Computer-Algebra. *Spektrum der Wissenschaft*, pages 98–102, März 1996. Cited on page(s) 98.

Schwarz, M. and Sharir, M. On the piano movers problem. In Schwarz, J., Hopcroft, J., and Sharir, M., editors, *Planning, Geometry and Complexity of Robot Motion*, pages 154–186. Ablex Publishing Corp., New Jersey, 1987. Cited on page(s) 233.

Sederberg, T. W. Applications to computer aided geometric design. In Cox, D. A. and Sturmfels, B., editors, *Applications of Computational Algebraic Geometry*, volume 53 of *Proceedings of Symposia in Applied Mathematics*, pages 67–89. AMS, Providence, 1998. Cited on page(s) 237, 240.

Sedgewick, R. *Algorithms*. Addison-Wesley, Reading, Massachusetts, 1983. Cited on page(s) 491.

Segal, Daniel. *Polycyclic Groups*. Cambridge University Press, Cambridge, 1983. Cited on page(s) 76, 459, 491.

Seidenberg, A. An elimination theory for differential algebra. *Univ. California Publications in Math.*, 3(2):31–65, 1956. Cited on page(s) 104.

Seidensticker, T. *Diploma thesis, University of Karlsruhe, unpublished*, 1998. Cited on page(s) 168.

Seiler, W. M. Completion to involution in AXIOM. In Calmet, J., editor, *Rhine Workshop on Computer Algebra. Proceedings*, pages 103–104, Karlsruhe, Germany, 1994a. Universität Karlsruhe. Cited on page(s) 264.

Seiler, W. M. Pseudo-differential operators and integrable systems in AXIOM. *Computer Physics Communications*, 79(2):329–340, April 1994b. Cited on page(s) 264.

Seiler, W. M. Applying AXIOM to partial differential equations. Internal Report 95-17, Universität Karlsruhe, Fakultät für Informatik, 1995. Cited on page(s) 103.

Seiler, W. M. Computer algebra and differential equations — an overview. Internal Report 97–25, Universität Karlsruhe, Fakultät für Informatik, 1997. Cited on page(s) 96.

Seiler, W. M. Indices and solvability for general systems of differential equations. In Ghanza, V. G., Mayr, E. W., and Vorozhtsov, E. V., editors, *Computer Algebra in Scientific Computing — CASC '99*, pages 365–385. Springer-Verlag, Berlin, 1999. Cited on page(s) 109.

Seiler, W. M. A combinatorial approach to involution and δ-regularity. Preprint Universität Mannheim, 2000. Cited on page(s) 103.

Semenov, A. L. Logical theories of one-place functions on the set of natural numbers. *Math. USSR Izvestiya*, 22(3):587–818, 1984. Cited on page(s) 138.

Seress, Akos. An introduction to computational group theory. *Notices of the American Mathematical Society*, 44:671–679, 1997. Cited on page(s) 66.

Seress, Akos. *Permutation Group Algorithms*. Cambridge University Press, Cambridge, 1999. in press. Cited on page(s) 71, 73, 490.

Serf, P. *The rank of elliptic curves over real quadratic number fields of class number 1*. PhD thesis, Universität des Saarlandes, Saarbrücken, 1995. Cited on page(s) 49.

Serpette, R., Vuillemin, J., and Hervé, J. C. BigNum: A portable and efficient package for arbitrary-precision arithmetic. Technical report, INRIA Report, Roquencourt, 1990. Cited on page(s) 12.

Shamash, J. The Poincaré series of a local ring. *J. Algebra*, 12:453–470, 1969. Cited on page(s) 419.

Shamus Software. M.I.R.A.C.L. multiprecision integer and rational class library. Shamus Software Ltd., Dublin 3, Ireland, 1998. 115 pages. Cited on page(s) 12.

Shanks, D. Class number, A theory of factorization and genera. In Lewis, D. J., editor, *Proceedings of the Symposion on Pure Mathematics XX, 1969 Number Theory Institute*, pages 415–440, Providence, 1971. AMS. Cited on page(s) 49.

Shannon, C. E. A mathematical theory of communication. *Bell Systems Technical Journal*, 27:379–423 and 623–656, 1948. Cited on page(s) 129.

Shary, S.P. Optimal Solutions of Interval Linear Algebraic Systems. *Interval Computations 2*, pages 7–30, 1991. Cited on page(s) 111.

Sherring, J., Head, A. K., and Prince, G. E. Dimsym and LIE: symmetry determination packages. *Mathematical and Computer Modelling*, 25(8-9):153–164, 1997. Cited on page(s) 468.

Shirkov, D. V., Rostovtsev, V. A., and Gerdt, V. P., editors. *IV. International Conference on Computer Algebra in Physical Research, Dubna, USSR, 22–26 May 1990*, Singapore, 1991. World Scientific Publishing. ISBN 981-02-0687-9. Cited on page(s) 486, 496, 548, 571.

Shokrollahi, M. A. Stickelberger codes. *Designs, Codes and Cryptography*, 9: 1–11, 1996. Cited on page(s) 129.

Shokrollahi, M. A. and Wasserman, H. Decoding algebraic geometric codes beyond the error correction bound. In *Proceedings of the 30th Symposium on Theory of Computing*, pages 241–248, 1998. Cited on page(s) 130.

Shor, Peter W. Scheme for reducing decoherence in quantum computer memory. *Physical Review A*, 52(4):R2493–R2496, 1995. Cited on page(s) 130.

Short, Mark W. *The Primitive Soluble Permutation Groups of Degree less than 256*, volume 1519 of *Lecture Notes in Mathematics*. Springer-Verlag, 1992. Cited on page(s) 389.

Shoup, V. A new polynomial factorization algorithm and its implementation. Technical report, Universität des Saarlandes, 1994. Cited on page(s) 406.

Shoup, V. NTL: A library for doing number theory. http://www.cs.wisc.edu/~shoup, 1998. Cited on page(s) 12.

Shparlinski, I. E. *Computational and algorithmic problems in finite fields*. Kluwer Academic Publishers, Dordrecht, 1992. Cited on page(s) 23, 491.

Siemens AG. *ARITHMOS, Benutzerhandbuch*. Siemens AG, Bibl.-Nr. U 2900-I-Z87-1, 1986. Cited on page(s) 111.

Siksek, S. Infinite descent on elliptic curves. *Rocky Mountain J. Math*, 25: 1502–1538, 1995. Cited on page(s) 49.

Silverman, J. H. *The Arithmetic of Elliptic Curves*. Springer-Verlag, Berlin-Heidelberg-New York, 1986. Cited on page(s) 49, 491.

Silverman, J. H. and Tate, J. *Rational Points on Elliptic Curves*. Springer-Verlag, Berlin-Heidelberg-New York, 1992. Cited on page(s) 48, 491.

Silverman, R. D. The multiple quadratic sieve. *Mathematics of Computation*, 48:329–339, 1987. Cited on page(s) 43, 44.

Simon, G. Communication interface between SAC-2 and lidia. In *Proc. of OpenMath Workshop 6*, Zurich, July 1996. Cited on page(s) 406.

Sims, C. C. How to construct a baby monster. In Durham II Durham II [1980], pages 339–345. Cited on page(s) 196.

Sims, C. C. *Abstract Algebra: A Computational Approach*. John Wiley&Sons, New York, 1984. Cited on page(s) 491.

Sims, C. C. *Computation with Finitely Presented Groups*, volume 48 of *Encyclopedia of Mathematics and Its Applications*. Cambridge University Press, Cambridge, 1994. Cited on page(s) 4, 58, 76, 78, 196, 459, 491.

Singer, Michael. Formal solutions of differential equations. *Journal of Symbolic Computation*, 10(1):59–94, 1990a. Cited on page(s) 96.

Singer, M., editor. *Differential Equations and Computer Algebra*. Computational Mathematics and Applications Series. Academic Press, New York, 1991a. Cited on page(s) 492.

Singer, Michael. Liouvillian solutions of linear differential equations with liouvillian coefficients. *Journal of Symbolic Computation*, 11:251–273, 1991b. Cited on page(s) 95.

Singer, M. F. Algebraic solutions of $n^{th}$ order linear differential equations. *Queens Papers Pure Appl. Math.*, 54, 1979. Cited on page(s) 97.

Singer, M. F., editor. *Differential Equations and Computer Algebra (CADE-90)*, Computational Mathematics and Applications, London, 1990b. Academic Press. Cited on page(s) 97, 486.

Singer, M. F. An outline of differential Galois theory. In Tournier, E., editor, *Computer Algebra and Differential Equations*. Academic Press, New York, 1990c. Cited on page(s) 97.

Singer, M. F. Testing reducibility of linear differential operators: a group theoretic perspective. *Appl. Alg. Eng. Comm. Comp.*, 7:77–104, 1996. Cited on page(s) 97.

Singer, M. F. Direct and inverse problems in differential Galois theory. In Bass, H., Buium, A., and Cassidy, P. J., editors, *Selected Works of E.R. Kolchin with Commentary*. AMS, 1999. Cited on page(s) 98.

Singer, M. F., editor. *Effective Methods in Algebraic Geometry*, volume 164 of *Journal of pure and applied algebra*, Amsterdam, 2001. North-Holland. Special Issue. Cited on page(s) 489.

Singer, M. F. and Ulmer, F. Galois groups of second and third order linear differential equations. *Journal of Symbolic Computation*, 16:1–36, 1993a. Cited on page(s) 97, 98.

Singer, M. F. and Ulmer, F. Liouvillian and algebraic solutions of second and third order linear differential equations. *Journal of Symbolic Computation*, 16: 37–73, 1993b. Cited on page(s) 97.

Singer, M. F. and Ulmer, F. Linear differential equations and products of linear forms. *J. Pure Appl. Alg.*, 117 & 118:549–564, 1997. Cited on page(s) 97.

Singer, M. F. and van der Put, M. *Galois Theory of Difference Equations*. Lecture Notes in Mathematics 1666. Springer, Berlin, 1997. Cited on page(s) 98, 491.

Sipcic, S. R. What is wrong with the way we teach mechanics and how to fix it. *Mathematica in Education and Research*, 4(4):5–13, 1995. Cited on page(s) 255.

Sipser, M. and Spielman, D. Expander codes. *IEEE Transactions on Information Theory*, 42:1710–1722, 1996. Cited on page(s) 129.

Sit, W. Y. An algorithm for solving parametric linear systems. *Journal of Symbolic Computation*, 13(4):353–394, April 1992. Cited on page(s) 264.

Skea, J. and et al. On–line invariant classification database, 1997. http://www.astro.queensu.ca/~jimsk/. Cited on page(s) 175.

Skea, J. E. F. *Applications of SHEEP*, 1994. http://www.can.nl/SystemsOverview/Special/Tensoranalysis/SHEEP/shpdrv.ps.gz. Cited on page(s) 175.

Skiena, S. S. *Implementing Discrete Mathematics: Combinatorics and Graph Theory with Mathematica*. Addison-Wesley, Reading, Massachusetts, 1990. Cited on page(s) 492.

Sköldberg, Emil. More on the homology of associative algebras. Preprint, Stockholm University, 1997. Cited on page(s) 210, 211.

Sloane, N. J. A. Error-correcting codes and invariant theory: New applications of a nineteenth-century technique. *Amer. Math. Monthly*, 84:82–107, 1977. Cited on page(s) 34.

Smale, S. Topology and Mechanics II. the planar n-body problem. *Inventiones mathematicae*, 11:45–64, 1970. Cited on page(s) 177.

Smart, N. P. $S$-integral points on elliptic curves. *Math. Proc. Comb. Phil. Soc*, 116:391–399, 1994. Cited on page(s) 49.

Smirnov, V. A. *Mod. Phys. Lett. A*, 3:381, 1988. Cited on page(s) 166.

Smirnov, V. A. *Comm. Math. Phys.*, 134, 1990. Cited on page(s) 166.

Smirnov, V. A. *Mod. Phys. Lett.*, A10:1485–1500, 1995. Cited on page(s) 168.

Smith, G. Computing global extension modules for coherent sheaves on a projective scheme. *Preprint*, 1998. Cited on page(s) 418.

Smith, Larry. *Polynomial Invariants of Finite Groups*. A. K. Peters, Wellesley, Mass., 1995. Cited on page(s) 33, 491.

Smith, Larry. Polynomial invariants of finite groups: A survey of recent developments. *Bull. Amer. Math. Soc.*, 34:211–250, 1997. Cited on page(s) 34.

Smolka, G., Nutt, W., Goguen, J. A., and Meseger, J. Order-sorted equational computation. In Aït-Kaci and Nivat [1989b], chapter 10. ISBN 0-12-046371-7. Cited on page(s) 136.

Socorro, J., Macias, A., and Hehl, F. W. Computer algebra in gravity: Reduce-Excalc programs for (non-)Riemannian spacetimes. I. *Computer Physics Communication*, 115:264–283, 1998. Also available at: http://xxx.lanl.gov/abs/gr-qc/9804068. Cited on page(s) 176.

Sofroniou, M. Symbolic derivation of Runge-Kutta methods. *Journal of Symbolic Computation*, pages 265–296, 1994. Cited on page(s) 108.

Soicher, L. H. On simplicial complexes related to the suzuki sequence graphs. In Liebeck, M. W. and Saxl, J., editors, *Groups, Combinatorics and Geometry*, number 165 in LMS Lecture Notes, pages 240–248. Cambridge University Press, 1992. Cited on page(s) 90.

Soicher, L. H. GRAPE: a system for computing with graphs and groups. In Finkelstein, L. and Kantor, W. M., editors, *Groups and Computation*, volume 11 of *DIMACS Series in Discrete Mathematics and Theoretical Computer Science*, pages 287–291. AMS, 1993. Also see http://www-gap.dcs.st-and.ac.uk/~gap/Share/grape.html. Cited on page(s) 91, 474.

Soicher, L. H. and McKay, J. Computing Galois groups over the rationals. *J. Number Theory*, 20:273–281, 1985. Cited on page(s) 47.

Soiffer, Neil. Mathematical Typesetting in Mathematica. In Levelt [1995], pages 140–149. Cited on page(s) 150.

Sokolov, V. V. and Wolf, T. A symmetry test for quasilinear coupled systems. *Inverse Problems*, 15:L5–L11, 1999. Cited on page(s) 190, 465, 467.

Soleng, H. H. *Tensors in physics: User's guide and disk for the Mathematica package Cartan. Version 1.2.* Scandinavian University Press, Oslo, 1993. Cited on page(s) 183, 491.

Soleng, H. H. *Tensors in Physics*. Scandinavian University Press, Oslo, 1996. Cited on page(s) 175.

Sommese, Andrew J., Verschelde, Jan, and Wampler, Charles W. Numerical decomposition of the solution sets of polynomial systems into irreducible components. *SIAM Journal of Numerical Analysis*, 38(6):2022–2046, 2001. Cited on page(s) 123.

Sorgatz, A. *Dynamische Module – Eine Verwaltung für Maschinencode-Objekte zur Steigerung der Effizienz und Flexibilität von Computeralgebra-Systemen*. MuPAD Reports. B.G. Teubner, Stuttgart, 1996. Cited on page(s) 331, 450.

Sorgatz, A. *Dynamic Modules – User's Manual and Programming Guide for MuPAD 1.4*. Springer-Verlag, Berlin-Heidelberg-New York, 1998. Cited on page(s) 326.

Sorgatz, A. and Wehmeier, S. Towards high-performance symbolic computing: Using MuPAD as a problem solving environment. *Mathematics and Computers in Simulation*, 49:235–246, August 1999. Cited on page(s) 330.

Specker, E. and Strassen, V., editors. *Komplexität von Entscheidungsproblemen*, volume 43 of *Lecture Notes in Computer Science*. Springer-Verlag, Berlin-Heidelberg-New York, 1976. Cited on page(s) 492.

Spielman, D. Linear-time encodable and decodable error-correcting codes. *IEEE Transactions on Information Theory*, 42:1723–1731, 1996. Cited on page(s) 129.

Springer, Tonny A. *Invariant Theory*. Number 585 in Lecture Notes in Math. Springer-Verlag, Berlin, Heidelberg, New York, 1977. Cited on page(s) 33, 491.

Stanley, Richard P. Invariants of finite groups and their applications to combinatorics. *Bull. Amer. Math. Soc.*, 1(3):475–511, 1979. Cited on page(s) 34.

Stauduhar, R. P. The determination of Galois groups. *Math. Comput*, 27:981–996, 1973. Cited on page(s) 47.

Stauffer, D. *Computer Simulation and Computer Algebra*. Springer-Verlag, Berlin-Heidelberg-New York, 1989. Cited on page(s) 491.

Stauffer, D., Hehl, F. W., Ito, N., Winkelmann, V., and Zabolitzky, J. G. *Computer Simulation and Computer Algebra*. Springer, 1993. Cited on page(s) 491.

Stauffer, D., Hehl, F. W., Winkelmann, V., and Zabolitzky, J. G. *Computer Simulation and Computer Algebra—Lectures for Beginners*. Springer-Verlag, Berlin-Heidelberg-New York, 1988. Cited on page(s) 491, 492.

Steeb, W. H. and Euler, N. *Nonlinear Evolution Equations and Painlevé Test*. World Scientific, Singapore, 1988. Cited on page(s) 101.

Steeb, W.-H. and Lewien, D. *Algorithms and Computation with Reduce*. Bibliographisches Institut, Mannheim, 1992. Cited on page(s) 492.

Steen, L. A. Quoted in J. Ferrini-Mundy and K. G. Graham: An overview of the calculus curriculum reform effort: Issues for learning, teaching, and curriculum development. *American Mathematical Monthly*, 98 (7):627–635, 1991. Cited on page(s) 254.

Steinhauser, M. *Ph. D. thesis, University of Karlsruhe, (Shaker Verlag, Aachen)*, 1996. Cited on page(s) 169.

Stephani, H. *Differential Equations: Their Solution Using Symmetries*. Cambridge University Press, Cambridge, 1989. Cited on page(s) 491.

Stephani, H. and Wolf, T. Spherically symmetric perfect fluids in shear-free motion - the symmetry approach. *Class. Quantum Grav.*, 13:1261–1271, 1996. Cited on page(s) 467.

Stetter, H. Stabilization of polynomial system solving with Groebner bases. In Küchlin [1997], pages 117–124. Cited on page(s) 242.

Stetter, Hans J. Matrix eigenproblems are at the heart of polynomial system solving. SIGSAM *Bulletin: Communications on Computer Algebra*, 30(4):22–25, December 1996. Cited on page(s) 122.

Stetter, Hans J. The nearest polynomial with a given zero, and similar problems. SIGSAM BULLETIN: *Communications on Computer Algebra*, 33(4): 2–4, December 1999. Cited on page(s) 116, 117, 118.

Stone, M. H. The theory of representations of Boolean algebras. *Trans. American Math. Society*, 40:37–111, 1936. Cited on page(s) 135.

Storjohann, A. Near optimal algorithms for computing Smith normal forms of integer matrices. In *International Symposium on Symbolic and Algebraic Computation, Zurich, Suisse*, pages 267–274. ACM Press, 1996. Cited on page(s) 39.

Storjohann, A. An $O(n^3)$ algorithm for the Frobenius normal form. In Gloor [1998], pages 101–104. ISBN 1-58113-002-3. Cited on page(s) 40.

Storjohann, A. *Algorithms for Matrix Canonical Form*. PhD thesis, Eidgenössiche Technische Hochschule ETH , Zürich, Switzerland, 2000. Cited on page(s) 39, 40.

Storjohann, A. and Labahn, G. Asymptotically fast computation of Hermite normal forms of integer matrices. In *International Symposium on Symbolic and Algebraic Computation, Zurich, Suisse*, pages 259–266. ACM Press, 1996. Cited on page(s) 39.

Storjohann, A. and Labahn, G. A fast Las Vegas algorithm for computing the Smith normal form of a polynomial matrix. *Linear Algebra and Appl.*, 253: 155–173, 1997. Cited on page(s) 39.

Storjohann, A. and Villard, G. Algorithms for similarity transforms. In *Proc. Seventh Rhine Workshop on Computer Algebra*, Bregenz, Austria, 2000. Cited on page(s) 40.

Strassen, V. Algebraic complexity theory. In van Leeuwen, J., editor, *Handbook of Theoretical Computer Science*, volume A, pages 635–672. Elsevier Science Publishers, Amsterdam, 1990. Cited on page(s) 126, 128.

Stroeker, R. J. and Kaashoek, J. F. *Discovering Mathematics with Maple*. Birkhäuser, Basel, 1999. Cited on page(s) 257, 492.

Stroeker, R. J. and Tzanakis, N. Solving elliptic diophantine equations by estimating linear forms in elliptic logarithms. *Acta Arith.*, 67:177–196, 1994. Cited on page(s) 49.

Strubbe, H. *Comp. Phys. Comm.*, 8:1, 1974. Cited on page(s) 165.

Sturm, Thomas. Reasoning over networks by symbolic methods. Technical Report MIP-9719, FMI, Universität Passau, D-94030 Passau, Germany, December 1997. Also in the AAECC Journal 10 (1999) 1, 79-96. Cited on page(s) 206.

Sturm, Thomas and Weispfenning, Volker. Rounding and blending of solids by a real elimination method. In Sydow, Achim, editor, *Proceedings of the 15th IMACS World Congress on Scientific Computation, Modelling, and Applied Mathematics (IMACS 97)*, volume 2, pages 727–732, Berlin, August 1997. IMACS, Wissenschaft & Technik Verlag, Berlin, 1997. ISBN 3-89685-552-2. Cited on page(s) 139, 206, 241.

Sturm, Thomas and Weispfenning, Volker. An algebraic approach to offsetting and blending of solids. Technical Report MIP-9804, FMI, Universität Passau, D-94030 Passau, Germany, May 1998a. Cited on page(s) 206.

Sturm, Thomas and Weispfenning, Volker. Computational geometry problems in Redlog. In Wang, Dongming, editor, *Automated Deduction in Geometry*, volume 1360 of *Lecture Notes in Artificial Intelligence (Subseries of LNCS)*, pages 58–86. Springer-Verlag, Berlin Heidelberg, 1998b. Cited on page(s) 57.

Sturm, Thomas and Weispfenning, Volker. Computational geometry problems in Redlog. In Wang, Dongming, editor, *Automated Deduction in Geometry*, volume 1360 of *Lecture Notes in Artificial Intelligence (Subseries of LNCS)*, pages 58–86. Springer-Verlag, Berlin Heidelberg, 1998c. ISBN 3-540-64297-8. Cited on page(s) 139, 206, 241.

Sturmfels, Bernd. *Algorithms in Invariant Theory*. Springer-Verlag, Wien, New York, 1993. Cited on page(s) 33, 34, 215, 491.

Sturmfels, Bernd. *Gröbner Bases and Convex Polytopes*, volume 8 of *University Lecture Series*. AMS, Providence, Rhode Island, 1995. Cited on page(s) 30.

Sudan, M. Decoding of Reed-Solomon codes beyond the error-correction bound. *Journal of Complexity*, 13:180–193, 1997. Cited on page(s) 130.

Swarttouw, R. F. CAOP: On-line version of the Askey-scheme. *http://www.riaca.win.tue.nl/Software/CAOP*, 1996. Cited on page(s) 215.

Systemics Ltd. Class java.math.biginteger java.math.bignum. http://tinf2.vub.ac.be/~dvermeir/java/cryptix/doc/java.math, 1996. 10 pages. Cited on page(s) 13.

Szmielew, Wanda. Elementary properties of abelian groups. *Fund. Math.*, 41: 203–271, 1954. Cited on page(s) 137, 138.

Takayama, N. An approach to the zero recognition problem by Buchberger's algorithm. *Journal of Symbolic Computation*, 14:265–282, 1992. Cited on page(s) 92.

Takayama, N. An algorithm for finding recurrence relations of binomial sums and its complexity. *Journal of Symbolic Computation*, 20:637–651, 1995. Cited on page(s) 92.

Tangora, M. C., editor. *Computers in Algebra*. Marcel Dekker, New York, 1988. Cited on page(s) 492.

Tangora, M. C., editor. *Computers in Geometry and Topology*. Marcel Dekker, New York, 1989. Cited on page(s) 492.

Tarasov, O. V. *Nucl. Phys. B*, 480:397, 1996a. Cited on page(s) 166.

Tarasov, O. V. *Phys. Rev. D*, 54:6479, 1996b. Cited on page(s) 166.

Tarski, Alfred. Arithmetical classes and types of boolean algebras. *Bull. AMS*, 55:64, 1949. Cited on page(s) 137, 138.

Tarski, A. *A decision method for elementary algebra and geometry*. University of California Press, Berkeley, 1951. (2nd edition, revised). Cited on page(s) 138, 491.

Taubeneder, A. Implementierung eines Entscheidungsverfahrens für Boolesche Algebren. Master's thesis, Universität Passau, FMI, März 1991. Cited on page(s) 139.

te Riele, Herman J. J. and et al. Factorization of a 512-bits RSA key using the number field sieve. Posting in `sci.crypt.research`, January 2000. Cited on page(s) 45.

Teitelbaum, J. Euclid's algorithm and the Lanczos method over finite fields. *Mathematics of Computation*, 67(224):1665–1678, October 1998. Cited on page(s) 37.

Tentyukov, M. and Fleischer, J. A feynman diagram analyser DIANA. *Comput. Phys. Commun.*, 132:124–141, 2000. Cited on page(s) 166.

Tertychniy, S. I. Searching for electrovac solutions to Einstein-Maxwell equations with the help of computer algebra system GRG-EC. E-print gr-qc/9810057 (1998) (Moscow: 1997), 1997. Cited on page(s) 185.

Tertychniy, S. I. Searching for electrovac solutions to Einstein–Maxwell equations with the help of computer algebra system GRG-EC. http://xxx.lanl.gov/abs/gr-qc/9810057, 1998. Cited on page(s) 176, 182, 183.

Tertychniy, S. I. and Obukhova, I. G. $GRG_{EC}$: Computer algebra system for applications to gravity theory. *SIGSAM Bulletin*, 31:6, 1997. Cited on page(s) 182, 183, 184.

Thallinger, G. H. and Stetter, H. Singular systems of polynomials. In Gloor [1998], pages 9–16. ISBN 1-58113-002-3. Cited on page(s) 242.

The Mathworks Inc. MATLAB 5, 1996. http://www.mathworks.com. Cited on page(s) 228.

The MathWorks Inc. MATLAB User's Guide, Version 5, 1997. Cited on page(s) 111.

The OpenMath Society. Openmath. http://www.openmath.org, 2001. Cited on page(s) 219.

The Unicode Consortium. *The Unicode Standard, Version 3.0*. Addison-Wesley, 2000. ISBN 0-201-61633-5. as ammended by Unicode Standard Annex 27: Unicode 3.1 http://www.unicode.org/reports/tr27/ and by Unicode Standard Annex 28: Unicode 3.2 http://www.unicode.org/reports/tr28. Cited on page(s) 156.

Theißen, Heiko. *Eine Methode zur Normalisatorberechnung in Permutationsgruppen mit Anwendungen in der Konstruktion primitiver Gruppen*. Dissertation, Rheinisch Westfälische Technische Hochschule, Aachen, Germany, 1997. Cited on page(s) 389.

Theobald, P. Berechnung der HNF dünnbesetzer ganzzahliger Matrizen. Ph.D. thesis, Technische Universität Darmstadt, 1999. Cited on page(s) 406.

Thomas, J. *Differential Systems*. American Mathematical Society, New York, 1937. Cited on page(s) 103, 491.

Tkachov, F. V. Rep. No. INR P-0332 (Moscow, 1984), 1984. Cited on page(s) 166.

Tkachov, F. V. *Int. J. Mod. Phys. A*, 8:2047, 1993. Cited on page(s) 166.

Todd, J. A. and Coxeter, H. S. M. A practical method for enumerating cosets of finite abstract groups. *Proc. Edinburgh Math. Soc.*, 5:26–34, 1936. Cited on page(s) 69.

Tommila, M. apfloat, A C++ high performance arbitrary precision arithmetic package. http://www.iki.fi/~mtommila/apfloat, 1998. 32 pages. Cited on page(s) 12, 13.

Topunov, V. L. Reducing systems of linear differential equations to a passive form. *Acta Appl. Math.*, 16:191–206, 1989. Cited on page(s) 102.

Tournier, E. *Solutions formelles d'équations différentielles*. PhD thesis, Université scientifique, médicale et technologique de Grenoble, France, 1987. Cited on page(s) 370.
Tournier, E., editor. *Computer Algebra and Differential Equations (CADE-88)*, Computational Mathematics and Applications, London, 1988. Academic Press. Cited on page(s) 97, 486.
Tournier, E., editor. *Computer Algebra and Differential Equations*. Academic Press, New York, 1990. Cited on page(s) 492.
Tournier, E., editor. *Computer Algebra and Differential Equations (CADE-92)*, London Mathematical Society Lecture Note Series 193, 1992. Cambridge University Press. Cited on page(s) 97.
Trager, B. Algebraic factoring and rational function integration. In *Proc. ACM Symposium on Symbolic and Algebraic Computation*, pages 219–226, New York, 1976. Association for computing machinery. Cited on page(s) 18.
Trager, B. *Integration of Algebraic Functions*. PhD thesis, EECS, MIT, Cambridge (USA), 1984. Cited on page(s) 95.
Tran, Quoc-Nam and Winkler, Franz. CASA reference manual (version 2.3). RISC Linz Report Series 97-33, Research Institute for Symbolic Computation, 4040 Linz, Austria, Europe, October 1997a. Cited on page(s) 357.
Tran, Q.-N. and Winkler, F. An overview of CASA—a system for computational algebra and constructive algebraic geometry. In Effelterre, T. V., Recio, T., and Winkler, F., editors, *The Symbolic and Algebraic Computation (SAC) Newsletter*, volume 2. Stichting CAN, Computer Algebra Nederland, Universiteit van Amsterdam, the Netherlands, 1997b. Cited on page(s) 357.
Trautman, A. *Differential Geometry for Physicists*. Stony Brook Lectures. Bibliopolis, Napoli, 1984. Cited on page(s) 182.
Traverso, C. Gröbner trace algorithms. In Gianni, P., editor, *ISSAC'88*, volume 358 of *LNCS*, pages 125–138. Springer, 1988. Cited on page(s) 30.
Traverso, Carlo. Hilbert functions and the Buchberger algorithm. *J. Symb. Computation*, 22(4):355–376, 1996. Cited on page(s) 29, 30.
Traverso, Carlo, editor. *Proceedings of the 2000 International Symposium on Symbolic and Algebraic Computation*, August 2000. ACM Press. Cited on page(s) 489, 529, 554, 559.
Tsantilis, E., Puntigam, R. A., and Hehl, F. W. A Quadratic Curvature Lagrangian of Pawłowski and Rączka : A Finger Exercise with MathTensor. In Hehl et al. [1996], pages 231–240. Also available at: http://xxx.lanl.gov/abs/gr-qc/9601002. Cited on page(s) 176.
Tsarev, S. P. An algorithm for complete enumeration of all factorizations of a linear ordinary differential operator. In Lakshman, Y. N., editor, *Proc. ISSAC '96*, 1996. Cited on page(s) 97.
Tschebotaröw, N. G. and Schwerdtfeger, H. *Grundzüge der Galoisschen Theorie*. Nordhoff, Groningen–Djakarta, 1950. Cited on page(s) 47.
Tsfasman, M. A. and Vladut, S. G. *Algebraic-Geometric Codes*. Kluwer Academic Publishers, Boston, 1991. Cited on page(s) 129, 491.
Tsfasman, M. A., Vladut, S. G., and Zink, Th. Modular curves, Shimura curves, and Goppa codes better than the Varshamov-Gilbert bound. *Math. Nachrichten*, 109:21–28, 1982. Cited on page(s) 129.

Turing, Alan M. Rounding errors in matrix processes. *Quarterly J. Mech. Appl. Math.*, 1:287–308, 1948. Cited on page(s) 116.

Ueberberg, J. *Einführung in die Computeralgebra mit Reduce, für Mathematiker, Informatiker und Physiker.* Bibliographisches Institut, Mannheim, 1992. ISBN 3-411-15781-X. Cited on page(s) 492.

Ufnarovski, V. Introduction to noncommutative Gröbner bases theory. In Buchberger and Winkler [1998], pages 259–280. Cited on page(s) 58, 61.

Uhl, J., Davis, B., and Porta, H. The Calculus&Mathematica World-Wide Web site, 2001. http://www-cm.math.uiuc.edu/whatis/. Cited on page(s) 254.

Ulmer, F. and Weil, J. A. Note on Kovaçic's algorithm. *Journal of Symbolic Computation*, 22:179–200, 1996. Cited on page(s) 97.

Ung, B.-C.-V. NCSF, a Maple Package for Non-Commutative Symmetric Functions. *Maple Technical Newsletter*, 3:24–29, 1996. Cited on page(s) 345.

Ung, B.-C.-V. Fonctions Symétriques Non Commutatives. Thesis 98–08, Université de Marne-la-Vallée, Institut Gaspard-Monge, CNRS, 1998. Cited on page(s) 346.

Uspensky, J. V. *Theory of Equations.* McGraw-Hill, New York, Toronto, London, 1948. Cited on page(s) 19, 491.

van de Ven, A. E. M. Two-Loop Quantum Gravity with the Computer Algebra Program FORM. In Hehl et al. [1996], pages 192–209. Cited on page(s) 176.

van den Dries, L. Remarks on tarski's problem concerning (R,+,.,exp). In Longi, G., Longo, G., and Marcja, A., editors, *Logic Colloquium*, pages 97–121, Amsterdam, u.a., 1982. North-Holland. Cited on page(s) 138.

van den Dries, L. A generalization of the tarski-seidenberg theorem and some nondefinability results. *Bulletin of AMS*, 15:189–193, 1986. Cited on page(s) 138.

van den Dries, L. Elimination theory for the ring of algebraic integers. *Journal für reine und angewandte Mathematik*, 288:189–205, 1988. Cited on page(s) 138.

van den Dries, Lou, Macintyre, Angus, and Marker, David. The elementary theory of restricted analytic fields with exponentiation. *Annals of Mathematics*, 140:183–205, 1994. Cited on page(s) 138.

van den Dries, L. and Miller, C. On the real exponential field with restricted analytic functions. *Israel Journal of Mathematics*, 85:19–56, 1994. Cited on page(s) 138.

van der Put, M. Galois theory of differential equations, algebraic groups and Lie algebras. *Journal of Symbolic Computation*, 28:441–472, 1999. Cited on page(s) 97.

van der Put, M. and Ulmer, F. Differential equations and finite groups. MSRI Preprint 1998-058, 1998. Cited on page(s) 98.

van der Waerden, B. L. *Moderne Algebra*, volume 1 and 2. Springer-Verlag, Heidelberg, 1931. Cited on page(s) 491.

van der Waerden, B. L. *Moderne Algebra, Zweiter Teil.* Springer-Verlag, Berlin, second edition, 1940. Cited on page(s) 52.

van der Waerden, B. L. *Modern Algebra, 2nd ed.* F. Ungar, New York, 1950. Cited on page(s) 383.

van der Waerden, B. L. *Modern Algebra, volumes 1 and 2*. Frederic Unger Publishing Co., 1966. Cited on page(s) 18.
van Hoeij, M. Factorization of differential operators with rational functions coefficients. *Journal of Symbolic Computation*, 24:537–562, 1997. Cited on page(s) 97.
van Hoeij, M. Factoring polynomials and the knapsack problem, 2001. Available from hoeij@math.fsu.edu. Cited on page(s) 26.
van Hoeij, M. and Weil, J. A. An algorithm for computing invariants of differential Galois groups. *J. Pure Appl. Alg.*, 117 & 118:353–379, 1997. Cited on page(s) 97.
van Hulzen, J. A., editor. *EUROCAL'83, Computer Algebra*, volume 162 of *Lecture Notes in Computer Science*, 1983. Springer. Cited on page(s) 485.
van Hulzen, J. A., Hulshof, B. J. A., Gates, B. L., and van Heerwaarden, M. C. A Code Optimization Package for REDUCE. In Gonnet [1989], pages 163–170. ISBN 0-89791-325-6. Cited on page(s) 151.
van Leeuwen, M. A. A, Cohen, A. M., and B., Lisser. LiE manual, manual for the software package LiE for Lie group theoretical computations. Technical report, CWI/CAN, 1992. Cited on page(s) 408, 411.
van Lint, J. and van der Geer, G. *Introduction to Coding Theory and Algebraic Geometry*. Basel: Birkhäuser, 1988. Cited on page(s) 300.
van Lint, J. H., Høholdt, T., and Pellikaan, R. Algebraic Geometry Codes. In Pless, V. S., Huffman, W. C., and Brualdi, R. A., editors, *Handbook of Coding Theory*, volume 1, pages 871–961. Elsevier Science Publishers, 1998. Cited on page(s) 129.
van Lint, J. H. and Wilson, R. M. *A Course in Combinatorics*. Cambridge University Press, 1992. Cited on page(s) 129.
van Ritbergen, T. and et al. *Int. J. Mod. Phys.*, C6:513, 1995. Cited on page(s) 166.
van Tilborg, H. C. A. *An Introduction to Cryptology*. Kluwer International Series in Engineering and Computer Science. Kluwer Academic Publishers, 1988. Cited on page(s) 129, 491.
Vardi, I. *Computational Recreations with Mathematica*. Addison-Wesley, Reading, Massachusetts, 1991. Cited on page(s) 492.
Vasconcelos, Wolmer V. *Computational Methods in Commutative Algebra and Algebraic Geometry*. Springer, 1998. ISBN ISBN 3-540-60520-7. Algorithms and Computation in Mathematics, Volume 2. Cited on page(s) 491.
Vaz, J. The use of Computer Algebra and Clifford Algebra in teaching Mathematical Physics. In et al, R. Ablamowicz, editor, *Clifford Algebras with Numeric and Symbolic Computations*, pages 33–55. Birkhauser, 1996. Cited on page(s) 180.
Veigneau, S. Calcul Symbolique et Calcul Distribué en Combinatoire Algébrique. Thesis 96–39, Université de Marne-la-Vallée, Institut Gaspard-Monge, CNRS, 1996. Cited on page(s) 345.
Veigneau, S. *ACE—an Algebraic Combinatorics Environment for the Computer Algebra System MAPLE*. Institute Gaspard Monge, Université de Marne-la-Vallé, 1997. Cited on page(s) 491.

Veigneau, S. ACE, an Algebraic Combinatorics Environment for the computer algebra system MAPLE. User's Reference Manual, Version 3.0 98–11, Université de Marne-la-Vallée, Institut Gaspard-Monge, CNRS, 1998. Cited on page(s) 345.

Veldhuizen, T. Blitz++ user's guide. tveldhui@extreme.indiana.edu, 1998. 107 pages. Cited on page(s) 13.

Veltman, M. Schoonship. *CERN preprint*, 1967. Cited on page(s) 165, 169.

Vermaseren, J. A. M. Symbolic manipulation with FORM. *Computer Algebra Netherlands, Amsterdam*, 1991. Cited on page(s) 165, 169, 170, 471, 492.

Verschelde, J. PHCpack: A general-purpose solver for polynomial systems by homotopy continuation. *ACM Transactions on Mathematical Software*, 25: 251–276, 1999. Cited on page(s) 179.

Vetter, E., Schönhage, A., and Grotefeld, A. *Fast Algorithms*. Bibliographisches Institut, Mannheim, Germany, 1994. Cited on page(s) 12, 491.

Vetterli, M. and Herley, C. Wavelets and filter banks: Theory and design. *IEEE, Trans. on ASSP*, 9:2207–2232, 1992. Cited on page(s) 228.

Viehl, M. *Analytical Investigation of the Compressible Navier-Stokes Equations for low Mach and Reynolds Numbers*. Dissertation, Darmstadt University of Technology, Technische Universität Darmstadt, 1989. Cited on page(s) 223.

Villard, G. Generalized subresultants for computing the Smith normal form of polynomial matrices. *Journal of Symbolic Computation*, 20(3):269–286, 1995. Cited on page(s) 39.

Villard, G. Computing Popov and Hermite forms of polynomial matrices. In *International Symposium on Symbolic and Algebraic Computation, Zurich, Suisse*, pages 250–258. ACM Press, 1996. Cited on page(s) 41.

Villard, G. Fast parallel algorithms for matrix reduction to normal forms. *Appli. Alg. Eng., Comm. Comp.*, 8(6):511–538, 1997a. Cited on page(s) 41.

Villard, G. Further analysis of Coppersmith's block Wiedemann algorithm for the solution of sparse linear systems. In Küchlin [1997], pages 32–39. Cited on page(s) 38.

Villard, G. A study of Coppersmith's block Wiedemann algorithm using matrix polynomials. Rapport de Recherche 975-I-M, Institut d'Informatique et de Mathématiques Appliquées de Grenoble, www.imag.fr, April 1997c. Cited on page(s) 38.

Villard, G. Block solution of sparse linear systems over GF($q$): the singular case. *SIGSAM Bulletin*, 32(4):10–12, 1998. Cited on page(s) 38.

Villard, G. Computing the Frobenius normal form of a sparse matrix. In *Proc. Third International Workshop on Computer Algebra in Scientific Computing*, Samarkand, Uzbekistan, 2000a. Cited on page(s) 40.

Villard, G. Processor efficient parallel solution of linear systems of equations. *Journal of Algorithms*, 35:122–126, 2000b. Cited on page(s) 38.

Vivet, M. *Expertise mathematique: CAMELIA, un logiciel pour raisonner et calculer*. Theses d'état, Université Paris IV, Paris, 1984. Cited on page(s) 142.

Vuillemin, J. Practical cellular dividers. *IEEE Trans. on Computers*, C-39: 605–614, 1990. Cited on page(s) 13, 17, 21, 23.

Vulcanov, D. Algebraic programming in the Hamiltonian treatment of an infationary model. *Int. Journal of Mod. Physics C*, 6(3):317–326, 1995. Cited on page(s) 180.

Vulcanov, D. and Ghergu, F. Use of computer facilities in teaching general relativity. http://xxx.lanl.gov/abs/physics/9812004, 1998. Cited on page(s) 176.

W3C Math Working Group. *Putting Mathematics on the Web with MathML.* World Wide Web Consortium, March 2002. http://www.w3.org/Math/XSL/. Cited on page(s) 160.

Wagon, S. *Mathematica in Action.* Freeman, W.H., San Francisco, 1990. Cited on page(s) 492.

Wagstaff, S. S. and Smith, J. W. Methods of factoring large integers. In *Number Theory, New York 1984-85*, volume 1240 of *Lecture Notes in Mathematics*, pages 281–303, Berlin-Heidelberg, 1987. Springer-Verlag. Cited on page(s) 16.

Walcher, S. On transformation into normal form. *J. Math. Anal. Appl.*, 180: 617–632, 1993. Cited on page(s) 106.

Wall, C. T. C. Resolutions for extensions of groups. *Proc. Cambridge Philos. Soc.*, 57:251–255, 1961. Cited on page(s) 211.

Walton, Jeremy and Dewar, Michael. See what I mean? Using Graphics Toolkits to Visualise Numerical Data. In Hege, Hans-Christian and Polthier, Konrad, editors, *Visualization and Mathematics*, pages 279–299. Springer Verlag, Heidelberg, 1997. Cited on page(s) 150, 263.

Wang, Dongming. Mechanical manipulation for a class of differential systems. *Journal of Symbolic Computation*, 12(2):233–254, August 1991. Cited on page(s) 264.

Wang, Dongming. An elimination method for polynomial systems. *Journal of Symbolic Computation*, 16(2):83–114, August 1993. Cited on page(s) 203.

Wang, Dongming. Reasoning about geometric problems using an elimination method. In Pfalzgraf, J., editor, *Automatic Practical Reasoning*, pages 147–185, Wien, 1995. Springer-Verlag. Cited on page(s) 32, 33, 201, 203.

Wang, Dongming, editor. *Automated Deduction in Geometry*, volume 1360 of *Lecture Notes in Artificial Intelligence*, 1997. Springer. Cited on page(s) 488.

Wang, Dongming. Gröbner bases applied to geometric theorem proving and discovering. In Buchberger, Bruno and Winkler, Franz, editors, *Gröbner Bases and Applications*, volume 251 of *LNS*, pages 281–301, Cambridge, UK, 1998. LMS, Cambridge University Press. ISBN 0-521-63298-6. Cited on page(s) 30.

Wang, Dongming. *Elimination Methods.* Springer WienNewYork, 2001. ISBN ISBN 3-211-83241-6. Texts and Monographs in Symbolic Computation. Cited on page(s) 491.

Wang, Dongming and Gao, Xiao-Shan. Geometry theorems proved mechanically using Wu's method—part on Euclidean geometry. Mathematics-Mechanization Research Preprints 2, Institute of Systems Science, Academia Sinicia, Beijing, China, November 1987. Cited on page(s) 32, 33.

Wang, P. S., editor. *1981 ACM Symposium on Symbolic and Algebraic Computation*, New York, 1981. Academic Press. Cited on page(s) 485, 557.

Wang, P. S. FINGER: A Symbolic System for Automatic Generation of Numerical Programs in Finite Element Analysis. *Journal of Symbolic Computation*, 2(3):305–316, 1986. Cited on page(s) 151.

Wang, P. S., editor. *ISSAC '92, International Symposium on Symbolic and Algebraic Computation, Berkeley, California*, New York, 1992. ACM. Cited on page(s) 487, 509, 523, 525, 532, 541, 558, 596, 617.

Wassermann, Alfred. Finding simple $t$-designs with enumeration techniques. *J. of Combinatorial Designs*, 6:79–90, 1998. Cited on page(s) 374.

Watanabe, Shunro. An experiment toward a general quadrature for second order linear ordinary differential equations by symbolic computation. In *Lecture Notes in Computer Science*, number 174, pages 13–22. Springer–Verlag, 1984. Proccedings of EUROSAM 1984, Cambridge, England. Cited on page(s) 283.

Watanabe, S. and Nagata, M., editors. *ISSAC '90, Proceedings of the International Symposium on Symbolic and Algebraic Computation, August 20–24, 1990, Tokyo, Japan*, New York, 1990. ACM. ISBN 0-89791-401-5,0-201-54892-5. Cited on page(s) 487.

Watkins, A. J. P. A new approach to mathematics for engineers. *International Journal for Mathematics, Education, Science and Technology*, 24(5):689–702, 1993. Cited on page(s) 251.

Watkins, A. J. P. *An Approach to Automating Mathematics Teaching and Learning using DERIVE, a Mathematical Assistant*. MPhil. Thesis, The University of Plymouth, 1994. Cited on page(s) 251.

Watt, S. M., editor. *Proceedings of the 1991 International Symposium on Symbolic and Algebraic Computation, ISSAC'91, July 15–17, 1991, Bonn, Germany*, New York, 1991. ACM. ISBN 0-89791-437-6. Cited on page(s) 487, 496, 499, 509, 512, 523, 532, 536, 568.

Watt, Stephen M., Broadbery, Peter A., Dooley, Samuel S., Iglio, Pietro, Morrison, Scott C., Steinbach, Jonathan M., and Sutor, Robert S. *AXIOM Library Compiler User Guide*. NAG Ltd, Oxford, 1994a. ISBN 1-85206-106-5. Cited on page(s) 265, 492.

Watt, Stephen M., Broadbery, Peter A., Dooley, Samuel S., Iglio, Pietro, Morrison, Scott C., Steinbach, Jonathan M., and Sutor, Robert S. A first report on the $A^\sharp$ compiler. In Giesbrecht [1994b], pages 25–31. Cited on page(s) 266.

Weber, H. *Lehrbuch der Algebra*, volume 1 - 3. Reprint Chelsea Publ. Comp., New York, 1961, 1898/1899/1908. Cited on page(s) 491.

Weber, K. The accelerated integer GCD algorithm. *ACM Transactions on Mathematical Software,*, 21:11–122, 1995. Cited on page(s) 11.

Wedeniwski, S. Piologie, eine exakte arithmetische Bibliothek in C++. Interner Bericht, Informatik WSI 96-35, Univeristt Tbingen, 1996. 214 pages. Cited on page(s) 12, 13.

Wegener, I. *The Complexity of Boolean Functions*. Wiley-Teubner, Stuttgart, 1987. Cited on page(s) 128, 491.

Wegschaider, K. *Computer Generated Proofs of Binomial Multi-Sum Identities*. PhD thesis, RISC-Linz, 1997. Cited on page(s) 93.

Weiglein, G. Results for precision observables in the electroweak Standard Model at two-loop order and beyond. *Acta Phys. Polon.*, B29:2735, 1998. Cited on page(s) 171.

Weiglein, G., Mertig, R., Scharf, R., and Böhm, M. Computer-algebraic calculation of two-loop self-energies in the electroweak Standard Model. In Perret-Gallix, D., editor, *New Computing Techniques in Physics Research 2*, page 617, Singapore, 1992. World Scientific. Cited on page(s) 170, 171.

Weiglein, G., Scharf, R., and Böhm, M. Reduction of general two-loop self-energies to standard scalar integrals. *Nucl. Phys.*, B416:606, 1994. Cited on page(s) 171.

Weileder, M. *Integration elementarer Funktionen durch elementare Funktionen, Errorfunktionen und logarithmische Integrale*. Dissertation, Fachbereich Mathematik der Ludwig-Maximilians-Universität München, München, 1990. Cited on page(s) 95.

Weispfenning, V. Aspects of quantifier elimination in algebra. In Burmeister, P., editor, *Universal Algebra and its Links with Logic, Algebra, Combinatorics and Computer Science, Proceedings of the 25. Arbeitstagung über Allgemeine Algebra, Darmstadt 1983*, volume 4 of *Research and Expositions in Mathematics*, pages 85–105, Berlin, 1984. Heldermann Verlag. ISBN 3-88538-204-0. Cited on page(s) 138.

Weispfenning, Volker. Admissible orders and linear forms. *ACM SIGSAM Bulletin*, 21(2):16–18, 1987. Cited on page(s) 29.

Weispfenning, V. The complexity of linear problems in fields. *Journal of Symbolic Computation*, 5(1):3–27, 1988. Cited on page(s) 54, 204, 206.

Weispfenning, V. Constructing universal Gröbner bases. In Huguet, L. and Poli, A., editors, *Applied Algebra, Algebraic Algorithms and Error-Correcting Codes, 5th International Conference, AAECC-5, Menorca, Spain, June 15–19, 1987, Proceedings*, volume 356 of *Lecture Notes in Computer Science*, pages 408–417, Berlin-Heidelberg-New York, 1989a. Springer-Verlag. ISBN 3-540-51082-6, 0-387-51082-6. Cited on page(s) 30, 427.

Weispfenning, V. Gröbner basis for polynomial ideals over commutative regular rings. In Davenport [1989], pages 336–347. ISBN 3-540-51517-8, 0-387-51517-8. Cited on page(s) 62.

Weispfenning, V. Comprehensive Gröbner bases. *Journal of Symbolic Computation*, 14(1):1–29, 1992a. Cited on page(s) 54, 179, 427.

Weispfenning, V. Finite Gröbner bases in non-Noetherian skew polynomial rings. In Wang [1992], pages 329–334. Cited on page(s) 62.

Weispfenning, Volker. Parametric linear and quadratic optimization by elimination. Technical Report MIP-9404, FMI, Universität Passau, D-94030 Passau, Germany, April 1994a. Cited on page(s) 139.

Weispfenning, Volker. Quantifier elimination for real algebra—the cubic case. In *Proceedings of the International Symposium on Symbolic and Algebraic Computation (ISSAC 94)*, pages 258–263, Oxford, England, July 1994b. ACM Press, New York, 1994. Cited on page(s) 206.

Weispfenning, Volker. Solving parametric polynomial equations and inequalities by symbolic algorithms. In *Proceeding of the workshop:"Computer Algebra in Science and Engineering", (Bielefeld, Aug'94*, pages 163–179, Singapore, 1995. World Scientific. Cited on page(s) 28, 57.

Weispfenning, Volker. Quantifier elimination for real algebra—the quadratic case and beyond. *Applicable Algebra in Engineering Communication and Computing*, 8(2):85–101, February 1997a. Cited on page(s) 54, 204, 206.

Weispfenning, Volker. Simulation and optimization by quantifier elimination. *Journal of Symbolic Computation*, 24(2):189–208, August 1997b. Special issue on applications of quantifier elimination. Cited on page(s) 57, 139.

Weispfenning, Volker. A new approach to quantifier elimination for real algebra. In Caviness and Johnson [1998], pages 376–392. Cited on page(s) 57, 139, 427.

Weispfenning, Volker. Mixed real-integer linear quantifier elimination. In Dooley [1999], pages 129–136. Cited on page(s) 138.

Weiss, J., Tabor, M., and Carnevale, G. The Painlevé property for partial differential equations. *J. Math. Phys.*, 24:522–526, 1983. Cited on page(s) 100.

Wellin, P. R. A letter from the Editor. *Mathematica in Education*, 1(1):1, 1991. Cited on page(s) 254.

Wess, J. and Zumino, B. Covariant differential calculus on the quantum hyperplane. *Nucl. Phys. B Proc. Suppl.*, 18:302–312, 1991. Cited on page(s) 212.

Wester, M. A review of CAS mathematical capabilities. *Computer Algebra Netherland Nieuwsbrief*, pages 41–48, December 1994. Cited on page(s) 283, 286.

Wester, Michael J. *Computer Algebra Systems*. John Wiley, Chichester, 1999. Cited on page(s) 261, 391, 485, 492, 593.

Wetzel, S. *Lattice Basis Reduction Algorithms and their Aplications*. PhD thesis, Universität des Saarlandes, 1998. Cited on page(s) 406.

Wicker, S. B. and Bhargava, V. K., editors. *Reed-Solomon Codes and their Applications*. IEEE Press, 1994. Cited on page(s) 491.

Widiger, A. Deciding degree-four-identities for alternative rings by rewriting. In Bronstein et al. [1998], pages 277–288. Cited on page(s) 64.

Wiedemann, D. Solving sparse linear equations over finite fields. *IEEE Transf. Inform. Theory*, IT-32:54–62, 1986. Cited on page(s) 37, 38, 40.

Wildanger, K. *Über das Lösen von Einheiten- und Indexformgleichungen in algebraischen Zahlkörpern mit einer Anwendung auf die Bestimmung aller ganzen Punkte einer Mordellschen Kurve*. Dissertation, Technische Universität Berlin, 1997. Cited on page(s) 400.

Wilf, H. S. and Zeilberger, D. An algorithmic proof theory for hypergeometric (ordinary and "$q$") multisum/integral identities. *Inventiones Math.*, 108:575–633, 1992a. Cited on page(s) 92.

Wilf, H. S. and Zeilberger, D. Rational function certification of multisum/integral/"$q$" identities. *Bulletin of the American Mathematical Society*, 80:148–153, 1992b. Cited on page(s) 92.

Wilkie, A. J. Model completeness results for expansions of the ordered field of real numbers by restricted. *Journal of American Mathematical Society*, 9:1051–1094, 1996. Cited on page(s) 138.

Wilkinson, James H. *The Perfidious Polynomial*, pages 1–28. Mathematical Association of America, 1984. Cited on page(s) 113.

Wille, R. Concept lattices and conceptual knowledge systems. In Lehmann [1992], pages 493–516. Cited on page(s) 218.
Wilson, R. A. A new construction of the baby monster and its applications. *Bull. Lond. Math. Soc.*, 25:431–437, 1993. Cited on page(s) 196.
Wilson, R. A. An Atlas of sporadic group representations. In Curtis and Wilson [1998], pages 261–273. Cited on page(s) 196.
Winkler, Franz. On the complexity of the Gröbner bases algorithm over $K[x, y, z]$. In Fitch, J., editor, *EUROSAM'84*, volume 174 of *LNCS*, pages 184–194. Springer, 1984. Cited on page(s) 29.
Winkler, Franz. A p-adic approach to the computation of Gröbner bases. *J. Symb. Computation*, 6:287–304, 1988. Cited on page(s) 30.
Winkler, F. *Polynomial Algorithms in Computer Algebra*. Springer Wien-NewYork, 1996. ISBN ISBN 3-211-82759-5. Texts and Monographs in Symbolic Computation. Cited on page(s) 491.
Winograd, Sh. *Arithmetic Complexity of Computations*. Society for Industrial and Applied Mathematics, Philadelphia, 1980. Cited on page(s) 491.
Wirsing, M. Structured algebraic specifications: A kernel language. *Theoretical Computer Science*, 42:123–249, 1986. Cited on page(s) 423.
Wirsing, M. Algebraic specification. In van Leeuwen, J., editor, *Handbook of Theoretical Computer Science*, volume B, pages 675–788. Elsevier Science Publishers, Amsterdam, 1990. Cited on page(s) 143.
Wirth, M. C. Symbolic vector and dyadic analysis. *SIAM J. Comput.*, 8:306–319, 1979. Cited on page(s) 222.
Wirth, N. *Programming in Modula*. Springer-Verlag, Berlin, Heidelberg, New York, 1985. Cited on page(s) 423.
Wolf, T. An efficiency improved program LIEPDE for determining lie-symmetries of PDEs. In *Modern Group Analysis: advanced analytical and computational methods in mathematical physics*, pages 377–385, Catania, Italy, 1993. Kluwer Academic Publishers. Cited on page(s) 465.
Wolf, T. Programs for applying symmetries of PDEs. In *Proceedings of ISSAC 95*, pages 7–15, Montreal, 1995. ACM Press. Cited on page(s) 466.
Wolf, T. The program crack for solving PDEs in general relativity. In Hehl et al. [1996], pages 241–258. Cited on page(s) 176.
Wolf, T. A comparison of four approaches to the calculation of conservation laws. preprint, 23 pages, 1998a. Cited on page(s) 189.
Wolf, T. A comparison of four approaches to the calculation of conservation laws. preprint, 1998b. Cited on page(s) 465, 467.
Wolf, T. Killing pairs in flat space. *Gen. Rel. & Grav.*, 30:123–129, 1998c. Cited on page(s) 467.
Wolf, T. Structural equations for killing tensors of arbitrary rank. *Comp. Phys. Comm.*, 115:316–329, 1998d. Cited on page(s) 188.
Wolf, T., Brand, A., and Mohammadzadeh, M. Computer algebra algorithms and routines for the computation of conservation laws and fixing of gauge in differential expressions. *Journal of Symbolic Computation*, 27:221–238, 1999. Cited on page(s) 189, 465, 467.

Wolf, T. and Gebot, G. Automatic symmetry investigation of space-time metrics, in Proceedings of the Journees Relativistes '93 in Brussels. *Int. J. of Mod. Phys. D*, 3(1):323–326, 1994. Cited on page(s) 188, 467.

Wolff, M., Richard, Ch. and Gloor, O. *Analysis Alive – Ein interaktiver Mathematik-Kurs.* Birkhäuser, 1998. ISBN 3-7643-5966-8. Book with CD-ROM. Cited on page(s) 259.

Wolfram, S. *Mathematica: A System for Doing Mathematics by Computer.* Addison-Wesley, Reading, Massachusetts, 1988. Cited on page(s) 168, 492.

Wolfram, S. *Mathematica: A System for Doing Mathematics by Computer.* Addison-Wesley, Reading, Massachusetts, second edition, 1991. Cited on page(s) 480, 492.

Wolfram, S. *Mathematica: Ein System für Mathematik auf dem Computer.* Addison-Wesley, Reading, Massachusetts, 1992. Cited on page(s) 492.

Wolfram, S. *The MATHEMATICA book.* Wolfram Media and Cambridge University Press, Champaign, IL and Cambridge, UK, third edition, 1996. Cited on page(s) 3, 492.

Wolfram, S. *Mathematica - a system for doing mathematics by computer.* Addison-Wesley, New York, 1998. Cited on page(s) 165, 492.

Worfolk, P. A. Zeros of equivariant vector fields: Algorithms for an invariant approach. *Journal of Symbolic Computation*, 17:487–511, 1994. Cited on page(s) 34, 107, 216.

Woronowicz, S. L. Differential calculus on compact matrix pseudogroups (quantum groups). *Comm. Math. Phys.*, 122:125–170, 1989. Cited on page(s) 212.

Wu, Wen-Tsun. Basic principles of mechanical theorem proving in elementary geometries. *Journal of Systems Sciences and Mathematical Sciences*, 4:207–235, 1984a. Cited on page(s) 32, 33, 201, 203, 205.

Wu, Wen-Tsun. Some recent advances in mechanical theorem-proving of geometries. *Contemporary Mathematics*, 29:235–241, 1984b. Cited on page(s) 32, 33.

Wu, Wen-Tsun. Basic principles of mechanical theorem proving in elementary geometry. *Journal of Automated Reasoning*, 2:219–252, 1986. Cited on page(s) 32, 33, 201, 203, 205.

Xiang, Jieh, editor. *Proceedings of the 6th Conference on Rewriting Techniques and Applications*, volume 914 of *Lecture Notes in Computer Science*, Berlin, 1995. Springer. Cited on page(s) 488.

Yee, E. Application of the decomposition method to the solution of the reaction-convection-diffusion equation. *Appl. Math. Comp.*, 56:1–27, 1993. Cited on page(s) 224.

Yeh, R., editor. *Current Trends in Programming Methodology IV: Data and Structuring.* Prentice Hall, Englewood Cliffs, N.J. 07632, 1978. Cited on page(s) 492, 543.

Yokoyama, K. A modular method for computing the galois group of polynomials. In Cohen, A. and Roy, M.-F., editors, *MEGA '96*, volume 117-118 of J.Pure Appl. Algebra, pages 617–636, 1997. Cited on page(s) 48.

Yoshiara, S. A classification of flag-transitive classical $c.C_2$-geometries by means of generators and relations. *Europ. J. Combinatorics*, 12:159–181, 1991. Cited on page(s) 90.

Yvinec, Brönnimann, H. and M. Efficient exact evaluation of signs of determinant. *Algorithmica*, 27:21–56, 2000. Cited on page(s) 114.

Zagier, D. Large integral points on elliptic curves. *Mathematics of Computation*, 48:425–436, 1987. Cited on page(s) 49.

Zahalak, G. I., Rao, P. R., and Sutera, S. P. Large deformations of a cylindrical liquid-filled membrane by a viscous shear flow. *J. Fluid Mech.*, 179:283–305, 1987. Cited on page(s) 222.

Zariski, O. Some open questions in the theory of singularities. *Bulletin of the American Mathematical Society*, 77:481–491, 1971. Cited on page(s) 199.

Zariski, O. and Samuel, P. *Commutative Algebra*, volume 1 and 2. Van Nostrand, Princeton, 1958,1960. Reprint Springer Verlag, New York, 1975/1979. Cited on page(s) 491.

Zassenhaus, H. On Hensel factorization I. *Journal of Number Theory*, 1:291–311, 1969. Cited on page(s) 25.

Zassenhaus, H. A real root calculus. In Leech, J., editor, *Computational Problems in Abstract Algebra, Proceedings of a Conference Held at Oxford Under the Auspices of the Science Research Council, Atlas Computer Laboratory, 29. Aug. to 2. Sept. 1967*, pages 383–392, Oxford, 1970. Pergamon Press. ISBN 0-08-012975-7. Cited on page(s) 56.

Zeilberger, D. A fast algorithm for proving terminating hypergeometric identities. *Discrete Mathematics*, 80:207–211, 1990a. Cited on page(s) 91.

Zeilberger, D. A holonomic systems approach to special function identities. *Journal of Computational and Applied Mathematics*, 32:321–368, 1990b. Cited on page(s) 91.

Zeilberger, D. The method of creative telescoping. *Journal of Symbolic Computation*, 11:195–204, 1991. Cited on page(s) 91, 215.

Zemanek, H. Dixit Algorizmi. In Ershov and Knuth [1981], pages 1–81. Cited on page(s) 1.

Zharkov, A. Yu. Solving zero-dimensional involutive systems. In Gonzalez-Vega, L. and Recio, T., editors, *Algorithms in Algebraic Geometry and Applications, MEGA-94*, pages 389–399, Basel, 1996. Birkhauser. Cited on page(s) 31.

Zharkov, A. Yu. and Blinkov, Yu.A. Involutive approach to investigating polynomial systems. *Math. Comp. Simul.*, 42:323–332, 1996. Cited on page(s) 103.

Zharkov, A. Yu. and Blinkov, Yu. A. Involution approach to solving systems of algebraic equations. In *Proc. IMACS'93*, pages 11–16, 1993. Cited on page(s) 31.

Zhevlakov, K. A., Slin'ko, A. M, Shestakov, I. P., and Shirshov, A. I. *Rings That Are Nearly Associative*. Academic Press, 1982. Cited on page(s) 347.

Zhi, Lihong and Wu, Wenda. Nearest singular polynomials. *Journal of Symbolic Computation*, 26(6):667–675, 1998. Special issue on Symbolic Numeric Algebra for Polynomials S. M. Watt and H. J. Stetter, editors. Cited on page(s) 117, 121.

Zhytnikov, V. V. GRG: Computer algebra system for differential geometry, gravitation, and field theory. version 3.2. (Moscow), 1997. Cited on page(s) 182.

Zimmer, H. G. *Computational Problems, Methods, and Results in Algebraic Number Theory*, volume 262 of *Lecture Notes in Mathematics*. Springer-Verlag, Berlin-Heidelberg-New York, 1972. Cited on page(s) 491.

Zimmer, H. G. Complete determination of all torsion groups of elliptic curves with integral absolute invariant over quadratic and pure cubic fields. In de Koninck, J. M. and Levesque, C., editors, *Number Theory. Proc. Intern. Conf., Univ. Laval*, pages 670–698, Berlin, 1989. W. De Gruyter. Cited on page(s) 49.

Zimmer, H. G. Torsion groups of elliptic curves over cubic and certain biquadratic number fields. *Contemporary Math.*, 174:203–220, 1994. Cited on page(s) 49.

Zimmer, H. G. SIMATH– a computer algebra system for number theoretic applications. In Küchlin [1997], pages 365–375. Cited on page(s) 45.

Zimmermann, P. Ga"ıa: A package for the random generation of combinatorial structures. *MapleTech*, 1:38–46, 1997. Cited on page(s) 220.

Zimmermann, Paul. Comparison of three public domain multiprecision libraries: BigNum, gmp and Pari. Paul.Zimmermann@inria.fr, 1998. 8 pages. Cited on page(s) 13.

Zippel, R. Probabilistic algorithms for sparse polynomials. In *EUROSAM' 79*, Lecture Notes in Computer Science 72, pages 216–226, Berlin-Heidelberg-New York-London-Paris-Tokyo-Hong Kong, 1979. Springer. Cited on page(s) 14.

Zippel, R. Simplification of expressions involving radicals. *J. Symb. Comput.*, 1:189–210, 1985. Cited on page(s) 18.

Zippel, R. Interpolating polynomials from their values. *Journal of Symbolic Computation*, 9(3):375–403, 1990. Cited on page(s) 384.

Zippel, R. *Effective Polynomial Computation*. Kluwer Academic Publishers, 1993. Cited on page(s) 18, 491.

Zippel, R. E. Rational function decomposition. In Watt, Stephen, editor, *International Symposium on Symbolic and Algebraic Computation*, pages 1–6, New York, July 1991. ACM. Cited on page(s) 28.

Zippel, R. E., editor. *Computer Algebra and Parallelism, Second International Workshop, Ithaca, USA, May 9–11, 1990, Proceedings*, volume 584 of *Lecture Notes in Computer Science*, Berlin-Heidelberg-New York, 1992. Springer-Verlag. ISBN 3-540-55328-2, 0-387-55328-2. Cited on page(s) 146, 486, 520.

Zölzer, U. *Digitale Audiosignalverarbeitung*. B.G. Teubner, 1996. Cited on page(s) 227.

# Subject Index

TEX, 158

A#
- compiler, 266
absolute irreducibility, 26
abstract data types, 146, 324, 422
ActiveX, 328
addition theorems, 214
adjoint curve, 357
admissible term order, 29
aircraft turbine, 226
al-Khorezmi, 1
Aldes
- programming language, 439
ALDES/SAC-2, 421, 425
Aldor
- compiler, 265
- language, 265
algebra, 1, 57
- completely separating, 368
- difference, 104
- differential, 104
- graded, 209
- Grassmann, 209
- hereditary, 369
- homological, 207
- non-commutative, 298
- repetitive, 369
- tubular, 369
algebra of solvable type, 58, 61
algebra, identities, 63
algebra, representation, 64
algebra, strucural issues, 63
algebraic closure of a field, 26
algebraic complexity theory, 125
algebraic curve, 301, 357
algebraic geometry, 197, 298, 300, 301, 356
algebraic number field, 426
algebraic number theory, 4
- Simath, 442
algebraic numbers
- representation, 18
algebraic set, 357
algebraic specification, 135, 218, 422, 437
algorithm, 1
- analysis, 220
algorithm design, 422
algorithm documentation, 422

alternating tensor power, 408
Amaya, 160
Amiga, 422
Analysis Alive, 256, 259
analysis of algorithms, 220
animation, 256, 258, 259
ANU
- $p$-Quotient Program, 459
- Nilpotent Quotient Program, 460
- Soluble Quotient Program, 460
apfloat, 12
application
- introduction, 163
- mathematics, 195
- physics, 163
applications
- relativistic gravitation theory, 469
approximate solutions, 109
approximation
- balanced, 477
- spline, 477
approximation of function, 477
APU system, 384
arbitrary domain polynomial, 426
arithmetic circuits, 25
arithmetic geometry, 298
Askey-Wilson scheme, 214, 215
assessment, 247, 248
asymptotics, 220
Atari, 422
atlas of finite groups, 196
atom, 242
audio signal processing, 227
Austrian experiment, 246, 250
automated theorem proving, 135
automatic theorem proving, 201
automorphism group, 299
- of $p$-group, 460
Axiom, 3, 261, 266, 325
- categories, 262
- domains, 262
- FORTRAN generation, 151
- GB system, 153
- NAG interface, 152, 263
- packages, 262

baby monster, 196

Bagdad, 1
basis reduction, 413
benchmarks, 9, 178
Bergman, 210
Berlin, 244
Bernina, 331
Betti number, 463
Bianchi identity, 173
bifurcation, 105, 179, 215, 216, 252
BigNum, 12
bilinear form, 475
Birkhoff-Poincaré-Witt theorem, 59
black box
– FoxBox system, 384
– ProtoBox system, 384
black box matrix, 37, 384
– LinBox system, 384
– WiLiSys system, 384
black box polynomial, 25
black box representation, 383
block design, 299
block vectors, 38
Boolean ring, 135
boson, 164
branching, 408
Brauer character tables
– of sporadic groups, 85
BUBBLES, 171
Buchberger's algorithm, 29, 60, 135, 229
Burnside group, 460
Burnside matrix, 388

C++, 146, 430
CABRI initiative, 178
CAD/CAM, 234
calculus, 94
Calculus&MATHEMATICA project, 254, 255
Cannes, 351
CAOP, 215
Carmichael numbers, 41
CASIO Algebra FX2.0, 3, 245
CASIO CFX-9970G, 3, 245
CASL, 218
Cataneo equation, 192
category, 325
CAVO group, 248
Cayley, 3
Cayley-Menger determinants, 242
celestial mechanics, 176

center manifold, 105, 106
central configurations, 176
certificates
– inconsistent linear systems, 38
chain complex, 207, 412
character
– highest weight module, 408
– symmetric group, 408
character table
– generic, 87, 196
– modular, 429
– of sporadic groups, 84
characteristic set, 32, 52, 104, 205
characters
– of groups, 388
– of Hecke algebras, 388
chemical reaction system, 243
chemistry, 242
Chinese remainder theorem, 30
Christoffel symbol, 172
Classi (Sheep package), 175
classical field theory, 182
classical group, 475
Clifford algebra, 180
CLN, 12
closures, 413
closures theorems, 202
CoCoa, 210
code, 305
CoFI, 218
coframe, 184
coherent sheaf, 417
cohomology, 207, 305
– coherent sheaf, 417
cohomology group, 87, 197
cohomology ring, 87
collision avoidance problem, 233
Columbia, 254
Common Algebraic Specification Language, 218
Common Framework Initiative, 218
commutative algebra
– CoCoA, 364
commutative polynomials, 426
comparison theorem, 208
COMPHEP, 166
compilation of user code, 341
complete intersection, 419
completion
– of differential equations, 102

complex, 207
complex functions, 258
complexity, 126, 474
- average-case analysis, 220
- matrix group algorithms, 474
comprehensive Gröbner basis, 31, 427
computation tree, 125
computational geometry, 127
computational group theory, 196, 253
computer aided design, 234
computer aided modelling, 234
computer algebra, 1
- benchmarks, 9
- in education, 7
- in Netherlands, 247
- in research, 7
- outlook, 6
- plugin, 8
- standard communication, 8
computer algebra curriculum, 3, 4, 255
computer algebra system
- for abstract tensor calculus, 174
- for component tensor calculus, 174
computer analysis, 109
computer labs, 250–252, 254, 255
computer-aided engineering, 331
condensation, 86
confluence, 132
conjecture
-body problem, 179
conjecture by Mumford, 198
connection relations, 214
conservation laws, 187
constructive mathematics, 7
coset table methods, 387
cosmology, 190
covariant derivative, 183
Crack, 176
creative telescoping, 91
CREP, 368
critial load, 226
critical pair, 29, 133
- additional, 62
cryptographic techniques, 8
cryptography, 303
- discrete logarithm, 304
crystallographic groups, 85, 354
curriculum, 3, 4, 244–248, 250, 252–257
- and assessment, 247, 248
- issues, 249

curvature, 183, 184
- irreducible decomposition, 183
curve
- adjoint curve, 357
- offset curve, 358
- projective, 200
cyclo-hexane, 242
cycloheptane molecule, kinematics of, 232
cylindric decomposition, 206

D-Gröbner basis, 426
Dagwood system, 384
debugger, 324
decision procedure, 137
decomposable structure, 220
decomposition
- functional, 233
- homogeneous bivariate functional, 233
decomposition matrix, 429
decomposition of algebraic sets, 357
definite hypergeometric summation, 215
Delaney symbol, 243
Derive, 244, 250–252
Descartes, 201
design, 473
Dickson's lemma, 61
difference equations, 98
differential algebraic equations, 108
differential calculus, 212
- non-commutative, 212
differential equations, 4, 96, 109
- linearization, 190
- ordinary, 97
- partial differential eequations, 187
- symmetries, 190, 468
differential Galois group, 97
differential Galois theory, 48, 97
differential geometry, 180, 182, 212
differential Gröbner basis, 188
digital signal processor, 228
dimension, 53, 357, 463
dimensional
- reduction, 469
- regularization, 469
Dimsym, 468
diophantine equation, 195
DIP, 421, 425
direct kinematics problem, 229
direction field, 252
directional derivatives, 468

discrete logarithm, 304
discriminant, 23
disequation, 203
distributed computing, 330
divided power algebra, 60
division free computation
– determinant, 37
Dixon's resultant, 358
dynamic loading of modules, 341
dynamic module, 326
dynamic system, 195
dynamical systems, 105
– bifurcation, 107, 215, 216
– Hamiltonian, 106
– reversible, 106

education, 2, 244
– CoCoA, 364
educational experiment, 247
efficiency, 423
efficient algorithms, 217
Einstein equation, solutions of
– classification, 175
– equivalence problem, 175
– finding new, 176
– Visualzation, 176
Einstein equations, 184
Einstein-Cartan theory, 184
electromagnetic field, 184
elementary particle physics, 164
elimination ideal, 463
elimination theory, 54
elliptic curve, 303, 306
– counting points, 405
– discrete logarithm, 304
elliptic curve method (ECM), 44, 405
elliptic curves, 42
– Simath, 443
elliptic curves by computer, 253
embedding problem, 48
endomorphism rings, 86
Engel identity, 460
engineering, 221
enumeration, 219
enveloping algebra, 59, 61
equation, 133
– diophantine, 195
– equivarinat, 216
equational specification, 133
equivarinat, 35, 215, 216

equivariant equation, 216
equivariants, 107
error estimation, 110
error-correcting code, 298–300
– minimum weight, 300
Euclid's algorithm, 127
exact division, 36
exact sequence, 207
examples
– character theory, 387
– group theory, 387
Excalc, 176
– checking Bianchi identity, 173
expressivness, 423
Ext
– global, 418
– total, 420
extensibility, 423
exterior algebra, 60, 212
exterior differential forms, 173
exterior form, 183

factor combination, 25
factorization, 4, 405
– CoCoA, 364
– black box polynomial, 384
– complex coefficients, 26
– greatest factorial, 92
– of differential operator, 97
– polynomial, 24, 26, 305
– polynomial over a finite field, 24
– polynomial over the rationals, 25
– polynomial via PDE, 26
– polynomial via root power sums, 26
– square-free, 92
factorized Gröbner basis, 426
fast arithmetic, 4
Fermat test, 41
Fermat's little theorem, 41
FeynArts, 166, 170, 469
FeynCalc, 166, 170
Feynman diagram, 164, 167, 169, 469
– in quantum gravity, 176
FFT, 14, 228, 430
FGB system, 153, 178
field, 298
– function field, 300
field extension, 18
field theory, 169
filter bank

Subject Index 627

- multirate, 228
- perfect reconstruction, 228
filter banks, 218
final assessment, 247
financial engineering, 255
finite differences, 108
finite elements, 108
finite field, 426
- discrete logarithm, 304
finite geometry, 305, 473
finite presentations, 459
finitely presented structure, 134, 437
floating point, 425
fluid
- ideal, 184
- spin, 184
FORM, 165, 174
- language, 170, 171, 472
formal power series, 15
FormCalc, 170, 469
formula books, 215
FORTRAN, 150
- interface, 176
Fourier transform, 127, 218, 227
FoxBox system, 384
fractal walk, 353
free loop space, 210
Free University Berlin, 244
freelip, 12
frequency range, 227
Freudenthal Institute, 248, 249
function
- algebraic, 94
- elementary, 94
- hypergeometric, 92
- rational, 94
- trigonometric, 94
functional decomposition, 233
functor
- Ext, 208
- Tor, 208
functors
- derived, 207

G-algebra, 58
Galois group, 47, 306
- baby monster as, 196
GAP, 3, 211, 475
- graphical interface, 389
Gap system, 385

garbage collection, 352
- mark-sweep, 351
gauge field theory, 164
gauge group, 184
Gaussian elimination, 36, 425
GB, 330
GB system, 153, 327
GCD
- black box polynomial, 384
- straight line program, 384
GEFICOM, 169
general linear group, 474
general relativity, 172, 184
- geodesic equations, 187
- killing vector and killing tensors, 187
generating function, 214, 219
- exponential, 220
- holonomic, 92, 220
generating system
- rigid, 196
generators and relations, 58
generic character table, 87
- of Iwahori-Hecke algebra, 88
- of Weyl group, 88
generic programming, 384
genericity, 146
genetic algebra, 58
Gentran, 151
genus, 357
geodesic, 184
geometric degree bound, 126
geometry, 438
- algebraic, 298
- algebraic geometry, 356
- arithmetic, 298
- incidence, 298
Givaro system, 384, 385
GL(d,q), 474
GMP, 12, 327, 385
Golay code, 299
Goldwasser-Kilian-Atkin algorithm, 43
Goppa code, 300
Gosper algorithm, 91, 215
GR (general relativity), 172
GRACE, 166
graph theory, 298
graphical interface
- for GAP, 389
graphics, 150
graphics calculator, 247–249, 251

graphs, 473
Grasmmannian variety, 416
Grassmann algebra, 209
gravity, 182
– field equations, 183
– gauge, 184
– general relativity, 184
– Kaluza-Klein, 184
– Lagrangian, 184
– metric-affine, 184
– theory, 183
gravity theory
– computer algebra system $GRG_{EC}$, 184
greatest common divisor, 23
Green functions, 469
grey box approach, 258
$GRG_{EC}$, 182, 184
– capabilities, 185
– language of problem specification, 186
– realization, 185
Gröbner basis, 4, 29, 51, 92, 104, 205, 209, 227, 305, 306, 353, 358, 412, 421, 425, 426, 463
– CoCoA, 364
– comprehensive, 54, 179
– differential, 188
– for modules, 463
– non-commutative, 60, 209
– reduced, 30
Gröbner basis conversion, 353
– fractal walk, 353
– Gröbner walk, 353
Gröbner factorizer, 464
Gröbner fan, 30
Gröbner walk, 30, 353, 358
group, 134, 298, 473
– automorphism group, 299
– cohomology, 207, 208
– Galois group, 306
– homology, 208
– invariant theory, 305
– matrix group, 299
– nilpotent, 459
– of order 256 and 512, 197
– permutation group, 299, 306
– soluble, 459
– solvable, 459
– supersolvable, 127
group algebra, 298
group characters, 388

group ring, 305
group theory, 4, 253
– classification of finite simple groups, 196
– combinatorial, 62
– computational, 196
– sporadic groups, 196
GRTensorII, 173, 176

hand-held computer algebra tools, 3, 244–249
harmonic analysis, 218
Hearn, Anthony C., 333
Hecke algebra, 298
Hensel lifting, 25, 36
Hermite polynomials, 214
Hfloat, 12
Higgs boson, 169
high energy physics, 469
high enery physics, 333
high school mathematics, 3, 245
Hilbert basis, 216
Hilbert function, 30
– CoCoA, 364
Hilbert series, 209, 463
Hilbert's 14th problem, 34
Hilbert's 16th problem, 105
homological algebra, 207
homotopy methods, 179
Hopf algebra, 212
Horner's rule, 126
hybrid method, 112
hypergeometric function, 92
hypergeometric representation, 214
hypergeometric summation, 91, 215

IBM, 226
ideal, 51
– differential, 104
– radical, 104
ideal membership problem, 30, 62
ideal of points
– CoCoA, 364
identities (binomial), 92
identities (hypergeometric), 92
Illustrated Mathematics, 256, 258
image compression, 218
implicitization, 235, 358
independent sets, 463
indeterminate, 2
indicial computation, 469

## Subject Index 629

inductive completion, 136, 438
initial model, 135
innovative curriculum, 246
integer arithmetic, 305
integrability
– complete, 99
integrability condition, 102
integration, 4, 94
interactive, 422
interactive book, 323, 327
interface
– numerical library, 108
InternetExplorer, 160
intersection, 463
INVAR, 35, 216
invariant polynomials, 427
invariant ring, 33
invariant theory, 33, 87, 179, 197, 216, 305
– modular, 34
invariants, 107, 215
inverse kinematics problem, 229
involutive basis, 31, 427
involutive division, 31
involutive system, 102
IRENA, 152
irreducibility, 4
irreducible module, 475
isomerism problem, 242
isomorphic $p$-groups, 460

Jacobi polynomials, 214
Jacoby identity, 58
Java, 8
Jordan algebra, 58

KANT V4, 3
KASH, 253
Kharitonov theorem, 122
kinematics, 229
Knuth-Bendix completion, 133, 438
Koszul resolution, 209
Krylov method, 37

laboratory activities, 250
Laguerre polynomials, 214
Lanczos method, 37
LaTeX, 150, 323
lattice, 299
– LLL reduction, 305, 306
libI, 12

library interface, 424
Lie algebra, 58, 106, 207, 298, 408, 460
– cohomology, 208
– finite dimensional, 60
– homology, 208
Lie brackets, 468
Lie derivative, 183
Lie group, 408
– reductive, 408
Lie symmetries, 468
Lie-Bäcklund symmetries, 468
LinBox system, 384
linear algebra, 298, 425
– black box matrix, 384
– matrix, 306
linear connection, 183, 184
linear representation, 217
LIP, 12
LISP, 423
– interpreter, 424
logic formulae, 427
LoopTools, 469
Lyapunov function, 107

Macaulay, 210
Macaulay2, 253, 411
Macsyma, 222
– Gentran, 151
Magma, 35, 211, 253, 295, 475
Maple, 3, 35, 150, 158, 160, 252, 259, 477
– FORTRAN generation, 151
marks
– table of, 388
MAS, 35, 421
MATAD, 169, 171
Mathematica, 3, 150, 160, 168, 254–256, 258, 314
– FORTRAN generation, 151
– in education and research, 254, 255
– InterCall, 152
mathematical concepts, 245–247, 249, 250, 252, 254–256, 259
mathematical dictionaries, 214
mathematical experiments, 248, 258, 259
mathematics curriculum, 3, 244–248, 250, 252–254, 256, 257
MathLie, 190
MathML, 150, 152, 154, 181, 256, 330
– content, 156, 157
– fine-grained level parallel markup, 158

- presentation, 154, 155
- top-level parallel markup, 158
MathOffice, 150
MathPlayer, 160
MathTensor
- checking Bianchi identity, 173
- Variational calculations, 176
MathView, 152
Matlab, 152, 228
matrix
- Hilbert, 38
- nilpotent, 414
- structured, 38
- Toeplitz, 38
- Vandermonde, 38
matrix group algorithms, 474
matrix group recognition, 87
matrix multiplication, 126
matrix representation, 84
meat-axe, 85
mechanism synthesis, 234
memory management, 146, 351
Mersenne numbers, 43
metric-affine gravity, 176
Miller-Rabin primality test, 42
MINCER, 166, 169
minimal surfaces, 258
minor expansion, 37
MIRACL, 12
modified equation, 108
Modula-2, 421, 423
modular invariant theory, 34
module, 298
- highest weight, 408
- representation, 412
modules of syzygies, 427
modulo AC, 133, 438
Moenck's algorithm, 91
molecular structure, 242
MOLGEN, 476
MOLiS, 476
monad algebra, 59
monster, 87, 196
Monte Carlo algorithm, 474
Mozilla, 160
MP, 12, 153, 330
MPEG, 228
MPFUN, 12
MTU, 226
multi-threading, 352

multiplication
- of integers, 127
- of polynomials, 127
MuPAD, 151, 321
- GB system, 153

NAG, 261
NAG Library, 152, 263, 327
National Research Council report, 254
Navier-Stokes, 223
neighborhood graph, 357
Netherlands, 247
Netscape, 160
Newman-Penrose formalism, 184
Newtonian $N$-body problem, 176
nilpotent group, 459
- infinite, 460
nilpotent matrix, 414
Noetherian ring, 61
non-commutative polynomials, 426
non-degeneracy condition, 203
nonmetricity, 184
normal form, 132, 463
- Birkhoff, 105
- singularity, 106
Norton's irreducibility criterion, 86
notebook, 322
NP-completeness, 128
NTL, 12, 327, 385, 430
Nullstellensatz, 51
number field, 306
number field sieve (NFS), 44
number theory
- Simath, 443
numeric and symbolic, 112, 216
- methods, 195
numerical analysis, 108

octonions, 425
Ohio State University, 254
one-loop integrals, 469
OpenGL, 150, 329
OpenMath, 151, 153, 158, 330
optical scalars, 184
orbifold, 243
orbit space reduction, 216
Ore algebra, 92
Ore extension, 62
orthogonal polynomials, 214
overhead viewscreen, 251

overloading, 325

p-adic, 48
p-group, 459
package
- combstruct, 220
packing, 305
Padé approximants, 166
Painlevé theory, 99
parallel computer algebra, 351
parallel robots, 330
parametrization, 358
Parcan, 351
PARI, 12
PARI-GP, 253, 431
parser, 424
partial differential equations
- linear, 468
partially ordered set, 368
particle physics, 165
patents, 160
path algebra, 59
PC, 422
permanent, 128
permutation group, 299, 306
- invariants, 35
- nearly linear methods, 387
- partition backtrack methods, 387
perturbation lemma, 210
perturbation theory, 469
physics
- elementary particle, 164
- high energy, 469
- particle, 165
pilot
- experiment, 249
- schools, 248
Piologie system, 12
plethysm, 408
Plymouth, 250
Pocklington's theorem, 42
Poincaré gauge gravity, 184
Poincaré map, 107
polarization, 64
polycyclic group, 459
polycyclic presentation, 459
polymorphism, 422
polynomial, 13
polynomial arithmetic, 305, 425
polynomial reduction, 29

polynomial rings, 412, 421
portability, 423
practical problems, 257
pretty printing, 425
primary decomposition, 53
primary ideal decomposition, 426
prime decomposition, 463
primitive matrix group, 475
problem solving, 245
- process, 257
- tool, 252
Profi project, 248
program verification
- abstract data types, 146
projective curve, 200
projective variety, 418
properties of identifiers, 325
ProtoBox system, 384
pseudo-division, 23
pseudo-random sequence, 305
Puiseux expansion, 358
PVM, 327, 330

QCD, 165, 167
QED, 165
qepcad, 204
QFT, 164
quadratic form, 368
quadratic sieve (QS), 44, 405
quantifier elimination, 53, 137, 353
quantum field theory, 164
quantum group, 212
quantum space, 212
quantum tunneling, 255
quaternions, 425
quiver, 87
- Auslander-Reiten, 64
- Gabriel, 64
quotient, 463
quotient methods, 387

radical, 52, 463
radical simplification, 18
Raleigh-Ritz, 226
random elements, 474
- pseudo random elements, 474
ranking, 102
Ratappr, 477
rational function, 426
real quantifier elimination, 427

real root counting, 427
RealSolving, 327, 330
recognition algorithm, 475
- classical groups, 474
recurrence relation, 92, 220
redlog, 205
Reduce, 222, 333
- Dimsym package, 468
- Gentran, 151
- imperative programming, 335, 337
- interfaces, 339
- IRENA, 152
- packages, 341
- rule oriented programming, 336
- scope, 151
- variational calculations, 176
reduced Gröbner basis, 30
reductive groups
- invariants, 35
renormalization, 171
representation theory, 218
representations
- Lie group, 408
RepTiles, 438
research
- impact, 4
resolution, 412
- CoCoA, 364
- free, 208
- Koszul, 209
- projection, 208
resolvent, 47
resonance, 100
resultant, 23
- Dixon's resultant, 358
- multivariate resultant, 358
resultant system, 52
rewrite rule, 132
Ricci tensor, 173
Riemann-Roch, 300
Riemannian curvature, 172
rigidity, 48
ring
- commutative, 298
ring of invariants, 33
ring theory, 427
Ritt's problem, 105
Ritt-Raudenbush theorem, 104
RLISP, 341
Roadmap-Algorithm, 234

Robotics, 229
Rodrigues formula, 214
root system, 408
roots
- complex, 26
- high precision, 26
Runge-Kutta method, 108

S-polynomial, 29
$S$-matrix elements, 469
SAC-1, 439
SAC-2, 439
SAC-2/ALDES, 421, 425
Saclib, 352, 385, 439
SAGBI basis, 31, 106
saturation, 62
SchGV, 12
SCHOONSCHIP, 4, 165, 261
SCHOONSHIP, 174
Schreier-Sims algorithm, 475
scientific computing, 2, 7
Scientific Workplace system, 150
Scilab, 330
secondary school, 247–250
semigroup, 298
Sheep, 174, 175
short vectors in lattices, 25
Shoup, Victor
- NTL, 430
signal processing, 218
signature, 133
Simath, 3, 442
simcalc user interface, 444
sine-cosine equation, 229
Singular, 3, 35, 210, 330
singular linear systems, 38
singularities, 216
singularity, 99, 357
Sister Celine's technique, 92
skew polynomial ring, 58
skills, 247
Solovay-Strassen primality test, 42
soluble group, 459
solvable group, 459
solvable polynomial rings, 426
spacetime, 183
- dimension, 183
- post-Riemannian, 184
- Riemannian, 184
spacetime metric, 184

sparse interpolation, 384
sparse representation, 25
specification language, 423
specification processor, 424
spinor, 183
– Dirac field, 184
squarefree decomposition, 24
standard basis, 32, 197, 463
standard model, 169
– minimal supersymmetric, 169
standardisation, 153
Stauduhar, 47
stereo isomerism problem, 242
Stewart platform, 229
straight-line program, 14, 54, 125, 383
– Dagwood system, 384
strategies, 426
stream cipher, 305
string rewriting, 62
– system, 134
string theory, 184
structure constants, 58
structures (combintorial), 219
subgroup lattice
– display, 389
summation, 4, 91
superconcentrator, 127
supergravity, 184
supersolvable group, 127
swapping algebra, 58
symbolic and numeric, 112, 216, 217, 227
– methods, 195
symbolic calculator, 249
symbolic integration, 4
symbolic summation, 4
SYMCON, 216
symmetric tensor power, 408
SYMMETRY, 216
symmetry, 107, 215, 216, 438
– generalized, 99
– Lie, 98, 468
– Lie-Bäcklund, 99, 468
– of differential equations, 468
– of geometric objects, 468
symmetry breaking, 438
symmetry theorem, 177
system
– discrete mathematics, 385
– free, 385
– group theory, 385

syzygy, 358, 463

table of marks
– of groups, 388
tangent cone, 357
– algorithm, 32
tangent space, 216, 357
techExplorer, 160
telecommunication management network, 221
tensor, 183
– density, 184
tensor calculus, 172
tensor decomposition, 475
tensor integral, 170
tensor product, 408
tensorial computation, 469
TERA project, 384
term order, 426
term rewriting, 422
term rewriting system, 132, 437
– complete, 133
– terminating, 132
TeX, 150
theorem by Saito, 197
theorem proving, 201, 453
Theorist, 152
thermodynamics, 190
threads of control, 352
TI-83, 251
TI-89, 3, 244–246, 251
TI-92, 3, 245, 249, 251, 252
tilings, 438
Todd-Coxeter algorithm
– algebras, 60
topology
– algorithmic, 243
torsion, 184
transfer matrix, 255
transformation
– coordinate, 184
– frame, 184
– spinor, 184
triangular sets, 179
triangular systems, 464
Turing machine, 128
tutoring software, 327
TwoCalc, 170

UDX, 330, 331

Unicode, 156
– basic multilingual plane, 158
unification, 135
universal enveloping algebra, 208
universal Gröbner basis, 30, 427
University of Illinois, 253, 254
University of Missouri-Columbia, 254
University of Plymouth, 250
unmixed radical, 463
Urbana-Champaign, 253, 254
US National Research Council report, 254
user interfaces, 152

variable, 2
– elimination, 235
variational calculations, 176
variety, 51, 133
vector, 183
vector enumerator, 85
vector space, 298
visualization, 256–259
VRML, 150, 263

wavelet, 218, 227
weighted projective line, 369
Weyl group, 408
white-box/black-box principle, 251, 258
WiLiSys system, 384
word problem, 62
World Wide Web Consortium (W3C), 158
Wu's algorithm, 32
WZ-method, 92

XHTML, 160
XLOOPS, 166
XML, 150, 152, 154

Yang-Mills field, 184

Zariski's multiplicity conjecture, 199
Zeilberger algorithm, 215
zero-dimensional ideal, 426
$z$-transform, 227

# Index of Authors' Contributions

Akers, Larry (SciComp Inc.), 440
Apel, Joachim (Leipzig), 57, 212, 375

Backelin, Jörgen (Stockholm), 349
Balfagón, A. (Ramón Llull, Spain), 480
Bauer, Christian (Mainz), 391
Baumann, Gerd (Ulm), 190
Becker, Eberhard (Dortmund), 55
Behnke, Kurt (Düsseldorf), 221
Belabas, Karim (Orsay), 431
Berry, John (Plymouth), 250
Beth, Thomas (Karlsruhe), 128, 161, 217, 461
Betten, Anton (Bayreuth), 372
Bonadio, Allan (San Francisco), 454
Boston, Nigel (Urbana-Champaign), 253
Breuer, Thomas (Aachen), 385
Bronstein, Manuel (Sophia Antipolis), 94
Buchberger, Bruno (RISC–Linz), 453
Buchmann, Johannes (Darmstadt), 41, 44, 403
Bündgen, Reinhard (Böblingen), 132, 437
Bürgisser, Peter (Paderborn), 125

Calmet, Jacques (Karlsruhe), 140
Cannon, John (Sydney), 295
Capani, Antonio (Genova), 364
Clausen, Michael (Bonn), 125, 227
Cohen, Arjeh (Eindhoven), 57
Cojocaru, Svetlana (Chisinau), 349
Conrad, Marc (Saarbrücken), 442
Corless, Robert M. (London, Ontario), 112

Degen, Wendelin (Stuttgart), 428
Delgado Friedrichs, Olaf (Bielefeld), 438
Dentzer, Ralf (Heidelberg), 16, 17
Dewar, Michael (Oxford), 150, 261
Di Crescenzo, Claire (Grenoble), 370
Dolzmann, Andreas (Passau), 201
Dräxler, Peter (Bielefeld), 57, 368
Dress, Andreas W. M. (Bielefeld), 242, 438
Drijvers, Paul (Utrecht), 247

Eck, Hagen, 469
Egner, Sebastian (Eindhoven), 128, 461

Fachgruppe Computeralgebra (Steering Committee of the German special interest group of computer algebra), 1
Fleischer, Jochem (Bielefeld), 164
Ford, David (Montreal), 348
Fortenbacher, Albrecht (Berlin), 142
Fowler, David (Lincoln), 254
Frink, Alexander (Mainz), 391

Gatermann, Karin (Berlin), 96, 215, 216
von zur Gathen, Joachim (Paderborn), 24
Gautier, Thierry (INRIA LMC-IMAG, Grenoble), 146
Geck, Meinolf (Lyon), 359
Geiselmann, Willi (Karlsruhe), 128
Gerdt, Vladimir P. (Dubna), 96
Gerhard, Jürgen (Paderborn), 321
Giesbrecht, Mark (London), 38
Giusti, Marc (Palaiseau), 13, 51
Gloor, Oliver (Bern, Switzerland), 256, 351
Gonzalez–Vega, Laureano (Cantabria), 234
Grabmeier, Johannes (Deggendorf), 6, 21, 57, 195, 226, 261
Gräbe, Hans-Gert (Leipzig), 463
Graham, Ted (Plymouth), 250
Grassl, Markus (Karlsruhe), 128
Grayson, Daniel R. (University of Illinois at Urbana-Champaign), 411
Greuel, Gert-Martin (Kaiserslautern), 54, 197, 445
Gutierrez, Jaime (Cantabria), 26

Hahn, Thomas (Karlsruhe), 469
Haible, Bruno (Gentilly, France), 464
Hantzschmann, Karl (Rostock), 109
Head, Alan (Melbourne), 411
Heckenberger, István (Leipzig), 212
Hehl, Friedrich W. (Cologne), 163, 172
Heinicke, Christian (Cologne), 172
Hemmecke, Ralf (Linz), 356
Hereman, Willy (Boulder), 96
Hillgarter, Erik (Linz), 356
Hiss, Gerhard (Aachen), 84, 196, 429
Homann, Karsten (Karlsruhe), 217
Hong, Hoon (NCSU, Raleigh), 146

Huang, Weiguang (Sydney, Australia), 451
Hulpke, Alexander (Columbus), 385
Huson, Daniel (Rockville, USA), 438

Jacobs, David P. (Clemson), 346
Jaén, X. (Catalana, Spain), 480
Jebelean, Tudor (RISC–Linz), 453
Jerie, Michael (Bundoora, Australia), 469
Jung, Françoise (Grenoble), 370

Kaltofen, Erich (Raleigh), 4, 26, 36, 112, 383
Kant, Elaine (SciComp Inc.), 440
Kemper, Gregor (Heidelberg), 33
Kerber, Adalbert (Bayreuth), 89, 452, 476
Klappenecker, Andreas (Karlsruhe), 161, 217
Klaus, Uwe (Leipzig), 375
Klioner, Sergei (St.Petersburg, Russia – Dresden, Germany), 469
Klüners, Jürgen (Heidelberg), 47
Koepf, Wolfram (Kassel), 2, 214, 244
Kohnert, Axel (Bayreuth), 452
Kotsireas, Ilias (Paris), 176
Kovács, Peter (Berlin), 229
Kozen, Dexter (Cornell), 26
Kreckel, Richard (Mainz), 391
Kredel, Heinz (Mannheim), 421
Küblbeck, Sepp, 469
Küchlin, Wolfgang (Tübingen), 434
Kurth, Frank (Bonn), 227
Kutzler, Bernhard (Linz, Austria), 271

Lambe, Larry A. (NJ, USA / Bangor, Wales), 207
Landau, Susan (Amherst, USA), 18
Laßner, Wolfgang (Senftenberg), 57
Laue, Reinhard (Bayreuth), 89, 372, 476
Laushnyk, Oksana (Lviv), 477
Lazic, Dejan (Karlsruhe), 128
Leedham-Green, C.R. (London, Great Britain), 474
van Leeuwen, M.A.A. (Poitiers), 408
Lescanne, Pierre (Vandœuvre-les-Nancy), 477
Lewis, Robert H. (Fordham University), 380
Loos, Rüdiger (Tübingen), 11, 143, 161
Lübeck, Frank (Aachen), 359

Lux, Klaus (Tucson), 363, 429

Macsyma Inc. (submitted by R. H. Berman), 294
Madlener, Klaus E. (Kaiserslautern), 57
Matzat, B. Heinrich (Heidelberg), 6, 47
Melenk, Herbert (Berlin), 243, 333
Mertig, Rolf (Amsterdam), 471
Müller-Quade, Jörn (Karlsruhe), 128, 161, 217
Müller, Volker (Yogyakarta), 41

Newman, M.F. (Canberra, Australia), 459
Nickel, Werner (Darmstadt, Germany), 459
Niemeyer, Alice C. (Perth, Australia), 459, 474
Niesi, Gianfranco (Genova), 364
Niklasch, Gerhard (München), 16, 17
Nörenberg, Rainer (Bielefeld), 368
Nückel, Armin (Karlsruhe), 161, 217

O'Brien, E.A. (Auckland, New Zealand), 459, 474
Obukhov, Yuri N. (Moscow), 182
Ollivier, François (Palaiseau), 96
Opgenorth, Jürgen (Aachen), 354

Pardo, L. M. (Palaiseau, Santander), 13, 51
Paule, Peter (Linz), 91
Pesch, Michael (Passau), 421
Pfahler, Thomas (Darmstadt), 41, 44, 403
Pfister, Gerhard (Kaiserslautern), 445
Pflügel, Eckhard (Grenoble), 370
Plesken, Wilhelm (Aachen), 354
Pohst, Michael E. (Berlin), 20, 21, 45, 50, 396
Popov, Bogdan (Lviv), 477
Praeger, Cheryl E. (Perth, Australia), 474
Prince, Geoff (Bundoora, Australia), 469
Püschel, Markus (Pittsburgh), 461
Punjani, Minaz (London), 441

Rabuka, Scott (Waterloo, Canada), 308
Randall, Curt (SciComp Inc.), 440
Rees, Sarah (Newcastle), 436
Reinert, Birgit (Kaiserslautern), 57
Robbiano, Lorenzo (Genova), 364
Roch, Jean-Louis (LMC-IMAG, Grenoble), 146

Roesner, Karl G. (Darmstadt), 221
Roggenbach, Markus (Bremen), 217
Roy, Marie-Françoise (Rennes), 19
Rump, Siegfried M. (Hamburg), 110

Saunders, B. David (Newark), 36
Schaefer-Lorinser, Frank (Darmstadt), 128, 161
Schmitt, Susanne (Saarbrücken), 442
Schönemann, Hans (Kaiserslautern), 445
Schreiner, Wolfgang (RISC-Linz, Austria), 146
Schrüfer, Eberhard (St. Augustin), 180
Schüler, Axel (Leipzig), 212
Schulz, Tilman (Aachen), 354
Schupp, Sibylle (RPI), 143
Schwarz, Fritz (St. Augustin), 96
Seiler, Werner M. (Mannheim), 96
Sharp, Jenny (Plymouth), 250
Shokrollahi, M. Amin (Bell Labs), 125, 128
Sims, Charles C. (New Brunswick), 65
Soicher, Leonard H. (London), 473
Sorgatz, Andreas (Paderborn), 321
Steinberg, Stan (SciComp Inc.), 440
Steinhauser, Matthias (Bern), 167
Stillman, Michael E. (Cornell University), 411
Storjohann, Arne (Zurich), 38
Strehl, Volker (Erlangen), 91, 219
Sturm, Thomas (Passau), 201

Takayama, Nobuki (Kobe), 392
Tertychniy, Sergey I. (Moscow), 184
Thomas, Wolfgang (Kiel), 348
Townend, Stewart (Plymouth), 250

Ufnarovski, Victor (Lund), 349
Ulmer, Felix (Rennes), 96
Unger, Bill (Sydney), 295

Veigneau, Sebastien (Noisy-le-Grand), 345
Vermaseren, Jos A.M. (Amsterdam), 171
Villard, Gilles (ENS Lyon), 38, 146

Waits, Bert K. (Columbus), 245
Wassermann, Alfred (Bayreuth), 372
Watkins, Anthony (Plymouth), 250
Watt, Stephen M. (London, Ontario), 112, 154, 265
Weiglein, Georg (Geneva), 169
Weispfenning, Volker (Passau), 4, 23, 28, 32, 51, 137, 201
Wildanger, Klaus (Berlin), 396
Windsteiger, Wolfgang (Linz, Austria), 314
Winkler, Franz (Linz), 356
Wolf, Thomas (London), 187, 465

Young, Bob (SciComp Inc.), 440

Zayer, Jörg (Saarbrücken), 44
Zimmer, Horst G. (Saarbrücken), 48